Neurotransmitters, Drugs and Brain Function

Neurotransmitters, Drugs and Brain Function

Edited by

R. A. Webster

Department of Pharmacology, University College London, UK

JOHN WILEY & SONS, LTD

Chichester · New York · Weinheim · Brisbane · Singapore · Toronto

Baffins Lane, Chichester,
West Sussex PO19 1UD, UK

National 01243 779777
International (+44) 1243 779777
e-mail (for orders and customer service enquiries): cs-books@wiley.co.uk
Visit our Home Page on: http://www.wiley.co.uk or http://www.wiley.com

Other Wiley Editorial Offices

John Wiley & Sons, Inc., 605 Third Avenue,
New York, NY 10158-0012, USA

WILEY-VCH Verlag GmbH, Pappelallee 3,
D-69469 Weinheim, Germany

John Wiley & Sons Australia, Ltd. 33 Park Road, Milton,
Queensland 4064, Australia

John Wiley & Sons (Asia) Pte, Ltd. 2 Clementi Loop #02-01,
Jin Xing Distripark, Singapore 129809

John Wiley & Sons (Canada), Ltd. 22 Worcester Road,
Rexdale, Ontario M9W 1L1, Canada

Library of Congress Cataloging-in-Publication Data

Neurotransmitters, drugs and brain function / edited by R. A. Webster
p. , cm.
Includes bibliographical references and index.
ISBN 0-471-97819-1
1. Neurotransmitters. 2. Neurotransmitter receptors. 3. Brain–Pathophysiology.
4. Psychopharmacology. I. Webster, R. A., Ph.D.
[DNLM: 1. Neurotransmitters–physiology. 2. Brain–drug effects. 3. Brain Chemistry–drug effects. 4. Synaptic Transmission–drug effects. QV 126 N4955 2001]
QP364.7 .N479 2001
612.8′042–dc21

2001024354

British Library Cataloguing in Publication Data

A catalogue record for this book is available from the British Library

ISBN 0 471 97819 1

Typeset in 10/12pt Times from authors' disks by Dobbie Typesetting Limited, Tavistock, Devon
Printed and bound in Great Britain by Biddles Ltd, Guildford and King's Lynn
This book is printed on acid-free paper responsibly manufactured from sustainable forestry, in which at least two trees are planted for each one used for paper production.

Contents

List of Contributors

David A. Brown, Alasdair J. Gibb, S. Clair Stanford, Anthony H. Dickenson, Mark Farrant, and Roy Webster, all contributors are from the Department of Pharmacology, University College London, Gower Street, London WC1E 6BT

Preface

This book is about neurotransmitters, the substances released from neurons to act on neurons. It covers what they do, how they do it and how their activity is involved in brain function and affected by drugs and disease.

After an overview of neurotransmitter systems and function and a consideration of which substances can be classified as neurotransmitters, section A deals with their release, effects on neuronal excitability and receptor interaction. The synaptic physiology and pharmacology and possible brain function of each neurotransmitter is then covered in some detail (section B). Special attention is given to acetylcholine, glutamate, GABA, noradrenaline, dopamine, 5-hydroxytryptamine and the peptides but the purines, histamine, steroids and nitric oxide are not forgotten and there is a brief overview of appropriate basic pharmacology.

How the different neurotransmitters may be involved in the initiation and maintenance of some brain disorders, such as Parkinson's disease, epilepsy, schizophrenia, depression, anxiety and dementia, as well as in the sensation of pain, is then evaluated and an attempt made to see how the drugs which are used in these conditions produce their effect by modifying appropriate neurotransmitter function (section C). The final section (D) deals with how neurotransmitters are involved in sleep and consciousness and in the social problems of drug use and abuse.

The contents are based on lectures given by the contributors, all of whom are experienced in research and teaching, in a neuropharmacology course for final-year BSc students of pharmacology, physiology, psychology and neuroscience at University College London. The text should be of value to all BSc students and postgraduates in those and related disciplines. Those studying medicine may also find it useful especially if working in neurology or psychiatry.

We have tried to make the book readable rather than just factual and so references have been kept to a minimum, especially in the early chapters on basic neuropharmacology and although more are given in the applied sections, they are selective rather than comprehensive.

Section A

BASIC ASPECTS OF NEUROTRANSMITTER FUNCTION

1 Neurotransmitter Systems and Function: Overview

R. A. WEBSTER

INTRODUCTION

> Analysis of Biological Function generally presumes that function at one level arises from the interactions of lower-level elements. It is often relatively straightforward to identify elements that may be involved and their individual interactions. Modern cell and molecular biology, in particular, is very efficient at identifying new molecules, and establishing which molecules 'talk to' one another. However, as the accumulation of such studies gradually reveals a complex network of interactions, its output — the biological function — becomes ever harder to understand and predict. The system is reduced to its elements, but it is not clear how to integrate it again. Yet this is the ultimate functional goal. (Brezina and Weiss 1997)

The molecules referred to are the neurotransmitters (NTs) and their receptors, found in the brain; the biological function is the activity of the brain itself. Our understanding of that must be the ultimate goal.

We have no such pretensions in this book but we do hope to help you to understand how neurotransmitters may be involved in brain function and more particularly how their activity is modified by disease and drugs. As the above quotation implies, this will mean considering the synaptic characteristics of each neurotransmitter, but before we do so, it is important to consider some more general and basic aspects of neurotransmitter function. Thus:

- What is a neurotransmitter and how did the concept of chemical transmission arise?
- Which substances are neurotransmitters? Can they be sensibly classified and how do we know they are transmitters?
- Which neurons and pathways use which neurotransmitters and how are they organised?
- How do neurotransmitters work? What effects do they have on neuronal activity?
- What is known about the receptors to which they bind?
- How are neurotransmitters released and how is this controlled?
- How can neurotransmitter function be modified?

Most of these points are considered in detail in subsequent chapters but some will be touched on collectively here.

Neurotransmitters, Drugs and Brain Function. Edited by R. A. Webster

CHEMICAL TRANSMISSION

We might start by considering what we understand by the term 'neurotransmitter'. According to the *Oxford English Dictionary* (2nd edition) it is:

> A substance which is released at the end of a nerve fibre by the arrival of a nerve impulse and by diffusing across the synapse or junction effects the transfer of the impulse to another nerve fibre (or muscle fibre or some receptor).

Other dictionaries carry similar definitions.

Based on this definition a neurotransmitter could be exemplified by actylcholine (ACh) released from motor nerves to excite and contract the fibres of our skeletal muscles. Indeed the synapses there, i.e. the junctions between nerve and muscle fibres, are anatomically and chemically geared to act as a fast relay station. Acetylcholine released rapidly from vesicles in the nerve terminal, on arrival of the nerve impulse, binds quickly with postsynaptic sites (receptors). When activated these open channels for sodium ions which pass through into the muscle fibre to depolarise its membrane and cause contraction. The whole process takes less than one millisecond and the ACh is rapidly removed through metabolism by local cholinesterase so that contraction does not persist and the way is cleared for fresh ACh to act. Anatomically there is a precise and very close relationship between the nerve ending and the muscle fibre at histologically distinct end-plates, where the receptors to ACh are confined. It is better than having the nerve directly linked to the muscle since the time lost through imposing a chemical at the synapse between nerve and muscle is insignificant and the use of a chemical not only facilitates control over the degree of muscle tone developed, but fortuitously makes it possible for humans to modify such tone chemically.

Blocking the destruction of ACh potentiates its effects while blocking the receptors on which it acts produces paralysis (neuromuscular blockade). Indeed it was the curare impregnated into the darts used by native South American hunters, so that they could paralyse and then easily kill their prey, that motivated Claude Bernand to investigate its actions at the end of the nineteenth century and so demonstrate the chemical sensitivity of excitable tissue that led to the concept of chemical transmission. He did a very simple experiment. He took a sciatic nerve gastrocnemious muscle preparation from a frog (not the actual quest of the hunters), placed the muscle in one dish of appropriate salt solution and extended the nerve into another. Not surprisingly, simple wire electrodes connected to an activated induction coil induced contractions of the muscle whether placed directly on the muscle or on the nerve to it. When, however, curare was added to the dish containing the muscle, direct stimulation of the muscle still induced a contraction, but activation of the nerve was ineffective. This was not due to any effect of curare on the nerve because when curare was added to the nerve rather than the muscle dish, stimulation of the nerve was still effective. Thus there had to be a chemically sensitive site on the muscle, where it was linked with the nerve, which was affected by the curare. This did not prove that a chemical had been released from the nerve but some years later (1916) Otto Loewi found that if he cannulated the ventricle of a frog's heart, isolated with its vagus nerve intact, then when this was stimulated not only did the heart slow, as expected, but if fluid withdrawn from the ventricle was subsequently reintroduced the heart slowed again. This suggested the release of a chemical from the vagus, which was made even clearer by allowing the fluid perfused through one frog heart to drip onto a second one and establishing that when the first heart was slowed by stimulating its vagus the fluid from it also slowed the second heart when that was reached. Loewi did not identify the chemical, which he called vagustoff, but it was later shown to be acetylcholine (ACh), the first identified neurotransmitter (and it was also found to transmit the neural stimulation of skeletal muscle, which had been blocked by curare in the experiments of Bernard).

Now this brings us to the first problem with the dictionary definition of a neurotransmitter because in the heart ACh is not transmitting an excitatory impulse between nerve and muscle, it is causing inhibition. There are also other differences. Its cardiac effect, change in rate, occurs much more slowly, has nothing to do with the direct opening of any ion channel and is not blocked by curare. Thus the sites on cardiac muscle that are chemically sensitive to ACh, its so-called receptors, are different from those for ACh on skeletal muscle. In fact they are blocked by a different poison, namely atropine (from *Atropa belladonna*, Deadly Nightshade). These observations raise two important issues. First, it is the receptor which ultimately determines the effects of a neurotransmitter and second, since only the excitatory effects of ACh at the neuromuscular junction fulfil the original definition of a neurotransmitter in transmitting excitation, either acetylcholine cannot be considered to be a neurotransmitter in the heart, despite its effects, or the definition of a neurotransmitter needs modifying. We will return to this topic at the end of the chapter.

Thus a neurotransmitter can clearly have more than one effect and a moment's consideration of what is involved for your nervous system in effecting the processes that enable you to turn the pages of this book and read and remember some of its contents will make you realise just how much the nervous system has to achieve and how many different parts of it have to be involved and functionally integrated. This is without considering whether you feel content, anxious, or depressed and how that can affect your concentration and ability to read and learn or even turn over the pages. Clearly such processes must involve many different neural pathways and types of neuron producing different effects and presumably requiring a number of different chemicals (neurotransmitters). The importance and variety of such chemicals is also emphasised from a look at drug usage and the study of how they work.

There are many drugs that affect the nervous system for good (antidepressants, analgesics, anticonvulsants) and bad (toxins, poisons, drugs of abuse) and although it would be naive to think that any drug has only one effect, i.e. that an anti-epileptic drug will never cause any sedation, their demonstrably different primary effects, coupled with the diversity of their chemical structures, suggest that not only are drugs affecting different areas of the brain but as they are likely to do this at chemical synapses there must be a number of different chemical synapses and chemicals, i.e. neurotransmitters.

NEUROTRANSMITTER CLASSIFICATION

The following substances, listed alphabetically, have been widely implicated and generally accepted as neurotransmitters in the central nervous system (CNS), although some, such as glutamate, are much more important than others, e.g. adrenaline. Some classification is appropriate and the simplest and most commonly used is that based on chemical structure with the substances grouped as follows:

	Chemical group	Examples
A	Choline ester	Acetylcholine (ACh)
B	**Monoamines**	
	Catechol	Dopamine (DA), noradrenaline (NA) (adrenaline)
	Indole	5-Hydroxytryptamine (5-HT, serotonin)
	Imidazole	Histamine (HIST)
C	**Amino acids**	
	Acidic	Glutamate (GLT)
	Basic	γ-Aminobutyric acid (GABA), glycine
D	**Peptides**	Enkephalins, endorphins, cholecystokinin, substance P (Many others have been implicated)
E	**Purines**	Adenosine triphosphate (ATP), adenosine

In addition to the above it is now clear that the following substances may have an important central action but whether they can be classified as true neurotransmitters is uncertain:

F	**Steroids**	Pregnenalone, dehydroepiandrosterone
G	**Nitric oxide**	(A gas but it is always in solution in the brain)
H	**Eicosanoids**	Prostaglandins

A glance at the structure of the classical neurotransmitters (Fig. 1.1) shows that apart from the peptides (D) (and purines, E), most of them are fairly simple chemicals. Some authors therefore divide them into small (e.g. A, B, C) and large (peptides, D) molecular NTs. Although we will see that peptides certainly have some properties different from other NTs, in that they rarely have a primary neurotransmitter function and usually just complement the actions of those NTs in groups A–C, to put them in a class of their own and group all the others together simply on the basis of molecular size is inappropriate and misleading since it elevates the peptides to a status that is neither proven nor warranted.

NEURONS: STRUCTURE AND ENVIRONMENT

The neurons from which NTs are released number more than 7 billion in the human brain. Each (Fig. 1.2) consists of a cell body, the soma or perikaryon, with one major cytoplasmic process termed the axon, which projects variable distances to other neurons, e.g. from a cortical pyramidal cell to adjacent cortical neurons, or to striatal neurons or to spinal cord motoneurons. Thus by giving off a number of branches from its axon one neuron can influence a number of others. All neurons, except primary sensory neurons with cell bodies in the spinal dorsal root ganglia, have a number of other, generally shorter, projections running much shorter distances among neighbouring neurons like the branches of a tree. These processes are the dendrites. Their

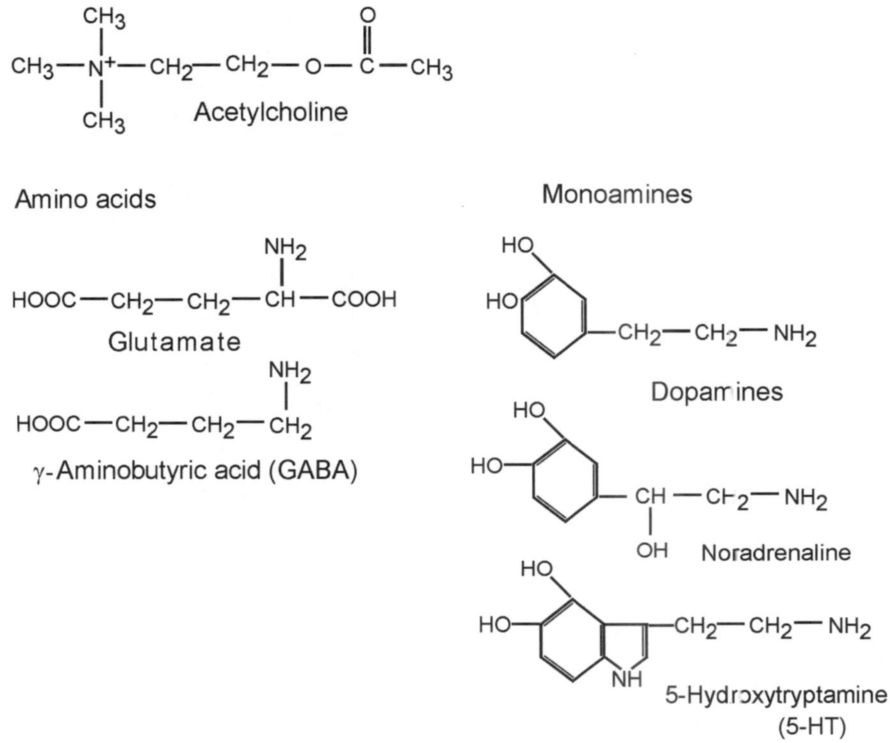

Peptides

Tyr–Gly–Gly–Phe–Met   Met-enkephalin

Arg–Pro–Lys–Pro–Gln–Gln–Phe–Phe–Gly–Leu–Met–NH_2 Substance P

Arg = arginine, Gly = glycine, Gln = glutamine, Leu = leucine, Lys = lysine,
Phe = phenylalanine, Pro = proline, Met = methionine

Figure 1.1 The chemical structures of the main neurotransmitters. The relatively simple structure of acetylcholine, the monoamines and the amino acids contrasts with that of the peptides, the simplest of which are the enkephalins which consists of five amino acids; substance P has eleven

absence from sensory, i.e. initiating, neurons immediately suggests that their function is associated with the reception of signals (inputs) from other neurons. Neuron cell bodies vary in diameter from 5 μm to 100 μm and axons from 0.1 μm to 10 μm, although these are enlarged at their terminal endings. Axons are generally surrounded by an insulating myelin sheath which is important for the propagation of action potentials generated in the neurons and gives the axons and the pathways they form a white colour which contrasts with the grey appearance of those areas of the CNS dominated by the presence of neuron cell bodies and their dendrites.

The axon terminals of one neuron synapse with other neurons either on the dendrites (axo-dendritic synapse) or soma (axo-somatic synapse). Synapses on another axon

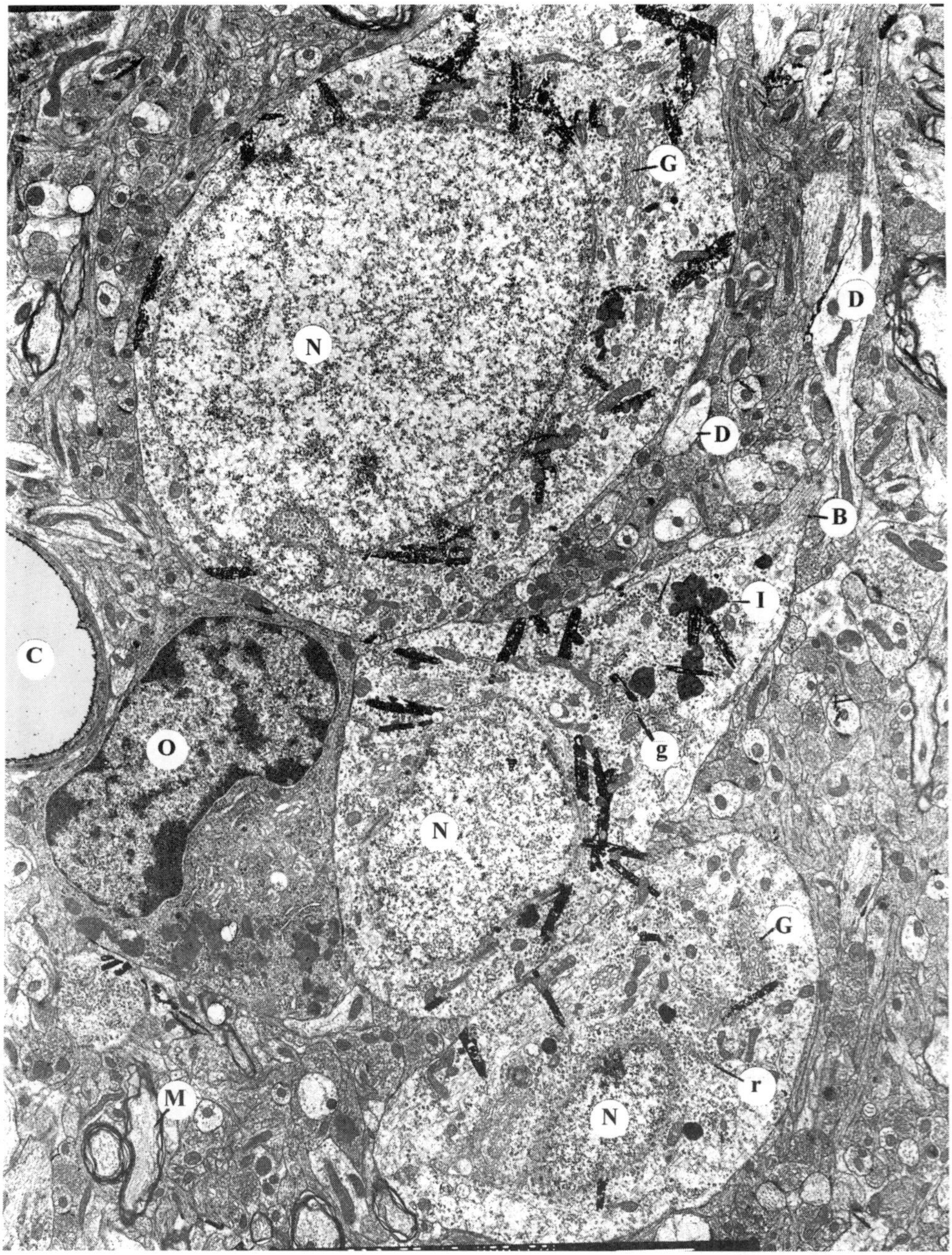

(a)

Figure 1.2 (a) Electron micrograph of three neuronal cell bodies in the anterior thalamic nucleus of the rat retrogradely labelled with horseradish peroxidase conjugated with cholera toxin B (dark bars) injected into the posterior cingulate cortex. N = nucleus of neurons, O = nucleus of oligodendrocyte, C = capillary, D = dendrite, G = Golgi apparatus, M = myelinated fibre, r = ribosome, l = lipofuscin pigment, g = granular endoplasmic reticulum. Picture kindly provided by Professor A. R. Lieberman (University College London). Reproduced by permission of Springer-Verlag GmbH & Co. KG, from Wang *et al.*, *Exp. Brain Res.* **126**: 369–382 (1999)

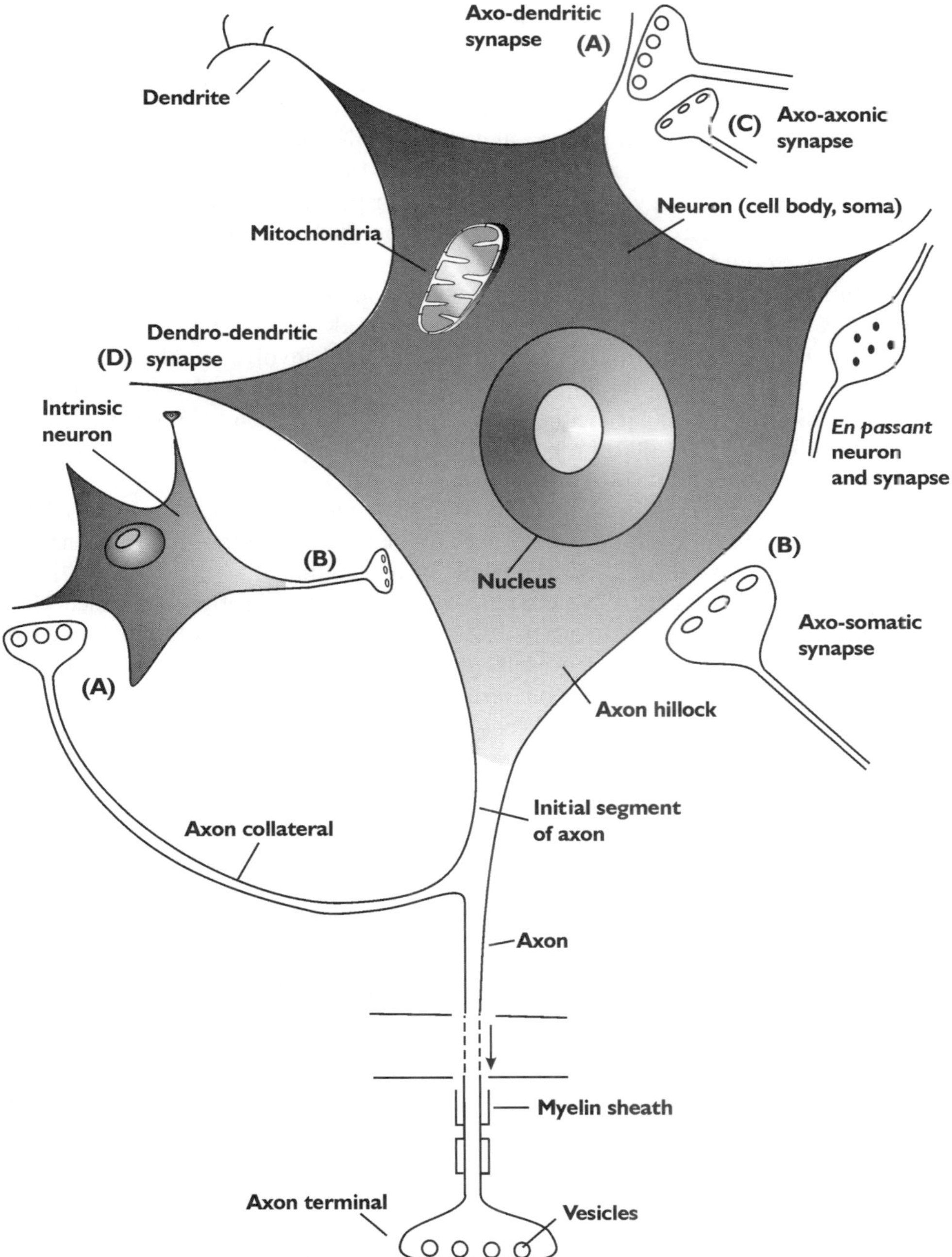

Figure 1.2 (b) Schematic representation of a neuron. The main features of a neuron are shown together with different synaptic arrangements (A) axo-dendritic, (B) axo-somatic, (C) axo-axonic and (D) dendro-dendritic. For more detail see section on 'Morphological correlates of synaptic function' and Fig. 1.7

terminal are also found (axo-axonal) and occasionally even between dendrites (dendro-dendritic) (see Fig. 1.2(b)). The morphology of synapses is considered later.

Like other cells, a neuron has a nucleus with genetic DNA, although nerve cells cannot divide (replicate) after maturity, and a prominent nucleolus for ribosome synthesis. There are also mitochondria for energy supply as well as a smooth and a rough endoplasmic reticulum for lipid and protein synthesis, and a Golgi apparatus. These are all in a fluid cytosol (cytoplasm), containing enzymes for cell metabolism and NT synthesis and which is surrounded by a phospholipid plasma membrane, impermeable to ions and water-soluble substances. In order to cross the membrane, substances either have to be very lipid soluble or transported by special carrier proteins. It is also the site for NT receptors and the various ion channels important in the control of neuronal excitability.

Microtubules (about 20 nm in diameter) and solid neurofilaments (10 nm) extend from the cell body into the axon and are found along its length, although not continuous. They give structure to the axon but are not involved in the transport of material and vesicles to the terminal, which despite its high level of activity does not have the facility for molecular synthesis possessed by the cell body. Such transport is considered to be fast (200–400 mm per day), compared with a slower transport (1 mm per day) of structural and metabolic proteins. Although axonal flow is mainly towards the terminal (ortho or anterograde) there is some movement (fast) of waste material and possibly information on synaptic activity back to the cell body (retrograde). The neuron is obviously very active throughout the whole of its length.

In addition to neurons the CNS contains various neuroglia (often just called glia). These can outnumber neurons by up to 10:1 in some areas and include star-like astrocytes with their long cellular processess which not only enable them to provide structural support for the nerve cells but also facilitate NT degradation and the removal of metabolites. Oligodendrites are glial cells which are involved in myelin formation and although they also have long processes, these are spirally bound rather than extending out as in the astocytes.

Neurons and glia are bathed in an ion-containing protein-free extracellular fluid which occupies less of the tissue volume (20%) in the brain than in other organs because of the tight packing of neurons and glia. In fact the whole brain is really suspended in fluid within its bony casing. The brain and spinal cord are covered by a thin close-fitting membrane, the pia mater and a thicker loose outer membrane, the dura mater. In the space between them, the subarachnord space, is the cerebrospinal fluid (CSF). This also flows into a series of ventricular spaces within the brain as well as a central canal in the cord and arises mainly as a secretion (ultra filtrate) of blood from tufts of specialised capillaries (the choroid plexus), which invaginate the walls of the ventricles. While the CSF is contiguous with the extracellular fluid within the brain and contributes to it, much of this fluid comes directly from the copious network of capillaries found throughout the brain. In fact neurons are never far from a capillary and their high metabolic rate means that despite contributing only 2% towards body weight, the nervous system receives 15% of cardiac output.

In most parts of the body, substances, other than large molecular ones like proteins, are filtered from the blood into the extracellular space through gaps between endothelial cells in the capillary wall. Such gaps are much narrower, almost non-existent, in brain capillaries and it is likely that any filtering is further reduced by the manner in which astrocytes pack around the capillaries. This constraint is known as the blood–brain barrier (BBB). It protects the brain from inappropriate substances, including all NTs and many drugs. To enter the brain as a whole is therefore almost as difficult for a substance as entering a neuron and again it has to be either very lipid soluble, when it can dissolve in and so pass through the capillary wall, or be transported across it.

NEUROTRANSMITTER FUNCTION

BASIC CIRCUITRY

In a classical neural pathway, such as that depicted in Fig. 1.3, neuron A must excite neuron B and at the same time inhibit neuron C in order to optimise the excitation of B. It could achieve this with one NT able to activate receptors linked to different events on B and C. Of course, neuron C would have other inputs, some of which would be excitatory and if the same NT was used it could activate the inhibitory mechanism on C as well. Also, the NT released from A might be able to stimulate as well as inhibit neuron C (Fig. 1.3(a)). Even the provision of separate receptors linked to excitation and inhibition would not overcome these problems since both would be accessible to the NT. One possible solution, used in the CNS, is to restrict the NT to the synapse at which it is released by structural barriers or rapid degradation. Also the inputs and receptors linked to excitation could be separated anatomically from those linked to inhibition and, in fact, there is electrophysiological and morphological evidence that excitatory synapses are mainly on dendrites and inhibitory ones on the soma of large neurons (Fig. 1.3(b)). Nevertheless, the problem of overlap would be eased if two NTs were released, one to activate only those receptors linked to excitation and another to evoke just inhibition, i.e. place the determinant of function partly back on the NT (Fig. 1.3(c)). This raises a different problem which has received much consideration. Can a neuron release more than one NT?

It was generally assumed that it cannot and this became known as Dale's Law. During his studies on antidromic vasodilation he wrote (1935) 'When we are dealing with two different endings of the same sensory neuron, the one peripheral and concerned with vasodilation and the other at a central synapse, can we suppose that the discovery and identification of a chemical transmitter at axon reflex dilation would furnish a hint as to the nature of the transmission process at a central synapse. The possibility has at least some value as a stimulus to further experiments'.

This it certainly has been and in the last few years much evidence has been presented to show that more than one substance (but not necessarily more than one conventional NT) can co-exist in one nerve terminal. This does not disprove Dale's Law (so called), since he was referring to 'a' not 'the' NT and to different endings of one neuron. In fact he was simply saying that if a neuron uses a particular transmitter at one of its terminals it will use it at another, although he did not add, irrespective of whether or not it uses more than one NT. This makes good sense especially since it is difficult to conceive how a neuron could achieve, let alone control, the release of different NTs from different terminals, unless the NTs were synthesised solely at the terminals independently of the cell body. In that way different substances might be released from different terminals of a neuron by arriving action potentials without the neuron having to do anything special

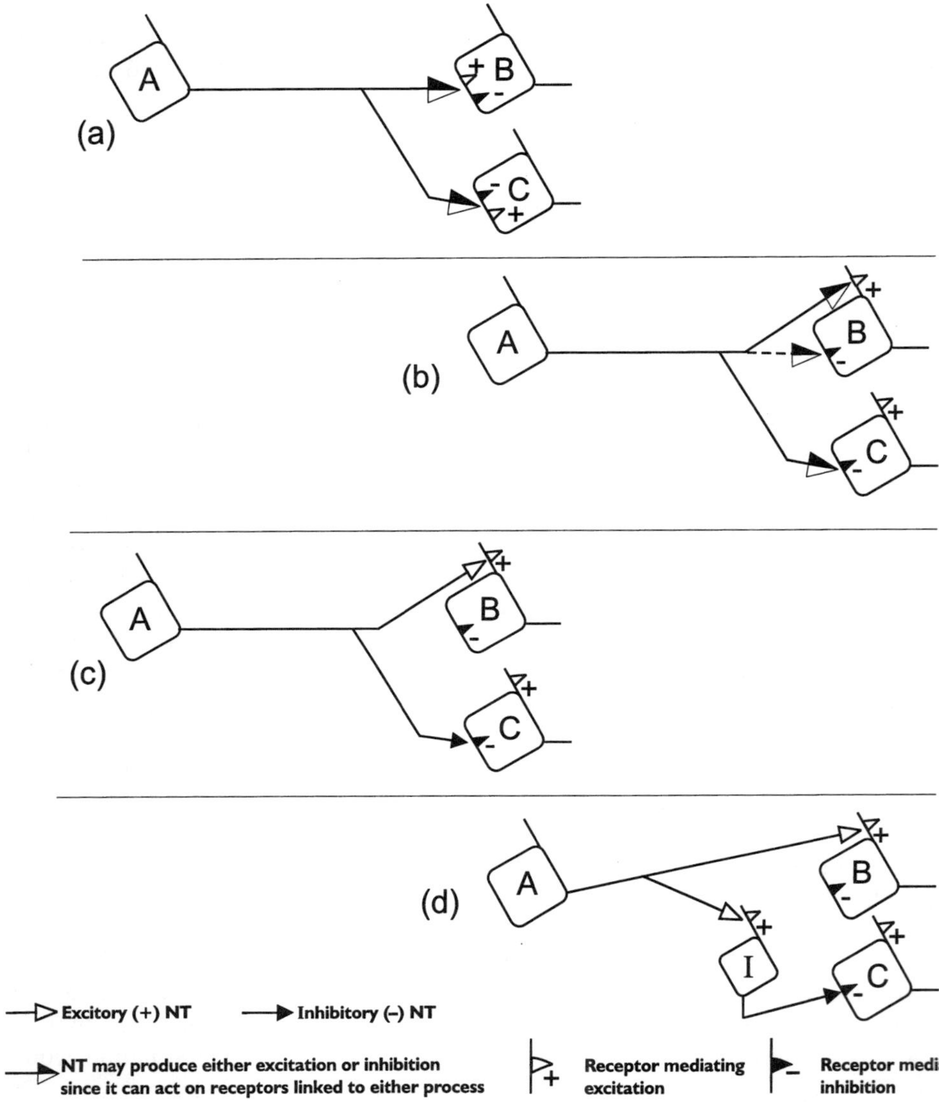

Figure 1.3 Some possible basic neurotransmitter–synaptic arrangements for the excitation and inhibition of different neurons. (a) The single NT activates neuron B and inhibits neuron C by being able to activate both excitatory and inhibitory receptors or, more probably, acting on one receptor linked to both events. There is potential, however, for the NT to activate any inhibitory receptors that may be on B or excitatory receptors on C. (b) The same NT is used as in (a) but the excitatory receptors are now only on dendrites and separated from the inhibitory receptors only on the soma. There is less chance of unwanted mixed effects. (c) Neuron A releases distinct excitatory and inhibitory NTs from its two terminals each acting on specific and morphologically separated receptors. But this depends on a neuron being able to release two NTs. (d) Neuron A releases the same NT from both terminals. It directly excites B but inhibits C through activating an inhibitory interneuron (I) which releases an inhibitory NT onto specific receptors on C. This last scheme (d) is clearly more functional and is widely used

to achieve it. Thus neuron A (Fig. 1.3) could then conceivably always release one NT at B and another at C or even two NTs at both but probably could not vary their release independently at different (or the same) synapses.

Fortunately there is another way in which one neuron can excite and inhibit different neurons using just one NT. Neuron A could excite B and inhibit C by the introduction of an inhibitory interneuron the activation of which by A, using the same excitatory NT as at B, automatically inhibits C (Fig. 1.3(d)). This form of inhibition is quite common in the CNS and in fact much inhibition is mediated by these so-called short-axon interneurons and a neuron may inhibit itself through feedback via an axon collateral synapsing onto an adjacent inhibitory short-axon interneuron (Fig. 1.2).

It might therefore be possible to set up a CNS with two NTs exerting fast excitatory and inhibitory effects through different receptors, situated on different parts of the neuron provided those were the only effects wanted. But this is not so. One neuron can receive hundreds of inputs and its activity and responsiveness is in fact balanced by such inputs producing different effects at differing speeds by using different NTs. So what are these different effects and how are they produced?

NEURONAL EXCITABILITY

POSTSYNAPTIC EVENTS

The neuronal membrane normally has a resting membrane potential around −70 mV (inside negative in respect of outside) with Na^+ and Cl^- concentrated on the outside and K^+ on the inside prepared to move down their concentration gradients when the appropriate ion channels are opened (Fig. 1.4). On arrival of an excitatory impulse the Na^+ channels are opened and there is an increased influx of Na^+ so that the resting potential moves towards the so-called equilibrium potential for Na^+ (+50 mV) when Na^+ influx equals Na^+ outflux but at −60 to −65 mV, the threshold potential, there is a sudden increase in Na^+ influx. This depolarisation leads to the generation of a propagated action potential. The initial subthreshold change in membrane potential parallels the action of the excitatory transmitter and is graded in size according to the amount of NT released. It is known as the excitatory postsynaptic potential (EPSP) and lasts about 5 ms.

An inhibitory input increases the influx of Cl to make the inside of the neuron more negative. This hyperpolarisation, the inhibitory postsynaptic potential (IPSP), takes the membrane potential further away from threshold and firing. It is the mirror-image of the EPSP and will reduce the chance of an EPSP reaching threshold voltage.

Such clear postsynaptic potentials can be recorded intracellularly with microelectrodes in large quiescent neurons after appropriate activation but may be somewhat artificial. In practice a neuron receives a large number of excitatory and inhibitory inputs and its bombardment by mixed inputs means that its potential is continuously changing and may only move towards the threshold for depolarisation if inhibition fails or is overcome by a sudden increase in excitatory input.

Not all influences on, or potentials recorded from, a neuron have the same time-course as the EPSP and IPSP, which follow the rapid opening of Na^+ and Cl^- ion channels directly linked to NT receptors. There are also slowly developing, longer lasting and smaller non-propagated (conditioning) changes in potential most of which appear to have a biochemical intermediary in the form of G-proteins linked to the activation (Gs) or

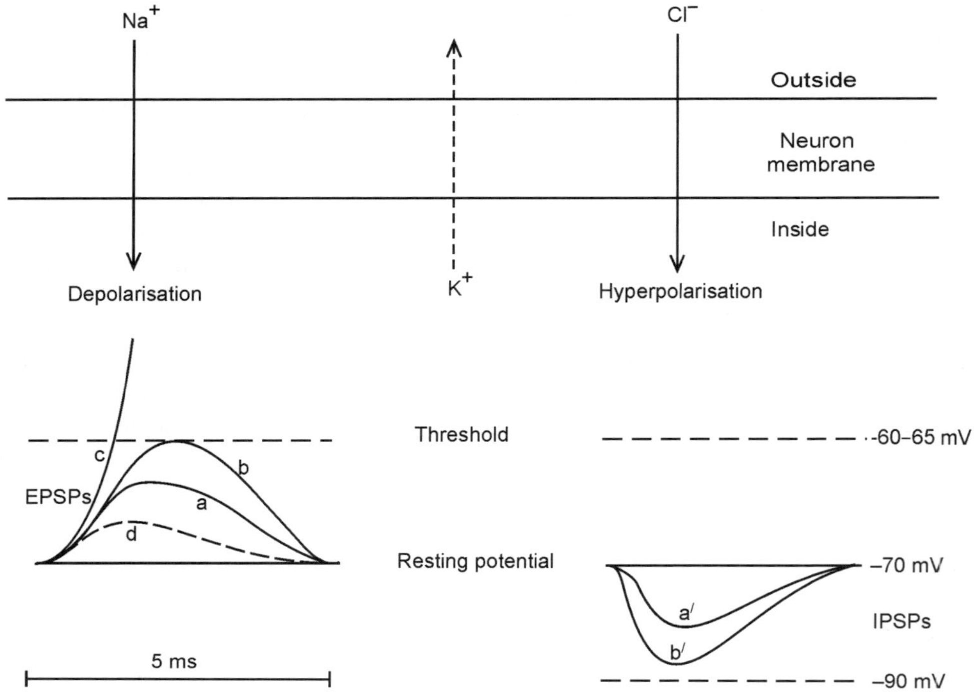

Figure 1.4 Ionic basis for excitatory postsynaptic potentials (EPSPs) and inhibitory postsynaptic potentials (IPSPs). Resting membrane potential (−70 mV) is maintained by Na^+ influx and K^+ efflux. Varying degrees of depolarisation, shown by different sized EPSPs (a and b), are caused by increasing influx of Na^+. When the membrane potential moves towards threshold potential (60–65 mV) an action potential is initiated (c). The IPSPs (a′b′) are produced by an influx of Cl^-. Coincidence of an EPSP (b) and IPSP (a′) reduces the size of the EPSP (d)

inhibition (Gi) of adenylate cyclase and cyclic AMP production or IP_3 breakdown (see Chapter 2). They can be excitatory (depolarising) or inhibitory (hyperpolarising) generally involving the opening or closing of K^+ channels. This can be achieved directly by the G-protein or second messenger but more commonly by the latter causing membrane phosphorylation through initiating appropriate kinase activity.

Thus the activity of a neuron can be controlled in a number of ways by NTs activating appropriate receptors (Fig. 1.5). Two basic receptor mechanisms are involved:

(1) **Ionotropic** Those linked directly to ion channels such as those for Na^+ (e.g. ACh nicotinic or some glutamate receptors) or Cl^- (GABA), involving fast events with increased membrane conductance and ion flux.

(2) **Metabotropic** Those not directly linked to ion channels but initiating biochemical processes that mediate more long-term effects and modify the responsiveness of the neuron. With these the first messenger of synaptic transmission, the NT, activates a second messenger to effect the change in neuron excitability. They are normally associated with reduced membrance conductance and ion flux (unless secondary to

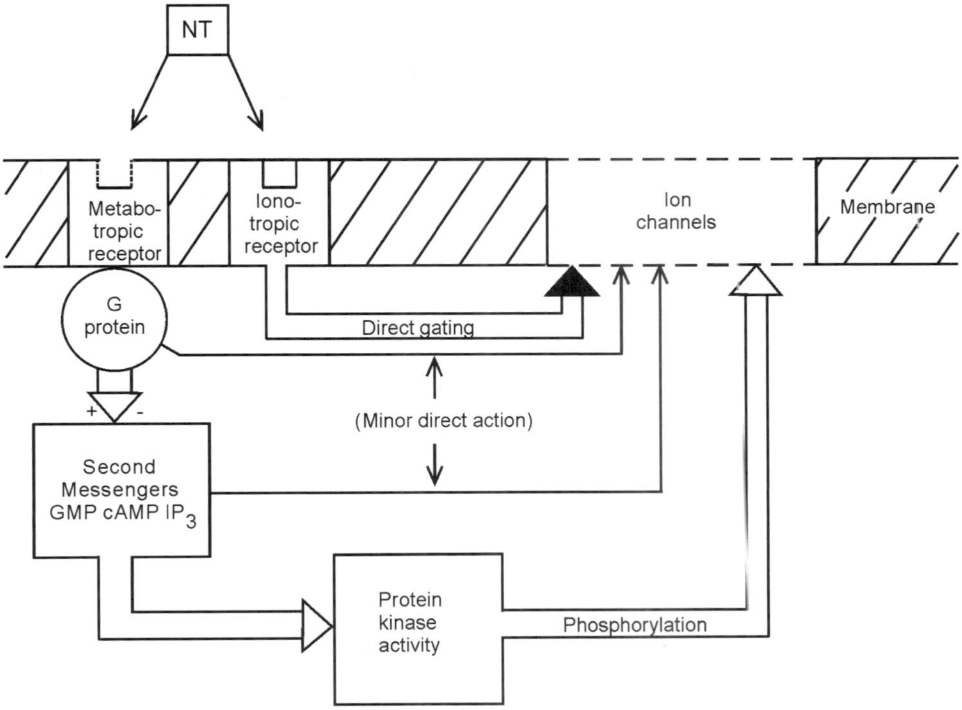

Figure 1.5 The degree of ion channel opening can be controlled (gated) either directly (ionotropic effect) or indirectly (metabotropic effect). In the former the neurotransmitter combines with a receptor that is directly linked to the opening of an ion channel (normally Na^+ or Cl^-) while in the latter the receptor activates a G-protein that can directly interact with the ion channel (most probably K^+ or Ca^{2+}) but is more likely to stimulate (Gs) or inhibit (Gi) enzymes controlling the levels of a second messenger (e.g. cAMP, GMP, IP_3). These in turn may also directly gate the ion channel but generally control its opening through stimulating a specific protein kinase that causes phosphorylation of membrane proteins and a change in state of the ion channel. The latter (metabotropic) effects may either open or close an ion channel (often K^+) and are much slower (100s ms to min) than the ionotropic ones (1–10 ms). A variety of neurotransmitters, receptors, second messengers and ion channels permits remarkably diverse and complex neuronal effects

an increased Ca^{2+} conductance) and may involve decreased Na^+ influx (inhibitory) or K^+ efflux (excitatory). Some amines (e.g. noradrenaline) may increase K^+ efflux (inhibitory).

These two basic mechanisms could provide a further classification for NTs, namely fast and slow acting, although one NT can work through both mechanisms using different receptors. The slow effects can also range from many milliseconds to seconds, minutes, hours or even to include longer trophic influences. What will become clear is that while one NT can modify a number of different membrane ion currents through different mechanisms and receptors, one current can also be affected by a number of different NTs. The control of neuronal excitability is discussed in more detail in Chapter 2.

PRESYNAPTIC EVENTS

So far we have assumed that a NT can only modify neuronal activity by a postsynaptic action. Recently, interest has also turned to presynaptic events. It has been known for many years that stimulation of muscle or cutaneous afferents to one segment of the spinal cord produces a prolonged inhibition of motoneuron activity without any accompanying change in conductance of the motoneuron membrane, i.e. no IPSP. Such inhibition is probably, therefore, of presynaptic origin and is, in fact, associated with a depolarisation of the afferent nerve terminals and a reduction in release of the excitatory NT. If it is assumed that the amount of NT released from a nerve terminal depends on the amplitude of the potential change induced in it, then if that terminal is already partly depolarised when the impulse arrives there will be a smaller change in potential and it will release less transmitter (Fig. 1.6). There is no direct evidence for this concept from studies of NT release but electrophysiological experiments at the crustacean neuromuscular junction, which has separate excitatory and inhibitory inputs, show that stimulation of the inhibitory nerve, which released GABA, reduced the EPSP evoked postsynaptically by an excitatory input without directly hyperpolarising (inhibiting) the muscle fibre. Certainly when GABA is applied to various *in vivo* and *in vitro* preparations (spinal cord, cuneate nucleus, olfactory cortex) it will produce a depolarisation of afferent nerve terminals that spreads sufficiently to be recorded in their distal axons.

Such presynaptic inhibition can last much longer (50–100 ms) than the postsynaptic form (5 ms) and can be a very effective means of cutting off one particular excitatory input without directly reducing the overall response of the neuron. How GABA can produce both presynaptic depolarisation and conventional postsynaptic hyperpolarisation by the same receptor, since both effects are blocked by the same antagonist, bicuculline, is uncertain (see Chapter 2) although an increased chloride flux appears to be involved in both cases. If nerve terminals are depolarised, rather than hyperpolarised by increased chloride flux, then their resting membrane potential must be different from (greater than) that of the cell body so that when chloride enters and the potential moves towards its equilibrium potential there is a depolarisation instead of a hyperpolarisation. Alternatively, chloride efflux must be achieved in some way.

This form of presynaptic inhibition must be distinguished from another means of attenuating NT release, i.e. autoinhibition. This was first shown at peripheral noradrenergic synapses where the amount of noradrenaline released from nerve terminals is reduced by applied exogenous noradrenaline and increased by appropriate (alpha) adrenoceptor antagonists. Thus through presynaptic (alpha) adrenoreceptors, which can be distinguished from classical postsynaptic (alpha) adrenoreceptors by relatively specific agonists and antagonists, neuronal-released noradrenaline is able to inhibit its own further (excessive) release. It is a mechanism for controlling the synaptic concentration of noradrenaline. This inhibition does not necessarily involve any change in membrane potential but the receptors are believed to be linked to and inhibit adenylate cyclase. Whether autoinhibition occurs with all NTs is uncertain but there is strong evidence for it at GABA, dopamine and 5-HT terminals.

There is also the interesting possibility that presynaptic inhibition of this form, with or without potential changes, need not be restricted to the effect of the NT on the terminal from which it is released. Numerous studies in which brain slices have been loaded with a labelled NT and its release evoked by high K^+ or direct stimulation show

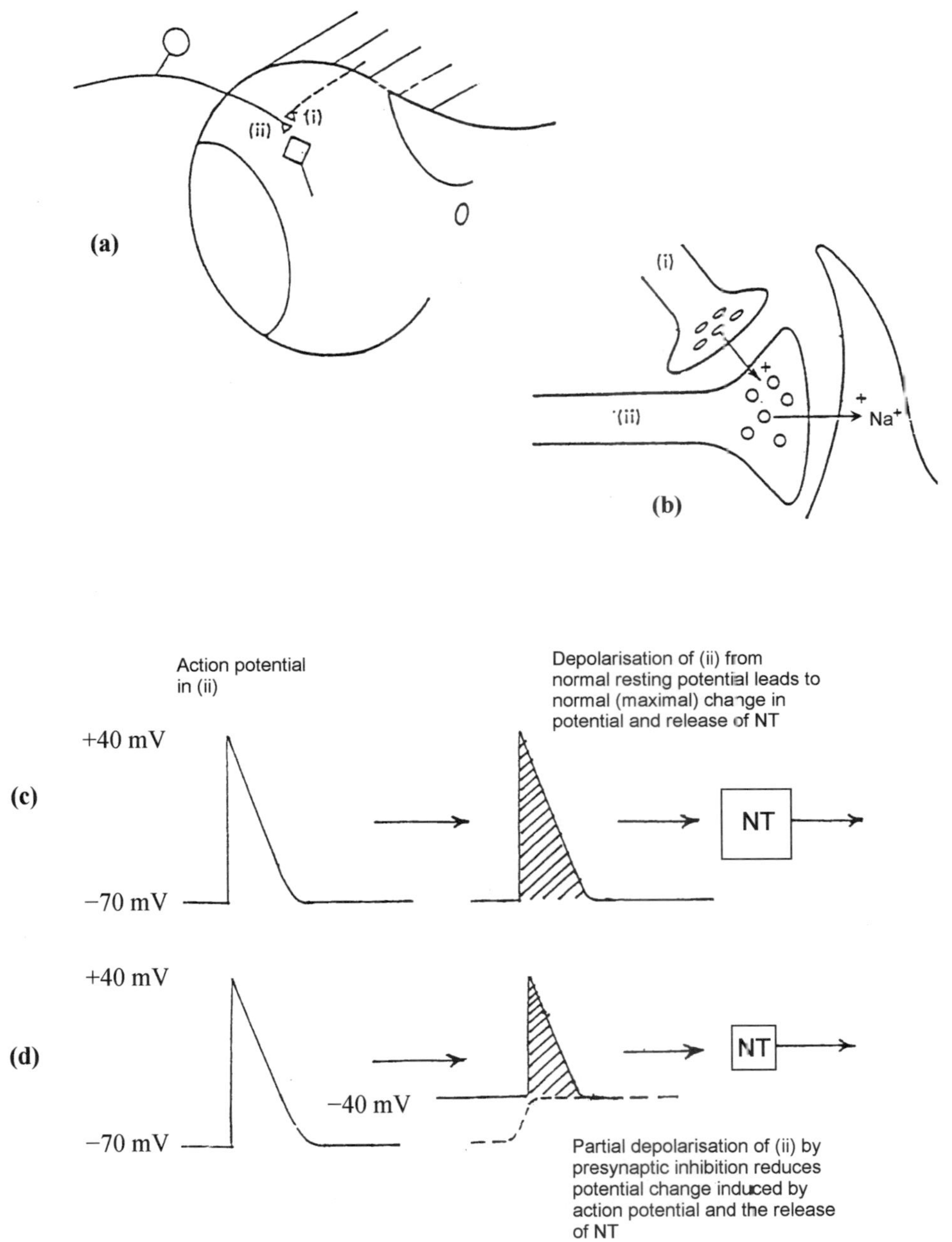

Figure 1.6 Presynaptic inhibition of the form seen in the dorsal horn of the spinal cord. (a) The axon terminal (i) of a local neuron is shown making an axo-axonal contact with a primary afferent excitatory input (ii). (b) A schematic enlargement of the synapse. (c) Depolarisation of the afferent terminal (ii) at its normal resting potential by an arriving action potential leads to the optimal release of neurotransmitter. (d) When the afferent terminal (ii) is already partially depolarised by the neurotransmitter released onto it by (i) the arriving acting potential releases less transmitter and so the input is less effective

that such release can be inhibited by a variety of other NTs. A noradrenergic terminal has been shown to possess receptors for a wide range of substances, so-called heteroceptors (see Langer 1981, 1997) and although this may be useful for developing drugs to manipulate noradrenergic transmission it seems unlikely that *in vivo* all of the receptors could be innervated by appropriate specific synapses or reachable by their NT. They may be pharmacologically responsive but not always physiologically active (see Chapter 4).

CONTROL OF SYNAPTIC NT CONCENTRATION

Having briefly discussed the presynaptic control of NT release it is necessary to consider how the concentration of a NT is controlled at a synapse so that it remains localised to its site of release (assuming that to be necessary) without its effect becoming too excessive or persistent.

Although one neuron can receive hundreds of inputs releasing a number of different NTs, the correct and precise functioning of the nervous system presumably requires that a NT should only be able to act on appropriate receptors at the site of its release. This control is, of course, facilitated to some extent by having different NTs with specific receptors so that even if a NT did wander it could only work where it finds its receptors and was still present in sufficient concentration to meet their affinity requirements. Normally the majority of receptors are also restricted to the immediate synapse.

Nevertheless, from release (collection) studies we know that enough NT must diffuse (overflow) to the collecting system, be that a fine probe *in vivo* or the medium of a perfusion chamber *in vitro*, to be detected. Thus one must assume that either the concentration gradient from the collecting site back to the active synaptic release site is so steep that the NT can only reach an effective concentration at the latter, or it is not unphysiological for a NT to have an effect distal from its site of release.

Released NT, if free to do so, would diffuse away from its site of release at the synapse down its concentration gradient. The structure of the synapse and the narrow gap between pre- and postsynaptic elements reduces this possibility but this means that there must be other mechanisms for removing or destroying the NT so that it, and its effects, do not persist unduly at the synapse but are only obtained by regulated impulse controlled release. In some cases, e.g. ACh, this is achieved by localised metabolising enzymes but most nerve terminals, especially those for the amino acids and monoamines, possess very high-affinity NT uptake systems for the rapid removal of released NT. In fact these are all Na^+- and Cl^--dependent, substrate-specific, high-affinity transporters and in many cases their amino-acid structure is known and they have been well studied. Transport can also occur into glia as well as neurons and this may be important for the amino acids. Of course, a further safeguard against an excessive synaptic concentration of the NT is the presence of autoreceptors to control its release.

Thus there are mechanisms to ensure that NTs neither persist uncontrollably at the synapse nor produce dramatic effects distal from it. Studies of glutamate release always show a measurable basal level (1–3 μM), although this may not all be of NT origin, and yet it is very difficult to increase that level even by quite intense stimulation. Whether this is a safeguard against the neurotoxicity caused by the persistent intense activation of neurons by glutamate (see Chapter 9), or just to ensure that neurons remain responsive to further stimulation is unclear, as is the mechanism by which it is achieved.

Despite the above precautions, it is still possible that NT spillover and extrasynaptic action may occur and indeed could be required in some instances. Thus the diffusion of glutamate beyond the synapse could activate extrasynaptic high-affinity NMDA or metabotropic receptors (Chapter 9) to produce long-lasting effects to maintain activity in a network. This may be important in long-term potentiation and memory effects. Crosstalk between synapses could also act as a back-up to ensure that a pathway functions properly (see Barbour and Hausser 1997).

MORPHOLOGICAL CORRELATES OF SYNAPTIC FUNCTION

Obviously different NTs have different synaptic actions and it is of interest to see to what extent there are morphological correlates for these differing activities.

As mentioned previously, an axon generally makes either an axo-dendritic or axo-somatic synapse with another neuron. Gray (1959) has described subcellular features that distinguish these two main types of synapse. Under the electron microscope, his designated type I synaptic contact is like a disk (1–2 μm long) formed by specialised areas of opposed pre- and postsynaptic membranes around a cleft (300 Å) but showing an asymmetric thickening through an accumulation of dense material adjacent to only the postsynaptic membrane. They are now often referred to as asymmetric synapses (Fig. 1.7). A type II junction is narrower (1 μm) with a smaller cleft (200 Å) and a more even (symmetric) but less marked membrane densification on both sides of the junction. In addition the presynaptic vesicles are generally large (300–600 Å diameter), spherical and numerous at the asymmetric type I synapse but smaller (100–300 Å), fewer in number and somewhat flattened or disk-like at the symmetric type II. Vesicles of varying shape can sometimes be found at both synapses, and while some differences are due to fixation problems, the two types of synapse described above are widely seen and generally accepted. They appear to be associated with fast synaptic events so that type I synapses are predominantly axo-dendritic, i.e. excitatory, and utilise glutamate while type II axo-somatic synapses are inhibitory generally utilising GABA, although the separation is not absolute. Asymmetric excitatory synapses outnumber GABA inhibitory symmetric synapses by up to 4:1, even though at such synapses there is usually only one actual synaptic junction whereas at the symmetrical inhibitory synapse there can be a number of such junctions — presumably to ensure adequate inhibitory control.

Unfortunately in routine EM (electron microscope) preparations one cannot identify the NT at individual synapses although structural features (shape of vesicle, asymmetric or symmetric specialisations) may provide a clue. At cholinergic synapses the terminals have clear vesicles (200–400 Å) while monoamine terminals (especially NA) have distinct large (500–900 Å) dense vesicles. Even larger vesicles are found in the terminals of some neuro-secretory cells (e.g. the neurohypophysis). One terminal can contain more than one type of vesicle and although all of them probably store NTs it is by no means certain that all are involved in their release.

Anatomical evidence can also be presented to support the concept of presynaptic inhibition and examples of one axon terminal in contact with another are well documented. These do not show the characteristics of either type I or II synapses but

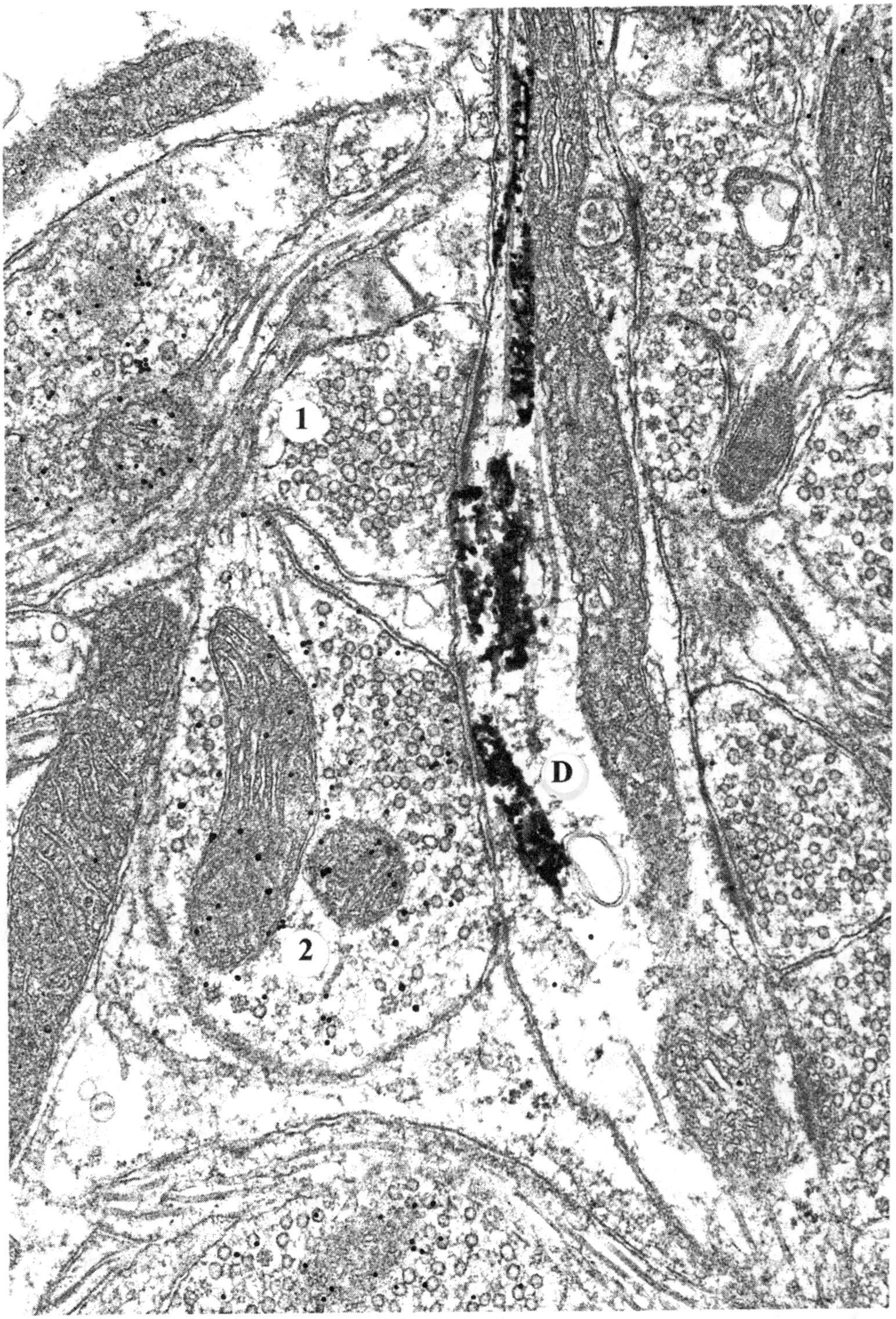

(a)

Figure 1.7 *Caption opposite*

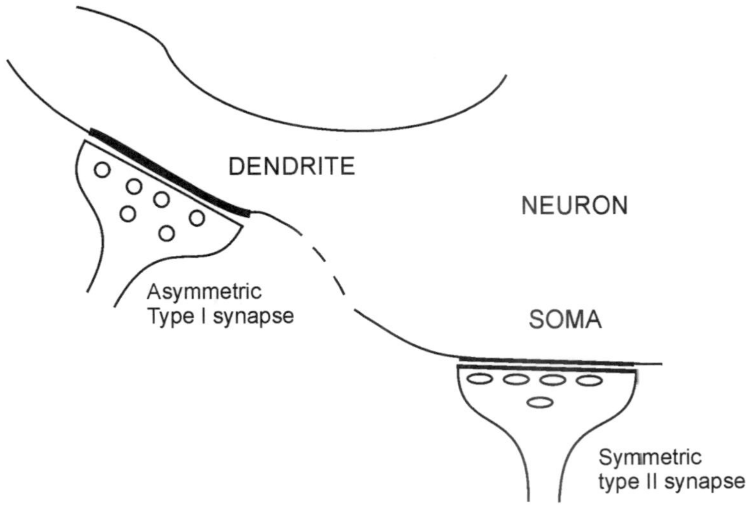

(b)

Figure 1.7 *continued* Different types of synaptic contact. (a) Symmetric and asymmetric synapses. The electromyograph from the anterior nuclear complex of the adult rat thalamus shows two terminals 1 and 2 establishing synaptic contact on the same dendrite. The electrodense material in the dendrite is a HRP reaction product and identifies the dendrite as that of a thalamocortical projection neuron (the tracer-cholera toxin B conjugated to HRP was injected into the cingulate cortex). Terminal 1 makes a prominent Gray type I (asymmetric) and terminal 2 a Gray type II (symmetric) synaptic contact. The latter is also labelled with gold particles indicating that despite the spherical vesicles obtained in the fixation procedure, it contains GABA since the material was immunoreacted with antibody against GABA (post-embedding immunogold method). The picture was kindly provided by Professor A. R. Lieberman (University College London). Reproduced from Wang *et al.*, *Brain Res. Bull.* **50**: 63–76 (1999) published by Elsevier Science. (b) Schematic representation of asymmetric (Gray type I) and symmetric (Gray type II) synapses. Asymmetric synapses are 1–2 μm long with a 30 nm (300 Å) wide cleft and very pronounced postsynaptic density. Presynaptic vesicles are round (30 nm diameter). Symmetric synapses are shorter (1 μm) with a narrower cleft (10–20 nm, 200 Å) and although the postsynaptic density is less marked it is matched by a similar presynaptic one. The presynaptic vesicles are more disk-like (10–30 nm diameter)

the shape of the presynaptic vesicle is of particular interest because even if the net result of activating this synapse is inhibition, the initial event is depolarisation (excitation) of the axonal membrane. This might suggest that the vesicles should be spherical but since the NT is GABA, normally an inhibitory transmitter, the vesicles could be flattened. Thus, does the type of synapse or the NT and its function determine the shape of the vesicle? Generally the vesicles at these axo-axonic synapses are flattened (or disk-like) but some have spherical vesicles and so while the situation is not resolved vesicle shape tends to be linked with the NT they house.

In the lateral superior olive, antibody studies have shown four types of axon terminal with characteristic vesicles (Helfert *et al.* 1992). Those with round vesicles contain glutamate, those with flattened vesicles have glycine, while large plemorphic vesicles contain glycine and GABA and small plemorphic ones only GABA. Interestingly when GABA and glycine were found in the same terminals in the spinal cord, the postsynaptic membrane had receptors to both NTs.

Dendro-dendritic synapses have also been described which show characteristic synaptic connections and we need to abandon the belief that one neuron can only influence another through its axon terminals. Dendro-dendritic synapses can also be reciprocal, i.e. one dendrite can make synaptic contact with another and apparently be both pre- and postsynaptic to it.

If NTs can have distal non-synaptic effects then nerve terminals that do not make definite synaptic connections could be apparent. In smooth muscle the noradrenergic fibres ramify among and along the muscle fibres apparently releasing noradrenaline from swellings (varicosities) along their length rather than just at distinct terminals. These fibres are termed *en passant* axons (see Fig. 1.2). In the brain many aminergic terminals also originate from *en passant* fibres but it seems that not all of them form classical synaptic junctions.

Monoamines can also be found in terminals at both symmetric and asymmetric synapses, but this may be partly because they co-exist with the classical transmitters glutamate and GABA. The fact that vesicular and neuronal uptake transporters for the monoamines can be detected outside a synapse along with appropriate postsynaptic receptors does suggest, however, that some monoamine effects can occur distant from the synaptic junction (see Pickel, Nirenberg and Milner 1996, and Chapter 6).

For further details on the concept of synaptic transmission and the morphology of synapses see Shepherd and Erulkar (1997) and Peters and Palay (1996) respectively.

NEUROTRANSMITTER ORGANISATION AND UTILISATION

In the periphery at the mammalian neuromuscular junction each muscle fibre is generally influenced by only one nerve terminal and the one NT acts on one type of receptor localised to a specific (end-plate) area of the muscle. The system is fitted for the induction of the rapid short postsynaptic event of skeletal muscle fibre contraction and while the study of this synapse has been of immense value in elucidating some basic concepts of neurochemical transmission it would be unwise to use it as a universal template of synaptic transmission since it is atypical in many respects.

In smooth muscle, by contrast, one sympathetic nerve fibre can influence a number of muscle fibres by releasing noradrenaline from varicosities along its length without there being any defined 'end-plate' junctions. The result of receptor activation is a slow change in potential and inactivation of the NT is initially by uptake and then metabolism. In other words, the NT function is geared to the slower phasic changes in tone characteristic of smooth muscle.

In the CNS there are many forms of neuronal organisation. One neuron can have many synaptic inputs and a multiplicity of NTs and NT effects are utilised within a complex interrelationship of neurons. There are also positive and negative feedback circuits as well as presynaptic influences all designed to effect changes in excitability and frequency of neuronal firing, i.e. patterns of neuronal discharge.

While we should try to exploit such differences between NT systems in developing drugs, rather than adopting a blanket concept of neurotransmission, it is still worth while trying to characterise different types of NT systems in the CNS in order to build up a functional framework and concept. The following patterns are suggested (see Fig. 1.8).

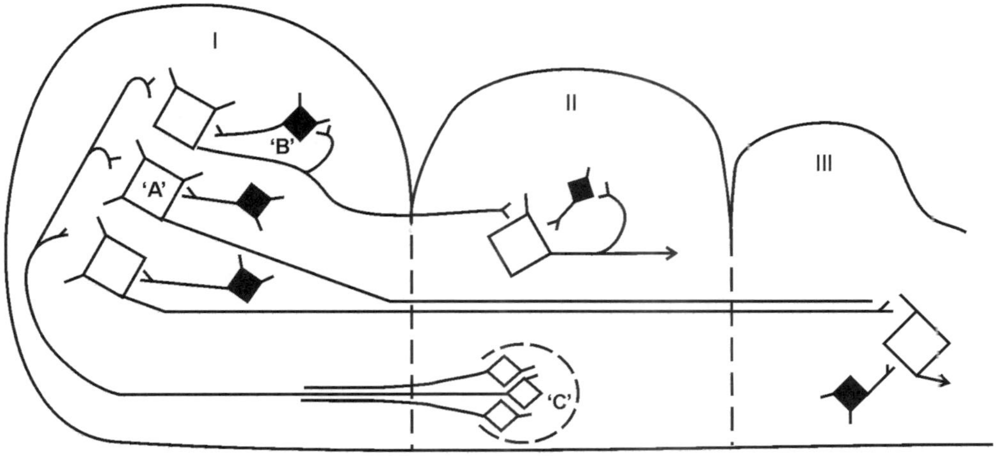

Figure 1.8 Some basic neuronal systems. The three different brain areas shown (I, II and III) are hypothetical but could correspond to cortex, brainstem and cord while the neurons and pathways are intended to represent broad generalisations rather than recognisable tracts. 'A' represents large neurons which have long axons that pass directly from one brain region to another, as in the cortico spinal or cortico striatal tracts. Such axons have a restricted influence often only synapsing on one or a few distal neurons. 'B' are smaller inter or intrinsic neurons that have their cell bodies, axons and terminals in the same brain area. They can occur in any region and control (depress or sensitise) adjacent neurons. 'C' are neurons that cluster in specific nuclei and although their axons can form distinct pathways their influence is a modulating one, often on numerous neurons rather than directly controlling activity, as with 'A'. Each type of neuron and system uses neurotransmitters with properties that facilitate their role

LONG-AXON (CLASSICAL) PATHWAYS (P1)

These include not only pathways with very long axons, such as the cortico-spinal and spino-thalamic tracts but also numerous shorter interconnecting systems, e.g. thalamus to cortex, etc. They may be regarded as the backbone of the CNS. The axons, especially the very long ones, show little divergence and have a relatively precise localisation, i.e. activation of particular motoneurons by stimulating a precise part of the motor cortex. Their influence on neurons is phasic and generally rapid with conventional EPSPs. Distinct axo-dendritic type I asymmetric synapses utilising glutamate acting on receptors (ionotropic) directly linked to the opening of N_a^+ channels are common and these systems form the basic framework for the precise control of movement and monitoring of sensation. Such pathways are well researched and understood by neuroanatomists and physiologists, but their localised organisation makes them, perhaps fortunately, somewhat resistant to drug action.

INTRINSIC CONTROLLING SYSTEMS (P2)

These are basically neurons whose cell body and axon terminals are both found in the same part of the CNS (Fig. 1.2). They are not concerned with transmitting information from one part of the CNS to another but in controlling activity in their own area. They can be excitatory but are more often inhibitory. They may act postsynaptically through conventional IPSPs (or slower potential changes) or presynaptically by modifying NT release. The former systems are generally thought to use amino acids as NTs, e.g. GABA or glycine acting on ionotropic receptors linked to Cl^- channels, while the latter systems may use GABA (presynaptic inhibition in cord) or peptides (enkephalin neurons). Excitatory interneurons may use ACh or an amino acid, like aspartate.

Since these interneurons exert a background control of the level of excitability in a given area or system their manipulation by drugs is of great interest (e.g. attemps to increase GABA function in epilepsy), especially if this can be achieved without adversely affecting important primary activity in the area. Although intrinsic neurons can only have a localised action they may be influenced by long-axon inputs to them and so incorporated into long pathway effects (Fig. 1.3(d)), such as the cortical inhibition of motoneurons.

MODULATING SYSTEMS (P3)

These have relatively long axons that originate from neurons that are grouped together in subcortical nuclei of perhaps a few hundred cell bodies but spread to vast areas of the brain and cord. The NTs, generally the monoamines noradrenaline, dopamine and 5-HT, are released at various sites along considerable lengths of the axon and distinct synaptic contacts may not always be seen. They may act either postsynaptically or presynaptically to produce slow changes in activity or modify NT release generally through secondary messenger systems. The tonic background influence of these systems and their role in behaviour have instigated the development and study of many drugs to manipulate their function. It also seems that the cholinergic input into the cortex from subcortical nuclei can also be included in this category (see Chapter 5).

Of course, while the identification of these distinct systems may be useful there are many neural pathways that would not fit easily into one of them. Thus some inhibitory pathways, such as that from the caudate nucleus to substantia nigra, utilising GABA, are not intrinsic neurons. The dopamine pathway from the substantia nigra to striatum may start from a small nucleus but unlike other monoamine pathways it shows little ramification beyond its influence on the striatum. The object of the above classification is not to fit all neural pathways and mechanisms into a restricted number of functional categories but again to demonstrate that there are different forms of neurotransmission.

CO-EXISTENCE (P4)

Although it may be argued that this is not a pattern of NT organisation but merely a feature of some (or possibly all) neurotransmitter systems, it justifies separate consideration. Since there is already good evidence for the existence of a fairly large number of different NTs, which it is assumed are released from their own specific neurons, and as they can produce a diversity of postsynaptic events one might consider the release of more than one NT from one terminal a somewhat unnecessary complication. Nevertheless since co-existence is established, its significance must be evaluated in respect of NT function and drug action. This is considered in more detail later (Chapter 12) but it is important to know which NTs co-exist and whether there is a definite pattern, i.e. does neurotransmitter A always occur with B and never with C and is the ratio A:B always the same? Also what effects do the NTs produce, how do they interact and are they both necessary for full synaptic transmission? The latter is a vital question for drug therapy based on NT replacement.

Thus it may be that a full understanding of how one NT works at a synapse will require knowledge of how that function depends on the actions of its co-released NT(s). It could unfold a whole new requirement and dimension to our understanding of synaptic physiology and pharmacology and the use of drugs. On the other hand, it may be of little significance in some cases for although cholinergic-mediated nicotinic and muscarinic responses as well as dopamine and peptide effects are observed in sympathetic ganglia, it is only nicotinic antagonists that actually reduce transmission, acutely anyway.

The brain could be likened to a television set in which the amino acids are providing the basic positive and negative power lines, while the other NTs (the multi-coloured wires) control the colour, contrast and brightness. All are required for a perfect picture but some are obviously more important than others.

FUNCTIONAL SYNAPTIC NEUROCHEMISTRY

To achieve their different effects NTs are not only released from different neurons to act on different receptors but their biochemistry is different. While the mechanism of their release may be similar (Chapter 4) their turnover varies. Most NTs are synthesised from precursors in the axon terminals, stored in vesicles and released by arriving action potentials. Some are subsequently broken down extracellularly, e.g. acetylcholine by cholinesterase, but many, like the amino acids, are taken back into the nerve where they are incorporated into biochemical pathways that may modify their structure initially but ultimately ensure a maintained NT level. Such processes are ideally suited to the fast transmission effected by the amino acids and acetylcholine in some cases (nicotinic), and complements the anatomical features of their neurons and the recepter mechanisms they activate. Further, to ensure the maintenance of function in vital pathways, glutamate and GABA are stored in very high concentrations (10 μmol/mg) just as ACh is at the neuromuscular junction.

By contrast, the peptides are not even synthesised in the terminal but are split from a larger precurser protein in the cell body or during transit down the axon. They are consequently only found in low concentrations (100 pmol/g) and after acting are broken down by peptidases into fragments that cannot be re-used. It is perhaps not surprising that they have a supporting rather than a primary role.

In between the above two extremes are the monoamines (1–10 nmol/g) which are preformed and stored in terminals but at much lower concentrations than the amino acids and when released are removed primarily by reuptake for re-use, or intraneuronal metabolism to inactive metabolites. Thus the appropriate synaptic organisation, biochemistry and receptor pharmacology of the NTs also varies in keeping with their function. It is often assumed, incorrectly, that the NTs found in the highest concentration are the most potent. In fact the opposite is true. Those like the amino acids while having high affinity for their receptors have low potency while the peptides found at much lower concentration have high potency but low affinity.

NEUROTRANSMITTER IDENTIFICATION

To achieve NT status a substance must fulfil three main criteria:

(1) *Presence*. It perhaps goes without saying that the proposed transmitter must be shown to be present in the CNS and preferably in the area and at the synapses where it is thought to act.
(2) *Release*. Stimulation of the appropriate nerves should evoke a measurable release of NT.
(3) *Identity of action*. The proposed NT must produce effects postsynaptically which are identical physiologically (appropriate membrane potential changes) and pharmacologically (sensitivity to antagonists) to that produced by neuronal stimulation and the relased endogenous NT.

These criteria should be regarded as guidelines rather than rules. As guidelines they provide a reasonable scientific framework of the type of investigations that must be undertaken to establish the synaptic role of a substance. As rigid rules they could preclude the discovery of more than one type of neurotransmitter or one form of neurotransmission. Nevertheless, the criteria have been widely employed and often expanded to include other features which will be considered as subdivisions of the main criteria.

PRESENCE

Distribution and concentration

It is generally felt that a substance is more likely to be a NT if it is unevenly distributed in the CNS although if it is widely used it will be widely distributed. Certainly the high concentration (5–10 μmol/g) of dopamine, compared with that of any other monoamine in the striatum or with dopamine in other brain areas, was indicative of its subsequently established role as a NT in that part of the CNS. This does not mean it cannot have an important function in other areas such as the mesolimbic system and parts of the cerebral cortex where it is present in much lower concentrations. In fact the concentration of the monoamines outside the striatum is very much lower than that of the amino acids but since the amino acids may have important biochemical functions that necessitate their widespread distribution, the NT component of any given level of amino acid is difficult to establish.

Nevertheless, useful information can be deduced from patterns of distribution. Glycine is concentrated more in the cord than cortex and in ventral rather than dorsal grey or white matter. This alone would be indicative of a NT role for glycine in the ventral horn, where it is now believed to be the inhibitory transmitter at motoneurons. GABA, on the other hand, is more concentrated in the brain than in the cord and in the latter it is predominantly in the dorsal grey so that although it is an inhibitory transmitter like glycine it must have a different pattern of activity.

Lesions in conjunction with concentration studies can also be useful. Section of dorsal roots and degeneration of afferent fibres produces a reduction in glutamate and substance P which can then be associated with sensory inputs. Temporary reduction of the blood supply to the cord causes preferential destruction of interneurons and a greater loss of asparate and glycine, compared with other amino acids and so links those amino acids with interneurons. Intrinsic neurons can also be destroyed through overactivity caused by kainic acid injections.

Subcellular localisation

A NT might be expected to be concentrated in nerve terminals and this can be ascertained since when nervous tissue is appropriately homogenised the nerve endings break off from their axons and surrounding elements and then reseal. Such elements are known as synaptosomes. They have been widely used to study NT release *in vitro* (Chapter 4) and some NT should always be found in them, at least if it is released from vesicles.

Synthesis and degradation

If a substance is to be a NT it should be possible to demonstrate appropriate enzymes for its synthesis from a precursor at its site of action, although peptides are transported to their sites of location and action after synthesis in the axon or distal neuronal cell body. The specificity of any enzyme system must also be established, especially if they are to be modified to manipulate the levels of a particular NT, or used as markers for it. Thus choline acetyltransferase (ChAT) may be taken as indicative of ACh and glutamic acid decarboxylase (GAD) of GABA but some of the synthesising enzymes for the monoamines lack such specificity.

After release there must be some way of terminating the action of a NT necessitating the presence of an appropriate enzyme and/or uptake mechanism. Such uptake processes can be quite specific chemically. Thus a high-affinity uptake (activated by low concentrations) can be found for glycine in the cord where it is believed to be a NT, but not in the cortex where is has no such action. This specific uptake can be utilised to map terminals for a particular NT, especially if it can be labelled, and also for loading nerves with labelled NT for release studies.

Of course, since CNS function depends on changes in the rate of neuronal firing, determined by a subtle balance between a number of different excitatory and inhibitory inputs, it may not always be necessary to destroy the NT rapidly. Excessive firing of a neuron may be controlled by activating a feedback inhibitory system or evoking presynaptic inhibition. There is also evidence for the release of the degrading enzyme together with NT at some purinergic (ATP) synapses (Kennedy *et al.* 1997) and possibly some cholinergic ones.

Pathways

If a substance (or its synthesising or degradative systems) can be demonstrated in particular neurons with a distinctive pattern of distribution, or bunched together into a well-defined nerve tract and/or nucleus, then this is not only good evidence for its role as a NT but it tells us something of its function. Indeed the distinct patterns of distribution of ascending monoamine pathways from brainstem nuclei could probably be considered as adequate evidence alone for their neurotransmitter role. In practical terms we can, of course, only study the release (and actions) of an endogenous NT if it can be evoked by stimulating an appropriate nerve pathway. Also the neurological and behavioural consequences of lesioning such pathways can tell us much about the functions of the NT. It is therefore useful to try to map NT pathways.

Receptors

If a NT is to be effective, there must be receptors for it to act on. Thus demonstrating the presence of receptors for the proposed NT at sites where it is found is further proof of its NT role. This could be done by recording some effect of the NT, e.g. change in neuronal firing, by establishing specific binding sites for it using it in a labelled form, or showing the presence of its receptor mRNA. Unfortunately a substance can bind to sites other than a receptor (e.g. uptake sites) and not all receptors are innervated.

RELEASE

A substance cannot be considered as a NT unless it is released. Unfortunately, although it may be possible to show the presence of a substance and some effect when it is applied directly to neurons its release may not be measurable for technical reasons. This is even more true if one strives for the ideal of demonstrating the release of an endogenous substance by physiological stimuli.

In the CNS access to the site of release is a major problem and attempts to achieve it have led to the development of a wide range of techniques of varying complexity and ingenuity or to short-cuts of dubious value (see Chapter 4). The feasibility of release studies in the CNS is to some extent dependent on the type of NT being studied. If we are dealing with a straightforward neural pathway with a number of axons going from A to B then by stimulating A and perfusing B we should be able to collect the NT. Unfortunately such arrangements are rare in the CNS and where they exist (e.g. corticospinal tract) it is not easy to perfuse the receiving (collecting) area. Sometimes the origin of a pathway is clear and easy to stimulate, e.g. NA fibres in the locus coeruleus, but fibre distribution in the cortex is so widespread that collection of sufficient amounts for detection can be very difficult, although current methods are beginning to achieve it.

These approaches are, in any case, only suitable for classical neurotransmitters. Those with slow background effects will probably not be released in large amounts. For such substances we require a measure of their utilisation, or turnover, over a much longer period of time. With NTs released from short-axon interneurons there are no pathways to stimulate and it becomes necessary to activate the neurons intrinsically by field stimulation, which is of necessity not specific to the terminals of the interneurons.

Apart from actually demonstrating release it is important to consider how NTs are released and whether they all need to be released in the same way, especially if they do different things. The variable time-courses of NT action referred to previously may require NTs to be released at different rates and in different ways, only some of which are achievable by, or require, vesicular mechanisms and exocytosis (see Chapter 4).

It should be remembered that with the possible exception of voltammetry when the monitoring electrode is sufficiently small to reach synapses, it is not the actual release of the NT that is being measured in perfusion studies. It is overflow. As discussed previously, most of any released NT is either physically restricted to the synapse or destroyed before it can diffuse away.

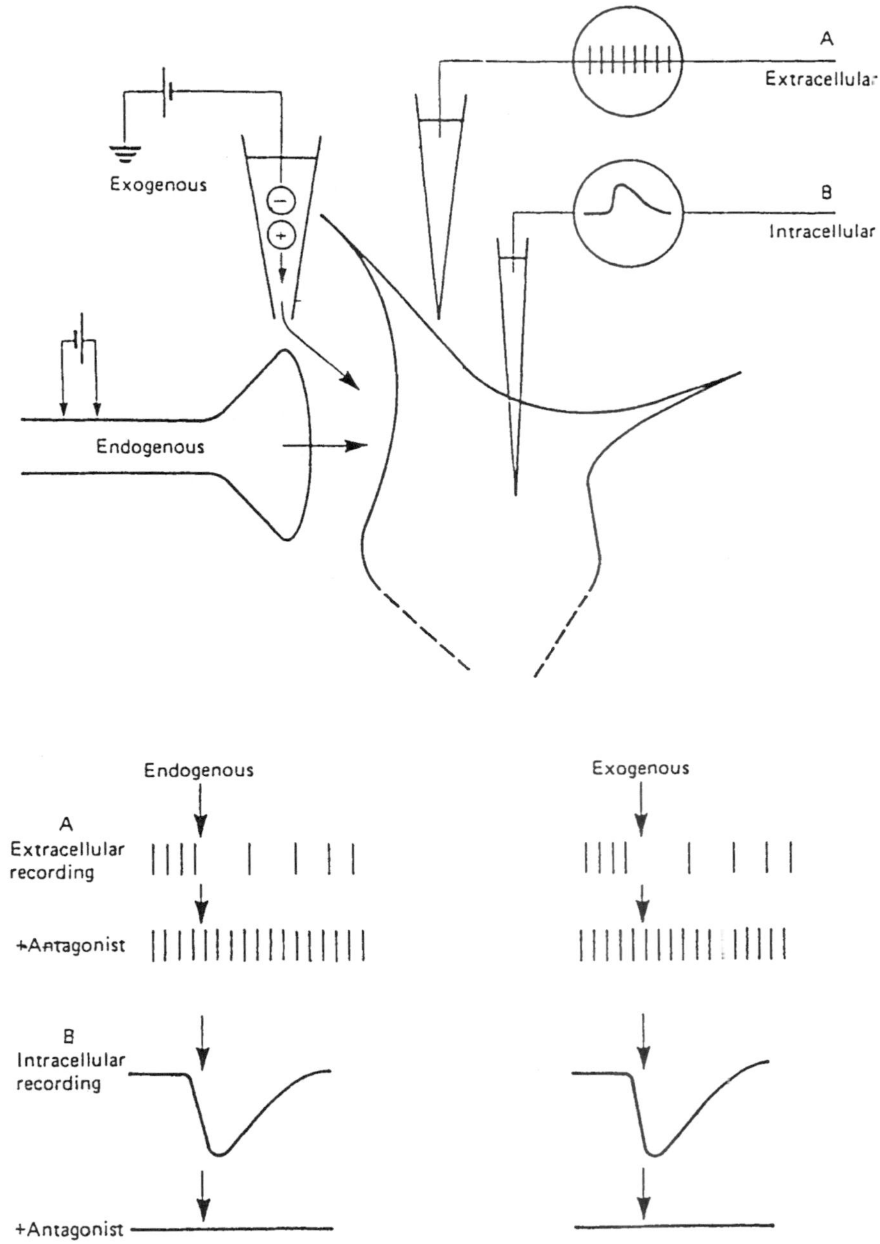

Figure 1.9 Comparison of the effects of an endogenously released and exogenously applied neurotransmitter on neuronal activity (identity of action). Recordings are made either of neuronal firing (extracellularly, A) or of membrane potential (intracellularly, B). The proposed transmitter is applied by iontophoresis, although in a brain slice preparation it can be added to the bathing medium. In this instance the applied neurotransmitter produces an inhibition, like that of nerve stimulation, as monitored by both recordings and both are affected similarly by the antagonist. The applied neurotransmitter thus behaves like and is probably identical to that released from the nerve

IDENTITY OF ACTION

Many people consider this to be the most important of all the criteria. Obviously a substance must have an effect of some kind if it is to be a NT but not all substances that have an effect on neurons need to be NTs. It may seem unnecessary to say this but the literature contains many accounts of the study of various substances on neuronal activity from which a NT role is predicted without any attempt to compare its effect with that of physiologically evoked (endogenous NT) effects. The importance of this safeguard is highlighted by the ease with which both smooth muscle and neurons will respond to a range of substances that are not released onto them as NTs. Thus the value of this criterion depends very much on the rigour with which it is applied and on its own is no more or less important than any other approach.

Ideally it should be shown that application of the proposed NT to a neuron, e.g. by iontopheresis (see Chapter 2), produces changes in membrane potential that are identical to and mediated by the same ionic mechanism as those produced by nerve stimulation and that the effects of both are equally overcome by an appropriate chemical antagonist. The basic system is outlined in Fig. 1.9. Clearly, changes in membrane potential can only be recorded if the neuron is large enough to take an intracellular electrode and even if it can be shown that the applied and released NT produce similar changes in membrane potential and share a common reversal potential and ionic mechanism this would not be so surprising, since the number of available ionic mechanisms is limited (i.e. both GABA and glycine produce hyperpolarisation by increasing chloride influx). Now that the properties of single ion channels can be recorded using modern patch-clamp techniques it will be necessary to show that application of the presumed NT produces identical changes in the frequency (n), degree (y, amount of current conducted) and duration (r) of channel opening to that achieved by synaptic activation. Unfortunately such a detailed analysis is presently only applicable to relatively simple systems with restricted innervations.

The use of antagonists is absolutely vital but even they can give false positives. Thus GABA, B-alanine and glycine all produce hyperpolarisation of cord motoneurons by increasing chloride influx but only GABA is unaffected by strychnine. Since strychnine abolishes inhibition in the cord, GABA cannot be the inhibitory NT but other features (distribution, release) had to be satisfied before glycine rather than B-alanine was shown to have that role.

It must be remembered that a substance can only be shown to be identical in its action with that of a particular endogenous NT if the latter's precise mode of action is clearly established and easily studied. Thus it may be relatively easy to consider those NTs mediating classical postsynaptic excitation through distinct potentical change but more difficult for NTs which function over a much longer time-course and possible without producing recordable potentical changes. Nevertheless they are still NTs. Or are they?

WHAT IS A NT?

The question is obviously an important one. Substances released from neurons are not always called neurotransmitters. Some of them are referred to as neuromodulators, neurohormones, neurotrophic factors or neurotoxins but since they all produce some effect on a neuron they could be said to have a transmitter role and justify the term neurotransmitter. On that basis every substance mentioned already and to be discussed further could be called a neurotransmitter. Even if we did try to distinguish between, say, a fast neurotransmitter and a slow neuromodulater effect, we have to realise that one substance can easily have both actions either at different synapses (ACh) or the same one (glutamate—ionotropic and metabotropic effects) and so could be both a neurotransmitter and a neuromodulater. If, however, the response they set up (transmit) has no reasonably quick and recordable effect, i.e. takes hours or days to develop (e.g. growth factors), or can actually kill the neuron (nitric oxide) it is difficult to conceive of them as neurotransmitters, however interesting they may be.

What this discussion does highlight, however, is that some modification is required to the standard dictionary definition of a neurotransmitter given in the introduction to this chapter, which sees a NT as a substance that transmits the impulse from one neuron to another neuron (or excitable cell). A more comprehensive definition of a NT might be

> A substance preformed, stored and then released from a neuron by a calcium dependent exocytotic mechanism activated by invading action potentials which induces a change in excitability and function of an adjacent neuron without entering the bloodstream.

This description would cover the classical NTs such as glutamate, GABA, ACh, DA, NA, and 5-HT as well as some peptides and ATP. That is irrespective of whether the effect produced by them is basic to the actual process of transmitting an impulse from one neuron to another, as with glutamate and ACh, rapidly inducing inhibition (GABA) or just making the neuron more or less responsive to other inputs (monoamines, peptides).

There is no room within the definition for nitric oxide, the prostaglandins and steroids mainly because they are not released in a controlled manner by neuronal activity and only the last are preformed. Thus if they are to be classified as NTs then the definition must be simplified so that a NT becomes:

> A substance produced in and released from a neuron to affect some aspect of neuronal function without being transported in the blood.

This would encompass those steroids synthesised in the CNS but not those entering it from the circulation (see Chapter 13).

While I feel that substances which only meet the abbreviated definition do not justify being called neurotransmitters, they will be treated as such in this text because this is the accepted practice and they are substances released from neurons to affect CNS function. Clearly, it is more important to distinguish between the different effects that a substance can produce when released from a nerve than to worry about what it is called. Nevertheless, it is unfortunate that the word neurotransmitter will inevitably be associated with the actual transmission of activity from one neuron to another and yet most of the substances we will be discussing do not actually do that.

REFERENCES

Barbour, B and Hausser, M (1997) Intersynaptic diffusion of neurotransmitter. *Trends Neurosci.* **20**: 377–384.

Brezina, V and Weiss, KR (1997) Analysing the functional consequences of transmitter complexity. *Trends Neurosci.* **11**: 538–543.

Dale, HH (1935) Pharmacology and nerve endings. *Proc. Roy. Soc. Med.* **28**: 319–332.

Gray, EG (1959) Axo-somatic and axo-dendritic synapses of the cerebral cortex. An electron-microscope study. *J. Anat.* **93**: 420–433. See also Gray, EG (1976) Problems of understanding the substructure of synapses. *Progr. Brain Res.* **45**: 208–234.

Helfert, RH, Juiz, JM, Bledsoe, SC, Bonneau, JM, Wenthold, RJ and Altschuler, RA (1992) Patterns of glutamate, glycine and GABA immunolabelling in four synaptic terminal classes in the lateral superior olive of the guinea pig. *J. Comp. Neurol.* **323**: 305–325.

Kennedy, C, Todorov, LD, Mihaylova-Todorova, S and Sneddon, P (1997) Release of soluble nucleotidases; a novel mechanism for neurotransmitter inactivation. *Trends Pharmacol. Sci.* **18**: 263–266.

Langer, SZ (1981) Presynaptic regulation of the release of catecholamines. *Pharmac. Rev.* **32**: 337–362.

Langer, SZ (1997) 25 years since the discovery of presynaptic receptors; present knowledge and future perspectives. *Trends Pharmacol. Sci.* **18**: 95–99.

Peters, A and Palay, SL (1996) The morphology of synapses. *J. Neurocytol.* **25**: 687–700.

Pickel, VM, Nirenberg, MJ and Milner, TA (1996) Ultrastructural view of central catecholaminergic transmission; immunocytochemical localization of synthesising enzymes, transporters and receptors. *J. Neurocytol.* **25**: 843–856.

Shepherd, GM and Erulkar, SD (1997) Centenary of the synapses from Sherrington to the molecular biology of the synapse and beyond. *Trends Neurosci.* **20**: 385–392.

2 Control of Neuronal Activity

D. A. BROWN

The task of nerve cells is to communicate. They do this using a combination of electrical signals (action potentials) and chemical signals (transmission). However, even the chemical signal has to be transduced to an electrical signal (the synaptic potential) in order to continue the process of communication from one neuron to another. Information is then coded in the frequency and pattern of action potential discharges. This chapter considers the question of how these electrical signals are generated and how their frequency and discharge patterns can be regulated.

THE RESTING STATE: ION GRADIENTS, PUMPS AND POTENTIALS

The electrical signals are carried by the movement of charged ions across the cell membrane. This makes use of the potential energy stored across the cell membrane in the form of *ionic gradients*. Concentration gradients for the principal ions across a typical nerve cell membrane are indicated in Fig. 2.1(a). The cell interior has a high concentration of K^+ ions and a low concentration of Na^+, Cl^- and Ca^{2+} ions relative to the exterior.

The ionic gradients themselves are generated by ion 'pumps' (carriers) (Fig. 2.1(b)). Thus, the Na^+/K^+ exchange pump (Na^+/K^+ ATPase) in the outer membrane generates the primary Na^+ and K^+ gradients across the cell membrane. Other pumps (a Ca^{2+} ATPase and/or a Na^+/Ca^{2+} exchange pump) generate a high concentration gradient for Ca^{2+} ions. These pumps consume energy (in the form of ATP). It has been estimated that about 40% of the oxygen consumption of the brain is used to drive the Na^+/K^+ exchange pump.

There is also an *electrical gradient* across the membrane. At rest, the normal value of this potential (E_{rest}) in most nerve cells is around -70 mV (inside $-$ve). In general, the ion pumps themselves are not directly responsible for this (though they can contribute, since they are not electroneutral). Instead, it is due primarily to the passive diffusion of K^+ ions back out of the cell down the chemical concentration gradient previously set up by the Na^+/K^+ exchange pump, leaving a small +ve charge deficit on the inside of the membrane. However, if K^+ were the only ion involved, then, from the K^+ concentration gradient, the *Nernst equation* predicts that the membrane potential should be about -90 mV:

Neurotransmitters, Drugs and Brain Function. Edited by R. A. Webster

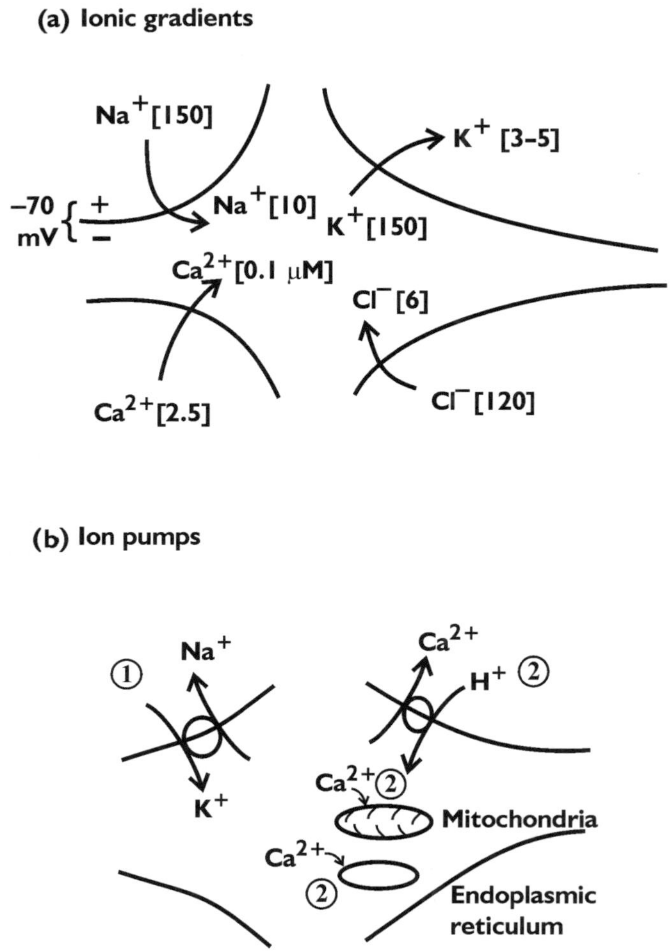

Figure 2.1 (a) Resting ionic gradients across a nerve cell membrane. Concentrations [ ] are in mM (except intracellular Ca^{2+}, in μM). Arrows show the direction of the electrochemical gradients for passive ionic movement. (b) Principal active ion pumps. (1): plasmalemmal Na^+/K^+ ATPase. (2) Ca^{2+} ATPases

$$E = RT/zF \ln([K^+]_{out}/[K^+]_{in})$$

This is the *equilibrium potential* for K^+ ions (E_K), i.e. the potential at which the electrical gradient pulling K^+ into the cell just balances the chemical concentration gradient forcing K^+ out of the cell. The 20 mV difference between E_{rest} and E_K is usually explained by assuming that the membrane is also slightly permeant to some other ion with a more positive equilibrium potential, such as Na^+. The membrane potential is then given by the *Goldman–Hodgkin–Katz (GHK)* or *constant-field equation*:

$$E = RT/zF \ln\{([K^+]_{out} + \alpha[Na^+]_{out})/([K^+]_{in} + \alpha[Na^+]_{in})\}$$

where α is the ratio of the permeability of Na^+ ions to that of K^+ ions (P_{Na}/P_K). The GHK equation then predicts a value of -70 mV for E_{rest} if the permeability of the membrane to Na^+ ions is about 4% of that to K^+ ions ($P_{Na}/P_K = 0.04$). However, it should be noted that, at this potential, although the fluxes of total cations in and out of the cell are equal, the cell will gradually accumulate Na^+ and lose K^+, which will have to be corrected by the Na/K exchange pump; since this involves energy expenditure, it is not a true equilibrium state. The GHK equation can be expanded to include terms for other ions, such as Cl^- ions, which can have a profound effect on the membrane potential under certain circumstances (e.g. during the activation of Cl^- channels by inhibitory neurotransmitters).

ION CHANNELS

In fact, a nerve cell membrane is not really 'permeant' to ions at all, in the sense that ions cannot diffuse across the lipid bilayer, since they cannot dissolve in the membrane lipids. Passive diffusion occurs entirely through *ion channels* — pore-forming membrane proteins. The resting potential may then be generated either by two sets of channels, one set permeant to K^+ and the other to Na^+ (or some other ion with a more positive equilibrium potential), with the former in the majority or opening more often; or a set of channels primarily permeant to K^+ but with some weak permeability to Na^+. Recently, a special class of 'resting' K^+ channels, the 'twin-pore' TASK channels, has been identified in some neurons, thus supporting the former proposition. Another class of K^+ channels that can contribute to the resting potential of neurons are inwardly-rectifying K^+ channels (Kir channels) — so-called because they conduct K^+ ions more readily into the cell than outwards.

Both TASK and Kir can be inhibited by certain neurotransmitters which act on receptors that couple to phospholipase C-activating G-proteins, such as acetylcholine, substance P and TRH. As a result, these transmitters can induce a sustained depolarisation of the receptive neurons (e.g. cerebellar granule cells and motor neurons by TASK-inhibition, basal forebrain neurons by Kir-inhibition). Several transmitters, acting on receptors coupled to other G-proteins of the Gi/Go family, are capable of *activating* another class of Kir channels (G-protein-gated inward rectifiers or GIRK channels, also known as Kir3), thereby *hyperpolarising* the neuron (and inhibiting it).

Other ion channels are closed at rest, but may be opened by a change in membrane potential, by intracellular messengers such as Ca^{2+} ions, or by neurotransmitters. These are responsible for the active signalling properties of nerve cells and are discussed below (see Hille 1992, for a comprehensive account). A large number of ion channels have now been cloned. This chapter concerns function, rather than structure, and hence does not systematically follow the structural classification.

THE ACTION POTENTIAL (Fig. 2.2)

This is the basic unit ('bit') of information processing in the nervous system. It is a transient electrical signal generated by the opening of *voltage-gated* Na^+ *channels*. These are normally shut at rest (or largely so), but are opened when the nerve cell membrane is depolarised by (e.g.) an excitatory transmitter. Since the entry of Na^+ ions further depolarises the membrane, so opening more Na^+ channels, the process becomes regenerative once the threshold potential is exceeded: this is the potential at which the rate of Na^+ entry exceeds the rate of K^+ efflux (and/or Cl^- entry). The membrane

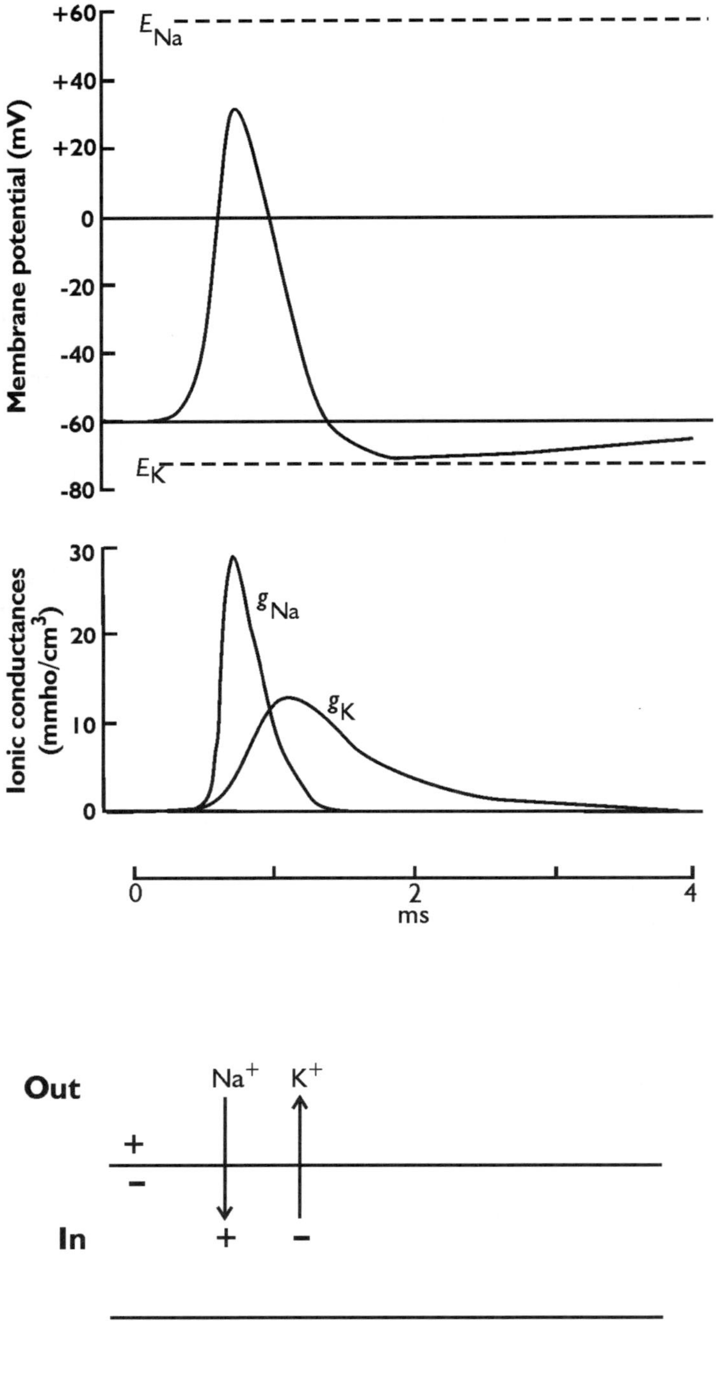

Figure 2.2 Ionic conductances underlying the action potential recorded from a squid axon. $g_{Na} = Na^+$ conductance; $g_K = K^+$ conductance. (Adapted from Hodgkin, AL and Huxley, AF (1952) *J. Physiol.* **117**: 500–544)

potential then moves transiently toward (but does not usually quite reach) the Na^+ equilibrium potential ($E_{Na} \sim +50$ to $+70$ mV; Fig. 2.2)—i.e. the membrane potential is reversed to inside-positive. *Repolarisation* results (in the first instance) from the *inactivation* of the Na^+ channels—that is, as the depolarisation is maintained, the channels close again (though at a slower rate than that at which they open). Recovery then requires that they progress back from the inactivated state to the resting closed state: this takes a little time, so the action potential becomes smaller and eventually fails during high frequency stimulation or during sustained depolarisation—a process of *accommodation*.

Local anaesthetics and some anti-epileptic drugs such as phenytoin and carbemazepine act by blocking Na^+ channels. Many of these have a higher affinity for the inactivated state of the Na^+ channel than for the resting or open states. Hence, by promoting inactivation, they selectively reduce high-frequency nerve impulses ('use-dependence'). This provides a rationale for the use of phenytoin and carbemazepine in controlling epileptic discharges.

In unmyelinated fibres (including the squid axon, where the ionic currents responsible for the action potential were first elucidated, see Fig. 2.2), and in unmyelinated regions of neurons, such as dendrites, somata and axon terminals, action potential repolarisation is accelerated by the delayed opening of additional voltage-gated K^+ channels—so-called *delayed rectifier* K^+ channels. These may be sustained or transient (inactivating) in kinetic behaviour. Since they take a few milliseconds to close as the potential recovers, in addition to hastening repolarisation, current flow through these channels leads to a transient after-hyperpolarisation ('undershoot') following each action potential. Where the action potential leads to the opening of voltage-gated Ca^{2+} channels (as in nerve terminals and neuron somata or dendrites—see below), the entry of Ca^{2+} also induces the rapid opening of large (100–200 pS in symmetrical high $[K^+]$) conductance ('BK') Ca^{2+}-activated K^+ channels, which can also accelerate action potential repolarisation. However, K^+ channels are normally absent from nodes of Ranvier and action potential repolarisation in myelinated fibres results solely from Na^+ channel inactivation. Thus, blocking K^+ channels with drugs such as tetraethylammonium or 4-aminopyridine (Fig. 2.3) does not affect conduction along myelinated fibres (though they can increase transmitter release, by prolonging the action potential in unmyelinated nerve terminals). They can also improve conduction in myelinated fibres following *demyelination* (e.g. in multiple sclerosis). This is because the action potential now has to be conducted along the demyelinated segments of the fibres (continuous conduction), instead of 'jumping' from node to node (saltatory conduction). (This is assisted by the spread of Na^+ channels from the nodes along the internodes after demyelination.) Since K^+ channels are normally present along internodal segments of myelinated fibres and the internodal Na^+ channel density is relatively low (even after demyelination), current through the K^+ channels tends to 'shunt' the Na^+ current and block internodal action potential conduction. Cooling the nerve has a similar effect to blocking K^+ channels: hence MS patients are very sensitive to temperature.

CALCIUM CHANNELS: TRANSMITTER RELEASE

When an action potential arrives at the axon terminal, it induces the release of a chemical transmitter. Transmitter release is a Ca^{2+}-dependent process (see Chapter 4) and requires a charge of Ca^{2+}. This is provided through the action potential-induced

Table 2.1 Types of calcium channel

Type	T	L	N	P/Q	R
α-subunit(s)	1G,H,I	1C,D	1B	1A	1E
Threshold[1]	Low	High	High	High	High
Inactivation	Fast	Slow	Moderate	None (P) Moderate (Q)	Fast
Location[2]	s/d	s/d	t,s/d	t,s/d	t,s/d
Blockers	Ni^{2+}	DHP[3]	ω-CTX-GVIA[4]	ω-Aga IVA[5]	
Main functions	Pacemaker	Spike	Transmitter release	Transmitter release	Transmitter release

[1]Low threshold around −60 mV; high threshold around −40 mV.
[2]s = soma; d = dendrites; t = axon terminals.
[3]Dihydropyridines.
[4]ω-Conotoxin GVIA.
[5]ω-Agatoxin IVA.

opening of *voltage-gated Ca^{2+} channels*. A variety of Ca^{2+} channels have been described, characterised by their kinetics, single-channel properties, pharmacology (especially sensitivity to different toxins) and molecular structure (Table 2.1). Those primarily responsible for transmitter release belong to the N (α1B), P/Q (α1A) and R classes (α1E). So far, no pharmacological agents capable of uniquely modifying Ca^{2+} channels involved in transmitter release have been described (other than polypeptide toxins). These, and other (L-type, T-type), Ca^{2+} channels are also variably present in neurons somata and/or dendrites, where they contribute to the regulation of neural activity in other ways (see below).

REGULATION OF Ca^{2+} CHANNELS BY NEUROTRANSMITTERS

N and P/Q channels are susceptible to inhibition by many neurotransmitters and extracellular mediators that act on receptors coupling to *Pertussis* toxin-sensitive G-proteins (primarily G_o)—for example, noradrenaline (via α2 receptors), acetylcholine (via M_2 and M_4 muscarinic receptors), GABA (via GABA-B receptors), opioid peptides (via μ/δ receptors) and adenosine (via A2 receptors) (see Fig. 2.4). Inhibition results from the release of the $\beta\gamma$ subunits of the trimeric ($\alpha\beta\gamma$) G-protein following its activation by the receptor. The $\beta\gamma$ subunit then binds to the Ca^{2+} channel in such a way as to shift its voltage sensitivity to more positive potentials, so that the channels do not open as readily during a rapid membrane depolarisation. This effect is 'reversible' in the sense that it can be temporarily reversed by applying a brief, strong depolarisation but then returns on rehyperpolarisation in the continued presence of free $\beta\gamma$ subunits (Fig. 2.4(a)). One interpretation of this is that the binding of the $\beta\gamma$ subunits is itself voltage-dependent. It should be noted that no 'second messenger' is necessary for this form of inhibition: instead, the $\beta\gamma$ subunit reaches, and binds to, a neighbouring Ca^{2+} channel when released from the activated G_o protein. (This G-protein is very abundant in nerve cell membranes.) As a result, the process of inhibition can be quite fast—within 30–50 ms following receptor activation. This is thought to provide the principal mechanism responsible for *presynaptic inhibition*, whereby neurotransmitters inhibit their own release (autoinhibition) during high-frequency synaptic transmission. This process can be replicated by applying exogenous transmitters or their analogues (see Fig. 2.4(b))

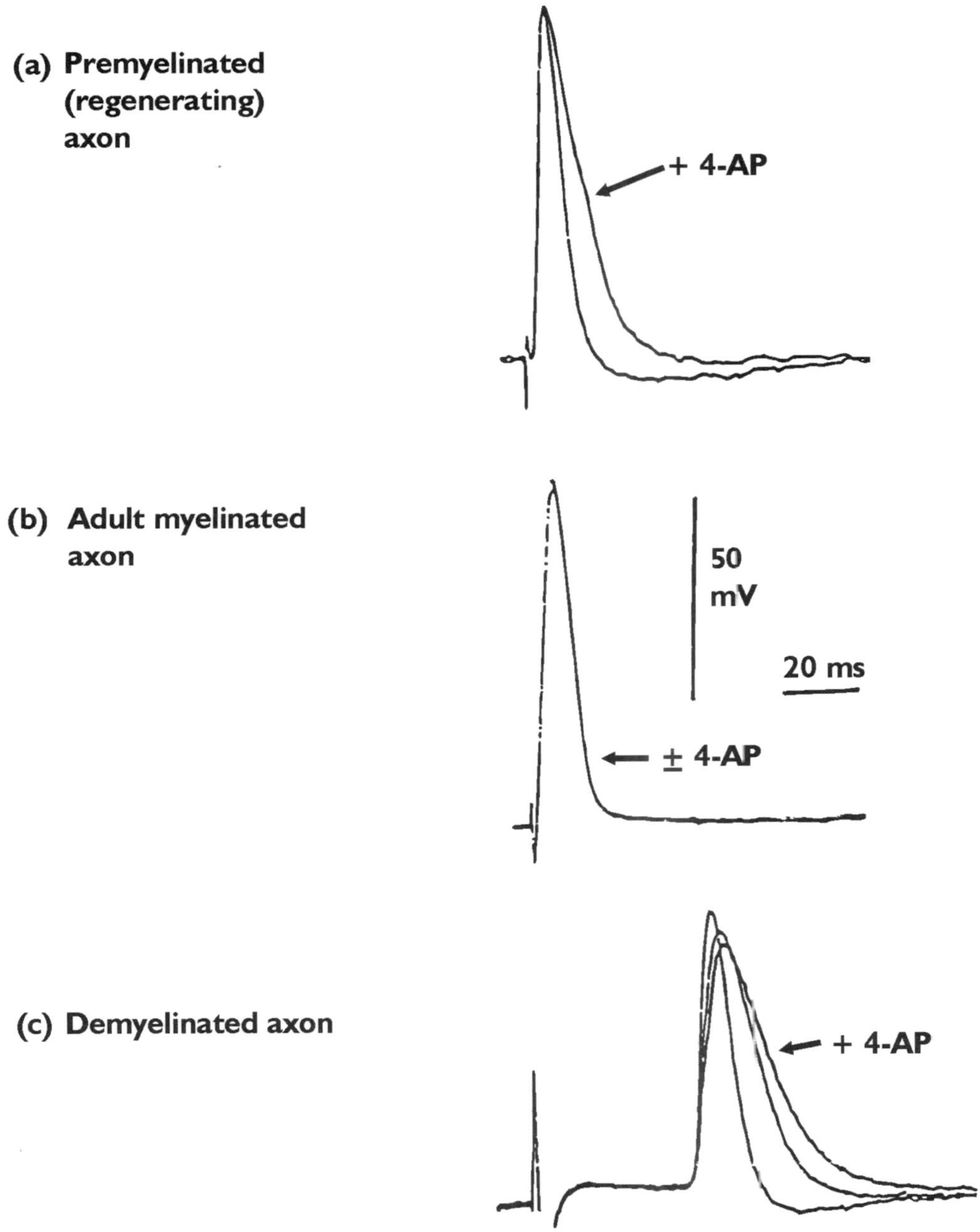

Figure 2.3 Role of K^+ channels in action potential repolarisation in mammalian axons, revealed using the K^+ channel blocking agent, 4-aminopyridine (4-AP, 0.5 mM). Records show intra-axonal recordings from (a) a regenerating sciatic nerve axon following nerve crush; (b) a normal sciatic nerve axon; and (c) a demyelinated ventral root axon after treatment with lysophosphatidylcholine. Note that 4-AP prolongs the action potential in (a) and (c) but not in (b). Thus, current through 4-AP-sensitive K^+ channels contributes to action potential repolarisation in premyelinated or demyelinated mammalian axons, whereas in normal myelinated axons repolarisation is entirely due to Na^+ channel inactivation. (Adapted from Fig. 2 in *Trends Neurosci* **13**: Black, JA *et al.* Ion Channel Organization of the Myelinated Fiber, p 48–54 (1990) with permission from Elsevier Science

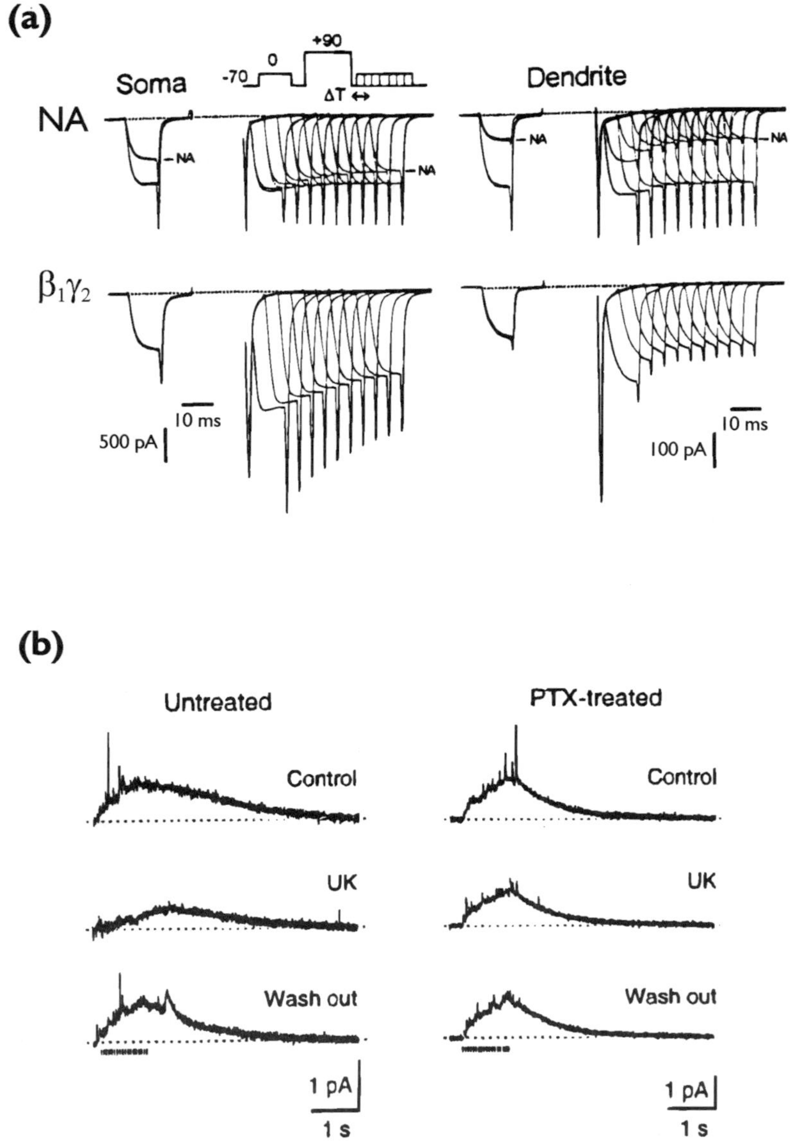

Figure 2.4 Noradrenergic inhibition of Ca^{2+} currents and transmitter release in sympathetic neurons and their processes. (a) Inhibition of currents through N-type Ca^{2+} channels by external application of noradrenaline (NA) or by over-expression of G-protein $\beta_1\gamma_2$ subunits, recorded from the soma and dendrite of a dissociated rat superior cervical sympathetic neuron. Currents were evoked by two successive 10 ms steps from -70 mV to 0 mV, separated by a prepulse to $+90$ mV. Note that the transient inhibition produced by NA (mediated by the G-protein G_o) and the tonic inhibition produced by the G-protein $\beta_1\gamma_2$ subunits were temporarily reversed by the $+90$ mV depolarisation. (Adapted from Fig. 4 in Delmas, P *et al.* (2000) *Nat. Neurosci.* **3**: 670–678. Reproduced with permission). (b) Inhibition of noradrenaline release from neurites of rat superior cervical sympathetic neurons by the α2-adrenoceptor stimulant UK-14,304, recorded amperometrically. Note that pretreatment with *Pertussis* toxin (PTX), which prevents coupling of the adrenoceptor to G_o, abolished inhibition. (Adapted from Fig. 3 in Koh, D-S and Hille, B (1997) *Proc. Natl. Acad. Sci. USA* 1506–1511. Reproduced with permission)

and suppressed by blocking the presynaptic receptors with antagonist drugs, which thereby selectively enhance the release of individual transmitters.

ION CHANNELS AFFECTING THE PATTERN AND FREQUENCY OF ACTION POTENTIAL DISCHARGES

The opening of Na^+ ion channels for the initiation of neuronal depolarisation and action potential generation, as described above, can be induced by excitatory neurotransmitters acting on receptors that are directly linked to cation channels. These include glutamate AMPA receptors (Chapters 3 and 10) and ACh nicotinic receptors (Chapter 6). The inhibitory neurotransmitter GABA has an opposing effect through receptors ($GABA_A$) that are directly linked to the opening of chloride channels, inducing an influx of Cl^- ions and subsequent hyperpolarisation (Chapters 1 and 11). There are, however, a number of other ion channels, generally for K^+ or Ca^{2+}, that have a more subtle controlling effect on neuronal activity. Their opening may be initiated by (or dependent on) preceding changes in membrane potential and ion flux, but they can be affected indirectly by various neurotransmitters, e.g. monoamines and peptides, acting on receptors linked to second messenger systems or more directly by various chemicals, some of which have clinical use. The role of these channels in controlling the overall activity of neurons is clearly important and needs to be considered.

'SLOW' K^+ CHANNELS AND ADAPTATION

The K^+ channels responsible for action potential repolarisation close fairly soon after repolarisation (usually within 5–10 ms). However, most nerve cells possess other K^+ channels which are opened during nerve cell discharges but which stay open much longer. These do not contribute much to the repolarisation of individual action potentials but instead affect the excitability of the neuron over periods of hundreds of milliseconds or even seconds.

Two principal types of channel having this effect have been identified and their properties are summarised in Table 2.2. The first type (*Ca^{2+}-activated K^+ channels* or K_{Ca} channels) are opened ('gated'), not by membrane voltage but by a rise in intracellular Ca^{2+} ion concentration. This means that they are activated by the Ca^{2+} influx through voltage-gated Ca^{2+} channels when these are opened during a somatic or dendritic action potential, or during trains of action potentials. They then close slowly as the intracellular Ca^{2+} concentration recovers, so producing a long-lasting after-hyperpolarisation (AHP) following an action potential or after trains of action potentials. The particular K_{Ca} channels thought to be responsible for the long AHP have a low conductance (~ 10 pS in symmetrical high [K^+] solution) so are called *'SK'* ('small-conductance K') channels: three variants of SK channel have now been cloned, SK1, 2 and 3. These are resistant to normal K^+ channel blocking agents such as tetraethylammonium or 4-aminopyridine, but can be selectively blocked (with varying affinities) by the bee-venom apamin or by certain quaternary ammonium compounds such as tubocurarine and derivatives therefrom.

The second type (*M-channels*) are voltage-gated, like delayed rectifier channels, but have a lower threshold (around -60 mV) and open 10–100 times more slowly when the

Table 2.2 'Slow' potassium channels

Type	Ca-activated	M-type
Descriptor	SK_{Ca}	K_M
Gene products	SK1–3	KCNQ2/3
Activated by	Intracellular Ca^{2+}	Voltage
Threshold	$>100\,nM\ [Ca^{2+}]_{in}$	$<-60\,mV\ V_m$
Blocked by	Apamin (SK2 > SK3 > SK1)	Linopirdine
Inhibited by	Acetylcholine[1]	Acetylcholine[1]
	Glutamate[2]	Glutamate[2]
	Noradrenaline[3]	Peptides[5]
	5-Hydroxytryptamine[4]	
Present in	Autonomic neurons	Sympathetic neurons
	Cortical pyramidal cells	Cortical pyramidal cells
	Hippocampal pyramidal cells	Hippocampal pyramidal cells
Function	Spike frequency adaptation	Spike frequency adaptation
		Membrane potential stabilisation

[1]Via m1, m3 muscarinic receptors.
[2]Via mGluR1,5 'metabotropic' glutamate receptors.
[3]Via β-adrenoceptors.
[4]Via 5-HT_2 receptors.
[5]Including bradykinin, angiotensin, substance P.

membrane is depolarised. They were originally called M-channels because they were inhibited by activating *M*uscarinic acetylcholine receptors. (This turns out not to be a very good definition since other channels can be inhibited by these receptors but the name has stuck.) It is now known that M-channels are composed of protein products of members of the KCNQ family of K^+ channel genes, mutations of which can give rise to certain forms of inherited epilepsy or deafness (depending where the proteins are expressed). M-channels, like SK channels, are generally resistant to common K^+ channel-blocking agents, but are selectively blocked by the 'cognition-enhancer' linopirdine and congeners thereof.

In spite of their different structure and gating mechanisms, these channels have quite a lot in common in functional terms. First, they both open and close slowly. (The SK_{Ca} channels open slowly because of the time taken to build up the required concentration of Ca^{2+} in the cell, and close slowly because it takes hundreds of milliseconds or seconds for the Ca^{2+} to be extruded. M-channels open and close slowly because of their slow intrinsic gating.) Second, although there are differences in their distribution among different types of neuron (e.g. M-channels are abundant in sympathetic neurons whereas SK_{Ca} channels are more important in the enteric neurons in the intestine), they also co-exist in many neurons (such as hippocampal and cortical pyramidal cells— including *human* cells). Third, they have rather similar effects when they open (see below). Fourth, they are both closed by some important neurotransmitters. Thus, acetylcholine (acting via muscarinic receptors) and glutamate (acting via metabotropic glutamate receptors) close both types of channel, but noradrenaline (acting via β-adrenoceptors) closes only the K_{Ca} channels. This effect makes an important contribution to the postsynaptic action of these transmitters, and is discussed further below.

Figures 2.5 and 2.6 shows some experimental records illustrating the function of these channels. Figure 2.5 illustrates the function of SK_{Ca} channels in a hippocampal pyramidal neuron. In the record marked 'control' in Fig. 2.5(b) the neuron was depolarised by injecting a 1-s long depolarising current. This makes it fire action

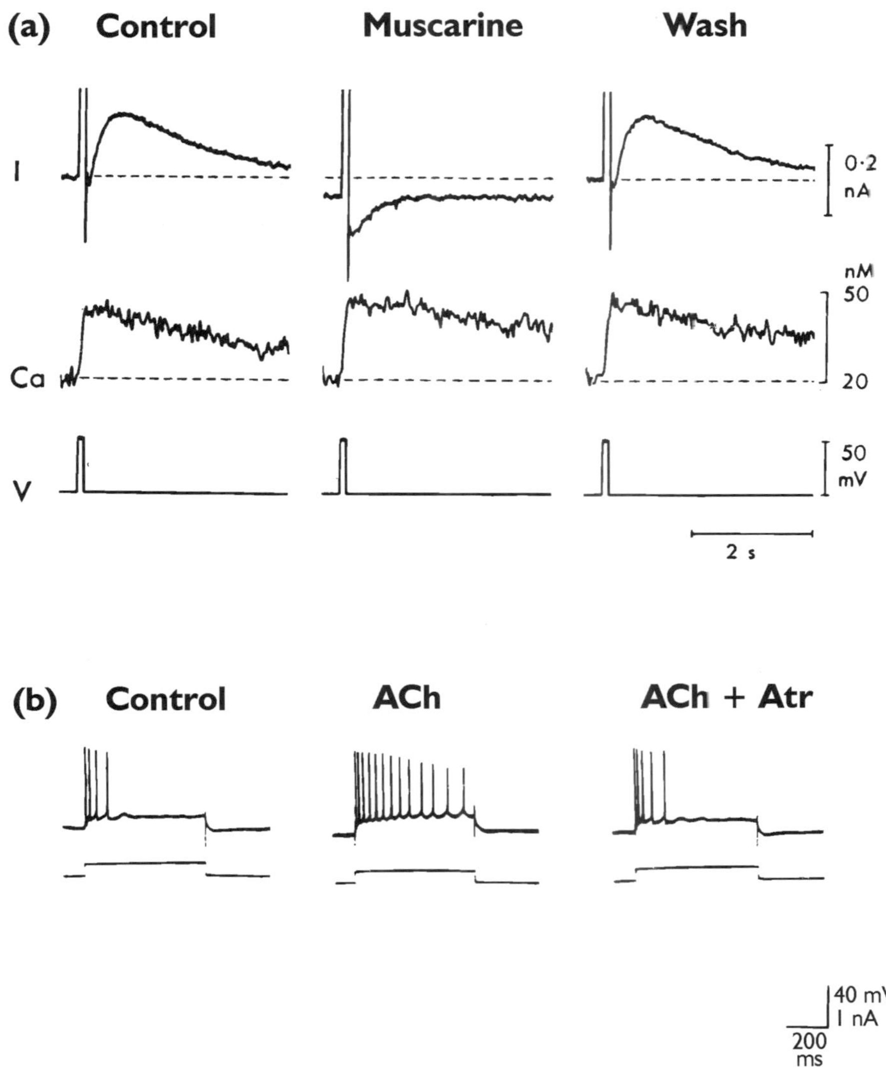

Figure 2.5 Effects of inhibiting SK_{Ca} Ca^{2+}-activated K^+ channels by stimulating muscarinic acetylcholine receptors (mAChRs) in rat hippocampal pyramidal neurons. (Micro-electrode recordings.) (a) Records showing SK_{Ca} current (*I*) and intracellular $[Ca^{2+}]$ transient (Ca) following a 50 ms depolarisation (*V*). The depolarisation opens voltage-gated Ca^{2+} channels. The resultant Ca^{2+} influx leads to a rise in intracellular $[Ca^{2+}]$ that (after a delay) activates the K_{Ca} current. The mAChR agonist muscarine (10 μM) does not affect the Ca^{2-} rise but inhibits the subsequent opening of the SK_{Ca} channels. (Adapted from Fig. 3 in Knopfel, T *et al.* (1990) *Proc. Natl. Acad. Sci. USA* **87**: 4083–4087. Reproduced with permission). (b) Records showing the effect of inhibiting the SK_{Ca} current on the firing properties of a hippocampal neuron. Under normal circumstances (control) the development of the SK_{Ca} current arrests action potential firing during tonic depolarisation induced by injecting 1-s depolarising current ('spike frequency-adaptation'). When the SK_{Ca} current is inhibited with acetylcholine (ACh, 200 μM) (see (a)) spike frequency-adaptation is reduced. This effect is reversed by adding 0.5 μM atropine, to block the mAChRs. Reprinted (adapted from Fig. 1) with permission, from *Acetylcholine Mediates a Slow Synaptic Potential in Hippocampal Pyramidal Cells*, Cole, AE and Nicoll, RA (1983) *Science* **221**: 1299–1301). American Association for the Advance of Science

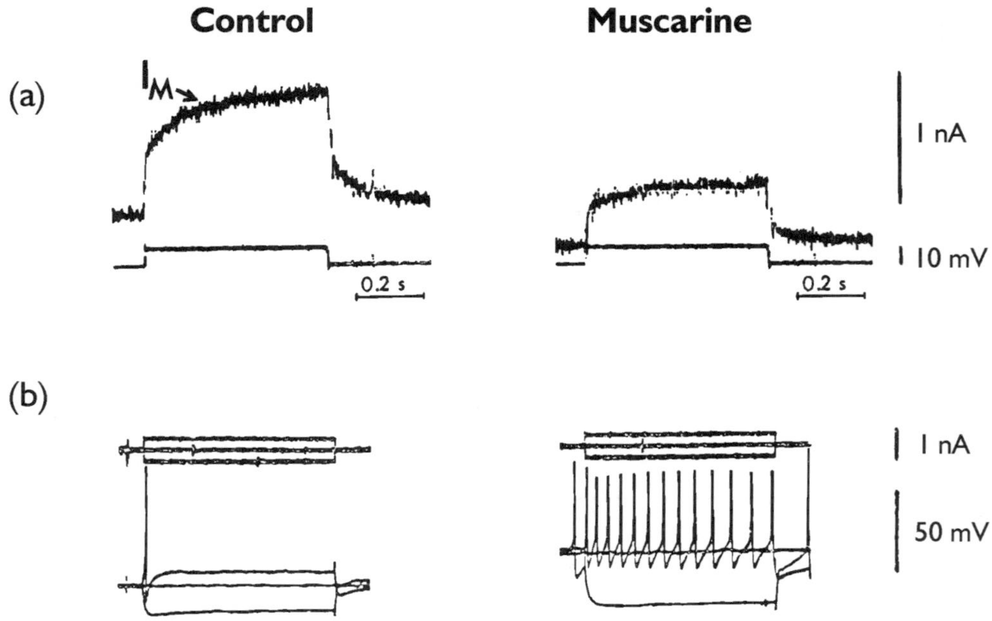

Figure 2.6 Effect of inhibiting M-type K^+ channels in rat superior cervical sympathetic neurons with the muscarinic acetylcholine-receptor (mAChR) stimulant, muscarine. Micro-electrode recordings from different neurons. (a). Current responses to +10 mV voltage steps from −50 mV holding potential. (b). Voltage responses to injecting depolarising and hyperpolarising currents from an initial resting potential of around −47 mV. Under control conditions, depolarisation produces a slow activation of the voltage-gated K^+ current, $I_{K(M)}$ ('I_M' in (a)); this raises the threshold for action potential generation so that the imposed depolarisation in (b) produces only a single action potential (i.e. this neuron, like that in Fig. 2.5, shows strong 'spike frequency-adaptation'). Muscarine strongly reduces $I_{K(M)}$; removal of this braking current now allows the neuron to fire a train of action potentials during the depolarising current injection. (Records in (a) A Constanti and DA Brown, unpublished; records in (b) adapted from Fig. 7 in *Intracellular Observations on the Effects of Muscarinic Agonists on Rat Sympathetic Neurones* by Brown DA and Constanti, A (1980) *Br. J. Pharmacol.* **70**: 593–608.) Reproduced by permission of Nature Publishing Group

potentials. However, the action potentials open Ca channels, so intracellular Ca^{2+} gradually rises as shown in Fig. 2.5(a), and this in turn opens SK_{Ca} channels to produce an outward (hyperpolarising) current. This current partly repolarises the cell and raises the threshold for action potential generation, so the action potential train in Fig. 2.5(b) dies out. The K_{Ca} channels were then inhibited with acetylcholine (or an analogue, muscarine). Now the SK_{Ca} channels cannot open, even though intracellular $[Ca^{2+}]$ still rises (Fig. 2.5(a)). This allows the action potential discharge to continue throughout the length of the depolarising current injection (Fig. 2.5(b)). Thus, the SK_{Ca} channels induce an *adaptation* of the action potential discharge to a maintained stimulus: this adaptation is lost when the SK_{Ca} channels are prevented from opening.

Figure 2.6 shows the effect of the M-channels on the action potential discharges of a rat sympathetic neuron during an equivalent (1-s) injection of depolarising current. (Hyperpolarising currents were also injected in this experiment, giving the downward voltage response.) This cell shows even stronger adaptation under normal circumstances ('control'), because the depolarisation itself is sufficient to open extra M-channels, even without the action potentials (Fig. 2.6(a)). When the opening of M-channels is inhibited by muscarine, this adaptation is again lost. Also note that muscarine has actually depolarised the cell—the level of membrane potential before injecting the current pulse has changed. This is because a few M-channels are open at the resting potential and actually contribute to the resting potential.

As mentioned above, M-channels and K_{Ca} channels co-exist in many neurons. This may seem odd, since Figs 2.5 and 2.6 suggest that they have the same effect. However, in practice, their effects are slightly different, depending on the pattern of stimulation, and in fact the two currents act synergistically—i.e. the effect of inhibiting both currents is far greater than the sum of inhibiting each individually. Their inhibition (separately or together) by neurotransmitters such as acetylcholine and noradrenaline removes a 'brake' on neural discharges and thereby induces a sustained increase in excitability. This is the prime mechanism underlying the arousal and attention-directing function of the ascending cholinergic and aminergic systems innervating the pyramidal cells of the cerebral cortex and hippocampus; the failure of this function, due to inadequate transmitter release, is thought to contribute to the cognitive deficits in such diseases as Alzheimer's disease.

SK_{Ca} and M channels are not the only K^+ channels regulated by transmitters. As noted above, transmitters can also close, or open, other K^+ channels that do not directly regulate excitability but instead determine the resting potential of the neuron, and hence depolarise or hyperpolarise the neuron.

Ca^{2+} CHANNELS: PLATEAU POTENTIALS AND PACEMAKING

As pointed out above, although the principal function of voltage-gated Ca^{2+} channels is to provide the charge of Ca^{2+} necessary for transmitter release, Ca^{2+} channels are also present on the somata and dendrites of most neurons. These include two classes of Ca^{2+} channel not involved in transmitter release—dihydropyridine-sensitive high-threshold L-type channels, homologous to the cardiac Ca^{2+} channels responsible for ventricular contraction and some pacemaking activity; and low-threshold, rapidly-inactivating T-type Ca^{2+} channels. These have multiple functions.

First, their opening during somato-dendritic action potentials provides the source of the increased intracellular [Ca^{2+}] required to open Ca^{2+}-activated K^+ channels—BK channels, to accelerate spike repolarisation, and SK channels, to induce spike-train adaptation and limit repetitive firing. The BK channels are activated (primarily) following entry of Ca^{2+} through L-type channels; the source of Ca^{2+} for *SK* channel activation varies with different neurons, and may be either through L-type or N-type channels.

Second, as in the ventricular muscle fibres of the heart, opening of L-type channels can generate sustained plateau potentials following the initial Na^{2+}-mediated action potential—for example, in the rhythmically firing neurons of the inferior olive (Fig. 2.7).

The T-type channels have a special 'pacemaking' function. This is well illustrated in thalamic relay neurons (Fig. 2.8). At resting potentials $\leqslant -60$ mV, these channels are inactivated and hence non-conducting (a voltage-sensitive closure process resembling Na^+ channel inactivation). Under these conditions, the relay neurons show sustained rhythmic firing when tonically depolarised. However, if the neurons are first hyperpolarised, T-channel inactivation is removed. Then, when the cells are depolarised, the T-channels open and generate a depolarising 'Ca^{2+} spike'. This in turn induces a rapid

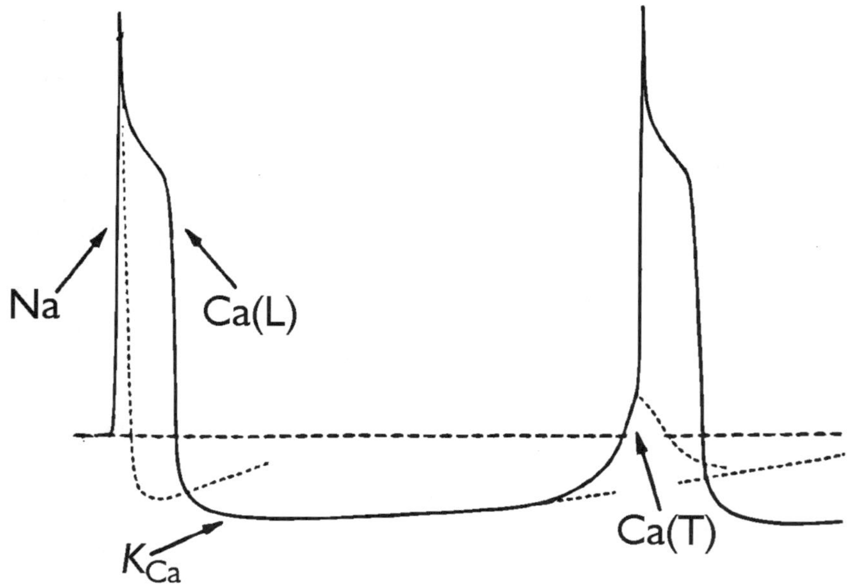

Figure 2.7 Oscillatory behaviour of guinea-pig inferior olivary neurons. The initial action potential, induced by the opening of conventional voltage-gated Na^+ channels, in turn opens voltage-gated L-type Ca^{2+} channels to produce a 'plateau potential'. The Ca^{2+} entry activates K_{Ca} channels, to produce a long-lasting (several hundred ms) after-hyperpolarisation. This de-inactivates the transient (T-type) Ca^{2+} channels (see Fig. 2.8). Hence, as the Ca^{2+} is extruded and the K_{Ca} current declines, the low-threshold T-type Ca^{2+} channels open, and the cell depolarises to reach the threshold for the Na^+ channel, giving a new action potential, and so on. The interval between the action potentials is 650 ms. (Adapted from Fig. 7 in Llinas, R and Yarom, Y (1981) *J. Physiol.* **315**: 569–584. Reproduced by permission of the Physiological Society)

'burst' of Na^+ action potentials. The burst is arrested first because the Na^+ channels inactivate, and then because the T-type Ca^{2+} channels inactivate. Both inactivation processes are removed when the cell hyperpolarises back again, so becoming available for another burst. As a result, the cells change their firing pattern from tonic firing to burst-firing simply dependent on membrane potential. This is thought to explain the switch between tonic firing in awake animals to burst-firing during slow-wave sleep. In the awake state, the neurons are maintained in a tonic state of depolarisation due to the release of neurotransmitters such as histamine and acetylcholine, which inhibit K^+ currents (see above), but hyperpolarise during slow-wave sleep when transmitter release diminishes — or when the receptors for the transmitters are blocked by anti-histamines or anti-cholinergic drugs. However, it should be emphasised that T-channels are quite widely distributed and their burst-inducing properties may also be important in some forms of epilepsy since they can be blocked by certain anti-epileptic drugs, such as ethosuximide.

Finally, entry of Ca^{2+} through somatic and dendritic Ca^{2+} channels activates calmodulin-dependent protein kinases to modulate transcription, and thereby plays a crucial role in certain components of neural development and plasticity.

Neither L nor T channels appear susceptible to the form of G-protein-mediated inhibition characteristic of N or P/Q channels. However, as in the heart cells, L-type Ca^{2+}

channels in the nervous system are susceptible to more indirect forms of modulation (both enhancement and inhibition) through receptor-mediated phosphorylation.

ANOTHER PACEMAKER CHANNEL: HYPERPOLARISATION-ACTIVATED CATION CHANNELS ('h-CHANNELS')

Another analogy with heart cells is the presence in many neurons of the cardiac pacemaker current I_h. (The neural current is sometimes dubbed I_Q.) This is a mixed cation current carried by channels that are permeant to both Na^+ and K^+ and which are opened by hyperpolarising the membrane — i.e. at potentials negative to the normal resting potential. It serves the same function as in the heart, to act as a pacemaker current. The way this works is illustrated in Fig. 2.9. As the membrane hyperpolarises (e.g. after an action potential, when K^+ currents are active), h-channels open to give an inward (depolarising) current (Fig. 2.9(a)). This leads to a slow depolarisation until the threshold for the T-type Ca^{2+} channels open, leading to a rapid depolarisation and spiking (Fig. 2.9(b)). The h-channels then switch off (because the cell is depolarised) and reopen during the subsequent hyperpolarisation. In this way sustained oscillations of membrane potential, leading to a steady rhythmic action potential discharge, can be maintained. The h-channels are blocked by low concentrations of Cs^+ ions, or by agents which block the cardiac current and slow the heart: such agents inhibit the neural membrane potential oscillations and discharges. Also like the cardiac pacemaker, the neural h-current is regulated by transmitters that activate adenylate cyclase, such as noradrenaline and 5-hydroxytryptamine: the cyclic AMP shifts the activation curve to more positive membrane potentials (by a direct action on the channels, not through phosphorylation), so accelerating the depolarisation and increasing the neural rhythm. Conversely, transmitters or mediators that inhibit adenylate cyclase, like enkephalins and adenosine, shift the activation curve to more negative potentials and slow rhythmic discharges.

RECORDING NEURAL ACTIVITY

RECORDING ION CHANNEL CURRENTS: THE PATCH-CLAMP

Currents through individual ion channels can be recorded using the *patch-clamp* technique (Fig. 2.10). A fine glass micro-electrode (tip diameter about 1 μm) filled with electrolyte solution is attached to the cell membrane by suction, forming a 'tight seal' (resistance 1 GΩ or more, i.e. 10^9 Ω), so that all current flowing through the channel enters the electrode. These currents are very small (a few picoamps, pA) so have to be amplified. The amplifier also incorporates a device for applying a potential to the pipette, so that the potential across the cell membrane at the tip of the pipette can be varied.

Figure 2.10(a) illustrates currents generated by K^+ ions flowing through an M-type K^+ channel in a ganglion cell membrane. By convention, the direction of current flow always refers to the direction in which +ve ions move. Thus, outward current is generated by +ve ions flowing out of the cell into the pipette (or −ve ions going the other way). Also by convention, outward current is depicted as an upward deflection in the recording. Note that the channel normally adopts one of two states — it is either

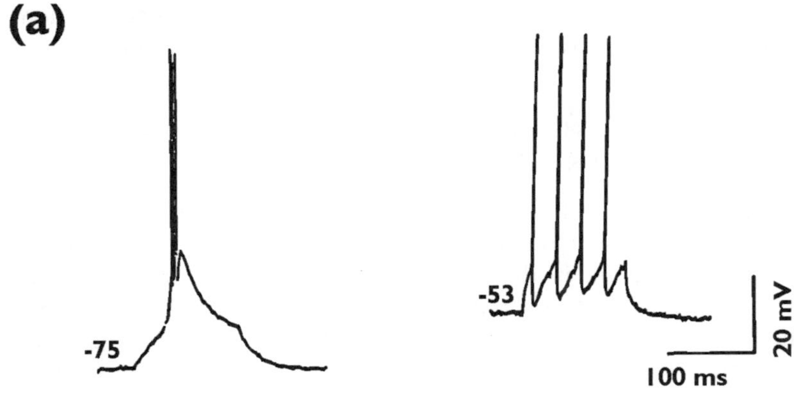

(a)
-75
-53
20 mV
100 ms

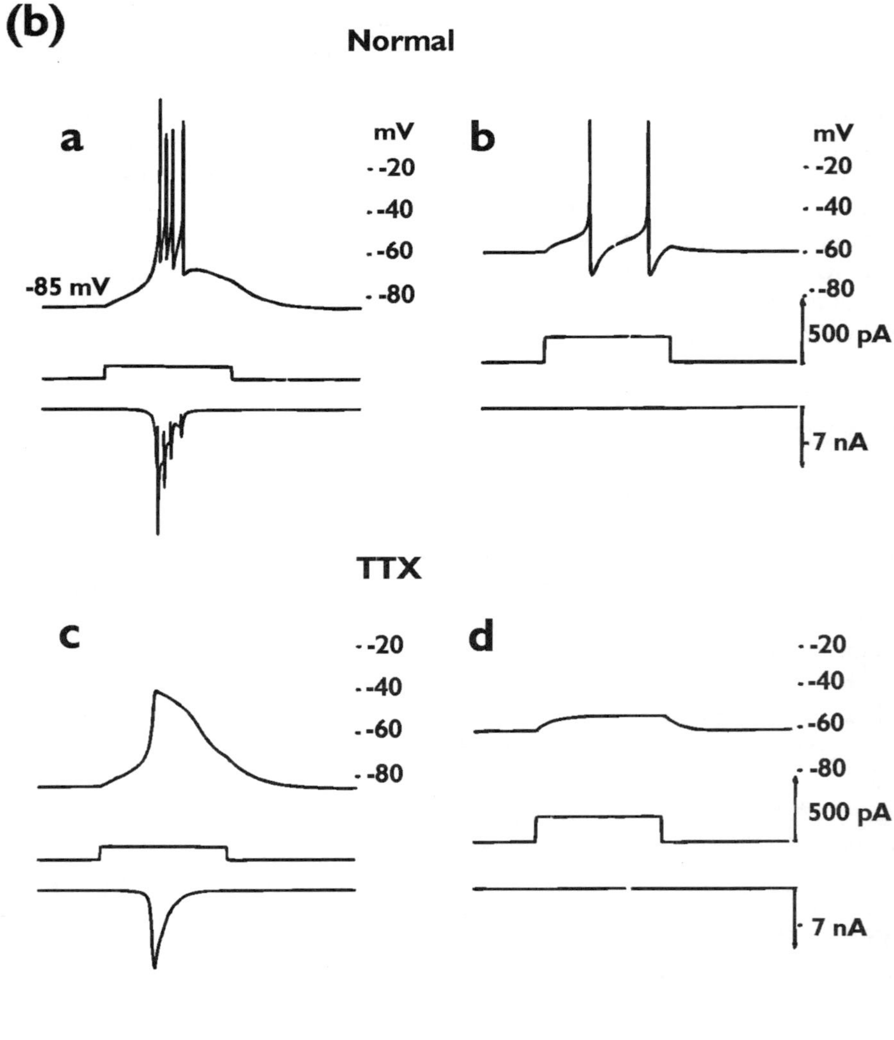

(b)
Normal
a
mV
-20
-40
-60
-85 mV
-80
b
mV
-20
-40
-60
-80
500 pA
7 nA
TTX
c
-20
-40
-60
-80
d
-20
-40
-60
-80
500 pA
7 nA

open or shut — but switches spontaneously between the two states. When it is shut, no current flows. When it is open, the current is fairly constant at any given potential. However, when the potential is changed, the amplitude of the channel current changes: this is because the current is given by Ohm's Law:

$$V = IR \text{ so } I = V/R, \text{ whence the single-channel current } i = g(V - E_K)$$

where g is the single-channel *conductance* (reciprocal resistance, units = Siemens, S), V is the membrane potential across the membrane patch and E_K is the equilibrium potential for K^+ ions. The conductance is normally constant, and is characteristic for the channel. Single-channel conductances are mostly within the range 2–100 picosiemens (pS): in this case, the conductance is about 8 pS with 2.5 mM [K^+] in the pipette solution.

This channel is voltage-sensitive — that is, its activity is increased when the membrane is depolarised. Thus, the channel opens very infrequently and for very short periods at −50 mV, but opens more frequently and for longer times at −30 mV. This activity is expressed by the *open probability* (P_o), that is, the probability that, at any given time, the channel is open (or, in other words, the proportion of time the channel spends in the open state). In the example illustrated, P_o was 0.02 at −50 mV and 0.27 at −30 mV.

FROM SINGLE-CHANNEL CURRENTS TO WHOLE-CELL CURRENTS

Figure 2.10(a) shows the activity of a single channel. There are many hundreds or thousands of such channels in the entire membrane of a single ganglion cell. The currents through all of these channels add up to give the *whole-cell current*. This can be recorded using the patch pipette by filling the pipette with a solution of similar ionic composition to that of the cytoplasm (i.e. with high [K^+] and low [Na^+] and [Ca^+]), then rupturing the membrane under the pipette tip (with pressure) once a seal has been established, or by adding an antibiotic ionophore such as nystatin or amphotericin to the pipette solution and letting these diffuse into the membrane under the pipette tip. In the former case, the solution in the pipette is in direct contact with the cytoplasm, so substances in the cytoplasm diffuse into the pipette and vice versa; nystatin and amphotericin conduct small ions such as Na^+ and K^+ across the cell membrane under the pipette tip, so providing good electrical contact with the cytoplasm, but do not permit total mixing of the two solutions. An older, but still useful, method is to insert one or more fine *micro-electrodes* filled with a strong K^+ solution into the cell and then let them seal into the membrane.

Figure 2.8 *(opposite)* Effects of the T-type Ca^{2+} current on the firing behaviour of guinea-pig thalamic relay neurons. (a) Dependence of firing behaviour on membrane potential. At a hyperpolarised potential (−75 mV), a current injection produces a brief burst of action potentials superimposed whereas at −53 mV the cell responds with a sustained train of action potentials. (Adapted from Fig. 2 in Jahnsen, H and Llinas, R (1984) *J. Physiol.* **349**: 205–226. Published for the Physiological Society by Cambridge University Press.) (b) Interpretation of the records in (a). Each record show voltage-trace (top), injected current pulse (middle) and T-type Ca^{2+} current (bottom). At the hyperpolarised potential (record **a**), the T-type Ca^{2+} current is de-inactivated ('primed'), so a depolarising current pulse opens T-channels to produce a 'Ca^{2+} spike' with superimposed Na^+ spikes. The Na^+ spike can be blocked with tetrodotoxin (TTX: record **c**), leaving a 'pure' Ca^{2+} spike and T-current. The T-current is transient and inactivates, so terminating the burst. At a depolarised potential (**b** and **d**), the T-channels are fully inactivated so depolarisation does not initiate a T-current (record **d**) and now evokes a train of Na^+ spikes instead of a burst (record **b**). (Computer simulation, adapted from Fig. 24 in *Electrophysiology of the Neuron* by Hugenaard and McCormick (1994). Published by Oxford University Press, New York — see Further Study)

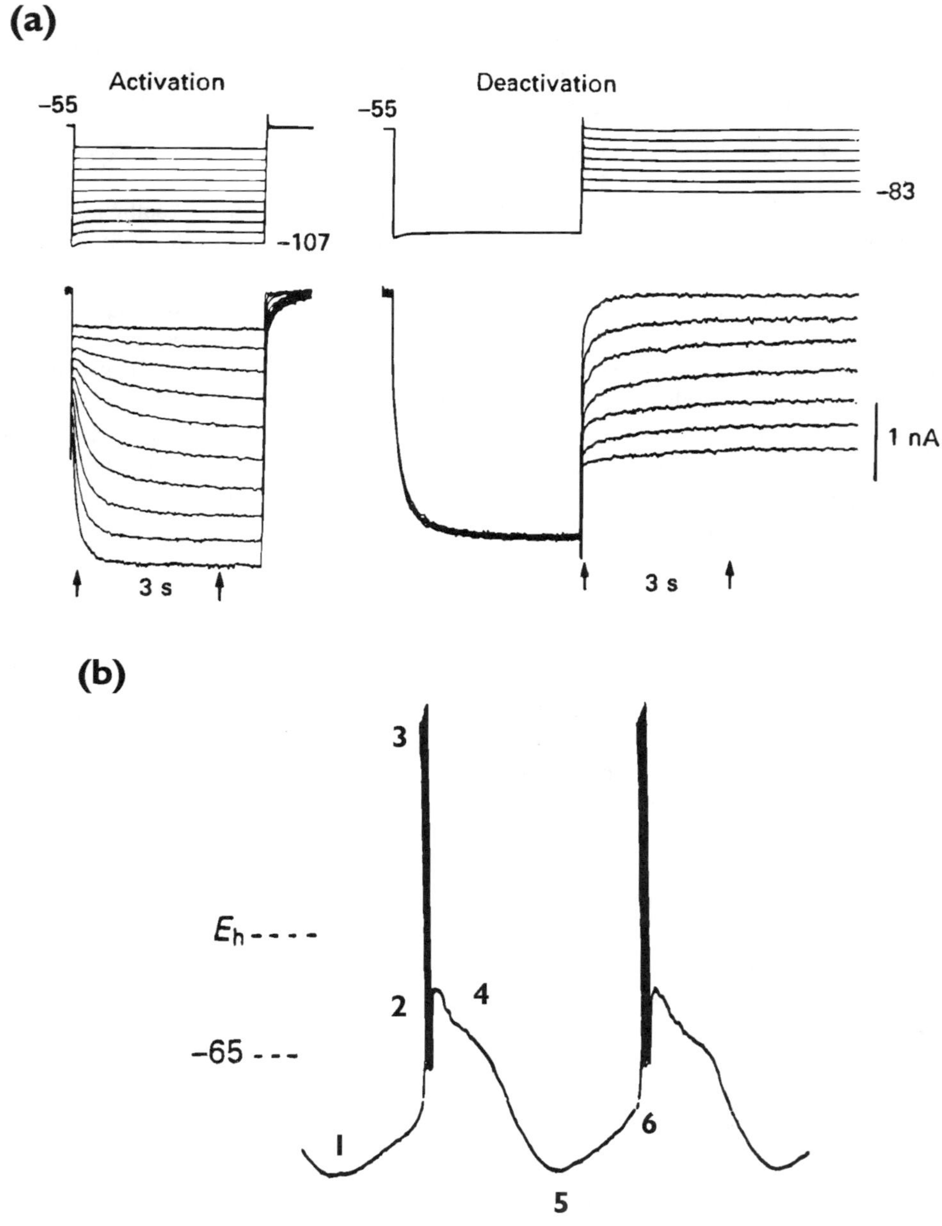

Figure 2.9 Hyperpolarisation-activated cation current I_h and its role in pacemaking in a guinea-pig thalamic relay neuron. (Adapted from Figs 2 and 14 in McCormick, DA and Pape, H-C (1990) *J. Physiol.* **431**: 291–318. Reproduced by permission of the Physiological Society.) (a) Records showing the time-dependent activation of the h-current by hyperpolarisation and its deactivation on repolarising. (b) Interpretation of rhythmic activity in a thalamic relay neuron. (1) The inter-spike hyperpolarisation activates I_h to produce a slowly rising 'pacemaker' depolarisation. (2) This opens T-type Ca^{2+} channels to give a more rapid depolarisation, leading to (3) a burst of Na^+ spikes (see Fig. 2.8). At (4) the depolarisation has closed (deactivated) the h-channels and has inactivated the T-channels. The membrane now hyperpolarises, assisted by outward K^+ current (5). This hyperpolarisation now removes T-channel in-activation and activates I_h (6), to produce another pacemaker potential

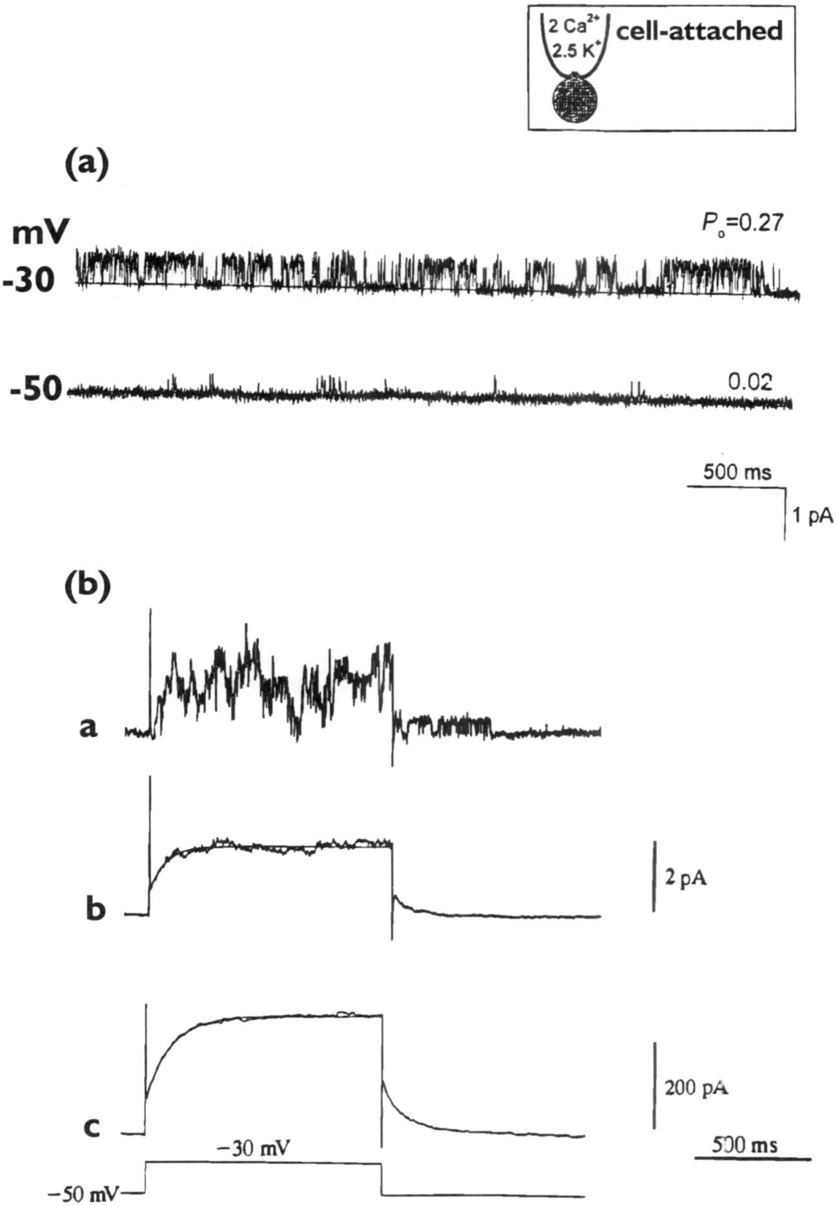

Figure 2.10 M-type K^+ channels: from single-channel currents to whole-cell currents. (a) Single-channel currents recorded from a dissociated rat sympathetic neuron using a cell-attached patch pipette held at estimated membrane potentials of −30 and −50 mV. (Adapted from Fig. 3 in Selyanko, AA and Brown, DA (1999) *Biophys. J.* **77**: 701–713. Reproduced with permission of the Biophysical Society.) (b) **a**: a cluster of single-channel openings recorded from a dissociated ganglion cell with a cell-attached patch pipette on stepping for 1 s from an estimated membrane potential of −50 mV to −30 mV. **b**: an averaged 'ensemble' current obtained on averaging the currents generated by 45 steps like that shown in **a**. **c**: mean whole-cell current recorded with a nystatin-perforated patch pipette during four steps from −50 to −30 mV. (Adapted from Fig. 3 in Selyanko, AA *et al.* (1992) *Proc. Roy. Soc. Lond. Ser. B* **250**: 119–125. Reproduced by permission of The Royal Society)

Figure 2.10(b) shows the relation between the activity of a small cluster of perhaps five individual M-channels recorded from a small patch of membrane with a cell-attached patch pipette (records **a** and **b**) and the whole-cell M-current recorded when the membrane patch under the electrode is permeabilised with nystatin B (record **c**), as seen when the membrane patch or the whole-cell membrane potential is suddenly stepped from −50 to −30 mV and back again. As predicted from Fig. 2.10(a), this depolarisation greatly increases the activity of the channels. However, they do not open instantly but instead take many milliseconds to open — that is, their voltage-gating is relatively slow compared to that of (say) a Na^+ channel. The time taken by any individual channel to assume its new level of open probability varies stochastically about a mean. This mean value is given by the time constant τ $(= 1/(1 - e))$. This can be estimated for a single channel, or for the small cluster of channels seen in Fig. 2.10(b), by repeating the depolarising step many times, then averaging the currents to give an ensemble current (record **b** in Fig. 2.10(b)). In this example, the average time-constant after 45 steps was 86 ms. The whole-cell current (record **c**) gives the current through all the channels in the cell membrane, so, since there are several hundred of them, the current is much larger (note that the current scale is 100 times larger) and one now sees an 'averaged' time-course after a single step (though in this experiment four steps were applied and averaged, to obtain a smoother trace). As one might expect, the time-course of the whole-cell current is quite similar to that of the ensemble of the currents through the small cluster of channels. (They may not be exactly the same, since individual channels in different parts of the cell may vary somewhat in behaviour, depending on their local environment.)

FROM CURRENT TO VOLTAGE

Currents through single channels and across the whole cell membrane are recorded under *voltage-clamp* — that is, the membrane potential is fixed. In a normal cell, however, the voltage is not fixed: the effect of the current is to change the voltage, and signals are normally seen as voltage signals. Figure 2.11 shows how the current through M-channels affects the membrane voltage. When the cell (a frog ganglion cell) was artificially hyperpolarised to −90 mV (left column) so that all of the M-channels were shut, very little current flowed when the voltage was changed (i.e. the membrane conductance was very low or its resistance was very high) (Fig. 2.11(a)). As a result, when a current was injected across the membrane (Fig. 2.11(b)), there was a large voltage change. (The time-course of this voltage change is dependent on the product of the membrane resistance and capacitance. Membrane capacitance is determined by the lipid composition of the membrane and is relatively constant at around 1 $\mu F/cm^2$ membrane.) However, when the cell was left to depolarise to its 'natural' level of (in this case) −46 mV (right-hand column), many M-channels were now open. A hyperpolarising step closes some of the channels, giving a slow decline in current, whereas depolarisation opened more, giving a slow increase in current — the gating of M-channels being characteristically slow, as shown in Fig. 2.10. So now when depolarising current is injected into the cell (bottom record), the membrane begins to depolarise as before but the depolarisation opens more M-channels, and the K^+ current through these extra M-channels hyperpolarises the membrane nearly back to where it started. Conversely, if one tries to hyperpolarise the membrane by injecting hyperpolarising current, the outward flux of K^+ ions diminishes as M-channels close, so the membrane

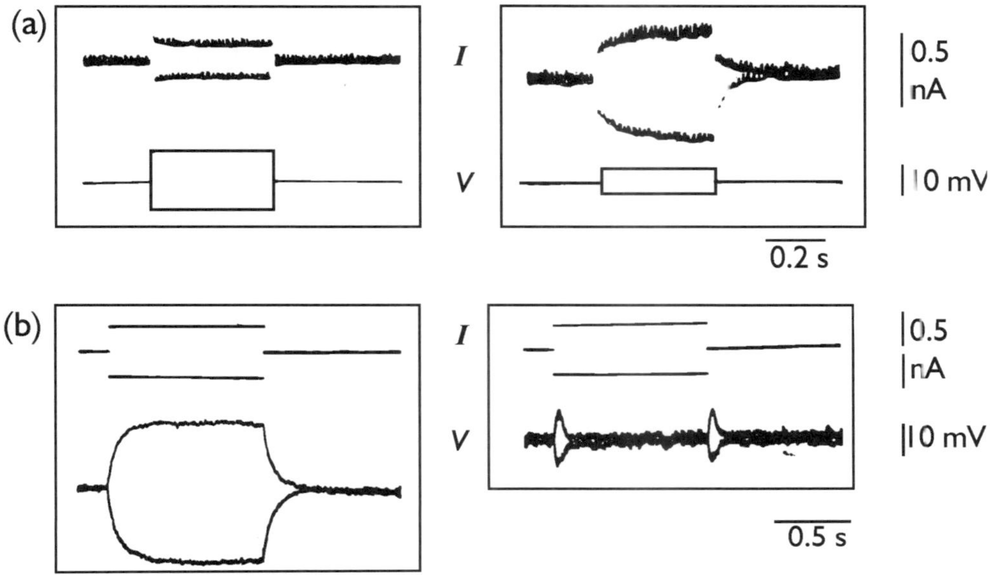

Figure 2.11 M-type K^+ currents: from voltage-clamp to 'current-clamp'. Recordings from frog sympathetic neurons. Upper traces in each record show currents (*I*), lower traces show voltage *V*). (a) Voltage-clamp records showing membrane currents evoked by 0.5-s voltage steps from holding potentials of −90 mV (left), where K_M channels are shut, and −46 mV (right), where K_M channels are open. (b) Voltage responses to 1-s current injections at the same two potentials observed when the voltage-clamp circuit is switched off. Note that the effect of activating the current is to severely reduce the voltage response to current injection. (Adapted from Fig. 6 in Brown, DA (1988) *Ion Channels*, Vol. 1 (Ed. Narahashi, T), Published by Plenum Press, New York, pp. 55–99)

re-depolarises back to where it started. Hence, because M-channels are voltage-sensitive, changes in voltage affect current through M-channels and changes in current through M-channels in turn affect voltage, in such a manner as to stabilise the membrane potential—a negative feedback effect. This is exactly the opposite effect to current through voltage-gated Na^+ channels: current through Na^+ channels depolarises the membrane and this increases the number of open Na^+ channels, so generating more depolarisation, to give positive feedback and hence generating the 'all-or-nothing' action potential.

Figure 2.12 shows another example of how current is converted into voltage—this time a synaptic current. The bottom trace shows a synaptic current recorded under voltage clamp at a preset voltage of −60 mV from a ganglion cell on giving a single shock to the preganglionic fibres. The synaptic current is generated by acetylcholine released from the preganglionic fibres, which opens nicotinic cation channels in the ganglion cell membrane to produce an inward cation current. The top trace shows what happens when the voltage-clamp circuit is switched off, to allow the membrane potential to change. The inward synaptic current now generates a depolarisation (the synaptic potential), which in turn initiates an action potential. This is exactly what synaptic potentials should do, of course, but no Na^+ current is seen under voltage clamp because the membrane potential is held below the threshold for Na^+ channel opening. This threshold is readily exceeded when the clamp circuit is turned off.

EXTRACELLULAR RECORDING

The action potential shown in Fig. 2.12 was recorded from inside the neuron with a micro-electrode. In many instances (particularly in humans!) it is neither convenient nor practicable to use intracellular recording. However, action potentials can still be recorded with extracellular electrodes, by placing the electrode near to the cell (Fig. 2.13). In this case, the electrode tip picks up the local voltage-drop induced by current passing into or out of the cell. Note that (1) the signal is much smaller than the full (intracellularly recorded) action potential and (2) it is essentially a differential of the action potential (because it reflects the underlying current flow, not the voltage change). Nevertheless, since neural discharges are coded in terms of frequency and pattern of

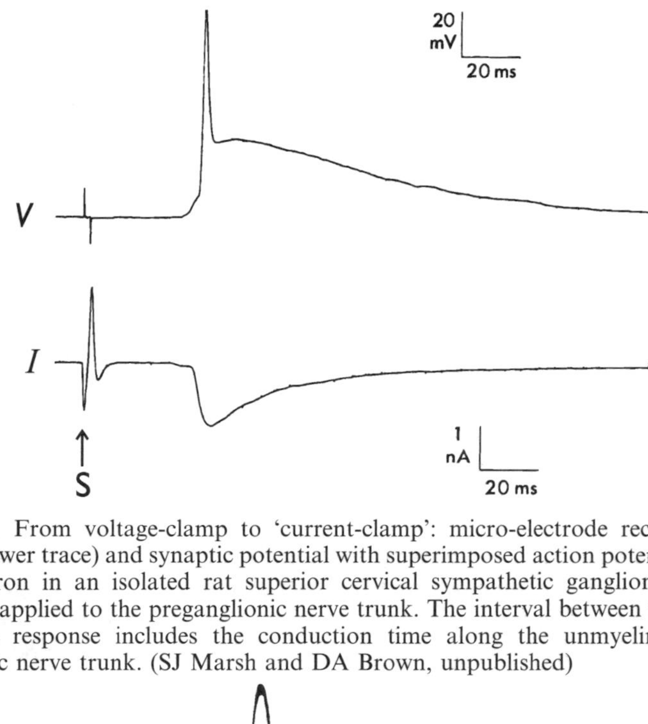

Figure 2.12 From voltage-clamp to 'current-clamp': micro-electrode recordings of synaptic current (*I*, lower trace) and synaptic potential with superimposed action potential (*V*, upper trace) from a neuron in an isolated rat superior cervical sympathetic ganglion following a single stimulus (S) applied to the preganglionic nerve trunk. The interval between the stimulus and the postsynaptic response includes the conduction time along the unmyelinated axons of the preganglionic nerve trunk. (SJ Marsh and DA Brown, unpublished)

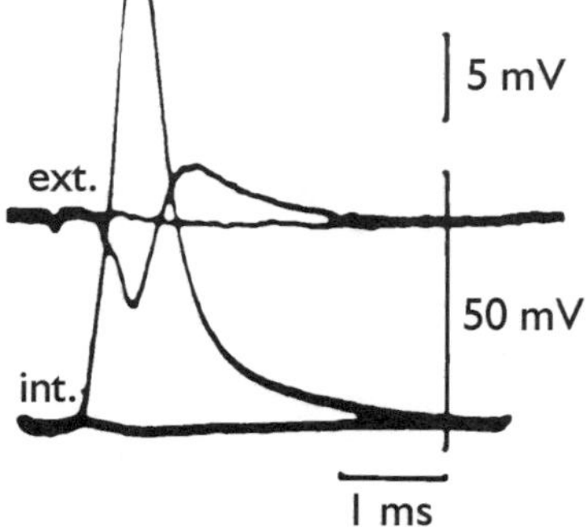

Figure 2.13 Relation between the action potential recorded intracellularly from a cat spinal motoneuron following antidromic stimulation (int.) and the local field potential recorded with an extracellular electrode (ext.). (Adapted from Terzuolo, AC and Araki, T (1961) *Ann. NY Acad. Sci.* **94**: 547–558). Published by NYAS

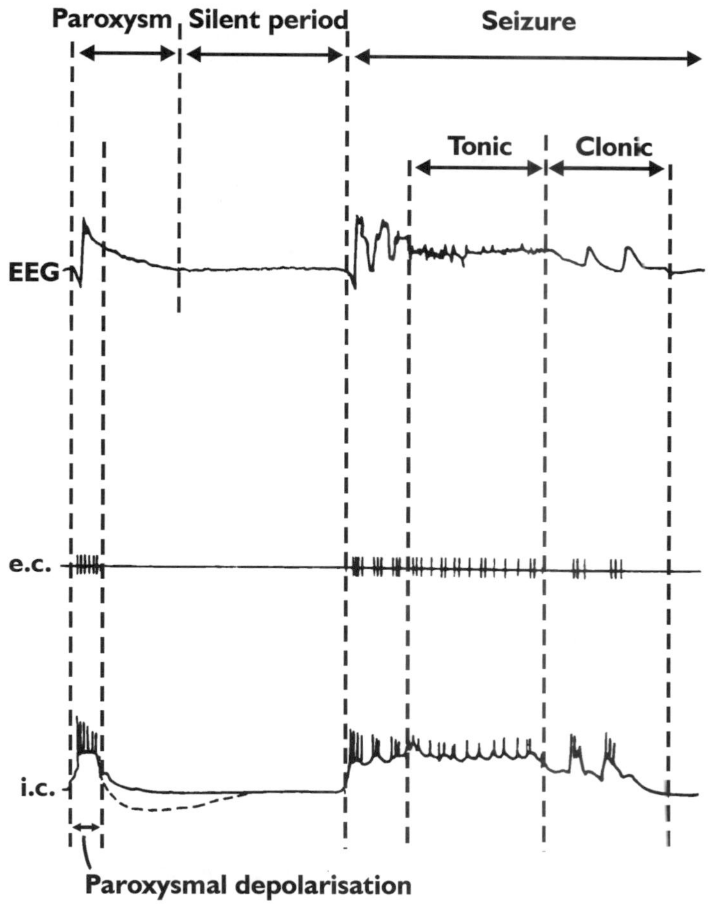

Figure 2.14 Relation between the EEG recorded from an epileptic focus on the surface of the cerebral cortex (EEG) and the activity of a single cortical neuron recorded extracellularly (e.c.) and intracellularly (i.c.) during an experimental epilepsy induced by topical application of penicillin. Note that the large EEG excursions correspond to the large (synchronised) depolarisations of the neuron, not to action potential discharges. (Adapted from *Brain Res.* **52**: Ayala, GF *et al.* Genesis of Epileptic Interictal Spikes. New Knowledge of Cortical Feedback systems suggests a Neurophysiological Explanation of Brief Paroxysms, 1–17 (1973) with permission from Elsevier Science)

action potential firing, such 'unit recording' provides the most convenient and useful method of studying neural activity in the intact nervous system.

Problems arise when the electrode is in contact with lots of cells. If these are firing asynchronously, the signals may cancel out so that individual action potentials become lost in the noise. This problem becomes less when the cells are made to discharge synchronously, by (for example) electrical stimulation. This is made use of to record *evoked potentials* with surface electrodes — for example, to measure conduction velocities along peripheral nerve trunks. Evoked potentials can also be recorded from the brain, via the scalp, along with the EEG (see below). However, the signals are very small (not surprisingly) so have to be averaged by computer. These are used to assess function of sensory systems or in evaluating the progress of demyelinating diseases.

THE EEG

This is a record of fluctuations in activity of large ensembles of neurons in the brain — primarily of the cortical pyramidal cells underneath the recording electrode. Unlike evoked responses, the EEG itself does not represent action potential activity: instead, it originates principally from summed synaptic potentials in pyramidal cell dendrites which (being longer-lasting) summate. However, as with extracellular recording in general, the strongest signal arises when activity of many neurons is synchronised. This happens (for example) in sleep, when large slow-wave activity is recorded: when the subject is woken, the EEG becomes *desynchronised*. Another instance of synchronised activity occurs in epilepsy (Figure 2.14) in which large numbers of neurons show a simultaneous depolarisation (the paroxysmal depolarising shift), again reflecting large underlying synaptic potentials.

FURTHER READING

Hille, B (1992) *Ionic Channels of Excitable Membranes*, 2nd edition, Sinauer Associates.

Kandel, ER, Schwartz, JH and Jessell, TM (2000) *Principles of Neural Science*, 4th edition, Prentice Hall.

Nicholls, JG, Martin, AR and Wallace, BG (1992) *From Neuron to Brain*, 3rd edition, Sinauer Associates.

REVIEWS ON SOME SPECIFIC ASPECTS

Brown, DA (1983) Slow cholinergic excitation — a mechanism for increasing neuronal excitability. *Trends Neurosci.* **6**: 302–307.

Brown, DA (2000) The acid test for resting potassium channels. *Curr. Biol.* **10**: R456–R459.

Dolphin, AC (1998) Mechanisms of modulation of voltage-dependent calcium channels by G proteins. *J. Physiol.* **506**: 3–11.

Hille, B (1994) Modulation of ion channel function by G protein-coupled receptors. *Trends Neurosci.* **17**: 531–536.

McCormick, DA (1989) Cholinergic and adrenergic modulation of thalamocortical processing. *Trends Neurosci.* **12**: 215–220.

Nicoll, RA (1985) The septo-hippocampal projection: a model cholinergic pathway. *Trends Neurosci.* **8**: 533–536.

Nicoll, RA (1988) The coupling of neurotransmitter receptors to ion channels in the brain. *Science* **241**: 545–551.

Pape, H-C (1996) Queer current and pacemaker. *Ann. Rev. Physiol.* **58**: 299–327.

Rogawski, MA (2000) KCNQ2/KCNQ3 K^+ channels and the molecular pathogenesis of epilepsy. *Trends Neurosci.* **23**: 393–398.

Vergara, C, Latorre, R, Marrion, NV and Adelman, JP (1998) Calcium-activated potassium channels. *Curr. Op. Neurobiol.* **8**: 321–329.

FURTHER STUDY

Hugenaard, J and McCormick, DA (1994) *Electrophysiology of the Neuron: an Interactive Tutorial* (Oxford University Press, New York). (This book is accompanied by a disk that allows the properties and functions of the different ion currents to be explored using a PC or Mac. Versions are also available through the web at *http://tonto.stanford.edu/eotn/*)

3 Neurotransmitter Receptors

ALASDAIR J. GIBB

THE RECEPTOR CONCEPT

All multicellular living things depend on chemical communication between cells to coordinate cellular behaviour. This communication is mediated by protein receptors. Cells send out messengers in the form of hormones and neurotransmitters. Receptors provide both **sensitivity** and **selectivity** in this system. The idea that there are specific receptors for hormones and drugs was developed by Erlich and Langley at the end of the nineteenth century, while Hill, Clark, Gaddum and Schild were pioneers in developing a quantitative understanding of the action of drugs. At that time, there was no evidence regarding the structural nature of receptors, although it was widely supposed they were proteins.

The value of receptors to higher animals becomes most obvious in considering the functioning of the central nervous system. The integration of sensory input, past experience and inborn instinct by the central nervous system in the generation of appropriate behavioural activity is only possible because of the specialised properties and diversity of neurotransmitter receptors in the nervous system which mediate signalling between neurons. It has long been recognised that a detailed knowledge of the neurotransmitter receptors in the brain is crucial to developing specific therapeutic approaches to correcting unwanted nervous system activity.

The aim of this chapter is to consider the structure, distribution and functional properties of neurotransmitter receptors in the brain in general and discuss the principles of how the action of drugs at these receptors can be studied. (*See relevant Chapters for detail of individual NT receptors.*)

IDENTIFICATION AND CLASSIFICATION OF NEUROTRANSMITTER RECEPTORS

EARLY PHARMACOLOGICAL STUDIES AND THE IMPACT OF MOLECULAR GENETIC TECHNIQUES

Traditionally receptors have been classified according to their pharmacology. Each neurotransmitter acts on its own family of receptors and these receptors show a high degree of *specificity* for their transmitter. Thus, the receptors on which acetylcholine (ACh) works do not respond to glutamate (or any other neurotransmitter) and vice versa. Diversity of neurotransmitter action is provided by the presence of multiple receptor subtypes for each neurotransmitter, all of which still remain specific to that neurotransmitter. This principle is illustrated by the simple observations outlined in

Neurotransmitters, Drugs and Brain Function. Edited by R. A. Webster

Chapter 1 which showed that since muscarine mimicked some of the actions of ACh (but not all) while nicotine mimicked the other actions of ACh, then ACh probably acted on two distinct types of receptors. The fact that atropine antagonised the muscarinic effects of ACh but not the nicotinic effects, while tubocurarine blocked the nicotinic effects provided firm evidence for this concept. These simple qualitative observations by Langley and others at the beginning of the twentieth century led to the development of more quantitative pharmacological methods that were subsequently used to identify and classify receptors. These methods were based on the use of both (1) agonist and (2) antagonist drugs:

(1) If a series of related chemicals, say noradrenaline, adrenaline, methyladrenaline and isoprenaline, are studied on a range of test responses (e.g. blood pressure, heart rate, pupil size, intestinal motility, etc.) and retain exactly the same order of potency in each test system, then it is likely that there is only one type of receptor for all four of these catecholamines. On the other hand, if, as Ahlquist first found in the 1940s, these compounds give a distinct order of potency in some of the tests, but the reverse (or just a different) order in others, then there must be more than one type of receptor for these agonists.
(3) If one set of these responses can be blocked (antagonised) by a drug that does not affect the other responses (e.g. propranolol blocks the increase in heart rate produced by adrenaline, but not the dilation of the pupil evoked by adrenaline) then this is good evidence that adrenoceptors in the pupil are not the same as those in the heart.

In fact, careful quantitative analysis of the order of activity of the agonists in each test, and of the precise potency of antagonists (see Chapter 5 for quantitative detail) has often successfully indicated, although rarely proved, the presence of subclasses of a receptor type (e.g. different muscarinic receptors). The affinity of receptors for selective antagonists determined using the Schild method was a mainstay of receptor classification throughout the second half of the twentieth century. Thus, a muscarinic receptor can be defined as a receptor with an affinity for atropine of around 1 nM and the M1 subtype of muscarinic receptor can be identified as having an affinity of around 10 nM for the selective antagonist, pirenzepine while muscarinic receptors in the heart (M2 subtype) are much less sensitive to pirenzepine block ($K_B \sim 10^{-7}$ M).

Classification of receptors according to agonist potency can be problematic because agonist potency depends partly on the density of receptors in the tissue and therefore use of selective antagonists has become a mainstay of receptor identification and classification. The development of radioligand binding techniques (see Chapter 5 for principles) provided for the first time a means to measure the density of receptors in a tissue in addition to providing a measure of the affinity of drugs for a receptor and allowed the relative proportion of different receptors in a tissue to be estimated.

These approaches to receptor identification and classification were, of course, pioneered by studies with peripheral systems and isolated tissues. They are more difficult to apply to the CNS, especially in *in vivo* experiments, where responses depend on a complex set of interacting systems and the actual drug concentration at the receptors of interest is rarely known. However, the development of *in vitro* preparations (acute brain slices, 'organotypic' brain slice cultures, tissue-cultured neurons and acutely dissociated neuronal and glial cell preparations) has allowed more quantitative pharmacological techniques to be applied to the action of drugs at neurotransmitter receptors while the development of new recording methods such as patch-clamp recording has allowed the study of drug action at central neurons to be made at ever more detailed levels.

Today we know not only that there is more than one type of receptor for each neurotransmitter, but we also know a great deal about the structural basis for the differences between receptor subtypes which are due to differences in the amino-acid sequence of the proteins which make up the receptor. How do we know this?

Finding the amino-acid sequence of a receptor protein has been approached in three main ways. The final aim of all three methods is to obtain a cDNA clone coding for the protein since the base sequence of this DNA allows the amino-acid sequence of the protein to be predicted:

(1) From purified receptor protein, obtain partial amino-acid sequence information which will allow molecular biologists to isolate the gene (or genes) coding for the receptor.
(2) cDNA library screening. From a receptor-rich tissue, isolate mRNA and create from this, a cDNA library. The library is then screened by, for example, functional expression in *Xenopus* oocytes or mammalian cell lines, for the proteins coded by the library. If positive expression is obtained, the library is subdivided until a single cDNA clone for the receptor is isolated.
(3) Homology screening. Using oligonucleotide probes based on known receptor sequences, search cDNA libraries for homologous sequences which may code for related receptors. The clones are then isolated and sequenced and used in expression studies to confirm the identity of the receptor.

The first tentative steps towards determining the structure of individual receptors were taken by protein chemists. A high-affinity ligand that binds specifically to the receptor (generally an antagonist) was identified by traditional pharmacological methods and attached to the matrix of an appropriate chromatography column. A tissue source, rich in receptors, is homogenised and the cell membranes disrupted with detergents to bring the membrane bound proteins into solution. This solution is then passed through the affinity column and the receptor of interest will stick to the column hence separating it from all the other proteins in the tissue. The receptor is then eluted from the column using a solution of ligand specific for the receptor. This strategy allowed isolation of the nicotinic acetylcholine receptor from the electric organ of the Californian ray (*Torpedo*). Almost 40% of the protein content of this tissue is ACh receptor. The isolation method used a snake toxin from the venom of the Taiwan banded krait (α-bungarotoxin) as the ligand of the affinity column and the purified receptor was eluted from the column using a high concentration of the competitive antagonist, tubocurarine. Following isolation of the protein, a partial N-terminus amino-acid sequence was obtained and from this sequence, oligonucleotide probes were made which were then used to screen a cDNA library to isolate a clone for the receptor. Since DNA sequencing is much faster than protein sequencing, the DNA sequence of the clone is then used to provide the amino-acid sequence of the receptor.

RECEPTOR MECHANISMS

It is often valuable to classify receptors according to their mechanism of action, because this is intimately related to structure. The neurotransmitter receptors in the brain are of two main types classified according to their structure and mechanism of action:

(1) Ion channel receptors
(2) G-protein-coupled receptors

The ion channel receptors are relatively simple in functional terms because the primary response to receptor activation is generated by the ion channel which is an integral part of the protein. Therefore, no accessory proteins are needed to observe the response to nicotinic AChR activation and the full functioning of the receptor can be observed by isolating and purifying the protein biochemically and reconstituting the protein in an artificial lipid membrane. In contrast, the G-protein-coupled receptors require both G-proteins and those elements such as phospholipase-C illustrated in Fig. 3.1, in order to observe the response to receptor activation (in this case a rise in intracellular calcium concentration resulting from the action of IP_3 on intracellular calcium stores).

Most receptors function as mediators of synaptic transmission between neurons. Figure 3.1 illustrates this for the case of a generic glutamatergic synapse. At this synapse glutamate is released from the presynaptic nerve terminal and acts on two different types of fast ionotropic glutamate receptors embedded in the postsynaptic membrane: AMPA receptors mediate an extremely rapid (within 1 ms) response to glutamate release resulting in a rapid depolarisation of the postsynaptic membrane (EPSP). On a slower time scale, the NMDA receptors mediate a slower EPSP which lasts over 100 ms, is carried partly by calcium ions and is voltage-dependent due to blocking of the NMDA receptor channel by Mg ions at negative membrane potentials. The AMPA receptor provides the depolarisation necessary to relieve the Mg block of the NMDA channel and so the calcium influx through the NMDA channel in effect provides a means to integrate synaptic activity mediated by the fast AMPA receptors. The synapse shown in Fig. 3.1 also illustrates that G-protein-coupled glutamate receptors may be located at the synapse, or perisynaptically and therefore can mediate slow synaptic transmission (on a time scale of 100 ms to seconds) whose characteristics will depend on the particular G-protein which is coupled to the metabotropic receptor. In this case, the receptor is coupled by G_q to phospholipase-C and results in IP_3 and diacylglycerol (DAG) production which in turn regulate intracellular calcium concentration and protein kinase-C activity. Thus, at any glutamatergic synapse in the brain there is the potential for a single neurotransmitter to generate fast and slow signals with particular characteristics which depend on the properties of the neurotransmitter receptors expressed in the target cell membrane.

RECEPTOR CLASSIFICATION IN THE POST-GENOMIC ERA

The definitive classification of receptors is by amino-acid sequence analysis. Since all properties of the receptor are determined by the amino-acid sequence of the protein this method has the final say. The explosion in use of molecular genetic techniques in the final decade of the twentieth century has led to the cloning and sequencing of the genes of all the known neurotransmitter receptors in the brain. From the gene sequence, the amino-acid sequence of the receptor protein can be inferred and hence a final classification of all receptors can be made. Ultimately, the human genome sequencing programme will mean that the amino-acid sequence of all human receptors will be known. Does this mean pharmacologists can now retire happy in the knowledge that all is now known that there is to know? Far from it! Gene cloning and sequencing has unveiled an increasingly vast diversity among receptor types which could barely have

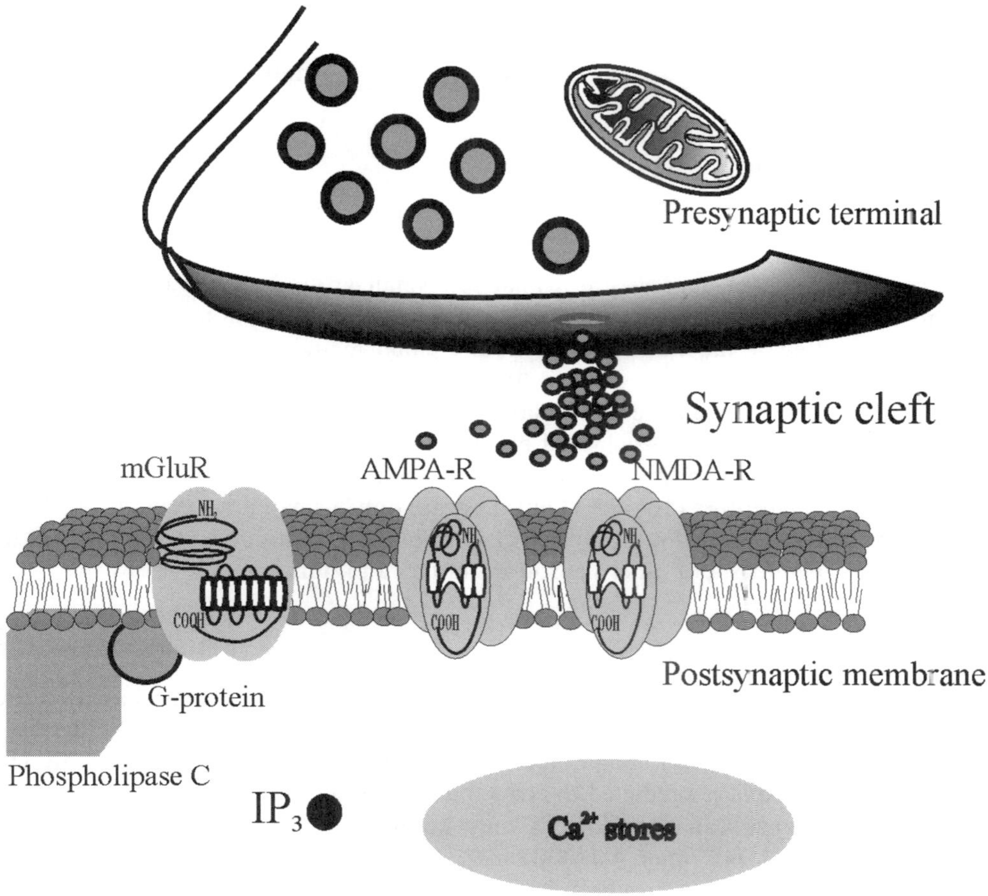

Figure 3.1 Schematic representation of a generic excitatory synapse in the brain. The presynaptic terminal releases the transmitter glutamate by fusion of transmitter vesicles with the nerve terminal membrane. Glutamate diffuses rapidly across the synaptic cleft to bind to and activate AMPA and NMDA receptors. In addition, glutamate may bind to metabotropic G-protein-coupled glutamate receptors located perisynaptically to cause initiation of intracellular signalling via the G-protein, G_q, to activate the enzyme phospholipase and hence produce inositol triphosphate (IP_3) which can release Ca^{2+} from intracellular calcium stores

been imagined by pharmacologists only 20 years ago. The properties and subtle functional differences between receptor subtypes can be studied in increasing detail utilising receptor expression systems such as *Xenopus* oocytes and clonal mammalian cell lines where single receptor populations at high density can be studied without the complications arising from the diversity of receptors present in brain tissue, or the difficulty of recording responses from receptors in the brain. Pharmacology has now entered the era of 'post-genomic' research in which the challenge is to utilise the diversity of receptor types revealed to us by gene cloning techniques in the development of subtype selective drugs. The hope is that if this diversity of receptor subtypes is matched by diversity of function in the brain, then subtype-selective drugs may provide the means to selective therapeutic agents with a minimum of side-effects for use in treating diseases of the brain.

The following sections of this chapter will consider some general and comparative aspects of receptor structure and function. More detailed material on these topics may be found in the relevant chapters on individual neurotransmitters.

ION CHANNEL RECEPTORS

GENERAL

The ligand-gated ion channels are a relatively small group of receptors responsible mainly for fast synaptic transmission at the neuromuscular junction, peripheral autonomic neuroeffector junctions, at autonomic ganglia, and at central synapses. Six different neurotransmitters are known to activate ligand-gated ion channel receptors (Table 3.1). A general principle in the nervous system is that only a few transmitters are used and diversity of effect is achieved by utilising a diversity of receptors. Except for glycine, all fast neurotransmitters have also been found to act at a diversity of G-protein-coupled receptors (Table 3.1) but these are only a small proportion of the huge array of G-protein-coupled receptors which exist in the nervous system and respond to monoamines, peptides and other neurotransmitters and hormones.

Subunit transmembrane topology

The ligand-gated ion channel receptors form three distinct super-families based on the number of times the receptor subunits are predicted to cross the cell membrane (Fig. 3.2). For the nicotinic acetylcholine receptors, $GABA_A$, $GABA_C$, 5-HT_3 and glycine receptors each subunit is predicted to cross the cell membrane four times (Fig. 3.2(a)). The exact transmembrane topology is only known with certainty for the nAChR (Unwin 1995). For the other 4-TM domain receptors (and for those in the 3-TM domain and 2-TM domain families) the transmembrane topology of each subunit has been inferred by analogy with the nAChR, from hydropathicity analysis of the subunit amino-acid sequence (about 20 hydrophobic amino acids are needed to form an alpha-helix long enough to span the cell membrane) and from experiments manipulating recombinant receptor subunits. All the 4-TM domain receptor subunits have both amino and carboxy terminals located on the outside of the membrane (Fig. 3.2(a)). The

Table 3.1 Fast neurotransmitters in the central nervous system

Transmitter	Ion channel receptors	G-protein-coupled receptors
Glutamate	Kainate AMPA NMDA	mGluR
GABA	$GABA_A$ $GABA_C$	$GABA_B$
Acetylcholine	Nicotinic	Muscarinic
5-HT	5-HT_3	5-HT
ATP	P2X	P2Y
Glycine	Gly-R	

Note:
Except for glycine, all fast transmitters act on both ion channel receptors and G-protein-coupled receptors. Within each receptor class, there may be several subtypes.

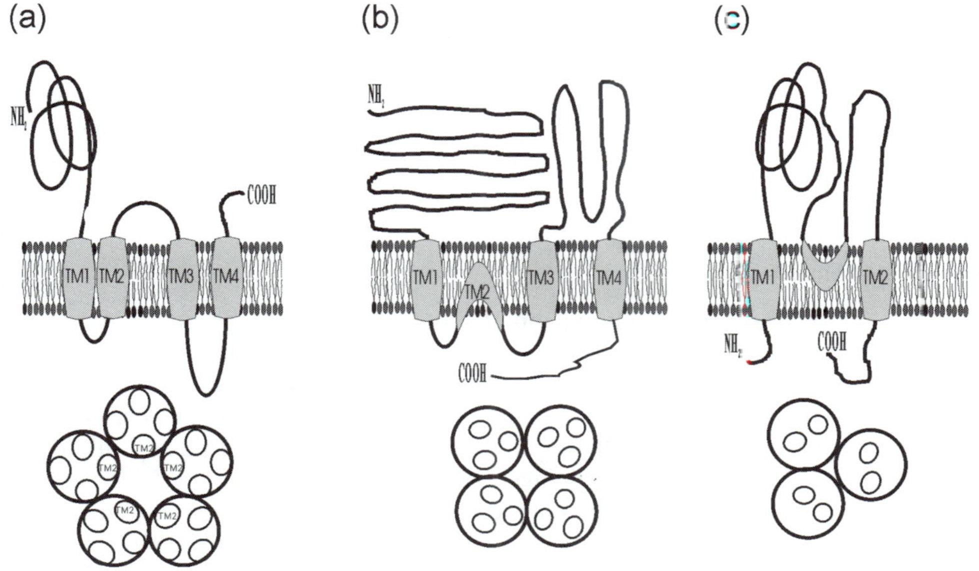

Figure 3.2 Transmembrane (TM) topology of the subunits of three different super-families of ion channel receptors denoted as 4-TM, 3-TM and 2-TM receptors. In (a), the topology of the 4-TM domain subunits is illustrated embedded in the cell membrane. Receptors in this class are the nicotinic acetylcholine receptors, $GABA_A$ and $GABA_C$ receptors, glycine receptors and 5-HT_3 receptors. Shown below is the likely pentameric stoichiometry of the 4-TM domain receptors with TM2 of each subunit lining the central ion channel. In (b), the transmembrane topology of the ionotropic glutamate receptors is shown. These have three true transmembrane domains. TM2 creates a pore-forming loop which penetrates into the cell membrane from the intracellular side. As illustrated below, the likely stoichiometry of the glutamate receptors is a tetramer. The exact contribution of TM1, TM3 or TM4 to forming the ion channel is uncertain. In (c), a subunit of the P2X ATP receptors is shown. These subunits cross the cell membrane only twice and the ion channel is probably formed by a short polypeptide loop entering the membrane from the outside. The exact stoichiometry of the P2X receptors is uncertain but current evidence supports the suggestion that they are trimers

agonist binding site is located in the amino terminal domain before the start of TM1 and the ion channel is formed by the TM2 domains of each subunit which come together to make up the complete receptor. Thus the amino acids in TM2 determine the ion conductance properties of the channel. For GABA and glycine receptors a Cl^- channel is formed while for the other ion channel receptors, the channel is largely cation non-selective for monovalent ions such as Na^+ and K^+ and is often also permeable to calcium. One of the key differences between different ion channel receptors for glutamate, ACh, 5-HT and ATP is in their relative permeability to calcium and this is controlled by the amino acids which line the ion channel.

The ionotropic glutamate receptors (kainate, AMPA and NMDA) are formed by subunits which are predicted to cross the cell membrane three times (Fig. 3.2(b)) with the TM2 region forming a loop into the membrane from the intracellular side. The ionotropic glutamate receptor subunits have a large extracellular amino terminal domain and a long intracellular carboxy terminal domain (Fig. 3.2(b)). In the glutamate receptor subunits the agonist binding site is formed by a 'clam-shell' structure where the amino terminal domain and the extracellular domain between TM3 and TM4 come together to create a pocket for agonist binding.

The P2X receptor subunits are unusual in having only two transmembrane domains with both the amino terminal and carboxy terminal located intracellularly. The ion channel is proposed by analogy with the structure of some potassium channels to be formed by a short loop which enters the membrane from the extracellular side (North and Surprenant 2000).

Subunit stoichiometry

The ion channel receptors are multi-subunit proteins which may be either homomeric (made up of multiple copies of a single type of subunit) or heteromeric (composed of more than one subunit type). These subunits come together after synthesis in the endoplasmic reticulum to form the mature receptor. Notice that stoichiometry is denoted by a subscript number. A receptor composed of two α and three β subunits is therefore denoted as having a stoichiometry of $\alpha_2\beta_3$. This can cause confusion when related subunits are given sequential numbers: $\beta1$, $\beta2$, $\beta3$, etc. The convention is therefore that subunits are numbered normally while stoichiometry is indicated by subscripts so that a pentamer of $\alpha4$ and $\beta3$ subunits might have a stoichiometry of $\alpha4_2\beta3_3$.

NICOTINIC RECEPTORS

All receptors in the 4-TM domain family are thought to form pentameric receptors in which five subunits (Fig. 3.2(a)) surround a central ion channel. Their structure has been most extensively studied in the case of the nicotinic acetylcholine receptor (analogous to the muscle endplate receptor) from *Torpedo* electroplaque (Unwin 2000) where there is now a detailed knowledge of the receptor in both resting and active conformations. The muscle receptor has a subunit stoichiometry of two α subunits, providing the agonist binding sites, and three other subunits (β, γ and δ). In adult muscle an ε subunit is present instead of the γ subunit which is found in the foetal-type receptor. The five subunits are arranged like the staves of a barrel around the central channel. Binding of ACh to the α subunits initiates a conformational change in the protein which, by causing rotation of all five TM2 domains lining the pore, opens the ion channel.

Diversity among neuronal nicotinic receptors is generated by having nine more different α subunits ($\alpha2$–$\alpha10$) and three further β subunits ($\beta2$–$\beta4$). These receptors are activated by nicotine and blocked by the antagonists hexamethonium, mecamylamine and trimetaphan, and the erythrina alkaloid dihydro-β-erythroidine. The neuronal nicotinic receptors are found in autonomic ganglia and in the brain may be either heteromeric (e.g. $\alpha4_2\beta2_3$) or homomeric ($\alpha7_5$). The $\alpha7$ receptor is likely to be the source of the α-bungarotoxin binding sites in the brain observed in autoradiograms of ^{123}I-α-bungarotoxin binding to brain sections (Clarke 1992) and α-bungarotoxin sensitive nicotinic receptors have been shown in a number of studies to stimulate transmitter release from nerve terminals such as dopaminergic terminals in the striatum and glutamatergic terminals in the cortex. Its main functional role may therefore be as a presynaptic receptor regulating transmitter release. It has a high affinity for nicotine and so may mediate some of the central effects of nicotine. Notice that a homomeric receptor has implications for the interpretation of functional studies since the number of agonist (and antagonist) binding sites on the receptor must equal the number of subunits. In the case of $\alpha7$ receptors, Hill coefficients (see Appendix) of around 1.8–2.0 have been observed suggesting that only any two of the five binding sites on the receptor need be occupied for efficient activation of the receptor.

GLYCINE RECEPTORS

Inhibitory glycine receptors with high affinity for the antagonist strychnine are predominantly found in the spinal cord and brainstem. Three different α subunits have been cloned ($\alpha1$–$\alpha3$) and a single β subunit ($\beta1$). Interestingly, the foetal-type of glycine receptor which is a homomer (the adult stoichiometry is likely to be $\alpha_3\beta_2$) has Hill coefficients nearer 3.0, suggesting three agonist sites need to be occupied for receptor activation.

GABA RECEPTORS

The GABA receptor subunits are one of the most diverse groups of ion channel receptor subunits in the brain. Six different α subunits ($\alpha1$–$\alpha6$), four β subunits ($\beta1$–$\beta4$), four γ subunits ($\gamma1$–$\gamma4$), an ε subunit, a π subunit, and three ρ subunits ($\rho1$–$\rho3$) have been found. This diversity of subunits is reflected in the complicated pharmacology of the GABA receptors. Functional GABA receptors can be formed by co-expression of any α subunit with any β subunit probably in the stoichiometry $\alpha_2\beta_3$. The agonist binding site is on the β subunit. However, these receptors, although activated by GABA and muscimol, potentiated by barbiturates and neurosteroids, and blocked by bicuculline and picrotoxin, lack classical benzodiazepine sensitivity which can be restored to the molecule by co-expressing a γ subunit with α and β.

The ρ subunits form $GABA_C$ receptors. These are activated by the weak agonist CACA, relatively insensitive to bicuculline, not sensitive to barbiturates or neurosteroids, and blocked by TPMPA. They have been described in the retina (where $\rho1$ is expressed) but relatively little is known of their function in the brain although they are widely expressed (Bormann 2000). For example, all three ρ subunits are found in cerebellar Purkinje cells. Native $GABA_C$ receptors may be homomeric pentamers of a single ρ subunit or heteromeric pentamers of more than one of the ρ subunits. $GABA_C$ receptor dose–response curves have Hill slopes around 3 or greater, unlike $GABA_A$ receptor dose–response curves where the Hill slope is usually around 2.

5-HT$_3$ RECEPTORS

5-HT$_3$ receptors are an exception to the general nomenclature of numbering subunits and denoting stoichiometry by subscripts. The 5-HT$_3$ receptor subtype is denoted by the subscript '3' and the two known subunits are referred to as 'A' and 'B'. When expressed alone, 5-HT$_3$-A subunits form functional homomeric receptors but these have a much smaller single channel conductance (less than 1.0 pS) and altered pharmacology compared to native 5-HT$_3$ receptors. Native 5-HT$_3$ receptors are likely to be pentameric heteromers of 5-HT$_3$-A and 5-HT$_3$-B subunits. 5-HT$_3$-B subunits, unlike 5-HT$_3$-A subunits, do not form functional receptors when expressed alone but when co-expressed with 5-HT$_3$-A subunits, the receptors formed have functional properties similar to native 5-HT$_3$ receptors. Synaptic transmission mediated by 5-HT$_3$ receptors has been described in only a few places in the brain, such as in the locus coeruleus and it has been suggested that most 5-HT_3 receptors are located presynaptically on nerve terminals where their calcium permeability may be important in potentiating transmitter release.

GLUTAMATE RECEPTORS

Glutamate receptors mediate fast synaptic transmission throughout the nervous system. They have been classified pharmacologically as three main types according to their sensitivity to three specific agonists: kainate, AMPA and NMDA. The NMDA receptors are unique among the ion channel receptors in requiring the binding of two separate agonists, glutamate and glycine in order to activate the receptor. Selective antagonists for both the glutamate binding site (AP5) and the glycine binding site (7-chloro-kynurenate) have been developed. AMPA and kainate receptors are often referred to as non-NMDA receptors and are blocked by the non-NMDA receptor antagonists CNQX and DNQX. Gene cloning has identified five kainate receptor subunits, four AMPA receptor subunits and six NMDA receptor subunits (Table 3.2).

Kainate receptors

Kainate receptors may be formed by homomeric combination of GluR5, GluR6 or GluR7 or by heteromeric combination of any of GluR5–7 with the kainate binding proteins, KA1 or KA2. KA1 and KA2 do not form functional homomeric receptors.

AMPA receptors

AMPA receptors can be formed by homomeric or heteromeric combinations of GluR1–4. However, most native AMPA receptors are more likely to be heteromeric assemblies

Table 3.2 Ion channel neurotransmitter receptor subunits

Acetylcholine	GABA	Glutamate
Muscle nicotinic AChR	*GABA*$_A$	*AMPA*
$\alpha1$ $\beta1$ γ ε δ	$\alpha1$ $\beta1$ $\gamma1$ δ ε π	GluR1
Neuronal nicotinic AChR	$\alpha2$ $\beta2$ $\gamma2$	GluR2
$\alpha2$ $\beta2$	$\alpha3$ $\beta3$ $\gamma3$	GluR3
$\alpha3$ $\beta3$	$\alpha4$ $\beta4$ $\gamma4$	GluR4
$\alpha4$ $\beta4$	$\alpha5$	*Kainate*
$\alpha5$	$\alpha6$	GluR5 KA1
$\alpha6$	*GABA*$_C$	GluR6 KA2
$\alpha7$	$\rho1$	GluR7
$\alpha8$	$\rho2$	*NMDA*
$\alpha9$	$\rho3$	NR1 (ξ) NR2A ($\varepsilon1$) NR3A
$\alpha10$		NR2B ($\varepsilon2$)
		NR2C ($\varepsilon3$)
		NR2D ($\varepsilon4$)

Notes:
Receptor subunits are numbered $\alpha1$, $\beta1$, etc. where there are more than one subunit of that type. Between different α subunits in any family there is around 60–70% amino-acid sequence homology whereas between α and β subtypes, for example, there is normally around only 40% sequence homology. Mouse NMDA receptor subunits are denoted by the Greek letters ξ and ε while rat and human are indicated as shown.

of GluR2 with one or more of the other AMPA receptor subunits. The main evidence for this came from the discovery that the mRNA for GluR2 is edited at the position coding for a glutamine residue in the TM2 region. mRNA editing produces an arginine residue and changes the functional properties of the receptor such that the receptor is no longer calcium permeable and the resulting synaptic current has a linear current–voltage relationship. This change in the GluR2 subunit dominates the functional properties of the receptor when heteromeric combinations of GluR subunits are expressed in oocytes or cell lines. Thus, wherever GluR2 subunits are expressed, the AMPA receptors so formed have a linear current–voltage relation and are Na^+ permeable but not Ca^{2+} permeable. This is the case at most synapses which have been investigated in the brain but interestingly, does not appear to be so at excitatory synapses onto interneurons where the AMPA receptors are probably heteromeric assemblies of GluR1, GluR3 and GluR4. mRNA for GluR2 is found to be almost 100% edited in the brain. The mRNAs for GluR5 and GluR6 are also edited at the equivalent position to that of GluR2 but the extent of editing varies between 50% and 90% in different brain regions. Thus kainate receptors are likely to be more heterogeneous than AMPA receptors in terms of their calcium permeability.

NMDA receptors

The NMDA receptors come in four main flavours depending on which NR2 subunit (or subunits) is involved in forming the receptor (Table 3.2). Functional NMDA receptors result from the co-assembly of one or more members of three different subfamilies of subunits: NR1, NR2 and NR3.

The NR1 family is composed of one subunit with nine different alternatively spliced variants. Block by NMDA channel blockers such as ketamine, MK-801 and phencyclidine is affected by which splice variant of the NR1 subunit is involved, probably because the NR1 splice variant affects the kinetics of channel activation (the effectiveness of any channel blocker being dependent on how much the channel is open). The glycine binding site is on the NR1 subunit and the glutamate binding site is on the NR2 and NR3 subunits.

The NR2 family is composed of four related subunits termed NR2A, NR2B, NR2C and NR2D and the NR3 family composed of two subunits termed NR3A and NR3B (Table 3.2). It is likely that the receptor is a tetramer composed of two NR1 subunits and two NR2 subunits or an NR2 plus an NR3 subunit. Expression of NR1 alone does not result in functional receptors, except in oocytes where it is now thought likely that the oocyte expresses a low level of native frog NR2 subunits. Expression of NR1 with an NR3 subunit in oocytes or cell lines does not result in functional receptors in contrast to expressing NR1 with an NR2 subunit which results in robust functional responses. An NR2 subunit is essential for function. None of the NR2 subunits form functional homomeric NMDA receptors.

Like the NR1 subunit, the NR2 subunits carry an asparagine residue in the pore-forming TM2 region (in an analogous position to the edited glutamine–arginine position of GluR2) that is important in regulating voltage-dependent Mg^{2+} block and calcium permeability of the channel and affects the affinity of non-competitive antagonists like ketamine. At present, subunit-selective antagonists have only been developed for NR2B-containing receptors. These antagonists such as ifenprodil and eliprodil act to inhibit NMDA receptors by increasing the proton sensitivity of the

NMDA receptors and have been widely investigated as possible neuroprotective agents for stroke and head injury patients (during stroke or mechanical injury pH falls which may result in enhanced effectiveness of this class of blocking drugs in the damaged region). However, the results of clinical trials with these drugs have so far been equivocal despite the known role of NMDA receptor activation in excitotoxic cell death. Compared to other glutamate receptor subunits, the NR2 subunits have the longest cytoplasmic carboxy terminal (>500 amino acids) which contains a conserved SXV sequence (where S is serine, X is any amino acid, and V is valine) that interacts with PSD-95, a postsynaptic anchoring protein and is thought to serve to anchor NMDA receptors at the synapse.

NR2 subunit expression is highly regulated during development and subunit segregation is observed between brain regions. In general, NR2B and NR2D are expressed early in development and NR2A and NR2C expression occurs in rats in the weeks after birth and continues into the adult in selected areas. In the adult cortex, hippocampus and thalamus NR2A and NR2B predominate while in the cerebellum and spinal cord, NR2C and NR2D are also expressed in the adult and their expression is specific to particular cell types.

Transgenic mice technology is increasingly used to investigate the functional significance of different receptor subunits to the development and behaviour of the whole animal. Mice lacking the $\varepsilon1$ (rat NR2A) subunit show apparently normal growth and mating behaviour, but an enhanced startle response and reduced hippocampal long-term potentiation (LTP) and spatial learning. Mice lacking the $\varepsilon2$ (rat NR2B) subunit die soon after birth showing impairment of suckling response, impairment of trigeminal neuronal pattern formation and impairment of hippocampal long-term depression (LTD). Mice lacking the $\varepsilon3$ (NR2C) subunit show apparently normal development and behaviour. Mice lacking the $\varepsilon4$ (NR2D) subunit show apparently normal growth and mating behaviour, but reduced spontaneous behavioural activity.

The details of NMDA receptor subunit expression during development are still only partly known at the single-cell level. Good examples of this can be found in the cerebellum where NMDA receptors have been investigated extensively by Cull-Candy and coworkers. In young animals, cerebellar granule cells express NR1 and NR2B. By three weeks old the rat granule cells express NR1, NR2A and NR2C. In contrast, cerebellar Golgi cells express NR1, NR2B and NR2D in both young and adult animals but it appears NR2D is not involved in synaptic transmission. In cerebellar Purkinje cells, young animals express NR1 and NR2D but in the adult, only NR1 is found and the cerebellar Purkinje cells are no longer responsive to NMDA, one of the few neurons in the brain to lack functional NMDA receptors.

The NR2 subunits confer diversity to the functional and pharmacological properties of NMDA receptors. They modulate properties such as the strength of Mg^{2+} block (NR2A and NR2B subunits are most sensitive and NR2C and NR2D are less sensitive to Mg^{2+}), glycine sensitivity (NR2D-containing receptors have the highest affinity for glycine) and kinetics of deactivation. NR2A-containing receptors have relatively rapid deactivation kinetics with a time constant of around 100 ms and mediate relatively fast synaptic currents, while NR2D-containing receptors have extremely slow deactivation kinetics with a time constant of around 4 s and have not yet been shown to be involved in synaptic transmission. The NMDA receptor single-channel properties have been shown to be dependent on the NR2 subunit. NR2A- and NR2B-containing receptors have a main single-channel conductance of 50 pS while NR2C- and NR2D-containing

Table 3.3 G-protein-coupled receptors

Receptor class	Neurotransmitter or hormone receptor
Amino acid receptors	Metabotropic glutamate and $GABA_B$ receptors
Monoamine receptors	Adrenoceptors, dopamine and 5-HT receptors, muscarinic and histamine receptors
Lipid receptors	Prostaglandin, thromboxane and PAF receptors
Purine receptors	Adenosine and ATP (P2Y) receptors
Neuropeptide receptors	Neuropeptide Y, opiate, cholecystokinin VIP, etc.
Peptide hormone receptors	Angiotensin, bradykinin, glucagon, calcitonin, parathyroid, etc.
Chemokine receptors	Interleukin-8
Glycoprotein receptors	TSH, LH/FSH, chorionic gonadotropin, etc.
Protease receptors	Thrombin

receptors have a main conductance level of 36 pS and a subconductance level of 18 pS.

G-PROTEIN-COUPLED RECEPTORS

The G-protein-coupled receptors generate a response by linking drug binding at the extracellular part of the receptor protein to activation of particular intracellular GTP binding proteins (G-proteins). The G-proteins act as transducers between the receptors and the effector systems. G-protein-coupled receptors represent the largest and most diverse class of membrane receptors: a super-family of receptors which ranges from rhodopsin to the odorant receptors (Table 3.3). These receptors are by far the biggest targets for drug development and particularly in the central nervous system they provide the opportunity to alter neuronal firing properties or release of transmitter from nerve terminals by inhibition of voltage-gated Ca^{2+} and K^+ channels, or by changing cAMP levels or by regulation of intracellular Ca^{2+} levels.

TRANSMEMBRANE TOPOLOGY AND TERTIARY STRUCTURE

X-ray crystallography has so far been unable to elucidate the structure of any membrane receptor. However, cryoelectron microscopy of two-dimensional arrays of receptors embedded in membrane lipid has produced electron density maps of both bacteriorhodopsin and bovine rhodopsin (Schertler, Villa and Henderson 1993). Bacteriorhodopsin is not a G-protein-coupled receptor but a proton pump while bovine rhodopsin is a true G-protein-coupled receptor. Both proteins are folded into seven helical bundles as illustrated in Fig. 3.3(a). Each of these seven alpha helices crosses the cell membrane giving rise to the fingerprint of the seven transmembrane (7-TM) domain receptors. The electron density map of rhodopsin can be compared with a helical-wheel diagram showing the amino acids of the 7-TM domains spiralling through the cell membrane (Fig. 3.3(b)). Molecular modelling suggests a particular packing arrangement for the alpha helices which is stabilised by a disulphide bridge between the second and third extracellular loops. The structure of other G-protein-coupled receptors is inferred to be similar to rhodopsin due to the presence of seven hydrophobic amino acid domains in all members of this receptor family so far sequenced.

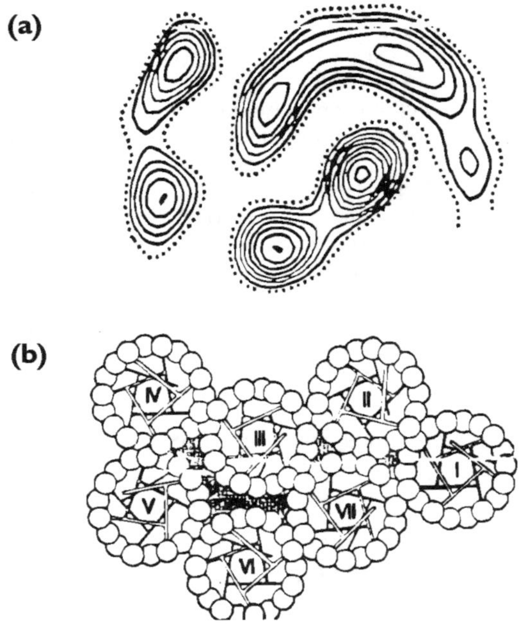

Figure 3.3 Molecular structure of G-protein-coupled receptors. In (a) the electron density map of bovine rhodopsin is shown as obtained by cryoelectron microscopy of two-dimensional arrays of receptors embedded in lipid membrane. The electron densities show seven peaks reflecting the seven α-helices which are predicted to cross the cell membrane. In (b) is shown a helical-wheel diagram of the receptor orientated according to the electron density map shown in (a). The diagram is seen as the receptor would be viewed from outside the cell membrane. The agonist binding pocket is illustrated by the hatched region between TM3, TM5 and TM6. (From Schertler *et al.* 1993 and Baldwin 1993, reproduced from Schwartz 1996). Reprinted with permission from *Textbook of Receptor Pharmacology*. Eds Foreman, JC and Johansen, T. Copyright CRC Press, Boca Raton, Florida

RECEPTOR ACTIVATION

Most structure–function information for the G-protein-coupled receptors has been inferred from molecular genetic experiments where single amino acids or groups of amino acids in the protein have been changed to investigate their role. The β-adrenoceptor was the first G-protein-coupled receptor to be cloned and a detailed picture of the receptor structure has emerged using the techniques of molecular biology combined with radioligand binding and classical pharmacology to study receptor function (see, for example, Lefkowitz *et al.* 1993). An outline structure of the β-adrenoceptor is shown in Fig. 3.4. Two main structural domains are recognised in all G-protein-coupled receptors:

(1) Ligand-binding domain
(2) G-protein-binding domain on the third intracellular loop

Ligand-binding domain

In the monoamine receptors the ligand-binding domain is located within the transmembrane helices. A pocket is formed between TM3, TM5 and TM6 where the agonist binds. A conserved aspartate residue in TM3 (Asp-113 in the β-adrenoceptor) and a

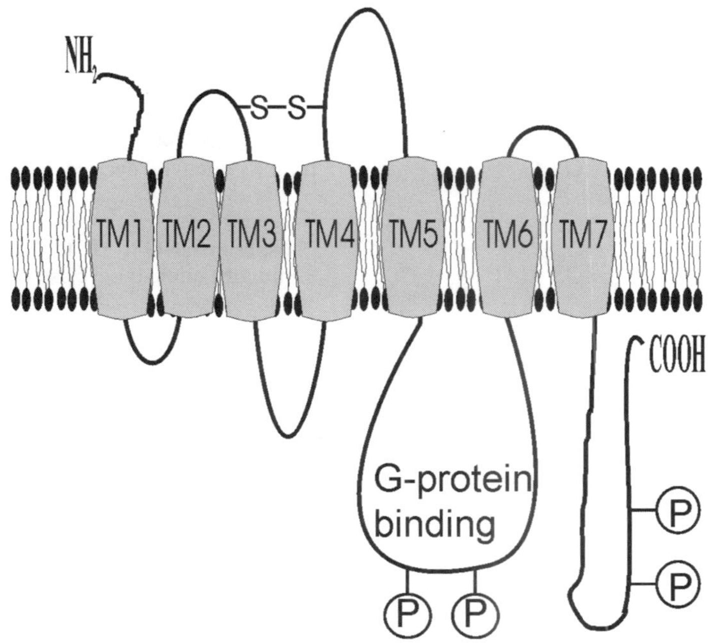

Figure 3.4 Transmembrane topology of a 7-TM domain G-protein receptor such as the β-adrenoceptor. Agonist binding is predicted to be within the transmembrane domains. The extracellular structure is stabilised by the disulphide bond joining the first and second extracellular loop. The third intracellular loop is the main site of G-protein interaction while the third intracellular loop and carboxy tail are targets for phosphorylation by kinases responsible for initiating receptor desensitisation

conserved phenylalanine in TM6 (Phe-290) and two serine residues in TM5 (Ser-204 and 207) are known to be crucial for agonist binding. Antagonists have been shown to have extra interaction points on TM4 and TM7 but are thought to largely share the same binding sites as the agonist and so can act by simple competition.

G-protein coupling

All rhodopsin-like G-protein-coupled receptors have a conserved arginine residue at the intracellular end of TM3 and this residue is thought to be crucial for G-protein activation. The third intracellular loop determines the class of G-protein activated by the receptor with the second intracellular loop and C-terminus also influencing G-protein binding in some cases. Four classes of G-protein are known:

(1) G_s—activates adenylyl cyclase (irreversibly activated by cholera toxin)
(2) G_i—inhibits adenylyl cyclase (inactivated by *Pertussis* toxin)
(3) G_q—activates phospholipase-C (not activated by *Pertussis* toxin or cholera toxin)
(4) G_o—inhibits voltage-dependent Ca^{2+} and K^+ channels (inactivated by *Pertussis* toxin)

Using chimaeric receptors it has been shown that swapping the third intracellular loop between receptors also swaps their G-protein selectivity. The G-protein-binding

Table 3.4 G-protein receptor-mediated responses

Receptor	G-protein activation	Effector mechanism
ACh-muscarinic		
M_1, M_3, M_5	G_q	Phospholipase-C to IP_3 to $[Ca^{2+}]_i$ regulation
M_2, M_4	G_o, G_i	Inhibits adenylyl cyclase, Ca^{2+}/K^+ channels
Histamine		
H_1, H_3	G_q	Phospholipase-C to IP_3 to $[Ca^{2+}]_i$ regulation
H_2	G_s	Stimulates adenylyl cyclase raising cAMP
Adrenoceptors		
α_1, α_2	G_o, G_i	Iinhibits adenylyl cyclase, Ca^{2+}/K^+ channels
β_1, β_2, β_3	G_s	Stimulates adenylyl cyclase raising cAMP
Opiate		
μ, δ, κ	G_o, G_i	Inhibits adenylyl cyclase, Ca^{2+}/K^+ channels
Dopamine		
D_1	G_s	Stimulates adenylyl cyclase raising cAMP
D_2, D_4	G_o, G_i	Inhibits adenylyl cyclase, Ca^{2+}/K^+ channels
D_3	G_q	Phospholipase-C to IP_3 to $[Ca^{2+}]_i$ regulation
$GABA_B$	G_o, G_i	Inhibits adenylyl cyclase, Ca^{2+}/K^+ channels
Glutamate-metabotropic		
mGluR1, 3, 5	G_q	Phospholipase-C to IP_3 to $[Ca^{2+}]_i$ regulation
mGluR2, 4, 6, 7	G_o, G_i	Inhibits adenylyl cyclase, Ca^{2+}/K^+ channels

domain and the carboxy terminal domain also contain a variable number of serine and threonine residues which are the target of protein kinases involved in receptor desensitisation. Different subtypes of G-protein-coupled receptor have evolved which couple to different G-proteins (Table 3.4) and it has been shown in heterologous expression systems that when a G-protein-coupled receptor is expressed at very high levels, the selectivity of G-protein coupling can be overruled and, for example, a receptor which normally couples to G_o can be made to activate G_i.

Receptor dimerisation

The $GABA_B$ receptors were the first G-protein-coupled receptors to be observed to form functional heterodimers (Bowery and Enna 2000) where two G-protein molecules come together to act as a dimer to enhance their combined response. A similar effect has recently also been described for dopamine and somatostatin receptors (Rocheville *et al.* 2000) and it is likely that this may occur with other G-protein-coupled receptors. The significance of this in terms of the pharmacology of the receptors is unclear, or indeed whether dimerisation affects mechanisms such as desensitisation.

7-TM DOMAIN RECEPTOR FAMILIES

Figure 3.5 shows diagrammatic illustrations of the transmembrane topology of the G-protein-coupled receptor families. Three main families have been identified:

(1) Rhodopsin-like 7-TM receptors
- Ligand binding within the transmembrane domains
- Monoamine, nucleotide and lipid receptors

(2) Glucagon, VIP and calcitonin family
- Ligand binding outside the transmembrane domains on cell surface

(3) Metabotropic glutamate receptors and chemosensor (Ca^{2+}) receptors
- Ligand binding on large extracellular N-terminus

Rhodopsin-like 7-TM receptors

By far the most studied family of the G-protein-coupled receptors are the rhodopsin-like receptors. These are also the largest group of receptors in number as they include receptors not only for the monoamines, nucleotides, neuropeptides and peptide hormones, but they also include the odorant receptors which number several hundreds of related receptors. These receptors have short N-termini, a conserved disulphide bridge between the TM2–TM3 and TM4–TM5 extracellular domains, and variable-length C-termini. In some cases the C-terminus is myristolyated which by tying the C-terminus to the cell membrane generates a fourth intracellular loop.

While the agonist binding domain is thought to be within the transmembrane domains for the monoamine and nucleotide receptors, neuropeptides are thought to bind close to the membrane surface on the extracellular domains of the receptor. It is still not clear whether non-peptide antagonists bind at the same or a different site on the receptor.

Glucagon, VIP and calcitonin family

These receptors are unlike the well-characterised rhodopsin-like family in that they have a large extracellular N-terminus and hormone binding seems to be dominated by this domain rather than the transmembrane domains. Receptors in this class include

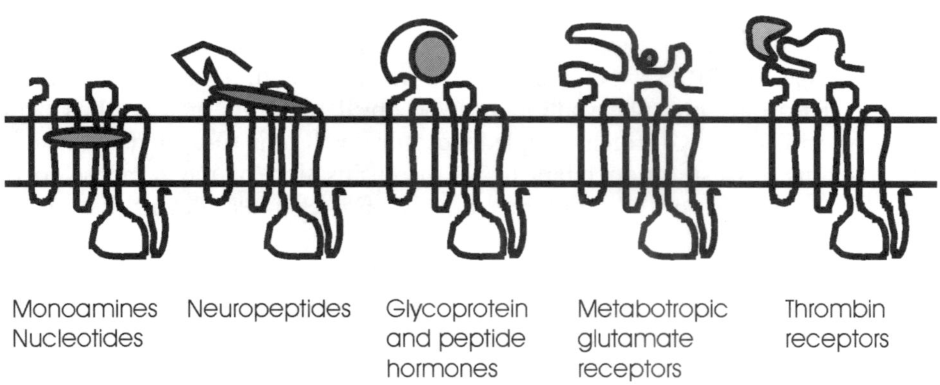

Figure 3.5 Ligand binding to G-protein-coupled receptors. Small ligands such as monoamines, nucleotides and lipids bind within the transmembrane domains while peptide and glycoprotein hormones bind outside the transmembrane region. Metabotropic glutamate receptors have agonist binding on the large N-terminal domain while the thrombin receptor is activated by cleavage of the N-terminal domain by thrombin (reproduced from Schwartz 1996). Reprinted with permission from *Textbook of Receptor Pharmacology*. Eds Foreman, JC and Johansen, T. Copyright CRC Press, Boca Raton, Florida

growth hormone-releasing hormone (GHRH), adrenocorticotrophic hormone-releasing factor and the neuropeptide, vasoactive intestinal polypeptide (VIP). All receptors in this family couple to G_s to stimulate adenylyl cyclase and hence raise cAMP levels.

Metabotropic glutamate receptors and chemosensor (Ca^{2+}) receptors

The metabotropic glutamate receptors are a distinct family of G-protein-coupled receptors which are homologous only to the Ca^{2+} sensors of the parathyroid and kidney. These receptors have an extremely large extracellular N-terminal domain of 500–600 amino acids (cf. 30–40 amino acids for the monoamine receptors). The binding of glutamate is thought to be within a 'clam-shell' region formed by two subdomains of the N-terminal region in an analogous manner to the binding of glutamate to the extracellular domain of the kainate, AMPA and NMDA receptor NR2 subunits.

Thrombin receptors

The thrombin receptor is unusual in that the receptor is activated by the enzymatic action of thrombin which cleaves the N-terminus of the receptor leaving the receptor constitutively active.

G-PROTEIN-COUPLED RECEPTOR DESENSITISATION

Desensitisation is a general phenomenon which serves as a negative feedback on receptor activity. The mechanisms of G-protein-coupled receptor desensitisation have been particularly investigated by Lefkowitz's group working on the β-adrenoceptors. In general, G-protein-coupled receptors desensitise following phosphorylation of specific serine and threonine residues on the third intracellular loop and on the COOH tail. These amino acids are targets for phosphorylation by protein kinases and specific G-protein receptor kinases (GRKs) which result in receptor desensitisation by inhibiting G-protein coupling, stabilising receptor *arrestin* binding, and controlling receptor internalisation. Two forms of desensitisation have been characterised: *homologous* and *heterologous*. Homologous desensitisation refers to desensitisation of the response to an agonist due to prior application of the same agonist. Homologous desensitisation is mediated by specific GRKs which are activated by the generation of free G-protein $\beta\gamma$ subunits. GRKs only phosphorylate agonist occupied receptors making their action relatively specific to the G-protein-coupled receptor involved in G-protein stimulation.

Heterologous desensitisation refers to the desensitisation of the response to one agonist by the application of a different agonist. For example, desensitisation of a response to adrenaline by application of 5-HT is mediated by protein kinase A or protein kinase C because these kinases can phosphorylate receptors which are not occupied by agonist. Phosphorylation disrupts the receptor–G-protein interaction and induces the binding of specific proteins, *arrestins* which enhance receptors internalisation via clathrin-coated pits. Thus desensitisation of G-protein-coupled receptors results in a decrease in the *number* of functional receptors on the cell surface.

CONSTITUTIVELY ACTIVE RECEPTORS

One current model of G-protein receptor activation is the **allosteric ternary complex** model of Lefkowitz and Costa. The agonist, receptor and G-protein must combine to form a *ternary* complex in order to generate a response. Thermodynamically, in this freely reversible mechanism receptors may occasionally adopt the active conformation in the absence of ligand and may then cause G-protein activation. This constitutive activity can be demonstrated by the effect of 'antagonists' which inhibit the unstimulated activity of the receptor and so are known as 'inverse agonists' (see Appendix). Constitutive activity has been produced by specific point mutations of the β-adrenoceptor where conversion of Ala-293 to glu-293 results in a tenfold increase in constitutive activity. In contrast, some antagonists ('pure antagonists') do not alter the constitutive activity of the receptor.

GENETIC DISEASE

Mutations producing constitutively active receptors have been found in several rare genetic diseases (Coughlin 1994) such as hyperthyroidism (TSH receptor), precocious puberty (LH receptor), and retinitis pigmentosa (rhodopsin). Clearly the ability to manipulate the degree of constitutive receptor activity using drugs could provide a therapeutic strategy in these diseases.

CONCLUSIONS

Neurotransmitter receptors have evolved as one of the key components in the ability of the central nervous system to coordinate the behaviour of the whole animal, to process and respond to sensory input, and to adapt to change in the environment. These same receptors are therefore ideal targets for drug action because of their central role in the activity of the nervous system. A rational approach to the development of new therapeutic strategies involving the action of drugs at receptors in the nervous system is based on knowledge of receptor structure, distribution and function.

APPENDIX

1 DERIVATION OF THE HILL COEFFICIENT (OR SLOPE) AS A DETERMINANT OF THE NUMBER OF BINDING SITES FOR AN AGONIST (NEUROTRANSMITTER) ON ITS RECEPTOR

The relationship between drug concentration and receptor occupancy

The simplest model for the reversible combination of a drug, A, with its receptor, R, is:

$$A + R \underset{k_{-1}}{\overset{k_{+1}}{\rightleftharpoons}} AR \qquad \text{(A3.1)}$$

The law of mass action states that the rate of a reaction is proportional to the product of the concentrations of the reactants. Thus the rate of the forward reaction is proportional to $[A][R] = k_{+1}[A][R]$, where k_{+1} is the *association rate constant* (with units of $M^{-1}\ s^{-1}$). Likewise, the rate of the backward reaction is proportional to $[AR] = k_{-1}[AR]$, where k_{-1} is the *dissociation rate constant* (with units of s^{-1}). At equilibrium, the rates of the forward and backward reactions will be equal so

$$k_{+1}[\mathrm{A}][\mathrm{R}] = k_{-1}[\mathrm{AR}]$$

Since receptors are embedded in cell membranes it is more convenient to consider the amount of receptor either free or bound as a proportion of the total receptor present. The proportion of receptors in state R (free) is

$$p_{\mathrm{R}} = [\mathrm{R}]/[\mathrm{R}]_{\mathrm{T}}$$

where $[\mathrm{R}]_{\mathrm{T}}$ is the total concentration of receptors and

$$p_{\mathrm{AR}} = [\mathrm{AR}]/[\mathrm{R}]_{\mathrm{T}}$$

is the proportion of receptors occupied by the drug. As the system is at equilibrium we can write

$$k_{+1}[\mathrm{A}][\mathrm{R}]/[\mathrm{R}]_{\mathrm{T}} = k_{-1}[\mathrm{AR}]/[\mathrm{R}]_{\mathrm{T}}$$

and hence

$$k_{+1}[\mathrm{A}]p_{\mathrm{R}} = k_{-1}p_{\mathrm{AR}}$$

Rearranging gives

$$[\mathrm{A}]p_{\mathrm{R}} = k_{-1}/k_{+1}p_{\mathrm{AR}}$$

$k_{-1}/k_{+1} = K_{\mathrm{A}}$ is defined as the *dissociation equilibrium constant*. This is often referred to as the *equilibrium constant*. Substitution for k_{-1}/k_{+1} gives

$$[\mathrm{A}]p_{\mathrm{R}} = K_{\mathrm{A}}p_{\mathrm{AR}}$$

Rearranging gives

$$p_{\mathrm{R}} = \{K_{\mathrm{A}}/[\mathrm{A}]\}\,p_{\mathrm{AR}}$$

In the simple two-state model shown in equation (A3.1) the receptor can only exist in either free or bound states and so

$$p_{\mathrm{R}} + p_{\mathrm{AR}} = 1$$

Substituting for p_{R} gives

$$\frac{K_{\mathrm{A}}}{[\mathrm{A}]}\,p_{\mathrm{AR}} + p_{\mathrm{AR}} = 1$$

Hence

$$p_{\mathrm{AR}} = \frac{[\mathrm{A}]}{K_{\mathrm{A}} + [\mathrm{A}]} \tag{A3.2}$$

This is known as the **Hill–Langmuir equation**. Its derivation assumes that the concentration of A does not change as drug receptor complexes are formed. In effect, the drug is considered to be present in such excess that the number of drug molecules in solution is many times greater than the number of receptor molecules available for the drug to bind to. It can be rearranged to

$$\frac{p_{\mathrm{AR}}}{1 - p_{\mathrm{AR}}} = \frac{[\mathrm{A}]}{K_{\mathrm{A}}}$$

and taking logs gives

$$\log\left(\frac{p_{AR}}{1-p_{AR}}\right) = \log[A] - \log K_A \tag{A3.3}$$

Hence a plot of log $(p_{AR}/(1-p_{AR}))$ against log[A] gives a straight line of unit slope. This is known as a **Hill plot**.

In practice, it is not often possible to directly measure p_{AR} except in radioligand binding experiments. In many experiments it is the relationship between agonist concentration [A] and percentage maximum response (y) which is measured (a dose–response curve) and the Hill plot is made by plotting

$$\log\left(\frac{y}{100-y}\right) = \log[A] - \log K_A \tag{A3.4}$$

The slope of this log–log plot is known as the **Hill coefficient** (n_H) or **Hill slope**. If the slope is 1 this may imply there is only one agonist binding site on the receptor, while a slope approaching 2 implies two binding sites. In practice, the slope of the line may be greater or less than unity and is rarely an integer. Factors which can affect the Hill slope are particularly the presence of more than one population of receptors with different affinities for the agonist contributing to the response ($n_H < 1$), occurrence of receptor desensitisation ($n_H < 1$), the presence of more than one agonist binding site on the receptor (as occurs with the ligand-gated ion channel receptors) where more than one site needs to be occupied for efficient activation of the receptor ($n_H > 1$), and the presence of spare receptors in the tissue ($n_H > 1$).

Concentration–response curves are often fitted *empirically* by the expression

$$y = y_{max}\frac{[A]^{n_H}}{[A]_{50}^{n_H} + [A]^{n_H}} \tag{A3.5}$$

where n_H is the Hill coefficient and y_{max} is the maximum response. $[A]_{50}$ is the concentration of A at which y is half maximal. Equation (A3.5) is known as the **Hill equation**. $[A]_{50}$ is sometimes denoted by K. However, the constant K obtained by fitting the Hill equation does not correspond to an equilibrium constant as defined above when deriving the Hill–Langmuir equation.

2 KINETICS AND MECHANISMS OF AGONIST ACTION

(a) Ion channel receptors

Agonist responses at ligand-gated ion channels and drug effects at ion channels are often more amenable to mechanistic investigation because the response (ionic current through open ion channels when measured with voltage or patch-clamp techniques) is directly proportional to receptor activation. This is a great advantage and has allowed electrophysiological techniques to be used to study ion channel activation and drug block of ion channels in great detail.

The first physically plausible mechanism for receptor activation was proposed by del Castillo and Katz (1957). They made the important distinction that agonist binding and channel opening of the AChR must occur as two separate steps:

$$A + R \underset{k_{-1}}{\overset{k_{+1}}{\rightleftharpoons}} AR \underset{k_{-2}}{\overset{k_{+2}}{\rightleftharpoons}} AR^* \tag{A3.6}$$

In the del Castillo and Katz model it is important to notice that the fraction of receptors occupied is the sum of both active (AR*) and occupied, but inactive (AR) receptors:

$$\begin{aligned} p_{\text{occ}} &= p_{\text{AR}} + p_{\text{AR*}}, = (1 + K_2)p_{\text{AR*}} \\ &= \frac{(1 + K_2)[\text{A}]}{K_1K_2 + (1 + K_2)[\text{A}]} \\ &= \frac{[\text{A}]}{\dfrac{K_1K_2}{1 + K_2} + [\text{A}]} \end{aligned}$$

This can be rewritten as

$$p_{\text{occ}} = \frac{[\text{A}]}{K_{\text{eff}} + [\text{A}]}$$

where K_{eff} is the **effective dissociation equilibrium constant**. Thus, most macroscopic estimates of the equilibrium constant for an agonist (radioligand binding, EC_{50} from the occupancy versus agonist concentration–response curve) the estimated equilibrium constant will depend on both affinity for the receptor and subsequent activation steps on the receptor.

(b) G-protein-coupled receptors

The simplest possible model for the action of such receptors is as follows:

$$\begin{aligned} A + R &\overset{K_A}{\rightleftharpoons} AR \\ AR + G &\overset{K_{ARG}}{\rightleftharpoons} ARG^* \end{aligned}$$

where the binding of a G-protein (G) to AR to form a ternary complex (ARG*) is described by the equilibrium constant K_{ARG}. However, recent results suggest that G-protein-coupled receptors (and potentially other receptors) can exist in the active state in the absence of agonist. These **constitutively active receptors** give rise to interesting new predictions for the shape of the dose–response curve and an alternative interpretation for the difference between agonists, partial agonists and antagonists (Lefkowitz *et al.* 1993; Jenkinson 1996). Suppose the receptors can isomerise spontaneously to and from an active form:

$$\underset{\text{(inactive)}}{R} \rightleftharpoons \underset{\text{(active)}}{R^*}$$

In principle, both states of the receptor could be occupied by a ligand, L:

$$\begin{array}{rcl} \text{(inactive) R} & \overset{K_A}{\rightleftharpoons} & \text{R*(active)} \\ + & & + \\ \text{L} & & \text{L} \\ K_{LR} \upharpoonleft\downharpoonright & & \upharpoonleft\downharpoonright K_{LR^*} \\ \text{(inactive) LR} & & \text{LR* (active)} \end{array}$$

If L combines only with R, then the presence of L will *reduce* the proportion of active receptors. L is said to be an **inverse agonist** or **negative antagonist**. If L combines with R* there will be an increase in active receptors and so L will behave as a conventional agonist. Where L has equal affinity for R and R*, it will not affect the fraction of receptors in the active state. However, it will reduce the binding of either a conventional or an inverse agonist, and so will behave as an antagonist.

REFERENCES

Baldwin, JM (1993) The probable arrangement of the helices in G-protein-coupled receptors. *EMBO J.* **12**: 1693–1703.

Bormann, J (2000) The 'ABC' of GABA receptors. *TIPS* **21**(1): 16–19.

Bowery, NG and Enna, SJ (2000) Gamma-aminobutyric acid(B) receptors: first of the functional metabotropic heterodimers. *J. Pharmacol. Exp. Ther.* **292**(1): 2–7.

Clarke, PB (1992) The fall and rise of neuronal α-bungarotoxin binding proteins. *TIPS* **3**(11): 407–413.

Coughlin, SR (1994) Expanding horizons for receptors coupled to G-proteins: diversity and disease. *Curr. Op. Cell Biol.* **6**: 191–197.

Del Castillo, J and Katz, B (1957) Interaction at endplate receptors between different choline derivatives. *Proc. Roy. Soc. Lond. Ser. B* **146**: 369–381.

Jenkinson, DH (1996) Classical approaches to the study of drug–receptor interactions. In *Textbook of Receptor Pharmacology* (Eds Foreman, JC and Johansen, T), CRC Press, Boca Raton, FL, pp. 159–185.

Lefkowitz, RJ, Cotecchia, S, Samama, P and Costa, T (1993) Constitutive activity of receptors coupled to guanine nucleotide regulatory proteins. *TIPS* **14**: 303–307.

Lukas, RJ, Changeux, JP, Le Novere, N, Albuquerque, EX, Balfour, DJ, Berg, DK, Bertrand, D, Chiappinelli, VA, Clarke, PB, Collins, AC, Dani, JA, Grady, SR, Kellar, KJ, Lindstrom, JM, Marks, MJ, Quik, M, Taylor, PW and Wonnacott, S (1999) International Union of Pharmacology. XX. Current status of the nomenclature for nicotinic acetylcholine receptors and their subunits. *Pharmacol. Rev.* **51**(2): 397–401.

North, RA and Surprenant, A (2000) Pharmacology of cloned P2X receptors. *Ann. Rev. Pharmacol. Toxicol.* **40**: 563–580.

Rocheville, M, Lange, DC, Kumar, U, Patel, SC, Patel, RC and Patel, YC (2000) Receptors for dopamine and somatostatin: formation of hetero-oligomers with enhanced functional activity. *Science* **288**(5463): 154–157.

Sautel, M and Milligan, G (2000) Molecular manipulation of G-protein-coupled receptors: a new avenue into drug discovery. *Curr. Op. Med. Chem.* **7**(9): 889–896.

Schertler, GF, Villa, C and Henderson, R (1993) Projection structure of rhodopsin. *Nature* **362**: 770–772.

Schwartz, TW (1996) Molecular structure of G-protein coupled receptors. In *Textbook of Receptor Pharmacology* (Eds Foreman, JC and Johansen, T), CRC Press, Boca Raton, FL, pp. 66–84.

Schwartz, TW and Rosenkilde, MM (1996) Is there a 'lock' for all agonist 'keys' in 7TM receptors? *TIPS* **17**: 213–216.

Unwin, N (1995) Acetylcholine receptor channel imaged in the open state. *Nature* **373**: 37–43.

Unwin, N (2000) Nicotinic acetylcholine receptor and the structural basis of fast synaptic transmission. *Phil. Trans. Roy. Soc. Lond. Ser. B* **355**: 1813–1829.

Vernier, P, Cardinaud, B, Valdenaire, O, Philippe, H and Vincent, J-D (1995) An evolutionary view of drug–receptor interaction: the bioamine receptor family. *TIPS* **16**: 375–381.

4 Neurotransmitter Release

S. C. STANFORD

INTRODUCTION

Release of transmitter from a neuron is triggered by the arrival of a propagated nerve impulse at its terminals. This wave of excitation causes the opening of voltage-gated Ca^{2+}-channels or mobilisation of Ca^{2+} from intracellular stores (e.g. the endoplasmic reticulum or 'calcisomes'). As a result, there is a phasic increase in free intracellular Ca^{2+}, probably to a concentration of about 0.1–1 mM, in regions of the terminal adjacent to the site of transmitter release (the 'active zone'). The subsequent fusion of neurotransmitter storage vesicles with the axolemma, together with the extrusion of their contents into the synapse, is thought to take about 100–200 μs; this cascade is therefore fast enough to effect rapid signalling between neurons.

While this chapter is concerned primarily with the neurochemical mechanisms which bring about and control impulse-evoked release of neurotransmitter, some of the methods used to measure transmitter release are described first. This is because important findings have emerged from studies of the effects of nerve stimulation on gross changes in transmitter release and intraneuronal stores. The actual processes that link neuronal excitation and release of transmitter from nerve terminals have been studied only relatively recently. The neurochemical basis of this stimulus–secretion coupling, which is still not fully understood, is described next. The final sections will deal with evidence that, under certain conditions, appreciable amounts of transmitter can be released through Ca^{2+}-independent mechanisms which do not depend on neuronal activation.

MEASUREMENT OF TRANSMITTER RELEASE

ESTIMATION OF TRANSMITTER TURNOVER *EX VIVO*

Until the development of sensitive assays and sophisticated collection techniques, release studies relied on measuring changes in the concentration of neurotransmitters in whole organs, or dissected brain regions, following nerve stimulation. However, under resting conditions, the transmitter content of any given organ or brain region is remarkably constant. The store of classical transmitters (monoamines and acetylcholine) in nerve terminals is rarely depleted by physiologically relevant rates of neuronal stimulation. This suggests that transmitter synthesis normally keeps pace with release. It follows that the rate at which the store of transmitter is replaced ('turnover rate') can be used to estimate its rate of release. Although this approach is rarely used nowadays, it is outlined here because it uncovered some important findings which are relevant to current studies of the regulation of transmitter release.

Neurotransmitters, Drugs and Brain Function. Edited by R. A. Webster

The turnover rate of a transmitter can be calculated from measurement of either the rate at which it is synthesised or the rate at which it is lost from the endogenous store. Transmitter synthesis can be monitored by administering [^{3}H]- or [^{14}C]-labelled precursors *in vivo*; these are eventually taken up by neurons and converted into radiolabelled product (the transmitter). The rate of accumulation of the radiolabelled transmitter can be used to estimate its synthesis rate. Obviously, the choice of precursor is determined by the rate-limiting step in the synthetic pathway: for instance, when measuring catecholamine turnover, tyrosine must be used instead of *l*-DOPA which bypasses the rate-limiting enzyme, tyrosine hydroxylase.

One limitation of this method is that the specific activity of the radiolabel is progressively diluted as the radiolabelled transmitter is released from neurons and replaced by that derived from unlabelled substrate. This method also assumes that there is no compartmentalisation of the terminal stores, yet there is ample evidence that newly synthesised acetylcholine and monoamines are preferentially released. An alternative approach is to monitor the rate at which the store of neurotransmitter is depleted after inhibition of its synthesis (Fig. 4.1). However, the rate of release of some neurotransmitters (e.g. 5-HT) is partly governed by their rate of synthesis and blocking synthesis blunts release.

It is already evident that the turnover rate of a transmitter is only a crude measure of its release rate. Further limitations are that there is appreciable intraneuronal metabolism of some neurotransmitters: notably, the monoamines. In such cases, turnover will overestimate release rate. Another problem, again affecting monoamines, is that some of the released neurotransmitter is taken back into the nerve terminals and recycled. This leads to an underestimate of release rate. Despite these drawbacks, studies of turnover rates uncovered some important features of transmitter release. In particular, they provided the first evidence for distinct functional pools of monoamines, acetylcholine and possibly other neurotransmitters: a 'release' pool, which could be rapidly mobilised for release, and a 'storage' or 'reserve' pool which had a slower turnover rate.

TECHNIQUES *IN VITRO*

In situ preparations

Many early studies of transmitter release depended on measuring its concentration in the effluent of a stimulated, perfused nerve/end-organ preparation. This technique is still widely used to study drug-induced changes in noradrenaline release from sympathetic neurons and the adrenal medulla. However, it is important to realise that the concentration of transmitter will represent only that proportion of transmitter which escapes into the perfusate ('overflow') (Fig. 4.2). Monoamines, for instance, are rapidly sequestered by uptake into neuronal and non-neuronal tissue whereas other transmitters, such as acetylcholine, are metabolised extensively within the synapse. Because of these local clearance mechanisms, the amount of transmitter which overflows into the perfusate will depend not only on the frequency of nerve stimulation (i.e. release rate) but also on the dimensions of the synaptic cleft and the density of innervation.

Synaptosomes

Synaptosomes are 'pinched-off' nerve terminals which become severed from the parent axon during gentle homogenisation of brain tissue and then subsequently reseal. They

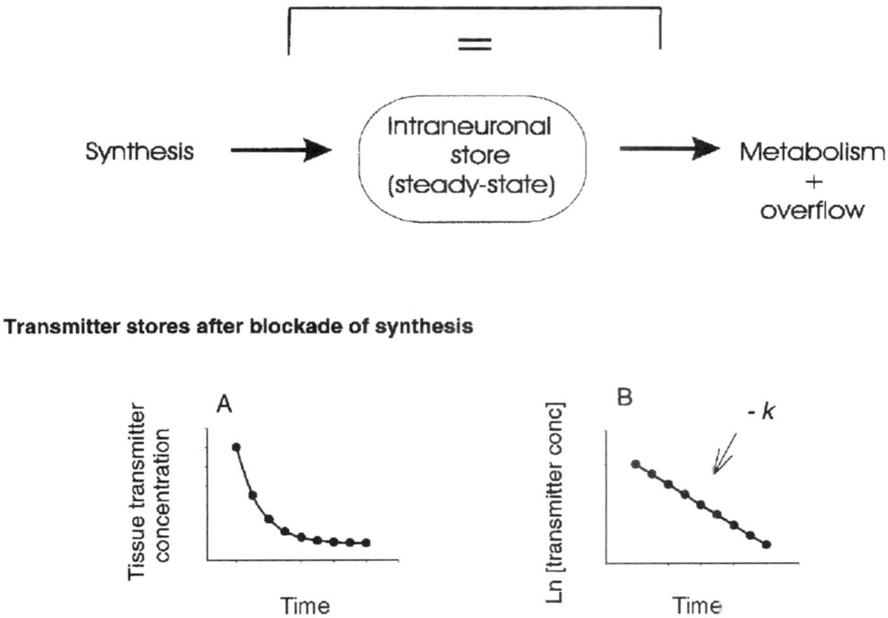

Figure 4.1 Turnover of classical neurotransmitters. At normal rates of neuronal activity, endogenous stores of neurotransmitter are maintained at constant (steady-state) levels, indicating that the supply of new neurotransmitter (through synthesis) meets the demand (determined by release and metabolism). Consequently, the rate of the depletion (A) of the endogenous store of transmitter after inhibition of its synthesis indicates turnover rate and is described by the equation:

$$[\mathrm{T}] = [\mathrm{T}]_0 \mathrm{e}^{-kt}$$

where [T] is the tissue concentration at time t; $[\mathrm{T}]_0$ is the transmitter concentration at time 0; and k is the rate constant for the efflux of transmitter. When plotted semi-logarithmically (B), the exponential decline in tissue stores of transmitter gives a straight line described by the equation:

$$\ln[\mathrm{T}] = \ln[\mathrm{T}]_0 - kt$$

At steady-state there is no net loss of transmitter from the system and so the rate of synthesis of transmitter equals the rate of its efflux. Thus:

$$\text{turnover rate (TOR)} = k \times \text{tissue transmitter concentration at time 0}$$

$$\mathrm{TOR} = k[\mathrm{T}]_0$$

For monoamines, turnover rate is only an approximate measure of release rate because of recycling of released transmitter and spontaneous metabolism of the endogenous store

are separated as a crude synaptosomal pellet from undisrupted tissue by differential centrifugation of the brain homogenates (Fig. 4.3). After resuspension of the synaptosomal pellet, they are perfused with artificial cerebrospinal fluid (aCSF) and the concentration of transmitter in the effluent used as an index of its release.

The main advantage of using synaptosomes is that they are free from any influence of the parent axon. Another is that, since the volume of extracellular space (the incubation medium) is functionally infinite, transmitter will not accumulate near the synaptosomes. This means that reuptake of released transmitter is unlikely to occur and that, under drug-free conditions, transmitter release will not be modified by activation of auto- or heteroceptors (see below).

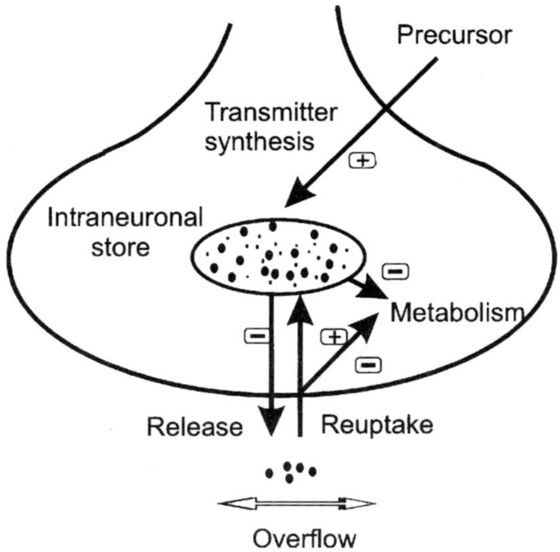

Figure 4.2 The intraneuronal stores of monoamines are maintained by synthesis from precursors taken in with the diet. The pool is depleted by release of transmitter and some spontaneous metabolism of intraneuronal transmitter. Released monoamines are inactivated by reuptake on membrane-bound transporters. Following reuptake, some transmitter might be recycled while the remainder is metabolised. Some transmitter escapes the reuptake process and overflows from the synapse in the extracellular fluid

A disadvantage of using synaptosomes is that they cannot be used to study transmitter release evoked by propagated nerve impulses, but the release, like that from intact neurons, is Ca^{2+}-dependent and K^{+}-sensitive. Pharmacological studies using synaptosomes have also provided evidence that the amount of transmitter that is released following their depolarisation is regulated by the activation of presynaptic receptors.

If the amount of transmitter released into the effluent is too small to be measured by a conventional neurochemical assay, the endogenous store can be 'tagged' by incubating the synaptosomes with radiolabelled transmitter (usually ^{3}H). After washing the synaptosomes, the release of ^{3}H into the perfusate is then measured using (extremely sensitive) liquid scintillation counting. This approach rests on the assumption that the radiolabelled transmitter is taken up only by the 'right' synaptosomes (e.g. that only those derived from noradrenergic neurons will take up [^{3}H]noradrenaline) which is not always the case. Another unjustified assumption is that the radiolabelled transmitter mixes freely with the endogenous (unlabelled) store. Finally, as it is the radiolabel, rather than the actual transmitter itself, which is being measured it must be established that the label remains attached to the neurotransmitter. For instance, [^{3}H]glutamate can be turned into [^{3}H]GABA or [^{3}H]glutamine and so care must be taken to ensure that any findings with ^{3}H refer to release of [^{3}H]glutamate rather than that of its metabolites.

Brain slices

These can be conventional cross-sectional slices up to 0.5 mm thick, or small cubes (0.1–0.5 mm on each plane) or pyramids of brain tissue. The main advantage of using slices is that, either by crudely passing currents across them or, in some cases, stimulating a

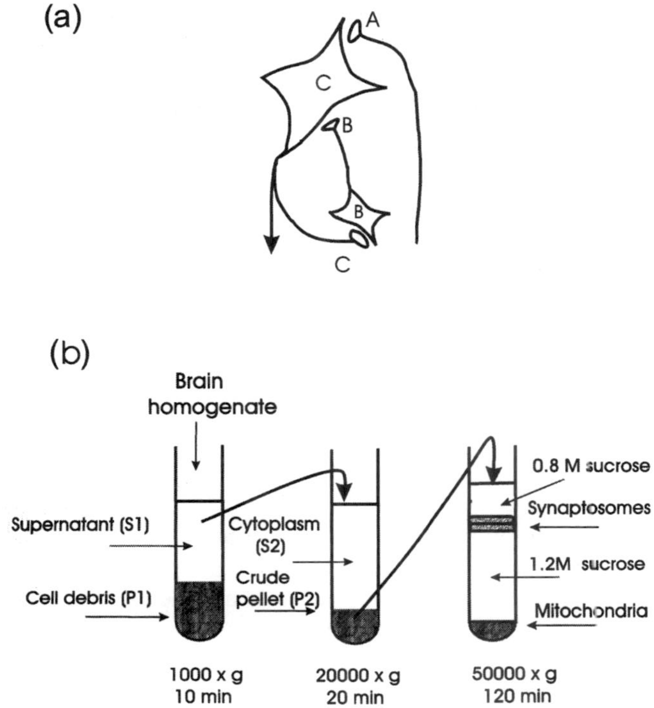

Figure 4.3 The distribution of neurotransmitter in different subcellular fractions. (a) Within the hypothetical brain area shown, where it is assumed that this neuronal arrangement is reproduced many times, the proportion of each neurotransmitter found in the synaptosomes, rather than the cell body (cytoplasmic) fraction, will vary considerably. On the assumption that, although concentrated in nerve terminals, the neurotransmitter will also be found in cell bodies and axons, it is likely that A will be almost entirely in synaptosomes, C mostly in the cytoplasm, while B will be more evenly divided. (b) Procedure for preparation and separation of synaptosomes. Brain tissue is homogenised and spun at $1000 \times g$ for 10 min to remove cell debris. The supernatant (S1) is spun at $20\,000 \times g$ for 10 min and the pellet (P2), containing synaptosomes and mitochondria, is spun through two layers of sucrose for 2 h at $50\,000 \times g$. This is 'sucrose density gradient centrifugation' and is based on the principle that an individual particle will settle in the zone of the sucrose gradient that has the same density. Thus, synaptosomes separate from other elements of the $20\,000 \times g$ pellet and settle at the interface between 0.8 M and 1.2 M sucrose

specific neuronal tract, they can be used to study impulse-evoked release of transmitter. Because the three-dimensional integrity of the tissue is maintained, they can also be used to study modulation of transmitter release by heteroceptors (see below).

One approach, and the first to be adopted, is to study transmitter release from slices which have been preloaded with radiolabelled transmitter. In these experiments, drug-induced changes in the release of transmitter is usually monitored using the 'double-pulse' technique. This involves comparing the effects of a test drug on the amount of transmitter released in response to a reference pulse and a second identical test pulse. If all the radiolabelled transmitter that overflows in the effluent is collected, and the amount which remains in the slice at the end of the experiment is also measured, it is possible to calculate not only how much radiolabelled transmitter was originally contained in the slice but also the effects of drugs on 'fractional release', i.e. the proportion of the store of radiolabelled transmitter which is released by nerve stimulation. As with synaptosomes, however, it cannot be assumed that incubation of slices in a medium containing radiolabelled transmitter results in its even distribution throughout the slice. Also, the problem of continuous dilution of the radiolabelled store with newly synthesised (unlabelled) transmitter must be borne in mind.

Modern sensitive chromatographic and voltammetric techniques now make it possible to estimate the release of unlabelled endogenous transmitter from slices of brain tissue (commonly the hippocampus and striatum) or spinal cord (Fig. 4.4). However, whatever analytical method is used, the thickness of the slice is paramount. It is important to maintain the balance between preserving the integrity of the tissue (the thicker the slice, the better) against maintaining tissue viability by perfusion with oxygenated aCSF (the thinner the slice, the better).

TECHNIQUES *IN VIVO*

The cortical cup

The cortical cup has been used for many years to monitor changes in transmitter release induced by physiological and pharmacological stimuli (Fig. 4.5). In the past, it was used most commonly to study release of amino acids and acetylcholine. More recently, it has

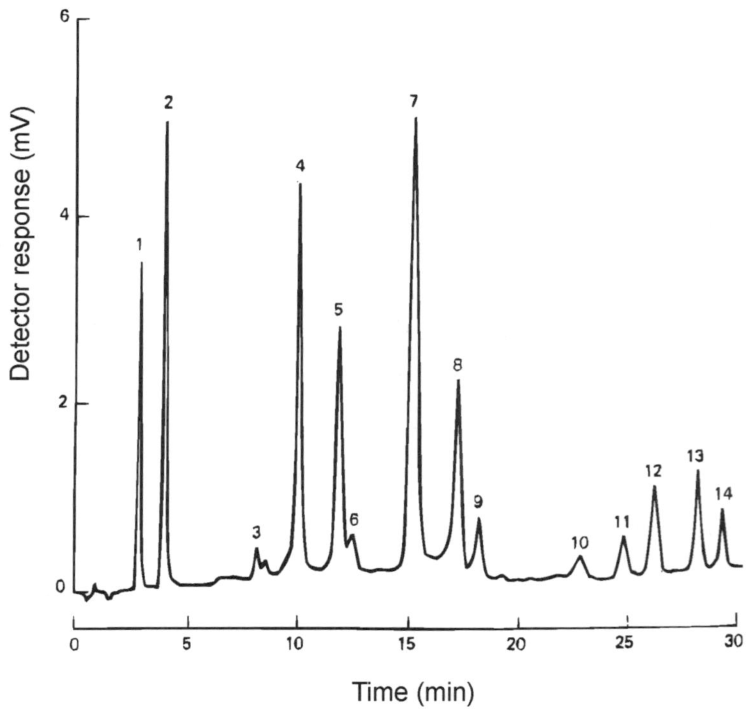

Figure 4.4 Release of amino acids from cortical slices exposed to 50 mM K^+. Measurements by HPLC and fluorescence detection after reaction of amino acids with *o*-phthalaldehyde: 1, aspartate; 2, glutamate; 3, asparagine; 4, serine; 5, glutamine; 6, histidine; 7, homoserine (internal standard); 8, glycine; 9, threonine; 10, arginine; 11, taurine; 12, alanine; 13, GABA; 14, tyrosine. Glutamate concentration is almost 1 pmol/μl which represents a release rate of 30 pmol/min/mg tissue

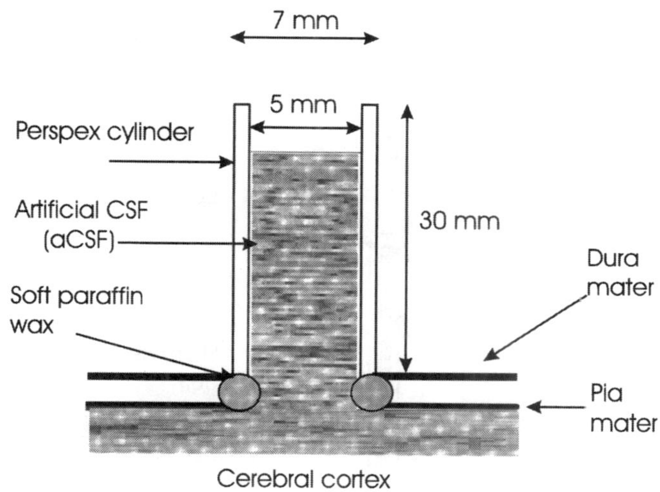

Figure 4.5 The cortical cup. The cup size obviously varies with the animal on which it is used. For the rat, the overall diameter is 6–7 mm and a typical flow rate is 50 μl/min

been adopted in studies of peptide and purine release as well (e.g. Phyllis, Smith-Barbour and O'Regan 1996). The cup comprises a 'well', constructed above a small hole drilled in the skull, which contains aCSF. Transmitters released from neurons near the surface of the brain accumulate in the cup and their concentration in the aCSF can be used as a crude index of release rate. The importance of this technique is that it enabled, for the first time, the monitoring of transmitter release in freely moving animals.

Microdialysis

This is a modification of the earlier push–pull cannula which could be used in anaesthetised animals only. The microdialysis probe which has an outside diameter of about 250 μm (Fig. 4.6) is implanted into the brain under anaesthesia and then subsequently perfused with aCSF. Solutes (including neurotransmitters) in the extracellular fluid of the brain diffuse down their concentration gradient into the probe. By taking samples of the effluent dialysate at regular intervals it is possible to monitor changes in transmitter release. This technique has been used for several years to study release of monoamines (e.g. Sharp, Umbers and Gartside 1997) but is now used to harvest acetylcholine and amino acids as well. Since the molecular cut-off of the dialysis membrane is in the region of 6–20 kDa (depending on the type of membrane used), this technique can also be used to measure release of some small neuropeptides (e.g. oxytocin and vasopressin).

One advantage of microdialysis is that it enables the study of transmitter release in specific brain areas or nuclei. To ensure its correct placement, the probe is implanted, under anaesthesia, by sterotaxic surgery. Another advantage is that the probe can be anchored in place with dental cement and experiments carried out later, in conscious freely moving animals once they have recovered from the anaesthetic. Indeed, comparison of results from studies carried out on both anaesthetised and freely moving subjects has revealed drug interactions with anaesthetics that can affect transmitter release: anaesthetic-induced changes in the regulation of noradrenaline release by α_2-adrenoceptors is a case in point. It is also possible to carry out long-term

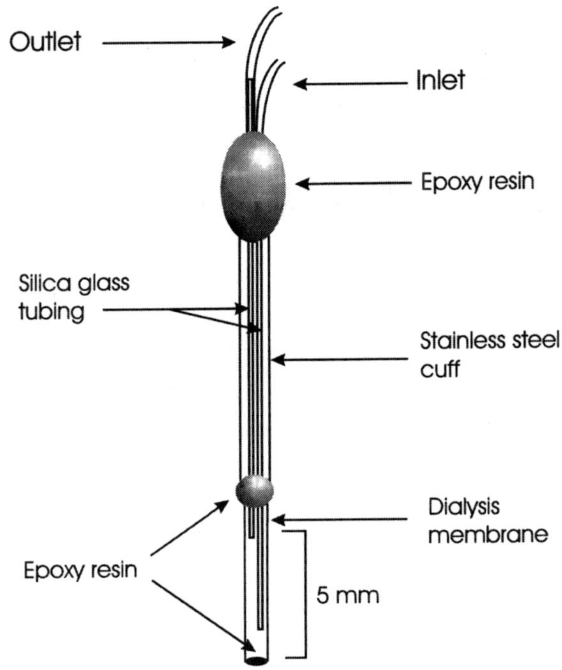

Figure 4.6 The tip of a microdialysis probe, expanded to show dialysis tubing around a steel cannula through the base of which fluid can flow out and then up and over the membrane. The length of membrane below the probe support can be altered (1–10 mm) to suit the size of the animal and the brain area being studied. Flow rates are normally below 2 μl/min

or repeated studies on the same animals but this requires a slight modification of the technique. In this case, a 'guide' cannula is first implanted, under anaesthesia, through which the microdialysis probe is inserted. Unfortunately, for a variety of reasons, each microdialysis probe can be used for only a few hours and so it has to be replaced each day. However, the presence of the guide cannula makes this a relatively straightforward process that requires only light sedation of the animal.

A further advantage of microdialysis is that, unlike the push–pull cannula or the cortical cup, the perfusion medium does not come into direct contact with the tissue being studied. This reduces damage caused by turbulence as well as enzymic degradation of the transmitter. For instance, acetylcholine, but not cholinesterase, will penetrate the probe membrane. Finally, because solutes will pass out of the probe, as well as into it, the probe can also be used for infusing ions (Fig. 4.7) or drugs into specific brain regions ('reverse' or 'retro-' dialysis) so that their local effects on transmitter release can be studied. This avoids many of the problems that arise when trying to determine the synaptic actions of drugs when these are administered systemically.

The rate at which the probes are perfused with aCSF is a compromise between the time required for the solutes in the CSF to reach equilibrium with those in the probe (the slower, the better) versus the ideal time-frame for studying changes in transmitter release (the shorter, the better). In general, flow rates of around 1–2 μl/min are used and the time which elapses between taking samples is determined by how much transmitter

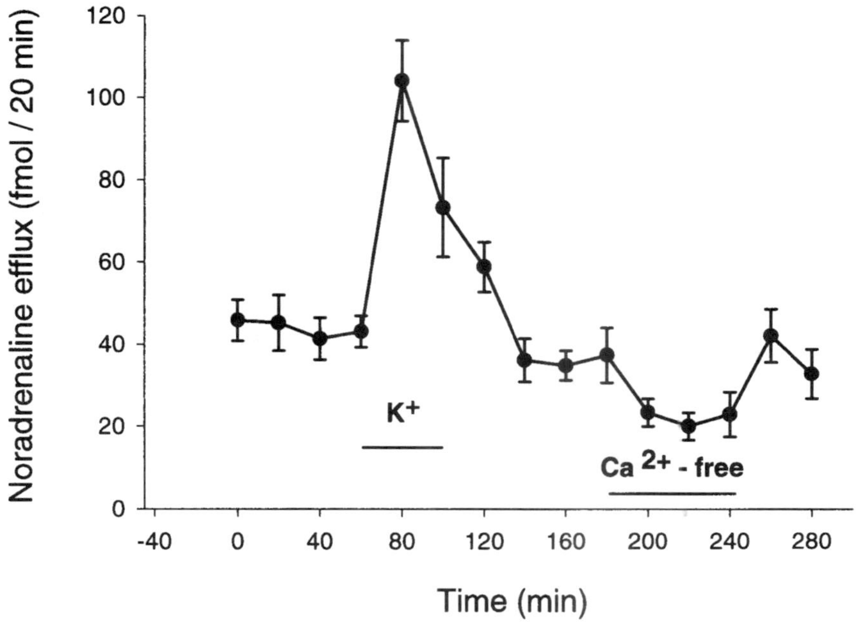

Figure 4.7 The effect of perfusion of the microdialysis probe with a medium containing a depolarising (80 mM) concentration of K^+, or Ca^{2+}-free medium, for the periods indicated by the bars. The graph shows efflux of noradrenaline in the frontal cortex of anaesthetised rats. Increasing the concentration of K^+ in the medium infused via the probe increases noradrenaline efflux whereas removing Ca^{2+} reduces it

is needed for analysis: i.e. by the sensitivity of the assay system. It is acknowledged that the solutes are not in equilibrium with the CSF outside the probe. In any case, the efficiency of the probe membrane limits the net influx (or efflux) of solutes to about 10–20% of the theoretical maximum. It should also be borne in mind that the microdialysis probe is not sampling the transmitter in the synapse: only that transmitter which escapes metabolism in, or reuptake from, synapses will migrate towards the probe. In the drug-free state, any change in the transmitter concentration in the dialysis samples is usually assumed to indicate a change in its rate of release from nerve terminals; this is supported by the spontaneous efflux of transmitters being Ca^{2+}-dependent and K^+-sensitive (Fig. 4.7). However, efflux does not always reflect release rate, especially if experimental interventions (e.g. infusion of monoamine uptake inhibitors) interfere with the clearance of transmitter from the synapse (Fig. 4.8).

Voltammetry

This can be carried out *in vitro* (in brain slices, cultured cell preparations) or *in vivo* and involves penetrating the experimental tissue with a carbon-fibre electrode of 5–30 μm in diameter (Fig. 4.9). This serves as an oxidising electrode and the Faradaic current generated by the oxidation of solutes on the surface of the electrode is proportional to their concentration. Obviously, only neurotransmitters which can be oxidised can be measured in this way so the technique is mainly limited to the study of monoamines and their metabolites. The amplitude of each peak on the ensuing voltammogram is a measure of solute concentration and individual peaks can be identified because different

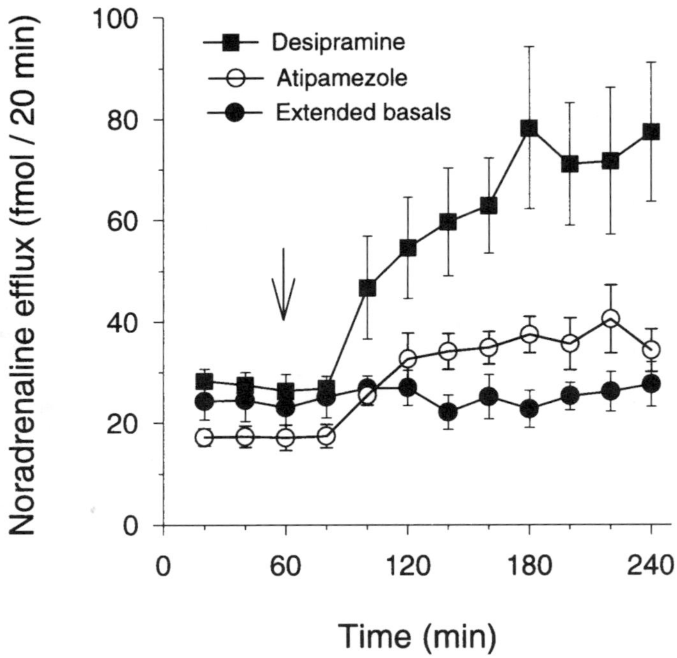

Figure 4.8 Noradrenaline concentration in dialysis samples from probes implanted in the rat frontal cortex. Spontaneous efflux of noradrenaline is stable throughout a 4 h sampling period ('extended basals') but is increased markedly when either the noradrenaline reuptake inhibitor, desipramine (5 μM), or the α_2-adrenoceptor antagonist, atipamezole (0.5 μM), is infused into the extracellular fluid via the microdialysis probe ('retrodialysis')

solutes oxidise at different potentials. Changes in the concentration of transmitters are monitored by rapid cycles of voltage scans (e.g. Palij and Stamford 1994). Since a complete scan takes only about 20 ms, the time resolution with voltammetry is much better than with microdialysis and is suitable for studying rapid, transient changes in transmitter release.

One difficulty with this method is that all oxidisable solutes in the external medium will be incorporated into the voltammogram and interfering peaks can be a problem. In fact, the concentration of monoamine metabolites and oxidisable solutes can be considerably greater than those of the parent amines which can be difficult to distinguish as a result. Ascorbic acid and uric acid are particularly problematic in this respect, although recent work suggests that an increase in the concentration of extracellular ascorbic acid could be a marker for the early phase of cerebral ischaemia. In general, voltammetry is most useful for measuring rapid (subsecond) changes in monoamine release. Under these circumstances, slower changes in the metabolites and other compounds do not interfere. Another problem is that the life of the electrode is limited by progressive 'poisoning' which diminishes its sensitivity. As a rule, voltammetric electrodes are best suited to 'acute' rather than 'chronic' measurements.

Biosensors

As the term suggests, the use of biosensors to measure transmitter release rests on exploiting a biological response which is proportional to the amount of transmitter in

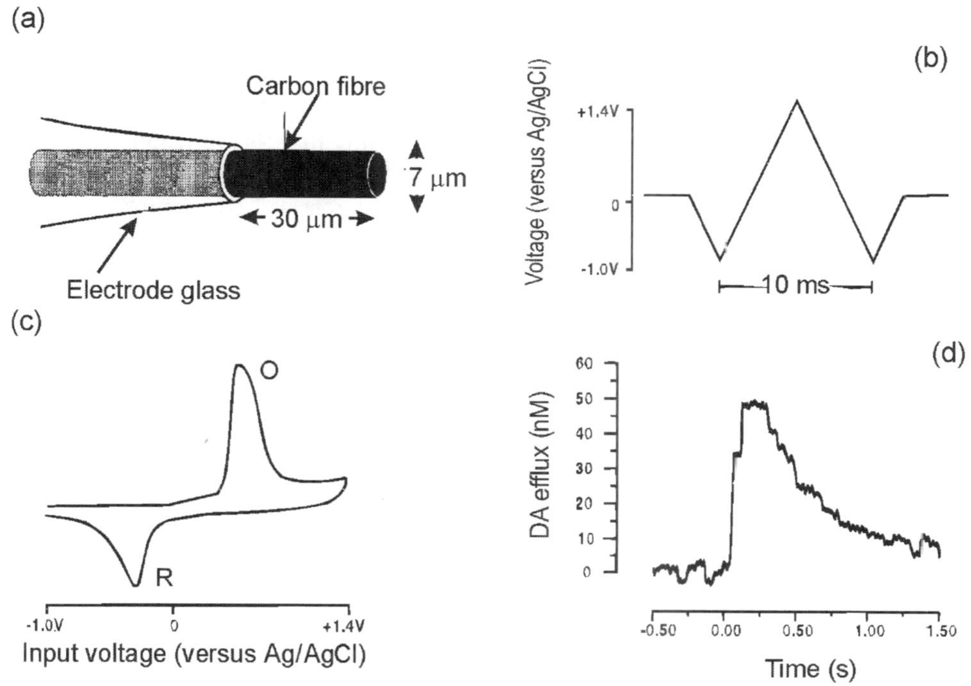

Figure 4.9 Basics of voltammetry. (a) The cut tip of the microelectrode surrounded by the glass insulation. (b) Input voltage waveform to the potentiostat (−1.0 to +1.4 V versus Ag/AgCl, 480 V/s). (c) Background current-corrected cyclic voltammogram of catecholamine obtained by plotting the Faradaic current against the input voltage waveform. The areas marked O and R are the oxidation and reduction currents. (d) Typical dopamine release and reuptake event following local electrical stimulation of a striatal slice. The trace is a plot of the oxidation peak height against time, calibrated for dopamine. (Figure and legend kindly supplied by J. A. Stamford)

the sample and which can be quantified. One of the earliest biosensors was the dorsal wall muscle of the leech which contracts in the presence of nM concentrations of acetylcholine. Others are the bioluminescent proteins, such as aequorin, which fluoresce in the presence of Ca^{2+}. Within a reasonable range, the fluorescence intensity is proportional to the cation concentration and so it can be used to monitor the increase in the intracellular concentration of Ca^{2+} during excitation of nerve terminals. More recently, biosensors have been developed which comprise electrodes coated with glucose oxidase or lactate oxidase. The activity of these enzymes generates a current that can be used to quantify the concentration of glucose and lactate on the surface of the electrode. This work is playing an important part in research on brain metabolism during neuronal activity. So far, these electrodes are used in 'on-line' assays of samples collected by microdialysis but might be adapted for measurements *in vivo* in the future.

WHERE DOES THE RELEASED TRANSMITTER COME FROM?

Two separate lines of research led to the proposal that transmitter released in response to neuronal excitation is derived from a vesicle-bound pool rather than from the neuronal cytoplasm. One piece of evidence came from electron microscopy which showed that nerve terminals were packed with vesicle-like organelles (Fig. 4.10). Using differential centrifugation, these vesicles were soon identified as the major storage sites for neurotransmitters. The second was electrophysiological evidence that the effect of neuronal release of acetylcholine on the postsynaptic membrane potential at the neuromuscular junction was quantal in nature, suggesting that this transmitter, at least, was released in discrete packets.

Early neurochemical investigations of the source of released transmitter measured noradrenaline release from chromaffin granules in the adrenal medulla. Chromaffin granules are considerably larger (250 nm diameter) than the storage vesicles in noradrenergic nerve terminals (40–100 nm) and so their experimental use avoided the constraint imposed by the low sensitivity of early assay techniques (see Winkler 1993). Yet, like noradrenergic neurons, the adrenal medulla is derived from the developing neural crest and noradrenaline release is activated by stimulation of preganglionic cholinergic neurons. Chromaffin granules therefore provide a useful model for processes involved in the storage and release of noradrenaline from neurons. Subsequent refinements of assays for noradrenaline enabled studies of noradrenaline release to be extended to stimulated sympathetic nerve/end-organ preparations. These experiments confirmed that noradrenaline was released from vesicle-bound packets of transmitter contained within the terminal vesicles. This is because its release was paralleled by the

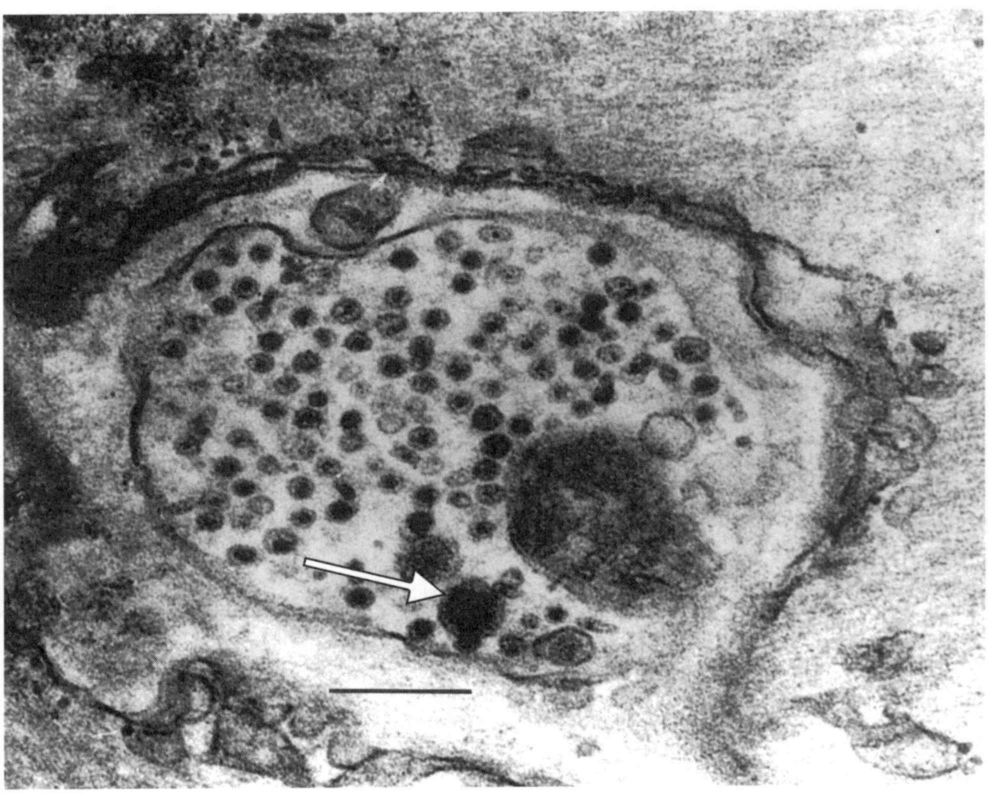

Figure 4.10 An electron micrograph of a terminal varicosity containing a large dense-core vesicle (LDCV), indicated by the arrow and many small synaptic vesicles (SSVs), some of which contain an electron dense core. Calibration mark: 250 nM. (Figure kindly supplied by M. Fillenz)

appearance of the proteins, dopamine-β-hydroxylase and chromogranins, which are found only in noradrenaline storage vesicles, whereas the cytoplasmic enzyme, lactate dehydrogenase, was not found in the perfusate.

Experiments of this kind have provided a great deal of evidence in favour of exocytotic release of vesicular noradrenaline. For example, by administering reserpine (which causes noradrenaline to leak out of the vesicles into the cytoplasm) together with an inhibitor of the enzyme monoamine oxidase (which will prevent metabolism of cytoplasmic noradrenaline), it is possible to redistribute the noradrenaline stored within nerve terminals because it leaks from the vesicles but is preserved within the neuronal cytoplasm. Under these conditions, the total amount of transmitter in the terminals is unchanged but impulse-evoked release rapidly diminishes.

Different evidence, mainly based on histological studies, suggested that acetylcholine is also released by vesicular exocytosis. Landmark experiments used a technique known as 'freeze–fracture' in which tissues are frozen rapidly during periods of intense transmitter release. It is then possible to fracture axolemma membranes in a way that separates their lipid bilayer. Electron microscopy reveals numerous pits in the membranes which are thought to reflect the vesicle/axolemma fusion pore of vesicles in the process of exocytosis. Subsequent studies, combining immunocytochemistry with electron microscopy, showed that proteins in the membranes of vesicles become incorporated into the axolemma during transmitter release. Furthermore, when neurons are stimulated in a medium containing an electron-dense marker, that does not penetrate the neuronal membrane, the marker later appears in vesicles inside the nerve terminals (Basbaum and Heuser 1979). This suggests that such markers are incorporated into the vesicles when they come into contact with the extracellular fluid during exocytosis. There is also some pharmacological evidence for exocytosis of acetylcholine. For instance, impulse-evoked release of this transmitter is prevented by the drug, vesamicol, which blocks uptake of acetylcholine from the cytoplasm into the terminal vesicles (Searl, Prior and Marshall 1991).

Although most evidence supports vesicular exocytosis of acetylcholine (see Ceccarelli and Hurlbut 1980), some researchers contest this view. An alternative suggestion is that an ATPase bound to the axolemma acts as a pore ('mediatophore'). According to this scheme, opening of the pore is triggered by an increase in the concentration of intracellular Ca^{2+} and allows gated release of aliquots of cytoplasmic acetylcholine. The vesicles are thought to serve merely as a reserve pool of transmitter and for sequestration of intracellular Ca^{2+} (Dunant 1994).

TRANSMITTER STORAGE VESICLES

Electron microscopy and biochemical techniques, such as sucrose density-gradient centrifugation (see Fig. 4.3), have enabled the characterisation of at least three types of vesicles in neurons. 'Large dense core vesicles' (LDCV: 75–100 nm), so-called because of their electron-dense core, contain mainly peptides and are thought to derive from the Golgi apparatus in the cell body. They are delivered to the terminals by fast axoplasmic transport and are the only type of vesicle to be found in axons (see Calakos and Scheller 1996). LDCVs are also found in nerve terminals but, in this part of the neuron, it is small synaptic vesicles (SSV: 40–50 nm) which predominate even though the proportion of LDCVs and SSVs varies from organ to organ (see Fillenz 1990). SSVs generally have an electron translucent core but, when gluteraldehyde is used as a fixative, the majority of SSVs in catecholamine-releasing neurons have an electron dense core and are known as 'small dense-core vesicles' (SDCVs) (Fig. 4.10). SSVs contain the classical transmitters (acetylcholine, monoamines and amino acids) which are mostly loaded in the nerve terminals.

Whether SSVs are derived from LDCVs has been a matter of considerable debate but differences in the protein markers in their membranes, notably synapsin (see below), makes this unlikely (see De Camilli and Jahn 1990). Also, electron microscopic evidence has shown that LDCVs, unlike SSVs, dock at the axolemma at sites remote from the active zone at the synapse. This suggests that they have different functions and regulatory processes which, since they contain peptides, agrees with the finding that their release requires higher frequencies of nerve stimulation than does that of the classical neurotransmitters. Whether this is because LDCVs and SSVs are confined to different regions of the nerve terminal, or have different affinities for, or a topographical relationship with, Ca^{2+} channels is unresolved (Sihra and Nichols 1993).

What happens to vesicles after exocytosis is controversial. Electron microscopy certainly shows that their membranes are recovered after fusion with the axolemma but precisely how this occurs is unresolved. One possibility is that they are retrieved intact from the active zone, immediately after release has taken place. Alternatively, they could become incorporated into, and mix with, the components of the axolemma but are reformed after sorting of the different membrane elements (see Kelly and Grote 1993). Recent studies of exocytosis from retinula cells of the *Drosophila* fly suggest that both these processes for membrane retrieval can be found within individual cells. These studies have shown that there is rapid recovery of vesicular membrane from the active zone. However, a second slower process exists which takes place at sites remote from the active zone and involves the formation of invaginations in the axolemma. This process is thought to precede endocytosis because the formation of these invaginations is followed by the appearance of tubular cisternae within the nerve terminal from which new vesicles bud-off (Koenig and Ikeda 1996). This finding raises the interesting question of whether these different processes lead to the formation of two different populations of synaptic vesicles with different release characteristics.

RELEASE VERSUS STORAGE POOL

Evidence from measurements of transmitter turnover has long suggested that not all transmitter is immediately available for release. This led to the concept of 'release' and 'storage' pools of neurotransmitter but explaining the physiological basis of these different pools has not been straightforward. Some evidence favours the view that the release pool of transmitter comprises those vesicles which are near to the 'active zone' where vesicles fuse with the axolemma and which is richly endowed with Ca^{2+} channels. The reserve pool would then comprise vesicles which are docked, more remotely, on the neuronal cytoskeleton. It is thought that vesicles move from one pool to the other as a result of the actions of protein kinases which effect cycles of phosphorylation/dephosphorylation of proteins, known as synapsins, which are embedded in the vesicle membranes.

SYNAPSINS

Synapsins comprise a family of phosphoproteins that are found only in association with SSVs. Although they account for only about 9% of the total vesicular membrane protein they probably cover a large proportion of their surface. So far, synapsins Ia, Ib, IIa, IIb and III, which are the products of different genes, have been identified.

Recent evidence suggests that, while synapsins might have a role in synaptogenesis, they also regulate the supply of vesicles to the release pool (Hilfiker *et al.* 1999). Experiments *in vitro* have shown that dephosphorylated synapsin I causes growth and bundling of actin filaments which are a major component of neuronal microfilaments. Such findings form the basis of the hypothesis that synapsin I forms a ternary complex with transmitter storage vesicles and the neuronal cytoskeleton, thereby confining vesicles to a reserve pool (Fig. 4.11; see Greengard, Benfenati and Valtorta 1994). All procedures which increase Ca^{2+}-dependent release of transmitter (e.g. electrical stimulation or increasing the concentration of K^+ in the external medium) induce phosphorylation of synapsin I. Phosphorylated synapsin dissociates from the vesicles and *F*-actin, reduces the number of vesicle anchoring sites, and so frees the vesicles to the release pool. Synapsin I is the substrate for several protein kinases but one of these, Ca^{2+}/calmodulin protein kinase II (CAM kinase II), is bound to vesicle membranes to some extent. It is thought that phosphorylation of synapsin I by CAM kinase II is triggered by the influx of Ca^{2+}. This process would enable synapsin to act as a regulator of the balance between the releasable and reserve pools of vesicles.

Much evidence supports this scheme. For example, neuronal depolarisation increases the amount of free synapsin in the cytosol and microinjection of CAM kinase II into the terminals of the squid giant axon or brain synaptosomes increases depolarisation-evoked transmitter release. By contrast, injection of dephosphorylated synapsin I into either the squid giant axon or goldfish Mauthner neurons inhibits transmitter release.

It has also been suggested that synapsin promotes vesicle clustering by a process which is not dependent on phosphorylation. It achieves this by forming cross-bridges between vesicles and by stabilising the membranes of the aggregated vesicles, thereby enabling them to cluster in the active zone without fusing with each other or the axolemma. When synapsin dissociates from the vesicles, as occurs during neuronal excitation, this membrane-stabilising action is lost. This would enable fusion of the membranes of vesicles, clustered near the active zone, with the axolemma. This scheme is supported by evidence that vesicles near the active zone have much lower concentrations of synapsin than those located more remotely (Pieribone *et al.* 1995). Further roles for synapsins are currently being investigated. For instance, it has been suggested that they might also regulate the kinetics of release, downstream of the docking process.

VESICULAR EXOCYTOSIS

Once vesicles detach from the cytoskeleton they are free to participate in the release process but our understanding of precisely how this is brought about is still sketchy, despite the wealth of information which has accumulated over recent years. What is clear is that it involves a complex cascade of regulatory processes focusing on proteins bound to vesicle membranes, the axolemma and some cytoplasmic factors (see Calakos

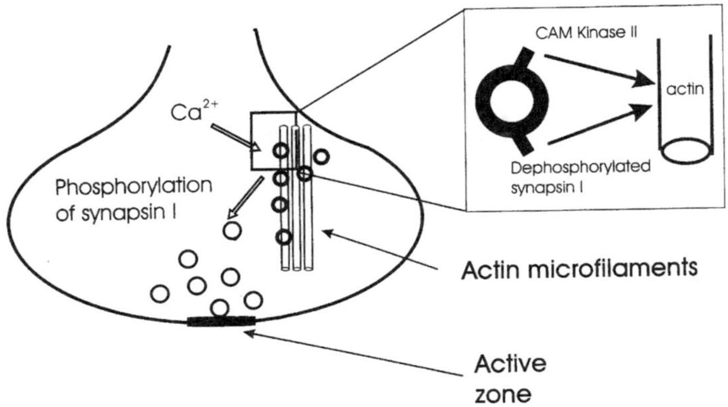

Figure 4.11 Dephosphorylated synapsin, associated with SSVs, is thought to form a heteromeric complex with CAM kinase II (also partially embedded in the vesicular membrane) and actin filaments. An increase in intracellular Ca^{2+} triggers phosphorylation of synapsin I which dissociates from the vesicular membrane. This frees the vesicles from the fibrin microfilaments and makes them available for transmitter release at the active zone of the nerve terminal

and Scheller 1996). The following sections will deal with those factors about which most is known and which are thought to have a prominent role in exocytosis. The extent to which this scheme explains release from large dense-cored vesicles is unclear, not least because these vesicles are not found near the active zone.

DOCKING AND FUSION

Because exocytosis is so rapid, it is believed that Ca^{2+} must trigger release from vesicles which are already docked at the active zone. The processes leading to docking and fusion of the vesicle with the axolemma membrane are thought to involve the formation of a complex between soluble proteins (in the neuronal cytoplasm) and those bound to vesicular or axolemma membranes. Much of this evidence is based on studies of a wide range of secretory systems (including those in yeast cells) but which are thought to be conserved in mammalian neurons.

From evidence collected to date, a scheme has emerged, known as the SNARE hypothesis (see Söllner and Rothman 1994) (Fig. 4.12). The soluble proteins referred to above include N-ethylmaleimide sensitive factor ('NSF', an ATPase) and SNAPs which comprise a family of 'soluble NSF attachment proteins'. Evidence, largely derived from studies of the Golgi apparatus, suggests that SNAPs have a general role in protein–protein interactions underlying membrane fusion. Proteins thought to act as SNAP receptors ('SNARES') are found both in the axolemma (known as 'target SNARES' or tSNARES) and vesicles (vSNARES). The vesicle protein, synaptobrevin (also known as 'vesicle associated membrane protein' or 'VAMP') is thought to act as a vSNARE and couples with the tSNARE proteins: syntaxin and SNAP-25 (synaptosomal associated protein: 25 KDa). It is envisaged that this complex of the two SNARES enables sequential binding of the soluble SNAPs and NSF. Subsequent hydrolysis of ATP by NSF enables dissociation of the complex and fusion of the membrane so that the vesicle contents can be discharged into the synapse.

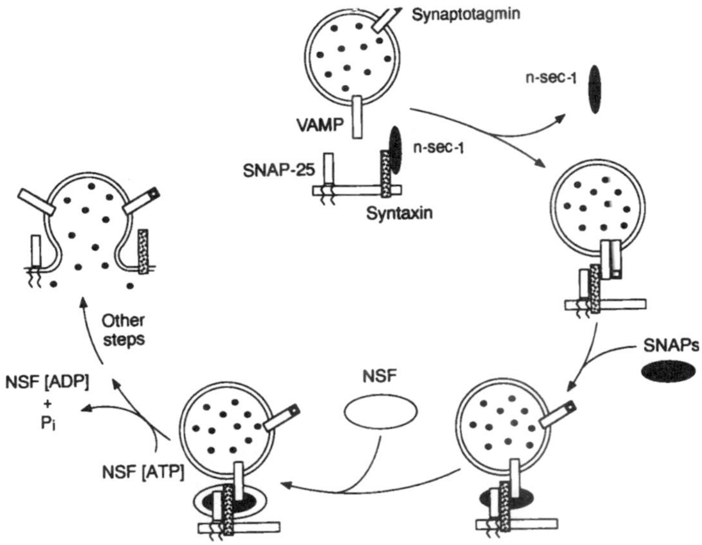

Figure 4.12 Hypothetical model of the action of SNAP receptors (SNARES) during vesicle docking, activation and fusion. It is thought that synaptic vesicle docking to the presynaptic plasma membrane requires the removal of a regulatory cytosolic protein, n-sec-1, from a VAMP (synaptobrevin) binding site on syntaxin. This results in the dissociation of synaptotagmin from the SNARE complex and binding of SNAPs and NSF, thus initiating fusion. The ATP hydrolysed by NSF results in disassembly of the SNARE complex. How the interconversion of these complexes occurs and which components trigger these processes is poorly understood. Proteins such as rab 3A, Ca^{2+} binding proteins and Ca^{2+} channels are likely to be involved. (From Söllner and Rothman 1994, page 346 with permission from Elsevier Science)

Much evidence supports a role for these proteins in exocytosis. For instance, injection of recombinant SNAP into the squid giant axon increases vesicular exocytosis. Also, membrane SNAP-25 and syntaxin are both targets for botulinum toxin while the vesicule protein, synaptobrevin, is a target for tetanus and botulinum toxins; both these toxins are well known for disrupting transmitter release.

How all these processes are influenced by Ca^{2+} is uncertain but another vesicle membrane-bound protein, synaptotagmin, is widely believed to effect this regulatory role (Littleton and Bellen 1995). Synaptotagmin has a single membrane-spanning domain with the NH_2–tail penetrating the vesicle and the COOH–tail extending into the cytoplasm. This tail binds Ca^{2+} and could enable synaptotagmin to act as a Ca^{2+}-sensor but, although it is found in adrenergic and sensory neurons, it appears to be absent from motor neurons.

Another protein, synaptophysin (p38), is the most abundant of the vesicle proteins and is found in the membranes of both SSVs and LDCVs. Its transmembrane structure resembles that of connexins which form gap junctions and has provoked the theory that neuronal excitation might cause synaptophysin to act as a fusion pore. There is no doubt that many other factors are involved in regulating the docking–fusion–extrusion process, including the Rab family of GTP-binding proteins and the Rab3 effectors, Rabphilin and Rim. For a detailed review of the role of all these factors in the exocytotic cycle, see Benfanati, Onofri and Giovedi 1999.

RECEPTOR-MEDIATED MODULATION OF Ca^{2+}-DEPENDENT TRANSMITTER RELEASE

Regulation of transmitter release does not rest solely on the frequency at which nerve impulses reach the terminals. Early experiments using stimulated sympathetic nerve/end-organ preparations *in situ*, or synaptosomes, indicated that release of [^{3}H]noradrenaline was attenuated by exposure to unlabelled, exogenous transmitter. This action was attributed to presynaptic adrenoceptors, designated α_2-adrenoceptors, which were functionally distinct from either α_1- or β-adrenoceptors. Later experiments have confirmed that α_2-adrenoceptors comprise a family of pharmacologically and structurally distinct adrenoceptor subtypes.

It is now generally accepted that there are receptors on nerve terminals which, when activated by released transmitter, attenuate its further release; these presynaptic receptors are known as 'autoreceptors'. However, there is some dissenting evidence. For instance, autoreceptors can only be synthesised in the cell bodies of neurons and are delivered to the terminals by axoplasmic transport. Yet α_2-adrenoceptors have not been found in either the cell bodies or axons of sympathetic nerves. Recently, it has been found that 'autoreceptor'-mediated modulation of transmitter release does not occur in cultured neurons unless they are co-cultured with target cells. Such findings fuel speculation that feedback inhibition of transmitter release might involve a transsynaptic mechanism. Nevertheless, many different types of neurons appear to have autoreceptors, including those that release acetylcholine (M_2), dopamine (D_2/D_3), GABA ($GABA_B$), 5-HT (5-HT_{1B} or 5-HT_{1D}) and histamine (H_3).

Electrophysiological studies in the CNS have exposed the presence of an α_2-autoreceptor with a different function. These are found on the cell bodies of noradrenergic neurons in the nucleus locus coeruleus of the brainstem. When activated, they depress the firing rate of noradrenergic neurons in the nucleus. This means that changes in the concentration of noradrenaline in the medium bathing these somatodendritic α_2-autoreceptors will modify the firing rate of central noradrenergic neurons. Other types of neurons have equivalent autoreceptors. Examples are: 5-HT_{1A} receptors on serotonergic neurons in the Raphé nuclei and $D_{2/3}$-autoreceptors on central dopaminergic neurons in the ventral tegmental area and substantia nigra.

Autoreceptor-mediated feedback control of transmitter release will obviously depend on enough transmitter accumulating in the synapse to activate the receptors. If the trains of stimuli are either too short, or their frequency too low, then transmitter release is not augmented by the administration of autoreceptor antagonists, implying that there is no autoreceptor activation (Palij and Stamford 1993). This is known as 'pseudo-one-pulse' stimulation. Conversely, at higher frequencies and long trains of stimulation, it becomes harder to inactivate the autoreceptors with antagonist drugs, presumably because of competition with increased concentrations of transmitter in the synapse.

Even more sophisticated control of neurotransmitter release is suggested by the possibility of 'heteroceptors'. These receptors are thought to be located on the terminals of, and to modulate transmitter release from, one type of neuron, but are activated by transmitter released from a different type of neuron (Laduron 1985). For example, noradrenaline has been proposed to modulate release of a wide range of transmitters (e.g. dopamine, 5-HT and glutamate) through activation of α_2-heteroceptors on the terminals of each of these different types of neuron. However, one factor that should be borne in mind is that most of the evidence for heteroceptors comes from studies of tissue slices with [^{3}H]preloaded transmitter stores. It is therefore hard to be certain that heteroceptors are actually located on the terminals of the [^{3}H]labelled neuron and to rule out the possibility that they form part of a polysynaptic loop. To avoid this problem, a few studies have used synaptosomes to test the effects of one transmitter on K^+-evoked release of another. This approach has shown that noradrenaline and histamine, at least, blunt [^{3}H]5-HT release from cortical synaptosomes. Whether the same is true for all the other interactions between noradrenaline and 5-HT release, as well as with other transmitters, remains to be seen.

Evidence suggests that co-transmitters in a terminal have their own autoreceptors and, in some cases, activation of their own presynaptic receptor can influence the release of the co-stored, classical transmitter. For instance, activation of P_{2Y}-autoreceptors by ATP is thought to affect the release of noradrenaline from sympathetic neurons. However, in other cases, feedback modulation of release of classical and their associated co-transmitters seems to have separate control mechanisms. This would suggest that either the two types of transmitter are concentrated in different nerve terminals or that mechanisms for regulating release target different vesicles located in different zones of the terminal (Burnstock 1990).

COUPLING RECEPTORS WITH EXOCYTOSIS

There are several ways in which activation of auto- or heteroceptors on nerve terminals could modify the amount of transmitter released by exocytosis. The fact that this will depend on the influence of second messengers is beyond doubt. What remains to be resolved is whether one mechanism is more important than the others, or whether this varies from tissue to tissue.

Taking α_2-adrenoceptors as an example, several possible mechanisms have been suggested (see Starke 1987). The first rests on evidence that these autoreceptors are coupled to a Gi (like) protein so that binding of an α_2-adrenoceptor agonist to the receptor inhibits the activity of adenylyl cyclase. This leads to a fall in the synthesis of the second messenger, cAMP, which is known to be a vital factor in many processes involved in exocytosis. In this way, activation of presynaptic α_2-adrenoceptors could well affect processes ranging from the docking of vesicles at the active zone to the actual release process itself.

Alternative mechanisms are equally likely. One possibility arises from evidence that activation of α_2-adrenoceptors reduces Ca^{2+} influx; this will have obvious effects on impulse-evoked exocytosis. In fact, the inhibition of release effected by α_2-adrenoceptor agonists can be overcome by raising external Ca^{2+} concentration. Finally, an increase in K^+ conductance has also been implicated: this would hyperpolarise the nerve terminals and render them less likely to release transmitter on the arrival of a nerve impulse. Any, or all, of these processes could contribute to the feedback inhibition of transmitter release. Similar processes could explain the effects of activation of other types of auto- or heteroceptors.

Ca^{2+}-INDEPENDENT RELEASE OF TRANSMITTER

CARRIER-MEDIATED RELEASE

It is now well established that transmitter in the cytoplasm of neurons can be released by a process which is not dependent on Ca^{2+}. For monoamines, this is best illustrated by the actions of amphetamine and its analogues. Studies of a range of substituted amphetamines, using cultured serotonergic neurons, have confirmed that this release is not prevented by either N-type or L-type Ca^{2+} channel blockers, or removal of Ca^{2+} from the incubation medium, or depletion of the vesicular pool of transmitter. The release is blocked by inhibitors of the axolemma 5-HT transporter on the axolemma that normally carries 5-HT back into the nerve terminals from the synaptic cleft. This suggests that amphetamine-induced release of 5-HT represents a reversed efflux of transmitter on the membrane-bound carrier (Rudnick and Wall 1992). Whether this process of reverse transport accounts for all the 5-HT which is released by amphetamine, or whether this drug has additional actions which affect transmitter release, remains unclear (see also Chapter 9).

Several other conditions can provoke this 'reverse pump' type of release. One is when the transmembrane ionic gradient is reversed. Experimentally this is achieved by reducing extracellular Na^{+}. Because the neuronal uptake of monoamines from the synapse by the transporter requires co-transport of Na^{+} and Cl^{-}, reversing the ionic gradient (so that the Na^{+} concentration is lower outside, than inside, the terminals) will drive the transporter in the 'wrong' direction. Such carrier-mediated release could explain the massive Ca^{2+}-independent release of noradrenaline during ischaemia which increases intracellular Na^{+} concentration and reduces intracellular K^{+}.

Amino acids might also be released in this way (see Attwell, Barbour and Szatkowski 1993). There is evidence that depolarisation of retinal horizontal cells and cultured type 2 astrocytes by glutamate increases intracellular $[Na^{+}]$ concentration sufficiently to drive the membrane transporter to carry GABA (together with Na^{+} and Cl^{-}) out of the neurons (Fig. 4.13). Glutamate release during ischaemia is also thought to involve such carrier-mediated transport. A similar process might also explain a glutamate-induced increase in glycine release from astrocytes in the hippocampus.

HETEROCARRIER-MEDIATED RELEASE

Finally, there is evidence that transporters for GABA are found on the terminals of neurons releasing other types of transmitters. Moreover, uptake of GABA by this 'heterocarrier' could result in co-transport of sufficient Na^{+} to depolarise the terminal and provoke exocytotic release of the resident neurotransmitter. This suggestion arises from findings that exposure of brain synaptosomes to GABA can trigger release of noradrenaline, dopamine and acetylcholine. This release is prevented by inhibitors of GABA uptake but not by GABA receptor antagonists or monoamine uptake blockers. Unlike the carrier-mediated release described above, this form of release is thought to be Ca^{2+}-dependent and to involve vesicular exocytosis. However, the contribution of this process to the physiological control of neurotransmission has not yet been resolved.

CONCLUSION

That impulse-evoked release of neurotransmitters depends on a Ca^{2+}-dependent extrusion from storage vesicles is beyond dispute. However, many details concerning the supply of vesicles that participate in this process, as well as the processes which regulate the docking and fusion of synaptic vesicles with the axolemma, remain uncertain. Nevertheless, it is clear that the amount of transmitter that is released in this

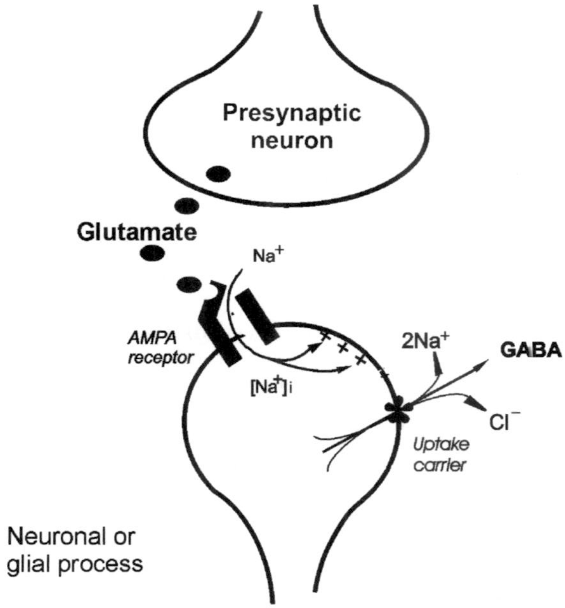

Figure 4.13 GABA release by reversed uptake ('retrotransport'). Depolarization of a neuronal, or glial cell process by glutamate, with a concomitant rise in $[Na^+]i$ reverses the operation of the GABA uptake carrier, raising [GABA]o. (Modified from Attwell, Barbour and Szatkowski 1993, with permission from the publisher Cell Press)

way is influenced by a network of auto- and heteroceptors. Activation of these receptors is coupled to the release process through their respective second messengers. It is also evident that vesicular exocytosis is not the only process which leads to release of transmitter from nerve terminals. Under certain conditions, axolemma-bound transporters can export transmitters from neurons or even evoke exocytosis. It seems that a range of processes contribute to release of neurotransmitters, all of which could have a vital role in the regulation of neurotransmission.

REFERENCES

Attwell, D, Barbour, B and Szatkowski, M (1993) Nonvesicular release of neurotransmitter. *Neuron* **11**: 401–407.

Basbaum, CB and Heuser, JE (1979) Morphological studies of stimulated adrenergic axon varicosities in the mouse vas deferens. *J. Cell Biol.* **80**: 310–325.

Benfanati, F, Onofri, F and Giovedi, S (1999) Protein–protein interactions and protein modules in the control of transmitter release. *Phil. Trans. Roy. Soc. Lond. B* **354**: 243–257.

Burnstock, G (1990) Noradrenaline and ATP as cotransmitters in sympathetic nerves. *Neurochem. Int.* **17**: 357–368.

Calakos, N and Scheller, RH (1996) Synaptic vesicle biogenesis, docking and fusion: a molecular description. *Physiol. Rev.* **76**: 1–29.

Ceccarelli, B and Hurlbut, WP (1980) Vesicle hypothesis of the release of quanta of acetylcholine. *Physiol. Rev.* **60**: 396–441.

De Camilli, P and Jahn, R (1990) Pathways to regulated exocytosis in neurons. *Ann. Rev. Physiol.* **52**: 625–645.

Dunant, Y (1994) Hormone and neurotransmitters release: four mechanisms of secretion. *Cell Biol. Int.* **18**: 327–336.

Fillenz, M (1990) *Noradrenergic neurons*, Cambridge University Press, Cambridge.

Greengard, P, Benfenati, F and Valtorta, F (1994) Synapsin I, an active-binding protein regulating synaptic vesicle traffic in the nerve terminal. In *Molecular and Cellular Mechanisms of Neurotransmitter Release* (Eds Starjne, L, Greengard, P, Grillner, S, Hokfelt, T and Ottoson, D), Raven Press, New York, pp. 31–45.

Hilfiker, S, Pieribone, VA, Czernik, AJ, Kao, H-T, Augustine, GJ and Greengard, P (1999) Synapsins as regulators of neurotransmitter release. *Phil. Trans. Roy. Soc. Lond. B* **354**: 269–279.

Kelly, RB and Grote, E (1993) Protein targeting in neurones. *Ann. Rev. Physiol.* **16**: 95–127.

Koenig, JH and Ikeda, K (1996) Synaptic vesicles have two distinct recycling pathways. *J. Cell Biol.* **135**: 797–808.

Laduron, PM (1985) Presynaptic heteroceptors in regulation of neuronal transmission. *Biochem. Pharmacol.* **34**: 467–470.

Littleton, JT and Bellen, HJ (1995) Synaptotagmin controls and modulates synaptic-vesicle fusion in a Ca^{2+} dependent manner. *Trends Neurosci.* **18**: 177–183.

Palij, P and Stamford, JA (1993) Real-time monitoring of endogenous noradrenaline release in rat brain slices using fast cyclic voltammetry. 2. Operational characteristics of the alpha2 autoreceptors in the bed nucleus of the stria terminalis, pars ventralis. *Brain Res.* **608**: 134–140.

Palij, P and Stamford, JA (1994) Real-time monitoring of endogenous noradrenaline release in rat brain slices using fast cyclic voltammetry. 3. Selective detection of noradrenaline efflux in the locus coeruleus. *Brain Res.* **634**: 275–282.

Phyllis, JW, Smith-Barbour, M and O'Regan, MH (1996) Changes in extracellular amino acid transmitters and purines during and following ischemias of different durations in the rat cerebral cortex. *Neurochem. Int.* **29**: 115–120.

Pieribone, VA, Shupliakov, O, Brodin, L, Hilfiker-Rothenfluh, S, Czernik, AJ and Greengard, P (1995) Distinct pools of synaptic vesicles in neurotransmitter release. *Nature* **375**: 493–497.

Rudnick, G and Wall, SC (1992) *p*-Chloroamphetamine induces serotonin release through serotonin transporters. *Biochemistry* **31**: 6710–6718.

Searle, T, Prior, C and Marshall, IG (1991) Acetylcholine recycling and release at rat motor nerve terminals studied using (−)-vesamicol and troxpyrrolium. *J. Physiol.* **444**: 99–116.

Sharp, T, Umbers, V and Gartside, SE (1997) Effect of a selective 5-HT reuptake inhibitor in combination with 5-HT1A and 5-HT1B receptor antagonists on extracellular 5-HT in rat frontal cortex *in vivo*. *Br. J. Pharmacol.* **121**: 941–946.

Sihra, TS and Nichols, RA (1993) Mechanisms in the regulation of neurotransmitter release from brain nerve terminals current hypothesis. *Neurochem. Res.* **18**: 37–58.

Söllner, T and Rothman, JE (1994) Neurotransmission: harnessing fusion machinery at the synapse. *Trends Neurosci.* **17**: 344–348.

Starke, K (1987) Presynaptic α-autoreceptors. *Rev. Physiol. Biochem. Pharmacol.* **107**: 73–146.

Winkler, H (1993) The adrenal chromaffin granule: a model for larger dense core vesicles of endocrine and nervous tissue. *J. Anat.* **183**: 237–252.

Section B

NEUROTRANSMITTERS AND SYNAPTIC TRANSMISSION

5 Basic Pharmacology and Drug Effects on Neurotransmitter Function

R. A. WEBSTER

INTRODUCTION

Drugs have the potential to bind to neurotransmitter receptors. Indeed that is the rationale for using them. If they combine with a receptor and activate it, they are called agonists but if they merely sit there to stop the NT, or another agonist drug, from binding to the receptor and producing an effect, they are called antagonists.

To bind to, i.e. have *affinity* for, a receptor, the drug, whether it is agonist or antagonist, must have the appropriate chemical structure but only agonists have the further ability, so-called *efficacy* (or *intrinsic activity*), to activate the receptor and the cellular mechanism to which it is linked. Ideally an antagonist has high (strong) affinity but zero efficacy. It is the chemical structure which determines to which receptor a drug combines and how specific it is in its action. Ideally a drug should only bind to one receptor type 1 but in reality very few, if any, are that specific, especially at high concentrations.

DRUG–RECEPTOR INTERACTIONS

The effect of an agonist drug whether it is measured as the ability to fire a neuron, inhibit an enzyme or reduce motor function, increases with the concentration of the drug and the number of receptors it occupies. In fact the magnitude of the response, like that of a chemical reaction, is proportional to the product of the concentration of the reactants, in this case the drug and its receptors, and as such obeys the law of Mass Action. Thus the rate at which a drug [D] combines with the receptor [R] to give occupied receptors (or drug receptor complexes [DR]), can be represented by its rate constant K_1 so that

$$[\mathrm{D}] + [\mathrm{R}] = K_1[\mathrm{DR}]$$

Since the drug–receptor interaction is reversible the drug also dissociates from the receptor at a rate K_2 when

$$[\mathrm{DR}] = K_2[\mathrm{D}] + [\mathrm{R}]$$

The equilibrium constant (K_A) for the reaction is thus given by

Neurotransmitters, Drugs and Brain Function. Edited by R. A. Webster

$$K_A = \frac{K_1}{K_2} = \frac{[D][R]}{[DR]}$$

When 50% of the receptors are occupied [R] and [DR], free and occupied receptors, must be equal and cancel out so that $K_A = D$, the concentration of drug required to bind to 50% of the receptors. Thus the lower the concentration of drug required to achieve this occupancy, the greater its affinity. Unfortunately its affinity does not necessarily reflect its potency in producing an effect.

The relationship between the dose (concentration) of a drug and the response it produces provides the so-called dose (or concentration) response curve (DRC). This is hyperbolic but is transformed to a sigmoid shape, which is linear over a large dose range, when the dose is plotted on a log scale (Fig. 5.1). Comparison of the concentrations of two or more drugs required to produce the same response is a measure of their relative potency. In Fig. 5.1 the dose–response curves for drugs A and B are one log unit apart and so A is ten times more potent than B. Often the potency of a drug is defined as the dose or concentration of drug required to produce 50% of the maximal response, i.e. the ED_{50} dose (or EC_{50} concentration). Drug C in Fig. 5.1 presents a different DRC. It obviously has agonist activity but since it cannot produce a maximal response it is known as a partial agonist. While such a property may seem unwanted the drug could still produce an adequate effect and avoid the danger of that becoming too great with increasing dose.

There are some points about the dose–response curve that justify consideration.

DOSE OR CONCENTRATION?

When a drug is administered to either humans or animals we obviously know the dose but not the concentration at its site of action. In this instance the relationship between the amount of drug and its effect really is a dose–response curve.

If a drug is added to an *in vitro* system in an organ or tissue bath then, provided the volume of the bathing solution is known, the concentration of drug can be calculated. Concentration is also known if a tissue is superfused with a prepared drug solution. In these instances, the response reflects drug concentration. Even then, the actual concentration of drug at the receptor site is not really known, since there can be a steep gradient between the concentration of drug in the medium and that at the actual receptor, especially if the drug is only in contact with the tissue for a short time. A proportion of most NTs is likely to be metabolised in, or taken up by, the tissue before reaching the receptor, although this is less likely with synthetic drugs.

POTENCY, AFFINITY AND EFFICACY

When looking at Fig. 5.1 it is pertinent to ask why drug A is more active than drug B. It could be that they are achieving the same response by acting through different receptors and that those targeted by A are either more numerous or better equipped to initiate the response. If they are both acting on the same receptor then obviously A has a more appropriate chemical structure to fit that receptor than B, but whether this has conferred on it a greater ability to combine with the receptor (affinity) or to activate it (efficacy) is unclear. It certainly should not be assumed that the EC_{50} is a measure of the affinity of the drug for the receptor. All responses are the result of a series of

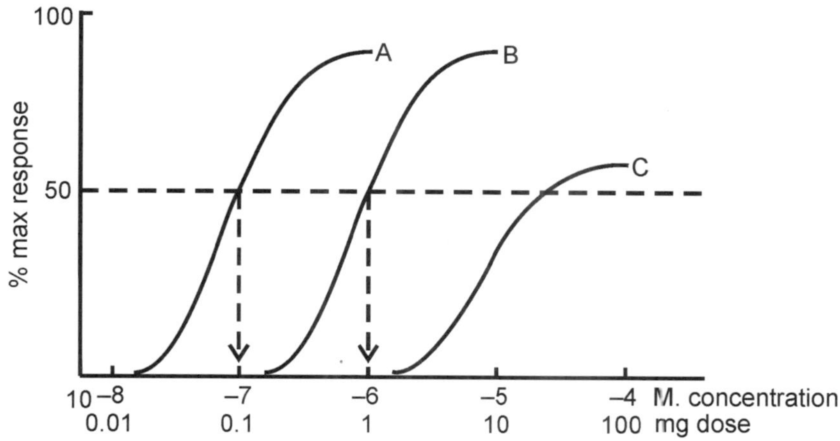

Figure 5.1 Dose (concentration) response curves for three drugs. Percentage response is plotted against log dose. The curves show that drug A achieves the same responses as drug B but at lower doses. A is in fact ten times more active than B since the same effect (e.g. 50% maximal response, ED_{50}) is obtained with 10^{-7} M (or 0.1 mg) A compared with 10^{-6} M (or 1.0 mg) B. Drugs A and B can both produce the maximal response and are full agonists. Drug C cannot produce a maximal response even at large doses and is known as a partial agonist

cellular events and, with the possible exception of studies on single-channel opening, not a direct measure of receptor occupancy. In any case, the efficacy of the drug must also be considered and since antagonists are devoid of that property their affinity and activity cannot be measured directly through a response (see below).

These problems can be overcome to some extent by using drugs labelled with a radioisotope (generally ^{3}H, ^{14}C or ^{125}I) and then directly determining the amount of label bound when the drug is incubated with samples of the appropriate tissue or, as with the nervous system, fragments of specially prepared isolated neuronal membranes that contain the receptors. Even this approach is not ideal since drugs will combine non-specifically with cellular elements other than the receptor. In practice this can be largely overcome (see Chapter 3). Experimentally, the test tissue is incubated with varying concentrations of the labelled drug (called ligand) until equilibrium is reached. The tissue is then separated from the incubation medium by filtration or centrifugation and dissolved in scintillation fluid which is measured for its radioactivity. This gives the total amount of drug bound, including specific binding to its receptors and any other non-specific tissue binding. The non-specific binding is estimated by running a parallel set of tissue samples incubated with medium containing both the labelled drug and an excess concentration of another unlabelled drug which binds to the same receptor. This should inhibit all the receptor binding of labelled drug. Any residual binding will be to non-specific sites (Fig. 5.2(a)). Subtraction of this non-specific binding from the total binding gives the specific receptor binding for the drug which is a saturable process. The relationship between the amount of ligand bound (B) and its concentration X can be represented, for a preparation where the total number of binding sites is $B_{\max}$, as

$$B = \frac{B_{\max} X}{X + K} \quad \text{where } K \text{ is the dissociation (affinity) constant}$$

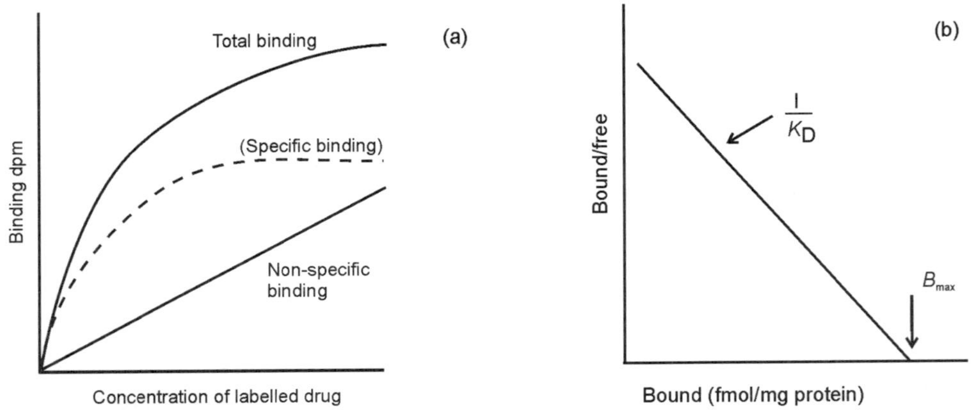

Figure 5.2 Measurement of specific saturable drug binding. (a) Plot of quantity of labelled drug bound against its increasing concentration in the bathing medium. Subtraction of non-specific from total binding gives the specific binding for the drug. (b) Scatchard plot of B/F (bound/free drug) against level of bound drug (B) gives a straight line the slope of which is $1/K_D$, while the intercept is B_{max}, the maximum number of that drug's binding sites, expressed as fmol/mg tissue protein. For experimental detail see text

Thus

$$\frac{B}{X} = \frac{B_{max}}{K} - \frac{B}{K}$$

If B/X is plotted against B (the Scatchard plot) it should give a straight line (Fig. 5.2(b)) with the slope ($1/K_D$) giving K and the intercept on the abscissa providing the maximal binding (B_{max}), expressed as fmol per mg tissue protein. The steeper the slope, the higher the affinity.

In many binding studies the relative abilities of a series of unlabelled drugs to displace a labelled ligand from a particular receptor is taken as a guide to their affinity for that receptor. This is normally represented as K_i, the concentration of drug required to displace half of the labelled ligand. Its accuracy depends on the chosen ligand only binding to the receptor it is intended to study and no other receptor. It must be emphasised that binding studies only measure the ability of a drug to combine with a receptor, they do not indicate whether it is an agonist or antagonist. Also compared with an antagonist the binding of an agonist may be affected in an uncertain manner by the change in state caused by the activation of the receptor.

DRUG ANTAGONISM

One drug can overcome the effect of another or reduce the activity of an endogenously released and active substance such as a neurotransmitter, either by competing with that substance for its receptor site (receptor antagonism) or stimulating a different receptor to induce an opposing effect (physiological or functional antagonism). The former may be regarded as true antagonism for in the latter case both drugs are actually agonists. It is epitomised by the use in asthma of beta adrenoceptor agonists like salbutamol to dilate bronchi that have been constricted by a cocktail of local mediators such as

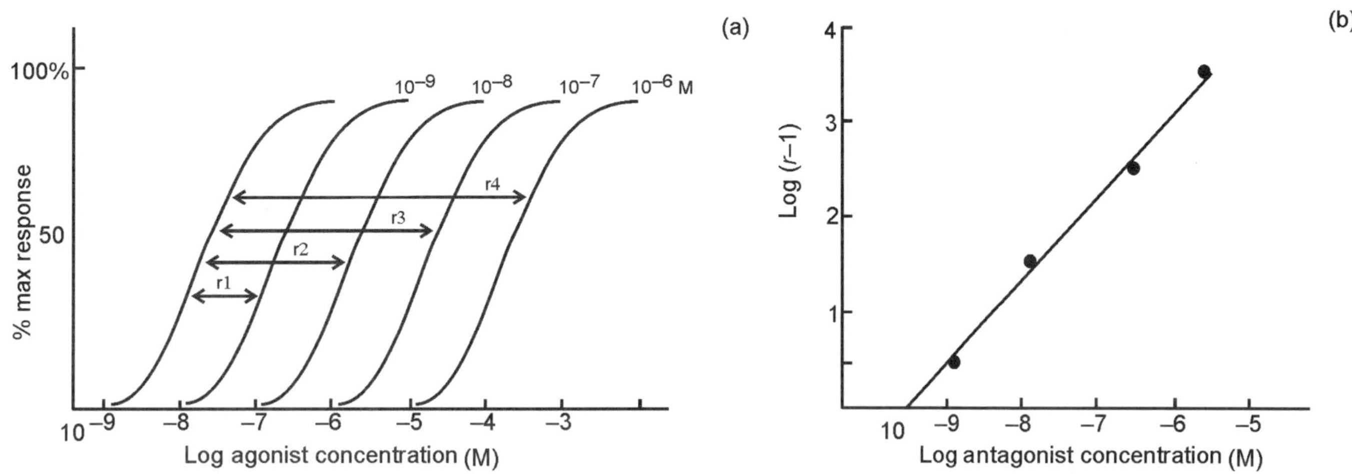

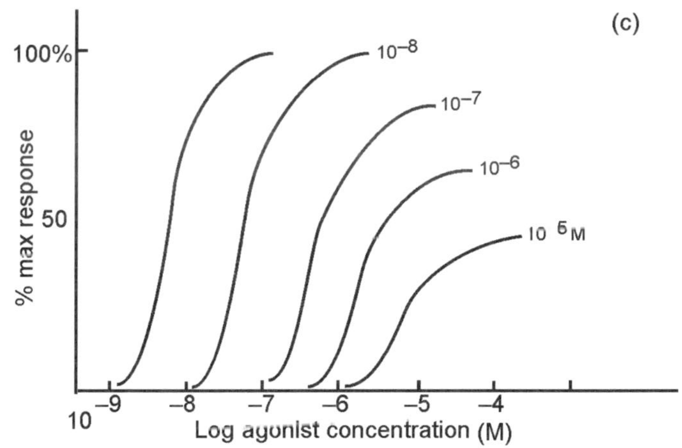

Figure 5.3 Drug antagonism. (a) Surmountable competitive antagonism. Dose–response curves are shown for an agonist alone and in the presence of increasing concentrations of the antagonist. The antagonism is surmountable because even in the presence of a large concentration of antagonist the agonist can still produce a maximal response and it is competitive because equal increments in the concentration of antagonist produces equal shifts in the DRCs. The shift is known as the dose ratio (*r*), and is the amount by which the dose of agonist must be increased in the presence of antagonist to produce the same response as in its absence. (b) A Schild plot. This shows dose ratio (*r*) measured as in (a) and plotted as $\log(r-1)$ against log concentration of antagonist. In true competitive antagonism the graph should have a slope of 1 and its intercept with the abscissa gives pK_B. The negative logarithm of this value is Schild's pA_2 measure of drug antagonism. In the example shown a K_B of 3.2×10^{-10} gives a pA_2 of 9.5. (c) Unsurmountable irreversible antagonism. In the presence of low concentrations of antagonist the agonist can still produce a maximum response because the antagonist does not occupy all the receptors and there are sufficient 'spare receptors' available for the agonist. As the concentration of antagonist is increased fewer spare receptors remain and since the antagonist does not dissociate from the receptors (as its binding is irreversible) the agonist is unable to produce a full response, i.e. the antagonism is unsurmountable

histamine, acetylcholine and kinins. In the CNS the inhibitory NT (GABA) could be regarded as the physiological antagonist of the excitatory NT (glutamate). When the agonist and antagonist compete for the same receptor the binding of the agonist and the response it produces are both reduced. Thus to obtain the same response in the presence, as in the absence of antagonist, the concentration of agonist must be increased and over a range of agonist concentrations this results in a parallel shift to the right in the position of its DRC (Fig. 5.3).

The degree of this shift, the amount by which the agonist concentration has to be increased in order to produce the same response in the presence as in the absence of the antagonist, is known as the dose ratio (r). The larger this ratio, the greater the shift in the DRC and the more potent is the antagonist. In fact if the antagonism is really competitive then the degree of shift of the DRC will be proportional to the increase in concentration of the antagonist used. From Fig. 5.3(a), it can be seen that increasing the antagonist concentration from 10^{-9} M to 10^{-8}, 10^{-7}, 10^{-6} M always produces the same tenfold increase in dose ratio (i.e. 10, 100, 1000). Also if the antagonism is competitive not only will the DRCs remain parallel but it should always be possible to restore the maximal response to the agonist by giving more of it, irrespective of the amount of antagonist present. This is known as reversible or surmountable competitive antagonism. Since both agonist and antagonist are continuously combining with and dissociating from the receptor the likelihood of either occupying it at any time will depend on their relative concentrations.

The dose ratio $r = (X_B/K_B) + 1$ where X_B is the concentration of antagonist and K_B its equilibrium constant. This can be expressed logarithmically as

$$\log(r - 1) = \log X_B - \log K_B$$

and a plot of $\log(r - 1)$ against $\log X_B$ (the Schild plot) should give a straight line with slope of 1, which intercepts the abscissa at the value $-\log K_B$ (pK_B) for the antagonist (Fig. 5.3(b)). This is frequently converted into a simple number by taking its negative logarithm, much as pH values represent hydrogen ion concentration, so that K_Bs of 10^{-7} or 3.2×10^{-7} mol/l become simply 7 or 6.5. This pA_2 value was defined by Schild as the negative logarithm of the molar concentration of antagonist required to give a dose ratio of 2. Thus the larger the pA_2 value, the smaller the concentration of antagonist needed and the greater its affinity and effectiveness. In practice full DRCs are rarely obtainable especially in studies on the CNS, or even necessary, provided that responses to two doses of agonist can be obtained at each concentration of antagonist. This will establish the position of the DRC and allow r to be calculated.

If the antagonist does not readily dissociate from the receptor, because it is bound firmly, then the agonist will not be able to displace it and restore a maximal response. At low concentrations of antagonist this may not be apparent. An agonist can often achieve a maximal response by activating only a small percentage of its receptors, so in the presence of low concentrations of a non-dissociating antagonist there may be sufficient spare receptors available for increased concentrations of the agonist still to achieve a maximal response. As the concentration of antagonist is increased, however, fewer unoccupied receptors are left and since the agonist cannot displace the antagonist a maximal response cannot be achieved (Fig. 5.3(c)). This is unsurmountable or irreversible competitive antagonism. It is still competitive because the drugs are competing for the same receptor. Sometimes an antagonist can inhibit the effect of an agonist not by occupying the same binding site on the receptor but some adjacent site or process necessary for the agonist's effect, such as the ion channel itself. This is non-competitive antagonism which may, or may not, be reversible, depending on the action of the antagonist. There are a number of drugs with this action on the glutamate NMDA receptors (Chapter 10).

PHARMACOKINETICS

This is generally accepted as the term to describe the absorption distribution and metabolism of a drug *in vivo* and it is these factors which determine how quickly and how much of the administered drug can actually reach its site of action (in the CNS) and be maintained there for the required time (see Fig. 5.4). Experimentally drugs are

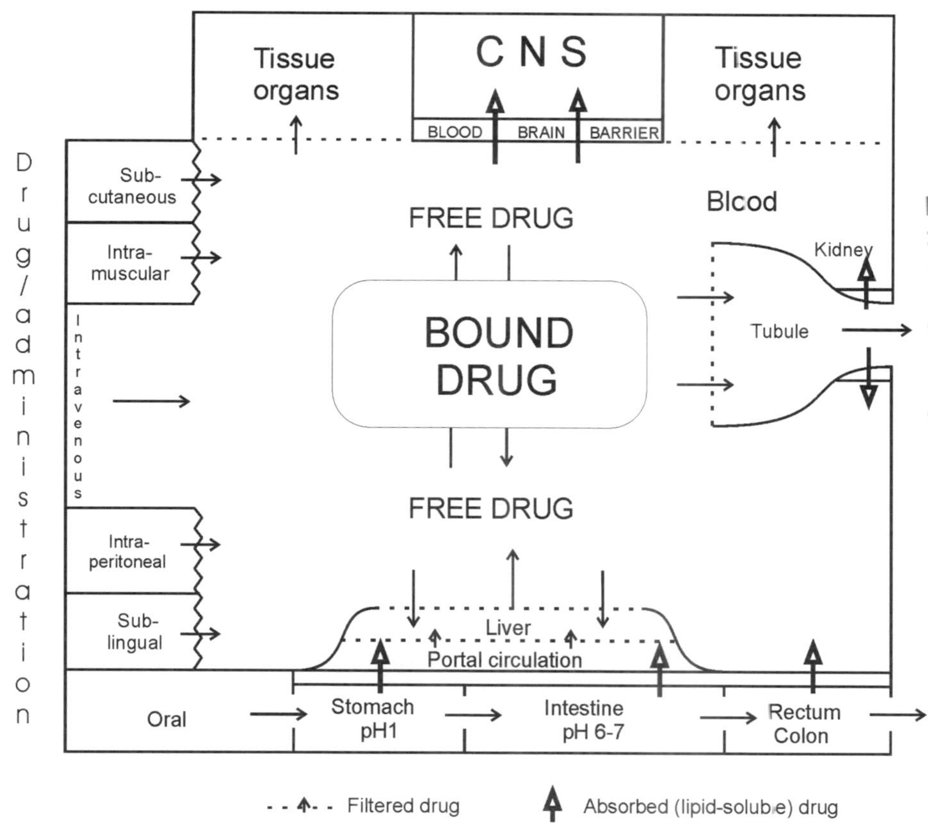

Figure 5.4 Pharmacokinetics. The absorption distribution and fate of drugs in the body. Routes of administration are shown on the left, excretion in the urine and faeces on the right. Drugs taken orally are absorbed from the stomach and intestine and must first pass through the portal circulation and liver where they may be metabolised. In the plasma much drug is bound to protein and only that which is free can pass through the capillaries and into tissue and organs. To cross the blood–brain barrier, however, drugs have to be in an unionised lipid-soluble (lipophilic) form. This is also essential for the absorption of drugs from the intestine and their reabsorption in the kidney tubule. See text for further details

usually given intravenously or intraperitonealy, but therapeutically most of them are taken orally.

The speed of onset of action of a drug depends primarily on how quickly it reaches the circulation. For this reason alone it is not surprising that intravenous administration produces the quickest response. Thereafter the rate and degree of absorption depends on the blood flow to the injected site and the surface area of vessels exposed to the drug. The response to an intramuscular injection in humans is quite rapid since our muscles are large and have a good blood supply. In laboratory animals muscle mass is small and so an intraperitoneal administration may be more effective because the drug solution can be given in relatively large volumes which disperse over a large surface area (the abdominal wall and intestinal surfaces).

Drugs taken orally are slow to act. Most are absorbed in the small intestine where the villi, which penetrate into the lumen, present a large surface area. Unfortunately in order to pass through the gut wall into the bloodstream the drug has to become dissolved in its cell's membranes and to achieve this it needs to be lipid-soluble.

Generally it is only the non-dissociated or unionised drug that is lipid-soluble and a drug's degree of ionisation depends on its dissociation constant (pK) and the pH of the environment in which it finds itself. For an acidic drug this is represented by the Henderson–Hasselbalch equation as

$$\mathrm{p}K - \mathrm{pH} = \log \frac{\text{conc-unionised drug } (Cu)}{\text{conc-ionised drug } (Ci)} \qquad \left(\mathrm{p}K - \mathrm{pH} = \log \frac{Ci}{Cu} \text{ for basic drug}\right)$$

Thus an acidic drug with a relatively low pK of 3 will be largely unionised (hundredfold) in the acidic environment (pH = 1) of the stomach since

$$3 - 1 = \log 2 = 100 = \frac{Cu}{Ci}$$

but in the more basic intestine it will be ionised, i.e.

$$3 - 6 = \log 3 = \frac{1}{1000} = \frac{Cu}{Ci}$$

It will then depend for its absorption on the large surface area of the intestine.

Drugs absorbed along the length of the gut do not enter straight into the general circulation but pass initially into the portal circulation to the liver where they may be subject to metabolism. In fact a high proportion of some orally administered drugs can be lost in this way without even reaching the main bloodstream but those given sublingually (under the tongue) or by suppository into the rectum bypass the portal system. Some drugs can also stimulate the production of microsonal-metabolising enzymes (e.g. phenobarbitone) in the liver and so increase the destruction of other drugs being taken at the same time.

Once in the blood most drugs will leave the circulation by being filtered through pores in the capillaries, provided they have a molecular weight below 6000, which is almost always the case, and are not bound to plasma protein (albumin) which is too large to be filtered. Although such binding, which commonly accounts for over 90% of plasma drug, does restrict movement, it also acts as a drug store. Unfortunately one drug can displace another from such binding and so elevate its free plasma concentration and create the potential for toxicity.

There are two sites in the body where a drug is not able to pass freely into the tissue. One is the placenta and the other the brain where the blood–brain barrier (see Chapter 1) is a formidable hindrance. Without pores in the capillaries a drug can only enter the CNS (or cross the placenta) by virtue of lipid solubility, as in the gut.

Since a drug is a foreign object, the body does its best to get rid of it. As the organ of excretion, the kidney has a copious blood supply and drugs are easily filtered through the glomerular capillaries into the kidney tubule and urine unless they are very large (e.g. hormones, heparin) or bound to plasma albumin. In fact most drugs would be rapidly lost if they were not so bound or showed sufficient lipid solubility to be reabsorbed through the wall of the kidney tubule back into the bloodstream. Thus a drug which is present in the unionised lipid-soluble form is more readily absorbed from the gut, can enter the CNS and is potentially longer acting as it will avoid excretion, unless it is rapidly metabolised.

To increase the chance of removing a drug, the body converts it into a water-soluble, ionised and so excretable form. This is generally a two-stage process involving initial metabolism (e.g. oxidation, reduction or hydrolysis) and then conjugation with something like glucuronic acid. The metabolite may occasionally be as, or more, active than the parent compound but is generally less so and can sometimes even be toxic.

The rate at which a drug is metabolised is generally proportional to its concentration (so-called first-order kinetics) but if there is an excess of drug and the metabolic process becomes saturated, then metabolism proceeds at a constant rate, the maximum possible, irrespective of concentration (zero-order kinetics). With some drugs, such as alcohol, this occurs even at low concentrations. The duration of action of a drug is represented by its half-life ($\frac{1}{2}t$) which is a measure of the time taken for its plasma concentration to fall by 50%. Obviously drugs with a short half-life have to be taken more frequently. To use a drug properly it is necessary to know not only what constitutes an effective plasma concentration but also how long that is maintained following dosage. This information can be obtained from pilot studies in humans but since there is considerable variation in how an individual responds to and metabolises a drug the effect of a drug can vary considerably between subjects.

This leads to the concept of therapeutic index. The potency of a drug is almost irrelevant. It is its specificity that matters. Thus if two drugs A and B are effective at the same dose in a patient, say 1 mg, but A produces toxic effects at 10 mg which are only seen with 500 mg of B then B is clearly a much safer drug than A, in that patient. The ratio of toxic to effective dose is the therapeutic index (TI). It is often expressed as

$$\frac{\text{toxic dose in 50\% of patients}}{\text{effective dose in 50\% of patients}}$$

In practice it is obviously difficult to actually determine the toxic and effective dose in 50% of treated patients in the same population but the concept of a maximum tolerated dose compared with an effective dose is of great importance.

DRUG EFFECTS ON NEUROTRANSMITTER FUNCTION

To establish the role of a particular NT experimentally it is necessary to modify its synaptic activity. This can be done most easily with drugs.

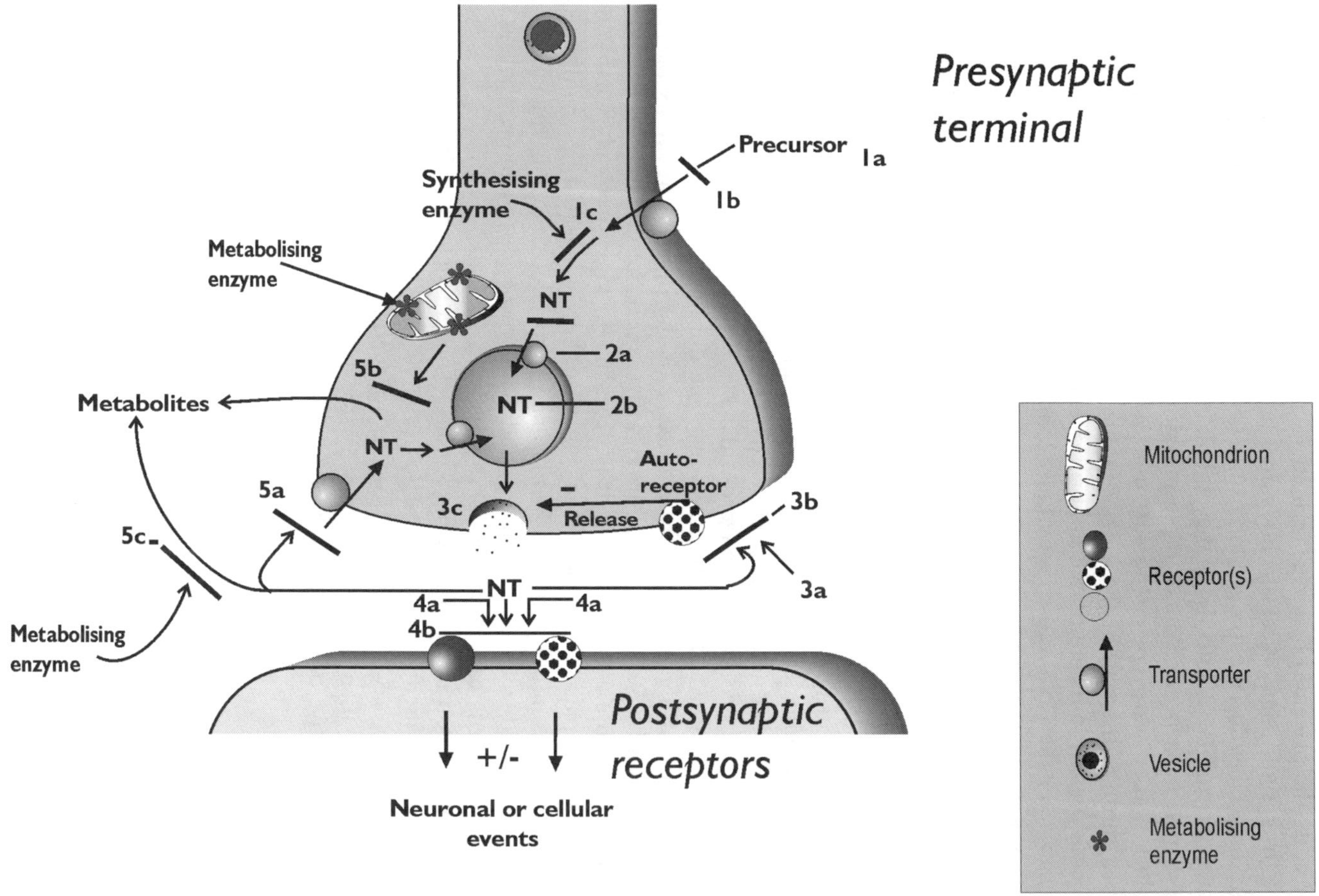

Figure 5.5 Diagrammatic representation of a synapse showing the sites at which drugs may act to increase or decrease the concentration and action of a neurotransmitter. Drugs can affect the synthesis (1), storage (2), release (3), action (4) and destruction (5) of the transmitter. The different ways in which they achieve this (a, b, c) are outlined in Table 5.1

Table 5.1 Drug modification of the different aspects of neurotransmitter function in synaptic transmission as illustrated in Fig. 5.5

	Aspect	Modification	Effect on NT function
1	Synthesis (i)	(a) Supplement precursor	+
		(b) Block precursor uptake	−
		(c) Inhibit synthesising enzyme	−
2	Storage (ii)	(a) Inhibit NT uptake into vesicle	− (+)
		(b) Inhibit NT binding in vesicle	− (+)
3	Release (exocytotic) (iii)	(a) Stimulation of negative autoreceptors	−
		(b) Blockade of autoreceptors	+
		(c) Disrupt release (exocytotic) mechanism	−
4	Action (iv)	(a) Mimic effect of NT on receptor	+
		(b) Block postsynaptic receptor	−
5	Destruction (v)	(a) Block uptake into neuron (and/or glia)	+
		(b) Inhibit intraneuronal metabolism	+
		(c) Inhibit extracellular metabolism	+

(+) = increased; (−) = decreased.

Notes:

(i) Synthesis may have multiple steps within the terminal process which provides more than one site for drug modification and one stage may even be within the vesicle. Providing extra synthesising enzyme is not a practical proposition but altering the availability of certain co-factors can have an influence (e.g. impaired vitamin V6 intake reduces GABA synthesis).

(ii) Drugs affecting storage (e.g. reserpine) are not likely to affect only one NT. Also although any reduction in the vesicular storage of NT will eventually lead to a reduction (−) in its normal exocytotic release by action potential, it is possible that the extra NT will build up sufficiently in the cytoplasm (if not metabolised) to diffuse out of the neuron or even induce its reverse transport out of the neuron through the membrane transporter that normally brings it in from the synapse (+).

(iii) Some terminals may also have receptors which augment release when activated by the NT as well as receptors for NTs other than that being released. Drugs that affect the actual release process, e.g. Ca^{2+} influx, are unlikely to have a specific effect on just one NT system unless concentrated in particular neurons by specific uptake.

(iv) Most NTs act on more than one postsynaptic receptor. This provides the opportunity to design drugs that will act specifically on just one of them although reproduction of the full effect of the NT may require the participation of more than one of its receptors (e.g. DA function in the basal ganglia). It should also be remembered that even if a drug is specific for just one NT receptor its effects will depend on how numerous and widely distributed that receptor is. This is particularly true of the much-used amino acids glutamate and GABA. Also many different functions could be linked to the same receptor (e.g. the numerous peripheral actions mediated by cholinergic muscarinic receptors).

(v) A NT may be taken up into glia (e.g. GABA) or even nerve terminals other than those from which it is released. As with synthesis, it is difficult to augment the action of the metabolising enzymes.

The concept that a drug is either an agonist or antagonist that acts at a receptor site is a simple one, especially if it is acting at the receptor for a NT that is linked either directly or indirectly through second messenger systems, to the control of ion channel opening and the excitability (discharge) of a neuron. The receptor or perhaps more precisely the receptive site can, however, also be part of an enzyme involved in the synthesis or metabolism of that NT, the transporter responsible for taking the NT (or its precursor) across the membrane of a storage vesicle or axon terminal or even the actual NT binding site within a vesicle. This means that a drug can modify the action of a NT and the function of the synapse where it acts in a number of ways. These will now be outlined in general and then covered in more detail for each particular NT in the following chapters (6–13).

At most synapses a conventional NT is synthesised from an appropriate precursor in the nerve terminal, stored in vesicles, released, acts on postsynaptic receptors and is then destroyed either by extraneuronal metabolism or intraneuronal metabolism after reuptake. Its release is triggered by invading action potentials and controlled by presynaptic autoreceptors. Although some NTs, e.g. peptides, are not synthesised in the terminal and others, e.g. NO, are formed on demand, some features of the 'typical' synapse, as described above, still apply to them. The mechanisms by which drugs may modify synaptic function through their effects on the synthesis, storage, release, action and destruction of a NT are shown in Fig. 5.5 and outlined in Table 5.1. They constitute what may be regarded as a template for how a drug may affect synaptic transmission.

Manipulating the activity of a NT in these ways helps to determine its function either at a synaptic level or in more general behavioural terms. Thus the clearest way of establishing the identity of the NT at a particular synapse is to ascertain which NT receptor antagonist blocks transmission there.

OTHER MEANS OF AFFECTING NEUROTRANSMITTER FUNCTION

Apart from the use of drugs the activity of a NT can be modified using one of the following approaches:

(1) Surgical or electrolytic lesions of a particular pathway or nucleus. Even with precise location and subsequent histological verification this approach rarely ablates all the axons (or neurons) in a pathway using the NT under study and can also affect other axons (or neurons).

(2) Toxins that gain access to a neuron through its uptake process and then destroy it in some way. This approach has been used mainly to destroy monoamine neurons with 5,6 or 5,7-dihydroxytryptamine targeting 5-HT neurons, 6-hydroxydopamine for dopamine (and to a lesser extent noradrenergic) neurons and MPTP for dopamine neurons (see Chapter 7). Only the latter is fully specific and effective systemically. The others need to be administered directly into the appropriate brain areas and while they may only affect the intended NT neurons, the injection may not affect all of them.

(3) Genetically bred animals in which the gene encoding a particular NT receptor, receptor subclass or synthesising enzyme has been knocked out. This approach becomes more realistic as we learn more of the subunit and amino-acid structure of receptors. While the effects will be more widespread than lesions it could have some advantage over drugs. It is difficult to design chemicals that are totally, or even adequately, specific as either agonists or antagonists for a particular subclass of receptor but if we know their structure then it should be possible to knock out a specified receptor subgroup.

6 Acetylcholine (ACh)

R. A. WEBSTER

INTRODUCTION

As outlined in Chapter 1, acetylcholine was the first neurotransmitter to be discovered and isolated. Its mode of action at peripheral synapses and in particular at the neuromuscular junction has been extensively studied and it is clearly the NT at all first synapses outside the CNS, whether this be at sympathetic or parasympathetic ganglia of the autonomic nervous system, the adrenal medulla or neuromuscular junctions in skeletal muscle. In these instances it transmits fast excitation through nicotinic receptors linked directly to the opening of Na^+ channels. At parasympathetic nerve endings, such as those of the vagus on smooth and cardiac muscle and secretary cells, as well as just those few sympathetic nerve endings to sweat glands, it is also the neurotransmitter. In these instances it has much slower excitatory or inhibitory effects mediated through muscarinic receptors utilising second messenger systems.

By contrast, the central actions of ACh are perhaps still less well understood than those of some more recently discovered NTs like dopamine and GABA. It does not appear to have a clear primary function but often an important supporting role. Attempts to understand its central actions were not encouraged by the knowledge that even those anticholinergic drugs that clearly cross the blood–brain barrier have few marked central effects and handicapped by the difficulty in measuring its release and turnover, or mapping its pathways.

Until the recent development of appropriate HPLC techniques capable of detecting pmol amounts (see Flentge *et al.* 1997) ACh could only be measured chemically by relatively lengthy and expensive procedures (e.g. gas chromotography), which were not always very sensitive, or by bioassays. Although the latter, using muscle preparations that responded to ACh, such as the dorsal muscle of the leech, the rectus abdominus of the frog or certain clam hearts, were reasonably sensitive they were tiresome and not easily mastered. Thus studies on the release and turnover of ACh have not been as easy as for the monoamines.

Similarly, cholinergic nerves could only be visualised indirectly. Staining for cholinesterase, the metabolising enzyme for ACh, gave some information on the location of cholinergic synapses, where it is found postsynaptically rather than in nerve terminals, but it is not specific to cholinergic nerves. Fortunately choline acetyltransference (ChAT), which acetylates choline in the synthesis of acetylcholine, is specific to cholinergic nerve terminals and its labelling by immunochemistry has much facilitated the mapping of cholinergic pathways (Fig. 6.7).

Neurotransmitters, Drugs and Brain Function. Edited by R. A. Webster

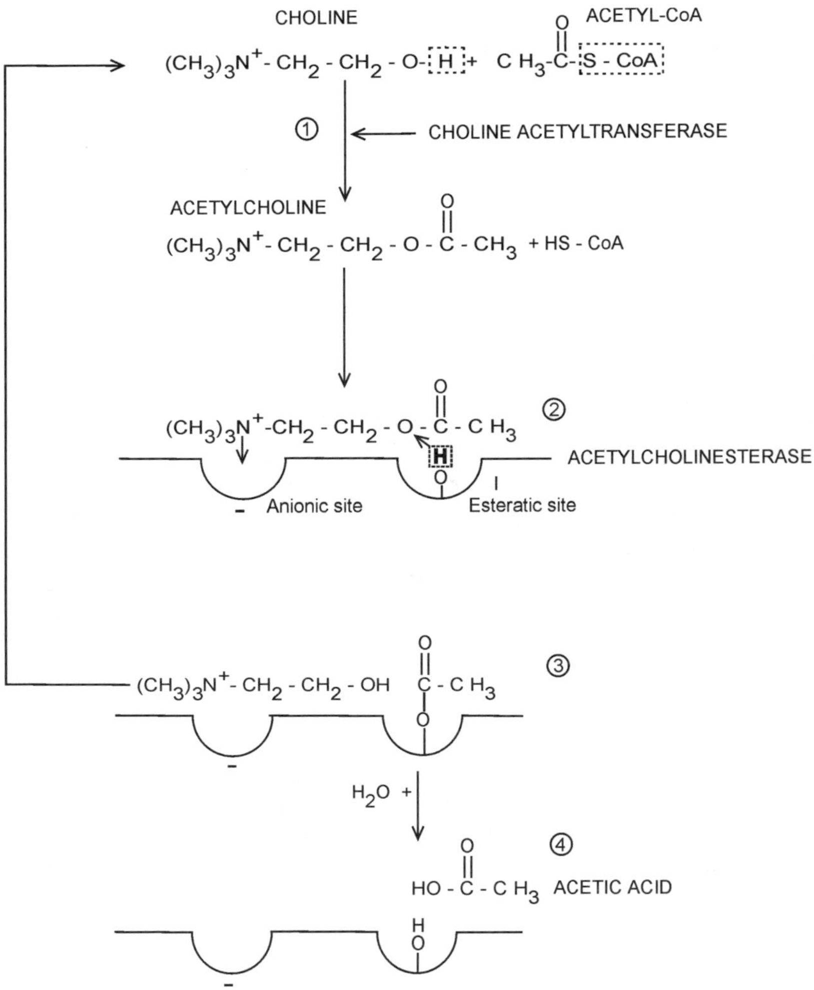

Figure 6.1 Synthesis and metabolism of acetylcholine. Choline is acetylated by reacting with acetyl-CoA in the presence of choline acetyltransferase to form acetylcholine (1). The acetylcholine binds to the anionic site of cholinesterase and reacts with the hydroxy group of serine on the esteratic site of the enzyme (2). The cholinesterase thus becomes acetylated and choline splits off to be taken back into the nerve terminal for further ACh synthesis (3). The acetylated enzyme is then rapidly hydrolised back to its active state with the formation of acetic acid (4)

In contrast to all this negativity, it must be acknowledged that more is known about the structure and function of cholinergic receptors and synapses, especially the nicotinic ones, than for the receptors of any other NT. It is unfortunate that nicotinic synapses are not very common in the CNS.

NEUROCHEMISTRY

The basic biochemistry of the synthesis and destruction of ACh is outlined in Fig. 6.1 and put into the context of the cholinergic synapse in Fig. 6.2.

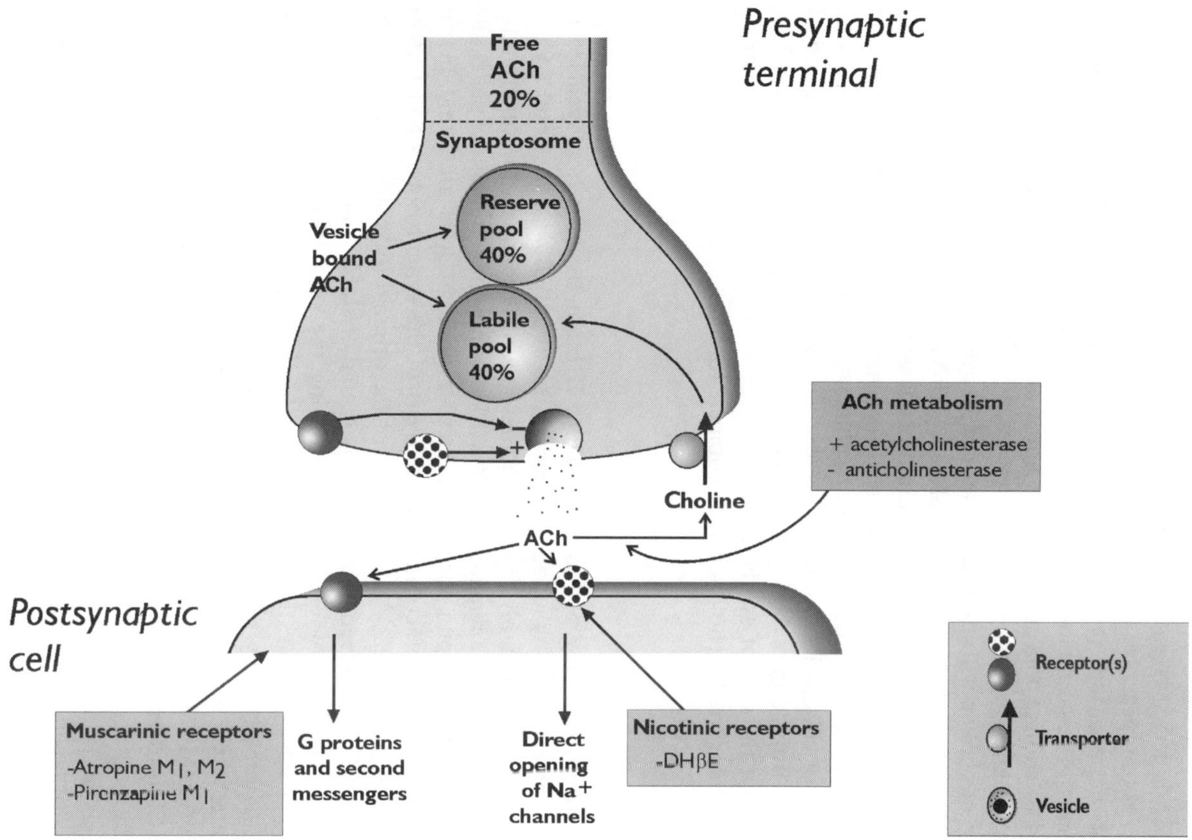

Figure 6.2 Diagrammatic representation of a cholinergic synapse. Some 80% of neuronal acetylcholine (ACh) is found in the nerve terminal or synaptosome and the remainder in the cell body or axon. Within the synaptosome it is almost equally divided between two pools, as shown. ACh is synthesised from choline, which has been taken up into the nerve terminal, and to which it is broken down again, after release, by acetylcholinesterase. Postsynaptically the nicotinic receptor is directly linked to the opening of Na^+ channels and can be blocked by compounds like dihydro-β-erythroidine (DHβE). Muscarinic receptors appear to inhibit K^+ efflux to increase cell activity. For full details see text

SYNTHESIS

Acetylcholine is synthesised in nerve terminals from its precursor choline, which is not formed in the CNS but transported there in free form in the blood. It is found in many foods such as egg yolk, liver and vegetables although it is also produced in the liver and its brain concentration rises after meals. Choline is taken up into the cytoplasm by a high-affinity ($K_m = 1$–$5\,\mu M$), saturable, uptake which is Na^+ and ATP dependent and while it does not appear to occur during the depolarisation produced by high concentrations of potassium it is increased by neuronal activity and is specific to cholinergic nerves. A separate low-affinity uptake, or diffusion ($K_m = 50\,\mu M$), which is linearly related to choline concentration and not saturable, is of less interest since it is not specific to cholinergic neurons.

The reaction of choline with mitochondrial bound acetylcoenzyme A is catalysed by the cytoplasmic enzyme choline acetyltransferase (ChAT) (see Fig. 6.1). ChAT itelf is synthesised in the rough endoplasmic reticulum of the cell body and transported to the axon terminal. Although the precise location of the synthesis of ACh is uncertain most of that formed is stored in vesicles. It appears that while ChAT is not saturated with either acetyl-CoA or choline its synthesising activity is limited by the actual availability of choline, i.e. its uptake into the nerve terminal. No inhibitors of ChAT itself have been developed but the rate of synthesis of ACh can, however, be inhibited by drugs like hemicholinium or triethylcholine, which compete for choline uptake into the nerve.

STORAGE AND RELEASE

ACh is not distributed evenly within the neuron. If brain tissue is homogenised in isotonic salt solution containing an anticholinesterase, about 20% of the total ACh is released into solution, presumably from cell bodies, and it is found in the supernatant fraction on centrifugation. The remaining 80% settles within the sedimenting pellet and if this is resuspended and spun through a sucrose gradient it is all found in the synaptosome (nerve ending) fraction. After analysis of the synaptosomes and further centrifugation about half of this ACh, i.e. 40% of the original total still remains in the spun-down pellet. This is referred to as the firmly bound or stable 'pool' of ACh since it is not subject to hydrolysis by cholinesterase during the separation procedure. The other half (again 40% or the original) is found in the supernatant and undergoes hydrolysis unless protected by anticholinesterases. While it is generally assumed that some of this latter ACh was always in the synaptosomal cytoplasm probably half of it (20% of the original) comes from disrupted vesicles. This mixture of vesicular and cytoplasmic ACh is called the labile pool and is probably the most important source of releasable ACh, and also where newly synthesised ACh is found. Thus in studies in which tissue has been incubated with labelled precursor choline not only is this pool (fraction) heavily labelled but since most of the released ACh is also labelled it is assumed to come from this pool. With the passage of time there is interchange of ACh between the labile and the so-called fixed pool and in the absence of adequate resynthesis, i.e. blockage of choline uptake, it is likely that ACh will be released from the latter source as well.

Morphological evidence has also been obtained for two distinct vesicles with one designated VP_1 (see Whittaker 1987; Zinnerman *et al.* 1993) being larger but less dense than the other (VP_2). It is the latter which are thought to be incompletely filled with

ACh and considered to be the vesicles in the labile releasable pool. The evidence for and the actual mechanism of the vesicular release of ACh, mostly gained from studies at peripheral synapses, has been covered in Chapter 4.

Apart from inhibiting the uptake of choline and hence its availability for ACh synthesis, with hemicholinium (see above), there are no drugs that directly affect the actual storage or release of ACh. Some experimental tools have, however, been used such as vesamicol, which appears to block the packaging of ACh into its vesicles and thus initiates the slow rundown of releasable vesicular ACh. Some toxins also inhibit ACh release.

Botulinum toxin produced by the anaerobic bacillus *Clostridium botulinum* is unbelievably toxic with a minimum lethal mouse dose of 10^{-12} g. Its occurrence in certain, generally preserved, foods leads to an extremely serious form of poisoning (botulism) resulting in progressive parasympathetic, motor and eventually respiratory paralysis and death. There are no antidotes and mortality is high. Despite this frightening profile, the toxin is finding increasing therapeutic use in relieving some forms of localised muscle spasm such as those of the eyelids (blepharospasm). Obviously it has to be injected directly into the muscle in carefully calculated small amounts. Provided this is achieved its firm binding and slow dissociation ensures a local effect that can last a number of weeks.

Beta-bungarotoxin, a protein in cobra snake venom, also binds to cholinergic nerves to stop ACh release while α-bungarotoxin (from the same source) binds firmly to peripheral postsynaptic nicotinic receptors. The combined effect ensures the paralysis of the snake's victim.

While there is no active neuronal uptake of ACh itself, cholinergic nerve terminals do possess autoreceptors. Since these are stimulated by ACh rather than by the choline, to which ACh is normally rapidly broken down, it is unlikely that they would be activated unless the synaptic release of ACh was so great that it had not been adequately hydrolysed by cholinesterase.

ACh is widely distributed throughout the brain and parts of the spinal cord (ventral horn and dorsal columns). Whole brain concentrations of 10 nmol g^{-1} tissue have been reported with highest concentrations in the interpeduncular, caudate and dorsal raphé nuclei. Turnover figures of 0.15–2.0 nmol g^{-1} min^{-1} vary with the area studied and the method of measurement, e.g. synthesis of labelled ACh from [^{14}C]-choline uptake or rundown of ACh after inhibition of choline uptake by hemicholinium. They are all sufficiently high, however, to suggest that in the absence of synthesis depletion could occur within minutes.

METABOLISM

Released ACh is broken down by membrane-bound acetylcholinesterase, often called the true or specific cholinesterase to distinguish it from butyrylcholinesterase, a pseudo- or non-specific plasma cholinesterase. It is an extremely efficient enzyme with one molecule capable of dealing with something like 10 000 molecules of ACh each second, which means a short life and rapid turnover (100 µs) for each molecule of ACh. It seems that about 50% of the choline freed by the hydrolysis of ACh is taken back into the nerve. There is a wide range of anticholinesterases which can be used to prolong and potentiate the action of ACh. Some of these, such as physostigmine, which can cross the blood–brain barrier to produce central effects and neostigmine, which does not readily do so, combine reversibly with the enzyme. Others such as the pesticide, disopropylphosphofluorate (DYFLOS), form an irreversible complex requiring the synthesis of new enzyme before recovery. Recently longer acting but reversible inhibitors such as tetrahydro aminoacridine have found some use in the therapy of Alzheimer's disease (Chapter 8). The manner in which acetylcholinesterase is thought to bind to and react with ACh and how drugs may inhibit it are shown in Fig. 6.3.

In addition to its vital role in the metabolism of ACh, acetylcholinesterase has been shown somewhat surprisingly to be released in the substantia nigra, along with DA,

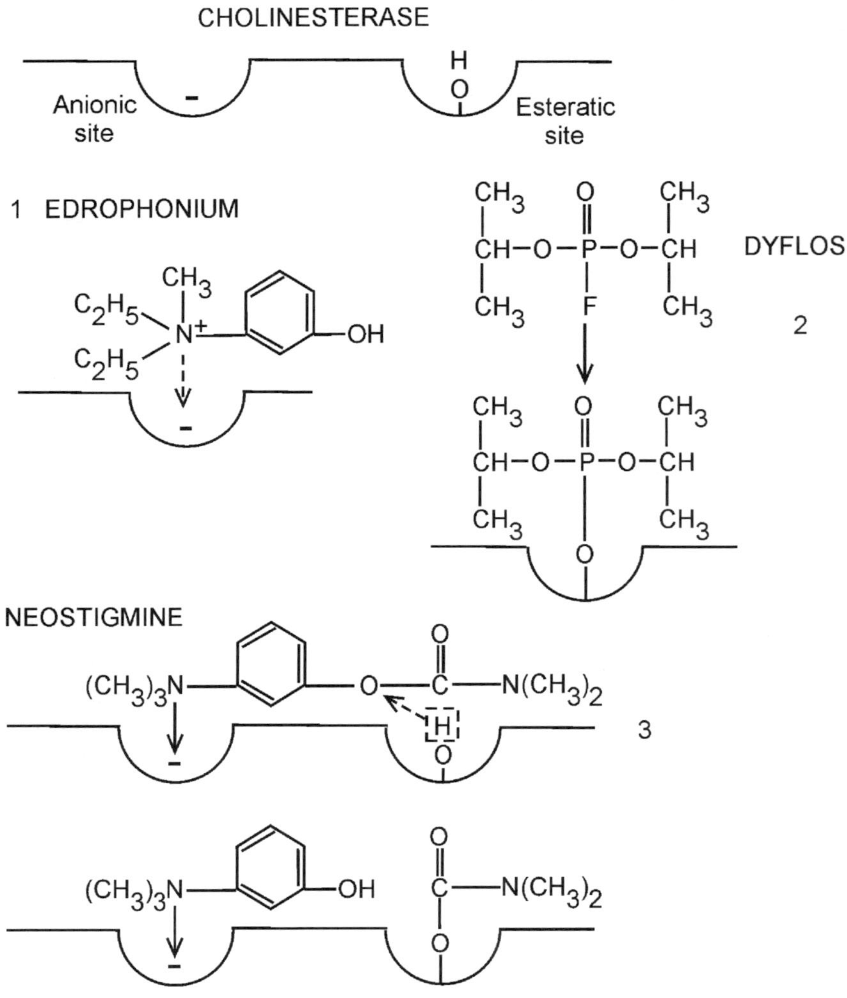

Figure 6.3 Modes of action of the anticholinesterase drugs. Cholinesterase, which has both an anionic and an ester site (Fig. 6.1), can be inhibited by drugs acting reversibly and irreversibly. Edrophonium is a short-acting inhibitor that binds reversibly with the anionic site (1) while DYFLOS reacts almost irreversibly with the esteratic site (2). Since hydrolysis of the enzyme is negligible new enzyme must be synthesised to overcome the effect of this very toxic compound. Clinically useful anticholinesterase like neostigmine have a medium duration of action ($\frac{1}{2}t = 1$ h). It binds to both sites on the enzyme (3) with the result that neostigmine itself is hydrolysed but the transfer of its carbamyl group to the enzyme's esteratic site produces a carbamylated enzyme which recovers by hydrolysis much more slowly (min) than after its acetylation by ACh (ms)

presumably from the soma and dendrites of DA neurons. Its function there is uncertain but purified preparation of the enzyme infused into the substantia nigra cause not only hyperpolarisation of the neurons, due to the opening of K^+ channels, but also a variety of motor effects in rats that are not related to its enzymatic activity and the turnover of ACh (see Greenfield 1991).

RECEPTORS

CLASSIFICATION AND STRUCTURE

As already mentioned, ACh acts on two distinct receptors: (a) nicotinic receptors, which mediate fast synaptic events and (b) muscarinic ones controlling much slower changes. This classification was originally based on the use of antagonists since atropine blocked only the slower events and curare only the fast ones. Their naming derives, perhaps unfortunately, from the fact that muscarine mimics the slow effects and nicotine the fast ones, initially anyway. As might be expected, the cholinergic receptors have been cloned and their structures established. In the CNS the muscarinic receptors outnumber the nicotinic possibly by 100:1, and, not surprisingly, they have been studied more extensively.

Nicotinic

Those receptors at the neuromuscular junction and in the electric organ of *Torpedo* have been studied much more than those in the CNS, but they all have similar characteristics. The peripheral receptor has four different protein subunits α, β, γ and δ but is pentameric with the α always doubled. In the CNS the receptors are less complex. Most have just two subunits α and β, but are again pentameric with 2α and 3β subunits situated around and forming the ion channel. Several variations of the α subunit, from $\alpha2$–$\alpha9$ ($\alpha1$ in periphery) and three in the β give the possibility for a number of different heteromeric receptors, although $\alpha4$ and $\beta2$ predominate and receptors with the configuration $\alpha4_2$ and $\beta2_3$ (2 of $\alpha4$ and 3 of $\beta3$) show the highest affinity for ACh. Homomeric assembled receptors of just $\alpha7$ subunits are also found. Each subunit folds into a four-transmembrane domain (m_1 m_2 m_3 m_4) with the m_3–m_4 loop linkage in the cytoplasm and the terminal amine and carboxyl groups extracellular (Fig. 6.4). (The accepted scheme for subunit and configuration numbering is outlined in Chapter 3).

The amino-acid sequence of each subunit is known and they are characteristic of a NT receptor that directly gates ion channels. Activation of the receptor requires 2 ACh molecules to combine with two α subunits. Pharmacologically it is not easy to distinguish between central and peripheral nicotinic receptors, let alone their variants. Those in the CNS are more like those found in peripheral ganglia than at the neuromuscular junction and are more readily blocked by dihydro-β-erythroidine than curare. Receptors containing $\alpha4\beta2$ subunits (the majority) are also not blocked by α-bungarotoxin but bind a shorter kappa or 'neuronal' bungarotoxin. Those receptors with the $\alpha7$ subunit, for which α-bungarotoxin has high affinity, will, however, bind that toxin.

Although drugs may not be able to distinguish between the subclasses of nicotinic receptor the last few years has seen the breeding of knock-out mice in which most of the

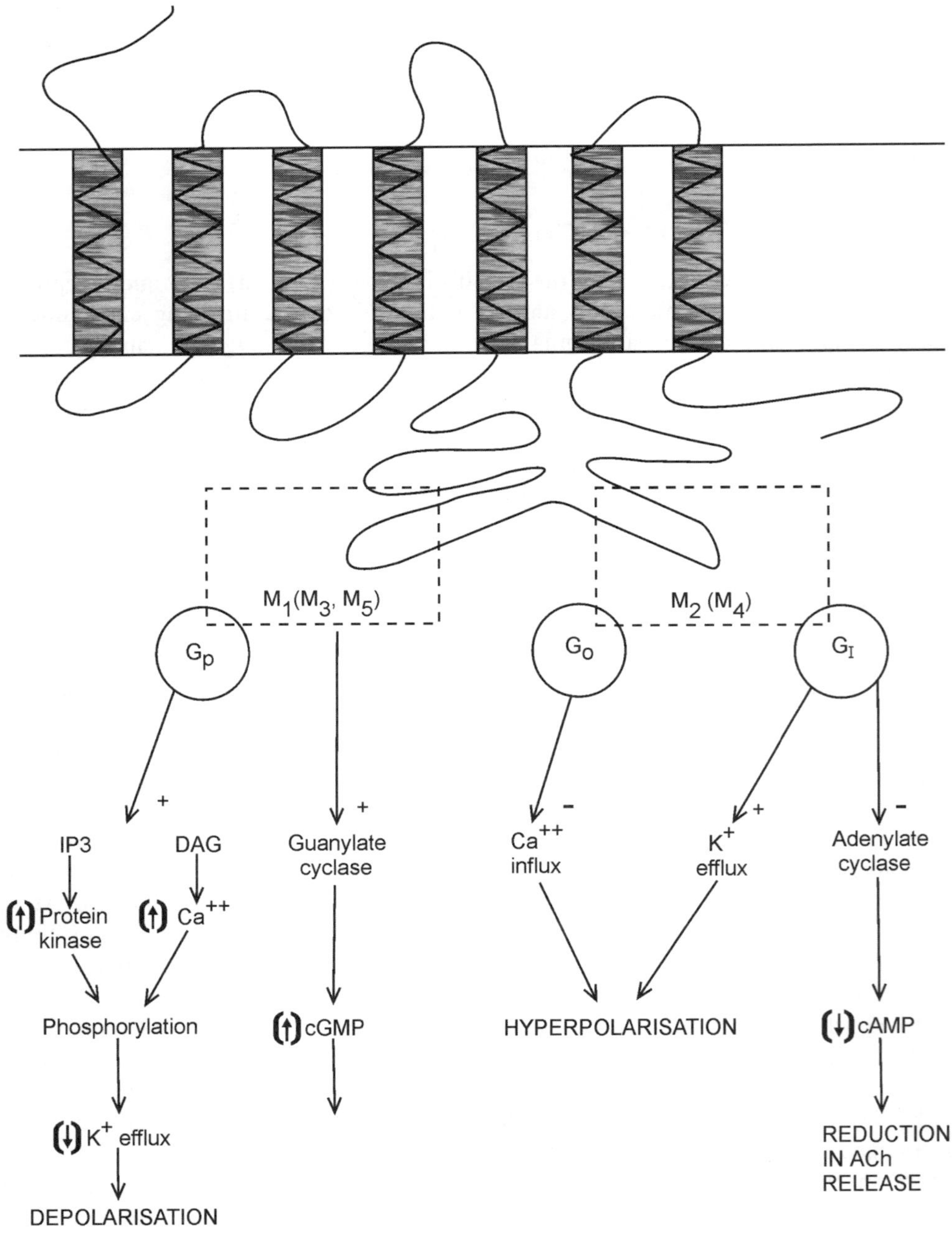

Figure 6.4 Schematic representation of the muscarinic receptor. All muscarinic receptors have seven transmembrane domains and the major difference between them is within the long cytoplasmic linkage connecting the fifth and sixth domains. This implies different G-protein connections and functions. Some possibilities are shown although the position of the M_1 and M_2 boxes is not intended to indicate their precise structural differences within the loop

mamalian nAChR subunits have been selectively deleted (see Cordero-Erausquin *et al.* 2000). While only those mice lacking subunits found mainly in peripheral nicotinic receptors (e.g. $\alpha3$ and $\beta4$) do not survive, others show little change in spontaneous behaviour but some reduced responses (less antinociception) to nicotine.

Muscarinic

Despite the wide variety of effects associated with the activation of muscarinic receptors on different peripheral organs it appeared that they were either identical or very similar because known antagonists, like atropine, were equally effective against all muscarinic responses. A decade ago, one drug, pirenzepine, was found to be a hundredfold more active against ACh-induced gastric acid secretion than against other peripheral muscarinic effects. The receptors blocked by pirenzapine became known as M_1 and all the others as M_2. Recently some differences between muscarinic M_2 receptors on heart (inhibitory) and those on exocrine glands (generally excitatory) became apparent through slight (fivefold) differences in the binding of some antagonist drugs (tools) such as AF-DX-116 and 4-DAMP. The former was more active on the receptors in the heart, accepted as M_2 receptors, while the glandular ones, blocked preferentially by 4-DAMP, became M_3. Molecular biology has since confirmed the existence of these three receptors and revealed (at the time of printing) two more — M_4 and M_5. The M_1 receptor mediates most of the central postsynaptic muscarinic effects of ACh while the M_2 is predominantly a presynaptic autoreceptor.

The structure of the muscarinic receptor is very different from that of the nicotinic. They are single-subunit proteins which belong to the group of seven transmembrane receptors (like adreno and dopamine receptors) typically associated with second messenger systems. The major difference between muscarinic receptors is in the long cytoplasmic linkage connecting the fifth and sixth transmembrane domain, suggesting different G-protein connections and functions. Thus M_1, M_3 and M_5 receptors are structurally similar and their activation causes stimulation of guanylate cyclase and an increase in cyclic GMP as well as inosotal triphosphate hydrolysis through an increase in G-protein (Gp) (Fig. 6.4).

The M_2 and M_4 receptors also show structural similarities. Through G-protein (G_1) they inhibit cyclic AMP production and open K^+ channels while activation of another G-protein (G_0) closes Ca^{2+} channels. The latter effect will cause membrane hyperpolarisation as will the G_1-induced increase in K^+ efflux. The reduction in cAMP production, although possibly leading to depolarisation, is more likely to explain the presynaptic reduction in ACh release associated with the M_2 receptor.

DISTRIBUTION

Cholinergic receptors should obviously be found where ACh is concentrated and cholinergic pathways terminate. Autoradiography with appropriately labelled ligands does in fact show M_1 receptors to be predominantly in the neocortex and hippocampus (where pathways terminate) and in the striatum where ACh is released from intrinsic neurons. By contrast, M_2 receptors are found more in the basal forebrain where ascending cholinergic pathways originate. Such a distribution is in keeping with the postsynaptic action of the M_1 receptor and the presynaptic cell body (autoinhibition) mediated effects of its M_2 counterpart. Unfortunately the ligands available for labelling are not sufficiently specific to use this technique to reliably distinguish M_1 from M_3 and M_5 receptors or M_2 from M_4. *In situ* hybridisation studies of receptor mRNA, which detects cell body receptors, is more sensitive and confirms the M_1 dominance in the neocortex, hippocampus and striatum with M_2 again in subcortical areas. Receptor mRNA for the M_3 is, like that for M_1, in the cortex and hippocampus but not in the striatum while that for M_4 is highest in the striatum and low in the cortex. Elucidation of the precise functional significance of such a distribution awaits the arrival of much more specific ligands for the receptor subsets. In their absence a more detailed analysis of the distribution of muscarinic and nicotinic receptors is not justified here but see Hersch *et al.* (1994), Levey *et al.* (1991), Wall *et al.* (1991) and Wess (1996).

Nicotinic receptors have been found and studied predominantly in the hippocampus, cerebral cortex and viatral tegmented area (VTA).

FUNCTION

Activation of nicotinic receptors causes the rapid opening of Na^+ channels and membrane depolarisation. This is a feature of cholinergic transmission at peripheral neuromuscular junctions and autonomic ganglia but while it is found in the CNS, it is not widely observed. Exogenously applied nicotinic agonists have been shown to directly excite neurons through somato-dendritic receptors in various brain regions while the excitatory response of GABA interneurons in the hippocampus and dopamine neurons in the VTA following some afferent stimulation is reduced by nicotinic antagonists (see Jones, Sudweeks and Yakel 1999). Nicotinic receptors also mediate the fast response of ACh released at the endings of collaterals from motoneuron axons to adjacent inhibitory interneurons (Renshaw cells) in the ventral horn of the spinal cord (see below).

More recently much interest has been directed towards presynaptic nicotinic receptors that have been shown to enhance the release of a number of NTs, i.e. ACh, DA, NA, glutamate and GABA, in perfused synaptosomes or slices from various brain regions, as well as DA into microdialysates of the striatum *in vivo*. Thus they can be hetero- and not just autoreceptors (see Wannacott 1997). Since activation of these receptors can actually evoke, and not just facilitate, NT release, they probably work directly on nerve terminals to increase Na^+ influx and initiate sufficient depolarisation to activate voltage-sensitive Ca^{2+} channels, although there could also be an influx of Ca^{2+} itself through the nicotinic gated channel. In fact the high permeability of some neuronal nicotinic receptors to Ca^{2+} ions provides an obvious mechanism for increasing transmitter release. Differences in the sensitivity of the presynaptic receptors to various agonists and antagonists indicate some heterogeneity but their relatively low affinity for nicotine (EC_{50} about 1 μM) and the absence of clear evidence for their innervation means that their physiological role remains uncertain. Their activation by exogenous agonists could, however, have interesting therapeutic applications such as an increase in ACh release in Alzheimer's disease and mAChRs have been found to be reduced in the cortex and hippocampus of such patients.

Although ACh does not have a primary excitatory role like glutamate in the CNS, it does increase neuronal excitability and responsiveness, through activation of muscarinic receptors. It achieves this in two ways, both of which involve closure of K^+ channels (see Chapter 2 and Brown 1983; Brown *et al.* 1996). The first is a voltage-dependent K^+ conductance called the M conductance, Gm or Im. It is activated by any attempt to depolarise the neuron, when the opening of the M-channel and the consequent efflux of K^+ counteracts the depolarisation and limits the generation of spikes. This current is inhibited by activation of muscarinic receptors and so ACh will tend to keep the neuron partially depolarised and facilitate repetitive firing and bust spiking. This slow cholinergic excitation in hippocampal neurons is shown in Fig. 6.5.

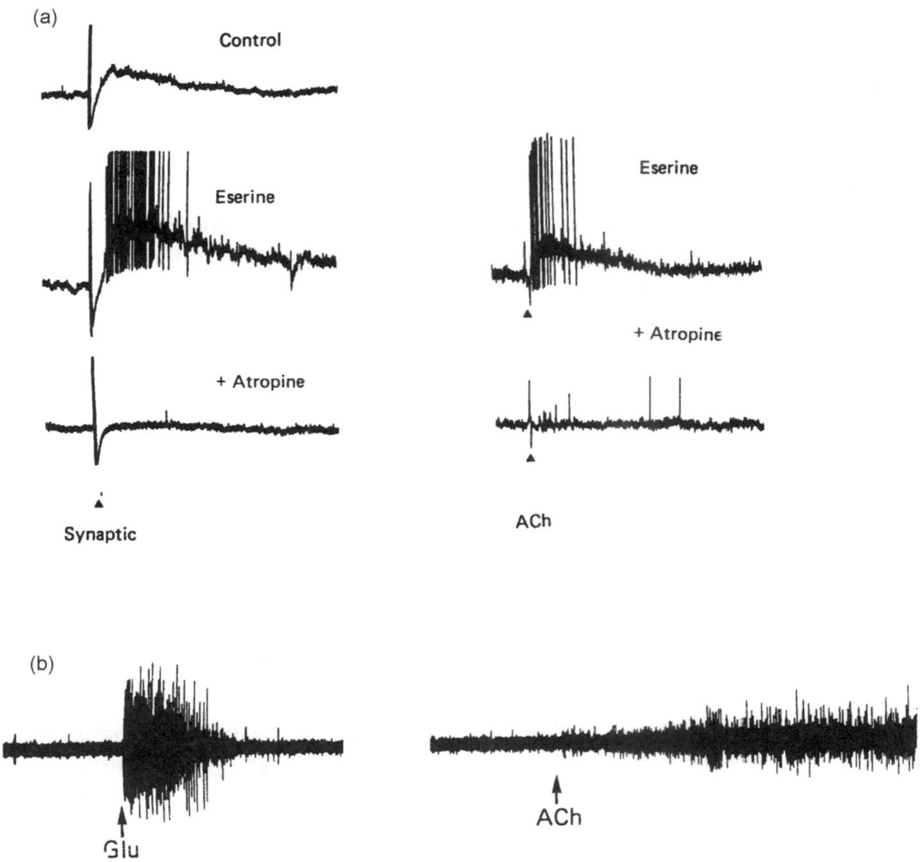

Figure 6.5 Illustrations of the slow excitatory effect of ACh. (a) Electrical stimulation (left-hand traces) of presumed cholinergic fibres (striatum oriens) in the rat hippocampal slice preparation (20 Hz for 0.5 s) induced a short latency epsp followed by an ipsp and a later slow epsp, recorded intracellularly in pyramidal neurons (control). The slow epsp was selectively potentiated by the anticholinesterase drug eserine (2 μM) with the generation of action potentials. This firing and the slow epsp, but not the fast epsp or ipsp, were eliminated by the muscarinic antagonist atropine (0.1 μM). The iontophoretic application of ACh (right-hand traces) in the presence of eserine only produced the slow epsp and superimposed firing which were also antagonised by atropine (resting membrane potentials 57–60 mV). These recordings show that the slow but not the fast epsp is cholinergic. (b) Extracellulary recorded multiunit response in layer V of the guinea pig anterior cingulate cortex slice preparation. Micropipette application of glutamate (10 ms of 1 mM) caused a rapid generation of action potentials up to 100 μV in amplitude while ACh (40 ms of 1 mM) only generated smaller (10–20 μV) potentials more slowly (50 ms) with larger ones (30–50 μV) appearing later. These results again demonstrate the slow excitatory effect produced by ACh compared with the larger and more rapid primary depolarisation of glutamate. ((a) reproduced with kind permission from the *Journal of Physiology* and from Cole and Nicoll 1984 and (b) from McCormick and Prince 1986a)

Studies of the hippocampal neurons have shown a second K^+ current which mediates the long-lasting after-hyperpolarisation following spiking. This is not voltage activated but is switched on by Ca^{2+} entry through channels opened during the initial depolarisation. It is inhibited by activation of muscarinic receptors and so its reduction will also lead to repetitive firing. ACh in fact seems to dampen the inbuilt brakes on cell firing (see also Chapter 2 and Figs 2.5 and 2.6).

M_1 and M_3 receptors mediate the excitatory effects and since this postspike hyperpolarisation is blocked by phorbol esters and is therefore presumably dependent on IP_3 production, one would expect it to be mediated through M_1 receptors (see above), especially as these are located postsynaptically. Unfortunately it does not appear to be affected by pirenzapine, the M_1 antagonist. By contrast, muscarinic inhibition of the M current is reduced by the M_1 antagonist but as it is not affected by phorbol esters is not likely to be linked to IP_3 production, an M_1 effect.

ACh can sometimes inhibit neurons by increasing K^+ conductance and although it has been found to hyperpolarise thalamic neurons, which would normally reduce firing, strong depolarisation may still make the cell fire even more rapidly than normal. This appears to be because the hyperpolarisation counters the inactivation of a low-threshold Ca^{2+} current which is then activated by the depolarisation to give a burst of action potentials (McCormick and Prince 1986b).

AGONISTS AND ANTAGONISTS

Many drugs bind to cholinergic receptors but few of them enter the brain and those that do are not noted for their effects.

AGONISTS

Some agonists, such as methacholine, carbachol and bethanecol are structurally very similar to ACh (Fig. 6.6). They are all more resistant to attack by cholinesterase than ACh and so longer acting, especially the non-acetylated carbamyl derivatives carbachol and bethanecol. Carbachol retains both nicotinic and muscarinic effects but the presence of a methyl (CH_3) group on the β carbon of choline, as in methacholine and bethanecol, restricts activity to muscarinic receptors. Being charged lipophobic compounds they do not enter the CNS but produce powerful peripheral parasympathetic effects which are occasionally used clinically, i.e. to stimulate the gut or bladder.

Pilocarpine, arecoline and, of course, muscarine itself are naturally occurring muscarinic agonists, while oxotremorine is a synthetic one, which, as its name implies, can cause muscle tremor through a central effect.

In view of the preponderance of muscarinic receptors in the CNS and the conceived need to augment the muscarinic actions of ACh in the treatment of Alzheimer's disease, much attention has been given recently to the synthesis of agonists that penetrate the blood–brain barrier, especially those that act specifically on M_1 receptors.

Few drugs, apart from nicotine itself, act specifically on nicotine receptors. One is methylcarbachol, which lacks the muscarinic effects of carbachol and another is dimethylphenylpiperazinium (DMPP), which appears to have some selectivity for the neuronal nicotinic receptor. Neither of them can cross the blood–brain barrier.

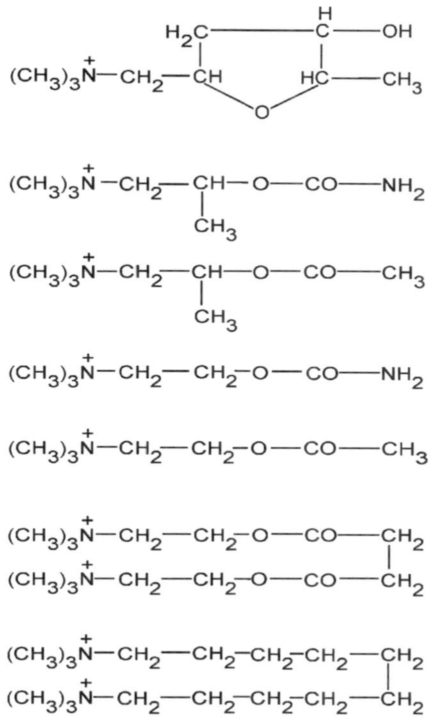

Figure 6.6 Structure of some cholinergic agonists and antagonists. Acetylcholine has the structure to activate both muscarinic and nicotinic receptors. Carbachol retains these actions but is longer acting because it lacks the terminal methyl group and is not so readily hydrolysed by cholinesterase (see Fig. 6.1). Methacholine with the methyl side chain lacks nicotinic activity but can be hydrolysed while bethanechol has a similar action but, like carbachol, is not easily hydrolysed. They show chemical similarities to muscarine. Suxamethonium is like two acetylcholine molecules joined together and has transient nicotinic activity at the neuromuscular junction before desensitising (blocking) those receptors. Decamethonium has a similar but much longer blocking action because, unlike suxamethonium, it is not hydrolysed by plasma cholinesterase

Although nicotine receptors are few in number, the proven ability to stimulate NT release may initiate the search for more effective centrally acting agonists.

ANTAGONISTS

The neuromuscular blocking action of the poison *d*-tubocurarine (curare) has been known for a century. It works by competing with ACh for binding to the nicotinic receptor. Others have been developed such as suxamethonium (succinylcholine) which is essentially two molecules of ACh joined together (Fig. 6.6). Perhaps not surprisingly, it is initially an agonist that causes a depolarisation of muscle fibres and actual twitching, before producing a depolarisation block of transmission. There are a large number of competitive antagonists apart from curare, such as gallamine, pancuronium and atracurium, while decamethonium works like suxamethonium as a depolarising agent. They can be used to produce neuromuscular block and skeletal muscle paralysis in surgery and prevent the damage to limbs that can occur in the electroconvulsive treatment of depression. An interesting pharmacological distinction between these two classes of neuromuscular blocking agents is that the effect of the competitive receptor blockers like curare can be overcome by increasing the concentration of ACh, which is achieved *in vivo* by giving an anticholinesterase, while the blocking action of the depolarising drugs is not reversed.

Drugs that block the nicotinic receptors on autonomic ganglia, such as hexamethonium, probably do so by actually blocking the Na^+ ion channel rather than the receptor. Generally these receptors appear to resemble the central ones more than those at the neuromuscular junction and dihydro-β-erythroidine is one drug that it is an effective antagonist in both ganglia and the CNS.

In contrast to the nicotinic antagonists and indeed both nicotinic and muscarinic agonists, there are a number of muscarinic antagonists, like atropine, hyoscine (scopolamine) and benztropine, that readily cross the blood–brain barrier to produce central effects. Somewhat surprisingly, atropine is a central stimulant while hyoscine is sedative, as least in reasonable doses. This would be the expected effect of a drug that is blocking the excitatory effects of ACh on neurons but since the stimulant action of atropine can be reversed by an anticholinesterase it is still presumed to involve ACh in some way. Generally these compounds are effective in the control of motion but not other forms of sickness (especially hyoscine), tend to impair memory (Chapter 18) and reduce some of the symptoms of Parkinsonism (Chapter 15).

DRUGS AND THE DIFFERENT MUSCARINIC RECEPTORS

While five different muscarinic receptors have now been distinguished, atropine and the other antimuscarinics discussed above show little specificity for any of them, although pirenzapine is most active at the M_1 receptor. Much effort has been expended in the search for more specific muscarinic agonists and antagonists and while a few compounds have emerged which, from binding studies at least, show some (but never dramatic) selectivity, the results have been somewhat disappointing. As M_1 receptors mediate the postsynaptic excitatory effects of ACh while M_2 cause autoinhibition of its release, then augmenting ACh activity requires an M_1 agonist coupled with an M_2 antagonist capable of crossing the blood–brain barrier as well as an M_1 antagonist that will not. Even then the peripheral effects of the M_2 antagonist such as dry mouth and blurred vision can be unpleasant. Such possible permutations of agonist and antagonists in the treatment of dementia are considered in more detail in Chapter 18.

CHOLINERGIC PATHWAYS AND FUNCTION

Three distinct and basic CNS neuronal systems were referred to in Chapter 1, namely: long-axon neurons, intrinsic short-axon neurons and those in brainstem nuclei with extensively branching and ramifying ascending axons. The ubiquitous nature of ACh as a NT is evidenced by it being employed as such in all three situations to some extent, although for the first it is mainly confined to the periphery where it is released from long-axon preganglionic fibres and somatic motor nerves to skeletal muscle. In the striatum it is released from intrinsic interneurons and in the cortex from the terminals of ascending axons from subcortical neurons in defined nuclei. See Fig. 6.7 for detail.

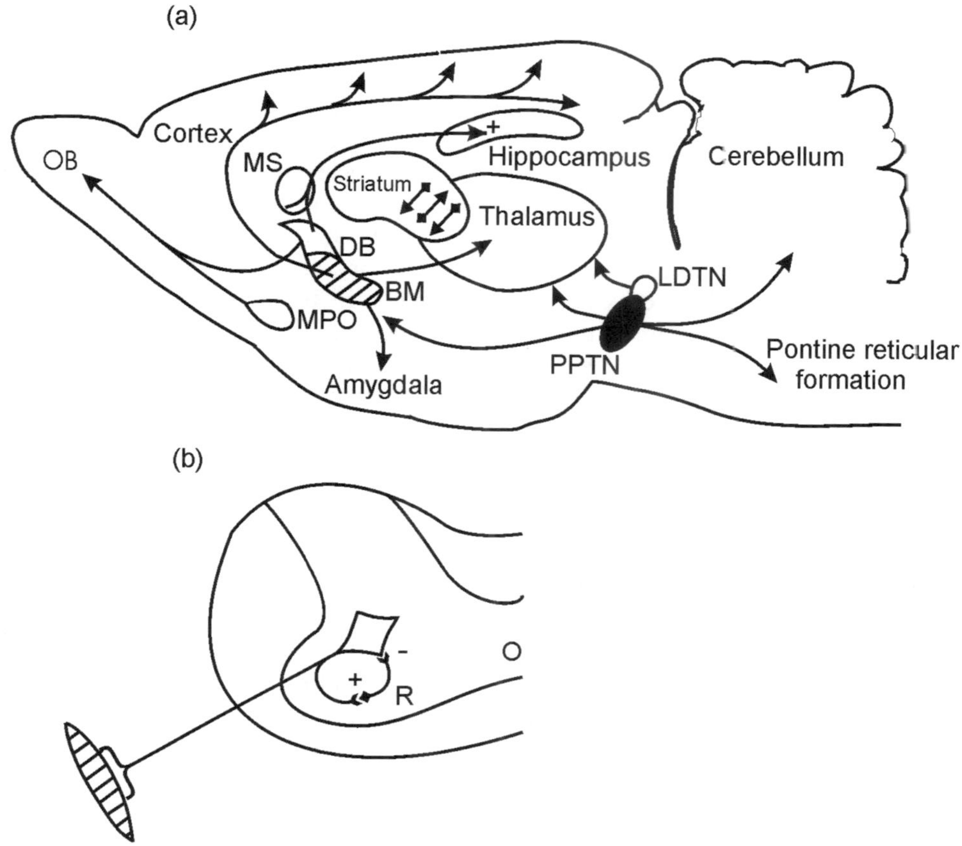

Figure 6.7 Cholinergic pathways. (a) Acetylcholine is found in intrinsic neurons within the striatum but the main pathways are the cortical projections from the nucleus basalis magnocellularis (BM) which also sends axons to the thalamus and amygdala. There are other projections from the medial septum (Ms) and the nucleus of the diagonal band, or diagonalis broco (DB), to the hippocampus and from the magnocellular preoptic nucleus (MPO) and DB to the olfactory bulb (OB). The DM and BM are sometimes referred to as the substantia inominata. Collectively all these nuclei are known as the magnocellular forebrain nuclei (FN). Other cholinergic nuclei are found more caudally in the tegmentum. The paramedian (or penduncυlo) pontine tegmental nucleus (PPTN) sends afferents to the paramedian pontine reticular formation and cerebellum but more importantly to the thalamus (lateral geniculate nucleus) and the more cephalic cholinergic neurons in MPO. Activation of neurons in PPTN during REM sleep gives rise to the PGO (ponto–geniculo–occipital) waves (see Chapter 22). There is a smaller lateral and dorsal tegmental nucleus (LDTN) with afferents projections like that of the PPTN, especially to the thalamus, but its role is less clear (see Woolf 1991). In the ventral horn of the spinal cord (b) ACh is released from collaterals of the afferent motor nerves to skeletal muscle to stimulate small interneurons, Renshaw cells (R), that inhibit the motoneurons

SPINAL CORD

Since ACh is the transmitter at the skeletal neuromuscular junction one might also expect it to be released from any axon collaterals arising from the motor nerve to it. Such collaterals innervate (drive) an interneuron (the Renshaw cell) in the ventral horn of the spinal cord, which provides an inhibitory feedback onto the motoneuron. Not only is ACh (and ChAT) concentrated in this part of the cord but its release from antidromically stimulated ventral roots has been demonstrated both *in vitro* and *in vivo*. Also the activation of Renshaw cells, by such stimulation, is not only potentiated by anticholinesterases but is also blocked by appropriate antagonists. In fact it illustrates the characteristics associated with both ACh receptors. Stimulation produces an initial rapid and brief excitation (burst of impulses), which is blocked by the nicotinic antagonist dihydro-β-erythroidine, followed, after a pause, by a more prolonged low-frequency discharge that is blocked by muscarinic antagonists and mimicked by muscarinic agonists. Thus in this instance although ACh is excitatory, as in other areas of the CNS, the activation of Renshaw cells actually culminates in inhibition of motoneurons. Pharmacological manipulation of this synapse is not attempted clinically and although administration of nicotinic antagonists that are effective at peripheral autonomic ganglia and can pass into the CNS, such as mecamylamine, may cause tremor and seizures, it cannot be assumed that this results from blocking cholinergic inhibition of spinal motoneurons.

STRIATUM

The concentration of ACh in the striatum is the highest of any brain region. It is not affected by de-afferentation but is reduced by intrastriatal injections of kainic acid and so the ACh is associated with intrinsic neurons. Here ACh has an excitatory effect on other neurons mediated through muscarinic receptors and is closely involved with DA (inhibitory) function. Thus ACh inhibits DA release and atropine increases it, although the precise anatomical connection by which this is achieved is uncertain and the complexity of the interrelationship between ACh and DA is emphasised by the fact that DA also inhibits ACh release. In view of the opposing excitatory and inhibitory effects of ACh and DA in the striatum and the known loss of striatal DA in Parkinsonism (see Chapter 15) it is perhaps not surprising that antimuscarinic agents have been of some value in the treatment of that condition, especially in controlling tremor, and that certain muscarinic agonists, like oxotremorine, produce tremor in animals.

CORTEX

Cholinergic neurotransmission has been most thoroughly studied in the cortex where the role of ACh as a mediator of some afferent input is indicated by the finding that undercutting the cortex leads to the virtual loss of cortical ACh, ChAT and cholinesterase. That it is not the mediator of the primary afferent input has been shown by the inability of atropine to block the excitatory effect of stimulating those pathways and the fact that such stimulation causes a release of ACh over a wide area of the cortex and not just localised to the area of their cortical representation (see Collier and Mitchell 1967). Indeed there have been many experiments which show that the release of ACh in the cortex is proportional to the level of cortical excitability, being increased by a variety of convulsants and decreased by anaesthesia (Fig. 6.8). The origins of this diffuse cholinergic input have been traced in the rat to the magnocellular forebrain nuclei (MFN) by mapping changes in cortical cholinesterase and ChAT after lesioning specific subcortical nuclei. The most important of them appears to be the nucleus basalis magnocellularis, similar to the nucleus of Maynert in humans, which projects predominantly to the frontal and parietal cortex and is thought to be affected

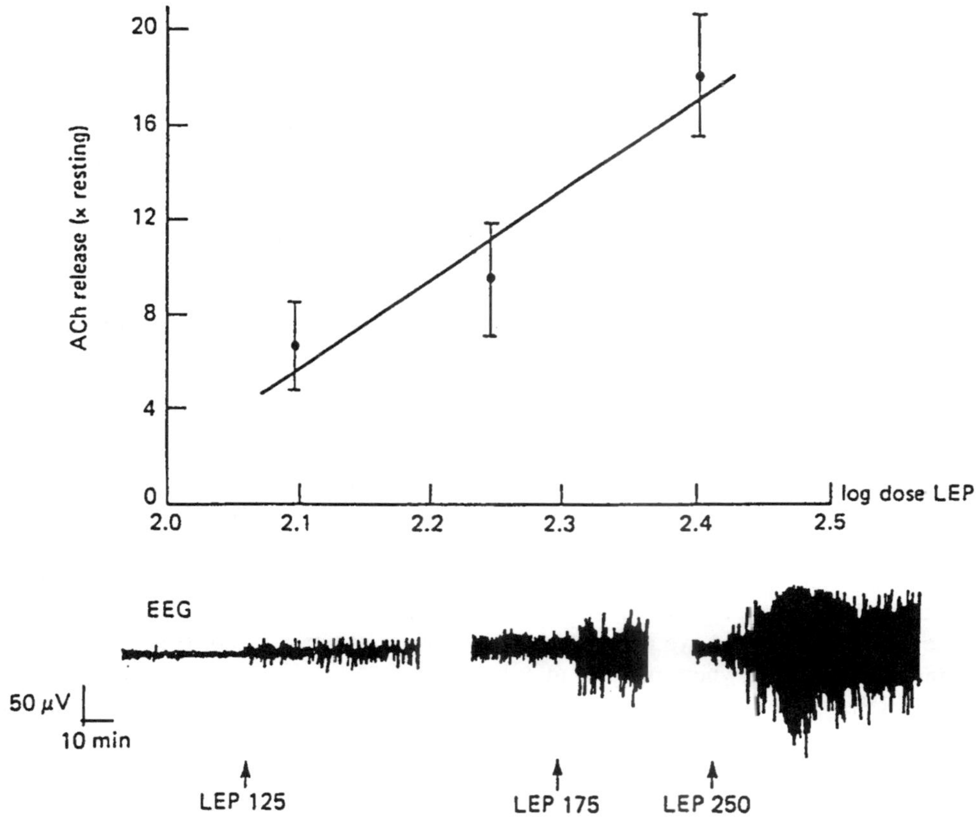

Figure 6.8 ACh release and cortical activity. Correlation between acetylcholine release and EEG activity after injections of leptazol (LEP mg kg^{-1} intravenously) into the urethane anaesthetised rat. ACh was collected in a cortical cup incorporating EEG recording electrodes. Mean values ±SE, $n = 6$ (unpublished data, but see Gardner and Webster 1977)

in Alzheimer's disease. This nucleus, together with the diagonal band, forms the substantia innominata and the dorsal neurons of this band also join with those in the medial septum to provide a distinct cholinergic input to the hippocampus (Fig. 6.7), which may play a part in memory function (see Chapter 18).

There is a second group of cholinergic neurons more caudally in the pontine tegmentum, the pendunculo pontine tegmental nucleus (PPPTN) and a smaller laterodorsal tegmental nucleus (LDTN). Their role in sleep and waking is discussed below and in Chapter 22.

Despite the excitatory effect of ACh in the cortex and its increased release during convulsive activity, antimuscarinic agents have only a slight sedative action (indeed, as emphasised above, atropine may cause excitation) and no anticonvulsant activity, except possibly in reducing some forms of experimentally induced kindling. ACh appears to exert a background excitatory effect on cortical function and while it may not directly stimulate the firing of pyramidal cells it will sensitise them to other excitatory inputs through its muscarinic activity.

AROUSAL AND SLEEP

Such a diffuse excitatory action of ACh in the cortex could fit it for a role in the maintenance of arousal and in fact the forebrain cholinergic nuclei, described above, appear to be innervated by the ventral part of the so-called ascending reticular system or pathway, which originates in a diffuse collection of brainstem neurons (see Chapter 22). If this pathway is lesioned the cortical EEG becomes quiescent but when stimulated it produces a high-frequency low-voltage desynchronised (aroused) EEG, which can be countered by antimuscarinic and potentiated by anticholinesterase drugs. Unfortunately this does not seem to apply to the actual behavioural arousal produced by such stimulation and suggests that ACh does not have a primary and certainly not a unique role in the maintenance of consciousness or sleep, although the firing of forebrain cholinergic neurons increases during the transition from sleep to waking. ACh does, however, feature prominently in one aspect of sleep behaviour.

As we relax in preparation for and pass into sleep, the active desynchronised 'awake' EEG characterised by the low-amplitude (5–10 μV) high-frequency (10–30 Hz) beta waves becomes progressively more synchronised giving larger (20–30 μV) and slower (8–12 Hz) alpha waves, and then even slower (1–4 Hz) and bigger (30–150 μV) delta waves. This so-called slow-wave sleep is interrupted at intervals of some 1–2 h by the break-up and desynchronisation of the EEG into an awake-like pattern. Since this is accompanied by rapid eye movements, even though sleep persists and can be deeper, the phase is known as rapid eye movement, REM or paradoxical, sleep. It is a time when dreaming occurs and when memory may be secured.

Such REM sleep may occur some four or five times during a night's sleep and can occupy 20% of sleep time. More importantly, for this discussion, it can be intensified by anticholinesterases and reduced by antimuscarinics and it is accompanied, and in fact preceded, by burst firing of a group of cholinergic neurons in the pedunculo pontine tegmental nucleus (PPTN). Neurons from this nucleus, which is quite distinct from the nucleus basalis, project to the paramedian pontine reticular formation, the thalamic lateral geniculate body and thus to the occipital cortex, all of which show increased activity during REM sleep to give PGO (ponto–geniculo–occipital) waves. Clearly sleep is not just a passive event and while cholinergic activity may be important in the production of REM sleep it does not appear to be responsible for turning it off or for actually inducing sleep. Many other NTs and neuronal networks come into this (see Chapter 22).

COGNITION AND REWARD

Not only is REM sleep a time for dreaming but it is also believed to be a time for the laying down (consolidation) of memory. This is only one observation among many that implicates ACh in the memory process. Certainly antimuscarinic drugs like atropine are well known to impair cognitive function in both animals and humans. In the former antimuscarinic drugs appear to impair both the acquisition and retention of some learned tasks, as in the Morris water maze. This involves placing a rat in a circular tank of water containing a stand with a platform just below the surface but which is not clearly visible because the vessel walls or water have been made opaque. Generally the rat quickly learns (2–3 trials) to identify the position of and swims to the platform. That ability is impaired by pretreatment with antimuscarinics which increase the number of trials (possibly tenfold) required before the animal swims directly to the platform and can increase the time to achieve it if given after the task has been learnt.

Perhaps the strongest evidence for the role of ACh in cognitive processes comes, however, from the finding that in Alzheimer's disease there is a reasonably selective loss of cholinergic neurons in the nucleus basalis and that augmenting cholinergic function with anticholinesterase and to some extent by appropriate muscarinic agonists can help to restore memory function in the early stages of the disease. How cholinergic function can facilitate the memory process is uncertain. It is generally thought that the laying down of memory is in some way dependent on the high-frequency discharge of hippocampal neurons in which long-term potentiation or LTP (the persisting potentiated response to a normal afferent input after a prior and short intense activation) plays an important part (see Chapter 18). Unfortunately while NMDA antagonists impair LTP, antimuscarinics do not. Of course, ACh will, by blocking K^+ efflux, increase the likelihood of neurons discharging repetitively.

While it is the muscarinic receptor which is primarily concerned with the cognitive effects of ACh it has recently been shown that part of the cholinergic septal input to the hippocampus innervates excitatory nicotinic receptors on GABA interneurons. Since these appear to synchronise the activity of the main hippocampal glutamate neurons their stimulation could influence hippocampal function and memory process (see Jones, Sudweeks and Yakel 1999). The fact that there is a cholinergic projection from the pedunculo pontine tegmental nucleus to the dopomine neurons of the ventral tegmental area (VTA) and that its excitatory effect is mediated through nicotinic receptors could also implicate them and so ACh, in the reward process. This is thought to be mediated in part through the mesolimbic and mesocortical dopamine pathways arising from the VTA and may offer an explanation for the addictive nature of nicotine and smoking.

REFERENCES AND FURTHER READING

Brown, DA (1983) Slow cholinergic excitation — a mechanism for increasing neuronal excitability. *Trends Neurosci.* **6**: 302–307.

Brown, DA, Gahwiler, BH, Marsh, SJ and Selyanko, AA (1986) Mechanisms of muscarinic excitatory synaptic transmission in ganglia and brain. *Trends Pharmacol. Sci.* **17**: Suppl, 66–71.

Caulfield, MP and Birdsall, NMJ (1998) Classification of muscarinic acetylcholine receptors. *Pharmacol. Rev.* **50**: 279–290.

Changeux, JP (1990) The nicotinic acetylcholine receptor: an allosteric protein prototype of ligand-gated ion channels. *Trends Pharmacol. Sci.* **11**: 485–492.

Changeux, JP, Bertrand, D, Corringer, PJ, Dehoene, S, Edelstein, S, Leno, C, Novere, N le, Marubio, L, Picciotto, M and Zoli, M (1998) Brain nicotinic receptors: structure and regulation, role in learning and reinforcement. *Brain Res. Rev.* **26**: 198–216.

Cole, AE and Nicoll, RA (1984) Characterization of a slow cholinergic post synaptic potential recorded *in vitro* from rat hippocampal pyramidal cells. *J. Physiol.* **352**: 173–188.

Collier, B and Mitchell, JF (1967) Release of ACh during consciousness and after brain lesions. *J. Physiol.* **188**: 83–98.

Cordero-Erausquin, M, Marubio, LM, Klink, R and Changeux JP (2000) Nicotinic receptor function: new perspectives from knockout mice. *Trends Pharmacol. Sci.* **21**: 211–218.

Eglen, RM and Watson, N (1996) Selective muscarinic receptor agonists and antagonists. *Pharmacol. Toxicol.* **78**: 59–78.

Flentge, F, Venema, K, Koch, T and Korf, J (1997) An enzyme-reactor for electrochemical monitoring of choline and acetylcholine. Applications in high-performance liquid chromatography, brain tissue, microdialysis and cerebral fluid. *Annal. Biochem.* **204**: 305–311.

Gardner, CR and Webster, RA (1977) Convulsant–anticonvulsant interactions on seizure activity and cortical acetylcholine release. *Eur. J. Pharmacol.* **42**: 247–256.

Greenfield, SA (1991) A noncholinergic action of acetylcholinesterase (AchE) in the brain. From neuronal secretion to the generation of movement. *Cell. Mol. Neurobiol.* **11**: 55–77.

Hersch, SM, Guten Kunst, CA, Rees, HD, Heilman, CT and Levey, AJ (1994) Distribution of M_1–M_4 muscarinic receptor proteins in the rat striatum. Light and electron microscopic immunochemistry using subtype-specific antibodies. *J. Neurosci.* **141**: 3351–3363.

Hulme, EC, Birdsall, NJM and Buckley, NJ (1990) Muscarinic receptor subtypes. *Ann. Rev. Pharm. Toxicol.* **30**: 633–673.

Jones, S, Sudweeks, S and Yakel, JL (1999) Nicotinic receptors in the brain: correlating physiology with function. *Trends Neurosci.* **22**: 555–562.

Levey, A, Kitt, CA, Simmonds, WF, Price, DL and Brann, MR (1991) Identification and localisation of muscarinic acetylcholine receptor proteins in brain with subtype specific antibodies. *J. Neurosci.* **11**: 3218–3226.

Lindstrom, JM (1997) Nicotinic acetylcholine receptors in health and disease. *Mol. Neurobiol.* **15**: 193–222.

McCormick, DA and Prince, DA (1986a) Mechanism of action of acetylcholine in the guinea-pig cerebral cortex *in vitro*. *J. Physiol.* **375**: 169–194

McCormick, DA and Prince, DA (1986b) Inhibitory effect of acetylcholine in the thalamus. *Nature* **319**: 402–405.

Wall, SJ, Yasada, RP, Hory, F, Flagg, S, Martin, BM, Ginns, EI and Wolfe, BB (1991) Production of antisera selective for M_1 muscarinic receptors using fusion proteins: distribution of M_1 receptors in rat brain. *Mol. Pharmacol.* **39**: 643–649.

Wannacott, S (1997) Presynaptic nicotinic ACh receptors. *Trends Neurosci.* **20**: 92–98.

Webster, HH and Jones, BE (1988) Neurotoxic lesions of the dorsalateral pontomesencephalic tegmentum-cholinergic cell area in the rat. II Effects upon sleep–waking states. *Brain Res.* **458**: 285–302.

Wess, J (1996) Molecular biology of muscarinic acetylcholine receptors. *Crit. Rev. Neurobiol.* **10**: 69–99.

Whittaker, VP (1987) Cholinergic synaptic vesicles from the electromotor nerve terminals of Torpedo: composition and life cycle. *Ann. NY Acad. Sci.* **493**: 77–91.

Woolf, NJ (1991) Cholinergic systems in mammalian brain and spinal cord. *Prog. Neurobiol.* **37**: 475–524.

Zinnerman, H, Volknandt, W, Wittich, B and Hausinger, A (1993) Synaptic vesicle lifecycle and synaptic turnover. *J. Physiol. (Paris)* **87**: 159–170.

7 Dopamine (DA)

R. A. WEBSTER

Dopamine (3:4 dihydroxyphenylethylamine), like noradrenaline and adrenaline, is a catecholamine and in addition to its independent neurotransmitter role in the CNS it is a precursor to noradrenaline (NA) in all central and peripheral noradrenegic neurons.

PATHWAYS

It became possible to visualise neurons which contained catecholamines when it was discovered that these amines reacted with formaldehyde vapour (later replaced by glyoxylic acid) to produce isoquinoline condensation products which emitted a bright-green fluorescence when visualised under ultra-violet light. This was distinguishable from the yellow fluorescence of 5-HT and could be separated from that for NA by appropriate pharmacological manipulations or adjustments to the microscopic techniques. Using this procedure, which is known as the Falk–Hillarp technique, Dahlstrom and Fuxe (1964) located and numbered nuclei in the hindbrain (pons medulla) in which either DA (A8–A12) or NA (1–7) was concentrated. The pathways were then established by axotomy since lesion of the axon is followed by loss of the NT and fluorescence at the neuron's terminals (destination of pathway) but not from its cell bodies (origin).

Most of the DA cell bodies (about 400 000) in the human brain are found in the A9 nucleus which forms the zona compacta (dorsal part) of the substantia nigra (SN), although a few cell bodies are found in the more ventral zona reticulata and in the zona lateralis as well (Fig. 7.1). A8 is lateral, caudal and somewhat dorsal to A9 and A10 whereas A10 is ventral to A9. Axons from A9 form the major contribution, together with some from A8, to the principal DA nigrostriatal pathway running to the striatum (caudate nucleus and putamen) and amygdala. This pathway is lateral to, but runs with, a more medial DA pathway, predominantly from A10, which innervates the nucleus accumbens and olfactory tubercle (mesolimbic pathway) as well as parts of the cortex (mesocortical system) such as the prefrontal and perirhinal cortex. The DA innervation to the anterior cingulate cortex also comes from A10 but with some axons from A9. There is in fact no clear divide between A9 and A10 and some overlap of their pathways. The DA mesolimbic tract and the noradrenergic bundles come together in the medial forebrain bundle before entering the cortex.

Neurotransmitters, Drugs and Brain Function. Edited by R. A. Webster

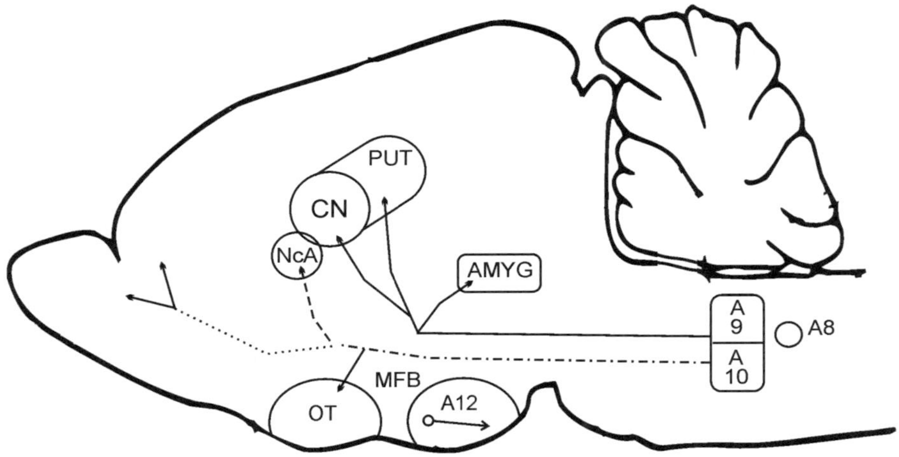

Figure 7.1 Dopamine neuronal pathways. AMYG, amygdala; CN, caudate nucleus; MFB, medial forebrain bundle; NcA, nucleus accumbers; OT, olfactory tubercle; PUT, putamen; SN, substantia nigra. For full details see text and Moore and Bloom (1978) and Lindvall and Bjorkland (1978)

A further totally separate DA pathway arises from A12 in the arcuate nucleus and forms the tuberoinfundibular tract in the median eminence to the pituitary gland for controlling prolactin release. This is partly achieved by DA being released into capillaries of the hypothalamic–hypophyseal portal system and then inhibiting the prolactin releasing cells (lactotrophs) of the anterior pituitary.

While the nigrostriatal pathways are ipsilateral some crossing occurs in fibres from the ventral tegmental A10 nucleus. These pathways are shown diagramatically in Fig. 7.1. Further details can be obtained from Moore and Bloom (1978) and Lindvall and Bjorkland (1978). The nuclei provide distinct loci for activating the dopamine systems for electrophysiological, release and behavioural studies and for their destruction by electrolytic lesion or injection of the toxin 6-hydroxydopamine (6-OHDA).

The concentration of DA in different brain areas of the rat is in keeping with the distribution of its pathways. It is concentrated in the striatum (10 μg/g), nucleus accumbens (5 μg/g) and olfactory tubercle (6 μg/g) but in the cortex there is much less (0.1 μg/g). Cells in the substantia nigra in humans and primates differ from those in other species in containing granules of the lipoprotein pigment called neuromelanin. The melanin granules are free in the cytoplasm and give the SN a distinctive dark colour. Cells in this nucleus can also have hyaline inclusion bodies, the Lewy bodies, which are not common normally but appear to increase dramatically in patients with Parkinsonism. In humans the SN neurons are very closely aligned to blood vessels which could make them readily influenced by blood-borne agents and might explain why they are vulnerable as in Parkinson's disease. Certainly they will require considerable biochemical back-up to maintain function in all their terminals.

NEUROCHEMISTRY

The biochemical pathways in the synthesis and metabolism of dopamine are shown in Fig. 7.2 and their position in the context of the dopamine synapse in Fig. 7.3.

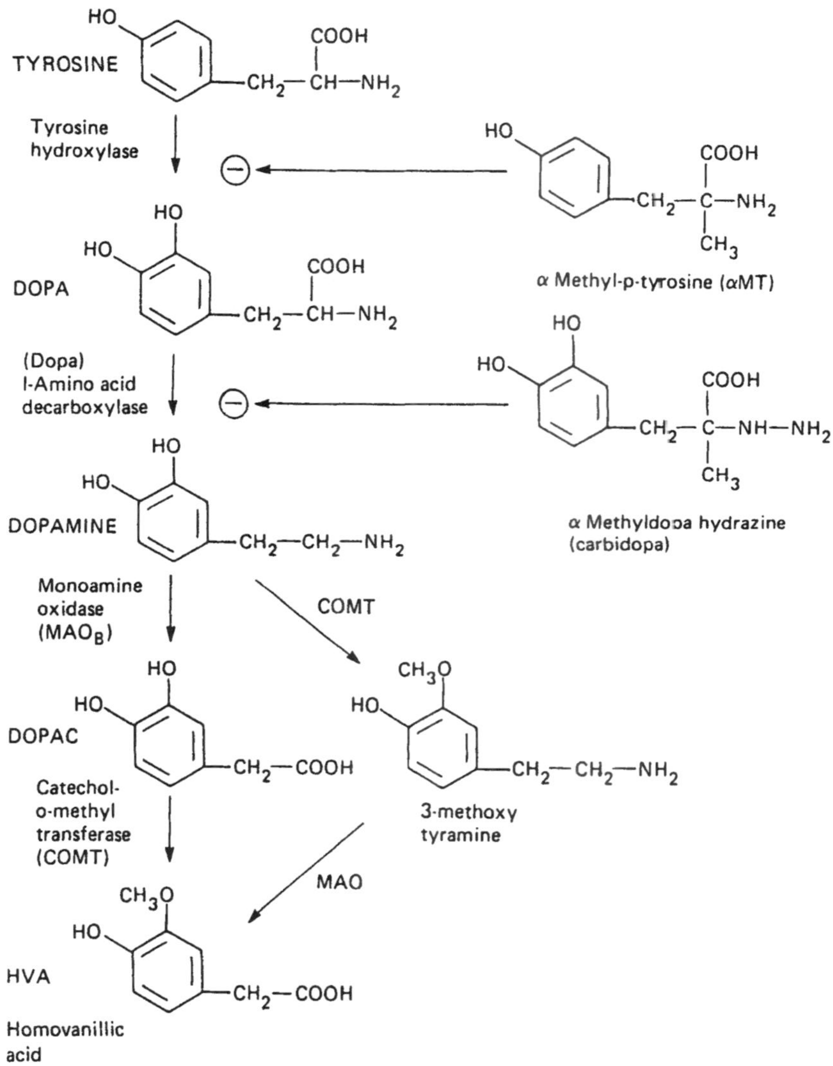

Figure 7.2 Biochemical pathways for the synthesis and metabolism of dopamine. (−) indicates drug inhibition of enzyme activity

SYNTHESIS

The synthesis and metabolism of DA are very similar to that of NA, even when it functions as a NT in its own right. Although both phenylalanine and tyrosine are found in the brain it is tyrosine which is the starting point for NA and DA synthesis. It appears to be transported into the brain after synthesis from phenylalanine (phenylalanine hydroxylase) in the liver rather than from phenylalanine found in the brain. Despite the fact that the concentration of tyrosine in the brain is high (5×10^{-5} M) very little body tyrosine (1%) is used for the synthesis of DA and NA.

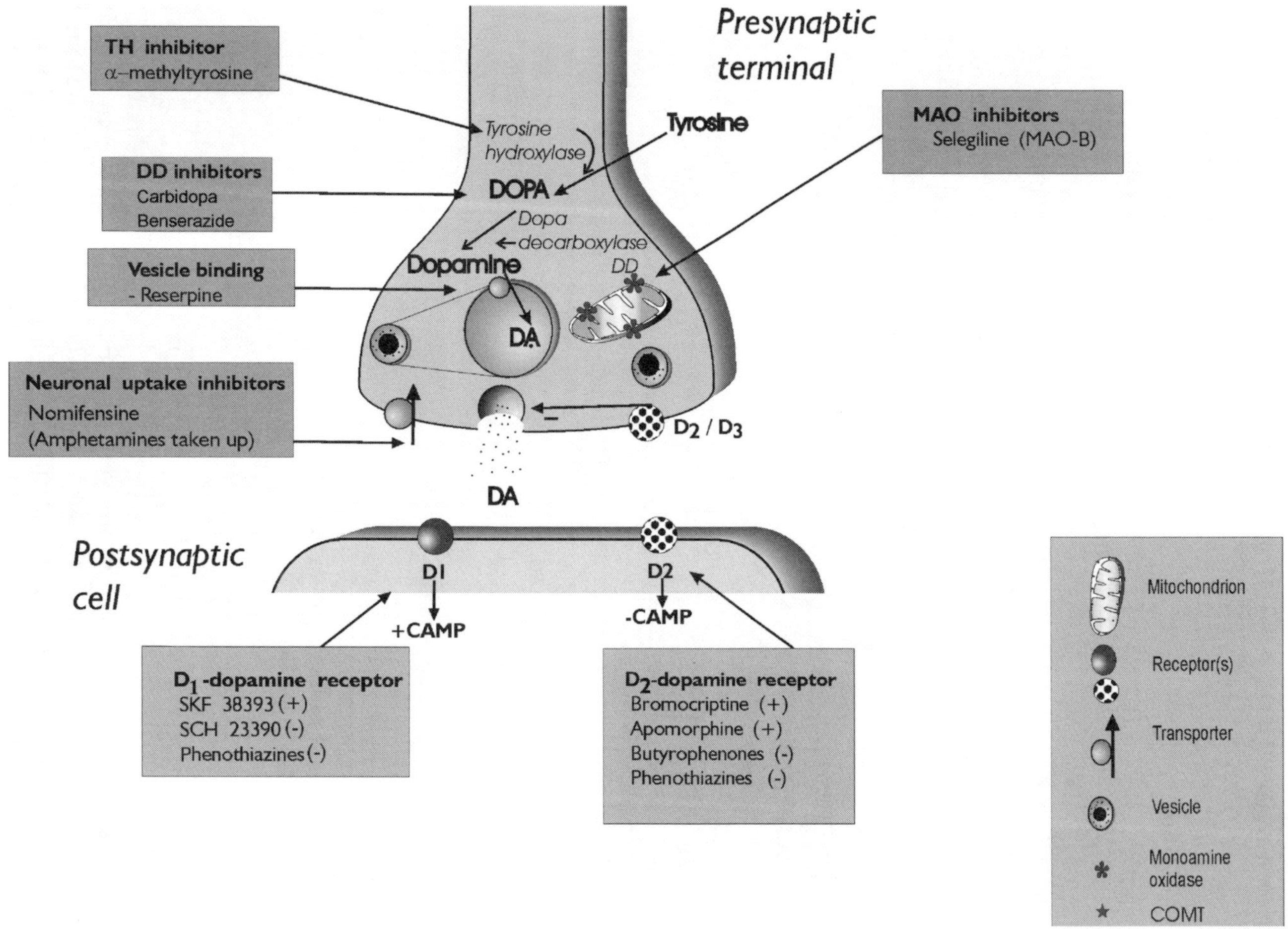

Figure 7.3 Diagrammatic representation of a dompaminergic synapse. (+) = stimulation, agonists; (−) = antagonism

Tyrosine hydroxylase

Tyrosine is converted to dopa by the cytoplasmic enzyme tyrosine hydroxylase. This is the rate-limiting step (K_m 5×10^{-6} M) in DA synthesis, it requires molecular O_2 and Fe^{2+} as well as tetrahydropterine (BH-4) cofactor and is substrate-specific. It can be inhibited by α-methyl-*p*-tyrosine, which depletes the brain of both DA and NA and it is particularly important for the maintenance of DA synthesis. Since the levels of tyrosine are above the K_m for tyrosine hydroxylase the enzyme is normally saturated and so it is not possible to increase DA levels by giving tyrosine.

Dopa decarboxylase

By contrast, the cytoplasmic decarboxylation of dopa to dopamine by the enzyme dopa decarboxylase is about 100 times more rapid (K_m: 4×10^{-4} M) than its synthesis and indeed it is difficult to detect endogenous dopa in the CNS. This enzyme, which requires pyridoxal phosphate (vitamin B6) as co-factor, can decarboxylate other amino acids (e.g. tryptophan and tyrosine) and in view of its low substrate specificity is known as a general L-aromatic amino-acid decarboxylase.

While a number of drugs, e.g. α-methyl dopa, inhibit the enzyme they have little effect on the levels of brain DA and NA, compared with inhibition of tyrosine hydroxylase and they also affect the decarboxylation of other amino acids. Some compounds, e.g. α-methyl dopa hydrazine (carbidopa) and benserazide, which do not easily enter the CNS have a useful role when given in conjunction with levodopa in the treatment of Parkinsonism (see Chapter 15) since the dopa is then preserved peripherally and so more enters the brain.

Controls of synthesis

It is possible to deplete the brain of both DA and NA by inhibiting tyrosine hydroxylase but while NA may be reduced independently by inhibiting dopamine β-hydroxylase, the enzyme that converts DA to NA, there is no way of specifically losing DA other than by destruction of its neurons (see below). In contrast, it is easier to augment DA than NA by giving the precursor dopa because of its rapid conversion to DA and the limit imposed on its further synthesis to NA by the restriction of dopamine β-hydroxylase to the vesicles of NA terminals. The activity of the rate-limiting enzyme tyrosine hydroxylase is controlled by the cytoplasmic concentration of DA (normal end-product inhibition), presynaptic dopamine autoreceptors (in addition to their effect on release) and impulse flow, which appears to increase the affinity of tyrosine hydroxylase for its tetrahydropteridine co-factor (see below).

METABOLISM

Just as the synthesis of DA and NA is similar so is their metabolism. They are both substrates for monoamine oxidase (MAO) and catechol-O-methyl transferase (COMT). In the brain MAO is found in, or attached to, the membrane of the intraneuronal mitochondria. Thus it is only able to deaminate DA which has been taken up into nerve endings and blockade of DA uptake leads to a marked reduction in the level of its deaminated metabolites and in particular DOPAC. The final metabolite, homovanillic acid (HVA), is one that has been both deaminated and O-methylated so it must be assumed that most of any released amine is initially taken back up into the nerve where it is deaminated and then subsequently O-methylated (Fig. 7.2). Certainly the brain contains much more DOPAC (the deaminated metabolite of DA) than the corresponding O-methylated derivative (3-methoxytyramine). It is possible, however, that the high levels of DOPAC, as found particularly in rat brain, partly reflect intraneuronal metabolism of unreleased DA and it is by no means certain that the metabolism of DA to HVA is always initially to DOPAC. Thus released DA that is not taken up into neurons is probably O-methylated initially by COMT.

O-methylation

It is generally accepted that COMT is an extracellular enzyme in the CNS that catalyses the transfer of methyl groups from S-adenylmethionine to the meta-hydroxy group of the catechol nucleus. Until recently the only inhibitors of this enzyme were pyragallol and catechol which were too toxic for clinical use. Now other inhibitors have been developed, e.g. entacapone and tolcapone, but these are used mainly to protect dopa (also a catecholamine) from O-methylation, in the treatment of Parkinson's disease (Chapter 15).

Deamination

Monoamine oxidase exists in two forms, MAO_A and MAO_B. The former is more active against NA and 5-HT than it is against DA, which is a substrate for both, even though, like β-phenylethylamine, it is more affected by MAO_B. It seems likely that MAO_B is the dominant enzyme in human brain and inhibitors of it, such as selegiline, have some value in the treatment of Parkinson's disease by prolonging the action of the remaining endogenous DA as well as that formed from administered levodopa.

Uptake

The removal of released DA from the synaptic extracellular space to facilitate its intraneuronal metabolism is achieved by a membrane transporter that controls the synaptic concentration. This transporter has been shown to be a 619 amino-acid protein with 12 hydrophobic membrane spanning domains (see Giros and Caron 1993). Although it has similar amino-acid sequences to that of the NA (and GABA) transporter, there are sufficient differences for it to show some specificity. Thus DA terminals will not concentrate NA and the DA transporter is blocked by a drug such as nomifensine which has less effect on NA uptake. Despite this selectivity some compounds, e.g. amphetamine and 6-OHDA (but not MPTP), can be taken up by both neurons. The role of blocking DA uptake in the central actions of cocaine and amphetamine is considered later (Chapter 23).

STORAGE

Most DA (up to 75%) is stored in vesicles like NA. This can be disrupted by the rauwolfia alkaloid, reserpine and by drugs like tetrabenazine. It should be emphasised that these drugs deplete the neurons of amines by stopping their incorporation into vesicles so that it leaks out and is deaminated. They do not cause an active release of amine.

RELEASE AND TURNOVER

Short-term control (autoreceptors)

As with many neurons (e.g. NA) there are presynaptic autoreceptors on the terminals of dopamine neurons whose activation attenuate DA release. Although most of these receptors appear to be of the D_2 type, as found postsynaptically, D_3 receptors are also found. It is possible that in addition to the short-term control of transmitter release they may also be linked directly to the control of the synthesising enzyme tyrosine hydroxylase. It seems that autoreceptors are more common on the terminals of nerves in the nigrostriatal (and possibly mesolimbic) than mesocortical pathway.

Autoreceptors are also found on the cell bodies of DA neurons, in the substantia nigra (A9) and ventral tegmentum (A10) where their activation leads to a reduction in cell firing. To what extent they are stimulated by endogenous DA is uncertain but systemic DA agonists certainly activate them to inhibit the neuron, and since DA antagonists alone can increase the firing of DA neurons that implies that the autoreceptors could be tonically active. This can have important implications, as we shall see later when considering the mode of action of DA antagonists in the treatment of schizophrenia (Chapter 17).

There is much evidence (e.g. Cheramy, Leviel and Glowinski 1981) from both *in vitro* and *in vivo* perfusion studies that DA is released from the dendrites of DA neurons in both A9 and A10 even though those dendrites do not contain many vesicles compared with axon terminals. The release and changes in it may also be slower and longer than that at axon terminals and the synaptic arrangement between the releasing dendrites and postsynaptic target is not clear. DA receptors also appear to be on neurons other than dopamine ones and on the terminals of afferent inputs to A9 (and A10). It seems that the activation of the DA neurons may partly be controlled by the effects of the dendritically released DA on such inputs.

Long-term control

Generally the concentration of DA remains remarkably constant irrespective of the level of neuronal activity. One reason for this is that nerve stimulation increases tyrosine hydroxylase activity and DA synthesis. It is thought that tyrosine hydroxylase can exist in two forms with low and high affinities for its tetrahydropteredine co-factor (BH-4) and that nerve traffic increases the high-affinity fraction.

Certainly the activity of tyrosine hydroxylase is greater in the DA neurons of the substantia nigra (17.5 nmol dopa synthesised/mg protein/h) than in the NA neurons of the locus coeruleus (4–5), as is the turnover of the amine itself (1.7 μg/h) compared with that of NA (1.0) (see Bacopoulus and Bhatnager 1977). In the caudate nucleus and nucleus accumbens the turnover of DA is even higher at 7.4 and 2–6 μg/g/h respectively.

NEUROTOXINS

The 6-hydroxylated form of DA, 6-hydroxydopamine (6-OHDA) is taken up into both DA and NA nerve terminals where it is readily oxidised to compounds that cause degeneration of the terminals over a period of days. To produce a central effect it must be administered directly into the brain by intracerebroventricular (icv) injection. NA terminals can be protected by prior injection of the NA uptake inhibitor desmethylimipramine. Alternatively, small amounts may be injected directly by stereotaxic techniques into particular DA nuclei where uptake takes place even if terminals are not present.

Recently much interest has centred on a very specific toxin for DA neurons. This is 1-methyl-4-phenyl-1,2,3,6-tetrahydropyridine (MPTP). It was discovered when a student, who was addicted to pethidine, tried to manufacture 1-methyl-4-phenyl-4-propionoxy-piperidine (MPPP) but took a short-cut in synthesis and produced MPTP. When he administered this to himself he developed Parkinsonism. MPTP destroys DA neurons. Again this process depends on the neuronal uptake mechanism, since MPTP itself is not the active material. It needs to be deaminated to MPP^+ which is then taken up by DA nerve terminals.

DOPAMINE RECEPTORS

CLASSIFICATION

The original discovery and classification of DA receptors was based on the results of three distinct studies:

(1) Stimulation of adenylate cyclase
(2) Ligand binding
(3) Inhibition of prolactin release

The adenylate cyclase discovered originally in bovine superior cervical ganglia, and then found in homogenates of rat striatum, was specific to DA, in that it was activated by other DA agonists like ADTN, but not greatly by NA or 5-HT. Some other drugs with established DA-like effects proved, however, to be either partial agonists (apomorphine) or ineffective (bromocriptine). Also while some neuroleptic (antipsychotic) drugs that are DA antagonists in behavioural studies, such as the thioxanthenes and phenothiazines, antagonised this effect with a relative potency that compared with their antipsychotic activity, other potent neuroleptics like the butyrophenones were relatively ineffective. Overall there was a poor correlation between antipsychotic activity and DA antagonism as measured by blockade of DA-induced cAMP production.

Ligand-binding studies, originally with [^{3}H] dopamine and [^{3}H] haloperidol but subsequently using [^{3}H] spiperone, demonstrated the existence of a specific binding site for them in membrane preparations from mammalian striatum. Displacement studies with a whole range of neuroleptic drugs also showed that not only was the rank order different from that for blocking the adenylate cyclase but also correlated much better with antipsychotic activity. Additionally DA agonists like bromocriptine, which were ineffective in increasing cAMP production, showed appropriate binding.

When tested on prolactin release in isolated mammatrophs of bovine anterior pituitary, apomorphine appeared a full agonist (inhibiting release) while antagonism of the inhibition of prolactin release by the neuroleptics showed a potency more similar to that for binding than for blocking cAMP production. Also the inhibition of prolactin release by DA was not accompanied by any change in intracellular cAMP and therefore was not linked to it.

Thus the establishment of two clear dopamine effects, one directly linked to stimulation of adenylate cyclase and the other inhibition of prolactin release, which was independent of adenylate cyclase stimulation but associated with distinct binding sites led to the concept, formulated by Kebabian and Calne (1979), that DA effects were mediated through two distinct receptors. One was linked to stimulation of adenylate cyclase (D_1) while the other (D_2) did not appear to be associated with the enzyme but had distinct binding sites. The justification for this classification was subsequently enhanced by the synthesis of two compounds, SKF 38393 and SCH 23390. The former activated the DA adenylate cyclase without affecting prolactin release or spiperone binding, i.e. it was a D_1 agonist, while the latter blocked the stimulation of adenylate cyclase, again without affecting prolactin release or binding. It was a D_1 antagonist. The basis for this early classification is shown in Table 7.1.

Although some subsequent pharmacological studies suggested that perhaps there could be a subdivision of both the D_1 and D_2 receptors, the paucity of appropriate agonists and antagonists (and indeed of test responses) precluded its justification until molecular biology took over. Cloning studies show that structurally there are two distinct groups of DA receptors, D_1 and D_2. There is a D_5 variant of D_1 as well as D_3 and D_4 forms of D_2. The D_1 and D_5 receptors are linked to activation of adenylate cyclase and the D_2 group to its inhibition, although this is not its main effect on neurons (see later).

Despite this profusion of receptors the D_1 and D_2 predominate (over 90% of total) and most known effects of DA, its agonists and antagonists are mediated through the D_2 receptor. Although the above nomenclature is now accepted it might have been better, as suggested by Sibley and Monsma (1992), to retain D_1 and D_2 to represent the two families and then subdivide them as D_{1A} for (D_1), D_{1B} for (D_5), then D_{2A} for (D_2), D_{2B} for (D_3) and D_{2C} for (D_4), even though variants of all five have been found. Their

Table 7.1 Evidence for the initial basic classification of D_1 and D_2 dopamine receptors

(a) **Discovery of specific DA stimulated striatal adenylate cyclase**
But (i) Effect only antagonised by some neuroleptics,
phenothiazines–YES
thioxanthenes–YES
butyrophenones–NO
(metoclopromide inactive)
(ii) Effect not reproduced by DA agonists like bromocriptine.

(b) **Binding studies: ^{3}H dopamine ^{3}H haloperidol**
Displacement studies with a wide range of neuroleptics showed good correlation with their clinical potency in schizophrenia.
DA agonists such as bromocriptine show binding.

(c) **Prolactin release from isolated mammatrophs from anterior pituitary**
DA agonists — bromocriptine, apomorphine and ADTN decrease release.
Blocked by neuroleptics — similar in effectiveness to their binding affinities (b).
Not linked to stimulation of adenyl cyclase.

Notes:
Studies with various agonists and antagonists showed that the effects on (a) differed in potency from both (b) and (c) and were thus associated with a receptor (D_1) different from that (D_2) linked to (b) and (c). See text for detail.

structure has been established in both rat and human brain and they are generally similar in the two. The human D_2 receptor shows a protein sequence which is 96% identical to that of the rat D_2 and although the similarity is only 91% between the human and rat D_1 receptor, it is 96% in the transmembrane region. It is differences in the amino-acid sequences in this region that primarily justify the classification into two groups (D_1 and D_2) rather than their total amino-acid number. Basically the D_1 (and D_5) receptors differ from the D_2 (D_3, D_4) in having a much shorter third cytoplasmic loop and a much longer intracellular C-terminus (Fig. 7.4), which appears to be a feature of receptors linked to G_s and the stimulation of adenylate cyclase. Based on amino-acid sequencing the D_3 receptor is only 53% homologous with the D_2 (but 75% in the transmembrane region) while with D_4 it is only 41% (56%). The D_5 receptor shows 50% homology with the D_1 rising to 80% in the transmembrane region. So-called short and long variants of the D_2 receptor (D_{2S} and D_{2L}) have also been discovered, differing by the presence or absence of a run of 29 amino acids in the third intracellular loop. For more detail see Sibley and Monsma (1992).

DISTRIBUTION AND MECHANISMS

The potential value of the discovery, classification and subdivisions of any NT receptors rests on the knowledge that

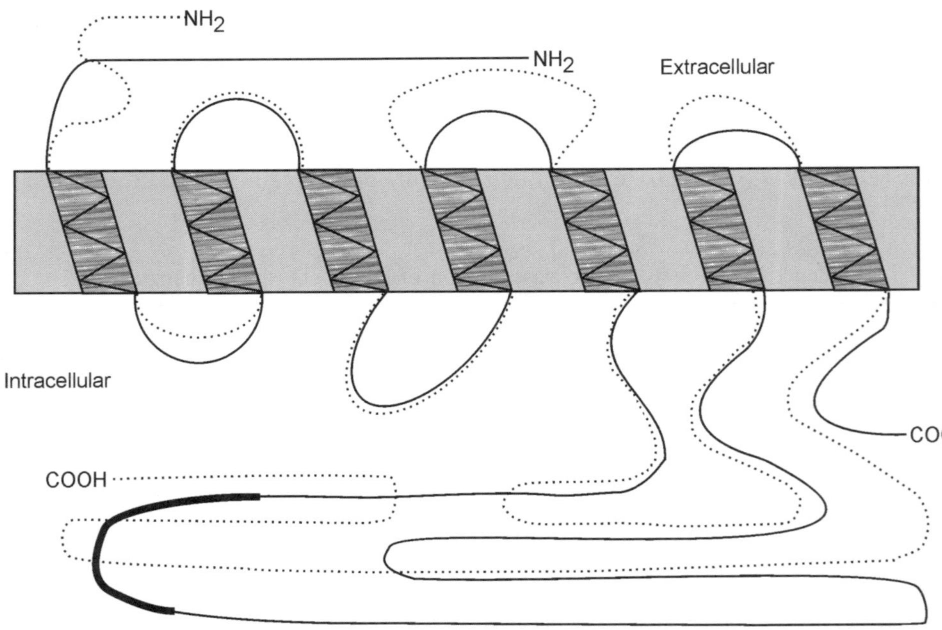

Figure 7.4 Comparative schematic representation of the D_1 (· · ·) and D (—) dopamine receptor. The figure attempts to highlight the major differences between extra- and intracellular loops, especially the intracellular loops between transmembrane sections 5 and 6 and the much longer C terminal of the D_1 compared with the D_2 receptor. It is based on the proposed topography of Sibley and Monsma (1992). The thickened length of the D_2 receptor represents the amino-acid sequence missing in the short form of the receptor. No attempt has been made to show differences in amino-acid sequencing or transmembrane topography

(1) Those receptors are linked to different cellular actions and/or are located in different brain regions or parts of the neuron so as to produce different functional effects.
(2) There are appropriate specific agonists or antagonists to establish and exploit those differences.

To some extent these requirements are cyclic since the establishment of different functions (1) depends on the availability of appropriate drugs (2). There is no shortage of drugs, especially antagonists, but since the main difference in structure between DA receptors is intracellular, rather than at the binding or recognition site, very specific drugs may be hard to produce. Since receptors can be expressed in cell lines the affinity of drugs for the different receptors can, however, be established, as can their cellular actions. Detection of appropriate mRNA also makes it possible to map the distribution of the receptors. The main characteristics of the DA receptors are summarised below and in Table 7.2.

D_1 receptor family

D_1 Highest expression in human striatum and nucleus accumbens and olfactory tubercle but also some in cortex and hypothalamus. In the striatum 50% of medium sized striato-nigral neurons, which also express substance P, express them. They are

Table 7.2 Dopamine receptor characteristics

Main class	D_1		D_2		
Subclass	D_{1A}	D_{1B}	D_{2A}	D_{2B}	D_{2C}
Named	D_1	D_5	D_2	D_3	D_4
No. amino acid (human)	446	477	414 (s) 443 (l)	440	387
DA affinity	Low	Moderate	Moderate	High	High
K_1 (nM) approx.	2000	200	599	20	20
<u>Effector</u>					
Activation of adenylate cyclase	↑	↑	↓	(↓)	(↓)
IP_3 turnover	(↑)	–	↑	–	–
Ca^{2+} influx	(↑)	–	↓	↓	↓
K^+ efflux	–	–	↑	–	–
Agonists	✓		✓	7-OHDPAT?	
Antagonists	✓		✓		Clozapine
Number	High	Low	High	Low	Low
<u>Distribution</u>					
Striatum	✓		✓		
Nuc. accumbens	✓		✓	✓ (Mainly)	
Frontal cortex					
Hippocampus		✓		✓	✓
Hypothalamus		✓		✓	(Midbrain)
Substantia nigra			✓	✓	
VTA (auto-receptors)			(✓)	(✓)	

Note:
S = short, L = long versions of D_2 receptor, ↑ = main effect observed, (↓) = some evidence of an effect. See Sibley and Monsma (1997), Sokoloff and Schwartz (1995) and Strange (1996).

linked primarily to stimulation of adenylate cyclase but also increase IP_3 turnover. They have a low micromolar affinity for DA ($K_1 \sim 2\,\mu M$).

D_5 Highest concentration in hippocampus and hypothalamus but much lower expression overall. Also linked to stimulation of adenylate cyclase but higher submicromolar affinity for DA ($K_1 \sim 200\,nM$). Also found in rat striatum and nucleus accumbens.

D_2 receptor family

D_2 Mostly in striatum, nucleus accumbens and olfactory tubercle but also on neuron cell bodies in substantia nigra and ventral tegmentum where they are the autoreceptors for locally (dendritic) released DA. The loss of specific D_2 antagonist binding in the striatum after lesions of the afferent nigro-striatal tract indicates their presynaptic autoreceptor role on terminals there. Other lesion studies have also established D_2 receptors on other inputs such as the cortico striatal tract.

As with D_1 receptors some 50% of striatal medium-sized cells contain them but they are different neurons as they co-express enkephalin rather than substance P. The importance of this difference in the therapy of Parkinsonism is taken up later (Chapter 15). D_2 receptors are also expressed on larger cells—probably cholinegic. Although linked to inhibition of adenylate cyclase (and IP_3 turnover) this is not their primary action. They increase K^+ conductance (hyperpolarise neurons) but also inhibit Ca^{2+} entry through voltage-sensitive channels, probably directly. When functioning as autoreceptors, these effects would also reduce DA release. The affinity for DA is slightly higher for the D_2 ($K_1 \sim 400\,nM$) than for D_1 receptors. No pharmacological differences have been established between the long or short forms of the D_2 receptor.

D_3 Much less abundant than D_2. Mainly in limbic regions (nucleus accumbens and olfactory tubercle) but also in hypothalamus. Some in caudate and cortex and also expressed on DA neurons in substantia nigra, presumably as autoreceptors. No effect on adenylate cyclase but inhibits Ca^{2+} entry (autoreceptor role). High affinity for DA ($K_1 \sim 25\,nM$).

D_4 Again very few in number compared with D_2 but located in frontal cortex, midbrain and amygdala. High affinity for DA ($K_1 \sim 20\,nM$) and a number of variants in humans.

Comparison of the K_1 values of various agonists and antagonists for the different receptors (Table 7.3) shows that whereas there are a number of drugs that readily distinguish between the D_1 and D_2 families and can be used to study their function, none distinguish between D_1 and D_5 and there is little to choose between D_2, D_3 and D_4 activities. Some differences that have been exploited are the low affinity of raclopride for D_4 receptors (compared with D_2 and D_3), the high affinity of clozapine and the benzamide derivative YM 43611 for the D_4 (cf. D_2, D_3) and that of 7-OH-DPAT for D_3. Since only the latter is an agonist, however, their value in establishing the roles of the D_3 and D_4 receptors is limited, although the high affinity of clozapine for D_4 receptors and their location in the frontal cortex has been considered, somewhat controversially, to be of significance in the aetiology and therapy of schizophrenia (see Chapter 17).

Table 7.3 (a) Dissociation constants (K_i) for various agonists and antagonists at the different dopamine receptors. (b) Indication of specificity for D_2 compared with D_1 receptors and between D_2 (i.e. D_2, D_3 and D_4) receptors

(a)

	D_1	D_5	D_2	D_3	D_4
Agonist					
Bromocriptine	440*	440*	8*	5*	290*
Quinpirole	1900	–	5	24*	30*
7-OH-DPAT	5000*	–	10	1*	550
SKF 38393	1	0.5*	150*	5000*	1000*
Apomorphine	0.7	–	0.7*	32*	4*
Antagonist					
Chloropromazine	90*	130*	3	4	35
Haloperidol	80*	100*	1	7*	2
Clozapine	170*	330*	230*	170*	21
Raclopride	18 000		1.8	3.5	2400
Spiperone	350*	3500*	0.06	0.6	0.08
S-Sulpiride	45 000	77 000	15*	13*	1000
YM 43611	10 000+	10 000+	43	11	2
SCH 23390	0.2*	0.3	1100*	800*	3000*

(b)

	Specific for D_2 family		Ratio to D_1 activity
Agonist	$D_2 = D_3 > D_4$	Bromocriptine	50
	$D_2 > D_3 = D_4$	Quinpirol	<400
	$D_3 > D_2 >> D_4$	7-OH-DPAT	<5000
Antagonist	$D_2 = D_3 >> D_4$	Raclopride	10 000
	$D_2 = D_4 > D_3$	Spiperone	6000
	$D_2 = D_3 > D_4$	S-Sulpiride	3000
	$D_4 > D_3 = D_7$	YM 43611	250
	Specific for D_1 family		**Ratio to D_2 activity**
Agonist	$D_1 = D_5$	SKF 38393	150+
Antagonist	$D_1 = D_5$	SCH 23390	1000

Note:
All values shown are taken from Seeman and Van Tol (1994) except for those for YM 43611 (Hidaka *et al.* 1996). Asterisked values are considered approximate.

SYNAPTIC EFFECTS

Because DA is very much localised to one brain area (striatum) and as there is such a pronounced DA pathway from the substantia nigra to the striatum it would be reasonable to assume that the effect of this pathway on striatal neuron activity is well established. Unfortunately this is not the case.

Over the years a large number of studies using extracellular recording in the striatum have shown that iontophoretic DA depresses 75–100% of all neurons responding to it, irrespective of whether spontaneous, excitatory amino acid-induced, or synaptic-evoked activity was being monitored. This inhibitory response is slow in onset (up to 15 s) and long in duration (possibly minutes). Stimulation of the substantia nigra can produce inhibition, excitation or mixed effects but it is possible, despite the high proportion of

DA neurons in this nucleus, that not all the effects are elicited by the release of DA. Most neuroleptics block the inhibitory effects of applied DA but some, e.g. haloperidol, are less active against SN-evoked inhibition. Generally these studies lacked specific agonists and antagonists used microintophoresis which is not really quantitative and with extracellular recording gave little information on the state of polarisation of the neuron.

Unfortunately the picture was not clarified by intracellular recordings from striatal neurons which, as these need to be large to take an electrode, are not necessarily typical (only 10%) of most striatal neurons innervated by DA afferents. Stimulation of the substantia nigra invariably produces a monosynaptic depolarisation in them that is blocked by haloperidol, but which may proceed to a hyperpolarisation, if the stimulus is strong enough. DA iontophoresced onto the same neuron may also cause depolarisation (Kitai, Sugimori and Kocsis 1976) but can still reduce its discharge. Mixed effects are often seen with DA and when it is infused in increasing concentrations into the striatum through a push–pull cannula it generally depresses extracellularly recorded cell firing but low concentrations can produce excitation or bimodal excitation–inhibition (Schoener and Elkins 1984). Voltammetry studies with an electrode that can also be used for recording neuronal firing have shown that increasing nigrostriatal stimulation induces not only an increase in DA release but also an inhibition of neurons (after some initial but variable excitation of large neurons), which outlasts the rise in extracellular DA. Thus the effects of endogenous DA appear to be critically dependent on its extracellular concentration and it may be that while synaptic effects can be excitatory, extrasynaptic ones are inhibitory. Some of this effect may also be indirect through reducing the release of excitatory NTs such as glutamate from cortico-striatal fibres or ACh from intrinsic neurons.

In view of the known cellular actions of DA, such as increased K^+ efflux and reduced Ca^{2+} currents associated with D_2 receptor activation in cell lines, inhibition would be the expected response to DA, especially as cyclic AMP, which is increased by D_1 receptor activation also inhibits striatal neurons. In fact although many DA synaptic effects are blocked by D_2 antagonists like haloperidol, the role of D_1 receptors should not be overlooked.

Iontophoretic studies on rat striatal neurons (Hu and Wang 1988) showed that while the release of DA by low currents facilitated glutamate-induced activation, high current efflux inhibited it. Although these effects were reduced by the D_2 antagonist haloperidol it was the D_1 agonist SKF 38393 which mimicked them, causing activation when released by low currents but inhibition at higher ones. Both effects were abolished by the D_1 antagonist SCH 23390. By contrast, the D_2 agonist quinpirol produced a less marked biphasic effect in which inhibition dominated.

A number of studies in fact show clear D_1 effects. Intracellular recording from striatal neurons in rat brain slices show a cAMP-mediated D_1-dependent (blocked by SCH 23390) suppression of a voltage-dependent sodium current which make the cell less responsive.

Repetition of some of these approaches using more modern techniques, e.g. whole cell and patch-clamp recording from dissociated striatal neurons, shows a similar mixed picture. An observed D_1-sensitive suppression of the sodium current and a shift of the inactivating voltage in a hyperpolarising direction, together with a depression of certain Ca^{2+} currents, would make the neuron less excitable. The D_2 effect in these measures is less clear with reports of both a depolarising and hyperpolarising shift of the

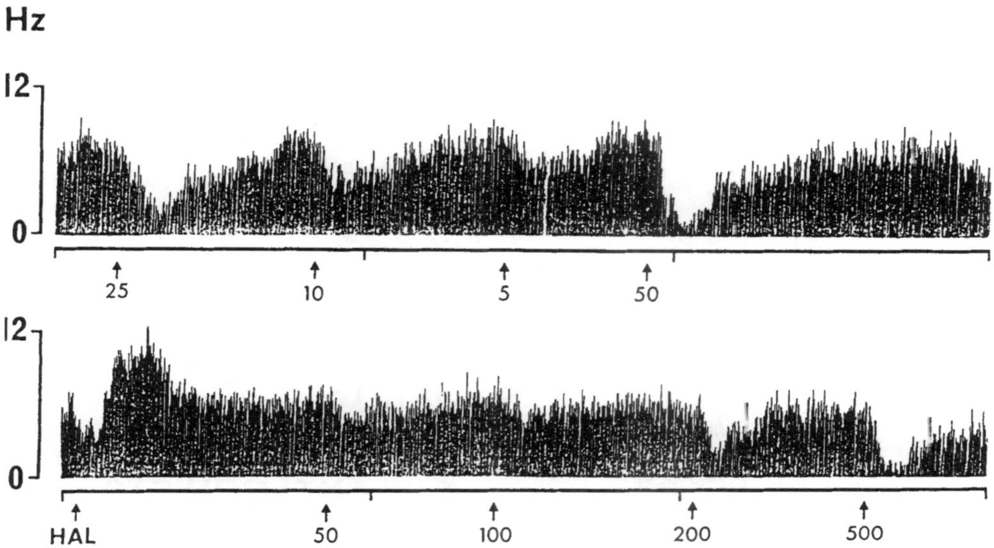

Figure 7.5 Rate recording of the dose-dependent inhibitory effects of apomorphine (μg/kg) on the spontaneous activity of a neuron in the medial prefrontal cortex of the halothane anaesthetised rat and its antagonism by haloperidol (HAL, 0.5 mg/kg). Time scale is 50 min intervals. Reproduced by permission from Dalley (1992)

inactivation curves and an increased opening of a potassium conductance (see Calabresi *et al.* 1987).

What is clear from all these experiments is that DA can have a bimodal effect depending on how much is applied or released, and which receptors are involved. Excitation is more common at low concentrations and inhibition at higher ones. What happens *in vivo* is not clear but *in vivo* voltammetry certainly suggests that the extracellular concentration of DA can be very high and this would favour the more commonly observed inhibition.

In other brain areas which receive a DA input, such as the nucleus accumbens and prefrontal cortex, it appears to be inhibitory and predominently D_2-mediated. This is clear from Fig. 7.5 which shows inhibition by apomorphine (mixed D_2, D_1 agonists) of the firing of neurons in the medial prefrontal cortex of the anaesthetised rat and its antagonism by the D_2 antagonist haloperidol.

These are, of course, extracellular recordings but more recent intracellular studies in both rat and guinea pig accumbens slices show that DA produces a D_2-mediated depolarisation and a D_1 hyperpolarisation which appear to be dependent on decreased and increased K^+ conductances respectively. This would certainly fit in with the belief that DA mediates the positive effects of schizophrenia by a D_2-mediated stimulation of the nucleus accumbens (see Chapter 17).

It is perhaps not surprising that DA produces such mixed effects. The D_1 receptor is primarily linked to the activation of adenylate cyclase and then protein kinase A. The response to its activation will therefore depend on the ion channels and other proteins modulated by the kinase which can vary from one neuron to another. Since the D_2 receptor is not so closely associated with just one G-protein, this gives it the potential for even more effects (see Greenhoff and Johnson 1997).

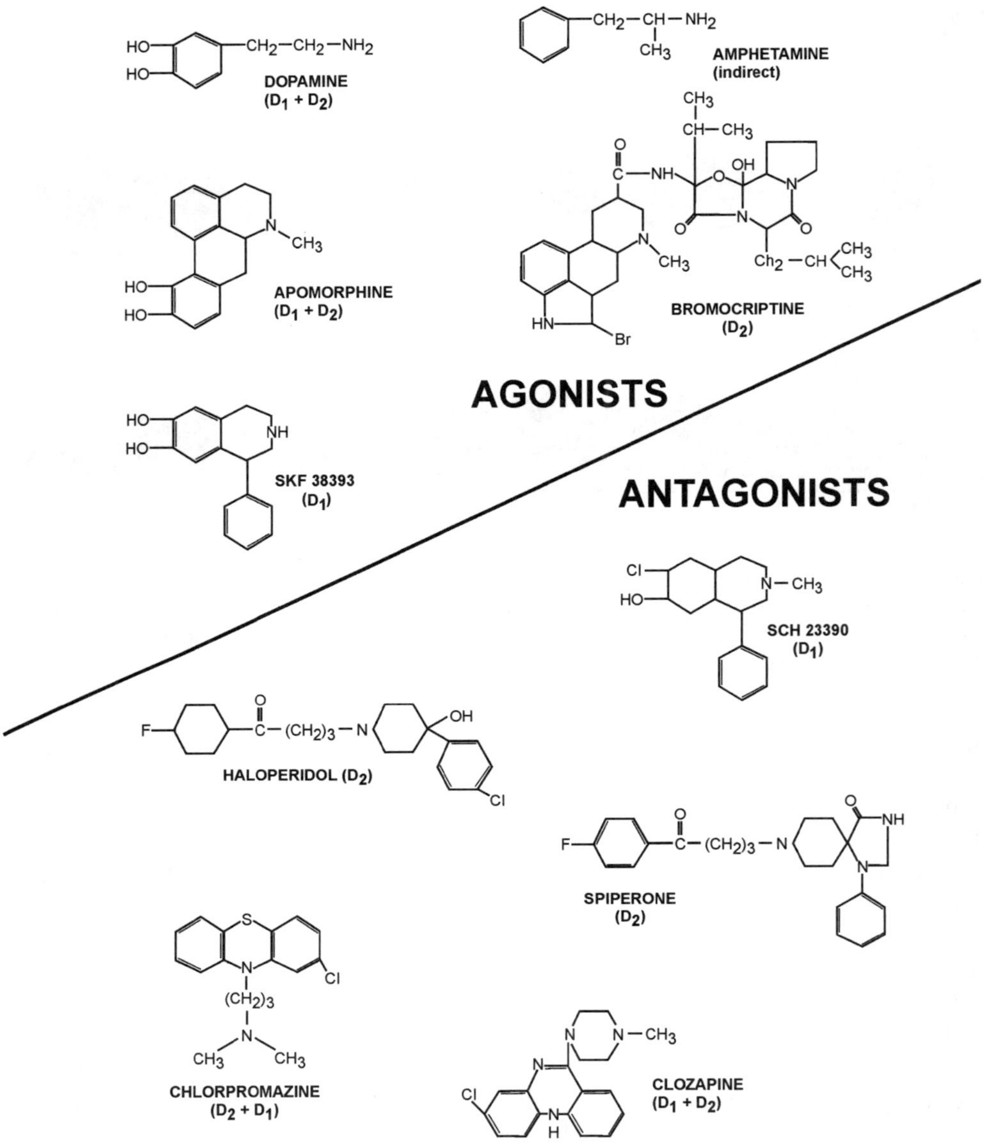

Figure 7.6 Chemical structures of some dopamine agonists and antagonists

PHARMACOLOGY OF THE DOPAMINE SYNAPSE

The sites of action of drugs affecting the dopamine synapse are indicated in Fig. 7.3. Those modifying the synthesis, storage, release, uptake and metabolism of DA have been covered above in the appropriate sections on neurochemistry. The actions and uses of agonists and antagonists are outlined in Table 7.4 and covered in detail in appropriate chapters. Their structures are given in Fig. 7.6.

AGONISTS

Those for the D_2 receptor (e.g. bromocriptine) have a particular value in the treatment of Parkinson's disease by reproducing the effects of the dopamine lost through degeneration of the nigrostriatal tract (Chapter 15). They are also used to reduce the undesirable effects of prolactinaemia (high plasma prolactin), such as amenorrhoea and galactorrhoea.

ANTAGONISTS

Again it is the D_2 compounds (e.g. chlorpromazine haloperidol) that generally have some benefit in and are the mainstay of therapy for schizophrenia although D_2 antagonism alone appears inadequate (Chapter 17). They also reduce dyskinesias such as those seen in Huntington's Chorea (Chapter 15). Some are used to control drug- and fever-induced vomiting and although any D_2 antagonist is effective, proclorperazine, metoclopramide and domperidone are more generally used. The latter two have fewer central effects since domperidone does not cross the blood–brain barrier while metaclopromide has the additional peripheral effect of increasing gastric emptying possibly by stimulating 5-HT receptors.

AMPHETAMINE

This is generally known as an indirectly acting sympathomimetic amine because peripherally it mimics activation of the sympathetic system although centrally it primarily reproduces the effect of DA. Chemically (Fig. 7.6) it is a phenylamine not a catecholamine and so it has no direct effect on any DA or NA receptor. It is, however, a substrate for the high-affinity DA (and NA) neural membrane transporter and is taken up into nerve terminals by that process. When released from the transporter into the cytoplasm its place is taken by DA which is then transported out (exchange diffusion). It is said to have released DA. In addition, amphetamine increases cytoplasmic DA by weakly reducing its uptake into vesicles. As a reasonably lipophilic compound amphetamine can enter the vesicles where being a weak base it takes up H^+ ions. This makes the vesicle interior less acidic and the reduction in pH gradient across the vesicle membrane appears to inhibit DA uptake into the vesicle. In addition, amphetamine is an inhibitor of MAO but it preferentially attacks the A form so its effect is greater on the breakdown of NA than of DA.

Most of the motor effects of amphetamine, especially stereotypy, are due to the release of DA as are its psychotic effects such as hallucinations. Its ability to mimic the action of DA in reward and reinforcement behaviour may contribute to its abuse potential (see Chapter 22) but its arousal (stimulant) properties also involve NA release.

CENTRAL FUNCTIONS

It is perhaps easier to identify some of the central functions of DA than that of the other monoamines because not only does it have distinctive central pathways associated with particular brain areas, but it has few peripheral actions. Also the actions of its antagonists reveal its central effects. These are summarised in Table 7.4.

Table 7.4 Summary of dopamine function

Function	Pathways	Effect of DA agonist	Effect of DA antagonist	Receptor
Control of motor function	Nigrostriatal tract from substantia nigra (A9)	Animals: Stereotypy. Rotation if one tract is lesioned Humans: Induces dyskinesias Effective in Parkinsonism	Animals: Catalepsy Humans: Reduces dyskinesias Induces Parkinsonism	Mainly D_2 some D_1
Initiation of behaviour	Mesolimbic pathway to nucleus accumbens from VTA (AIO)	Animals: Increases locomotor activity and intracranial self-stimulation	Animals: Decreases activity and self-stimulation	D_2
	Mesocortical pathways to prefrontal cortex from VTA (AIO)	Humans: Hallucinations, psychoses (reward, reinforcement)	Humans: Reduces positive symptoms of schizophrenia	
Control (inhibition) of prolactin release	Tuberoinfundibular tract from A12 in the arcuate nucleus of the median eminence to pituitary	Humans: Hypoprolactaemia	Humans: Hyperprolactaemia Galactorrhoea Amenorrhoea	D_2
Emesis	No distinct pathway DA receptors in chemoreceptor pathway zone	Vomiting	Anti-emetic (not motion sickness)	D_2

DA antagonists are anti-emetic, elevate plasma prolactin and have major motor and behavioural effects. Thus DA must be involved in the initiation of vomiting, the secretion of prolactin and control of motor and behavioural activity. Its role in emesis and as the prolactin release inhibitory factor have been adequately covered above. Its motor and behavioural function will now be considered.

MOTOR ACTIVITY

People with Parkinson's disease show a specific degeneration of the nigrostriatal tract so DA must be linked in some way to the control of motor function. It is also known that an imbalance of DA function on the two sides of the rat brain, either by stimulation or lesion of one SN, causes off-line or rotational movement (Ungerstadt and Arbuthnott 1970). This is best shown some days after 6-OHDA lesion of one substantia nigra and its nigrostriatal pathway when systemic apomorphine (DA agonist) causes animals to turn away from the lesioned side (contraversive), presumably

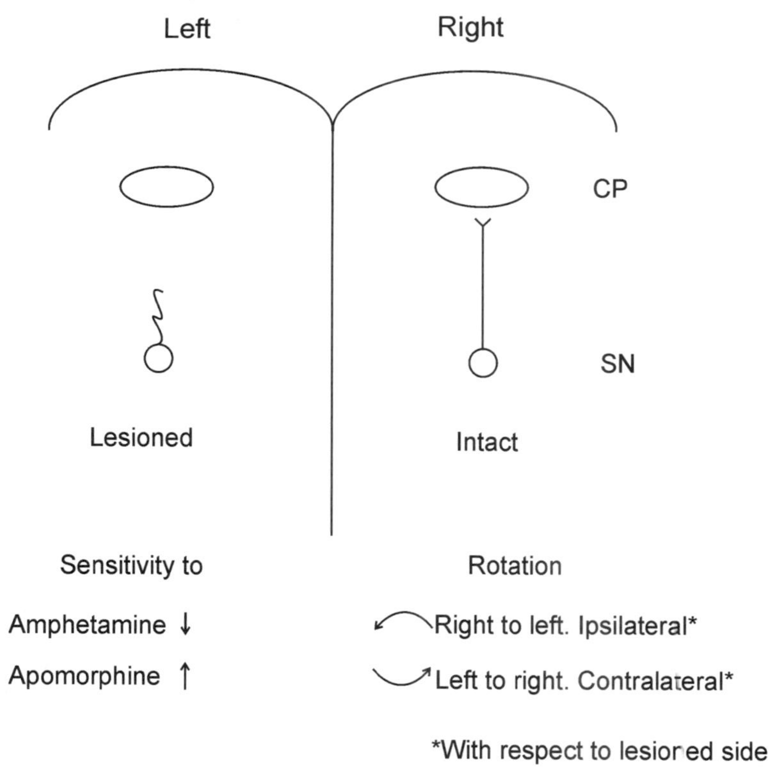

Figure 7.7 Dopamine-induced rotation in the rat in which one (left) nigrostriatal dopamine pathway from the substantia nigra (SN) to the caudate putamen (CP) has been lesioned by a prior injection (14 days) of 6-hydroxydopamine. Amphetamine, an indirectly acting amine, releases DA and so can only act on the right side. Since the animal moves away from the dominating active side it induces ipsilateral rotation (i.e. towards the lesioned side). By contrast, the development of postsynaptic supersensitivity to DA on the lesioned side ensures that apomorphine, a directly acting agonist, is actually more active on that side and so the animal turns away from it (contralateral rotation)

because the denervated striatum has become supersensitive and therefore more responsive than the control side to the DA agonist. Conversely, the indirectly acting amine amphetamine promotes movement towards the lesioned side (ipsiversive) because it can only release DA in the intact striatum (Fig. 7.7). Thus animals move away from the side with the most responsive and active striatum. These drugs also produce other motor activity including increased locomotion and a so-called 'stereotype' behaviour in which rats sniff avidly around the cage and spend much time licking and rearing. It appears that stereotypy is due to activation of the nigrostriatal pathway as it is absent after lesion of the substantia nigra and follows apomorphine and amphetamine injection into the striatum, whereas locomotor responses to amphetamine are reduced by lesions to A10 and can be induced by its injection into the nucleus accumbens.

Another indication of the importance of DA in motor control is the observation that in humans its precursor levodopa, and DA agonists like bromocriptine, not only overcome the akinesia of Parkinsonism but in excess will actually cause involuntary movements, or dyskinesia (Chapter 14). Also it is well known that DA antagonists like chlorpromazine and haloperidol produce Parkinsonian-like symptoms in humans (and catalepsy in animals) and, as indicated above, reduce the dyskinesia of Huntington's Chorea. Thus DA seems to sit on a knife edge in the control of motor function (Fig. 7.8).

PSYCHOSES

The main use clinically of DA antagonists is in the treatment of schizophrenia (Chapter 17) and the control of mania. Since psychotic symptoms are also a side-effect of levodopa therapy in Parkinsonism and as amphetamine causes hallucinations and schizophrenic-like symptoms in humans, presumably by releasing DA, it appears that DA also has an important part to play in the control and induction of psychotic symptoms. It is possible that the role of DA in psychosis is mediated primarily through the mesolimbic and mesocortical pathways and its control of motor function through the striatum, and there is evidence that the neurons from which these pathways arise have different characteristics. Although there is some overlap between the various DA nuclei in respect of the location of the cell bodies of the neurons that give rise to the different DA pathways, neurons can be identified by antidromic activation of their terminal axons in the appropriate projection areas. Recordings from neurons so identified show that they have differing firing patterns. Those cells innervating the prefrontal cortex fire at a much higher rate (9.7 Hz) than those to the cingulate and piriform cortex (5.9 or 4.3) and the striatum (3.1). Unlike the NA neurons they are also remarkably little affected by the state of the animal, i.e. its wake–sleep cycle. The cells in A10 which form the mesocortical pathway are also less easily inhibited by DA agonists suggesting that they probably have fewer autoreceptors. Unfortunately it seems that the DA postsynaptic receptor is the same at both sites so it has been difficult to divorce the antischizophrenic from the extrapyramidal-inducing activity of DA antagonists (see Chapter 16).

REWARD AND REINFORCEMENT

We expect reward to be pleasurable and it is assessed in animals by their willingness to seek and approach something, such as a lever linked to either food dispensing or brain

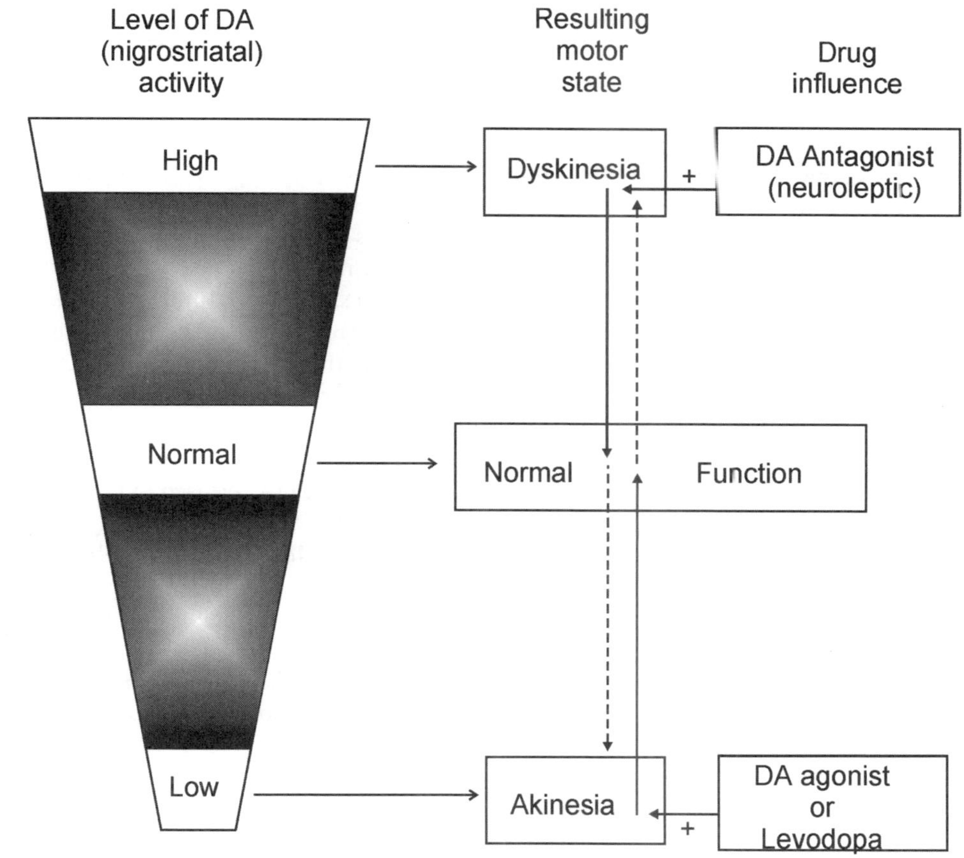

Figure 7.8 Dopamine and motor function. When nigrostriatal dopamine activity is normal so is motor function. Any reduction in this DA activity, as in Parkinson's disease, results in reduced motor activity, i.e. akinesia. By contrast, too much DA activity, as in Huntington's Chorea, produces abnormal motor function, i.e. dyskinesia. The latter may be controlled by neuroleptic drugs (DA antagonists) but they can swing the balance in DA activity sufficiently to produce akinesia (Parkinsonism). DA agonists (and levodopa) may overcome akinesia but can induce DA overactivity and dyskinesia (peak dose effect) (see Chapter 15)

stimulation. Reinforcement is the manner in which one event (stimulus) strengthens the likelihood of its repetition, i.e. repeated lever pressing for a pleasurable reward. In this instance, of course, reinforcement is rewarding but it need not be. Reward and reinforcement are considered by some to be the basis of motivation.

In 1954, Olds and Milner first described the effects of intracranial self-stimulation (ICSS). Rats with electrodes implanted in certain brain regions appeared to find the stimulation mediated through them to be rewarding (pleasurable) and so would seek out whatever part of their surroundings they associated with it. In addition, such self-stimulation reinforced the animal's inclination to indulge in other activity such as pressing a lever for a food reward. Since it was thought that the brain pathways and NTs mediating ICSS could also be responsible for more natural pleasurable rewards such as food, drink and sex their identification generated much interest.

The brain area most commonly, but not uniquely, associated with ICSS is the medial forebrain bundle (MFB). This includes the axons of noradrenergic as well as dopaminergic neurons but it appears to be the ventral tegmentum area and the A10 DA neurons innervating the nucleus accumbens (and prefrontal cortex) which is most active, as evidenced by 2-deoxyglucose autoradiography during self-stimulation. In fact effective, i.e. rewarding, self-stimulation through electrodes in the VTA is accompanied by DA release in the nucleus accumbens (Fiorino *et al.* 1993). The threshold current (or frequency) of stimulation for the initiation of ICSS is lowered by amphetamine and raised by DA antagonists while the rate of lever pressing in response to a particular level of ICSS is potentiated or reduced by the same respective procedures. Animals also learn to press a lever to initiate the administration (injection) of certain drugs in preference to obtaining food or water and will continue this to a point of intoxication. Dopamine uptake blockers and D_2 (or mixed) agonists are all strongly sought in self-administration trials and it became generally accepted that DA was paramount in mediating the reinforcing effects not only of ICSS but also of drug abuse and sex.

As a result of these observations it has been suggested that DA released in the nucleus accumbens is important in motivation by linking reward (especially when it is food) with the motor activity required to achieve it (Mogenson, Jones and Yim 1980). It is difficult, however, to distinguish a pure behavioural role for DA in actually initiating the sense of reward and motivation from its undisputed part in facilitating the motor response necessary to obtain the reward, e.g. a lever press in rats.

Salamone, Cousins and Snyder (1997) in fact suggest that the function of DA in the nucleus accumbens should not be described by terms such as motivation, reinforcement and reward. Rather it should be considered to mediate the higher-order motor and sensory processes that are important for the activation of aspects of motivation and responsiveness to conditioned stimuli.

DOPAMINE RECEPTORS, FUNCTION AND SYNERGISM

With the discovery of five distinct DA receptors within two major families, one might hope that the effects of the different DA pathways would be mediated through different receptors. Unfortunately this is not the case. As indicated above, it is generally the D_2 receptor that is important. Thus only D_2 antagonists have anti-emetic activity and only D_2 agonists, like bromocriptine, reduce plasma prolactin levels. In schizophrenia it is again the D_2 antagonists that are effective, although D_1 and D_4 receptors have been implicated (see Chapter 16) while in Parkinson's disease the symptoms can be alleviated by D_2 but not D_1 agonists if they are given alone. In this condition, however, some D_1 stimulation may augment the effect of the D_2 agonists (Chapter 14), suggesting a synergism between the two receptors. This synergism has been observed in both electophysiological studies on striatal neuron activity and some animal behaviours.

The electrophysiological studies of DA function in the striatum, reported above, suggest some similarities between D_1- and D_2-mediated effects. A clear synergism has been seen in fact, on the so-called type I DA neurons in the ventral tegmentum *in vitro* (Momiyama, Naoyuki and Saso 1993). The D_2 agonist quinpirole produced hyperpolarisation ($\uparrow K^+$ conductance) and reduced-action potential numbers but the D_1 agonist SKF 38393 had no effect alone. When given with quinpirole, however, it increased the ability of the D_2 agonist to raise the threshold for action potential

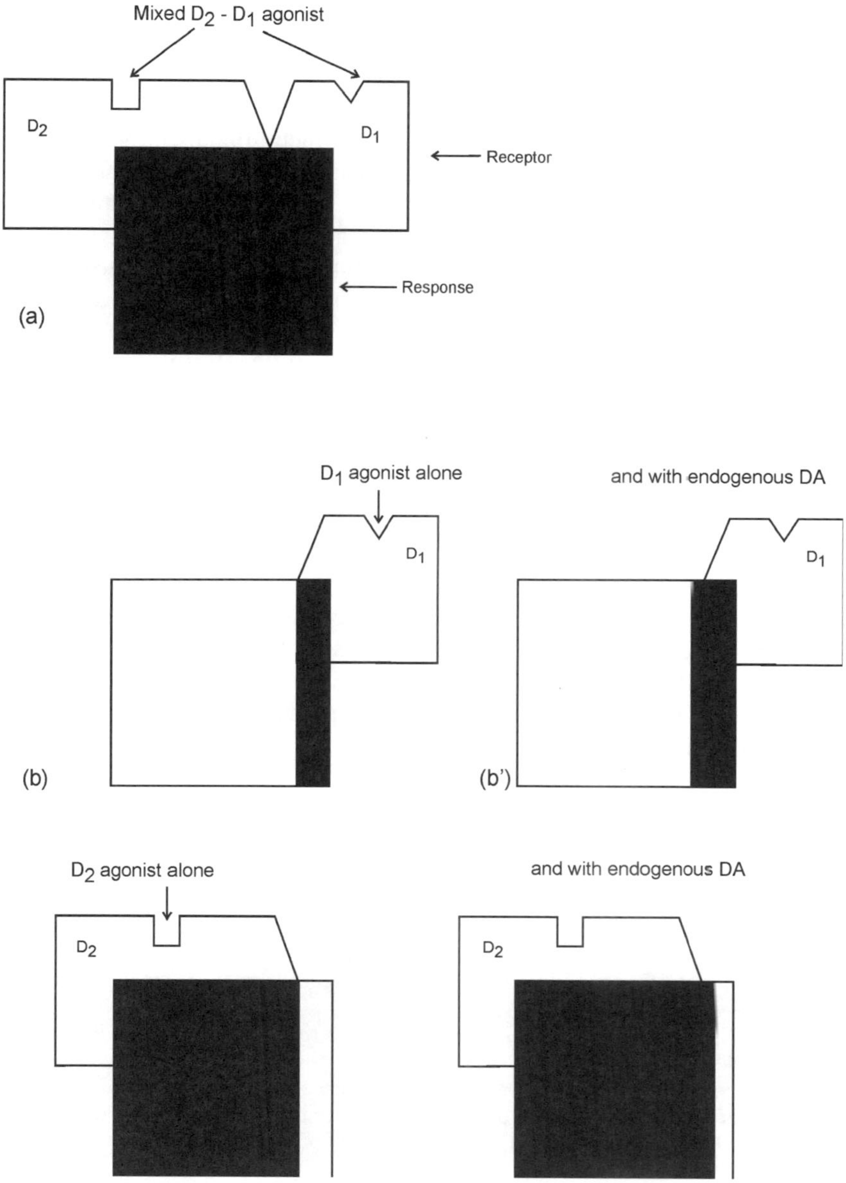

Figure 7.9 Synergism between dopamine D_1 and D_2 receptor activation. The level of behavioural and motor response to dopamine agonists is shown by the extent of shading in the response box. A mixed agonist, including DA itself, can produce a full response (a) by activating both D_1 and D_2 receptors. A D_1 agonist alone (b) has little effect compared with a D_2 agonist alone (c). The effect of the D_1 agonist is greater in the presence of endogenous DA acting synergistically on D_2 receptors (b′) and so can be partly reduced by a D_2 antagonist that has no D_1 receptor activity. While the effect of the D_2 agonist can also be augmented by DA acting on the D_1 receptor (c′) this increase is less marked and a D_1 antagonist has little effect on D_2 activity. The need to activate, or block, both the D_1 and D_2 family of DA receptors in order to reproduce or elimate full endogenous DA activity has implications in the treatment of Parkinson's disease and schizophrenia

generation and reduced firing and this potentiation was abolished by the D_1 antagonist SCH 23390.

Behaviourally, a D_1 agonist like SKF 38593 has few effects alone in rats apart from inducing grooming and some sniffing. By contrast, the D_2 agonist bromocriptine produces a marked behavioural stereotypy during which animals move avidly around the cage sniffing, gnawing, digging and then rearing. This is even more pronounced with a mixed D_2 and D_1 agonist like apomorphine. Surprisingly the activity of the D_1 agonist was severely attenuated by a D_2 antagonist (haloperidol) while the D_1 antagonist SCH 23390 slightly reduced the stereotypy of the D_2 agonist bromocriptine (Waddington 1989). Also, the effect of bromocriptine was increased by addition of the D_1 agnonist.

On this evidence it appears that a D_1 agonist is only fully effective if endogenous DA is present to act on D_2 receptors while a D_2 agonist also requires, although not to the same extent, some DA to act on D_1 receptors. Of perhaps more importance is the fact that a full DA effect depends on the activation of both D_1 and D_2 receptors even though the latter is dominant (Fig. 7.9).

At a time when pharmacological endeavour is aimed at producing drugs with limited and specific receptor profiles, the possibility that more than one receptor needs to be activated in order to replicate an action of an endogenous NT like DA is disturbing. Its significance in the therapy of Parkinson's disease is considered in Chapter 15. It must be remembered, however, that despite the links between D_1- and D_2-receptor mediated effects and the equality in their number, no D_1 agonist or antagonist produces or blocks any of the main known effects of DA either in humans or animals, whereas their D_2 counterparts are active.

REFERENCES AND FURTHER READING

Bacopoulus, NG and Bhatnager, RK (1977) Correlation between tyrosine hydroxylase activity and catecholamine concentration or turnover in brain regions. *J. Neurochem.* **29**: 631–643.

Calabresi, P, Mercuri, N, Stanzione, P, Stefani, A and Bernard, G (1987) Intracellular studies on the dopamine-induced firing inhibition of neostriatal neurons *in vivo*: evidence for D_1 receptor involvement. *Neurosciences* **20**: 757–771.

Cheramy, A, Leviel, V and Glowinski, J (1981) Dendritic release of dopamine in the substantia nigra. *Nature* **289**: 537–542.

Dahlstrom, A and Fuxe, K (1964) Evidence for the existence of monoamines containing neurons in the central nervous system. 1. Demonstration of monoamines in the cell bodies of brainstem neurons. *Acta Physiol. Scand.* **62** (Suppl 232): 1–55.

Dalley, JW (1992) Typical and atypical neuroleptic drug effects on dopamine and other neurotransmitter functions in rodents. PhD thesis, University of London, p. 125.

Fiorino, DF, Coury, A, Fibiger, HC and Phillips, AG (1993) Electrical stimulation of reward sites in the ventral tegmentum area increases dopamine transmission in the nucleus acumbers of the rat. *Behav. Brain Res.* **55**: 131–141.

Giros, B and Caron, MG (1993) Molecular characterization of the dopamine transporter. *Trends Pharmacol. Sci.* **14**: 43–49.

Greenhoff, J and Johnson, SW (1997) Electrophysiological effects of dopamine receptor stimulation. In *The Dopamine Receptors* (Eds Neve, KA and Neve, RL), Humana Press, Totowa, NJ, pp. 267–304.

Hidaka, K, Tada, S, Matsumoto, M, Ohmori, J, Tasaki, Y, Nomura, T, Usuda, S and Yamaguchi, T (1996) *In vitro* pharmacological profile of YM-43611, a novel D_2-like receptor antagonist with high affinity and selectivity for dopamine D_3 and D_4 receptors. *Brit. J. Pharmacol.* **117**: 1625–1632.

Hu, XT and Wang, RY (1988) Comparison of effects of D_1 and D_2 dopamine receptor agonists on neurons in the rat caudate putamen; an electrophysiological study. *J. Neurosci.* **8**: 4340–4348.

Kebabian, JW and Calne, DB (1979) Multiple receptors for dopamine. *Nature* **277**: 93–96.

Kitai, ST, Sugimori, M and Kocsis, JC (1976) Excitatory nature of dopamine in the nigro-caudate pathway. *Exp. Brain Res.* **24**: 351–363.

Lindvall, O and Bjorkland, A (1978) Organisation of catecholamine neurons in the rat central nervous system. In *Handbook of Psychopharmacology*, Vol. 9 (Eds Iversen, LL, Iversen, SD and Snyder, SH), Plenum, New York, pp. 139–231.

Mogenson, G, Jones, D and Yim, CY (1980) From motivation to action; functional interface between the limbic system and the motor system. *Progr. Neurobiol.* **14**: 69–97.

Momiyama, T, Naoyuki, T and Saso, M (1993) A mechanism underlying dopamine D_1 and D_2 receptor-mediated inhibition of dopaminergic neurons in the ventral tegmental area *in vitro*. *Br. J. Pharmacol.* **109**: 933–940.

Moore, RY and Bloom, FE (1978) Central catecholamine neuron systems, anatomy and physiology of the dopamine system. *Ann. Rev. Neurosci.* **1**: 129–169.

Neve, KA and Neve, RL (Eds) (1996) The Dopamine Receptor, Humana Press, New York.

Salamone, JD, Cousins, MS and Snyder, BJ (1997) Behavioural functions of nucleus accumbens dopamine; empirical and conceptual problems with the anhedonia hypothesis. *Neurosci. Behav. Rev.* **21**: 341–359.

Schoener, EP and Elkins, DP (1984) Neuronal response to dopamine in rat neostriatum. A push–pull perfusion study. *Neuropharmacology* **25**: 611–616.

Schultz, W (1997) Dopamine neurons and their role in reward mechanisms. *Curr. Opin. Neurobiol.* **7**: 191–197.

Seeman, P and Van Tol, HHM (1994) Dopamine receptor pharmacology. *Trends Pharacol. Sci.* **13**: 61–72.

Sibley, DR and Monsma Jr, FJ (1992) Molecular biology of dopamine receptors. *Trends Pharmacol. Sci.* **13**: 61–70.

Sokoloff, P and Schwartz, JC (1995) Novel dopamine receptors half a decade later. *Trends Pharmacol. Sci.* **16**: 270–274.

Strange, PG (1994) Dopamine D_4 receptors: curiouser and curiouser. *Trends Pharmacol. Sci.* **15**: 317–319.

Strange, PG (1996) Dopamine receptors: studies on structure and function. *Adv. Drug Res.* **28**: 313–351.

Ungerstadt, U and Arbuthnott, GW (1970) Quantitative recording of rotational behaviour in rats after 6-hydroxy dopamine lesions of the nigrostriatal dopamine system. *Brain Res.* **24**: 485–493.

Waddington, JL (1989) Functional interactions between D_1 and D_2 dopamine receptor systems: their role in the regulation of psychomotor behaviour, putative mechanisms and clinical relevance. *J. Psychopharm.* **3**: 54–63.

8 Noradrenaline

S. C. STANFORD

INTRODUCTION

Noradrenaline, 3,4-dihydroxyphenylethanolamine (Fig. 8.1), is released from terminals of noradrenergic neurons in the brain, from most postganglionic sympathetic neurons and from chromaffin cells in the adrenal medulla. Its role in the periphery (the so-called 'sympathoadrenal system') has been evident for nearly a century, but its function in the brain is much harder to define. This chapter will describe recent developments in our understanding of the neurochemistry and pharmacology of noradrenergic neurons and adrenoceptors as well as outlining theories to explain how changes in central noradrenergic transmission might influence behaviour.

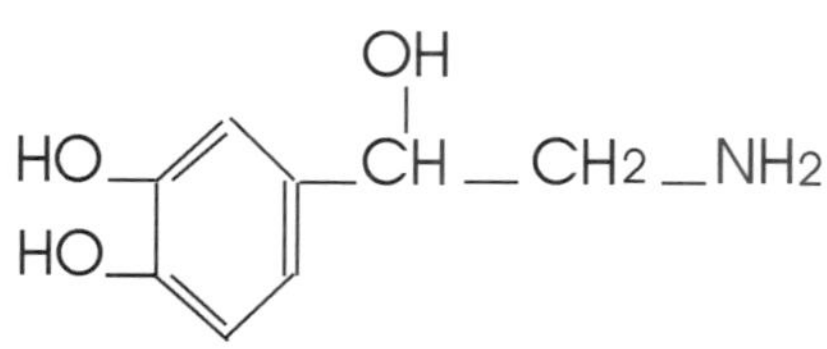

Figure 8.1 The chemical structure of noradrenaline

PATHWAYS WITHIN THE CNS

The cell bodies of central noradrenergic neurons are all clustered within two bilateral groups of nuclei (numbered A1 to A7) in the brainstem (Fig. 8.2). These comprise the locus coeruleus complex and the lateral tegmental nuclei. The locus coeruleus proper (nucleus A6) has attracted most interest because it accounts for approximately 45% of all the noradrenergic neurons in the brain. In the rat, there are only about 1500 noradrenergic cell bodies in the locus coeruleus of each hemisphere but their neurons branch extensively and project throughout the neuraxis. Retrograde tracing has shown that over 50% of neurons within the locus coeruleus innervate both the cortex and the cerebellum, for instance. The majority, if not all, of these fibres are thought to be retained ipsilaterally. The density of innervation varies from brain region to brain region and this is reflected, to some extent, by regional variation in tissue noradrenaline content (Table 8.1).

Neurotransmitters, Drugs and Brain Function. Edited by R. A. Webster

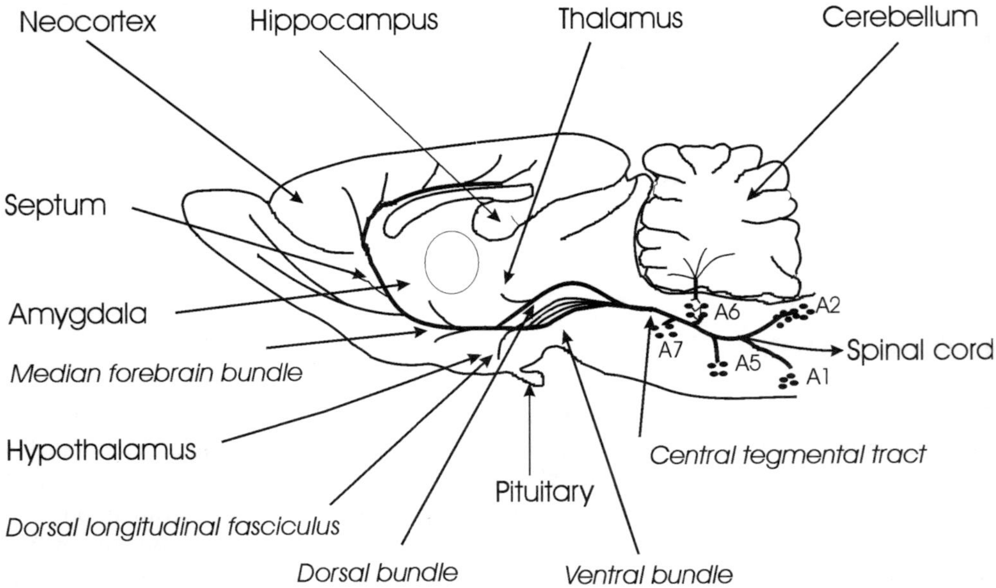

Figure 8.2 The distribution of noradrenergic neurons in the brain. The cell bodies are clustered in nuclei (A1–A7) in the pons/medulla regions of the brainstem and their axons project both rostrally and caudally to most regions of the neuraxis. The major nucleus is the locus coeruleus (A6)

The activity of noradrenergic neurons within the locus coeruleus is governed by two major afferent systems: a GABAergic (inhibitory) input from the nucleus prepositus hyperglossi and an (excitatory) glutamatergic projection from the nucleus paragingantocellularis (Aston-Jones *et al.* 1991). However, dendrites of neurons with cell bodies lying within the locus coeruleus can extend into the area surrounding the nucleus (the pericoerulear region) and could well be influenced by other neurotransmitters and neuromodulators.

Many brain areas are innervated by neurons projecting from both the locus coeruleus and the lateral tegmental system but there are exceptions (Fig. 8.3). The frontal cortex, hippocampus and olfactory bulb seem to be innervated entirely by neurons with cell bodies in the locus coeruleus whereas most hypothalamic nuclei are innervated almost exclusively by neurons projecting from the lateral tegmental system. The paraventricular nucleus (and possibly the suprachiasmatic nucleus, also) is an exception and receives an innervation from both systems.

Table 8.1 The concentration of noradrenaline in different brain regions (μg/g wet weight of tissue)

Cortex	0.1
Hippocampus	0.25
Hypothalamus	0.2
Pons medulla	0.35

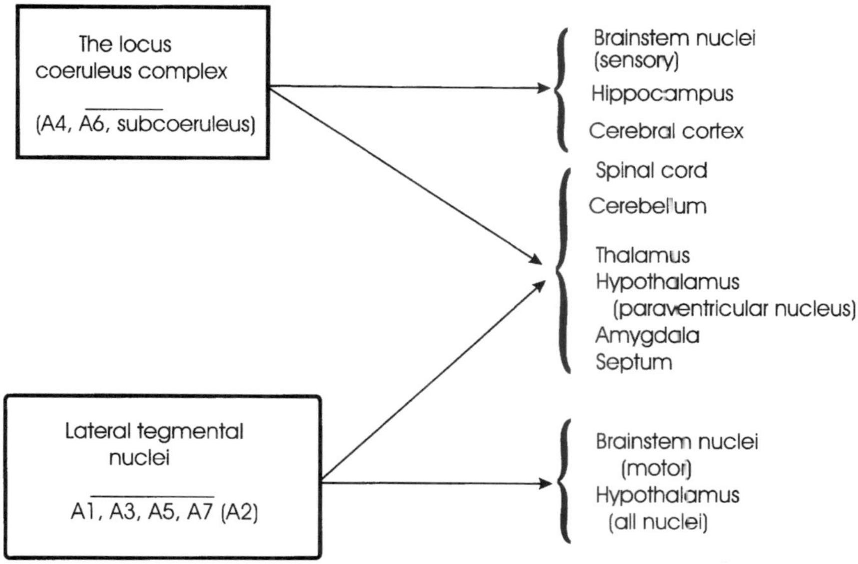

Figure 8.3 Brain areas receiving a prominent noradrenergic innervation. Most brain regions are innervated by neurons projecting from both the locus coeruleus and the lateral tegmental system. However, the frontal cortex, hippocampus and olfactory bulb are innervated exclusively by neurons with cell bodies in the locus coeruleus. With the exception of the paraventricular nucleus (and possibly the suprachiasmatic nucleus) hypothalamic nuclei are innervated by neurons projecting from the lateral tegmental system

The extensive branching and widespread distribution of noradrenergic neurons within the CNS has long been cited as evidence that this is a spatially and functionally diffuse neuronal system. This view was reinforced by an early report that few of these neurons formed specialised synaptic contacts. This fostered the impression that the locus coeruleus represents a 'switch' which, when activated, causes noradrenaline to be discharged from neurons, throughout the brain, in a non-selective manner. However, it is now known that, in the cortex at least, over 90% of the noradrenergic nerve terminals form specialised synaptic contacts with postsynaptic elements (Papadopoulos and Parnavelas 1991). There is also evidence that neurons in different zones of the locus coeruleus are morphologically distinct (at least six different types of noradrenaline-containing cells have been identified) and project to different brain regions or brain systems. In fact, neurons from different noradrenergic nuclei even innervate different types of neuron in the terminal field but, although it is certain that different noradrenergic nuclei have different functions, little is known about their physiological specialisations, largely because of their extensive reciprocal connections. All this evidence (reviewed in Stanford 1995) challenges the view that the central noradrenergic system operates in a non-selective manner.

NEUROCHEMISTRY OF NORADRENALINE

The effects of drugs on the synthesis, storage, release and destruction of noradrenaline, summarised in Fig. 8.4, are discussed in the following sections.

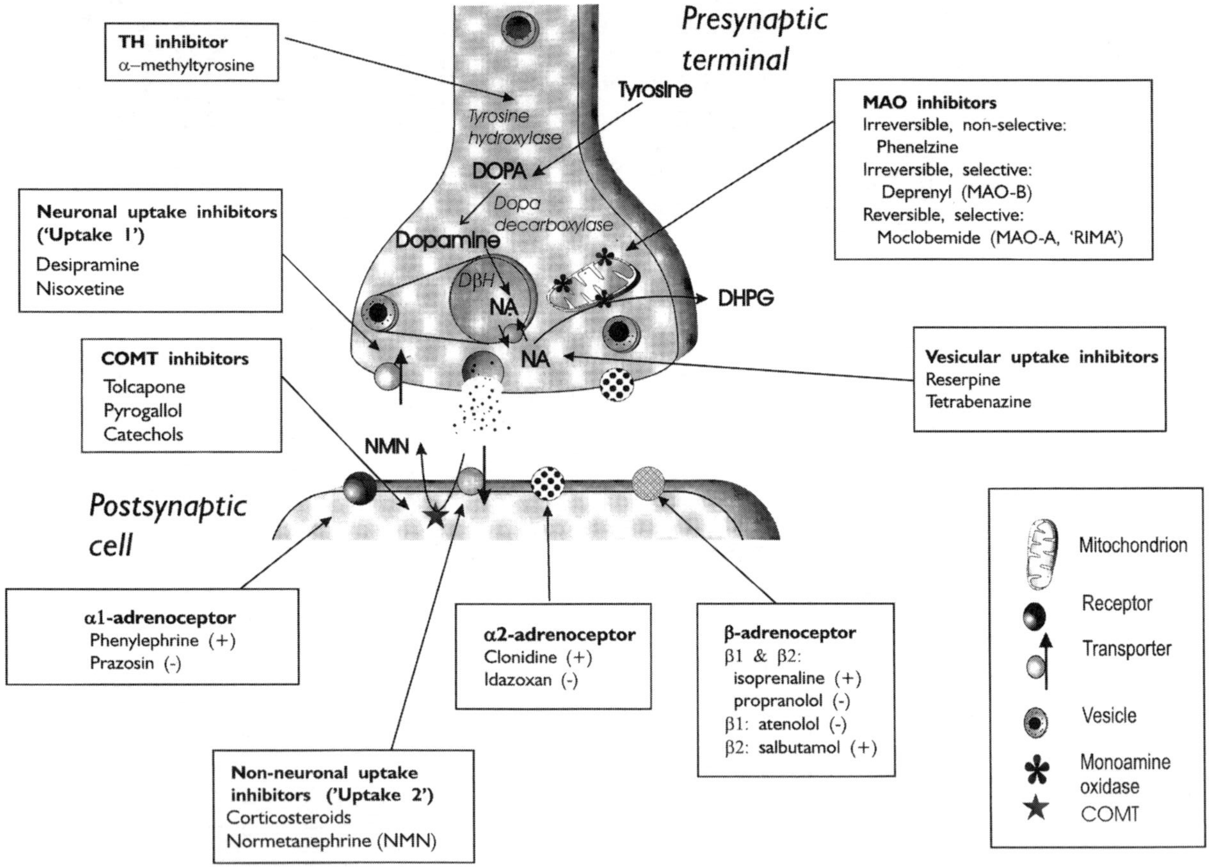

Figure 8.4 The site of action of drugs that modify noradrenergic transmission

SYNTHESIS

The pathway for synthesis of the catecholamines dopamine, noradrenaline and adrenaline, illustrated in Fig. 8.5, was first proposed by Hermann Blaschko in 1939 but was not confirmed until 30 years later. The amino acid *l*-tyrosine is the primary substrate for this pathway and its hydroxylation, by tyrosine hydroxylase (TH), to *l*-dihydroxyphenylalanine (*l*-DOPA) is followed by decarboxylation to form dopamine. These two steps take place in the cytoplasm of catecholamine-releasing neurons. Dopamine is then transported into the storage vesicles where the vesicle-bound enzyme, dopamine-β-hydroxylase (DβH), converts it to noradrenaline (see also Fig. 8.4). It is possible that *l*-phenylalanine can act as an alternative substrate for the pathway, being converted first to *m*-tyrosine and then to *l*-DOPA. TH can bring about both these reactions but the extent to which this happens *in vivo* is uncertain. In all catecholamine-releasing neurons, transmitter synthesis in the terminals greatly exceeds that in the cell bodies or axons and so it can be inferred

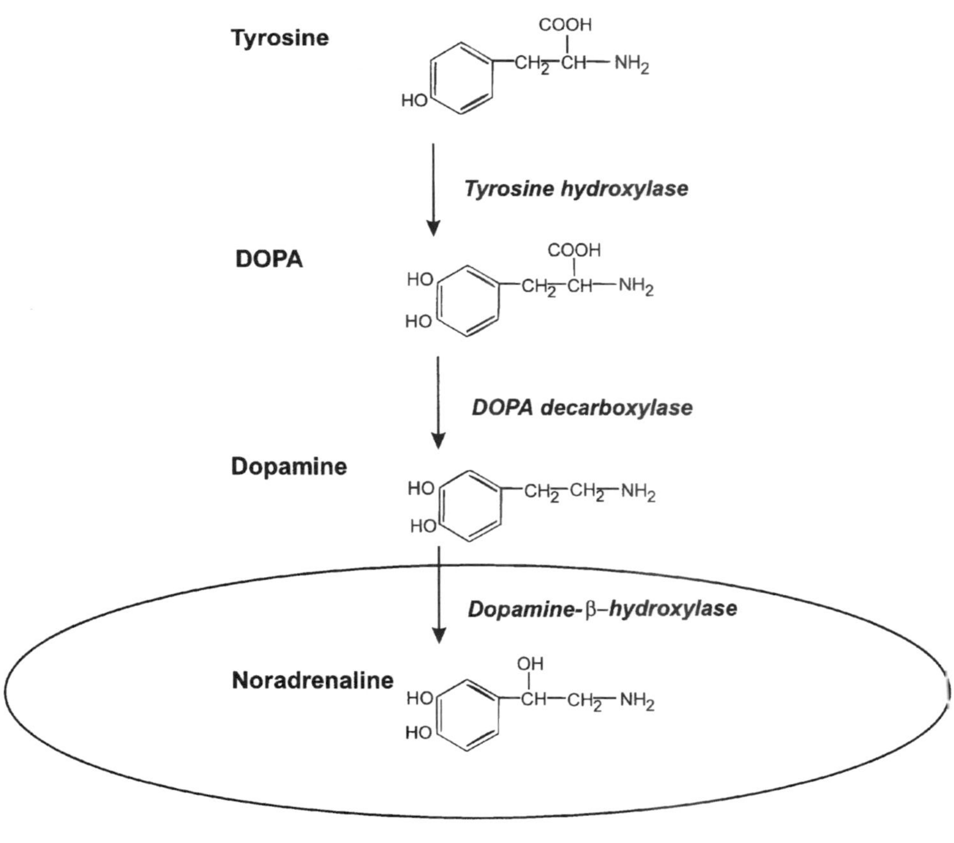

Figure 8.5 The synthetic pathway for noradrenaline. The hydroxylation of the amino acid, tyrosine, which forms dihydroxyphenylalanine (DOPA) is the rate-limiting step. Conversion of dopamine to noradrenaline is effected by the vesicular enzyme, dopamine-β-hydroxylase (DβH) after uptake of dopamine into the vesicles from the cell cytosol

that all the factors (enzymes and storage vesicles) which are vital for this process undergo axoplasmic transport after their assembly in the cell body.

It was recognised as early as the 1960s that conversion of tyrosine to *l*-DOPA was the rate-limiting step in the synthesis of noradrenaline. This emerged from experiments showing that incubation of tissues with high concentrations of tyrosine had no effect on the rate of synthesis of noradrenaline, whereas incubation with high concentrations of *l*-DOPA or dopamine increased it. More evidence came from experiments showing that the rate of conversion of [^{3}H]tyrosine to [^{3}H]DOPA was increased if the sympathetic nerves innervating the test tissue were stimulated, whereas stimulation of nerves innervating tissues incubated in a medium containing [^{3}H]DOPA did not accelerate synthesis of [^{3}H]dopamine or [^{3}H]noradrenaline.

Because the enzymes, DOPA decarboxylase and DβH, have a high affinity for their substrates, neither *l*-DOPA nor dopamine accumulate in noradrenergic nerve terminals under normal conditions. Exceptionally, DβH can become the rate-limiting enzyme, such as during *l*-DOPA treatment of Parkinson's disease which greatly increases the intraneuronal pool of dopamine. It is thought that DβH can also be rate-limiting during periods of intense or prolonged impulse-evoked release of noradrenaline. This is because a high release rate compromises the supply of vesicles that not only store and release noradrenaline but are also the site of its synthesis after uptake of dopamine from the cytosol. Evidence suggests that vesicular uptake of dopamine is reduced after periods of intense neurotransmission; this results in its accumulation in the cytosol until new vesicles are delivered to the terminals.

TH is a mixed-function oxidase with an absolute requirement for reduced pterin co-factor, 6*R*-tetrahydrobiopterin, molecular oxygen and Fe^{2+}. The K_m of the enzyme for its substrate is thought to be well below tissue concentrations of tyrosine and so the enzyme is probably normally about 80% saturated. This makes it unlikely that the supply of tyrosine limits enzyme activity and synthesis of noradrenaline under normal circumstances. However, it was clear from the earliest studies of noradrenergic neurons that the synthesis of noradrenaline was increased during neuronal activity, whether this is induced pharmacologically (e.g. by blockade of presynaptic α_2-adrenoceptors which inhibit release of noradrenaline) or by physiological stimuli (e.g. cold exposure or hypoglycaemia). Such findings suggested that synthesis and release are coupled in some way.

It is now known that regulation of this enzyme involves multiple mechanisms affecting both rapid, transient changes in enzyme activity and long-latency, long-lasting changes in enzyme synthesis involving increased TH gene transcription. The factors controlling the synthesis of noradrenaline have been studied more, and are better understood, than those of most other neurotransmitters and therefore justify detailed consideration.

Regulation of tyrosine hydroxylase activity

Short-term

At first, it was thought that control of TH activity depended on inhibition by its end-product, noradrenaline, which competes with the binding of co-factor. According to this scheme, release of noradrenaline would diminish end-product inhibition of the enzyme and so ensure that synthesis is increased to replenish the stores. When the neurons are quiescent, the opposite would occur: i.e. intraneuronal accumulation of noradrenaline would automatically blunt synthesis. Much evidence was deemed to support this view. For instance, when metabolic breakdown of cytoplasmic noradrenaline was prevented by treatment with an inhibitor of the enzyme, monoamine oxidase (MAO: see below), the rate of synthesis of [^{3}H]noradrenaline from [^{3}H]tyrosine was markedly reduced (Neff and Costa 1966).

However, as early as the 1970s, it was obvious that end-product inhibition of TH could not be the main factor regulating the rate of noradrenaline synthesis. Clearly, the hydroxylation of tyrosine takes place in the cytoplasm and so it must be cytoplasmic noradrenaline that governs enzyme activity. Yet, it is vesicle-bound transmitter that undergoes impulse-evoked release from the neuron. Also, when neurons are releasing noradrenaline, its reuptake from the synapse is increased and, even though some of this transmitter ends up in the vesicles, or is metabolised by MAO, there should be a transient increase in the concentration of cytoplasmic noradrenaline which would *increase* end-product inhibition of TH.

To overcome these difficulties it was suggested that there was a small 'strategic pool' of cytoplasmic noradrenaline that inhibited the activity of TH. Nevertheless, even this small pool was eventually ruled out as a regulator of TH. This followed *in vitro* experiments investigating the effects of addition of reduced pterin co-factor on the activity of the enzyme derived from the vas deferens. It was predicted that the activity of enzyme derived from control tissues would be increased by the addition of co-factor *in vitro* whereas that enzyme derived from stimulated tissues should not increase because the TH would already have been maximally activated by endogenous co-factor during nerve stimulation. In fact, co-factor increased noradrenaline synthesis in both instances, suggesting that noradrenaline synthesis depended primarily on factors that directly activate TH, rather than on removal of end-product inhibition. The extent to which end-product inhibition of TH contributes to the regulation of its activity under physiologically relevant (e.g. drug-free) conditions remains uncertain.

The first clue to the processes which normally regulate TH activity came from experiments showing that electrical stimulation of sympathetic neurons increased the affinity of this enzyme for its co-factor and reduced its affinity for noradrenaline (for detailed reviews of this topic see Zigmond, Schwarzschild and Rittenhouse 1989; Fillenz 1993; Kaufman 1995; Kumar and Vrana 1996). Several lines of investigation showed that activation of TH was in fact paralleled by its phosphorylation and it was this process that accounted for the changes in the enzyme's kinetics (Table 8.2).

Table 8.2 The effects of phosphorylation of tyrosine hydroxylase on enzyme kinetics (based on Kaufman 1995)

	K_m for co-factor	V_{max}
cAMP-dependent protein kinase (PKA) (pH 6.0)	↑↑	No change
cAMP-dependent protein kinase (PKA) (pH 7.0–7.4)	↓	↑↑
Ca^{2+}-calmodulin dependent protein kinase (CAM-PK II)	No change	↑
Ca^{2+}-phospholipid dependent protein kinase (PKC) (pH 7.0)	↓↓	No change
cGMP-dependent protein kinase (PKG) (pH 6.0)	↓↓	↑
cAMP-independent protein kinase	No change	↑

It is now known that such phosphorylation is activated by several protein kinases, including Ca^{2+}/phospholipid-dependent protein kinase (PKC), which reduces its K_m for co-factor, and cGMP-dependent protein kinase. These factors phosphorylate different sites on the enzyme, although some are shared by different kinases. In rat TH, serine residues, Ser^8, Ser^{19}, Ser^{31} and Ser^{40}, have been identified as targets and Ser^{40} seems to be a common target for all the kinases. It is thought that this site produces a conformational change in the enzyme that reduces its affinity for catecholamines. All these regulatory sites reside on the N-terminus of the enzyme, whereas it is the C-terminus that comprises the catalytic site. In addition to all these changes, phosphorylation of the enzyme changes the pH optimum for maximal enzyme activity and so the kinetics of this enzyme depend on the pH of the incubation medium to some extent.

In the periphery, some of the primary triggers for these processes have been identified. Acetylcholine seems to be one such factor because stimulation of preganglionic nerves *in vivo* increases enzyme activity. However, nicotinic and muscarinic receptor antagonists do not completely prevent this increase. The residual activation is attributed to peptides of the secretin–glucagon subgroup, including VIP and secretin; both these peptides activate cAMP synthesis. Purinergic transmitters could also be involved.

Finally, recent findings suggest that, in humans, four different mRNAs for TH are produced from a single gene. The translation products of these mRNAs differ in their amino-acid sequence in the N-terminal domain, rather than the catalytic C-terminus, and are likely to differ in their kinetics and susceptibility to different protein kinases. Regional differences in the distribution of these enzyme isoforms suggest that they might differ functionally, a possibility that is being explored currently.

Long-term

The first reports that TH activity could be altered without changes in its kinetics came from studies of the adrenal medulla of rats in which catecholamine release was stimulated by exposure of rats to a cold environment. The increase in enzyme activity was prevented by protein synthesis inhibitors, suggesting that it was due to an increase in TH gene transcription rather than activation of existing enzyme. Since then, physiological and pharmacological stimuli that increase demand on the transmitter store have consistently been shown to trigger such induction of TH enzyme. Increased TH protein has also been detected in noradrenergic cell bodies of sympathetic ganglia and the locus coeruleus. At all these sites, as in the adrenal medulla, the increase is evident after about 24 h. However, changes in the terminals take several days to appear, presumably because of the time required for axoplasmic transport of the enzyme.

The signal for increased synthesis of TH protein in the adrenal gland certainly depends on an intact cholinergic innervation. Moreover, in the denervated gland, the increase induced by perfusion with exogenous acetylcholine is prevented by nicotinic antagonists. However, nicotinic antagonists do not completely prevent the increase in glands with an intact cholinergic innervation. These findings suggest that activation of nicotinic receptors by ACh is normally only partly responsible for the increase. Other factors now known to regulate TH gene transcription include glucocorticoids and nerve growth factor (NGF). Although details are far from clear, protein kinases (especially PKA), diacyl glycerol and Ca^{2+} are all thought to be crucial intracellular messengers for increased gene transcription. It should also be borne in mind that enzyme induction is not limited to TH: the same stimuli also increase DβH synthesis but less is known about factors mediating this process.

STORAGE

In common with other classical transmitters, noradrenaline is stored in vesicles that accumulate in the terminal varicosities. This was first shown by experiments that combined sucrose density–gradient centrifugation of tissue homogenates (see Fig. 4.3) with electron microscopy and assay of the noradrenaline content of the different layers of the gradient. These studies confirmed that the noradrenaline-rich layers of the gradient coincided with those layers in which the vesicles were clustered. This suggested that the vesicles were the major storage site for noradrenaline within the nerve terminals. Further studies examined the effects of ligation or cooling the axons of sympathetic neurons for several days. Electron micrographs of the zone around the obstruction showed that the vesicles accumulated on the side nearest the cell body, confirming that they were assembled in the cell body and transported to the terminals by anterograde axoplasmic transport. The life cycle of these vesicles was discussed in more detail in Chapter 4.

The concentration of noradrenaline in the vesicles is thought to be in the region of 0.1–0.2 M and it is estimated that there is a concentration gradient, in the order of 10^4–10^6-fold, driving the transmitter out of the vesicles towards the cytoplasm. The vesicular compartmentalisation of noradrenaline is made possible by its active uptake on vesicular monoamine transporters (VMATs) and its subsequent binding, in an osmotically inert matrix, within the vesicles. One obvious function of these transporters is thus to protect and conserve the releasable vesicular pool of transmitter. However, it is thought that they also protect neurons from potentially toxic effects of an excess of cytoplasmic noradrenaline and also maintain a concentration gradient favouring noradrenaline reuptake from the synapse (see below).

Uptake of noradrenaline into the vesicles depends on an electrochemical gradient driven by an excess of protons inside the vesicle core. This gradient is maintained by an ATP-dependent vesicular H^+-triphosphatase. Uptake of one molecule of noradrenaline into the vesicle by the transporter is balanced by the counter-transport of two H^+ ions (reviewed by Schuldiner 1998). It is thought that either binding or translocation of one H^+ ion increases the affinity of the transporter for noradrenaline and that binding of the second H^+ actually triggers its translocation.

Reserpine irreversibly inhibits the triphosphatase that maintains the proton gradient and so it depletes neurons of their vesicular store of transmitter. This explains why restoration of normal neuronal function rests on delivery of new vesicles from the cell bodies. Some amphetamine derivatives, including methylenedioxymethamphetamine (MDMA), are also substrates for the transporter and, as a result, competitively inhibit noradrenaline uptake. Another way of inhibiting the transporter is by dissipation of the pH gradient across the vesicular membrane: *p*-chloroamphetamine is thought to act in this way.

Much of the early work on these transporters was carried out on the chromaffin granules of the bovine adrenal medulla. These studies revealed the transporter to be a polypeptide of 80 kDa. However, two VMATs have now been characterised and these are the products of different genes. Evidence suggests that both have 12

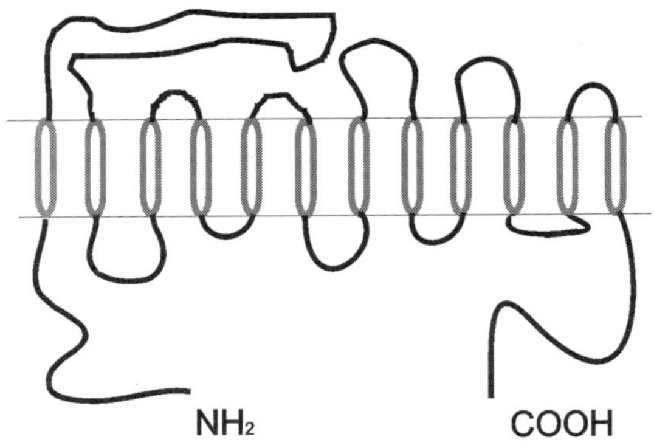

Figure 8.6 Schematic diagram of the proposed structure of the vesicular monoamine transporter. There are 12 transmembrane segments with both the N- and C-termini projecting towards the neuronal cytosol. (Based on Schuldiner 1998)

transmembrane-spanning domains (TMDs) with a large hydrophobic loop, facing the vesicular core, between TMD1 and TMD2. Both the N- and C-termini project towards the neuronal cytosol (Fig. 8.6). There are species differences, but VMAT1 and VMAT2 differ in their distribution. In fact, the expression of these proteins in individual cells might be mutually exclusive. They also differ in their sensitivity to the reversible uptake inhibitor, tetrabenazine, and their affinity for substrates such as amphetamine and histamine. Only VMAT2 binds histamine and tetrabenazine and this protein consistently binds amines with a higher affinity than does VMAT1. In the rat, VMAT1 is found in non-neuronal tissue, including the adrenal medulla, whereas VMAT2 is found in neurons, only. In other species, the distribution is not so distinct, with mRNA for VMAT2 being reported in the adrenal medulla as well as the brain.

RELEASE

Studies of release of noradrenaline from sympathetic neurons provided the first convincing evidence that impulse (Ca^{2+})-dependent release of *any* transmitter depended on vesicular exocytosis. Landmark studies carried out in the 1960s, using the perfused cat spleen preparation, showed that stimulation of the splenic nerve not only led to the detection of noradrenaline in the effluent perfusate but the vesicular enzyme, DβH, was also present. As mentioned above, this enzyme is found only within the noradrenaline storage vesicles and so its appearance along with noradrenaline indicated that both these factors were released from the vesicles. By contrast, there was no sign in the perfusate of any lactate dehydrogenase, an enzyme that is found only in the cell cytosol. The processes by which neuronal excitation increases transmitter release were described in Chapter 4.

While the amount of noradrenaline released from the terminals can be increased by nerve stimulation, it can be increased much more by drugs, like phenoxybenzamine, which block presynaptic α-adrenoceptors. These receptors are normally activated by increased noradrenaline in the synapse and trigger a feedback cascade, mediated by second messengers, which blunts further release of noradrenaline. These presynaptic autoreceptors play an important part in ensuring that transmitter stores are conserved and preventing excessive stimulation of the postsynaptic cells.

Pharmacological characterisation of this receptor revealed that it was unlike classic α-adrenoceptors found on smooth muscle. In particular, receptors modulating noradrenaline release have a higher affinity for the agonist, clonidine, and the antagonist, yohimbine. This distinctive pharmacology led to the subdivision of α-adrenoceptors into the α_1- and the α_2-subtypes. Although the latter is the subtype responsible for feedback inhibition of noradrenaline release, the majority of α_2-adrenoceptors are actually found postsynaptically in some brain regions. There is still some debate over the identity of the subtype of α_2-adrenoceptors responsible for feedback inhibition of transmitter release. However, most studies agree that the $\alpha_{2A/D}$-subtype has the major role, although the α_{2B}- and α_{2C}-subtypes might contribute to this action. Species differences in the relative contributions of these different receptors are also possible.

It is α_{2A}-adrenoceptors that are found on cell bodies of noradrenergic neurons in the locus coeruleus. These receptors are activated by noradrenaline released from branches ('recurrent collaterals') of noradrenergic neurons projecting from the locus coeruleus and inhibit neuronal firing (Cederbaum and Aghajanian 1976). α_2-Adrenoceptors in the brain thus depress noradrenaline release through two distinct processes: inhibition of the release process following activation of terminal autoreceptors and depression of neuronal firing following activation of receptors on the cell bodies.

The exact process(es) by which α_2-adrenoceptors blunt release of transmitter from the terminals is still controversial but a reduction in the synthesis of the second messenger, cAMP, contributes to this process. α_2-Adrenoceptors are negatively coupled to adenylyl cyclase, through a *Pertussis* toxin-sensitive Gi-like protein, and so their activation will reduce the cAMP production which is vital for several stages of the chemical cascade that culminates in vesicular exocytosis (see Chapter 4). The reduction in cAMP also indirectly reduces Ca^{2+} influx into the terminal and increases K^+ conductance, thereby reducing neuronal excitability (reviewed by Starke 1987). Whichever of these release-controlling processes predominates is uncertain but it is likely that their relative importance depends on the type (or location) of the neuron.

α_2-Adrenoceptors are not the only receptors to modulate noradrenaline release. In the periphery and CNS (Murugaiah and O'Donnell 1995) activation of presynaptic β-adrenoceptors has the opposite effect: i.e. it augments release of noradrenaline. The increase in cAMP production resulting from activation of these receptors is an obvious explanation for how this might occur. The precise role of these receptors in regulation of noradrenaline release *in vivo* is uncertain because noradrenaline has a relatively low affinity for these receptors. However, one suggestion is that, in the periphery, they are preferentially activated by circulating adrenaline which has a relatively high affinity for these receptors. This activation could enable circulating adrenaline to augment neuronal release of noradrenaline and thereby effect a functional link between these different elements of the sympathoadrenal system. However, the extent to which this actually happens is uncertain as is a physiological role for β-adrenoceptors in regulation of noradrenaline release in the brain.

Noradrenaline release might also be modulated by receptors on noradrenergic nerve terminals that are activated by other neurotransmitters ('heteroceptors'). Unfortunately, most studies of this type of modulation have been carried out in tissue slices and so it is not possible to rule out the possibility that 'heteroceptors' are in fact part of a polysynaptic loop and that they influence noradrenaline release only indirectly. Nevertheless, there is some evidence from studies of hippocampal synaptosomes that activation of muscarinic, $GABA_B$ or adenosine (A_1) receptors depresses noradrenaline release while activation of $GABA_A$ receptors increases it.

A further possible mechanism, that would enable different types of neurons to modify noradrenaline release, is suggested by recent *in vitro* studies of brain slices. These have revealed that noradrenaline release is increased when the slices are superfused with a solution containing GABA. This release is prevented by an inhibitor of GABA uptake but unaffected by the presence of $GABA_A$ receptor antagonists, such as bicuculline. There is no doubt that this form of release depends on vesicular exocytosis because it is Ca^{2+}-dependent, sensitive to tetrodotoxin and, like impulse-dependent release, it is attenuated by α_2-adrenoceptor agonists (see above). Since uptake of GABA by GABA transporters on noradrenergic nerve terminals ('heterocarriers') involves co-transport of Na^+ ions into the terminal (Fassio *et al.* 1996) it is possible that this uptake increases Na^+ influx enough to depolarise the terminals and trigger exocytotic release of noradrenaline. The extent to which this process occurs under normal physiological conditions *in vivo* remains to be seen.

NEURONAL REUPTAKE OF NORADRENALINE

In common with other monoamines, the actions of released noradrenaline are terminated by its rapid reuptake from the synaptic cleft. This uptake process relies on membrane-bound noradrenaline transporters which are glycoproteins closely related

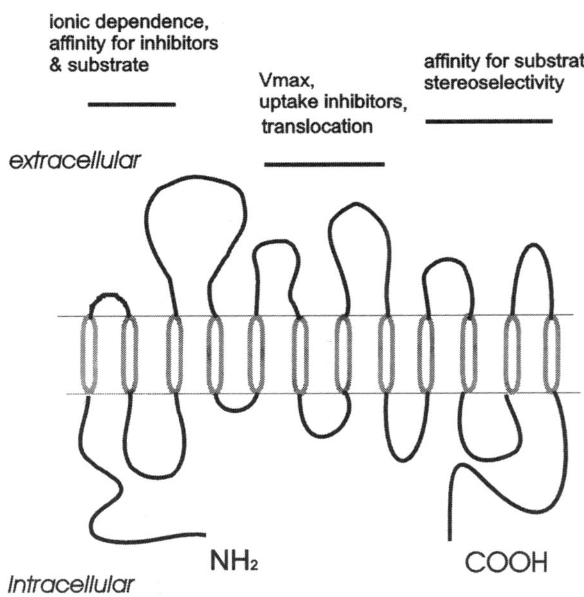

Figure 8.7 Schematic diagram of the proposed structure of the noradrenaline neuronal transporter showing the 12 transmembrane, hydrophobic domains with the N- and C-termini projecting towards the cell cytoplasm. Binding domains for specific ligands are thought to be within regions indicated by the solid bars. (From Stanford 1999, reproduced with permission)

to the transporters for several other transmitters (e.g. GABA and 5-HT). All these transporters have 12 hydrophobic transmembrane domains (TMDs), a large hydrophyllic loop between TM3 and TM4, and intracellular N- and C-termini. The hypothetical structure of the noradrenaline transporter is illustrated in Fig. 8.7. Because co-transport of both Cl^- and Na^+ is required for the uptake of noradrenaline, this is regarded as one of the family of Na^+/Cl^- transporters.

Exactly how this transporter carries noradrenaline across the neuronal membrane is not known but one popular model proposes that it can exist in two interchangeable states. Binding of Na^+ and noradrenaline to a domain on its extracellular surface could trigger a conformation change that results in the sequential opening of outer and inner channel 'gates' on the transporter. This process enables the translocation of noradrenaline from the extracellular space towards the neuronal cytosol.

So far, only one noradrenaline transporter has been cloned. Point-mutation and splicing studies indicate that different zones of the transporter determine its substrate affinity and selectivity, ionic dependence, V_{max}, and the binding site for uptake inhibitors such as desipramine (Povlock and Amara 1997). Because the cloned transporter is a target for the reuptake inhibitor, desipramine, it is thought to reflect the native transporter in the brain and peripheral tissues. However, in the periphery, two native reuptake processes (neuronal uptake, 'uptake$_1$' and extraneuronal uptake, 'uptake$_2$') have been recognised for over 30 years and recently, a third, desipramine-insensitive uptake site has been found in hepatocytes. These are quite distinct uptake mechanisms because they have different substrate affinities and antagonist sensitivities. As yet, few studies have investigated the possibility that more than one uptake process exists in the brain but since two mRNAs for noradrenaline transporters have been isolated from brain tissue (Pacholczyk, Blakely and Amara 1991) there could be more than one transcription factor. Also, the so-called 'extraneuronal transporter' for noradrenaline, responsible for 'uptake 2', has recently been found on glial cells in the brain (Russ *et al.* 1996). At the very least, intracellular messengers could modify substrate affinity of the transporter, by causing its phosphorylation or glycosylation (Bönisch, Hammermann and Brüss 1998), and so markedly affect its function. Whether or not there are different gene products, splice variants, or posttranslational changes, it has been suggested that abnormal distributions of functionally distinctive noradrenaline transporters could underlie some psychiatric and neurological disorders.

METABOLISM

After reuptake into the cytosol, some noradrenaline may be taken up into the storage vesicles by the vesicular transporter and stored in the vesicles for subsequent release (see above). However, it is thought that the majority is broken down within the cytosol of the nerve terminal by monoamine oxidase (MAO; EC1.4.3.4). A second degradative enzyme, catechol-*O*-methyl transferase (COMT; EC2.1.1.6), is found mostly in non-neuronal tissues, such as smooth muscle, endothelial cells or glia. The metabolic pathway for noradrenaline follows a complex sequence of alternatives because the metabolic product of each of these enzymes can act as a substrate for the other (Fig 8.8). This could enable one of these enzymes to compensate for a deficiency in the other to some extent.

MAO is bound to the outer membrane of mitochondria and is responsible for the oxidative deamination of noradrenaline. There are two isoforms of this enzyme, MAO-A

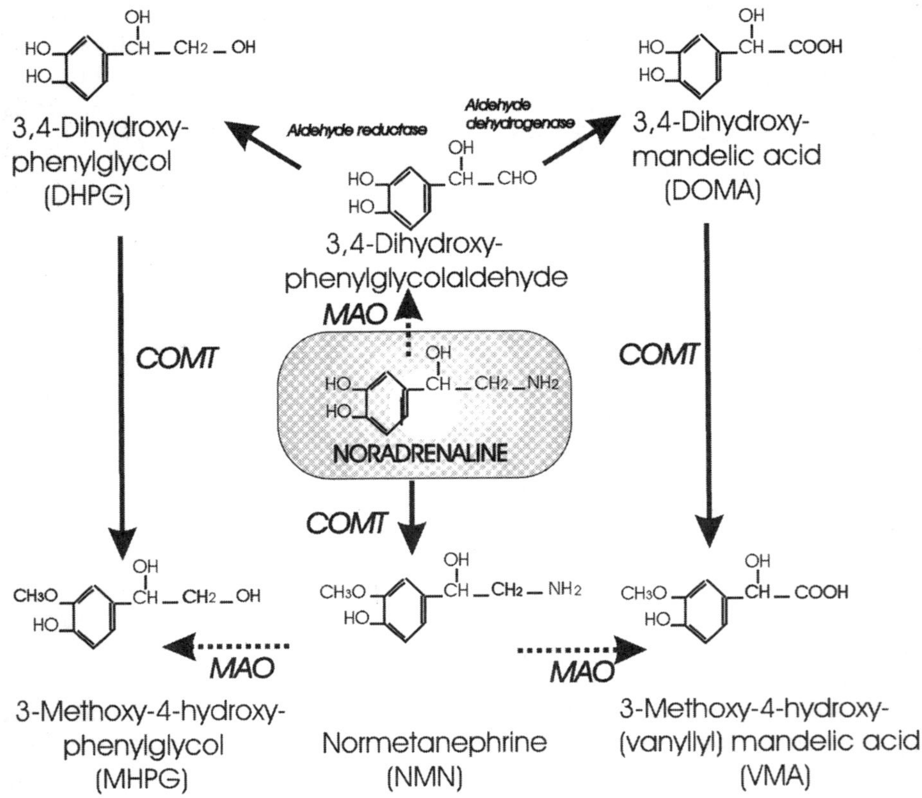

Figure 8.8 The metabolic pathway(s) for noradrenaline. MAO is responsible for the oxidative deamination of noradrenaline derivatives while COMT *O*-methylates noradrenaline. Most intraneuronal metabolism involves MAO while COMT is mainly found extraneuronally. However, both these enzymes can act on each other's products, yielding a complex cocktail of metabolites. The reasons for this complex network of metabolites are not known

and MAO-B, which hybridise to different cDNAs and are encoded by different genes on the X chromosome. MAO-A is the more important *in vivo* because it preferentially metabolises noradrenaline. However, *in vitro*, MAO-B will metabolise noradrenaline at high substrate concentrations. MAO probably also has an important role in development: a genetic deficiency of MAO-A causes some mental retardation and a tendency to bouts of aggression. MAO-B deficiency has no overt effects in the phenotype but a deficiency of both enzymes causes severe mental retardation and behavioural problems (Lenders *et al.* 1996). Of course, some of these abnormalities could be due to disruption of the metabolism of other monoamines, such as tyramine, which are also substrates for MAO.

Certainly, such a complex system for metabolism of noradrenaline (which is shared with the other catecholamines) strongly suggests that its function extends beyond that of merely destroying transmitter sequestered from the synapse. However, as yet, little is known about the regulation of this pathway and any influence it might have on noradrenergic transmission. One crucial, additional role for MAO appears to be the regulation of the intraneuronal stores of noradrenaline. Its predominantly intraneuronal location would suggest that its primary function is to ensure that there is always a low concentration of cytoplasmic noradrenaline. What can happen when the concentration of cytosplasmic noradrenaline is increased is illustrated by amphetamine. This drug causes a rise in the cytoplasmic noradrenaline and results in increased binding of this transmitter to the cytoplasmic side of the transporter which then carries it out of the neuron. Importantly, this form of noradrenaline release ('retrotransport') is independent of neuronal activation or intracellular Ca^{2+}.

By maintaining low concentrations of cytoplasmic noradrenaline, MAO will also regulate the vesicular (releasable) pool of transmitter. When this enzyme is inhibited, the amount of noradrenaline held in the vesicles is greatly increased and there is an increase in transmitter release. It is this action which is thought to underlie the therapeutic effects of an important group of antidepressant drugs, the MAO inhibitors (MAOIs) which are discussed in Chapter 20.

Because of their lack of selectivity and their irreversible inhibition of MAO, the first MAOIs to be developed presented a high risk of adverse interactions with dietary tyramine (see Chapter 20). However, more recently, drugs which are selective for and, more importantly, reversible inhibitors of MAO-A (RIMAs) have been developed (e.g. moclobemide). These drugs are proving to be highly effective antidepressants which avoid the need for a tyramine-free diet.

Further interest in MAO has been aroused as a result of recent research on drugs with an imidazole or imidazoline nucleus (Fig. 8.9). Although many of these compounds are potent and selective α_2-adrenoceptor ligands (e.g. the agonist, clonidine, or the antagonist, idazoxan), not all the binding of these compounds is explained by their high affinity for α_2-adrenoceptors. It is now known that many of these drugs have their own binding sites that are now classified as imidazoline (I-) receptors. One of these, the so-called I_2-receptor, has been found on MAO-B but there is general agreement that the I_2-receptor is not the same as the catalytic site on the MAO enzyme. Instead, it is thought that the I_2-receptor is an allosteric modulator of the catalytic site on MAO which, when activated, reduces enzyme activity. So far, the function of this

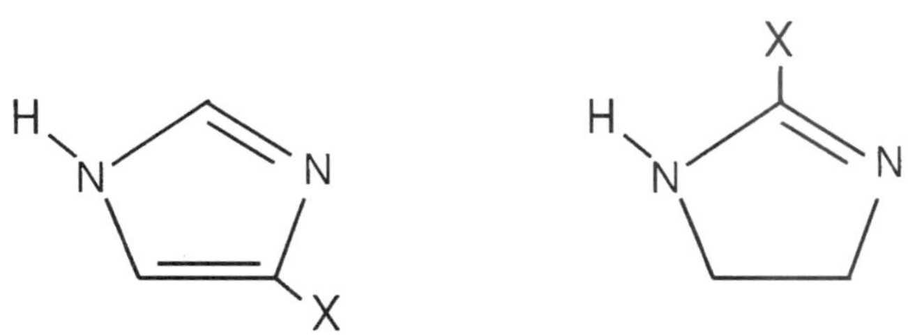

Imidazoles

Atipamezole
Cimetidine
Histamine
Medetomidine

Imidazolines

Clonidine
Idazoxan
Monoxidine
p-Aminoclonidine
Phentolamine

Figure 8.9 The chemical structure of imidazole and imidazoline, together with some well-known derivatives

receptor is unknown but it has been suggested that a dysfunction of I_2-receptors could contribute to neurodegenerative disorders such as dementia and Parkinson's disease.

There is also some evidence for subtypes of COMT but this has not yet been exploited pharmacologically. Certainly, the majority of COMT is found as soluble enzyme in the cell cytosol but a small proportion of neuronal enzyme appears to be membrane bound. The functional distinction between these different sources of COMT is unknown. COMT inhibitors also exist (e.g. pyrogallol), mostly as catechol derivatives, but so far, most have proved to be highly toxic. Only recently have drugs been developed which are selective for COMT; one of these agents, tolcapone, is used currently in treatment of Parkinson's disease (see Chapter 15).

NORADRENERGIC RECEPTORS

The division of adrenoceptors into α- and β-types emerged some 50 years ago and was based on the relative potencies of catecholamines in evoking responses in different peripheral tissues. Further subdivision of β-adrenoceptors followed characterisation of their distinctive actions in the heart (β_1), where they enhance the rate and force of myocardial contraction and in the bronchi (β_2), where they cause relaxation of smooth muscle. The binding profile of selective agonists and antagonists was the next criterion for classifying different adrenoceptors and this approach is now complemented by molecular biology. The development of receptor-selective ligands has culminated in the characterisation of three major families of adrenoceptors (α_1, α_2 and β), each with their own subtypes (Fig. 8.10). All these receptors have been cloned and belong to the superfamily of G-protein-coupled receptors predicted to have seven transmembrane domains (Hieble, Bondinell and Ruffolo 1995; Docherty 1998).

The α_1-subgroup is broadly characterised on the basis of their high affinity for binding of the antagonist, prazosin, and low affinity for yohimbine but they seem to be activated to the same extent by catecholamines. There are at least three subtypes which for historical reasons (Hieble, Bondinell and Ruffolo 1995) are now designated α_{1A}, α_{1B} and α_{1D}. α_{1A}-Adrenoceptors are distinguished by their selective antagonists tamsulosim

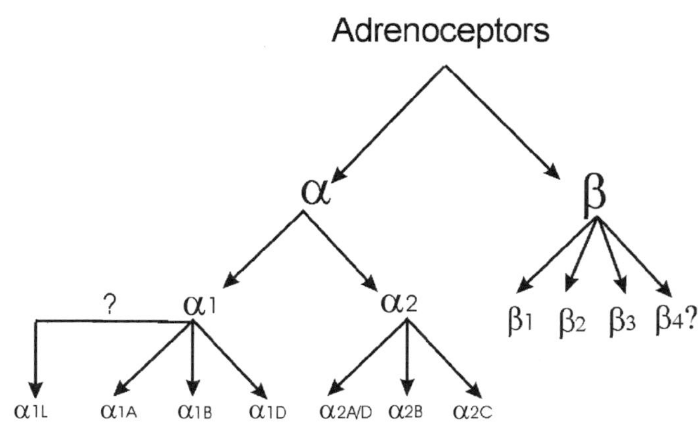

Figure 8.10 Subdivisions of α- and β-adrenoceptor families

and WB4101 but, whereas α_{1B}- and α_{1D}-adrenoceptors have ligands that distinguish them from α_{1A}-adrenoceptors, and from each other, these ligands bind to other transmitter receptors and so they are not really selective. An alternative classification (also based on sensitivity to prazosin) characterised two classes of receptor: α_{1H} and α_{1L} receptors. Whereas those classified as α_{1H} seem to overlap with α_{1A}, α_{1B} and α_{1D} receptors (and are now regarded as the same), there is no known equivalent of the α_{1L} receptor. Although it is still tentatively afforded the status of a separate receptor, it has been suggested that it is an isoform of the α_{1A} subtype (Docherty 1998).

All α_1-adrenoceptors are coupled to the $G_{q/11}$ family of G-proteins and possibly other G-proteins as well. When activated, they increase the concentration of intracellular Ca^{2+} through the phospholipase C/diacyl glycerol/IP_3 pathways (Ruffolo and Hieble 1994) but other routes have been suggested too. These include: direct coupling to Ca^{2+} (dihydropyridine sensitive and insensitive) channels, phospholipase D, phospholipase A2, arachidonic acid release and protein kinase C. Their activation of mitogen-activated protein (MAP) kinase suggests that they also have a role in cell proliferation. All three subtypes are found throughout the brain but their relative densities differ from one region to another. A detailed review of the classification of α_1-adrenoceptors is to be found in Zhong and Minneman (1999).

α_2-Adrenoceptors all have a high affinity for yohimbine (although there are species differences) and are negatively coupled to a *Pertussis* toxin-sensitive $G_{i/o}$-protein in some tissues whereas, in others, they appear to be insensitive to this toxin. Their activation inhibits target cell activity, resulting from reduced cAMP production, an increase in K^+ current and a reduced Ca^{2+} current. However, stimulatory effects of α_2-adrenoceptors have also been reported, although the underlying mechanisms are unclear. Paradoxically, the different receptor subtypes are characterised by their affinity for prazosin: the α_{2A}-subtype (found in human platelets) has a low affinity for this ligand, while the α_{2B}-subtype (isolated from neonatal rat lung) has a higher affinity. The α_{2C}-adrenoceptor, first isolated from opossum kidney (OK) cells, is distinguished by its characteristic relative affinities for yohimbine and prazosin. There is also functional evidence for an α_{2D}-adrenoceptor. This has not been granted the status of a separate subtype, partly because it has not been possible to produce a distinctive receptor clone, and it is now regarded as the rodent homologue of the human α_{2A}-subtype. (Bylund *et al.* 1994).

It is the $\alpha_{2A/D}$-adrenoceptor that predominates in the locus coeruleus and this subtype seems to be responsible for reducing neuronal excitability and transmitter release. Strangely, immunocytochemical studies suggest that most α_{2C}-receptors are intracellular. The explanation for this finding and its functional implications are as yet unknown but it could reflect differences in intracellular trafficking of different receptor subtypes.

Contrasting with the α_2-adrenoceptors, β-adrenoceptors activate cAMP synthesis and are coupled to G_s-proteins. The β_1 and β_2 subtypes were distinguished in the 1960s but the β_3-adrenoceptor has been characterised only recently, largely on the basis of its low affinity for the antagonist, propranolol. Unlike β_1- and β_2-adrenoceptors, this subtype is not found in the brain but probably has an important role in lipolysis by mobilising triglyceride stores and promoting thermogenesis (Giacobino 1995).

In the brain, autoradiography has shown that β_1- and β_2-adrenoceptors have quite distinct distributions. Thus, approximately 60% of the β-adrenoceptors in the neocortex are of the β_1-subtype while, in the cerebellum, it is the β_2-subtype that predominates. As yet, the functional implications of this uneven distribution are unclear and await the development of more subtype selective agents. However, unlike the α-adrenoceptor families, the affinity of catecholamines for β-adrenoceptors differs markedly: noradrenaline is a relatively weak agonist at the β_2-subtype whereas it is more potent than adrenaline at β_3-receptors.

Electrophysiological studies of the β-adrenoceptor have produced complex findings. In cardiac tissue, their activation leads to an increase in Ca^{2+} conductance and so they are regarded as excitatory receptors. β-Adrenoceptor activation in cortical pyramidal cells causes an increase in excitability mediated by a reduction of a Ca^{2+}-activated K^+ current. A different response is evoked in thalamic relay neurons where these receptors cause depolarisation and an increase in input conductance by resetting a hyperpolarisation-induced cation current. In the dentate gyrus their activation causes an increase in voltage-dependent Ca^{2+} currents through opening of Ca^{2+} channels.

Because of these disparate findings, it is difficult to assign particular electrophysiological changes to each of the adrenoceptors let alone to noradrenaline, more generally. A shortage of selective ligands aggravates this problem. Another difficulty concerns the uncertain location of the receptors responsible for initiating any changes. In tissue slices, the target receptors could be located on interneurons, rather than mediating direct axo-somatic interactions, for instance. The net effect of receptor activation could also depend on the underlying tonic activity of the target cell as well as the influence of other neurotransmitters that converge on the same G-protein.

Despite these obstacles, it has been suggested that the overall effect of interactions between noradrenaline and its receptors could be to increase the excitability and responsiveness of the target cells. This could make an important contribution to the governance of arousal and selective attention (McCormick, Pape and Williamson 1991). Another, similar suggestion is that noradrenergic transmission increases the signal-to-noise ratio of cell responses to incoming stimuli: i.e. it reduces the basal activity of target cells but increases their response to excitatory inputs. This is the so-called 'Kety hypothesis' (reviewed by Harley 1987).

WHAT IS THE FUNCTION OF NORADRENALINE IN THE BRAIN?

Because central noradrenergic pathways are so diffuse, and the synaptic effects of noradrenaline have a comparatively slow time-course, these neurons could have a wide range of functions, depending on the brain region being targeted and the neurobiological status of the individual. In general terms, however, it is agreed that noradrenergic neurons influence arousal. This encompasses not only the sleep/waking cycle (see Chapter 22) but also more specific activities, such as selective attention and vigilance (Aston-Jones *et al.* 1994). Indeed, depression and anxiety, both of which are relieved by drugs that modify noradrenergic transmission, can be regarded as arousal disorders. Yet, despite nearly 40 years of research, it is still uncertain whether an increase in noradrenergic transmission contributes to unpleasant emotional responses to environmental stimuli (e.g. fear and anxiety) or whether its main role is to ameliorate the emotional impact of such stimuli (i.e. contributes to 'coping').

Many electrophysiological studies have shown that single-unit activity of noradrenergic neurons in the locus coeruleus is increased by sensory stimuli. Effective stimuli range from those causing physical discomfort (e.g. footshock) and interoceptive cues (e.g. hypoglycaemia) to certain arousing environmental stimuli (e.g. the approach of the experimenter). On the basis of this evidence, it has been suggested that central noradrenergic neurons could form part of an 'alarm system'. This would be consistent with the attenuation of the neuronal response on repeated presentation of the test stimulus, the presumption being that this change underlies behavioural habituation.

The precise features of environmental stimuli that provoke increased noradrenergic transmission are unclear but recent experiments using *in vivo* microdialysis suggest that neither 'novelty' nor the 'aversiveness' of the stimulus alone is responsible (McQuade and Stanford 2000). Electrophysiological studies suggest that it could be the 'salience' (i.e. its significance or relevance to the individual), or change in salience, of a stimulus that is the key factor and that increased noradrenergic transmission in the brain mediates changes in selective attention.

Even if this turns out to be the case, it is likely that noradrenergic neurons in different brain regions make different contributions to this process. This complication is suggested by the results of a recent microdialysis study in which release of noradrenaline in response to the sound of a buzzer alone was provoked after repeated

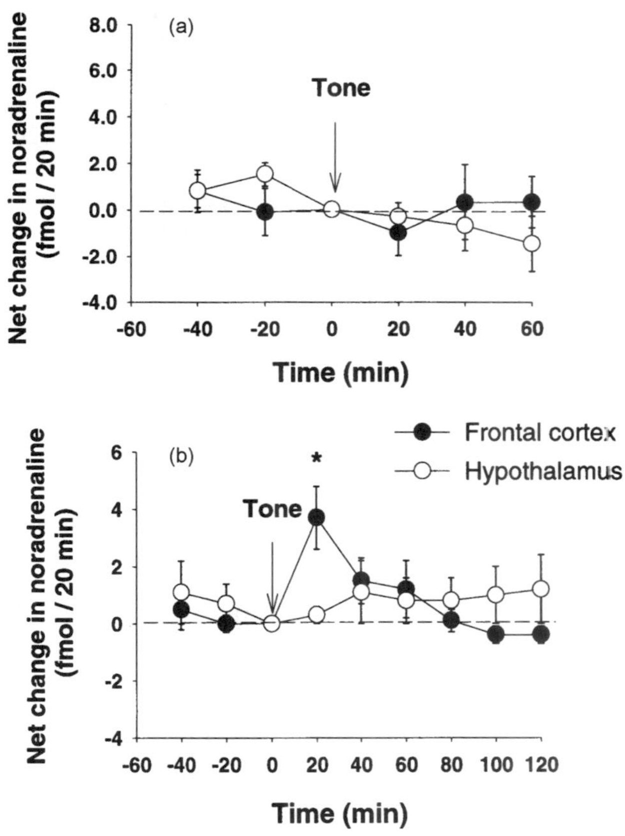

Figure 8.11 Noradrenaline efflux, measured by microdialysis, in the rat frontal cortex and hypothalamus. (a) Repeated exposure to a tone, alone, has no effect on noradrenaline efflux in either brain region. (b) After repeated pairing of the tone with transfer of the rat to a brightly lit (aversive) arena, the sound of the tone alone triggers a significant (*: $P < 0.05$, cf. last basal sample) increase in noradrenaline efflux in the frontal cortex, but not the hypothalamus. (Based on a figure from McQuade and Stanford 2000)

pairing of this normally neutral stimulus with transfer of the rats to a brightly-lit novel arena. This adaptive change occurred in the frontal cortex but not the hypothalamus suggesting that only noradrenergic neurons innervating the former brain region (i.e. those arising from the locus coeruleus) show adaptive changes in response to a change in the salience of an environmental stimulus (McQuade and Stanford 2000) (Fig. 8.11).

Another concept is that noradrenergic transmission influences the emotional impact of a given stimulus, i.e. individuals' ability to 'cope'. One obvious possibility is that inadequate noradrenergic transmission explains depression, whereas moderate activity provokes attentive interest that is vital for appropriate cognitive function, and excessive noradrenergic activation culminates in anxiety or agitation. Evidence supporting this single axis for central noradrenergic function/dysfunction is discussed in Chapters 19 and 20.

It is equally possible that the role and consequences of central noradrenergic transmission depend on the type or severity of the stimulus and individual differences in the neurobiological coding of behaviour. This would mean that the optimal behavioural response to a given environmental stimulus requires a specific increase in noradrenergic transmission. The optimal response could be determined genetically or by the individuals' previous experience of that stimulus, or both. Deviation of the response, in either direction (i.e. either under- or overactivity), would then result in a deficit in 'coping' (Fig. 8.12(a)). However, it is also possible to envisage disruption of this neurochemical coding of behaviour in the ways illustrated in Figs 8.12(b) and 8.12(c). If there is a shift of the curve to either the right or the left, then the noradrenergic response that would be optimal in normal subjects now produces a suboptimal coping response. In the case of a shift to the left, a reduction in noradrenergic transmission would be required to restore optimal coping whereas for a shift to the right, an increase would be required.

This hypothetical scheme means that there are two possible sources of mismatch that could account for an abnormal behavioural response to a given stimulus and result in an 'inability to cope'. One is that the underlying coding is correct but it is the noradrenergic response evoked by the stimulus that is inappropriate. A second is that the amplitude of the noradrenergic response to arousing stimuli is normal but the underlying coding is not.

Several findings support this model. For instance, an early report suggested that there is a positive correlation between the density of (postsynaptic) β-adrenoceptors in rat cortex and behavioural resistance to a mild environmental stress (novelty and frustration) but a negative correlation between these parameters when the stress is intensified (Stanford and Salmon 1992). More recently, it has been proposed that the phasic response of neurons in the locus coeruleus (which governs 'attentiveness') depends on their tonic activity (which determines arousal). Evidence suggests that the relationship between these two parameters is described by a bell-shaped curve and so an optimal phasic response is manifest only at intermediate levels of tonic activity (Rajkowski *et al.* 1998).

Obviously, it is extremely unlikely that noradrenergic transmission is the sole factor to determine the behavioural response to even simple environmental stimuli. Indeed, a bell-shaped dose–response curve immediately suggests the intervention of one or more additional factors (neurotransmitters?). Such interactions with other neurotransmitters could well define the relationship between noradrenergic transmission and the coding of the coping response.

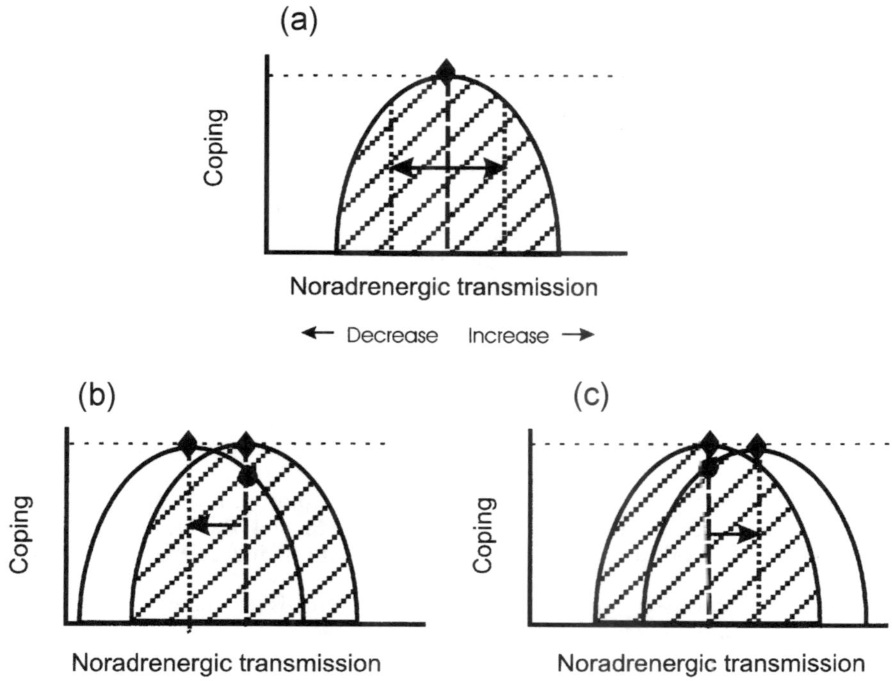

Figure 8.12 Schematic diagram showing the hypothetical relationship between noradrenergic transmission and an individual's ability to 'cope' with aversive environmental stimuli. (a) Optimal coping is attained when the brain rallies a specific noradrenergic response which is determined either genetically and/or by previous experience of the stimulus. Either a reduction or an increase in noradrenergic transmission produces a functional mismatch and diminishes coping. (b) The hatched area depicts the normal relationship between changes in noradrenergic transmission and coping with aversive stimuli (as illustrated in (a)). In these normal subjects, optimal coping is attained when the noradrenergic response to a specific stimulus corresponds to that marked (◆). If there is a leftward shift of the curve that describes the neurochemical coding of coping, then the (predetermined) noradrenergic response that would be optimal in normal individuals now produces suboptimal coping (●). One remedy for such a dysfunction is to reduce noradrenergic transmission so as to restore optimal coping. Similarly, in the case of a rightward shift of the coping curve (c), a predetermined noradrenergic response to a specific stimulus, that would be optimal in normal individuals, will again produce suboptimal coping (●). This time, the remedy is to increase noradrenergic transmission. In both (b) and (c) an alternative way to restore optimal coping would be to reverse the shift in the noradrenergic transmission/coping curve. This could explain the changes in mood that occur after chronic administration of drugs that cause long-latency changes in neurochemical factors that influence noradrenergic transmission (see Chapters 19 and 20)

SUMMARY

Much remains to be learned about the neurochemical regulation of noradrenergic transmission and even more research is required before we can define the role(s) of this neurotransmitter in the brain. Nevertheless, it is evident that these neurons are a crucial component of the network of monoamine influences on the limbic system and that they are capable of both short- and long-term adaptive changes that will influence emotion, motivation, cognition and many other aspects of behaviour.

REFERENCES

Aston-Jones, G, Shipley, MT, Chouvet, G *et al.* (1991) Afferent regulation of locus coeruleus neurons: anatomy, physiology and pharmacology. *Prog. Brain Res.* **88**: 47–73.

Aston-Jones, G, Rajkowski, J, Kubiak, P and Alexinsky, T (1994) Locus coeruleus neurons in monkey are selectively activated by attended cues in a vigilance task. *J. Neurosci.* **14**: 4467–4480.

Bönisch, H, Hammermann, R and Brüss, M (1998) Role of protein kinase C and second messengers in regulation of the norepinephrine transporter. *Adv. Pharmacol.* **42**: 183–186.

Bylund, DB, Eikenberg, DC, Hieble, JP *et al.* (1994) IV. International Union of Pharmacology Nomenclature of Adrenoceptors. *Pharmacol. Rev.* **46**: 121–136.

Cederbaum, JM and Aghajanian, GK (1976) Noradrenergic neurons of the locus coeruleus: inhibition by epinephrine and activation by the alpha-antagonist piperoxane. *Brain Res.* **112**: 413–419.

Docherty, JR (1998) Subtypes of functional α_1 and α_2-adrenoceptors. *Eur. J. Pharmacol.* **361**: 1–15.

Fassio, A, Bonanno, G, Fontana, G, Usai, C, Marchi, M and Raiteri, M (1996) Role of external and internal calcium on heterocarrier-mediated transmitter release. *J. Neurochem.* **66**: 1468–1474.

Fillenz, M (1993) Short-term control of transmitter synthesis in central catecholaminergic neurones. *Prog. Biophys. Molec. Biol.* **60**: 29–46.

Giacobino, JP (1995) Beta 3-adrenoceptors: an update. *Eur. J. Endocrinol.* **132**: 377–385.

Harley, CW (1987) A role for norepinephrine in arousal, emotion and learning: limbic modulation by norepinephrine and the Kety hypothesis. *Prog. Neuro-Psychopharmacol. Biol. Psychiat.* **11**: 419–458.

Hieble, JP, Bondinell, WE and Ruffolo, RR (1995) Alpha- and beta-adrenoceptors: from the gene to the clinic. 1. Molecular biology and adrenoceptor subclassification. *J. Med. Chem.* **38**: 3415–3444.

Kaufman, S (1995) *Tyrosine Hydroxylase. Advances in Enzymology and Related Areas of Molecular Biology*, Vol. 70 (Ed. Meister, A), John Wiley, New York.

Kumar, SC and Vrana, KE (1996) Intricate regulation of tyrosine hydroxylase activity and gene expression. *J. Neurochem.* **67**: 443–462.

Lenders, JWM, Eisenhofer, G, Abeling, NGGM *et al.* (1996) Specific genetic deficiencies of the A and B isoenzymes of monoamine oxidase are characterized by distinct neurochemical and clinical phenotypes. *J. Clin. Invest.* **97**: 1010–1019.

McCormick, DA, Pape, HC and Williamson, A (1991) Actions of norepinephrine in the cerebral cortex and thalamus: implications for function of the central noradrenergic system. *Prog. Brain Res.* **88**: 293–305.

McQuade, R and Stanford, SC (2000) A microdialysis study of the noradrenergic response in rat frontal cortex and hypothalamus to a conditioned cue for aversive, naturalistic environmental stimuli. *Psychopharmacology* **148**: 201–208.

Murugaiah, KD and O'Donnell, JM (1995) Facilitation of noradrenaline release from rat brain slices by beta-adrenoceptors. *Naunyn–Schmiedebergs Arch. Pharmacol.* **351**: 483–490.

Neff, NH and Costa, E (1966) The influence of monoamine oxidase inhibition on catecholamine synthesis. *Life Sci.* **5**: 951–959.

Pacholczyk, T, Blakely, RD and Amara, SG (1991) Expression cloning of a cocaine- and antidepressant-sensitive human noradrenaline transporter. *Nature* **350**: 350–353.

Papadopoulos, GC and Parnavelas, JG (1991) Monoamine systems in the cerebral cortex: evidence for anatomical specificity. *Prog. Neurobiol.* **36**: 195–200.

Povlock, SL and Amara, SG (1997) The structure and function of norepinephrine, dopamine and serotonin transporters. In *Neurotransmitter Transporters: Structure, Function, and Regulation* (Ed. Reith, MEA), Humana Press, Totowa, NJ, pp. 1–28.

Rajkowski, J, Kubiak, P, Ivanova, S and Aston-Jones, G (1998) State-related activity, reactivity of locus coeruleus neurons in behaving monkeys. *Adv. Pharmacol.* **42**: 740–744.

Ruffolo, RJ and Hieble, JP (1994) α-Adrenoceptors. *Pharmac. Ther.* **61**: 1–64.

Russ, H, Staust, K, Martel, R, Gliess, M and Schomig, E (1996) The extracellular transporter for monoamine transmitters exists in cells derived from human central nervous system glia. *Eur. J. Neurosci.* **8**: 1256–1264.

Schuldiner, S (1998) Vesicular neurotransmitter transporters. In *Neurotransmitter Transporters: Structure, Function, and Regulation* (Ed. Reith, MEA), Humana Press, Totowa, NJ, pp. 215–240.

Stanford, SC (1995) Central noradrenergic neurones and stress. *Pharmac. Ther.* **68**: 297–342.

Stanford, SC (1999) SSRI-induced changes in catecholaminergic transmission. In *Selective Serotonin Reuptake Inhibitors (SSRIs): Past, Present and Future* (Ed. Stanford, SC), RG Landes Co., Austin, TX, pp. 147–170.

Stanford, SC and Salmon, P (1992) β-Adrenoceptors and resistance to stress: old problems and new possibilities. *J. Psychopharmacol.* **6**: 15–19.

Starke, K (1987) Presynaptic α-autoreceptors. *Rev. Physiol. Biochem. Pharmacol.* **107**: 73–146.

Zhong, H and Minneman, KP (1999) α_1-Adrenoceptor subtypes *Eur. J. Pharmacol.* **375**: 261–276.

Zigmond, RE, Schwarzschild, MA and Rittenhouse, AR (1989) Acute regulation of tyrosine hydroxylase by nerve activity and by neurotransmitters via phosphorylation. *Ann. Rev. Neurosci.* **12**: 451–461.

9 5-Hydroxytryptamine

S. C. STANFORD

INTRODUCTION

The idoleamine, 5-hydroxytryptamine (5-HT), like the catecholamines, dopamine and noradrenaline, is found in both the periphery and the brain. In the cardiovascular system, it causes marked vasoconstriction and it is from this action that its alternative name, 'serotonin', derives. This chapter will concentrate on the aspects of 5-HT transmission which have made the greatest advances in recent years, particularly those for which some important and interesting questions remain unanswered. Although this material will obviously focus on 5-HT in the brain, the neurochemical mechanisms that regulate 5-HT transmission, such as its synthesis and inactivation, will apply generally to 5-HT-containing cells in the periphery (e.g. enterochromaffin cells in the gut and neurons of the myenteric plexus). All these processes, together with some well-known drugs that affect them, are summarised in Fig. 9.1.

DISTRIBUTION IN THE CNS

As with the other monoamines, the distribution of 5-HT-releasing neurons in the brain was first characterised in the 1960s using the Falck–Hillarp histochemical technique whereby 5-HT is converted to a compound that is fluorescent under ultra-violet light. This showed that the cell bodies of 5-HT neurons aggregate around the midline of the upper brainstem, forming distinct clusters (or nuclei) (Fig. 9.2). Since then, 5-HT neurons have been found in the noradrenergic locus coeruleus and the area postrema as well. Yet, despite this relatively restricted distribution of cell bodies, their processes project more or less throughout the whole neuraxis. For a detailed review of this topic, see Jacobs and Azmitia (1992) but an outline of key features is given here.

The clusters of 5-HT cell bodies (the so-called Raphé nuclei) were originally perceived as forming nine separate nuclei (designated B1–B9) but current nomenclature has reclassified these to some extent so that, currently the nuclei incorporate cell bodies from more than one of those described originally (Table 9.1). Despite these changes, all these nuclei are still regarded as forming two major groups.

The so-called 'inferior' group (B1–B4) projects mainly to brainstem nuclei, the head nuclei of some cranial nerves and the spinal cord. This means that these neurons are well placed for serving a key role in regulation of motor activity, autonomic function and nociception. In addition, there are numerous interconnections between the different

Neurotransmitters, Drugs and Brain Function. Edited by R. A. Webster

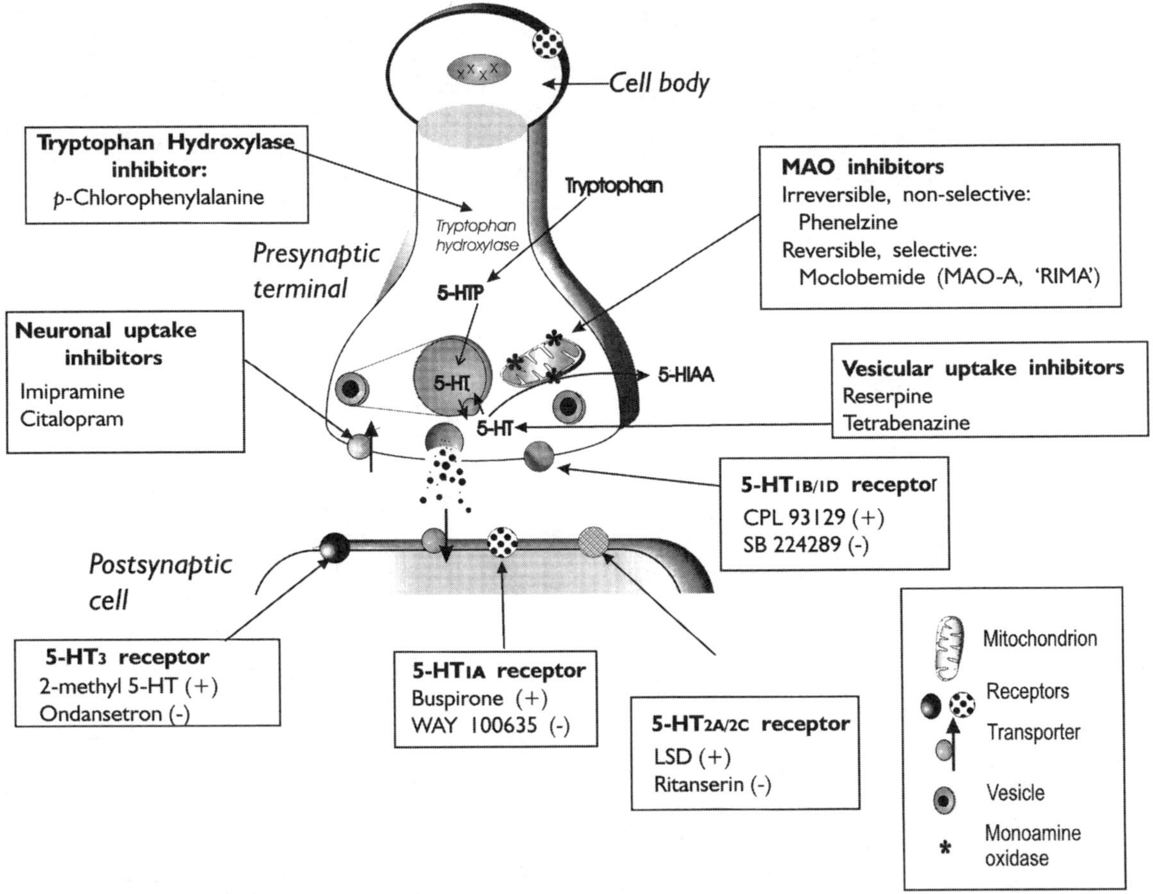

Figure 9.1 The distribution of 5-HT neurons in the brain. The cell bodies are clustered in nuclei (B1–B7) in the pons/upper medullary regions of the brainstem. The rostral cluster ('superior group') project mainly to forebrain areas while the caudal ('inferior') group projects mainly to the medulla and spinal cord. Collectively, these neurons innervate most regions of the central nervous system

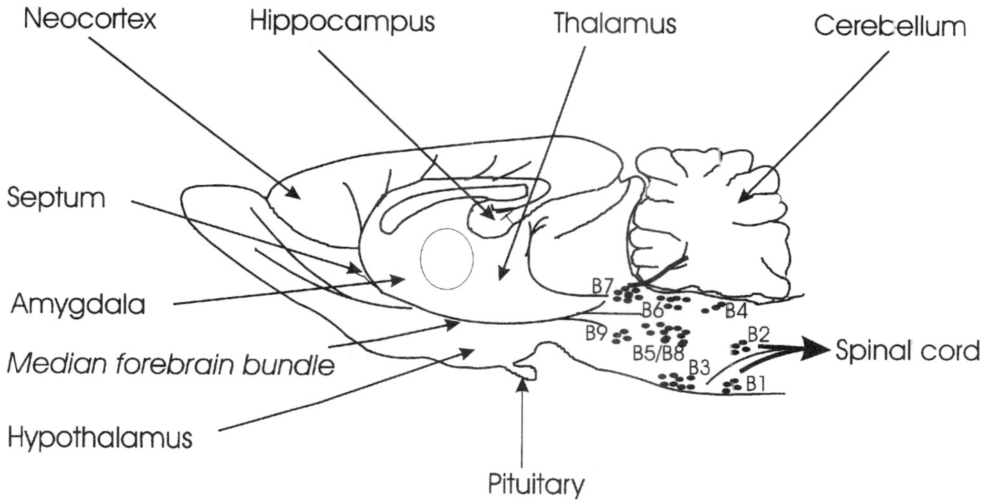

Figure 9.2 The site of action of drugs that modify 5-HT transmission

5-HT nuclei, as well as connections with other monoaminergic nuclei in the brainstem which are also implicated in these physiological functions.

Neurons in the 'superior' group (B5–B9) project rostrally, generally ipsilaterally, in six fibre tracts. The most prominent of these is the median forebrain bundle which contains both myelinated and unmyelinated 5-HT fibres. These mainly innervate limbic and sensory areas of the forebrain. Although extensive branching of the neuronal processes results in a considerable overlap in the terminal axonal fields of the different nuclei, there is evidence for some topographical organisation of the areas to which different nuclei project (Fig. 9.3). For instance, whereas fibres emanating from the dorsal Raphé nucleus (DRN) are the major source of 5-HT terminals in the basal ganglia and cerebellum, neurons in the median Raphé nucleus (MRN) provide the major input to the hippocampus and septum.

There is also some evidence for morphological differences between DRN and MRN neurons which could impinge on their function. Thus, the terminals of neurons from the DRN are relatively fine, unmyelinated, branch extensively and seem to make no specialised synaptic contacts, suggesting *en passant* release of 5-HT (type I). In contrast,

Table 9.1 The main subdivisions of 5-HT nuclei in the brain

Superior		Inferior	
B6 and B7	Dorsal Raphé nucleus	B1 and B4	Nucleus Raphé pallidus
B5 and B8	Median Raphé nucleus	B2	Nucleus Raphé obscurus
B8	Caudal linear nucleus	B3	Nucleus Raphé magnus
B9	Nucleus prosupralemniscus	B1 and B3	Neurons of the lateral paragigantocellular nucleus and the intermediate reticular nuclei
			Cells in the area postrema

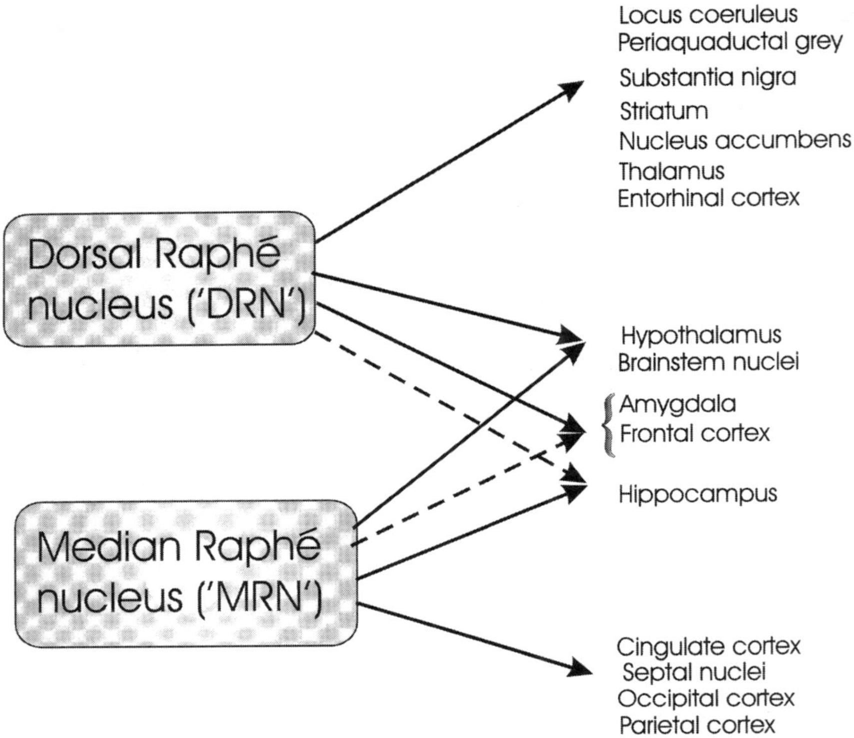

Figure 9.3 Brain regions to which neurons in the dorsal and median Raphé nuclei project. Some areas are innervated by neurons from both nuclei (e.g. hypothalamus) whereas others are innervated predominantly by either the MRN (e.g. the hippocampus) or the DRN (e.g. the amydgala)

those from the MRN are broader, often myelinated, with large varicosities, and they seem to form specialised synaptic contacts, suggesting targeted release of 5-HT (type II). The existence of co-transmitters, especially substance P, thyrotropin releasing hormone (TRH) and enkephalin, gives further options for functional specialisation of different neurons but, as yet, the distribution of these peptides within different nuclei has provided no specific clues as to how this might occur. In any case, species differences in the distribution of co-transmitters is a confounding factor.

In short, although the 5-HT system seems to have a rather non-specific influence on overall brain function, in terms of the brain areas to which these neurons project, there is clearly much to be learned about possible functional and spatial specialisations of neurons projecting from different nuclei.

SYNTHESIS

The first step in the synthesis of 5-HT is hydroxylation of the essential amino acid, tryptophan, by the enzyme tryptophan hydroxylase (Fig. 9.4). This enzyme has several features in common with tyrosine hydroxylase, which converts tyrosine to *l*-DOPA in

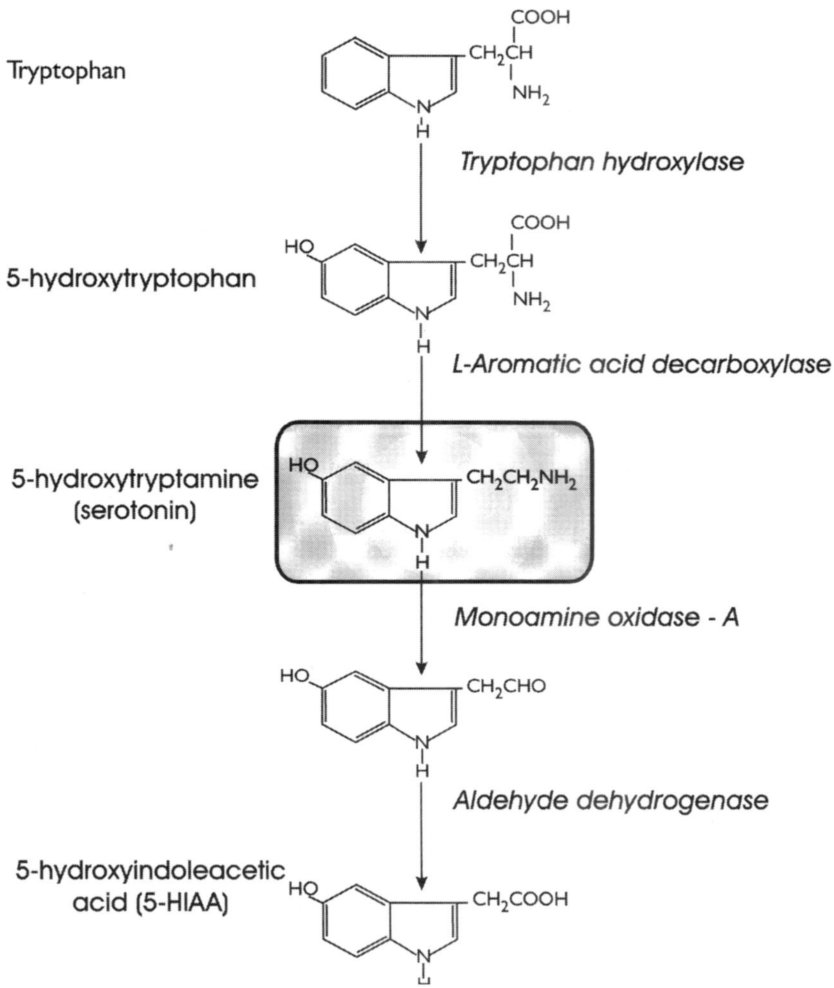

Figure 9.4 The synthesis and metabolism of 5-HT. The primary substrate for the pathway is the essential amino acid, tryptophan and its hydroxylation to 5-hydroxytryptophan is the rate-limiting step in the synthesis of 5-HT. The cytoplasmic enzyme, monoamine oxidase (MAO_A), is ultimately responsible for the catabolism of 5-HT to 5-hydroxyindoleacetic acid

the noradrenaline synthetic pathway. First, it has an absolute requirement for O_2 and the reduced pterin co-factor, tetrahydrobiopterin. Second, hydroxylation of tryptophan, like that of tyrosine, is the rate-limiting step for the whole pathway (reviewed by Boadle-Biber 1993) (see Chapter 8). However, unlike the synthesis of noradrenaline, the availability of the substrate, tryptophan, is a limiting factor in the synthesis of 5-HT. Indeed, the activated form of tryptophan hydroxylase has an extremely high K_m for tryptophan (50 μM), which is much greater than the concentration of tryptophan in the brain (10–30 μM). This means that not only is it unlikely that this enzyme ever becomes saturated with its substrate but also that 5-HT synthesis can be driven by giving extra tryptophan.

This influence of tryptophan availability on the rate of synthesis of 5-HT has some interesting implications. First, it predicts that a dietary deficiency of tryptophan could lead to depletion of the neuronal supply of releasable 5-HT. Indeed, this has been confirmed in humans to the extent that a tryptophan-free diet can cause a resurgence of depression in patients who were otherwise in remission (see Chapter 20). In contrast, a tryptophan-high diet increases synthesis and release of 5-HT. In fact, when given in combination with other drugs that augment 5-HT transmission (e.g. an MAO inhibitor or a 5-HT reuptake inhibitor), tryptophan can cause a life-threatening delirium known as the 'serotonin syndrome' (Gillman 1999).

Transport of tryptophan across the blood–brain barrier and neuronal membranes relies on a specific carrier for large neutral amino acids (LNAAs). Thus, although an increase in the relative concentration of plasma tryptophan, either through dietary intake or its reduced metabolism in a diseased liver, increases its transport into the brain, other LNAAs (such as leucine, isoleucine or valine) can compete for the carrier. This process forms the basis of an intriguing theory linking the intake of carbohydrates in the diet with an individual's mood. It is known that consumption of carbohydrates increases secretion of insulin which, in addition to its well-known glucostatic role, promotes uptake of LNAAs by peripheral tissues. However, it seems that tryptophan is less affected by insulin than the other LNAAs in this respect and so its relative concentration in the plasma increases, thereby increasing its transport into the brain (see Rouch, Nicolaidis and Orosco 1999). The resulting increase in synthesis and release of 5-HT is claimed to enhance mood. Although this scheme is rather controversial, it has been suggested as an explanation for the clinical improvement in some patients, suffering from depression or premenstrual tension, when they eat carbohydrates. It has also been suggested to underlie the carbohydrate-craving experienced by patients suffering from Seasonal Affective Disorder (Wurtman and Wurtman 1995).

Not a great deal is known about factors that actually activate tryptophan hydroxylase. In particular, the relative contribution of tryptophan supply versus factors that specifically modify enzyme activity under normal dietary conditions is unknown. However, removal of end-product inhibition of tryptophan hydroxylase has been firmly ruled out. Also, it has been established that this enzyme is activated by electrical stimulation of brain slices, even in the absence of any change in tryptophan concentration, and so other mechanisms are clearly involved.

So far, it has been established from *in vitro* studies that the enzyme undergoes phosphorylation, a process that changes the conformation of the enzyme protein and leads to an increase in its activity. This involves Ca^{2+}/calmodulin-dependent protein kinase II and cAMP-dependent protein kinase which suggests a role for both intracellular Ca^{2+} and enzyme phosphorylation in the activation of tryptophan hydroxylase. Indeed, enzyme purified from brain tissue innervated by rostrally projecting 5-HT neurons, that have been stimulated previously *in vivo*, has a higher activity than that derived from unstimulated tissue but this increase rests on the presence of Ca^{2+} in the incubation medium. Also, when incubated under conditions which are appropriate for phosphorylation, the K_m of tryptophan hydroxylase for its co-factor and substrate is reduced whereas its V_{max} is increased unless the enzyme is purified from neurons that have been stimulated *in vivo*, suggesting that the neuronal depolarisation *in vivo* has already caused phosphorylation of the enzyme. This is supported by evidence that the enzyme activation caused by neuronal depolarisation is blocked by a Ca^{2+}/calmodulin protein kinase inhibitor. However, whereas depolarisation alone increases enzyme V_{max}, it does not appear to affect the enzyme K_m and so a firm link between neuronal depolarisation and enzyme phosphorylation has yet to be established.

The apparent reliance of enzyme activation on phosphorylation and intracellular Ca^{2+} gives a clue as to how the rate of 5-HT synthesis might be coupled to its impulse-evoked release. Certainly, the impulse-induced increase in intracellular Ca^{2+}, and/or activation of the G protein-coupled receptors that govern synthesis of cAMP, could modify the activity of tryptophan hydroxylase. Indeed, this could explain why activation of either somal 5-HT_{1A} autoreceptors in the Raphé nuclei (which depress the firing rate of 5-HT neurons) or terminal 5-HT_{1B} autoreceptors (which depress 5-HT release) can reduce the production of cAMP and attenuate 5-HT synthesis.

The product of the hydroxylation of tryptophan, 5-hydroxytryptophan, is rapidly decarboxylated to 5-HT by a specific decarboxylase enzyme. This is generally thought to be a soluble enzyme which suggests that 5-HT is synthesised in the cytoplasm, before it is taken up into the storage vesicles. If this is the case, then considerable losses might be incurred from its metabolism by monoamine oxidase before it reaches the storage vesicles. Indeed, this could explain why 5-HT turnover seems to greatly exceed its rate of release.

The high affinity of the decarboxylase enzyme for its substrate (10 μM in the brain) makes it unlikely that this stage could ever become rate-limiting for the pathway as a whole. Nevertheless, the K_m for this enzyme is considerably higher than tissue concentrations of 5-hydroxytryptophan and so, again, supply of this substrate is likely to be a crucial factor.

Finally, as with the noradrenergic system, there is evidence for long-term changes in the rate of synthesis of 5-HT that are triggered by prolonged changes in its rate of release. These can be traced to the rate of gene transcription and the ensuing synthesis of enzyme protein ('enzyme induction'). It has even been shown that mRNA for tryptophan hydroxylase shows a daily rhythm in cultured eye-cups maintained in the dark. Again, not a great deal is known about the underlying control mechanisms but the synthesis of tryptophan hydroxylase, at least, is increased by exposure of 5-HT neurons *in vivo* to the growth factor, brain-derived neurotrophic factor ('BDNF; Siuciak *et al.* 1998). Steroid hormones also seem to modulate tryptophan hydroxylase gene transcription but research in this area is confounded by the variation in this effect across different tissues and different hormones, with both increases and decreases being reported.

STORAGE

As with the other monoamines, 5-HT is found primarily in storage vesicles (30–35 nm diameter) where 'serotonin-binding proteins' (SBPs) have also been identified. These seem to form a macromolecular complex with 5-HT. In fact, three such proteins have now been characterised, but only one of them, 45 kDa SBP, appears to be secreted into the synapse along with 5-HT. Whether they serve any role other than forming an osmotically inert storage matrix for 5-HT is unknown.

In other respects the storage of 5-HT resembles that of noradrenaline with its uptake by vesicles resting on energy-dependent, 'vesicular monoamine transporters' (VMATs) (see Chapter 8). Functional disruption of this transporter, either through competitive inhibition (e.g. by methylenedioxymethamphetamine (MDMA, 'Ecstasy')) or dissipation of the pH gradient across the vesicle membrane that drives the uptake of 5-HT (e.g. by MDMA, reserpine and *p*-chloroamphetamine) explains why these compounds deplete the vesicular pool of 5-HT.

RELEASE

Impulse-evoked release of 5-HT, like that of noradrenaline, is subject to fine control by a system of autoreceptors, in particular 5-HT_{1A} receptors on the cell bodies of neurons in the Raphé nuclei and 5-$HT_{1B/1D}$ receptors on their terminals. Because these are all $G_{i/o}$ protein-coupled receptors, their activation reduces the synthesis of cAMP so that 5-HT_{1A} agonists (or 5-HT itself) decrease neuronal excitability and the firing of Raphé neurons whereas activation of 5-$HT_{1B/1D}$ receptors seems to disrupt the molecular cascade that links the receptor with transmitter release (see Chapter 4).

There is some evidence that receptors for other neurotransmitters on 5-HT nerve terminals also modify release of 5-HT. These include nicotinic receptors (increase release from striatal synaptosomes), α_{2A}-adrenoceptors (depress cortical release) and H_3-receptors (cortical depression). Because changes in 5-HT release on activation of these receptors is evident in synaptosomal preparations, it is likely that these are true ‘heteroceptors’.

Finally, the actions of the so-called ‘5-HT releasing agent’, *d*-fenfluramine, which is well known for its anorectic effects, should be mentioned here. This compound inhibits 5-HT uptake but its metabolite, *d*-norfenfluramine, increases 5-HT release as do high doses of *d*-amphetamine. It is important to realise that this 5-HT release is independent of nerve impulses and the action of such compounds rests on their effects on the 5-HT transporters on the storage vesicles and terminal membrane. Once these drugs have been taken up into 5-HT neurons by the transporter, they cause 5-HT to leak out of its storage vesicles and, ultimately, to be extruded from the neuron by retrotransport (see below and Chapter 4 for further details).

Until recently, *d*-fenfluramine was used to control appetite, in preference to *d*-amphetamine, because it has a lower affinity for the catecholamine transporter and so its uptake into noradrenergic and dopaminergic neurons is much less than that of amphetamine. This is thought to explain why, at anorectic doses, this compound lacks the psychotropic effects and dependence-liability that are real problems with *d*-amphetamine. Unfortunately, despite this therapeutic advantage, this compound has had to be withdrawn from the clinic because of worries that it might cause primary pulmonary hypertension, valvular heart disease and even long-term neuropathy.

INACTIVATION

As with other monoamines, the actions of 5-HT are terminated by its reuptake from the synapse by another member of the family of Na^+/Cl^--dependent transporters. The 5-HT transporter has many features in common with its catecholamine equivalent (described fully in Chapter 8; see Fig. 8.7), including its presumed 12 transmembrane-spanning domains. However, the cloned 5-HT transporter has a K_m for 5-HT of about 450 nM whereas its K_i for both noradrenaline and dopamine is some ten thousand-fold greater (Povlock and Amara 1997) which means that it is relatively selective for uptake of 5-HT. The uptake process itself requires the inward co-transport of one Na^+ ion and one Cl^- ion while K^+ (or H^+) is carried in the opposite direction. The energy required to maintain the ionic gradients that drive this process is provided by a Na^+/K^+-dependent ATPase.

As might be expected, mRNA for the 5-HT transporter is found in high concentrations in the Raphé nuclei but it is also found in other brain regions. Whether this means that non-5-HT neurons can synthesise this protein is unknown but there is some evidence that it is synthesised in astrocytes, at least. One complication is that there are multiple forms of mRNA for the 5-HT transporter, but there is, as yet, no evidence for transporter subtypes in the CNS. However, it must also be remembered that 5-HT transporters are found in the peripheral tissues, notably platelets, mast cells, the placental brush-border and adrenal chromaffin cells and it is possible that these are not all identical.

Inhibitors of 5-HT uptake include the tricyclic antidepressants and the selective serotonin reuptake inhibitors (SSRIs, which are discussed in detail in Chapter 20) as well as compounds like cocaine and *d*-amphetamine. Because cocaine is not transported into the neuron it is thought to bind to a site on the transporter protein. This has a negative allosteric effect on the protein and prevents binding of 5-HT to its domain. It has even been suggested that there could be an endogenous ligand for this site which regulates 5-HT uptake. By contrast, *d*-amphetamine is transported into neurons and so acts as a competitive inhibitor of 5-HT reuptake. Other inhibitors that are transported into the neuron, and which are thought to bind to the same site, include *p*-chloroamphetamine, MDMA and fenfluramine.

Of course, all these transporter inhibitors release 5-HT (see above) but exactly how they do this is uncertain. One suggestion is that, because they can also penetrate the cell membrane directly, they recycle continuously through their active transport into the cell and passive outward diffusion. This is thought gradually to dissipate the ionic gradient that is needed for the transporter to take up 5-HT into the neuron and so culminates in the outward transport of 5-HT into the synapse (Rudnick 1997). This action is compounded by their disruption of the vesicular transporters (VMATs) since, once they gain access to the neuron, they diminish the proton gradient required for the VMATs to function properly, possibly because they are weak bases (see Chapter 8; Fig. 8.6). This leads to leakage of 5-HT into the cytoplasm where its ensuing increased concentration ensures that a large pool is available for its retrotransport into the synapse.

Recent evidence indicates that the 5-HT transporter is subject to post-translational regulatory changes in much the same way as neurotransmitter receptors (Blakeley *et al.* 1998). Protein kinase A and protein kinase C (PKC), at least, are known to be involved in this process. Phosphorylation of the transporter by PKC reduces the V_{max} for 5-HT uptake and leads to sequestration of the transporter into the cell, suggesting that this enzyme has a key role in its intracellular trafficking. Since this phosphorylation is reduced when substrates that are themselves transported across the membrane bind to the transporter (e.g. 5-HT and *d*-amphetamine), it seems that the transport of 5-HT is itself linked with the phosphorylation process. Possibly, this process serves as a homeostatic mechanism which ensures that the supply of functional transporters matches the demand for transmitter uptake. By contrast, ligands that are not transported (e.g. cocaine and the selective serotonin reuptake inhibitors (SSRIs)) prevent the inhibition of phosphorylation by transported ligands. Thus, such inhibitors would reduce 5-HT uptake both by their direct inhibition of the transporter and by disinhibition of its phosphorylation (Ramamoorthy and Blakely 1999).

In platelets, depletion of intracellular Ca^{2+} reduces 5-HT transport and this points to calmodulin as another endogenous regulator and its antagonists do inhibit 5-HT uptake. In contrast, activation of adenosine (A3) receptors seems to upregulate the transporter, possibly through the PKG, NO/cGGP pathway.

Glycosylation sites have also been identified on the transporter and recent findings suggest that the sex steroids, estradiol and testosterone increase transcription of the transporter gene and, in turn, the density of transporters in the DRN but not the MRN (McQueen *et al.* 1999). Although it is not yet clear whether this involves a direct effect on transporter gene expression, this finding does suggest that transporters associated with these two groups of neurons are subject to different control mechanisms.

Many studies have attempted to show changes in 5-HT uptake or the density of transporters either in depressed patients, or after treatment with antidepressants. Most have found a reduction in the density of uptake sites, labelled with the tricyclic reuptake inhibitor, [^{3}H]imipramine, in depression. However, there appears to be no change in the density of uptake sites when these are labelled with the selective serotonin reuptake inhibitor, [3H]paroxetine. Another problem is that, even in studies showing a reduction in transporter density, there are no consistent changes in 5-HT uptake. An intriguing suggestion that could account for this anomaly is that there are 'spare' transporters. More recently, research has been directed towards a search for genetic polymorphisms of the 5-HT transporter gene that might account for disorders including depression, bipolar disorder, anxiety, substance abuse and autism. So far, no certain links with either the expression of, or vulnerability to, any disorder have emerged.

One drug that seems to cause quite marked, long-term changes in 5-HT transporter function is MDMA. Single-photon emission-computed tomography (SPECT) which provides an image of the binding of [^{123}I]2β-carbomethoxy-3β-(4′-iodophenyl)-tropane (β-CIT) to the transporters in the (living) human brain shows that this is greatly reduced (and, in some cases, totally absent) in subjects who claim to use MDMA (Semple *et al.* 1999). Interestingly, fenfluramine, another 5-HT-releasing agent, does not seem to have this effect. It has been suggested that loss of transporters in users of MDMA is due to the death of 5-HT neurons and that this is evidence for its neurotoxic effects. This toxicity is thought to be mediated by the formation of quinones and then free radicals from the metabolites of MDMA, although there are alternative explanations (see Sprague, Everman and Nichols 1998) and some individuals still dispute that this drug is actually neurotoxic in humans. At the very least, there is accumulating evidence for long-term deficits in cognitive and neuroendocrine function in users of MDMA and, of even greater concern, it is not known whether these are reversible.

METABOLISM

5-HT is metabolised primarily by MAO to 5-hydroxyindoleacetic acid (5-HIAA) (Fig. 9.4). *In vitro*, 5-HT is the preferred substrate for the MAO_A, rather than the MAO_B isoenzyme (see Chapter 8) and this appears to be the case *in vivo* since MAO_A, but not MAO_B, knock-out mice have increased concentrations of 5-HT in the brain. Obviously, because of its indole nucleus, 5-HT is not a substrate for the enzyme COMT which metabolises the catechol derivatives, dopamine and noradrenaline. However, other metabolic products of 5-HT are theoretically possible and one, 5-hydroxytryptophol, which results from the reduction of its intermediate metabolite, 5-hydroxyindoleacetaldehyde, instead of oxidation to 5-HIAA, has been identified in the brain.

The comparatively straightforward link between 5-HT and its primary metabolite, 5-HIAA, encouraged many researchers to use changes in the ratio of tissue concentrations of 5-HIAA and 5-HT as an index of the rate of release of 5-HT *ex vivo*. However, it has been clear for some time that the majority of 5-HT is metabolised in the cytoplasm by MAO before it is released from 5-HT nerve terminals. Consequently, the reliability of the 5-HIAA:5-HT ratio as an index of transmitter release is rather dubious, although it could be used as an acceptable measure of MAO activity. In any case, the development of *in vivo* microdialysis means that changes in the concentration of extracellular 5-HT can now be monitored directly which, under drug-free conditions, provides a far more reliable indication of any changes in the rate of release of 5-HT.

RECEPTORS

Over the last 20 years, the development of receptor-selective ligands, coupled with advances in molecular biology, has resulted in the number of 5-HT receptors increasing from a modest two (identified by Gaddum and Picarelli in 1957) to the 14 recognised to date (Table 9.2). These form seven distinct families which, with the exception of the 5-HT_3 receptor, are all G protein-coupled with seven transmembrane-spanning domains. Apart from 5-ht_{1E}, 5-ht_{1F}, 5-ht_5 and 5-ht_6 subtypes, for which genes have been identified, even though the native receptor protein remains elusive (hence their lower-case nomenclature), all are expressed in the CNS.

All the native 5-HT receptors characterised so far are found postsynaptically, with respect to 5-HT terminals, and some are located presynaptically where they regulate the firing rate of 5-HT neurons and/or release of transmitter from their terminals. There is also evidence that some regulate the release of other transmitters in the terminal field and so could act as 5-HT heteroceptors. For instance, 5-HT_{1B} receptor agonists inhibit K^+-evoked release from synaptosomes preloaded with either [^{3}H]dopamine, [^{3}H]noradrenaline, [^{3}H]prolactin or [^{3}H]glutamate. Apart from regulating neuronal firing and transmitter release, activation of certain 5-HT receptors with selective ligands causes specific behavioural or physiological changes (Table 9.3) but, in some cases, these can vary from species to species. There is also some evidence that 5-HT_{1A} receptors, at least, might influence gene expression and neurogenesis and so they could have far-reaching effects on brain function.

Essential features of the different receptor subtypes are highlighted here and, except where indicated, references to specific points can be found in the definitive review of this subject by Barnes and Sharp (1999).

5-HT_{1A}

Although the distribution of these receptors is widespread in the brain, they are found postsynaptically in high concentrations in the hippocampus, septum and amygdala and also on cell bodies of 5-HT neurons in the Raphé nuclei. They are negatively coupled, via $G_{i/o/z}$ proteins, to adenylyl cyclase such that their activation reduces production of cAMP. In turn, this leads to an increase in K^+ conductance and hyperpolarisation of

Table 9.2 The characteristics of 5-HT receptors

Family	Actions	Subtype	Second messenger	Selective/preferential agonist	Selective antagonist	Location
5-HT_1	Hyperpolarisation	5-HT_{1A}	$G_{i/o/z} \rightarrow \downarrow cAMP$	8-OH-DPAT	WAY 100635	Pre- and postsynaptic and heteroceptor
		$5\text{-HT}_{1B/D}$	$G_{i/o} \rightarrow \downarrow cAMP$	CP 93129	SB 224289	Pre- and postynaptic and heteroceptor
		5-HT_{1D}	$G_{i/o} \rightarrow \downarrow cAMP$	N/A	BRL 15572	Pre- (?) and postsynaptic and heteroceptor
		5-ht_{1E}	$G_{i/o}$ (?)$\rightarrow \downarrow cAMP$	N/A	N/A	
		5-ht_{1F}	$G_{i/o}$ (?)$\rightarrow \downarrow cAMP$	LY 334370	N/A	
5-HT_2	Depolarisation	5-HT_{2A}	$G_{q/11:}$ activates phospholipase C, $\downarrow gK^+$ and $\uparrow[Ca^{2+}]_{in}$. Also activates PLA2 and arachidonic acid pathway	CP 94253	MDL 100907	Postynaptic and heteroceptor
		5-HT_{2B}	$G_{q/11:}$ activates phospholipase C, $\downarrow gK^+$ and $\uparrow[Ca^{2+}]_{in}$. Also activates PLA2 and arachidonic acid pathway	BW 723C86	SB 204741	
		5-HT_{2C}	$G_{q/11:}$ activates phospholipase C, $\downarrow gK^+$ and $\uparrow[Ca^{2+}]_{in}$. Also activates PLA2 and arachidonic acid pathway	MK 212	SB 242084	Postsynaptic and heteroceptor
5-HT_3	Depolarisation		Ligand-gated Na^+/K^+ channel	2-methyl 5-HT	Ondansetron	Postsynaptic and heteroceptor
5-HT_4			$G_s \rightarrow \uparrow cAMP$	SDZ 216454	GR 113808	Postsynaptic and heteroceptor
5-ht_5	?		G protein-coupled?	N/A	N/A	?
5-ht_6	Depolarisation		$G_s \rightarrow \uparrow cAMP$	N/A	SB 271046	?
5-HT_7	Depolarisation		$G_s \rightarrow \uparrow cAMP$	N/A	SB 258719	?

Table 9.3 Behavioural and physiological responses affected by 5-HT receptors

	5-HT_{1A}	5-HT_{1B}	5-HT_2	5-HT_3	5-HT_4	5-HT_7
Anxiety/panic	✓		✓	(✓?)		
Cognition					(✓?)	
Food intake	✓	✓	✓	✓		
Hallucinations			✓			
Mood	✓		✓			
Nausea/vomiting				✓		
Obsessive behaviour			✓			
Pain		✓	✓	✓		
Psychosis			✓			
Sexual function		✓	✓			
Sleep/circadian rhythms		✓	✓			✓
Thermoregulation	✓	✓	✓			

the host cell resulting in the inhibition of firing of cells in the Raphé nuclei and reduced release of 5-HT from their terminals. The importance of this action, as a possible explanation for the delay in the therapeutic effects of those antidepressants that increase the concentration of extracellular 5-HT, is discussed in Chapter 20.

There is some evidence that pre- and postsynaptic receptors do not respond in exactly the same way to drug challenges and it has even been suggested that they are not identical. For instance, the drug BMY 7378 behaves as an agonist at presynaptic 5-HT_{1A} receptors but has a low intrinsic activity at the postsynaptic site where it acts as an antagonist. However, there is as yet insufficient evidence to claim that there are subtypes of this receptor and, in any case, differences in the receptor reserve at pre- and postsynaptic sites could well explain some of the apparently conflicting findings.

Given that the firing rate, and hence the release of 5-HT, is greater in awake animals than in those that are asleep (see Chapter 22), it is not surprising that the effects of 5-HT_{1A} antagonists on neuronal firing and 5-HT release are more evident in behaving, conscious subjects than in those that are anaesthetised. Indeed, this should be borne in mind when perusing the literature on this subject. However, in terms of their gross physiological effects in awake animals, consistent findings are that 5-HT_{1A} receptor agonists induce hypothermia and increase food intake. They also reduce anxiety and, so far, this is the only action to be exploited clinically. Even so, only one such compound, buspirone, is licensed for use in the clinic and it is still not known whether its anti-anxiety effect is mediated by activation of pre- or postsynaptic 5-HT_{1A} receptors (see Chapter 19).

Another well-known agonist at these receptors is lysergic acid diethylamide (LSD) and, for several years it was thought that this explained its hallucinogenic effects. However, this drug is a non-selective ligand that also binds to $5\text{-HT}_{2A/2C}$ receptors. Although activation of 5-HT_{1A} receptors by LSD seems to have some effects on motor activity, this site can be ruled out as being responsible for its hallucinogenic effects. This is not least because neither buspirone, which is also an agonist of these receptors, nor reserpine, which diminishes 5-HT transmission, have any hallucinogenic actions in humans. In fact, experimental preclinical models strongly indicate that 5-HT_{1A} agonists could be beneficial in treatment of both the positive and negative symptoms of schizophrenia. For instance, they increase the concentration of extracellular dopamine in the frontal cortex but diminish apomorphine-induced stereotypy in rats.

Importantly, they achieve this without inducing extrapyramidal side-effects or increasing prolactin secretion, which are real problems with neuroleptics. These results have been borne out by preliminary clinical trials of buspirone, used in combination with neuroleptics, and several novel 5-HT$_{1A}$ agonists (e.g. BSF 190555) are currently under development for this clinical application (Meltzer 1999).

5-HT$_{1B}$

These receptors are found postsynaptically, mostly in the basal ganglia, striatum and frontal cortex, but they are also thought to lie on 5-HT nerve terminals where their activation reduces release of 5-HT. However, there seem to be regional differences in the extent to which this population of receptors is tonically activated by extracellular 5-HT and so the literature describing the effects of 5-HT$_{1B}$ antagonists on 5-HT release is somewhat confusing. 5-HT$_{1B}$ receptors are also found in the Raphé nuclei and their antagonism increases release of 5-HT in the DRN. Because of the dearth of 5-HT$_{1B}$ selective ligands, the gross physiological effects that result from activation of these receptors is largely uncharacterised. However, it is possible that they contribute to the regulation of circadian rhythms by blunting 5-HT release in the suprachiasmatic nucleus of the hypothalamus and, in mice in which this receptor has been 'knocked-out', the observed quiescence suggests that their activation increases locomotor activity (see also Chapter 22). Activation of these receptors could also contribute to the anti-migraine effects of sumatriptan, a non-selective 5-HT$_{1B/1D}$ agonist (but see below).

An interesting development in this area is the possibility that there could be an endogenous ligand, for these receptors: 5-HT-moduline. This is a tetrapeptide that is released from neurons and is claimed to be the first allosteric modulator of a G protein-coupled receptor to be identified so far. Functionally, 5-HT-moduline behaves like a 5-HT$_{1B}$ antagonist and so increases terminal release of 5-HT (Massot *et al.* 1998) and it is thought that this could be an important step in maintaining a sustained increase in release of 5-HT during stress.

5-HT$_{1D}$

Probably the most notable feature of this receptor is the confusion arising from its classification and nomenclature! Soon after characterisation of the 5-HT$_{1D}$ receptor, which was found in certain species (e.g. the human) it was determined that this was in fact a variant of the 5-HT$_{1B}$ receptor which had already been found in other species (e.g. the rat). These receptors were therefore regarded as species variants and came to be described as the 5-HT$_{1B/1D}$ subtype. Since then, another 5-HT$_1$ receptor subtype has been identified and current nomenclature dictates that this is the (new) 5-HT$_{1D}$ receptor.

So far, little is known about this novel 5-HT$_{1D}$ receptor but, in the rat and human, its mRNA is found, albeit in low concentrations, in the basal ganglia, nucleus accumbens, hippocampus, frontal cortex and Raphé nuclei. It is negatively coupled to adenylyl cyclase and is possibly located presynaptically, on both the 5-HT neuronal cell body and terminals, but this has yet to be confirmed.

Interest in this receptor has been generated by the possibility that its activation accounts for the anti-migraine effects of the non-selective 5-HT$_{1B/1D}$ agonist, sumatriptan. The exact process(es) that account for this action are unresolved but favoured possibilities include vasoconstriction of cerebral arteries and/or blockade of neurogenic pain and inflammation generated by vascular afferents within the trigeminal nucleus. Because mRNA for the 5-HT_{1B} receptor is not found in this nucleus in humans, activation of the 5-HT_{1D} receptor is thought to be responsible for these effects.

5-ht_{1E} AND 5-ht_{1F}

Although the existence of further 5-HT_1 receptor subtypes with atypical pharmacology was suspected from radioligand binding studies, the 5-ht_{1E} and 5-ht_{1F} receptors have been characterised mainly from the identification of the genes that encode receptors with properties similar to the rest of the 5-HT_1 family. Radioligand binding and the distribution of mRNA for the 5-ht_{1E} receptor suggest that it is found mainly in the striatum but it also seems to be present in the amygdala, frontal cortex and globus pallidus, to some extent. However, the lack of specific ligands has prevented their pharmacological characterisation. As a result of these limitations, they are currently assigned only 'lower-case' status. Recently, the anti-migraine drug, sumatriptan, has been found to bind to 5-ht_{1F} receptors with an affinity similar to that for 5-$HT_{1B/1D}$ receptors and so they might have a role in migraine.

5-HT_{2A}

These receptors are mainly found in the cortex and basal ganglia. They are coupled to phospholipase A and have an excitatory effect on the host cell as a result of the ensuing reduction in K^+ conductance. The many well-known agonist ligands for this receptor include DOI (1-(4-iodo-2,5-dimethoxyphenyl)-2-aminopropane hydrochloride), DOM (2,5-dimethoxy-4-methylamphetamine) and LSD. In rodents, these compounds cause a characteristic 'head twitch' and increase motor activity and, in humans, they are all hallucinogens. Preclinical studies reinforce the view that 5-HT_{2A} (and possibly 5-HT_{2C}) receptor activation underlies the hallucinogenic effects of these compounds (Krebs-Thomson, Paulus and Geyer 1998). This is consistent with recent evidence that all the atypical neuroleptics, such as clozapine, risperidone and olanzepine, act as antagonists at this receptor, an action that could well contribute to their therapeutic effects in schizophrenia.

In addition to their psychotropic effects, activation of 5-HT_{2A} receptors induces hyperthermia, which could explain this dangerous action of MDMA. Finally, an unusual feature of these receptors is that they are downregulated by prolonged exposure to antagonists, as well as agonists. The reason for this is uncertain but it could suggest that drugs which hitherto have been regarded as antagonists are, in fact, inverse agonists.

5-HT_{2B}

The use of labelled antibodies suggests their presence in the amygdala, septum, hypothalamus and cerebellum. However, little is known about these receptors, mainly because of the shortage, until recently, of selective ligands, their low density and the limited distribution of their mRNA in the brain.

5-HT_{2C}

Again, conventions for nomenclature have somewhat confused the status of this receptor which was first known as a 5-HT_{1C} receptor. However, as information accrued from cloning studies, pharmacological characterisation, and discovery of its second messenger system, it became evident that this receptor shared the characteristics of the 5-HT_2 receptor family, rather than those of a 5-HT_1 receptor. The switch in classification from a 5-HT_{1C} to a 5-HT_{2C} receptor explains the gap in the 5-HT_1 receptor family.

The 5-HT_{2C} receptor was first found in the choroid plexus, where it is thought to regulate the formation of CSF, but it has since been found in cortical and limbic areas as well as the basal ganglia. In the choroid plexus, at least, its actions seem to be mediated by activation of phospholipase C with a resulting depolarisation of the host cell. Like the 5-HT_{2A} subtype, 5-HT_{2C} receptors are downregulated by prolonged exposure to antagonists (inverse agonists?) as well as agonists. The discovery that 5-HT_{2C} receptor mRNA is subject to posttranslational changes suggests that there could be several different isoforms of this receptor and it cannot be assumed that they are functionally the same.

As far as can be certain, given the lack of selective ligands, their activation elsewhere in the brain is thought to culminate in reduced locomotor activity and hyperthermia. However, interest in these receptors as possible therapeutic targets is fostered by evidence that their agonists, such as *m*CPP, appear to be profoundly anxiogenic (see Chapter 19) and reduce food intake (see below).

5-HT_3

These receptors are quite different from any other monoamine receptor in that they are not coupled to G proteins. Instead, they comprise a pentameric complex of subunits that incorporates an ion channel. This is selective for the cations Na^+ and K^+ which, when opened, leads to depolarisation of the host cell. 5-HT_3 receptors are found at high concentrations in the brainstem and area postrema. However, they are also found elsewhere in lower concentrations, notably in the cortex, amygdala and hippocampus, where they are thought to be associated mainly with GABAergic neurons.

Interestingly, their function is modified by many agents that allosterically modify $GABA_A$ receptor function (e.g. barbiturates and steroids). A further parallel with the $GABA_A$ receptor is that there could well be differences in subunit composition of 5-HT_3 receptors such that different heteromeric complexes form receptors which are functionally distinct.

These receptors are best known for their stimulation of dopamine release. Indeed, it is attenuation of dopamine release in the area postrema by the 5-HT_3 receptor antagonist, ondansetron, that is thought to explain its anti-emetic effects. However, they are also thought to influence release of other neurotransmitters including GABA, acetylcholine and noradrenaline; they are even thought to increase somatodendritic release of 5-HT in the Raphé nuclei. So far, despite vigorous attempts to find other clinical applications for ondansetron, none has proved convincing.

5-HT_4

This receptor is positively coupled to adenylyl cyclase which results in closure of K^+ channels, culminating in an increase in excitability of the host cell and a delay in repolarisation. Its density is high in the basal ganglia and the nucleus accumbens but studies of the pharmacology of this receptor may well be complicated by the discovery that it has several 'splice' variants (four in humans). So far, the literature on its behavioural effects is somewhat inconsistent but agonists of this receptor are being explored as possible cognitive enhancers.

5-ht$_5$ AND 5-ht$_6$

The existence of these receptors was predicted by cDNA sequence analysis and their protein products have been studied in cell expression systems. 5-ht$_5$ receptor mRNA is found in the cortex, hippocampus, olfactory bulbs and cerebellum but the native receptor has still not been characterised and so it has been assigned only 'lower-case' nomenclature. However, studies using antibodies generated against these receptors have shown that they are present on glial cells and investigations of cloned receptors suggest that they are negatively coupled to $G_{i/o}$ proteins and reduce activation of adenylyl cyclase

In contrast, the 5-ht$_6$ receptor is positively coupled to G_s proteins and increases adenylyl cyclase activity. Again, the native 5-ht$_6$ receptor has not been characterised but *in situ* hybridisation suggests that its mRNA is present in the amygdala, nucleus accumbens, striatum, cortex and olfactory tubercle. Many antipsychotic agents and some antidepressant drugs show high-affinity binding to this receptor where they act as antagonists but it remains to be seen whether this contributes to their therapeutic profile. The recent development of selective antagonists for 5-ht$_6$ receptors could help to answer this question but, so far, the most promising findings are that their antagonists increase seizure threshold and could turn out to be beneficial in the treatment of epilepsy.

5-HT$_7$

Again, these receptors are positively coupled to adenylyl cyclase through a G_s protein. However, at least three splice variants are expressed in human tissue and the impact of these different isoforms on the function of these receptors is not known. Radioligand binding and the distribution of 5-HT$_7$ receptor mRNA suggest that the density of these receptors is high in the thalamus, hippocampus and hypothalamus where (in the suprachiasmatic nucleus) they are thought to synchronise circadian rhythms with the light cycle (see Chapter 22).

WHY ARE THERE SO MANY RECEPTORS?

It is obvious that strenuous efforts have been invested in the research of 5-HT receptors and, in particular, in the development of receptor-selective agonists and antagonists. All this has been done in the hope that it might be possible to control a specific 'switch' in the brain that governs a particular aspect of 5-HT function and which would be beneficial therapeutically. A further ambition is that, by avoiding activation of other 5-HT receptors, the risk of any unwanted side-effects would be eliminated. Of course, it is equally possible that reduction in non-specific receptor interactions could actually unmask some side-effects.

This approach has worked to some extent in that the 5-HT_{1A} agonist, buspirone, is an acknowledged anxiolytic drug that lacks some of the problems associated with benzodiazepines. Also, the benefits of the 5-HT_3 antagonist, ondansetron, in relieving nausea, with minimal side-effects, are undisputed. However, in other respects, this approach to drug development has been disappointing. This is probably because all the 5-HT receptors have a wide and overlapping distribution in the brain and it would be naive to suppose that any physiological response relies exclusively on the activation of any single 5-HT receptor.

What the overall physiological consequences of either an increase or decrease in 5-HT transmission in any brain region might be is beyond the scope of this chapter. However, it is certain that the diverse cocktail of 5-HT receptors in every brain region gives scope for flexibility and refinement in the 5-HT response that would not be possible if there were only the two receptors identified by Gaddum. This flexibility applies not only to the qualitative features of the response but also its duration. Another dimension of sophistication is added by the different affinities of 5-HT for each of its receptors and differences in their rates of desensitisation. An interesting discussion of how all these variables could affect overall 5-HT transmission in the brain can be found in Uphouse (1997).

WHAT DOES 5-HT DO IN THE BRAIN?

The final challenge is to define the function of 5-HT in the brain. This is not easy because the actions of drugs that target specific receptors leads us to believe that 5-HT helps to regulate: mood, anxiety, sleep, body temperature, appetite, sexual behaviour, movement, intestinal motility, cardiovascular function (central and peripheral) and nociception, at least. While a detailed explanation of the physiology of each of these functions is not possible here, and many are covered in appropriate chapters of this book, two topics are of particular interest. One is the general role of 5-HT during the waking state: this is discussed below because, in the light of recent discoveries, we might have to modify the currently accepted view. A second is the role of 5-HT in feeding, a subject to which this chapter has referred to some extent already. This will be covered here because the regulation of body weight is becoming an increasingly important research area, reflecting the growing concern about the serious health problems linked with obesity.

THE ROLE OF 5-HT IN THE WAKING STATE

One puzzle concerning 5-HT transmission in the brain, and a defining feature of 5-HT neurons in the DRN and MRN, is their slow, rhythmic firing rate of 1–2 spikes/s. In fact, this 'clock-like' discharge is even maintained *in vitro*. Certainly, electrophysiological studies *in vivo* have shown that neurons in the DRN do respond to environmental stimuli but, unlike noradrenergic neurons, they do not seem to have a role in homeostasis or the response to aversive stimuli. This is deduced from findings that the single-unit response of neurons in the DRN is not affected by a range of aversive stimuli such as environmental heat or systemic pyrogens; drug-induced changes in systemic blood pressure or glucoregulatory challenge. They are even apparently unaffected by a variety of painful stimuli. In all these cases, the activity of these neurons during the stimulus is no greater than that expressed during active waking (Jacobs and Azmitia 1992). In fact, it is generally thought that the only consistent change in response of these neurons is that reflecting changes in the sleep–waking cycle such that these neurons are maximally active during waking but can become totally quiescent during rapid eye movement (REM) sleep (see Chapter 22).

As a result of such findings, it has been suggested that 5-HT neurons in the brain are concerned merely with regulation of motor responses. Specifically, that serotonergic transmission serves to coordinate target cell responses by adjusting their excitability to match the animal's general level of arousal. In so doing, they could be responsible for gating motor output and coordinating homeostatic and sensory function (Jacobs and Azmitia 1992; Jacobs and Fornal 1999). This would be consistent with evidence that increases in the firing rate of neurons in the DRN precede an increase in arousal (see Chapter 22). This could mean that the frequency of discharge codes the state of arousal and primes target cells for forthcoming changes in the motor response to sensory inputs. It has even been claimed that 5-HT neurons projecting to the primary visual cortex are involved more in the interpretation of movement in the visual field than its qualitative features.

Evidence deemed to support this theory comes from the discovery of a population of neurons in the DRN that, unlike the majority, do not show any increase on waking and some may even reduce their firing during orientation to environmental stimuli. However, they do increase their activity during vegetative motor behaviours involving oral–buccal movements (chewing, grooming). Some are even active during anticipation of food, suggesting that they are capable of developing responses to conditioned environmental cues.

To some extent, this proposal is supported by microdialysis studies of changes in 5-HT efflux in the terminal fields of 5-HT neurons. For instance, increased 5-HT efflux in the striatum, induced by immobilisation of rats, occurs only during the period of increased motor activity that follows the animals' release (Takahashi *et al.* 1998). A single swim stress also fails to increase 5-HT efflux in the medial prefrontal cortex of rats.

However, evidence that challenges the 'motor output' theory is that a second bout of swim stress does increase 5-HT efflux (Petty *et al.* 1997). Indeed, it would be interesting to know whether this reflects any long-term influence of 5-HT-moduline on regulation of 5-HT release (see above). Also, 5-HT efflux is increased in the Raphé nuclei and their terminal fields after handling of rats (Adell, Casanovas and Artigas 1997), hypoglycaemic shock (Vahabzadeh, Boutelle and Fillenz 1995) and even during a conditioned fear response in which animals actually freeze (i.e. when their motor response is suppressed; Yoshioka *et al.* 1995). Another finding that is difficult to reconcile with 5-HT simply governing motor responses is that its efflux in the brain is increased by exposure of rats to inescapable, uncontrollable stress during which animals develop 'learned helplessness' (see Chapter 20; Petty *et al.* 1994) and yet the striking feature of this behaviour is that there is a deficit in animals' motor activity. Finally, there is compelling evidence that 5-HT transmission in the amygdala affects emotional, rather than merely motor, components of anxiety (see Chapter 19).

Overall, it remains to be seen whether or not changes in the release of 5-HT in the terminal field parallel changes in the firing rate of neurons in the Raphé nuclei. Certainly the network of hetero- and presynaptic receptors, described above, could make it feasible to adjust 5-HT release in the terminal field despite the 'clock-like' firing rate of these neurons. However, this question might be resolved by the recent discovery of a subpopulation of 5-HT neurons in the Raphé nuclei that is affected by aversive stimuli. Thus, in brain slices, taken from rats that had previously experienced stress *in vivo*, the 'stress hormone', corticotropin-releasing factor (CRF), increased the neuronal firing rate of neurons in the region of the caudal DRN/rostral MRN. It is proposed that these (presumed) 5-HT neurons form a distinct mesocorticolimbic group that, unlike neurons in other zones of the DRN and MRN, contribute to the stress response *in vivo* (Lowry *et al.* 2000). If so, this could explain the neuronal origin of the increase in 5-HT efflux in forebrain areas during stress and could suggest that certain 5-HT neurons have an important role in the stress response rather than merely governing motor activity.

THE ROLE OF 5-HT IN GOVERNING FOOD INTAKE

Obviously, regulation of food intake depends on many neurotransmitters and hormones but this final section will outline the role played by central 5-HT transmission in this process. It had been the belief for some time that increased 5-HT transmission in the brain reduces food intake (Blundell 1977) and this certainly explains the satiety in rats that follows infusion of 5-HT into the paraventricular nucleus (PVN) of the hypothalamus. However, recent studies using microdialysis have found that 5-HT efflux in the lateral hypothalamus is itself increased by food intake, suggesting the existence of a feedback control system. In fact, because the increase in 5-HT efflux is greater in genetically obese rats than in their lean counterparts, it has been proposed that there is a deficiency in the 5-HT inhibition of food intake in obesity.

The increase in food intake induced by 5-HT_{1A} receptor agonists is entirely consistent with the view that increased 5-HT transmission in the brain reduces food intake (hypophagia) since the activation of 5-HT_{1A} autoreceptors in the Raphé nuclei would ensure a reduction in 5-HT release from the nerve terminals. Of course, this explanation does not tackle the question of what role, if any, is served by postsynaptic 5-HT_{1A} receptors and, in fact, recent studies suggest that links between 5-HT_{1A} receptor activation and food intake are far more complex than originally proposed. For instance, in microdialysis studies of the lateral hypothalamus, 5-HT efflux was reduced, as expected, by systemic administration of the 5-HT_{1A} agonist, 8-OH-DPAT, in freely feeding, but not food-deprived, rats. The finding that 8-OH-DPAT reduces food intake in fasted pigs but increases it in satiated animals is even more difficult to explain. Clearly, more research is needed before these apparently incongruous findings can be reconciled.

So far, the role(s) of other 5-HT receptors in feeding seem to be more consistent and it is generally agreed that activation of 5-HT_{1B} or 5-$HT_{2A/2C}$ receptors reduces food intake. It has even been reported that stress-induced hypophagia is ameliorated by 5-HT_{2A} antagonists and that this indicates a link between abnormal 5-HT_{2A}-receptor mediated transmission in the PVN and anorexia nervosa (see Samanin and Grignaschi 1996).

A link between 5-HT transmission and hypophagia is further reinforced by the pharmacology of drugs that reduce food intake ('anorectic' agents), such as *d*-fenfluramine or sibutramine. Both these compounds have actions that should lead to increased synaptic concentrations of 5-HT in the brain, albeit through different mechanisms (see below), but whether or not this actually explains the anorectic actions of *d*-fenfluramine is controversial.

Certainly, infusion of *d*-fenfluramine into the region of the PVN reduces food intake by increasing satiety. Moreover, microdialysis studies have shown that systemic administration of this drug causes a rapid but short-lived increase in the extracellular concentration of 5-HT in this brain region. This increased efflux is due to the effects of *d*-fenfluramine on the vesicular transporter that leads to impulse-independent extrusion of 5-HT into the synapse by the terminal membrane transporter. Although this action fits well with the hypophagic effects of 5-HT, the increased 5-HT release is possibly not responsible for the effects of *d*-fenfluramine. This doubt has emerged from studies such as those showing that neither a neurotoxic lesion of 5-HT neurons in the brain, nor depletion of neuronal transmitter stores by administration of the synthesis inhibitor, *p*CPA, prevents the reduction in food intake caused by *d*-fenfluramine (see Curzon, Gibson and Oluyomi 1997). Instead, a direct effect of this compound and/or its metabolite, norfenfluramine, on 5-HT receptors has been proposed.

Another drug that reduces food intake through advancing satiety is the non-selective 5-HT/noradrenaline reuptake inhibitor, sibutramine. However sibutramine also increases thermogenesis by increasing metabolism in brown adipose tissue (BAT). This action, which contributes to loss of body weight, is thought to be mediated by an increase in activation of sympathetic afferents to BAT by central 5-HT neurons because, like sibutramine's effect on satiety, this action is prevented by administration of a peripheral ganglion blocking agent.

An important distinction between the effects of sibutramine and *d*-fenfluramine is highlighted by microdialysis studies (Heal *et al.* 1998). These show that the rate of increase in 5-HT efflux in the region of the PVN, after administration of sibutramine, is slow, progressive and long-lasting. This is because it relies on the accumulation of extracellular 5-HT following the inhibition of its reuptake after impulse-dependent release. This time-course contrasts with the rapid and transient increase in 5-HT efflux which results from the fenfluramine type of impulse-independent release from nerve terminals. In fact, this rapid increase in 5-HT release is thought to underlie the serious adverse side-effects of *d*-fenfluramine that have led to its withdrawal from the clinic.

Although the above findings confirm that 5-HT efflux is increased by anorectic drugs, the question remains as to how this leads to a reduction in food intake and, in particular, what 5-HT receptors are involved? Unfortunately, despite the development of selective 5-HT receptor ligands, this question remains unanswered. Nevertheless, it seems that *d*-fenfluramine and sibutramine increase satiety in different ways because 5-HT$_{2A/2C}$ receptor antagonists, such as ritanserin, do not block this effect of *d*-fenfluramine but they do inhibit that of sibutramine. In fact, it is still not at all certain which 5-HT receptor(s) mediate the effects of *d*-fenfluramine. Some studies suggest that 5-HT$_{1B}$ receptors are responsible but evidence for this is inconsistent. Another possibility is that it is in fact the metabolite of *d*-fenfluramine, norfenfluramine, which increases satiety through its potent activation of 5-HT$_{2C}$ receptors. Although this remains unconfirmed, 5-HT$_{2C}$ knock-out mutant mice are less sensitive to the effects of this drug on satiety than are their wild-type counterparts.

A final, important distinction between sibutramine and *d*-fenfluramine is that the actions of the former, but not the latter, rest on its modification of both 5-HT and noradrenergic transmission. Thus, the reduction in food intake by sibutramine is partially blocked by α_1- or β_1-adrenoceptor antagonists as well as 5-HT$_{2A/2C}$ or 5-HT$_{2B/2C}$ antagonists. In fact, there appears to be a synergistic interaction between these two transmitter systems. This is illustrated by a study of the effects of the selective serotonin reuptake inhibitor, fluoxetine, and the selective noradrenaline uptake inhibitor, nisoxetine on food intake and BAT thermogenesis. Using doses of these drugs that were ineffective on either measure when given alone, they did increase both satiety (Jackson *et al.* 1997) and BAT thermogenesis when given in combination. The 'hard-wiring' of this 5-HT/noradrenaline interaction is currently being researched.

Clearly, this is a complex physiological system, involving multiple feedback and feedforward pathways that interacts with other 'food-factors' (e.g. NPY, leptin, orexins, corticosteroids and other monoamines) and it will take a great deal of research to unravel all these networks. However, the progress that has been made so far underlines the crucial role played by 5-HT in this process and reinforces the view that drugs targeting specific 5-HT receptors could turn out to be effective anti-obesity agents that lack the adverse effects of 5-HT releasing agents.

REFERENCES

Adell, A, Casanovas, JM and Artigas, F (1997) Comparative study in the rat of the actions of different types of stress in the release of 5-HT in the raphe nuclei and forebrain areas. *Neuropharmacology* **36**: 735–741.

Barnes, NM and Sharp, T (1999) A review of central 5-HT receptors and their function. *Neuropharmacology* **38**: 1083–1152.

Blakely, RD, Ramamoorthy, S, Schroeter, S, Qian, Y, Apparsundaram, S, Galli, A and DeFelice, LJ (1998) Regulated phosphorylation and trafficking of antidepressant-sensitive serotonin transporter proteins. *Biol. Psychiatry* **44**: 169–178.

Blundell, JE (1977) Is there a role for serotonin (5-hydroxytryptamine) in feeding? *Int. J Obesity* **1**: 15–42.

Boadle-Biber, MC (1993) Regulation of serotonin synthesis. *Prog. Biophys. Molec. Biol.* **60**: 1–15.

Curzon, G, Gibson, EL and Oluyomi, AO (1997) Appetite suppression by commonly used drugs depends on 5-HT receptors but not on 5-HT availability. *Trends Pharmacol. Sci.* **18**: 21–25.

Gaddum, JH and Picarelli, ZP (1957) Two kinds of tryptamine receptor. *Brit. J. Pharmacol.* **12**: 323–328. (Reproduced in *Brit. J. Pharmacol.* **120** (Suppl): 134–139.)

Gillman, PK (1999) The serotonin syndrome and its treatment. *J. Psychopharmacol.* **13**: 100–109.

Heal, DJ, Cheetham, SC, Prow, MR, Martin, KF and Buckett, WR (1998) A comparison of the effects on central 5-HT function of sibutramine hydrochloride and other weight-modifying drugs. *Brit. J. Pharmacol.* **125**: 301–308.

Jackson, HC, Needham, AM, Hutchins, LJ, Mazurkiewicz, SE and Heal, DJ (1997) Comparison of the effects of sibutramine and other monoamine reuptake inhibitors on food intake in the rat. *Brit. J. Pharmacol.* **121**: 1758–1762.

Jacobs, BL and Azmitia, EC (1992) Structure and function of the brain serotonin system. *Physiol. Rev.* **72**: 165–229.

Jacobs, BL and Fornal, CA (1999) Activity of serotonergic neurons in behaving animals. *Neuropsychopharmacology* **21**: 9S–15S.

Krebs-Thomson, K, Paulus, MP and Geyer, MA (1998) Effects of hallucinogens on locomotor and investigatory activity and patterns: influence of 5-HT_{2A} and 5-HT_{2C} receptors. *Neuropsychopharmacology* **18**: 339–351.

Leibowitz, SF, Alexander, JT (1998) Hypothalamic serotonin in control of eating behavior, meal size, and body weight. *Biol. Psychiatry* **44**: 851–864.

Lowry, CA, Odda, JE, Lightman, SL and Ingram, CD (2000) Corticotropin-releasing factor (CRF) increases the *in vitro* firing rates of serotonergic neurones in the rat dorsal raphé nucleus: evidence for selective activation of a topographically organised mesolimbocortical system. *J. Psychopharmacol.* **14** (Suppl 2): A72.

Massot, O, Rousselle, JC, Grimaldi, B, Cloez-Tayarani, I, Fillion, MP, Plantefol, M, Bonnin, A, Prudhomme, N and Fillion, G (1998) Molecular, cellular and physiological characteristics of

5-HT-moduline, a novel endogenous modulator of 5-HT1B receptor subtype. *Ann. NY Acad. Sci.* **861**: 174–182.

McQueen, JK, Wilson, H, Sumner, BEH and Fink, G (1999) Serotonin transporter (SERT) mRNA and binding site densities in male rat brain affected by sex steroids. *Molec. Brain Res.* **63**: 241–247.

Meltzer, HY (1999) The role of serotonin in antipsychotic drug action. *Neuropsychopharmacology* **21** (Suppl 2): 106S–115S.

Petty, F, Kramer, G, Wilson, L and Jordan, S (1994) *In vivo* serotonin release and learned helplessness. *Psychiatry Res.* **52**: 285–293.

Petty, F, Jordan, S, Kramer, GL, Zukas, PK and Wu, J (1997) Benzodiazepine prevention of swim-stress induced sensitization of cortical biogenic amines: an *in vivo* microdialysis study. *Neurochem. Res.* **22**: 1101–1104.

Povlock, SL and Amara, SG (1997) The structure and function of norepinephrine, dopamine and serotonin transporters. In: *Neurotransmitter Transporters.* Humana Press, New York, pp 1–28.

Ramamoorthy, S and Blakely, RD (1999) Phosphorylation and sequestration of serotonin transporters differentially modulated by psychostimulants. *Science* **285**: 763–766.

Rouch, C, Nicolaidis, S and Orosco, M (1999) Determination, using microdialysis, of hypothalamic serotonin variations in response to different macronutrients. *Physiol. Behav.* **65**: 653–657.

Rudnick, G (1997) Mechanism of biogenic amine neurotransmitter transporters. In: *Neurotransmitter Transporters.* Humana Press, New York, pp 1–28, 73–100.

Samanin, R and Grignaschi, G (1996) Role of 5-hydroxytryptamine receptor subtypes in satiety and animal models of eating disorders. In *Drug Receptor Subtypes and Ingestive Behaviour* (Eds Cooper, SJ and Clifton, PG), Academic Press, London, pp. 39–58.

Semple, DM, Ebmeier KP, Glabus MF, O'Carroll, RE and Johnstone, EC (1999) Reduced in vivo binding to the serotonin transporter in the cerebral cortex of MDMA ('ecstasy') users. *Br. J. Psychiatry* **175**: 63–69.

Siuciak, JA, Clark, MS, Rind, HB, Whittemore, SR and Russo, AF (1998) BDNF induction of tryptophan hydroxylase mRNA levels in the rat brain. *J. Neurosci. Res.* **52**: 149–158.

Sprague, JE, Everman, SL and Nichols, DE (1998) An integrated hypothesis for the serotonergic axonal loss induced by 3,4-methylenedioxymethamphetamine. *Neurotoxicology* **19**: 427–442.

Stock, MJ (1997) Sibutramine: a review of the pharmacology of a novel anti-obesity agent. *Int. J. Obesity* **21** (Suppl 1): S25–S29.

Takahashi, H, Takada, Y, Nagai, N, Urano, T and Takada, A (1998) Extracellular serotonin in the striatum is increased after immobilisation stress only in the nighttime. *Behav. Brain Res.* **91**: 185–201.

Uphouse, L (1997) Multiple serotonin receptors: too many, not enough, or just the right number? *Neurosci. Biobehav. Rev.* **21**: 679–698.

Vahabzadeh, A. Boutelle, MG and Fillenz, M (1995) Effects of changes in rat brain glucose on serotonergic and noradrenergic neurons. *Eur. J. Neurosci.* **7**: 175–179.

Wurtman, RJ and Wurtman, JJ (1995) Brain serotonin, carbohydrate-craving, obesity and depression. *Obesity Res.* **3**: 477S–480S.

Yoshioka, M, Matsumoto, M, Togashi, H and Saito, H (1995) Effects of conditioned fear stress on 5-HT release in the rat prefrontal cortex. *Pharmacol. Biochem. Behav.* **51**: 515–519.

10 Amino Acids: Excitatory

A. H. DICKENSON

INTRODUCTION

For many years the amino acid glutamate (Fig. 10.1) has been known to play the major role in the transmission of excitatory signals via long axonal projections of neurons in the central nervous system. In fact, of the billions of long-axon neurons in the central nervous system, the majority use glutamate as their principal transmitter as do excitatory intrinsic neurons. A large proportion of peripheral sensory fibres conveying touch- and pain-related information contain glutamate and aspartate as do visual, auditory and other sensory afferent fibres. This is also the case for neurons in the CNS linking different areas of the brain and spinal cord. Due to the metabolic role of glutamate and the fact that it is the precursor for GABA, the inhibitory amino-acid, precise localisation studies have been fraught with difficulties. However, both release studies and, more importantly, electrophysiological recordings have shown that glutamate functions as a transmitter at many synapses. In the case of C-fibres, the co-existence of glutamate with peptides such as substance P and/or CGRP would make it highly likely that a noxious stimulus releases both peptides and excitatory amino acids from the afferent nociceptive fibres. Here the coincident actions of glutamate in concert with peptides have a functional importance that is discussed later.

While aspartic acid (aspartate) is also found in the CNS and has excitatory effects on neurons, little is known of its precise location and action although it may be released from intrinsic neurons and hippocampal pathways. It will not be discussed further.

NEUROCHEMISTRY

Due to the major role of glutamate, not only as a component of proteins but also as a key step in intermediate metabolism, the production and metabolism of the amino acid are compartmentalised in neurons. It may be that the transmitter pool of glutamate uses the amino acid from any source given that it can be produced from such diverse origins as glucose, aspartate, glutamine and oxoglutarate. Once release occurs there are high-affinity uptake sites in both terminals and glia that remove the transmitter from the synaptic cleft (Fig. 10.2).

These points have important functional implications. While neuronal glutamate may come from glucose via pyruvate, the Krebs cycle and transamination of alpha-oxoglutamate, it seems likely that most of the transmitter originates from the deamination of glutamine. After release, the high-affinity uptake sites (transporters)

Neurotransmitters, Drugs and Brain Function. Edited by R. A. Webster

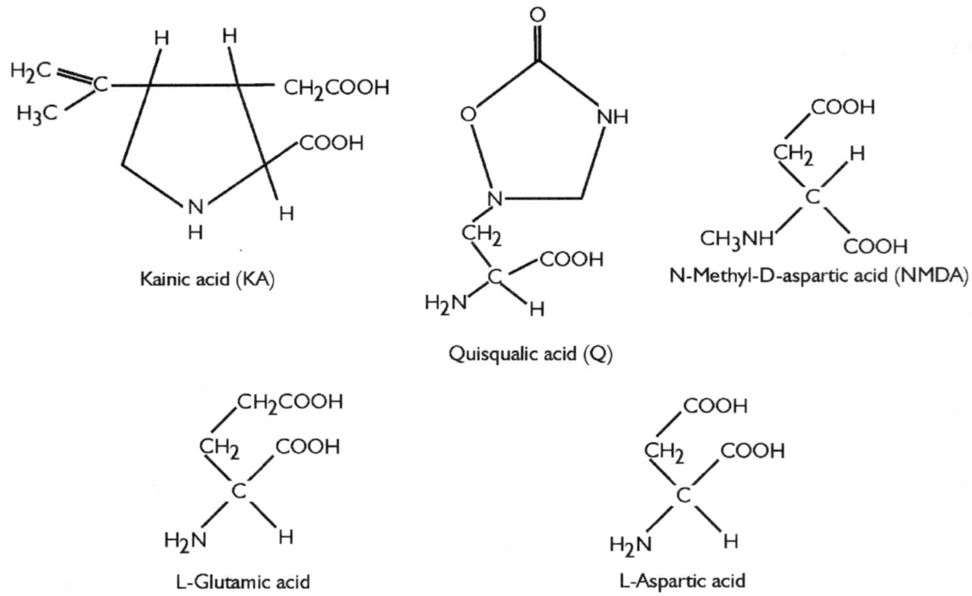

Figure 10.1 Structures of the transmitters and synthetic agonists at the various receptors for glutamate

remove glutamate from the synapse, partly back into the nerve terminal or more probably into adjacent glial cells. In the latter, it is converted by glutamine synthetase into glutamine which then passes back into the extracellular fluid (the CSF levels are high, of the order of 0.5 mM) to be taken up by the glutamate nerve terminal. Here it is deaminated to neurotransmitter available glutamate by mitochondrial glutaminase. Thus there is a conservation rather than a net synthesis of glutamate. This complex but very general biochemical process provides very little opportunity for drug modification of glutamate synthesis or metabolism.

Unlike other transmitter systems, there are no obvious mechanisms for dampening glutamate release. Presynaptic autoreceptors for glutamate are mostly of the kainate type (see below) and appear to act as positive rather than negative influences on further release of the amino acid. Although poorly characterised at present, inhibitory autoreceptors of the metabotropic type of receptors may act to inhibit release of glutamate.

RECEPTORS—STRUCTURE AND FUNCTION

The extensive early literature on the idea of important excitatory roles for amino acids originated in work in the 1950s by Curtis and his group in Australia from iontophoretic studies which showed powerful excitatory effects of a number of dicarboxylic amino acids on a variety of CNS neurons. Thus despite any knowledge at that time of the receptors or the availability of selective antagonists, important roles were proposed for these transmitters in neuronal function. These early predictions have been confirmed by more recent approaches using selective agonists and antagonists which have now allowed the separation of the

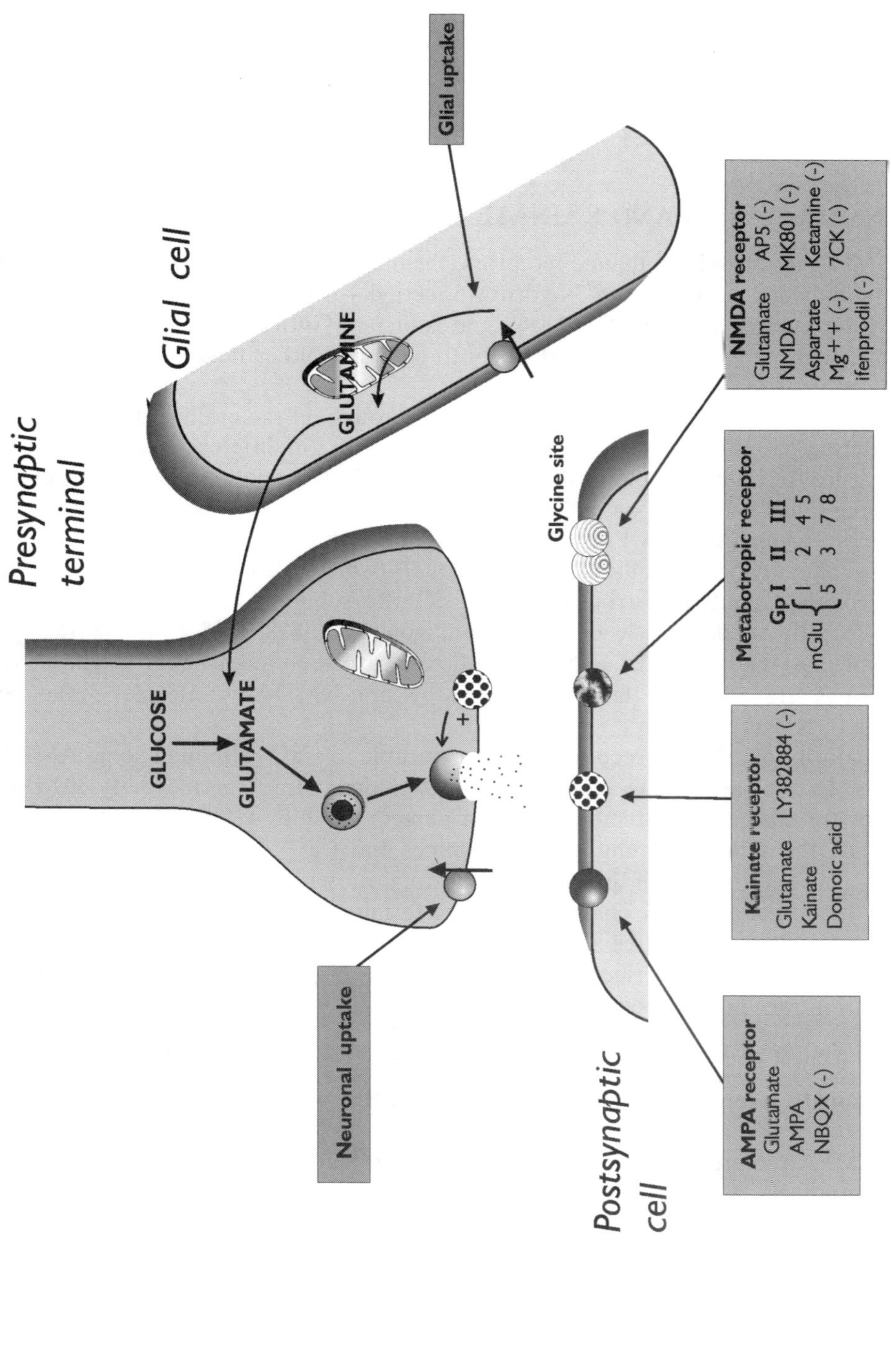

Figure 10.2 Site of action of drugs affecting glutamate synapses

receptors for glutamate into four main types. Block of a physiological response by an antagonist is good evidence for a functional role of any transmitter in CNS events and this has now been achieved for glutamate in many areas of the CNS.

The nomenclature for the glutamate receptors is confusing. Originally, the receptors were called N-methyl-D-aspartate (NMDA) and non-NMDA with the latter later subdividing into quisqualate and kainate. Now, the accepted classification is into AMPA, kainate, NMDA and metabotropic. This latter class of receptor is further divided into three groups (I, II and III) containing at least two subtypes. Figure 10.1 shows the agonists at the receptors.

NON-NMDA—AMPA AND KAINATE RECEPTORS

Non-NMDA ionotropic glutamate receptors (the majority sodium channel containing) can be subdivided into α-amino-3-hydroxy-5-methyl-4-isoxazole propionate (AMPA) (comprising cloned subunits GluR1–4) and kainate (GluR5–7, KA1–2) preferring receptors, with native receptors most likely to comprise either homo- or heteromeric pentamers of these subunits.

mRNA coding for GluR1–4 subunits is found throughout the brain and spinal cord, with differing patterns of expression of GluR2 and GluR1 in different regions. There is also evidence for both presynaptic AMPA and particularly kainate-preferring receptors comprising GluR5 subunits on neuronal terminals in various areas of the CNS.

There are competitive AMPA receptor antagonists (Fig. 10.3) of which NBQX (6-nitro-7-sulphamoylbenzo(f)quinoxaline-2,3-dione) (which also displays micromolar affinity for the kainate-preferring GluR5 and GluR6 subunits3) is the most selective, and the recently developed selective GluR5 antagonist LY382884 [3S,4aR,6S,8aR-6-(4-carboxyphenyl)-methyl-1,2,3,4,4a,5,6,7,8,8a-deca-hydroisoquinoline-3-carboxylic acid]. These drugs are allowing the roles played by non-NMDA ionotropic glutamate receptors to be gauged.

The majority of AMPA receptors are impermeable to Ca^{2+}, although some AMPA receptors, as well as kainate receptors, have significant Ca^{2+} permeability. AMPA receptors are multimeric assemblies of four cloned subunits, GluR1–4, but it is the absence of the GluR2 subunit that determines the Ca^{2+} permeability of AMPA receptors, since editing out of this subunit following transcription into mRNA results in the introduction of a positive charge in the pore-forming region (Q/R site), which is not present in GluR1,3 or 4, 9 (see also AMPA receptors in Chapter 3). AMPA receptors lacking GluR2 have Ca^{2+} permeability ratios up to $P_{Ca}/P_{Na} = 3$. Since calcium is such a ubiquitous intracellular messenger any receptor that allows this ion to enter neurons is likely to be important in plasticity in the CNS. In a similar manner to the AMPA receptor, RNA editing in the pore region at the Q/R site controls the Ca^{2+} permeability of the kainate receptor subunits GluR5 and GluR6, with significant levels of the unedited (Ca^{2+} permeable) version of these receptors present in the adult CNS.

The AMPA receptor subunits are all found within many regions of the CNS but in differing numbers, and, in the spinal cord, have differing lamina distributions. GluR1 and GluR2 are generally the most abundant AMPA receptor subunits with lower levels of GluR3 and GluR4. The majority of AMPA receptors allow Na^+ to enter neurons and thus in most areas of the CNS studied, the initial stage in excitatory synaptic transmission is a fast-depolarising response due to the release of glutamate and subsequent activation of the AMPA receptor.

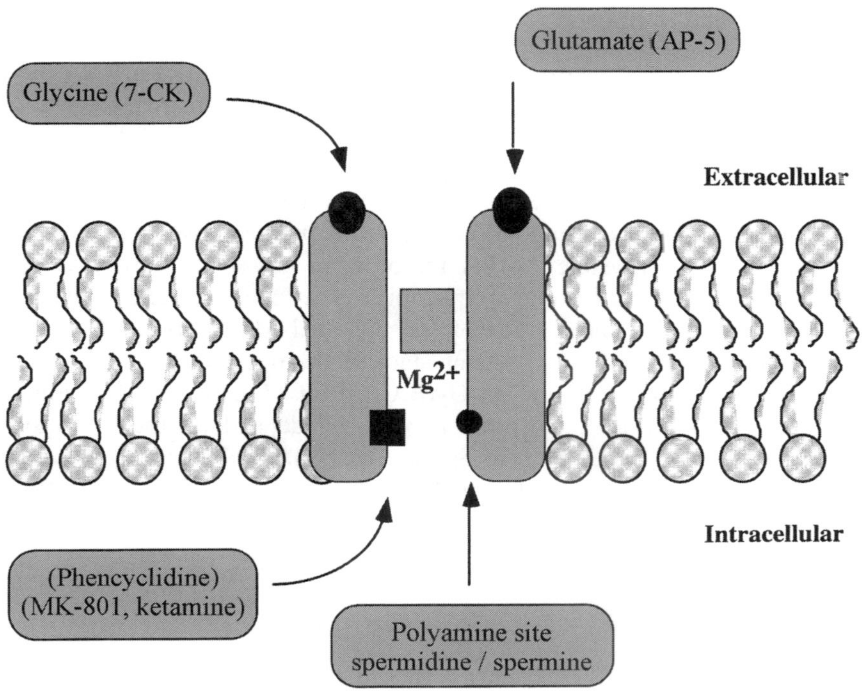

Figure 10.3 Structure of the NMDA receptor–channel complex. The receptor has a complicated structure and this is highlighted by the presence of many pharmacologically distinct binding sites through which the receptor activity can be modulated. The channel associated with the receptor is blocked by Mg^{2+} at resting potential (−70 mV). Receptor activation requires the removal of this Mg^{2+} block (voltage-gated) as well as the binding of glutamate and the co-agonist, glycine (ligand-gated). The different binding sites (glutamate, phencyclidine, polyamine, glycine) are illustrated, and together with antagonists which act at the various sites (in parentheses). The polyamine site is an intracellular site which modulates the affinity of other agonists and antagonists

mRNA coding for the kainate receptor subunits GluR5 and GluR7 is also found in isolated neurons in the CNS although many kainate GluR5 receptors are thought to be located presynaptically on terminals of neurons that release glutamate. Kainate receptors are therefore thought to be excitatory autoreceptors that enhance the release of glutamate. It could be predicted that the widespread distribution of AMPA receptors precludes the use of antagonists at this receptor in therapy since adverse effects are highly likely. By contrast, the kainate receptor might be an interesting target since its functional role will be linked to the level of glutamate release. Thus, antagonists at this receptor should reduce excessive glutamate release while having less effect on more normal functional synapses.

The role of the kainate receptor system in the brain is at an early stage since there are as yet few pharmacological tools to study its function. However, mutations in the kainate receptor genes have been made in mice and there is a GluR6 kainate receptor knock-out mouse. Kainate binding is absent in areas of the brain which normally have high levels such as the hippocampus. Here, in normal animals kainate receptors mediate a postsynaptic current which is absent in the GluR6 knock-out mouse. The mice have reduced motor activity but can learn maze tasks. The knock-out mouse is resistant to kainate-induced seizures.

Studies have shown that neurons expressing high levels of GluR1 mRNA but lacking GluR2 are found in the superficial laminae of the spinal cord, an area where nociceptive primary afferents terminate, suggesting that a subpopulation of AMPA receptors in this region may have significant Ca^{2+} permeability. Calcium-permeable non-NMDA receptors have been demonstrated in spinal cord slices using kainate-induced cobalt loading. Studies performed using cultured neurons *in vitro* have suggested that Ca^{2+} entry through Ca^{2+}-permeable AMPA receptors in the spinal cord may provide a mechanism for the strengthening of transmission at synapses and enhancement of nociceptive transmission. Other studies have suggested a link between Ca^{2+}-permeable AMPA receptors and inhibitory systems since in the dorsal horn of the spinal cord many of these receptors are found on GABA neurons. Clearly, the functional role of Ca^{2+}-permeable non-NMDA receptors *in vivo* will depend on their location in the integrated circuitry of the CNS. Joro Spider Toxin (JSTx) has been reported to be a selective blocker of Ca^{2+}-permeable non-NMDA responses evoked by AMPA/kainate rather than those evoked by NMDA and so will be a useful tool for studying the roles of these receptors.

NMDA RECEPTORS

Much attention has been focused on the role of the N-methyl-D-aspartate (NMDA) receptor for glutamate, activation of which produces slow prolonged neuronal depolarisation. Thus unlike the AMPA receptor, it is not responsible for the fast transmission of excitation nor the initiation of impulses but has been shown to be critical for maintaining excitatory responses such as the manifestation of wind-up in spinal cord, long-term potentiation in the hippocampus, epileptiform activity and in neuroexcitotoxicity. Mechanisms of central amplification of a nociceptive input have been suggested to underlie aspects of the enhanced spinal transmission of nociceptive messages in protracted pain states, and in this case there is good clinical evidence to support the concepts that have arisen from animal studies.

The NMDA receptor has a heteromeric structure composed of two subunit types; NR1 and NR2, the latter having four subunits (NR2A–NR2D) (Fig. 10.2). Molecular genetic techniques have demonstrated that native NMDA receptors are likely to be composed of a combination of the NR1 subunit (which can exist in eight different splice variants) and one or more of the four NR2 subunits which are the main determinants of functional diversity among the NMDA receptors (see Chapter 3 for further details). It has been shown that there are distinct developmental and spatial expression patterns of NMDA receptor NR1 subunit splice variants and NR2 receptor subunits in the CNS.

Although the exact subunit stoichiometry is not yet known for any NMDA receptor, heterologous expression studies suggest that they are likely to be tetramers composed of two NR1 subunits and two NR2 subunits providing the possibility for considerable structural diversity of NMDA receptors. The subtypes have been partially mapped in the CNS and show differing regional distributions. As the subunit composition imparts different physiological characteristics to the receptor, this would imply that different functional roles of the NMDA receptor could be separated—at present there are only antagonists for the NR2B subtypes. The NMDA receptor is a non-specific cation channel in that both sodium and calcium enter, but the latter ion appears to be the predominant factor in the alteration of neuronal activity. This is not simply due to the large amounts of calcium that enter neurons and thus the degree of excitability that ensues but also simply that many intracellular pathways are calcium dependent.

The NMDA receptor is a complex entity. Functional modulation of the receptor can be achieved through actions at various recognition sites including the primary transmitter site, the site where glutamate binds (competitive), the phencyclidine (PCP) site (uncompetitive) situated in the channel of the receptor, the polyamine modulatory site and the strychnine-insensitive glycine site where glycine is a required co-agonist with glutamate (Fig. 10.3). Potentially, there are several ways in which the effect of released glutamate can be antagonised through NMDA receptor blockade. Numerous studies have investigated the potential use of antagonists acting through the different recognition sites. However, due to the ubiquitous nature of the receptor, it has often been difficult to achieve therapeutic effects at the target organ, in the absence of adverse side-effects. Prototypical antagonists for the various sites are shown in Fig. 10.3 — namely AP5 for the receptor, MK-801 for the channel (although the clinically used drugs ketamine and memantine also act at this site) and then there are other agents such as 7-chlorokynurante for the associated glycine site.

Alterations in the transmission of neuronal information via NMDA receptors arise due to two main factors, the first being that the calcium influx through the channel produces large depolarisations, and the second due to the unique profile of the receptor–channel complex, which requires various conditions for operation, and therefore is not necessarily involved in synaptic transmission at all times and under all circumstances. The release of the excitatory amino acids is obviously needed but in addition, glycine is required as a co-agonist, and this is of pharmacological and therapeutic interest, as antagonists of this site can produce inhibitions of NMDA-mediated events. The latter condition would appear to be ever present, due to the levels of glycine available in the brain and spinal cord. However, for the action of glycine on NMDA receptors to take on a clear physiological role, the concentration of glycine present at the synapse must normally be kept below saturating levels. Although it is not exactly clear if this occurs, it is thought that a glycine transporter is involved, whose distribution closely matches that of NMDA receptors in the CNS, so the glycine concentration present at glutamate synapses may be regulated by glycine uptake. The key role of glycine in activation of the receptor is borne out by the ability of antagonists at this site to produce inhibitions of NMDA-mediated transmission. Finally, an induced depolarisation of the neuron to relieve the resting voltage-dependent magnesium block of the channel is a prerequisite for activation of the complex. For these reasons, the NMDA receptor–channel complex is not a participant in 'normal' synaptic transmission, but when the correct conditions are achieved the complex will rapidly become activated and add a powerful depolarising or excitatory drive to synaptic transmission.

The NMDA receptor is an ionotropic receptor coupled to a cation channel, which is blocked by physiological levels of Mg^{2+} at the resting membrane potential — the sensitivity to magnesium block depends on the subunit composition as does the glycine sensitivity. The channel is blocked in a voltage-dependent manner so the receptor can only operate after sufficient repeated depolarisation. In the spinal cord, the removal of the Mg^{2+} block is mediated by peptides, including tachykinins, which are co-released with glutamate. After a brief acute stimulus, pain transmission from C-fibres is largely mediated by the action of glutamate on AMPA/kainate receptors. When the stimulus is sustained or its intensity is increased, however, the action of substance P on NK-1 receptors produces sufficient membrane depolarisation so that the Mg^{2+} block can now be removed and the NMDA receptor activated. These events underlie central hyperexcitability and result in a significant amplification of the response. Substance P therefore plays an important role in this instance in recruiting NMDA receptors and contributes to the cascade of events leading to the enhancement and prolongation of the neuronal response. In other CNS areas, the NMDA receptor may be allowed to participate in synaptic events by glutamate acting on AMPA receptors. How AMPA/kainate receptors provide an excitatory drive of sufficient length to remove the block of the NMDA receptor channel while being fast ionotropic receptors is unclear. However, different subtypes of the NMDA receptor have differing sensitivity to both glycine and magnesium and the particular channel openings vary in both amplitude and duration. Thus regional specific conditions may control the receptor and determine its properties.

The NMDA receptor is therefore unique in that it is not simply ligand-gated but also voltage-gated due to the channel block imparted by magnesium. No other receptor requires two ligands (e.g. glutamate and glycine) for receptor activation.

METABOTROPIC RECEPTORS

This fourth type of receptor for glutamate (mGluRs), so named as they are members of the seven transmembrane-spanning family, is the least well understood. The poor understanding of this class of receptor stems from the fact that there are eight receptors in the class which fall into three groups, divided by sequence homology, effector mechanisms and, to some extent, their pharmacology. We are presently lacking sufficiently potent and selective antagonists at all these metabotropic glutamate receptors to probe their roles. They are coupled through G-proteins to potassium and calcium channels and while Group I (mGluR 1 and 5) receptors interact with IP3 systems, both Group II and III inhibit adenyl cyclase. Thus broadly, the Group I receptors are therefore excitatory and Groups II and III are inhibitory. There is some evidence for both pre- and postsynaptic locations of all groups of receptors. Functionally, the mGluRs have been implicated in memory, pain, anxiety and neurodegeneration with few specific details due to the lack of antagonists.

FUNCTIONAL ROLES

EPILEPSY

There is much evidence that both the initiation and maintenance of epileptic seizures involves the release of glutamate even though there is clear evidence that reduced GABA function may be equally impaired. Drugs that block NMDA receptors are anticonvulsant experimentally whereas the clinically effective antiepileptic drug lamotrigine reduces glutamate release as part of its action (see Chapter 16). As yet no NMDA receptor antagonists have been tested clinically.

PAIN

The excitatory amino acids are found in most sensory fibres of both large- and small-diameter fibres and, in the latter, they are co-localised with peptides such as substance P. The co-existence of these two transmitters suggests that they are released together in response to a noxious stimulus and hence contribute to the transmission of pain. While AMPA receptors are activated in response to brief acute stimuli and are involved in the fast events of pain transmission, NMDA receptors are only activated following repetitive noxious inputs, under conditions where the stimulus is maintained (for more details see Chapter 21). NMDA receptors have been implicated in the spinal events underlying 'wind-up', whereby the responses of dorsal horn neurons are significantly increased after repetitive C-fibre stimulation despite the constant input. Thus the activation of this class of receptors brings about a marked increase in neuronal excitability and is responsible for the amplification and prolongation of neuronal responses in the spinal cord. Substantial evidence exists for the involvement of NMDA receptors in various pathological pain states. Studies have demonstrated the effectiveness of NMDA receptor antagonists in animal models of inflammation, neuropathic pain, allodynia and ischemia. Both pre- and postsurgical administration of antagonists were shown to be effective, suggesting that the induction and maintenance of these ongoing pain states are dependent on NMDA receptor-mediated events.

Neuropathy may produce a prolonged activation of NMDA receptors, due to a sustained afferent input to the spinal cord, and this may result in a relatively small but continuous increase in the extracellular level of glutamate. As ketamine is a licensed drug which use-dependently blocks the NMDA receptor channel, the positive effects of this drug in patients with neuropathic pain (although not without side-effects) would strongly suggest that the NMDA receptor is as important in sensory processing of painful events in humans as suggested by the animal work.

MEMORY

It is generally accepted that long-term potentiation (LTP) is a key event in the processes that lead to the laying down of memories in the brain. LTP is a long-lasting enhancement of synaptic effectiveness that follows certain types of tetanic electrical stimulation to input pathways into the hippocampus. Although much of the work has been based on the hippocampus, where it was first documented, LTP has also been described in areas such as cortex, amygdala and spinal cord. The consensus would be that synaptic activation during high-frequency stimulation triggers a series of intracellular events that lead to the expression of synaptic potentiation with the release of glutamate being the first step. This persistent increase in synaptic efficacy is thought then to be critical for memory, presumably the acquisition. Much of the vast literature is based on electrophysiology, mainly *in vitro*, and so despite the conceptual appeal of LTP, the functional studies find it much harder to link LTP with memory. From a large clinical literature, the hippocampus appears to be a key structure in memory, and blocking glutamate receptors causes reduced memory-like behaviour in animals. Also the more recent description of activity-dependent long-term depression (LTD) could be associated with the processes of forgetting. However, this may be overly simple since LTD also occurs in the cerebellar cortex and might contribute to the motor aspects of learning in animals. LTD is reversibly blocked by NMDA receptor antagonists which suggests that postsynaptic Ca^{2+} entry through the NMDA receptor channel is critical for LTD induction.

Nearly all mechanistic studies of LTP have been carried out in the CA1 region of hippocampal slices, where Schaffer collateral/commissural fibres make monosynaptic contacts with the dendrites of CA1 pyramidal cells. It is generally accepted that the

Table 10.1 Classification of NMDA receptors

	Metabotropic	Ionotrophic		
		NMDA	AMPA	Kainate
Channel		Sodium calcium	Sodium*	Sodium*
Endogenous agonist	Glutamate	Glutamate	Glutamate	Glutamate
Other agonists	Quisqualate	NMDA	AMPA Kainate	Kainate
Antagonists	L-AP3	MK-801 Memantine Ketamine Dextrophan CPP 7-CK (glycine site)	CNQX NBQX	LY382884

Abbreviations: N-methyl-D-aspartate (NMDA), alpha-amino-3-hydroxy-5-methyl-4-isoxazole-propionic acid (AMPA), L(+)-2 amino-3-phosphonopropionic acid (L-AP3), 6-cyano-7-nitroquinoxaline (CNQX), 2,3-dihydroxy-6-nitro-7-sulfamyl-benzo-f-quinoxaline (NBQX), 3-(2-carboxypiperazin-4-yl)-propyl-1-phosphonic acid (CPP), 7 Chlorokynureic acid (7-CK).

*Some AMPA and kainate receptors are calcium permeable. Some of the antagonist structures are shown in Fig. 10.4.

induction of LTP at this synapse requires activation of postsynaptic NMDA receptors by synaptically released glutamate during adequate postsynaptic depolarisation. This results in the relief of the voltage-dependent block of the NMDA receptor channel by Mg^{2+}. Ca^{2+} then enters the postsynaptic neurons or dendrite as a necessary and perhaps sufficient trigger for LTP. Although the NMDA receptor channel may be the critical entry point for the Ca^{2+} involved in triggering LTP, activating voltage-dependent calcium channels during NMDA receptor blockade can also cause an increase in synaptic efficacy. Once induction of LTP has occurred, the maintenance of LTP is then non-NMDA receptor dependent, favouring the idea of intracellular mechanisms as key factors.

It has also been suggested that activation of mGluRs enhances NMDA receptor-mediated LTP and there is also good evidence that switching off GABA mechanisms is also a prerequisite. One of the most controversial areas in the study of the mechanisms of LTP has been the search for a so-called retrograde messenger, a factor that is released from the postsynaptic neuron and diffuses back across the synapse to modify neurotransmitter release from the presynaptic terminal. The necessity for the existence of such a messenger was first suggested by the finding that LTP was associated with an increase in the concentration of glutamate in perfusates. Although there is data that the candidate retrograde messenger could be arachidonic acid, most recent work indicates the gas NO (nitric oxide), although there is almost as much evidence against as for this molecule (see Chapter 13). Postsynaptic NMDA receptors are involved in the induction of both LTP and both forms of plasticity appear to need retrograde messengers, and use common intracellular events—what occurs when these mechanisms converge will then determine whether neurons become potentiated or depressed.

On the basis of the events that occur in pain and LTP, it is easy to see how the actions of glutamate relate to the excessive firing of neurons—as yet no NMDA receptor antagonist has been tested in human epilepsy.

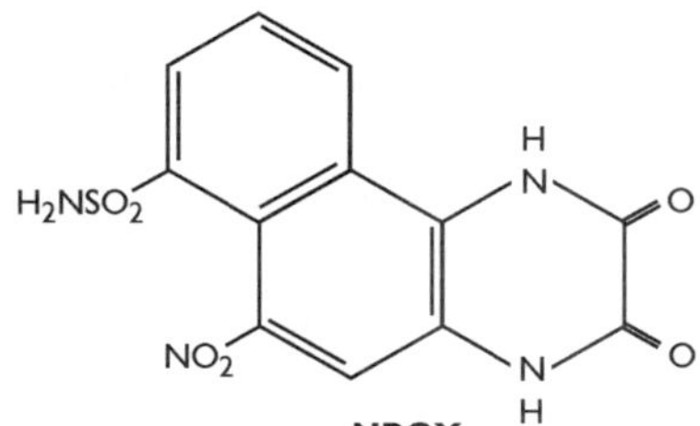

NBQX

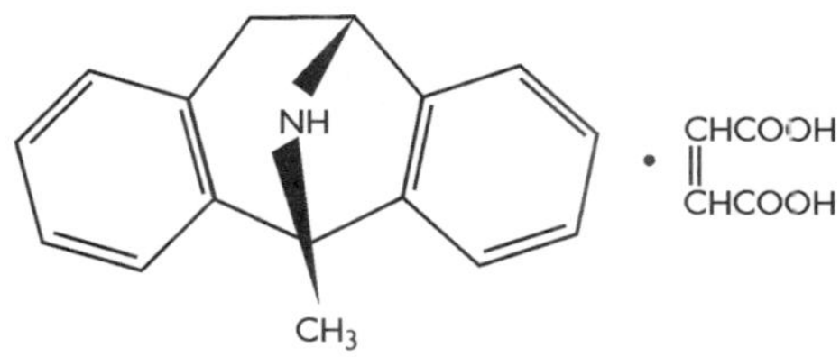

MK801 (dizocilpine)

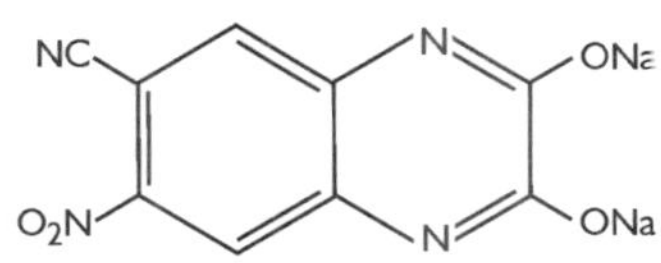

CNQX

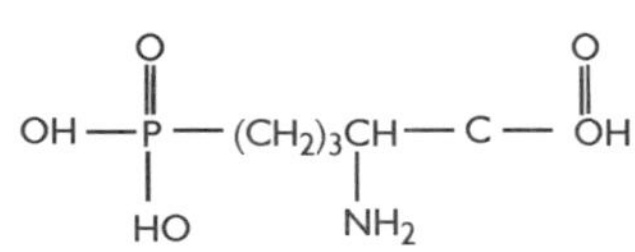

AP-5

Figure 10.4 Structures of some antagonists at the various receptors for glutamate. CNQX is an AMPA antagonist but NQQX has greater selectivity. AP-5 is an NMDA receptor antagonist while MK-801 blocks the NMDA receptor channel (non-competitive)

EXCITOTOXICITY

The final issue relating to the function of the NMDA receptor is excitotoxicity. Briefly, the depolarisation may drive neurons into a state where large quantities of calcium enter the neuron. For this to occur, the release of glutamate would have to be excessive and in this context, cerebral ischaemic episodes are thought to disrupte the reuptake of glutamate into neurons and glia. The consequent influx of calcium, if excessive, can bring water into the neuron as a result of the cation entry. These osmotic changes can then lead to swelling and damage to the cell, although if the neuronal activity is reduced, then the osmotic stress is reversible. A second delayed phase of neuronal damage then occurs in that intracellular signalling is driven by the high calcium levels leading to permanent destruction of the neuron. A number of culprits have been identified, including activation of kinases, phospholipases leading to the generation of arachidonic acid and free radicals, nitric oxide synthase and also lipases and proteases.

The overactivation of glutamate receptors is therefore thought to be a key initial step in the neuronal and glial cell loss following cerebral vascular accidents. Despite this, the trials of NMDA receptor antagonists in patients with brain ischemia had so far been disappointing with poor efficacy and marked side-effects. Both factors could be improved by targeting NMDA receptor subtypes but it may be that AMPA and kainate receptors also have key roles in excitotoxicity. Another issue is that even when NMDA receptors are blocked the influx of calcium through voltage-gated calcium channels may induce neuronal damage.

It is unclear as to what extent events such as these are responsible for the cell death seen in neurological disorders like Parkinson's, Huntington's and Alzheimer's diseases. However, a combination of motoneuron disease, dementia and a Parkinson-like syndrome was possibly triggered by a constituent of the cyclad seed, used in Guam in times of famine for which the most likely candidate appears to be an excitatory amino-acid agonist. Certainly, there is evidence for a defect in mitochondrial energy metabolism in PD which may lead to neuronal depolarisation and an easier removal of the voltage-dependent Mg^{2+} block of the NMDA receptor. The resulting excessive neuronal excitation may contribute to nigrostriatal cell death. Whatever the case, once PD is established, the corticostriatal and subthalamofugal pathways, that use glutamate, are overactive. In MPTP-treated monkeys both NMDA and non-NMDA antagonists have efficacy. In rats, NR2B antagonists stimulate motor function.

Finally, AIDS dementia has parallels with cerebral ischemia or stroke and again the key mechanism appears to involve overactivation of glutamate receptors, in particular the NMDA receptor, followed by excessive influx of calcium and the generation of free radicals.

Clearly there is much therapeutic potential for drugs acting on glutamate systems but much more progress is needed.

DEVELOPMENT

Glutamate receptor expression is developmentally regulated and glutamate-mediated neurotransmission is generally enhanced in the immature brain at ages when certain glutamate receptors are transiently overexpressed. Furthermore, receptor subunit composition differs compared to the adult. Glutamate receptors play a critical role in neuronal plasticity and activity-mediated growth during brain development and yet the immature brain is more vulnerable than the adult to excitotoxic neuronal injury, suggesting that the functional state of glutamate receptors modifies the response of the brain to injury. In view of the general role of glutamate in the development and plasticity of connections in the immature CNS it is perhaps surprising that as yet we know little of its function apart from a pivotal role in the organisation of sensory pathways. In the CNS, NMDA receptor binding is much more restricted in the adult than in young animals although in broad terms, affinity appears the same as in the adult. In the immature hippocampus, NMDA EPSPs are much greater in amplitude and significantly less sensitive to Mg^{2+} although glycine modulation appears the same as in the adult. In the spinal cord, during the first postnatal week, NMDA and NMDA induced elevations of $[Ca^{2+}]_i$ are markedly elevated, gradually declining to adult levels although the AMPA response or resting $[Ca^{2+}]_i$ do not show these developmental changes. The neonatal brain represents a unique problem because any therapy based on glutamate receptors will somehow have to avoid adverse effects on the physiological roles of these receptors in plasticity and synaptic development.

FURTHER READING

Burnashev, N (1996) Calcium permeability of glutamate-gated channels in the central nervous system. *Curr. Opin. Neurobiol.* **6**: 311–317.

Colingridge, G and Singer, W (1990) Excitatory amino acid receptors and synaptic plasticity. *Trends Pharm. Sci.* **11**: 290–296.

Conn, PJ and Pin, J-P (1997) Pharmacology and functions of metabotropic glutamate receptors. *Ann. Rev. Pharmacol.* **37**: 205–237.

Daw, NW, Stein, PSG and Fox, K (1994) The role of NMDA receptors in information processing. *Ann. Rev. Neurosci.* **16**: 207–222.

Dickenson, AH (1990) A cure for wind-up: NMDA receptor antagonists as potential analgesics. *Trends Pharm. Sci.* **11**: 307–309.

Dingledine, R *et al.* (1999) The glutamate receptor ion channels. *Pharmacol. Rev.* **51**: 7–61.

Frerking, M and Nicoll, RA (2000) Synaptic kainate receptors. *Curr. Opin. Neurobiol.* **10**: 342–351.

Gegelashvili, G and Schousboe, A (1997) High affinity glutamate transporters: regulation of expression and activity. *Mol. Pharmacol.* **52**: 6–15.

Hollman, M and Heinemann, S (1994) Cloned glutamate receptors. *Ann. Rev. Neurosci.* **17**: 31–108.

Lerma, J (1997) Kainate reveals its targets. *Neuron* **19**: 1155–1158.

Malinow, R, Mainen, ZF and Hayashi, Y (2000) LTP mechanisms: from silence to four-lane traffic. *Curr. Opin. Neurobiol.* **10**: 352–357.

McGaugh, JL and Izquierdo, I (2000) The contribution of pharmacology to research on the mechanisms of memory formation. *Trends Pharm. Sci.* **21**: 208–210.

Turrigiano, GG and Nelson, SB (2000) Hebb and homeostasis in neuronal plasticity. *Curr. Opin. Neurobiol.* **10**: 358–364.

11 Amino Acids: Inhibitory

M. FARRANT

INTRODUCTION

Amino acids are the most widely used neurotransmitters in the central nervous system (CNS). Just as glutamate is the major excitatory neurotransmitter in the CNS (Chapter 10), two other amino acids, GABA (γ-aminobutyric acid) and glycine, are the principal inhibitory neurotransmitters. GABA and glycine are found throughout the CNS, but while GABA serves as a neurotransmitter in virtually every brain area, glycine fulfils this role predominantly, although not exclusively, in the caudal portions of the brain, the spinal cord, and the retina. This chapter describes the basic neurochemistry and pharmacology of GABAergic and glycinergic synapses.

GABA

The non-protein amino acid GABA is found in nearly all pro- and eukaryotic organisms, including plants. Its presence in mammalian brain was described 50 years ago, and the progress towards its subsequent acceptance as an important central neurotransmitter has been well documented (Roberts 1986). Over the years, numerous neurochemical, immunocytochemical and electrophysiological studies, as well as more recent investigations made possible by the cloning of GABA receptors, GABA transporters and the enzymes responsible for GABA synthesis, have confirmed the presence of GABAergic synapses throughout the CNS. Many of these synapses arise from local circuit interneurons, which are extremely diverse in both form and function (e.g. Freund and Buzsaki 1996; Gupta, Wang and Markram 2000). In addition, GABA is found in a number of neurons that contribute to identified pathways between brain regions. The majority of these are associated with the basal ganglia and include, for example, projections from the striatum to the globus pallidus and substantia nigra as well as projections from the globus pallidus and substantia nigra zona reticulata to several brain areas. Outside the basal ganglia, one of the best-known GABAergic projection neurons is the Purkinje cell of the cerebellar cortex.

GABA is an 'inhibitory' neurotransmitter in as much as its principal action is to cause membrane hyperpolarisation, thus reducing neuronal activity (see below). The large number and widespread distribution of GABAergic synapses has led to the idea that the nervous system is highly restrained. While this is undoubtedly an over-simplification, the importance of such inhibition for normal brain function is illustrated by experimental or pathological situations in which blocking or impairing the action of

Neurotransmitters, Drugs and Brain Function. Edited by R. A. Webster

GABA results in uncontrolled neuronal firing or seizures. GABA-mediated inhibition, however, does not act solely as a 'brake' but serves a variety of roles that far exceed the simple suppression of excitability. For example, tonic inhibitory input can transform the underlying firing pattern of a target cell and, by changing its electronic properties, alter both its temporal and spatial integration of excitatory inputs and hence the way information is processed. Moreover, individual neurons should not be considered in isolation. In networks of neurons, inhibitory connections may be organised to provide negative feedback (recurrent inhibition) and so lead to oscillatory behaviour. By controlling the precise timing of firing in multiple target cells inhibitory interneurons may also synchronise activity among neuronal populations and even enhance the effect of excitatory inputs. GABAergic neurons may also contact other GABAergic neurons, producing their ultimate effects through a process of disinhibition. Finally, it is important to note that, while GABA is classically viewed as an inhibitory neurotransmitter, under certain circumstances (described below) it can cause overt excitatory actions, and that this would appear to be its principal role in the developing nervous system.

NEUROCHEMISTRY OF GABA

SYNTHESIS AND CATABOLISM OF GABA

The synthesis and metabolism of GABA is closely linked with that of glutamate and the citric acid or tricarboxylic acid (TCA) cycle (Fig. 11.1). GABA is produced by the decarboxylation of glutamate, a reaction catalysed by the enzyme glutamic acid decarboxylase (GAD). GAD is found in several non-neuronal tissues (including ovary and pancreas) but within the CNS it is a specific marker of GABAergic neurons, where it is present in the cytoplasm as both soluble and membrane-bound forms, principally in the axon terminals. Labelling with antibodies against GAD has thus proved a particularly valuable technique for the identification of these neurons and their synaptic boutons. The breakdown of GABA occurs as a transamination reaction catalysed by the mitochondrial enzyme 4-aminobutyrate aminotransferase (GABA transaminase; GABA-T). In this process the amino group from GABA is transferred onto the TCA cycle intermediate α-ketoglutarate, producing glutamate and succinic semialdehyde. The latter is in turn converted by the enzyme succinic semialdehyde dehydrogenase (SSADH) into succinate, which re-enters the TCA cycle. This synthesis and catabolism of GABA is often referred to as the 'GABA-shunt', as it acts as a shunt of the normal TCA pathway from α-ketoglutarate to succinate. Other potential routes of GABA production have been described — involving deamination and decarboxylation reactions from putrescine, spermine, spermidine and ornithine — but the vast majority of GABA is generated by means of the GABA-shunt.

GABA-T and SSADH are also present in the mitochondria of glial cells and are responsible for the degradation of GABA recovered from the extracellular space (see below). In this case the glutamate formed from the action of GABA-T is converted into glutamine by the cytosolic enzyme glutamine synthetase (GS). Glial glutamine serves as an important precursor for both neuronal glutamate and GABA. It is transported from glia into neurons where the mitochondrial enzyme phosphate-activated glutaminase (PAG) converts it back into glutamate. This neuronal glutamate can then be converted to GABA, either directly or following metabolism via the TCA cycle. The interconversion of glutamate and α-ketoglutarate is achieved by two further groups of

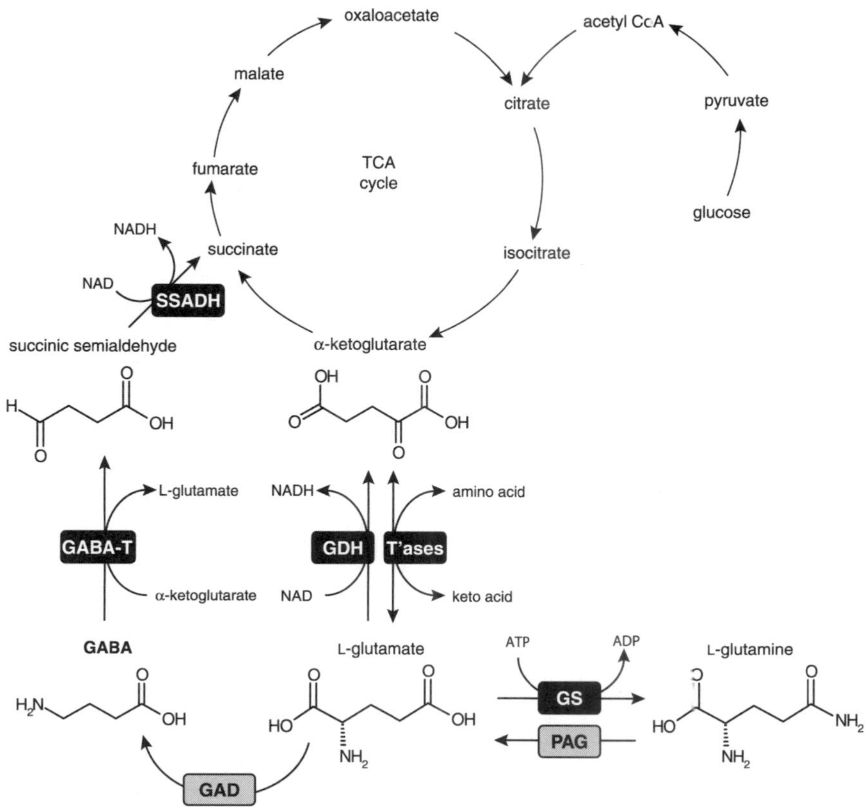

Figure 11.1 Enzymes responsible for the synthesis and metabolism of GABA. Enzymes responsible for the syntheisis (GAD) and metabolism (GABA-T and SSADH) of GABA, and their relationship to the TCA cycle and the amino acids glutamate and glutamine. Precursor glutamate is derived from glutamine by phosphate-activated glutaminase (PAG) and from α-ketoglutarate by aminotransferases, including aspartate and alanine aminotransferases (T'ases) and GABA-T. In glia, glutamate can be converted to glutamine by glutamine synthetase (GS). Other abbreviations are given in the text. Dark-grey boxes denote enzymes present in both neurons and glia, light-grey boxes denote enzymes present only in neurons

enzymes found in the mitochondria of both neurons and glia: the multi-enzyme complex glutamate dehydrogenase (GDH), and several aminotransferases (including aspartate and alanine aminotransferases) whose action is analogous to that of GABA-T. Fig. 11.2 shows the pathways of GABA metabolism in the context of a GABAergic synapse.

Regulation of GAD

Of key importance in the synthesis of GABA is the short-term regulation of GAD activity. Increasing the availability of glutamate does not lead to an increase in the production of GABA, suggesting that GAD may normally be saturated with its substrate. Instead, the control of GAD activity is intimately linked to the enzyme's requirement for the co-factor pyridoxal-5′-phosphate (PLP; a form of vitamin B_6) (Martin and Rimvall 1993). GAD exists in two states; an inactive apoenzyme (apoGAD) lacking the co-factor and active

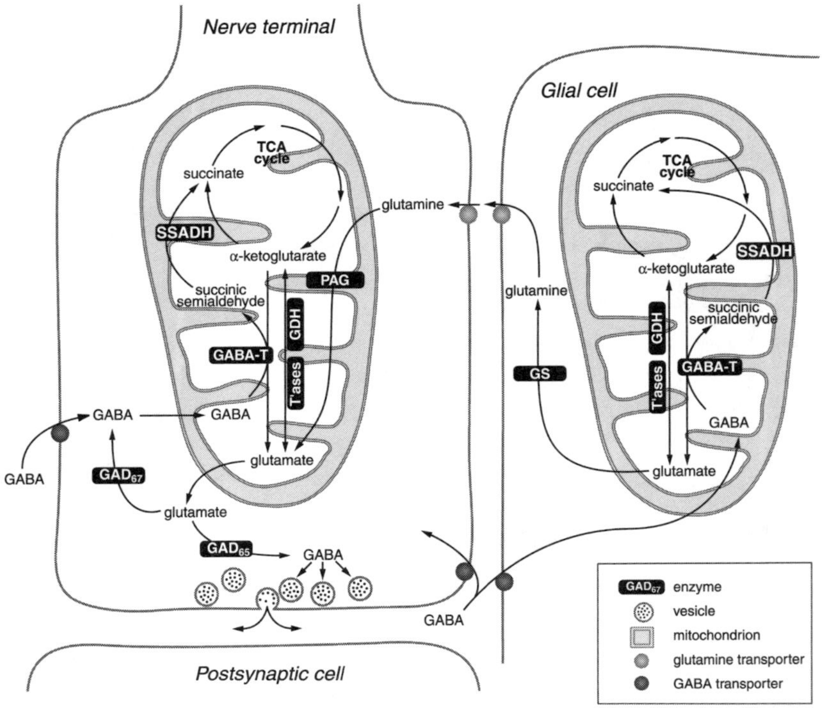

Figure 11.2 Pathways for GABA metabolism. GABA is synthesised in nerve terminals by GAD. GABA produced by both GAD_{67} and GAD_{65} can be used as a neurotransmitter but GAD_{65} is preferentially associated with synaptic vesicles. Synaptically released GABA is recovered into neurons and glia by GABA transporters (not shown is the possible release of GABA by reversal of these transporters). In both neurons and glia, GABA is degraded in mitochondria by GABA-T. Glutamine produced in glial cells is exported to neurons and converted to glutamate (after Soghomonian and Martin 1998)

holoenzyme (holoGAD) complexed with PLP. During the synthetic process GAD can undergo cycles of interconversion between these states. As illustrated in Fig. 11.3, in the primary reaction sequence active holoGAD combines with glutamate to form a complex, which, after decarboxylation, yields an intermediate product that is rapidly converted to GABA and free holoGAD. The intermediate product can also undergo an alternative reaction to produce succinic semialdehyde (SSA) and pyridoxamine-5′-phosphate (PMP) which, unlike PLP, dissociates from GAD to leave inactive apoenzyme, requiring additional PLP to be reactivated.

Traditionally, two processes have been considered important with respect to the regulation of GAD. First, GABA may promote conversion of GAD from its active to its inactive state, and so cause feedback inhibition of GABA synthesis. Second, ATP appears to inhibit, while inorganic phosphate promotes, the reactivation of GAD by PLP. During periods of increased neuronal activity, when the consumption of ATP increases, a rise in the level of phosphate should stimulate the conversion of inactive to active GAD, thereby increasing GABA synthesis. More recently, it has been suggested that soluble and membrane-bound forms of GAD may be differentially regulated. The soluble form of GAD is activated by a phosphatase that causes dephosphorylation while the membrane-bound form is activated following phosphorylation by a vesicular protein kinase (Hsu *et al.* 1999).

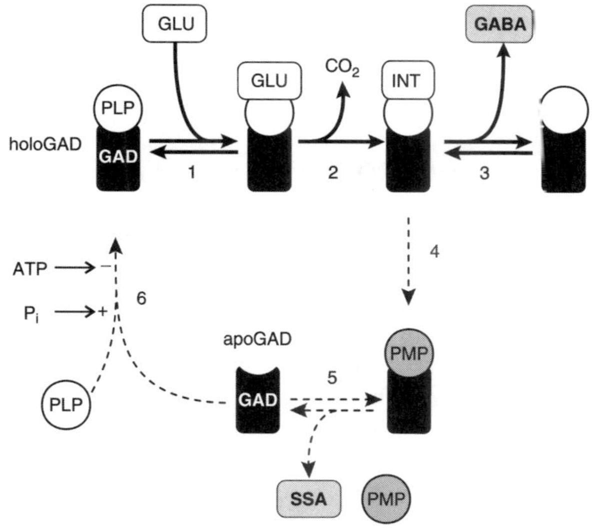

Figure 11.3 Regulation of GAD during the synthesis of GABA. Active GAD (GAD-PLP) combines with glutamate (1) to form a complex (GAD-PLP–GLU). After decarboxylation (2) this yields GABA and GAD-PLP (3). The intermediate product (GAD-INT) can undergo an alternative reaction (4) to produce succinic semialdehyde (SSA) and pyridoxamine-5′-phosphate (PMP). PMP dissociates from GAD (5) leaving inactive enzyme, which requires additional PLP to be reactivated (6), a process that is affected by ATP and inorganic phosphate

Two isoforms of GAD

In addition to the inactive and active states of GAD, there are two distinct forms of the enzyme. The two isoforms, GAD_{67} and GAD_{65}, named for their respective molecular masses (~67 and ~65 kDa), are encoded by separate, independently regulated genes, *GAD1* and *GAD2* (Erlander *et al.* 1991). GAD_{67} and GAD_{65} differ substantially in their amino-acid sequence, their interaction with PLP, their kinetic properties, and their regulation (Soghomonian and Martin 1998). Individual cells contain both forms of GAD but the ratio of the two differs among different neuronal populations. GAD_{65} is located preferentially in nerve terminals, both in the cytosol and as a membrane-bound form closely associated with synaptic vesicles into which the newly synthesised GABA is accumulated (see below). GAD_{65}, unlike GAD_{67}, is not saturated with PLP and forms the majority of the apoenzyme present in brain (about half of the total GAD). This has led to the view that a proportion of GAD_{65} exists as a pool of inactive enzyme, ready to combine with PLP in response to cellular signals for increased GABA synthesis.

Further insights into the role of GAD isoforms in the synaptic release of GABA have been provided by the techniques of gene manipulation. Mice lacking the GAD_{67} gene have a greatly reduced level of brain GABA (Asada *et al.* 1997). The neurological significance of this reduction is difficult to ascertain: GABA appears essential for the normal development of the palate and one consequence of the reduced production in GABA in these mice is a cleft palate that is responsible for their death soon after birth. In contrast, mice lacking the GAD_{65} gene show only a modest reduction in total brain GABA but exhibit spontaneous seizures and a greater susceptibility to chemical convulsants (Asada *et al.* 1996; Kash *et al.* 1997). In these mice basal GABAergic transmission is normal but GABA release during sustained synaptic activation is reduced (Obata *et al.* 1999; Tian *et al.* 1999). Together these results suggest that GAD_{67} is responsible for the synthesis of most brain GABA, but GAD_{65} is intimately involved in synthesis of GABA required for the refilling of the releasable pool of synaptic vesicles.

Inhibitors of GAD

Several drugs are known to inhibit GAD, either directly or through interaction with the co-factor PLP. The largest group of inhibitors are the hydrazides, such as isoniazid, semicarbazide and thiosemicarbazide. These are carbonyl-trapping agents that react with the aldehyde group of PLP; as many other enzymes use PLP as a co-factor, these agents are not specific for GAD. Two other agents, allylglycine and 3-mercaptopropionic acid, are competitive inhibitors of GAD. In general, GAD inhibitors reduce the level of GABA in the brain and cause seizures in experimental animals that, in the case of the hydrazides, can be overcome by application of vitamin B_6, the precursor of PLP. Similarly, in humans an inherited defect in pyridoxine metabolism is characterised by a low concentration of GABA in the cerebrospinal fluid, and intrauterine or neonatal seizures that also respond to treatment with vitamin B_6. These findings support the notion that maintained synthesis of GABA is an important factor in the control of overall brain excitability.

STORAGE OF GABA

Within nerve terminals, GABA, like other classical non-peptide neurotransmitters, is stored in synaptic vesicles into which it is accumulated by active transport. The uptake of GABA from the cytosol (where it is present at a concentration of a few millimolar) into the vesicle lumen (where it may reach several 100 millimolar) is dependent on a vesicular protein that transports cytosolic GABA in exchange for lumenal protons. The proton electrochemical gradient that drives this uptake is generated by a H^+-ATPase located in the vesicle membrane. Like vesicular transporters for monoamines and acetylcholine, the 'GABA transporter' recognises more than one substrate, and can also transport glycine (see below).

A gene (*unc-47*) encoding a vesicular GABA transporter was first identified from experiments on the simple nervous system of the nematode worm *C. elegans*. Mammalian homologues were subsequently cloned from rat and mouse; these were named VGAT (for vesicular GABA transporter; McIntire *et al.* 1997) or VIAAT (for vesicular inhibitory amino acid transporter; Sagne *et al.* 1997), respectively. These essentially identical clones have sequences predicting proteins of approximately 520 amino acids with ten transmembrane domains and, when expressed in mammalian cell lines, confer vesicular GABA and glycine transport. Immunohistochemical studies showed that VGAT/VIAAT is concentrated not only in the terminals of GABAergic neurons but also in those of neurons known to use glycine as a neurotransmitter (Gasnier 2000). As yet, no specific blockers or modulators of VGAT/VIAAT activity have been identified.

UPTAKE OF GABA

Once released from a vesicle, GABA molecules are able to activate receptors located on the pre- or postsynaptic membrane before rapidly diffusing out of the synaptic cleft. The ultimate removal of GABA from the extracellular space, and the maintenance of a low extracellular GABA concentration (low micromolar), is achieved by the high-affinity Na^+- and Cl^--dependent uptake of GABA into both GABAergic neurons and glial cells. Like the accumulation of GABA into vesicles, this is a secondary active transport mechanism, but in this case GABA uptake is coupled to the movement of Na^+ down its electrochemical gradient into the cell.

Drugs which block the uptake of GABA may be beneficial in conditions of reduced GABA function, as they are likely to prolong the action of synaptically released GABA (Thompson and Gahwiler 1992). The uptake of GABA is inhibited by a variety of simple GABA analogues, including nipecotic acid, β-alanine, 2,4-diaminobutyric acid (DABA), *cis*-3-aminocyclohexane-carboxylic acid (ACHC), 4,5,6,7-tetrahydroisoxazolo [4,5-c]pyridin-3-ol (THPO) and guvacine (Fig. 11.4), but as most of these do not cross the blood–brain barrier they have been of experimental interest only. In early studies, a number of compounds were suggested to preferentially inhibit GABA uptake into neurons (DABA and ACHC) or glia (β-alanine and THPO), while others were clearly non-selective (nipecotic acid and guvacine).

Cloned GABA transporters

This simple distinction between glial and neuronal uptake has required revision following the molecular cloning of a family of four Na^+- and Cl^--dependent GABA transporters, each encoded by a different gene: GAT-1, GAT-2, GAT-3 and BGT-1 (reviewed by Palacin *et al.* 1998). The nucleotide sequence of GAT-1 predicts a protein of 599 amino acids with a presumed structure containing twelve membrane-spanning regions. The transport of each GABA molecule into the cell is coupled to the movement of 2 Na^+ and 1 Cl^-. All the GABA transporters share a similar structure, with approximately 50% amino acid identity. GAT-1 appears to be mainly neuronal in origin as its mRNA is found in neurons and it is inhibited more effectively by neuronal than by glial uptake inhibitors. Nevertheless, immunohistochemical studies suggest some expression in glial cells. GAT-2 is found in cells of the ependyma and arachnoid membrane surrounding the brain and may play a role in the regulation of GABA in cerebrospinal fluid (it is also found in other tissues such as liver). GAT-3 is present in brain, principally in glia, but also in some neurons. BGT-1 was isolated from kidney and transports the osmolyte betaine as well as GABA (hence betaine/GABA transporter). It is present in the brain but its precise location and role are unclear.

In parallel with the identification of distinct transporters for GABA there has been continued interest in the development of selective blockers of these transporters and the therapeutic potential that could result from prolonging the action of synaptically released GABA. It has been known for a long time that certain pro-drugs of nipecotic acid (e.g. nipecotic acid ethyl ester) are able to cross the blood–brain barrier and are effective anticonvulsants in experimental models of epilepsy. More recently, several different systemically active lipophillic compounds have been described that act selectively on GAT-1, GAT-2 or GAT-3 (Fig. 11.4). Of these, tiagabine (gabitril), a derivative of nipecotic acid that acts preferentially on GAT-1, has proved clinically useful in cases of refractory epilepsy.

METABOLISM OF GABA

Once recovered into GABAergic nerve terminals or glia, GABA is metabolised to succinic semialdehyde and then to succinate. As detailed above, these reactions are catalysed by GABA-T and SSADH, respectively. The actions of these two enzymes are closely linked. Aminotransferase reactions are reversible but GABA-T breaks down GABA, rather than producing it, because the irreversible action of SSADH rapidly oxidises the product SSA to succinate (Fig. 11.1). SSA may also be reduced by the enzyme succinic semialdehyde reductase (SSAR) to form γ-hydroxybutyric acid (GHB).

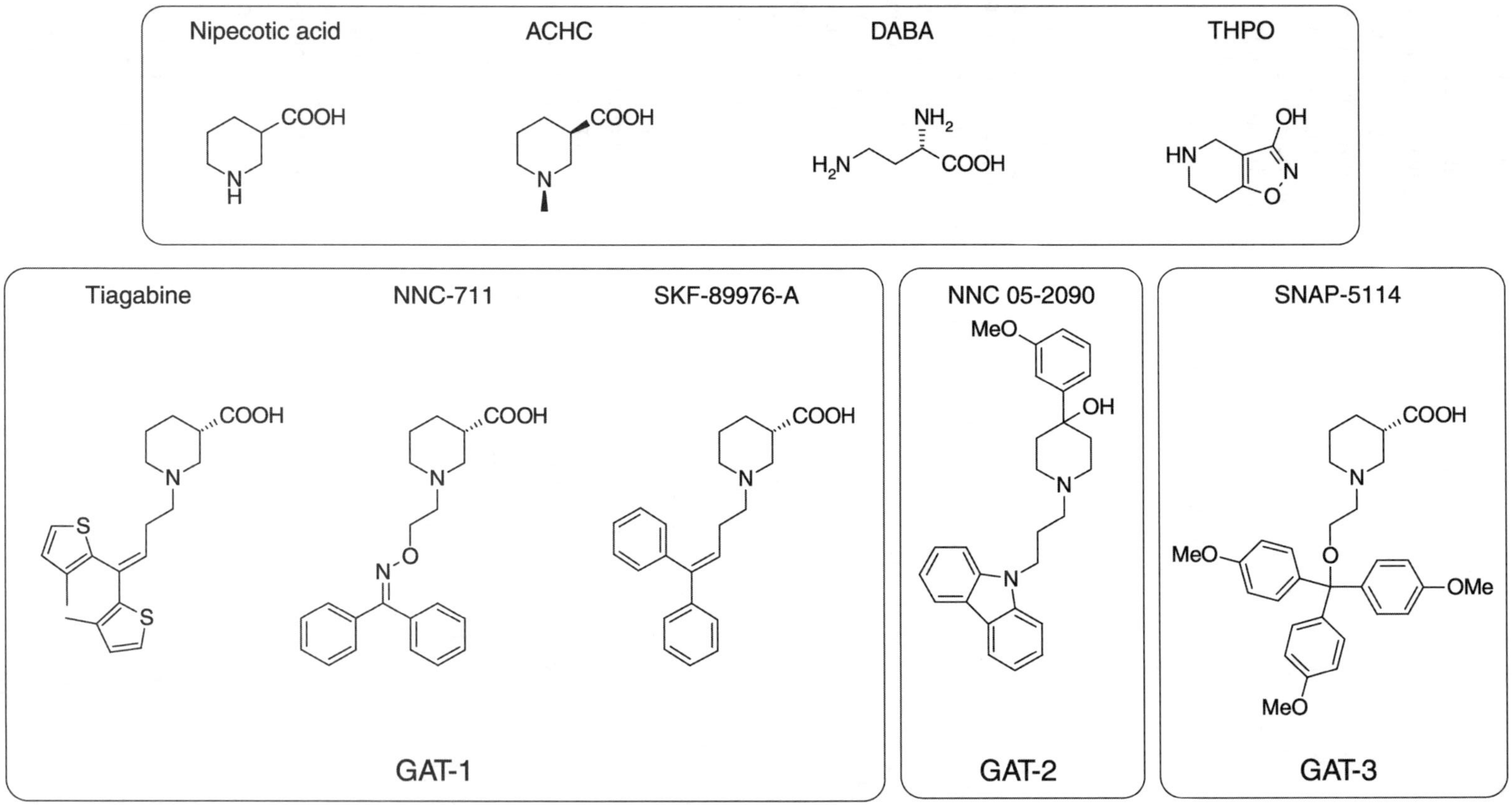

Figure 11.4 Blockers of GABA transport.The upper panel shows the structure of several GABA analogues that inhibit GABA transport into neurons or glia (see text). The lower panels show more recently developed compounds that exhibit selectivity for various cloned GABA transporters

Inhibitors of GABA-T

Inhibition of GABA-T leads to an elevation of brain GABA and, presumably because of an enhanced presynaptic availability of the transmitter, this has an anticonvulsant effect. Inhibitors of GABA-T include aminooxyacetic acid, 5-amino-1,3-cyclohexadienenecarboxylic acid (gabaculine), γ-vinyl GABA (vigabatrin) and 2-propylpentanoic acid (valproate). The first two are PLP antagonists and are of experimental interest only. Vigabatrin is an irreversible inhibitor of GABA-T and has been used clinically as an anticonvulsant. Valproate is a widely used anticonvulsant but it is not clear to what extent inhibition of GABA-T contributes to its therapeutic properties, as it also inhibits SSADH and SSAR, and inhibits Na^+ currents, thus limiting neuronal firing.

GABA RECEPTORS

The actions of GABA are mediated by receptors belonging to three distinct classes, termed $GABA_A$, $GABA_B$ and $GABA_C$. $GABA_A$ and $GABA_C$ receptors form membrane channels (ionotropic receptors) and their activation leads to an increased permeability to chloride (Cl^-) and bicarbonate (HCO_3^-) ions. $GABA_B$ receptors belong to the family of G-protein-coupled receptors (metabotropic receptors) and can modify the activity of the enzyme adenylate cyclase, suppress transmitter release by directly inhibiting calcium channels or hyperpolarise postsynaptic cells by directly activating potassium channels.

$GABA_A$ RECEPTORS

$GABA_A$ receptors are the most widely expressed of the GABA receptors in the CNS and are found at the vast majority of GABAergic synapses. Binding of two molecules of GABA to the receptor causes the opening of an integral transmembrane anion channel (Bormann, Hamill and Sakmann 1987). As the Cl^- permeability of the channel is approximately five times that of HCO_3^-, under most circumstances the net flux is dominated by Cl^-. Because of this the amplitude and direction of GABA-gated currents, and the resultant transmembrane potential changes, are determined largely by the sign and magnitude of the difference between the membrane potential (V_m) and the chloride equilibrium potential (E_{Cl}).

If Cl^- were passively distributed across the neuronal membrane E_{Cl} would equal V_m. However, neurons possess a variety of transport mechanisms for extrusion or uptake of Cl^- (Kaila 1994). The value of E_{Cl} is dictated by the net result of these chloride-extruding or accumulating mechanisms. Mature central neurons tend to maintain a low intracellular Cl^- through the activity of a Cl^--extruding K^+/Cl^- co-transporter (KCC2). Thus, in many neurons, E_{Cl} is more negative than V_m (although variable, typical values would be −70 and −65 mV, respectively). Under these circumstances, an increase in chloride conductance (g_{Cl}) leads to an influx of Cl^- that results in membrane hyperpolarisation (a movement towards E_{Cl}). This is the classic GABA-mediated inhibitory postsynaptic potential (IPSP). The IPSP transiently (tens of milliseconds) moves the membrane potential to a more hyperpolarised value, away from the threshold for action potential initiation (Fig. 11.5). In cells in which the net Cl^- extrusion is less (because KCC2 or other Cl^--extruding mechanisms are less active), E_{Cl} may be very close to, or slightly positive to, V_m. The effect of GABA is still inhibitory, as the increase in g_{Cl} tends to hold the membrane potential close to E_{Cl}, thus making it more difficult to trigger an action potential. This is often referred to as 'silent' or 'shunting' inhibition, and may play a role at both pre- and postsynaptic sites.

Neurons also possess mechanisms for Cl^- accumulation. One such is a $Na^+/K^+/2Cl^-$ co-transporter (NKCC1). Activity of NKCC1 can result in E_{Cl} being positive to V_m, such that an increase in g_{Cl} leads to chloride exit from the cell and membrane depolarisation. GABA may still be inhibitory, as long as E_{Cl} is below the threshold for spike initiation. This is thought to be the mechanism underlying one form of $GABA_A$ receptor-mediated presynaptic inhibition, originally identified in primary afferents to spinal neurons and best characterised in nerve terminals of the pituitary (Zhang and Jackson 1995). Here, modest depolarisation of the nerve terminal promotes inactivation of Na^+ channels, limiting the invasion of the action potential and reducing the effectiveness with which it triggers Ca^{2+} influx and transmitter release. NKCC1 is also present in developing neurons, prior to the expression of KCC2. As a result, the hyperpolarising or shunting inhibition, typical of the postsynaptic action of GABA in the adult, is absent. Instead, E_{Cl} is substantially more positive than V_m and GABA may produce sufficient depolarisation to activate voltage-dependent Ca^{2+} channels or trigger action potentials. Thus, in some situations in the developing nervous system GABA may function effectively as an excitatory neurotransmitter. The shift during development to an inhibitory action reflects the increased expression of the Cl^--extruding KCC2 (Rivera *et al.* 1999). One interesting example of the flexibility of this system is seen in cells of the suprachiasmatic nucleus of the hypothalamus (a brain region involved in the generation of circadian rhythms). Here, GABA is inhibitory during the night but excitatory during the day. This seems to result from diurnal variations in E_{Cl} such that during the night the GABA reversal potential is negative to V_m while during the day it is more positive than V_m so that GABA depolarises the cell above spike threshold (Wagner *et al.* 1997).

$GABA_A$ RECEPTOR PHARMACOLOGY

$GABA_A$ receptors are classically defined by their sensitivity to the antagonist bicuculline. Other widely used antagonists include SR-95531 (gabazine), picrotoxin, and cage convulsants such as TBPS (*t*-butylbicyclophosphorothionate). Like bicuculline, SR-95531 is a competitive antagonist acting at the GABA binding site while picrotoxin and TBPS are non-competitive, acting at a site which may be more closely associated with the chloride ion channel. Of these antagonists, only bicuculline is selective, as the others also act at $GABA_C$ receptors (see below). Similarly, no truly selective $GABA_A$ agonist is known. Muscimol is commonly used to activate $GABA_A$ receptors but it is also a partial agonist at $GABA_C$ receptors, while THIP (4,5,6,7-tetrahydroisoxazolo[5,4-c]pyridin-3-ol) is a partial agonist at $GABA_A$ receptors and an antagonist at $GABA_C$ receptors (Chebib and Johnston 2000). In addition to these agonists and antagonists, an enormous range of structurally diverse compounds can affect $GABA_A$ receptors. The most important of these are the benzodiazepines, barbiturates, neuroactive steroids and various general anaesthetics. The various sites of action of these drugs are depicted schematically in Fig. 11.6.

Benzodiazepines

Benzodiazepines such as chlordiazepoxide (Librium) and diazepam (Valium) were discovered in the early 1960s and found to have clinically important anxiolytic,

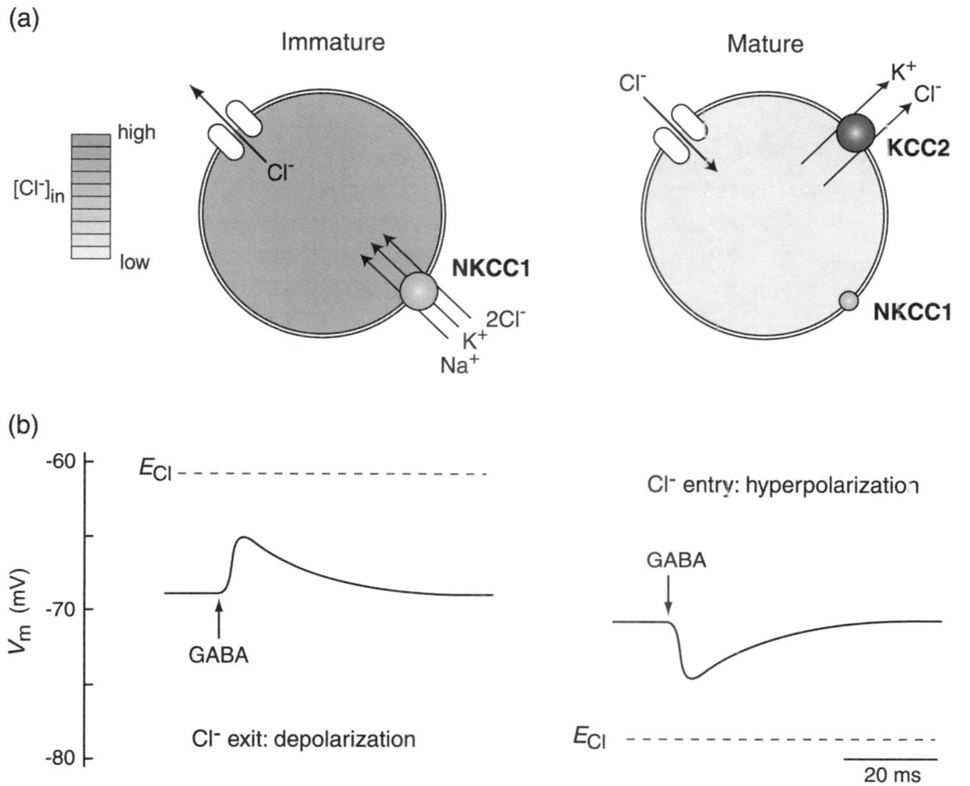

Figure 11.5 Chloride distribution and the $GABA_A$ response. The change in membrane voltage (V_m) that results from an increase in chloride conductance following activation of $GABA_A$ receptors is determined by the resting membrane potential and the chloride equilibrium potential (E_{Cl}). (a) Immature neurons accumulate Cl^- via NKCC, while mature neurons possess a Cl^--extruding transporter (KCC2). (b) In immature neurons $GABA_A$ receptor activation leads to Cl^- exit and membrane depolarisation while in mature neurons the principal response is Cl^- entry and hyperpolarisation. This is the classic inhibitory postsynaptic potential (IPSP)

anticonvulsant and muscle relaxant properties. It was not until the late 1970s, however, that their mode of action was established. Then, in studies of cultured neurons, benzodiazepines were found to increase the conductance change produced by GABA, shifting the GABA concentration–response curve to the left. They had no effect on conductance in the absence of GABA. At approximately the same time, specific high-affinity binding sites for benzodiazepines were identified in the CNS using radiolabelled compounds. A significant correlation between the binding affinity of various benzodiazepines and their potency in behavioural tests suggested that these sites mediated the central effects of benzodiazepines. A close association between benzodiazepine binding sites and the GABA receptor was indicated by the discovery of increased binding of benzodiazepines in the presence of GABA and increased binding of GABA in the presence of benzodiazepines.

The term 'benzodiazepine' refers to a specific chemical structure. Numerous benzodiazepine-receptor ligands exist which have different structures. These include the β-carbolines (e.g. methyl-6,7-dimethoxy-4-ethyl-β-carboline 3-carboxylate; DMCM), triazolopyridazines (e.g. CL 218872), imidazopyridines (e.g. zolpidem), and pyrazoloquinolinones (e.g. CGS 8216). In experimental animals these compounds produce

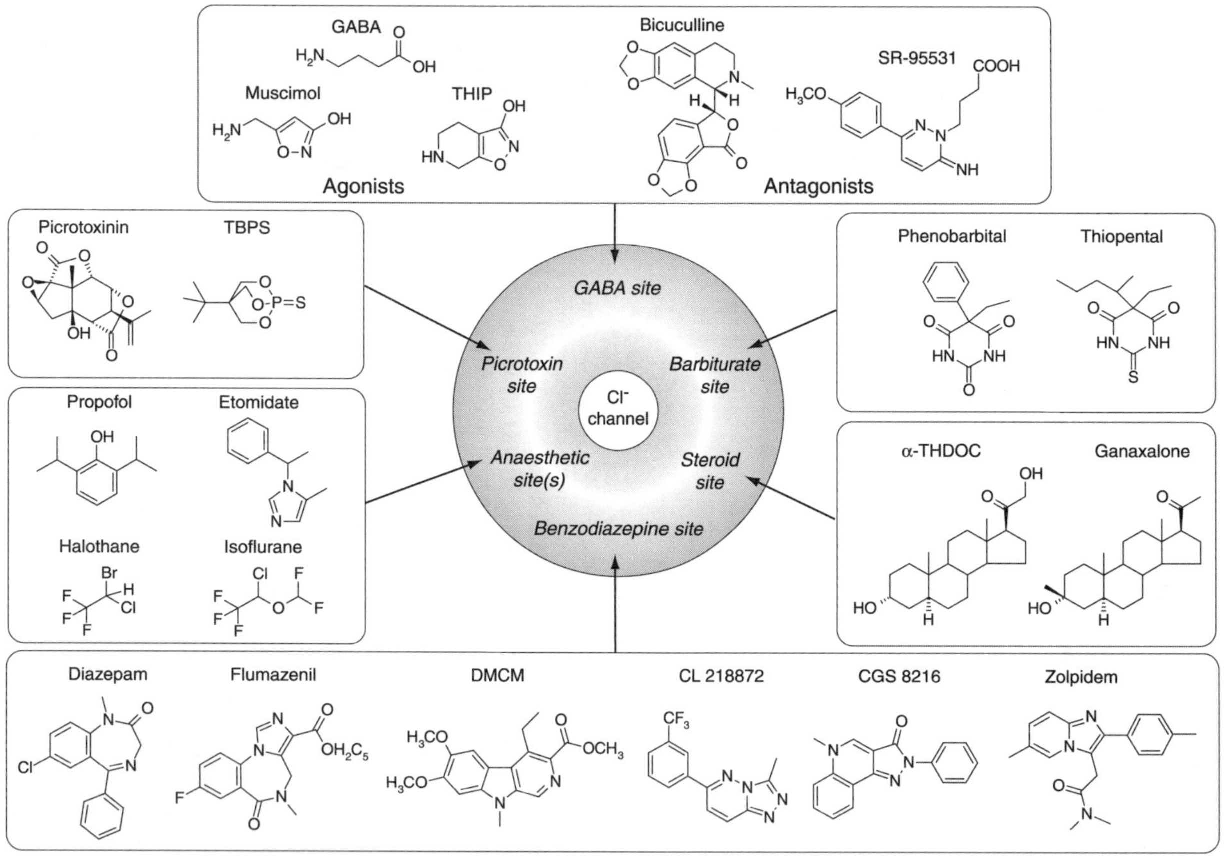

Figure 11.6 Schematic representation of the $GABA_A$ receptor complex. Examples of the many structurally diverse compounds that act at different sites on the receptor (see text for details). Picrotoxinin, the active component of picrotoxin, and TBPS act as non-competitive antagonists. The barbiturates, steroids and anaesthetics are positive allosteric modulators, as are the benzodiazepine site ligands shown, with the exception of DMCM (negative allosteric modulator) and flumazenil (benzodiazepine site antagonist)

effects ranging from anticonvulsant and anxiolytic (benzodiazepine-like) to proconvulsant/convulsant and anxiogenic. Between these two extremes are compounds, such as the imidazodiazepine flumazenil, that display only a limited degree of intrinsic activity but which are capable of antagonising the effects of the clinically useful benzodiazepines as well as those of the convulsant ligands or so-called 'inverse agonists' (negative allosteric modulators). All the compounds appear to act at the same or overlapping sites on the receptor complex. In studies of $GABA_A$ receptor single-channel currents, anxiolytic benzodiazepines, such as diazepam, increase the response to GABA but do not generally change the conductance of individual Cl^- channels. Instead they increase the affinity of the receptor for GABA and, in steady-state experiments, increase the frequency of channel opening, in a manner equivalent to increasing the concentration of GABA. At GABAergic synapses such compounds prolong the decay of the postsynaptic current and may also increase its peak amplitude. Inverse agonists such as DMCM reduce the response to GABA by decreasing the frequency of channel opening.

Barbiturates

Like benzodiazepine agonists, barbiturates possess sedative, anxiolytic and anticonvulsant properties. Although certainly not their only site of action, sedative barbiturates, such as phenobarbitone or the clinically used intravenous anaesthetic thiopentone, cause a marked potentiation of GABA responses. Unlike benzodiazepines, barbiturates increase the time for which GABA-activated channels are open and increase the length of bursts of openings. At higher concentrations barbiturates can activate Cl^- channels even in the absence of GABA. Neither effect is due to an action at the benzodiazepine binding site, as they are not blocked by the benzodiazepine antagonist flumazenil.

Steroids

It has been known since the 1940s that steroids of the pregnane series have anaesthetic properties. These early studies led to the development of the intravenous anaesthetic althesin, the active component of which is alphaxolone (3α-hydroxy-5α-pregnan-11,20 dione). This compound has barbiturate-like actions and is also able to directly activate Cl^- channels in the absence of GABA. These properties are shared by several endogenous steroids (synthesised in the brain or adrenal glands), the most potent being the reduced metabolite of progesterone, 3α-hydroxy-5α-pregnan-20-one (allopregnanolone or 3α,5α-THP) and the reduced metabolite of dexoycorticosterone, 3α,21-dihydroxy-5α-pregnan-20-one (allotetrahydrodeoxycorticosterone or α-THDOC). Particular interest in these compounds stems from the fact that they may act as endogenous modulators of $GABA_A$ receptors and their levels are altered by stress as well as during the menstrual cycle and pregnancy. For example, during menstruation decreasing levels of progesterone result in a decline in the production of allopregnanolone. Recently it has been demonstrated that such an abrupt decline (akin to drug withdrawal) can cause changes in the properties of $GABA_A$ receptors that may underlie the symptoms associated with premenstrual syndrome, including increased susceptibility to seizures and insensitivity to benzodiazepine agonists. Steroids appear to act at a distinct site on the GABA-receptor complex, as flumazenil does not block their action, and the Cl^- currents they evoke directly can be potentiated by barbiturates (and vice versa). The 3β-methyl-substituted synthetic analogue of allopregnanolone, ganaxolone (3α-hydroxy-3β-methyl-5α-pregnan-20-one) is less easily metabolised than its endogenous parent compound, allowing activity following oral administration, and is currently under investigation as an anticonvulsant.

Anaesthetics

Steroids, such as alphaxolone, and barbiturates, such as thiopentone, represent only two classes of the many structurally diverse molecules found to induce general anaesthesia. Although a number of these clearly have actions on a range of targets, including glycine, 5-HT_3, nicotinic and glutamate receptors, all, with the exception of the dissociative anaesthetic ketamine, have an effect on $GABA_A$ receptors at relevant concentrations. For example, the intravenous anaesthetic agents propofol, propanidid and etomidate markedly enhance responses to GABA (apparently by prolonging bursts of Cl^- channel openings) and are capable of directly evoking Cl^- currents. The currents produced by these agents at high doses, as well as those caused by steroids and barbiturates, are blocked by bicuculline, indicating that they are due to activation of the Cl^- channel associated with the $GABA_A$ receptor. It is also now clear that volatile anaesthetics such as halothane and isoflurane as well as alcohols (including ethanol), rather than having non-specific membrane-disrupting actions, owe at least some of their properties to a potentiation of GABA responses, through a direct interaction with sites on $GABA_A$ receptors.

STRUCTURE OF $GABA_A$ RECEPTORS

Over the past decade or so significant advances have been made in our understanding of the structure of the $GABA_A$ receptor, which is now known to be formed by the assembly of multiple subunit proteins. In 1987 two subunits of the receptor, designated α and β, were cloned (Schofield *et al.* 1987). Following on from this work, 16 mammalian subunits encoded by distinct genes have now been identified. These genes encode proteins of approximately 450–550 amino acids (depending on the species) which, according to their sequence similarities, have been grouped into seven families—α, β, γ, δ, ε, π and θ (Barnard *et al.* 1998; Bonnert *et al.* 1999). The α, β and γ families contain multiple isoforms ($\alpha1$–$\alpha6$, $\beta1$–$\beta3$ and $\gamma1$–$\gamma3$) and in a number of cases additional complexity is generated by alternative mRNA splicing.

The subunits share varying degrees of sequence identity but have a similar predicted tertiary structure. This consists of four membrane-spanning α-helices (M1–M4), a large extracellular N-terminal region, a large intracellular domain between M3 and M4 and a short extracellular C-terminal portion (Fig. 11.7). The highest degree of conservation is in the transmembrane regions and the greatest variation in the intracellular loop between M3 and M4. The extracellular domain contains potential N-linked glycosylation sites and a β-loop formed by a disulphide bridge between two cysteine residues. The intracellular loops of β and γ subunits contain sites for phosphorylation by a variety of protein kinases, including cAMP-dependent protein kinase, cGMP-dependent protein kinase, protein kinase C, Ca^{2+}/calmodulin-dependent protein kinase and tyrosine kinase, which may be important in the regulation of receptor function. These general features are very similar to those of two other ligand-gated ion channels, the nicotinic acetylcholine receptor and the glycine receptor (see below) and there is a considerable degree of sequence homology among these proteins. By analogy with the nicotinic acetylcholine receptor, it is thought that the $GABA_A$ receptor is formed by the assembly of five subunits around a central ion channel, with the M2 region of each subunit forming the lining of the channel (Fig. 11.7).

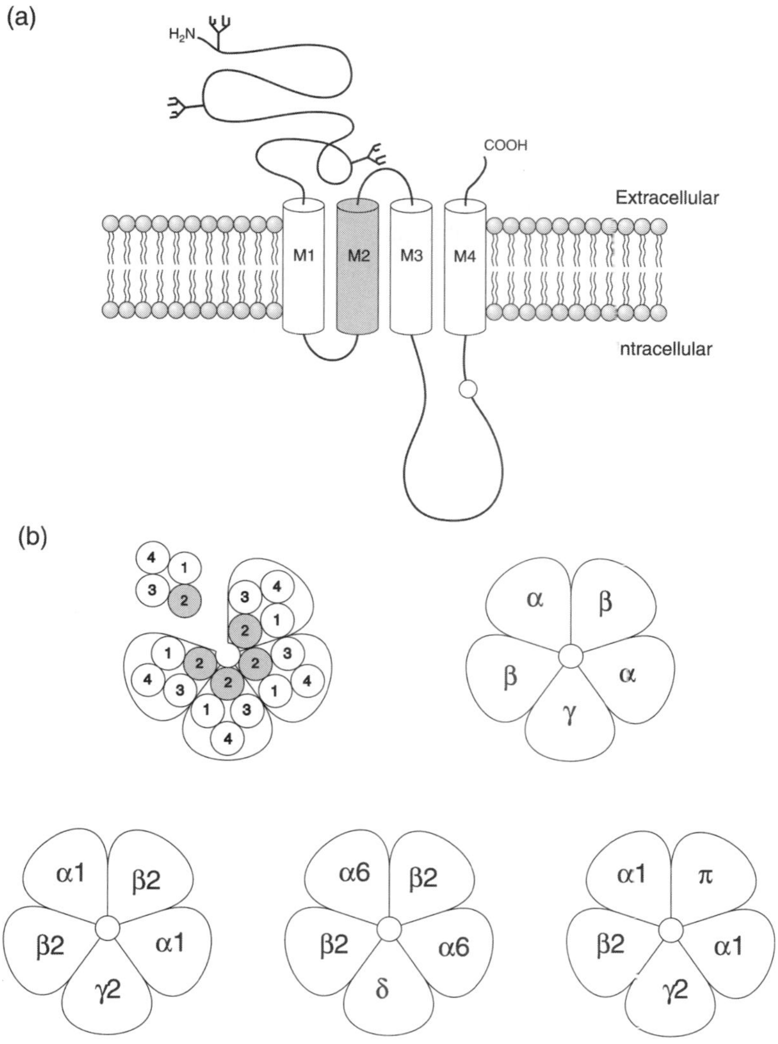

Figure 11.7 Presumed arrangement of $GABA_A$ receptor subunits to form a receptor–channel complex. (a) Diagrammatic representation of an individual subunit with four transmembrane regions, extracellular sites for glycosylation and a site for phosphorylation on the intracellular loop between M3 and M4. (b) Association of five subunits to form a central ionophore bounded by the M2 region of each subunit. The suggested stoichiometry of the most widely expressed form of receptor is 2α, 2β and 1γ. Shown below are the possible subunit combinations of one such benzodiazepine-sensitive receptor together with a benzodiazepine-insensitive receptor in which the γ subunit is replaced by a δ, and a π-containing receptor with four different subunit types

Subunit combinations and receptor function

Expression studies in *Xenopus* oocytes or transfected cell lines originally suggested that functional GABA-activated chloride channels could be formed by receptor subunits of each class in isolation. However, much better expression occurs with two or more subunit types in combination and it is likely that most native receptors contain at least three different subunits. Co-expression of α and β subunits results in the assembly of functional receptors that can be activated by GABA and are sensitive to the antagonists bicuculline and picrotoxin and show modulation by barbiturates. But only when a γ subunit is expressed in conjunction with an α and a β subunit is benzodiazepine binding and potentiation of GABA seen. As benzodiazepines do not bind to γ subunits alone, it is likely that the conformation of the receptor is appropriate for benzodiazepine binding only when all three subunit types are present.

The large number of cloned subunit proteins makes it clear that $GABA_A$ receptors themselves must be diverse. An illustration of this diversity is provided by the pharmacology of benzodiazepine ligands. Even before the existence of $GABA_A$ receptor subunits was recognised, variations in the binding of radiolabelled drugs to native benzodiazepine receptors had led to the suggestion that not all GABA receptors were the same. Two types of benzodiazepine receptor were postulated — BZI and BZII. These had similar affinity for agonists such as diazepam and antagonists such as flumazenil, but BZI receptors showed a higher affinity for triazolopyridazines (e.g. CL 218872) and β-carbolines (e.g. β-CCM). It is now clear that the molecular basis for these differences resides in the variety of α subunits. Thus, while γ subunits are required for benzodiazepine binding, the precise nature of this interaction depends on the type of α subunit present. Heteromeric recombinant receptors ($\alpha\beta\gamma$) containing an $\alpha 1$ subunit exhibit BZI-type pharmacology, receptors containing $\alpha 2$, $\alpha 3$ or $\alpha 5$ subunits exhibit BZII pharmacology, while receptors containing $\alpha 4$ or $\alpha 6$ subunits have a low affinity for both benzodiazepines and β-carbolines. Studies involving site-directed mutagenesis of the various subunits have narrowed down even further the precise amino acid residues responsible for these differences in benzodiazepine pharmacology, as well as those involved in the binding of GABA. Altogether, such data suggest that GABA molecules bind at the interface of α and β subunits while benzodiazepines bind at the interface of α and γ subunits.

The complexity afforded by different α, β and γ subunits is increased further by the existence of the δ, ε, θ and π subunits. The δ subunit preferentially associates with the $\alpha 4$ and $\alpha 6$ subunits. Receptors containing this subunit are unusual in having a particularly high affinity for GABA and muscimol and a reduced sensitivity to benzodiazepines and neurosteroids. The most recently cloned subunits, ε, θ and π, are the least well understood. The sequence of the ε subunit is most closely related to that of the γ subunits but studies in recombinant expression systems show that it assembles with α and β subunits to form receptors that are insensitive to benzodiazepines and show altered sensitivity to anaesthetics (pregnanolone, pentobarbital and propofol). The θ subunit is most closely related to the β subunits; it coassembles with α, β and γ subunits to form receptors with a low affinity for GABA, although other subunit combinations (notably $\alpha\beta\theta\varepsilon$ or $\alpha\theta\varepsilon$) have been suggested. The ε and θ subunits have a fairly restricted pattern of expression that includes the hypothalamus and brainstem nuclei such as the locus coeruleus. The sequence of the π subunit is most closely related to that of the β subunits. Unlike the other $GABA_A$ subunits it is principally found in peripheral tissues, including lung, thymus, prostate and particularly the uterus.

Heterogeneity of native $GABA_A$ receptors

Given that the pharmacological and biophysical properties of recombinant $GABA_A$ receptors have been shown to depend critically on their subunit composition, much effort has been directed towards understanding the assembly of native receptors. This could provide a rational basis for the design of compounds able to interact with specific subpopulations of $GABA_A$ receptors in different brain regions that may be involved in different aspects of brain function. Clearly, many hundreds of different receptor types could arise from the assembly of 16 different subunits into a pentameric structure. However, numerous studies, involving the use of subunit-specific antibodies to localise or to purify receptor populations, have suggested that the restricted distribution and preferential assembly of these subunits results in the generation of no more than a dozen favoured receptor types. Of these, the most common receptor type is composed of $\alpha 1$, $\beta 2$ and $\gamma 2$ subunits. Several lines of evidence suggest that the most likely stoichiometry of these receptors is 2α, 2β and 1γ (although assemblies containing 2α, 1β and 2γ have also been described). As indicated above, in less widely expressed assemblies, the δ or ε subunits can substitute for the γ subunit, while the π and θ subunits may co-assemble with α, β *and* γ subunits.

Advances in our understanding of the functional significance of $GABA_A$ receptor heterogeneity have also come from studies of mice lacking specific subunit genes or expressing altered receptor subunits. To date, mutant mice have been generated that lack the $\alpha 1$, $\alpha 6$, $\beta 2$, $\beta 3$, $\gamma 2$ or δ subunits (Rudolph *et al.* 2001; Sur *et al.* 2001). In the case of the $\gamma 2$ subunit deletion, neurons cultured from newborn mice show a complete lack of sensitivity to benzodiazepines (Gunther *et al.* 1995). By introducing a histidine residue (instead of the normal arginine) at position 101 in the $\alpha 1$ subunit of mice—making receptors containing this subunit insensitive to benzodiazepines—it has also been possible to determine which of the various effects of benzodiazepines are mediated by $\alpha 1$-containing receptors and which by receptors containing $\alpha 2$, $\alpha 3$ or $\alpha 5$ subunits. This approach showed that $\alpha 1$-containing $GABA_A$ receptors are involved in the sedative and amnesic actions of benzodiazepines (McKernan *et al.* 2000; Rudolph *et al.* 1999). Complementary experiments have shown that the anxiolytic actions of benzodiazepines are mediated by $\alpha 2$-containing receptors and the muscle-relaxant actions by $\alpha 2$- and $\alpha 3$-containing receptors (Rudolph *et al.* 2001; Crestani *et al.* 2001).

$GABA_B$ RECEPTORS

$GABA_B$ receptors are found in both the peripheral nervous system and CNS. They were first identified in the late 1970s, during studies of noradrenaline release from axon terminals of sympathetic post-ganglionic fibres in rat atria. GABA was found to reduce the evoked release of transmitter but this action was not blocked by the conventional antagonists bicuculline and picrotoxin. The effect of GABA was mimicked not by muscimol but by the compound (*R*)-4-amino-3-(4-chlorophenyl)butanoic acid (baclofen), a GABA analogue that has no effect on GABA receptors linked to Cl^- channels. To distinguish between the established receptors and the newly identified bicuculline-insensitive receptors the terms $GABA_A$ and $GABA_B$ were introduced (reviewed in Bowery 1993).

$GABA_B$ RECEPTOR PHARMACOLOGY

Baclofen and the phosphinic analogue of GABA, 3-aminopropyl phosphinic acid (APPA), selectively activate $GABA_B$ receptors. The first antagonists identified were 2-OH-saclofen and phaclofen, the sulphonic and phosphonic acid analogues of baclofen, respectively (Kerr and Ong 1995). In recent years, a number of more potent and systemically active antagonists have been developed. These include CGP 54626 and GCP 62349 (Fig. 11.8), the high affinity of the latter proving instrumental in the eventual cloning of $GABA_B$ receptors (see below). As yet, no modulatory compounds of the type described for $GABA_A$ receptors have been identified.

$GABA_B$ RECEPTOR MECHANISMS

Depending on cell type and the location of the receptor on neurons, $GABA_B$ receptors act via G-proteins to affect the activity of either Ca^{2+} channels, K^+ channels or adenylate cyclase (Bowery and Enna 2000). For example, in dorsal root ganglion (DRG) neurons baclofen was found to inhibit the Ca^{2+}-dependent phase of the DRG action potential, an effect attributed to block of voltage-activated Ca^{2+} currents. A similar action on presynaptic Ca^{2+} channels was presumed to underlie the block of neurotransmitter release by baclofen. This has now been demonstrated by recording Ca^{2+} currents from presynaptic terminals directly (Takahashi, Kajikawa and Tsujimoto 1998). In the CNS such presynaptic $GABA_B$ receptors are found on terminals releasing a variety of transmitters, as well as on GABAergic terminals themselves, where they may act as autoreceptors.

In several neuronal types baclofen was also shown to cause a postsynaptic hyperpolarisation due to the opening of K^+ channels (E_K is more negative than V_m). These postsynaptic $GABA_B$ receptors can be activated by synaptically released GABA and give rise to a delayed and long-lasting (hundreds of milliseconds) hyperpolarisation which, following as it does the $GABA_A$ receptor-mediated fast IPSP, is often referred to as the 'late IPSP' (Dutar and Nicoll 1988). In general, both types of $GABA_B$ response can be blocked by the same antagonists or by treatment with *Pertussis* toxin (which blocks the activation of G-proteins of the $G_{i/o}$ class). However, at some sites presynaptic effects of baclofen are not blocked by all agents, indicating that multiple types of $GABA_B$ receptor may exist (see below).

STRUCTURE OF $GABA_B$ RECEPTORS

$GABA_B$ receptors long resisted attempts at expression cloning of the type used to identify $GABA_A$ receptor subunits, partly because the requirement for G-protein coupling to ion channels made functional assays in cell lines or oocytes difficult to devise. Recently, however, a $GABA_B$ receptor ($GABA_BR1$) was successfully isolated from a rat cDNA library by screening transfected cells with a high-affinity radiolabelled $GABA_B$ receptor antagonists, binding of which does not require the presence of G-proteins (Kaupmann *et al.* 1997). Two isoforms of the receptor were identified which differed only in the length of their amino-terminals—these were termed $GABA_BR1a$ (960 amino acids) and $GABA_BR1b$ (844 amino acids). Additional isoforms, both termed $GABA_BRIc$, have also been identified in rat (Pfaff *et al.* 1999) and human (Ng *et al.* 2001). Each protein has a predicted structure consisting of a large N-terminal and seven transmembrane domains, similar to metabotropic glutamate receptors.

The cloned receptors, when expressed in cell lines and studied by radioligand binding assays, showed some of the expected pharmacology of $GABA_B$ receptors. However, the affinity of agonists was much lower than seen with native receptors and not all expected coupling to effector systems could be demonstrated (possibly because of inappropriate or inefficient linkage to G-proteins). Subsequently, a second receptor protein, $GABA_BR2$, was identified (Jones *et al.* 1998; Kaupmann *et al.* 1998; White *et al.* 1998), and shown to interact with $GABA_BR1$, through the coiled regions of their intracellular C-termini, to form fully functional heterodimers (Fig. 11.8). $GABA_BR1$

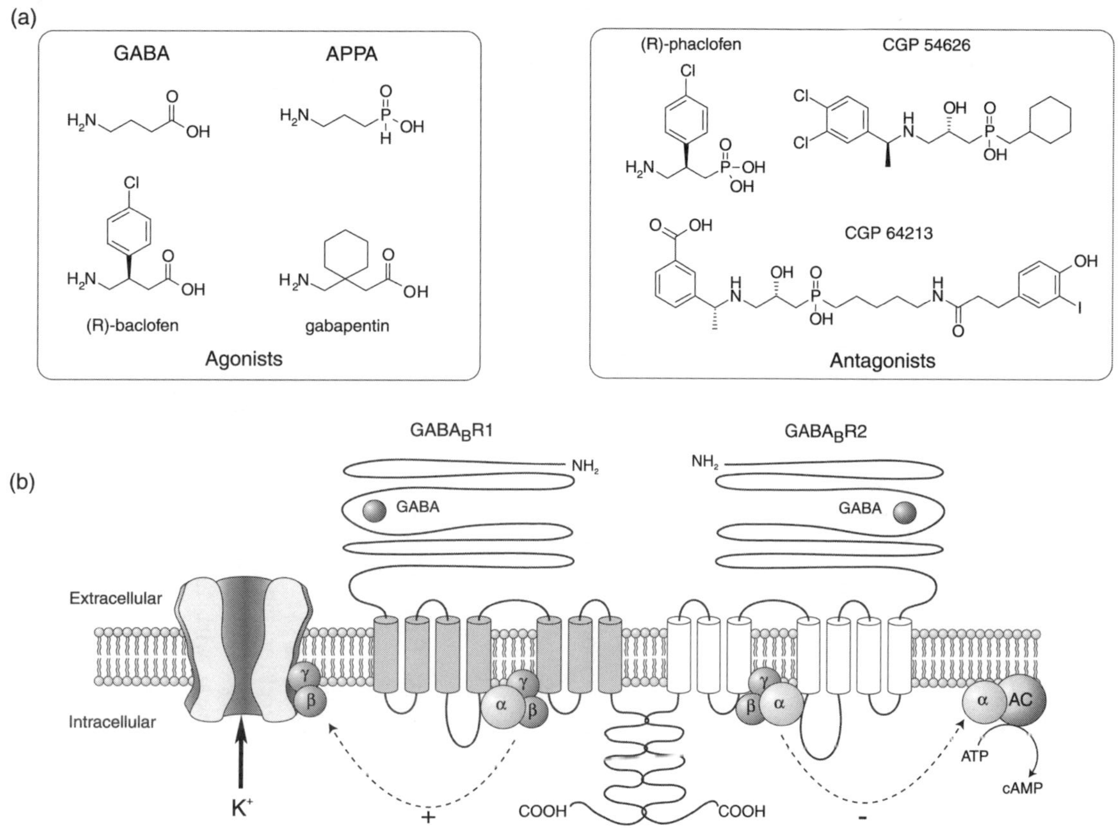

Figure 11.8 Pharmacology and structure of $GABA_B$ receptors. (a) Various $GABA_B$ agonists and antagonists described in the text. (b) A $GABA_B$ receptor shown as a dimer containing one copy of $GABA_BR1$ and one copy of $GABA_BR2$, joined by their coiled intracellular carboxy-terminals. The large extracellular amino-terminals are the proposed sites of GABA binding. A G-protein is shown linked to each of the $GABA_BR$ proteins. The activated β/γ subunits trigger the opening of a K^+ channel while the α subunit is shown inhibiting the activity of adenylate cyclase (AC). At presynaptc sites the β/γ subunits would inhibit a Ca^{2+} channel (after Bowery and Enna 2000)

and $GABA_BR2$ are found in areas of the brain known to contain $GABA_B$ receptors, including hippocampus, cerebellum and cortex, but some differences in their distribution suggest that in certain cases homomeric $GABA_BRs$ may be functional or that dimerisation may occur with other unidentified $GABA_BRs$ (Bowery and Enna 2000). The existence of structurally and pharmacologically distinct pre- and postsynaptic $GABA_B$ receptors is supported by the recent demonstration that gabapentin, an anticonvulsant GABA analog, is a selective agonist for postsynaptic $GABA_BR1a/R2$ heterodimers coupled to K^+ channels (Ng *et al.* 2001).

$GABA_C$ RECEPTORS

Early studies of the action of GABA and its analogues on spinal neurons revealed that the depressant action of one of these, *cis*-4-aminocrotonic acid (CACA), was not blocked by bicuculline. Several analogues of GABA shared the same properties and did not interact with the then newly described $GABA_B$ receptors. In 1984, the term $GABA_C$ was introduced to distinguish this third class of GABA receptor (Johnston 1996). Like $GABA_A$ receptors, $GABA_C$ receptors activate anion channels permeable to Cl^- (and HCO_3^-) and the responses are similarly governed by the distribution of Cl^- across the neuronal membrane.

$GABA_C$ RECEPTOR PHARMACOLOGY

$GABA_C$ receptors are defined by their insensitivity to bicuculline and their activation by conformationally restricted analogues of GABA such as CACA and (+)-CAMP (1*S*,2*R*-2-(aminomethyl)cyclopropanecarboxylic acid). They are blocked by picrotoxin but can be selectively antagonised by TPMPA (1,2,5,6-tetrahydropyridin4-ylphosphinic acid). Unlike $GABA_A$ receptors, they are not affected by benzodiazepines, barbiturates or anaesthetics (Barnard *et al.* 1998; Bormann 2000; Chebib and Johnston 2000).

STRUCTURE OF $GABA_C$ RECEPTORS

The best evidence for the existence of functional $GABA_C$ receptors and the clearest indication as to their molecular identity comes from work on the retina. Expression of retinal mRNA in *Xenopus* oocytes produces GABA-gated chloride channels with conventional $GABA_A$ receptor pharmacology as well as channels with characteristics of $GABA_C$ receptors (i.e. blocked by picrotoxin but not bicuculline). The basis of this distinction was made clear with the cloning from a retinal cDNA library of a new GABA receptor subunit termed ρ (Cutting *et al.* 1991). To date, three ρ subunits have been identified in mammals ($\rho1$–$\rho3$). Originally classed as $GABA_A$ subunits, with which they have ~35% sequence identity, they are now accepted as a distinct group of subunits, forming the basis of the relatively simple, and evolutionarily older, $GABA_C$ receptors. Unlike subunits of the $GABA_A$ receptor, ρ subunits form fully functional homomeric receptors and do not co-assemble with α or β subunits. These homomeric receptors are similar to native $GABA_C$ receptors, in that they are activated by GABA and CACA, blocked by picrotoxin and TPMPA but not bicuculline, and unaffected by barbiturates, benzodiazepines or anaesthetics (Fig. 11.9). Receptors formed from ρ subunits have a higher affinity for GABA than many $GABA_A$ receptors formed from $\alpha\beta\gamma$ combinations, have a lower single-channel conductance and produce currents that decay more slowly after removal of GABA.

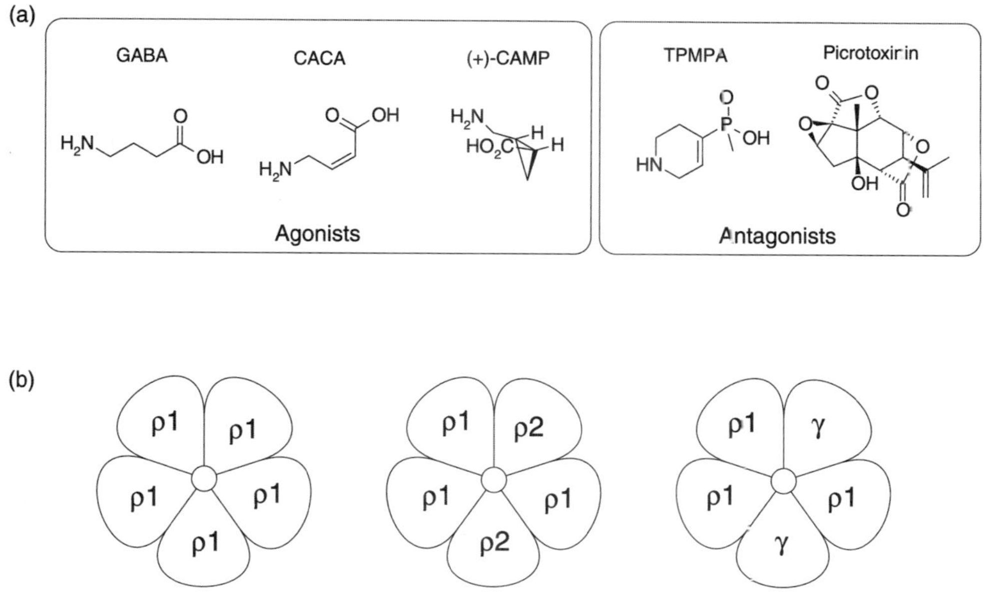

Figure 11.9 $GABA_C$ receptor pharmacology and structure. (a) Various $GABA_C$ agonists and antagonists described in the text. Picrotoxinin is the active component of picrotoxin and also acts at $GABA_A$ receptors. (b) Presumed subunit structures of $GABA_C$ receptors. The receptors can form as homomeric assemblies of ρ subunits but native receptors may be heteromeric assemblies of ρ subunits (e.g. $\rho 1$ and $\rho 2$) or may contain both ρ and γ subunits

In the retina, electrophysiological data indicate that $GABA_C$ receptors are present on horizontal and bipolar cells, and ρ subunits have been localised to subsets of synapses formed by amacrine cells onto the axon terminals of rod bipolar cells. Activation of these presynaptic receptors inhibits glutamate release from the bipolar cells. However, the true molecular composition of native $GABA_C$ receptors is still under investigation. While homomeric receptors formed from ρ subunits share many features of retinal $GABA_C$ receptors, a number of discrepancies have been noted in the details of ion permeability, single-channel conductance and channel open time (Wotring, Chang and Weiss 1999). Thus, it has been suggested that native $GABA_C$ receptors may be composed of heteromeric assemblies of ρ subunits or, in certain cases, that such assemblies may also contain a $\gamma 2$ subunit (Qian and Ripps 1999). All three ρ subunits have been identified in brain, but their precise location and the functional significance of this expression is unclear. In particular, the basis of $GABA_C$ receptor-like responses seen, for example, in the spinal cord, cerebellum, optic tectum and hippocampus is yet to be determined.

GLYCINE

Glycine is the simplest of all amino acids. It is involved in many metabolic pathways, is an essential component of proteins, and is found throughout the brain. A neurotransmitter role for glycine was first identified in the spinal cord, where it was found to be differentially distributed between dorsal and ventral regions and shown to cause hyperpolarisation of motoneurons (Werman *et al.* 1967). This inhibitory action of glycine is distinct from its subsequently identified role as a co-agonist at NMDA-type glutamate receptors (Chapter 10), and is mediated by receptors that share many features with $GABA_A$ receptors (see below). Glycine-mediated neurotransmission plays a key role in spinal cord reflexes, mediating reciprocal and recurrent inhibition of motoneurons by Renshaw cells, and is important in motor control and sensory pathways. Glycine receptors are also found in higher brain centres including the hippocampus, cortex and cerebellum.

NEUROCHEMISTRY OF GLYCINE

SYNTHESIS AND CATABOLISM OF GLYCINE

The details of glycine metabolism within neural tissue are poorly understood, and it is unclear to what extent neurons depend on *de novo* synthesis or uptake of glycine. Two enzymes are important in glycine metabolism; serine hydroxymethyltransferase (SHMT), which is thought to be present in the mitochondria of both neurons and glia, and the four-enzyme complex known as the glycine cleavage system (GCS), present in glia. SHMT catalyses the interconversion of L-serine and glycine while GCS catalyses the breakdown of glycine. Within neurons the action of SHMT leads to the conversion of L-serine to glycine, while in glia the coupling of SHMT and GCS results in the conversion of glycine to L-serine (Verleysdonk *et al.* 1999). The L-serine derived from glycine may be further metabolised, or released from glial cells to be taken up into neurons, forming a cycle analogous to the glutamine–glutamate cycle shown in Fig. 11.2. Glycine can also be formed by the action of aminotransferases (such as alanine-glyoxylate transaminase or glycine transaminase), in which the amino group from a donor amino acid is transferred onto glyoxlate, producing glycine and a keto acid.

STORAGE OF GLYCINE

Glycine, like GABA, is stored in synaptic vesicles. As described above, it seems likely that a common transport mechanism (VIAAT) is responsible for the accumulation of both amino acids. This lack of absolute specificity in the vesicular transporter means that the 'phenotype' of a neuron (GABAergic or glycinergic) is dictated by the relative concentrations of GABA and glycine in the cytosol. This will be determined by the expression of the respective biosynthetic enzymes and plasma membrane transporters. In certain cases neurons may release both GABA *and* glycine, which have been packaged into the same vesicles (Jonas, Bischofberger and Sandkuhler 1998; O'Brien and Berger 1999). The extent and significance of such co-release is unclear, but its effects will obviously depend on the types of pre- and postsynaptic receptors present at the synapse. Possible benefits of co-release may stem from the different kinetic properties of $GABA_A$ and glycine receptors, the ability to activate $GABA_B$ receptors or the modulatory action of glycine at NMDA receptors.

UPTAKE OF GLYCINE

Glycine is removed from the extracelluar space by high-affinity uptake into neurons and glia. Five glycine transporters have been identified in the CNS of mammals. All are members of the Na^+- and Cl^--dependent family transporters and are encoded by two independently regulated genes, *GLYT1* and *GLYT2*. Three GLYT1 isoforms (1a, b and c) and two GLYT2 isoforms (2a and b) are generated by alternative splicing (reviewed by

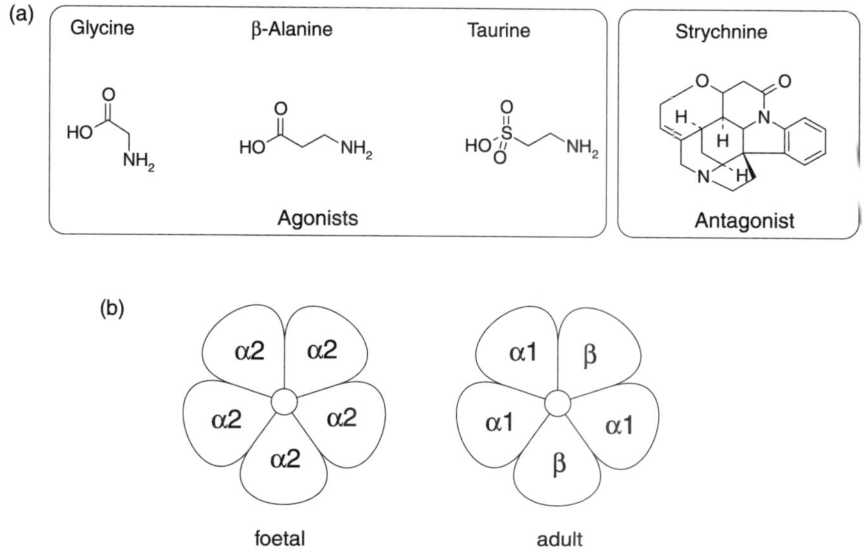

Figure 11.10 Glycine receptor pharmacology and structure. (a) Amino acids that act as agonists at glycine receptors, and strychnine a competitive antagonist. (b) Subunit composition of foetal and adult glycine receptors in the spinal cord. The receptors are shown with a pentameric assembly but the α and β subunits are distinct from those that form $GABA_A$ receptors. Picrotoxin is also an effective glycine antagonist and in recombinant systems is selective for homomeric receptors

Palacin *et al.* 1998). GLYT2 is found in neurons and GLYT1 is found predominantly in glia. The distribution of the transporters with respect to glycine receptors has led to the suggestion that both transporters are associated with glycinergic synapses, while GLYT1 may also regulate extracellular glycine levels at glutamatergic synapses and hence affect the activity of NMDA receptors. Relatively few selective blockers of glycine uptake have been described. GLYT1 isoforms are inhibited by sarcosine (*N*-methyl glycine) and various lipophillic derivatives of sarcosine, including NFPS (*N*[3-(4′-fluorophenyl)-3-(4′-phenyl-phenoxy)propyl]sarcosine) and ORG 24598 (Bergeron *et al.* 1998; Roux and Supplisson 2000). GLYT2a is inhibited by ORG 26176 but not by sarcosine.

GLYCINE RECEPTORS

Glycine receptors can be activated by a range of simple amino acids including glycine, β-alanine and taurine, and are selectively blocked by the high-affinity competitive antagonist strychnine (Fig. 11.10). Glycine receptors were originally isolated from spinal cord membranes on the basis of strychnine binding, and found to be composed of two membrane-spanning polypeptides (termed α and β) and an associated cytoplasmic protein (gephyrin). To date, four α subunit genes (α1–4) and a single β subunit gene have been identified, with several additional variants of the α1 and α2 isoforms produced by alternative splicing (reviewed by Kuhse, Betz and Kirsch 1995; Rajendra, Lynch and Schofield 1997). The α and β subunits are formed from approximately 420 and 470 amino acids, respectively, are similar in structure to $GABA_A$ subunits, and likewise form pentameric receptors with a central ion channel permeable to Cl^- and HCO_3^-. In recombinant expression systems the α subunits give rise to functional homomeric receptors or co-assemble to form heteromeric receptors. The β subunit is only incorporated into receptors when co-expressed with α subunits. Native receptors in the adult spinal cord contain 3 $\alpha 1$ and 2 β subunits whereas neonatal receptors are homomeric receptors formed from $\alpha 2$ subunits. The cytoplasmic protein gephyrin is not needed for the formation of functional receptors but plays an important role in the clustering of both glycine and $GABA_A$ receptors (Moss and Smart 2001).

Glycine receptors in the postsynaptic membrane, like $GABA_A$ receptors, most commonly generate a hyperpolarizing IPSP. In the brainstem, glycine receptors have also been shown to be present on presynaptic terminals, where they induce a small depolarisation that activates Ca^{2+} channels and *increases* neurotransmitter release (Turecek and Trussell 2001). This differs from the action of presynaptic $GABA_A$ receptors described above, where the depolarisation induced is sufficient to inactivate Na^+ channels and *decrease* neurotransmitter release. Unlike $GABA_A$ receptors, glycine receptors are inhibited by some steroids, unaffected by benzodiazepines and are relatively insensitive to barbiturates. However, native and recombinant glycine receptors are positively modulated by a wide range of general anaesthetics, including diethyl ether, halothane, isoflurane, chloral hydrate, brometone and trichloroethylene.

REFERENCES

Asada, H, Kawamura, Y, Maruyama, K, Kume, H, Ding, R, Ji, FY, Kanbara, N, Kuzume, H, Sanbo, M, Yagi, T and Obata, K (1996) Mice lacking the 65 kDa isoform of glutamic acid decarboxylase (GAD65) maintain normal levels of GAD67 and GABA in their brains but are susceptible to seizures. *Biochem. Biophys. Res. Commun.* **229**: 891–895.

Asada, H, Kawamura, Y, Maruyama, K, Kume, H, Ding, RG, Kanbara, N, Kuzume, H, Sanbo, M, Yagi, T and Obata, K (1997) Cleft palate and decreased brain γ-aminobutyric acid in mice lacking the 67-kDa isoform of glutamic acid decarboxylase. *Proc. Natl. Acad. Sci. USA* **94**: 6496–6499.

Barnard, EA, Skolnick, P, Olsen, RW, Mohler, H, Sieghart, W, Biggio, G, Braestrup, C, Bateson, AN and Langer, SZ (1998) International Union of Pharmacology. XV. Subtypes of γ-aminobutyric acidA receptors: classification on the basis of subunit structure and receptor function. *Pharmacol. Rev.* **50**: 291–313.

Bergeron, R, Meyer, TM, Coyle, JT and Greene, RW (1998) Modulation of *N*-methyl-D-aspartate receptor function by glycine transport. *Proc. Natl. Acad. Sci. USA* **95**: 15730–15734.

Bonnert, TP, McKernan, RM, Farrar, S, le Bourdelles, B, Heavens, RP, Smith, DW, Hewson, L, Rigby, MR, Sirinathsinghji, DJ, Brown, N, Wafford, KA and Whiting, PJ (1999) θ, a novel γ-aminobutyric acid type A receptor subunit. *Proc. Natl. Acad. Sci. USA* **96**: 9891–9896.

Bormann, J (2000) The 'ABC' of GABA receptors. *Trends Pharmacol. Sci.* **21**: 16–19.

Bormann, J, Hamill, OP and Sakmann, B (1987) Mechanism of anion permeation through channels gated by glycine and γ-aminobutyric acid in mouse cultured spinal neurones. *J. Physiol. (Lond).* **385**: 243–286.

Bowery, NG (1993) $GABA_B$ receptor pharmacology. *Ann. Rev. Pharmacol. Toxicol.* **33**: 109–147.

Bowery, NG and Enna, SJ (2000) γ-aminobutyric acid(B) receptors: first of the functional metabotropic heterodimers. *J. Pharmacol. Exp. Ther.* **292**: 2–7.

Chebib, M and Johnston, GA (2000) GABA-Activated ligand gated ion channels: medicinal chemistry and molecular biology. *J. Med. Chem.* **43**: 1427–1447.

Crestani, F, Low, K, Keist, R, Mandelli, M, Mohler, H and Rudolph U (2001) Molecular targets for the myorelaxant action of diazepam. *Mol. Pharmacol.* **59**: 442–445.

Cutting, GR, Lu, L, O'Hara, BF, Kasch, LM, Montrose-Rafizadeh, C, Donovan, DM, Shimada, S, Antonarakis, SE, Guggino, WB, Uhl, GR *et al.* (1991) Cloning of the γ-aminobutyric acid (GABA) rho 1 cDNA: a GABA receptor subunit highly expressed in the retina. *Proc. Natl. Acad. Sci. USA* **88**: 2673–2677.

Dutar, P and Nicoll, RA (1988) A physiological role for $GABA_B$ receptors in the central nervous system. *Nature* **332**: 156–158.

Erlander, MG, Tillakaratne, NJ, Feldblum, S, Patel, N and Tobin, AJ (1991) Two genes encode distinct glutamate decarboxylases. *Neuron* **7**: 91–100.

Freund, TF and Buzsaki, G (1996) Interneurons of the hippocampus. *Hippocampus* **6**: 347–470.

Gasnier, B (2000) The loading of neurotransmitters into synaptic vesicles. *Biochimie* **82**: 327–337.

Gunther, U, Benson, J, Benke, D, Fritschy, JM, Reyes, G, Knoflach, F, Crestani, F, Aguzzi, A, Arigoni, M, Lang, Y *et al.* (1995) Benzodiazepine-insensitive mice generated by targeted disruption of the γ 2 subunit gene of γ-aminobutyric acid type A receptors. *Proc. Natl. Acad. Sci. USA* **92**: 7749–7753.

Gupta, A, Wang, Y and Markram, H (2000) Organizing principles for a diversity of GABAergic interneurons and synapses in the neocortex. *Science* **287**: 273–278.

Hsu, CC, Thomas, C, Chen, W, Davis, KM, Foos, T, Chen, JL, Wu, E, Floor, E, Schloss, JV and Wu, JY (1999) Role of synaptic vesicle proton gradient and protein phosphorylation on ATP-mediated activation of membrane-associated brain glutamate decarboxylase. *J. Biol. Chem.* **274**: 24366–24371.

Johnston, GA (1996) $GABA_C$ receptors: relatively simple transmitter-gated ion channels? *Trends Pharmacol. Sci.* **17**: 319–323.

Jonas, P, Bischofberger, J and Sandkuhler, J (1998) Corelease of two fast neurotransmitters at a central synapse. *Science* **281**: 419–424.

Jones, KA, Borowsky, B, Tamm, JA, Craig, DA, Durkin, MM, Dai, M, Yao, WJ, Johnson, M, Gunwaldsen, C, Huang, LY, Tang, C, Shen, Q, Salon, JA, Morse, K, Laz, T, Smith, KE, Nagarathnam, D, Noble, SA, Branchek, TA and Gerald, C (1998) $GABA_B$ receptors function as a heteromeric assembly of the subunits $GABA_BR1$ and $GABA_BR2$. *Nature* **396**: 674–679.

Kaila, K (1994) Ionic basis of $GABA_A$ receptor channel function in the nervous system. *Prog. Neurobiol.* **42**: 489–537.

Kash, SF, Johnson, RS, Tecott, LH, Noebels, JL, Mayfield, RD, Hanahan, D and Baekkeskov, S (1997) Epilepsy in mice deficient in the 65-kDa isoform of glutamic acid decarboxylase. *Proc. Natl. Acad. Sci. USA* **94**: 14060–14065.

Kaupmann, K, Huggel, K, Heid, J, Flor, PJ, Bischoff, S, Mickel, SJ, McMaster, G, Angst, C, Bittiger, H, Froestl, W and Bettler, B (1997) Expression cloning of $GABA_B$ receptors uncovers similarity to metabotropic glutamate receptors. *Nature* **386**: 239–246.

Kaupmann, K, Malitschek, B, Schuler, V, Heid, J, Froestl, W, Beck, P, Mosbacher, J, Bischoff, S, Kulik, A, Shigemoto, R, Karschin, A and Bettler, B (1998) $GABA_B$-receptor subtypes assemble into functional heteromeric complexes. *Nature* **396**: 683–687.

Kerr, DI and Ong, J (1995) $GABA_B$ receptors. *Pharmacol. Ther.* **67**: 187–246.

Kuhse, J, Betz, H and Kirsch, J (1995) The inhibitory glycine receptor: architecture, synaptic localization and molecular pathology of a postsynaptic ion-channel complex. *Curr. Opin. Neurobiol.* **5**: 318–323.

Martin, DL and Rimvall, K (1993) Regulation of γ-aminobutyric acid synthesis in the brain. *J. Neurochem.* **60**: 395–407.

McIntire, SL, Reimer, RJ, Schuske, K, Edwards, RH and Jorgensen, EM (1997) Identification and characterization of the vesicular GABA transporter. *Nature* **389**: 870–876.

McKernan, RM, Rosahl, TW, Reynolds, DS, Sur, C, Wafford, KA, Atack, JR, Farrar, S, Myers, J, Cook, G, Ferris, P, Garrett, L, Bristow, L, Marshall, G, Macaulay, A, Brown, N, Howell, O, Moore, KW, Carling, RW, Street, LJ, Castro, JL, Ragan, CI, Dawson, GR and Whiting, PJ (2000) Sedative but not anxiolytic properties of benzodiazepines are mediated by the $GABA_A$ receptor alpha1 subtype. *Nat. Neurosci.* **3**: 587–592.

Moss, SJ and Smart, TG (2001) Constructing inhibitory synapses. *Nat. Rev. Neurosci.* **2**: 240–250.

Ng, GY, Bertrand, S, Sullivan, R, Ethier, N, Wang, J, Yergey, J, Belley, M, Trimble, L, Bateman, K, Alder, L, Smith, A, McKernan, R, Metters, K, O'Neill, GP, Lacaille, JC and Hebert TE (2001) γ-Aminobutyric acid type B receptors with specific heterodimer composition and postsynaptic actions in hippocampal neurons are targets of anticonvulsant gabapentin action. *Mol. Pharmacol.* **59**: 144–152.

O'Brien, JA and Berger, AJ (1999) Cotransmission of GABA and glycine to brain stem motoneurons. *J. Neurophysiol.* **82**: 1638–1641.

Obata, K, Fukuda, T, Konishi, S, Ji, FY, Mitoma, H and Kosaka, T (1999) Synaptic localization of the 67,000 mol. wt isoform of glutamate decarboxylase and transmitter function of GABA in the mouse cerebellum lacking the 65,000 mol. wt isoform. *Neuroscience* **93**: 1475–1482.

Palacin, M, Estevez, R, Bertran, J and Zorzano, A (1998) Molecular biology of mammalian plasma membrane amino acid transporters. *Physiol. Rev.* **78**: 969–1054.

Pfaff, T, Malitschek, B, Kaupmann, K, Prezeau, L, Pin, JP, Bettler, B and Karschin, A (1999) Alternative splicing generates a novel isoform of the rat metabotropic $GABA_BR1$ receptor. *Eur. J. Neurosci.* **11**: 2874–2882.

Qian, H and Ripps, H (1999) Response kinetics and pharmacological properties of heteromeric receptors formed by coassembly of GABA rho- and gamma 2-subunits. *Proc. Roy. Soc. Lond. B Biol. Sci.* **266**: 2419–2425.

Rajendra, S, Lynch, JW and Schofield, PR (1997) The glycine receptor. *Pharmacol. Ther.* **73**: 121–146.

Rivera, C, Voipio, J, Payne, JA, Ruusuvuori, E, Lahtinen, H, Lamsa, K, Pirvola, U, Saarma, M and Kaila, K (1999) The K^+/Cl^- co-transporter KCC2 renders GABA hyperpolarizing during neuronal maturation. *Nature* **397**: 251–255.

Roberts, E (1986) GABA: the road to neurotransmitter status. In *Benzodiazepine/GABA receptors and chloride channels: structural and functional properties* (Eds Olsen, R and Venter, CJ), Alan R. Liss, New York, pp. 1–39.

Roux, MJ and Supplisson, S (2000) Neuronal and glial glycine transporters have different stoichiometries. *Neuron* **25**: 373–383.

Rudolph, U, Crestani, F, Benke, D, Brunig, I, Benson, JA, Fritschy, JM, Martin, JR, Bluethmann, H and Mohler, H (1999) Benzodiazepine actions mediated by specific γ-aminobutyric $acid_A$ receptor subtypes. *Nature* **401**: 796–800.

Rudolph, U, Crestani, F and Mohler H (2001) $GABA_A$ receptor subtypes: dissecting their pharmacological functions. *Trends Pharmacol. Sci.* **22**: 188–194.

Sagne, C, El Mestikawy, S, Isambert, MF, Hamon, M, Henry, JP, Giros, B and Gasnier, B (1997) Cloning of a functional vesicular GABA and glycine transporter by screening of genome databases. *FEBS Lett.* **417**: 177–183.

Schofield, PR, Darlison, MG, Fujita, N, Burt, DR, Stephenson, FA, Rodriguez, H, Rhee, LM, Ramachandran, J, Reale, V, Glencorse, TA *et al.* (1987) Sequence and functional expression of the GABA A receptor shows a ligand-gated receptor super-family. *Nature* **328**: 221–227.

Soghomonian, JJ and Martin, DL (1998) Two isoforms of glutamate decarboxylase: why? *Trends Pharmacol. Sci.* **19**: 500–505.

Sur, C, Wafford, KA, Reynolds, DS, Hadingham, KL, Bromidge, F, Macaulay, A, Collinson, N, O'Meara, G, Howell, O, Newman, R, Myers, J, Atack, JR, Dawson, GR, McKernan, RM, Whiting, PJ and Rosahl TW (2001) Loss of the major $GABA_A$ receptor subtype in the brain is not lethal in mice. *J. Neurosci.* **21**: 3409–3418.

Takahashi, T, Kajikawa, Y and Tsujimoto, T (1998) G-Protein-coupled modulation of presynaptic calcium currents and transmitter release by a $GABA_B$ receptor. *J. Neurosci.* **18**: 3138–3146.

Thompson, SM and Gahwiler, BH (1992) Effects of the GABA uptake inhibitor tiagabine on inhibitory synaptic potentials in rat hippocampal slice cultures. *J. Neurophysiol.* **67**: 1698–1701.

Tian, N, Petersen, C, Kash, S, Baekkeskov, S, Copenhagen, D and Nicoll, R (1999) The role of the synthetic enzyme GAD65 in the control of neuronal γ-aminobutyric acid release. *Proc. Natl. Acad. Sci. USA* **96**: 12911–12916.

Turecek, R and Trusell, LO (2001) Presynaptic glycine receptors enhance transmitter release at a mammalian central synapse. *Nature* **411**: 587–590.

Verleysdonk, S, Martin, H, Willker, W, Leibfritz, D and Hamprecht, B (1999) Rapid uptake and degradation of glycine by astroglial cells in culture: synthesis and release of serine and lactate. *Glia* **27**: 239–248.

Wagner, S, Castel, M, Gainer, H and Yarom, Y (1997) GABA in the mammalian suprachiasmatic nucleus and its role in diurnal rhythmicity. *Nature* **387**: 598–603.

Werman, R, Davidoff, RA and Aprison, MH (1967) Inhibition of motoneurones by iontophoresis of glycine. *Nature* **214**: 681–683.

White, JH, Wise, A, Main, MJ, Green, A, Fraser, NJ, Disney, GH, Barnes, AA, Emson, P, Foord, SM and Marshall, FH (1998) Heterodimerization is required for the formation of a functional $GABA_B$ receptor. *Nature* **396**: 679–682.

Wotring, VE, Chang, Y and Weiss, DS (1999) Permeability and single channel conductance of human homomeric rho1 $GABA_C$ receptors. *J. Physiol. (Lond.)* **521**: Part 2, 327–336.

Zhang, SJ and Jackson, MB (1995) $GABA_A$ receptor activation and the excitability of nerve terminals in the rat posterior pituitary. *J. Physiol. (Lond.)* **483**: 583–595.

12 Peptides

A. H. DICKENSON

INTRODUCTION

The status of amino acids such as glutamate and GABA as neurotransmitters is well established and widely accepted. However, when peptides are being considered as transmitters, views tend to be more diverse. The definition of a peptide is a chain of amino acids which does not exceed 30 amino acids in length, the arbitrary cut-off before the molecule becomes a protein, which is too bulky to be stored, released and interact with a receptor molecule. One problem with considering peptides as transmitters is that many of the peptides active in the CNS have additional roles elsewhere in the body such as somatostatin controlling insulin and glucagon release and substance P and bradykinin acting on the vasculature. Nevertheless, it is clear that signalling molecules can have roles in many places in the body so there is no reason why a transmitter substance can act as a hormone via the vasculature on a distant site as well as at closer range when released from a nerve terminal to act on an adjacent neuron.

The realisation that peptides can function as neurotransmitters has increased the number of NTs in the brain by at least 20! The increasing number of synthetic agonists and antagonists for the peptide receptors means that function can now be probed and novel therapeutic targets are achieved. It cannot be ignored that the therapeutic effects of morphine and its antagonist naloxone arise from an ability to act on a receptor that is there for the functional effects of endogenous opioid peptide systems. This chapter will consider the life history of a peptide transmitter, comparing it to the classical transmitters such as the excitatory and inhibitory amino acids, acetylcholine and the monoamines and then briefly review the main groups of peptides and their receptors and some of the possible functional aspects of peptides in the CNS. En route, the principle of co-existence that started with the finding of a peptide co-habiting with acetylcholine in parasympathetic neurons will be extended to the CNS — the old 'rule' that a neuron made and used a single transmitter is unlikely to be the case for many brain and spinal cord neurons.

NEUROCHEMISTRY

PEPTIDES — PRODUCTION AND RELEASE

The release (calcium-dependent) and receptor actions of peptides resemble those of the 'classical' transmitters, the receptors being seven transmembrane-spanning receptors

Neurotransmitters, Drugs and Brain Function. Edited by R. A. Webster

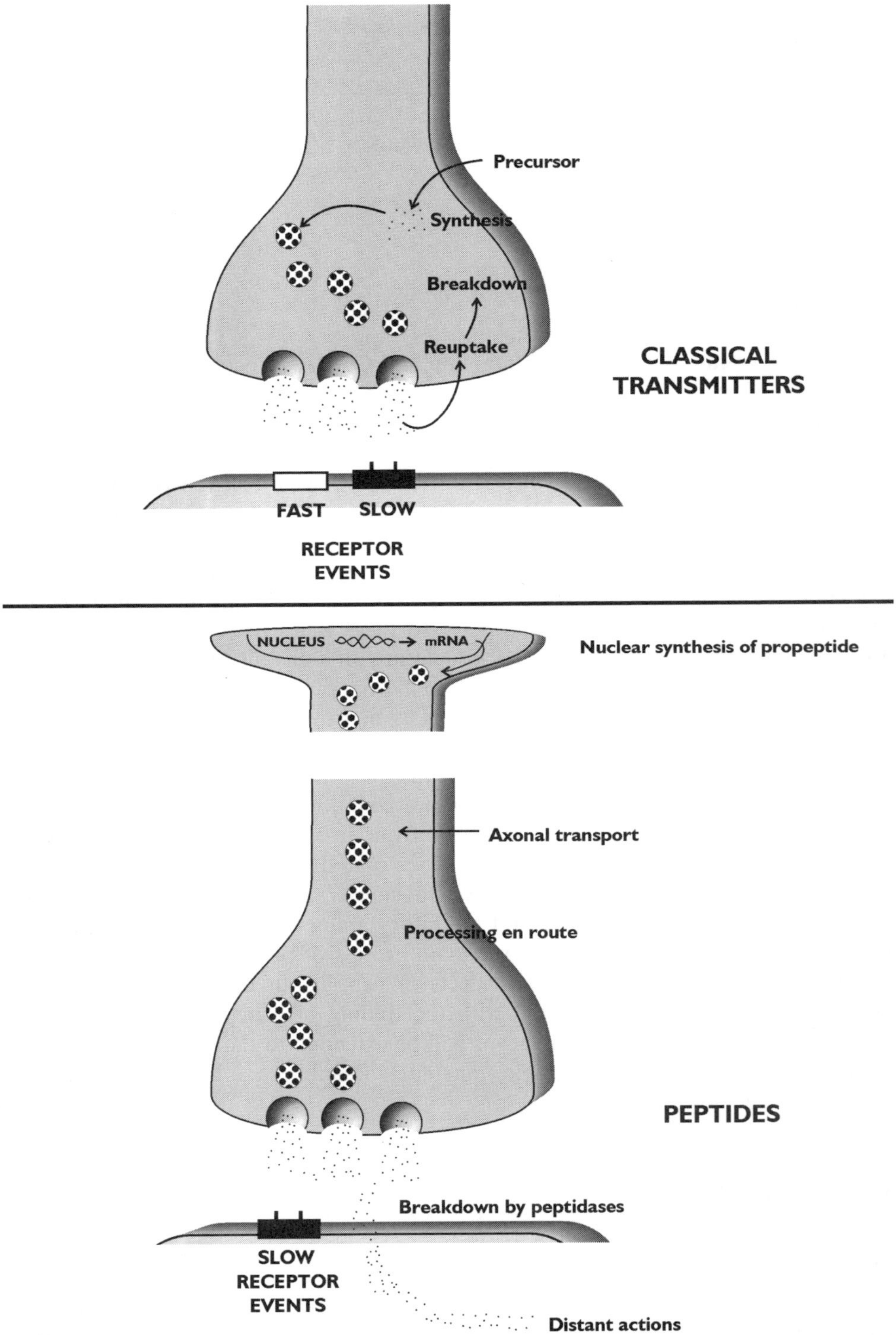

Figure 12.1 A comparison of the production, release and fate of 'classical transmitters' such as glutamate and the monoamines and a peptide. (Most neurons use both classical and peptide transmitters)

coupled to second messengers. As yet there is no evidence that a peptide acts on a receptor that is ionotropic—although it may yet turn out that a peptide receptor can couple to an ion channel, the data points to peptides acting as 'slow' transmitters.

A major difference between the 'classical' transmitters and peptides is that the production of a peptide is quite different since the synthesis of a peptide is in the form of a huge precursor of about 300 amino acids which is produced in the nucleus of the cell and then transported to the terminal being processed en route (Fig. 12.1). The prepropeptide is produced by translation in ribosomes and so occurs only in cell bodies or dendrites while the 'classical transmitters' are produced at the terminal via a short series of enzymatic steps from a simple precursor. The study of the production of the propeptides have revealed a series of principles in that:

- Some propeptides lead to the production of different, in terms of receptor affinities, peptides (substance P and neurokinin A act on neurokinin 1 and 2 receptors, respectively).
- Some propeptides produce multiple copies of similar peptides (met-enkephalin and leu-enkephalin act on the same delta opioid receptor).

The whole process of production of a peptide is sluggish simply because the size of the precursor is so great. Once produced the precursor is packaged into vesicles and then transported down the axon to the terminal. Axonal transport is generally a slow process in that mm–cm/day is rarely exceeded. Thus in a long axon the arrival of the peptide at the release site at the terminal will not be quick. While the precursor is being transported it is processed further by peptidases within the vesicles that cleave the larger parent molecule into smaller fragments. This process continues until the active peptide(s) is produced.

It is easy to speculate that in an active neuron with a rapid firing pattern, the continued release of a peptide may eventually lead to depletion of the peptide occurring. This has been shown in the peripheral nervous system. If this also happens in the CNS it would provide a mechanism whereby the release and resultant receptor effects of a transmitter no longer match the firing pattern and demands of the neuron and so could contribute to long-term adaptations of neurons by a reduction in the time over which a peptide is effective.

The release of some peptides may differ from that of other transmitters, depending on the firing rate of the neurons. The large vesicles needed to store a peptide may need a greater rate of depolarisation for membrane fusion and release of the contents. In the salivary gland the release of vasoactive intestinal polypeptide requires high-frequency stimulation whereas acetylcholine is released by all stimuli. Due to the complexities and problems of access to CNS synapses it is not known if the same occurs here but there is no reason why this should not. In sensory C-fibres a prolonged stimulus appears to be a prerequisite for the release of substance P.

BREAKDOWN

A peptide, once released, is not subject to reuptake like most transmitters, but is broken down by membrane peptidases. There are no known peptide transporters so that reuptake and re-use are not likely. The peptidases are predominantly membrane bound at the synapse and many are metalloproteases in that they have a metal moiety, most often zinc, near the active site. These enzymes are generally selective for particular amino-acid sequences so that one peptidase may cleave a number of peptides if the amino-acid sequences overlap. A number of peptidases are found in the vasculature, including aminopeptidases and angiotensin-converting enzyme and any peptide with an acidic amino acid near the amino-terminal end of the peptide will be degraded after systemic administration.

At a central synapse, the termination of action of a peptide relies on these peptidases. Thus, if there is considerable release at any one time, the peptide may saturate the enzyme(s) and so metabolism will not keep pace with release.

Thus the peptide could escape the synapse where it was released and then diffuse through the tissue. The peptide may then act at sites distant from the neuron that released it, and these sites will be determined simply by receptors for the particular peptide. Consequently, volume transmission or non-synaptic effects may be important. This has been shown for lutenising hormone releasing factor (LHRH) in sympathetic ganglion cells where the peptide can act on neurons over distances of many hundreds of microns. In the CNS, the areas of spinal cord where neurokinins can be detected increase markedly when a prolonged intense peripheral stimulus is applied, suggesting saturation of peptidases allowing the intact transmitter to move through the tissue. Finally, one of the neurokinins, substance P, is found with calcitonin gene-related peptide (CGRP) in sensory neurons that terminate in the spinal cord. These peptides potentiate each other's actions by competing for the same peptidase. When CGRP occupies the active site the actions of substance P are greater since there is less degradation.

MANIPULATION OF ENDOGENOUS PEPTIDES

The only approach so far has been centred on the opioids. The enkephalins are rapidly degraded by membrane-bound peptidases. The synthesis of peptidase inhibitors has been a successful strategy so that kelatorphan, a mixed peptidase inhibitor, inhibiting at least two of the important breakdown enzymes (aminopeptidase N/M and neutral endopeptidase) affords almost complete protection to the enkephalins. The spinal application of the inhibitor produces a reduction of nociceptive responses of cells with the pool of enkephalins protected by the inhibitor likely to be derived from both a segmental release and from descending pathways activated by the stimulus. The recent reports of a systemically active mixed peptidase inhibitor, RB101, is the next stage towards the clinical application of this novel approach to pain relief. Interestingly, the side-effect profile of RB101 appears to be unlike that of morphine in that it does not cause physical and psychological dependence, suggesting that the receptor activation produced by endogenous opioids is unlike that of exogenous drugs. This is to be expected as the peptide transmitter release and consequent receptor activation will only occur following physiological events—the inhibitors only act when the peptide is released. The potential problems with this approach are that of the two enzymes involved (aminopeptidase N/M and neutral endopeptidase), neither is selective for the enkephalins and inhibition of the former could also increase levels of angiotensin, endothelin, CCK and substance P, among others.

PLASTICITY IN PEPTIDES

The processing of the peptide en route from the cell body to the terminal can result in different products being released from the same gene product. Since the precursor is

Table 12.1

	Inflammation	Neuropathy
Glutamate levels	Normal	Normal
Substance P	Upregulation	Downregulation
CCK	Downregulation	Upregulation
Neuropeptide Y	Normal	Upregulation
Galanin	Normal	Upregulation

produced in the nucleus, gene induction and suppression can change the peptide content of neurons. Nerve growth factor appears important in this, as does NMDA-induced calcium influx. There is much evidence for an induction of early onset protoncogenes in neurons elicited by neuronal activity and *c-fos* and *c-jun* are protein markers of these events. When a gene is switched on or off after neuronal activity then some peptides will always be present in neuronal systems and others appear as a result of damage and/or dysfunction to neurons. Thus the pharmacology of a neuron will change as a consequence of pathological changes. This is best illustrated by consideration of sensory C fibres after peripheral inflammation or nerve damage, two conditions that commonly contribute to pain in patients (Table 12.1).

CO-EXISTENCE OF PEPTIDES WITH OTHER TRANSMITTERS

The nature of a peptide makes it easy to raise specific antibodies to amino-acid sequences unique to that particular peptide. On this basis peptides can be mapped accurately in neuronal tissues. When this type of analysis was applied to autonomic neurons, there was hardly a neuron that did not contain a peptide in addition to noradrenaline or acetylcholine. As the peripheral nervous system is easily accessible and end-points are simple to measure (salivation, blood pressure, etc.) the functional consequences of this co-existence have been demonstrated. Thus, VIP and acetylcholine interact to produce the full integrated salivation and vasodilatation that occur following stimulation of the parasympathetic supply to the salivary glands. Stimulation of the nerve causes secretion and vasodilatation — the former is muscarinic since it is blocked by atropine. When administered exogenously the same pattern is seen: ACh causes secretion and VIP dilates. When co-applied, both secretion and vasodilatation are potentiated and this may partly be due to VIP increasing muscarinic binding of ACh, an effect that has also been seen in the cortex. Noradrenaline and neuropeptide Y also have similar interactive effects on the vasculature. Thus, the vasoconstriction produced by both sympathetic stimulation and exogenous noradrenaline is enhanced in the presence of NPY. Co-existence of more than one neurotransmitter in a neuron is likely to be common in CNS neurons but only a small proportion of the billions of neurons have been investigated.

The consequences of co-existence are considerable since the two or more transmitters can:

- Synergise postsynaptically
- Oppose postsynaptically
- Alter release
- Alter breakdown

The combinations of co-existence within a single neuron that have been seen in the CNS are:

(1) Different peptides from the same gene product (met and leu enkephalin, substance P and neurokinin A). The former two act on the same receptor, the delta opioid receptor, whereas the latter act on different receptors, the neurokinin 1 and 2 receptors. Despite this, the receptors for the neurokinins produce the same direction of effect, a slow depolarisation, even though their distribution differs.

(2) Two different peptides (TRH and enkephalin, substance P and CCK) coming from different precursors. These combinations are intriguing in that TRH is excitatory yet enkephalins are inhibitory—complex postsynaptic effects can be envisaged. Substance P is excitatory and CCK acts on two receptors, A and B, with the former being the predominant CNS form.

(3) Co-existence of one or more peptides with a non-peptide transmitter within a single neuron. For example, dopamine and CCK are found together in some but not all of the monoamine neurons and there is a complex interaction involving mutual control of release. This could have implications for therapy for Parkinson's disease, psychoses, addiction, etc., all areas where dopamine is manipulated while the co-existing peptide has, as yet, not been attacked. In another instance, 5-HT and substance P can be found together in supraspinal pathways projecting to the spinal cord. Here the inhibitions produced by activation of these pathways are due to 5-HT yet drug-induced depletion of the monoamine leads to excitatory effects on sensory neurons that are likely to be due to the peptide. Furthermore, as in the periphery, noradrenaline and neuropeptide Y co-exist in some neurons and in C-fibres glutamate and substance P are found together—in this case, 90% of substance P-containing cells have glutamate alongside. Interestingly, as described in Chapter 10, the long slow peptide depolarisations elicted on release of the peptide allow glutamate to activate the NMDA receptor by removal of the Mg^{2+} block. Here substance P plays a permissive role determining which receptor(s) the amino acid can activate.

(4) Different non-peptides, e.g. GABA and 5-HT, are found in some of the descending monoamine neurons forming bulbospinal pathways, as well as substance P and enkephalin. This case of four peptides is the most seen thus far and it suffices to say that in theory, the target neuron could be exposed to information via two fast ionotropic receptors ($GABA_A$ and $5\text{-}HT_3$) and a mix of slow inhibitory and excitatory effects via the remaining 5-HT receptors, the NK1 receptor, the $GABA_B$ receptor and the delta opioid receptor.

Figure 12.2 shows some of the ways in which the co-existence of peptides with classical transmitters can interact at a synapse.

FUNCTION OF PEPTIDES

The peptides will now be considered individually in some detail. It must be noted that the large molecular size of the peptides means that they are even less likely to cross the blood–brain barrier than classical transmitters and the instability of peptides means that full functional studies require non-peptide agonists and antagonists. Whereas nature has provided morphine and medicinal chemists have made naloxone, tools are lacking for many other peptides.

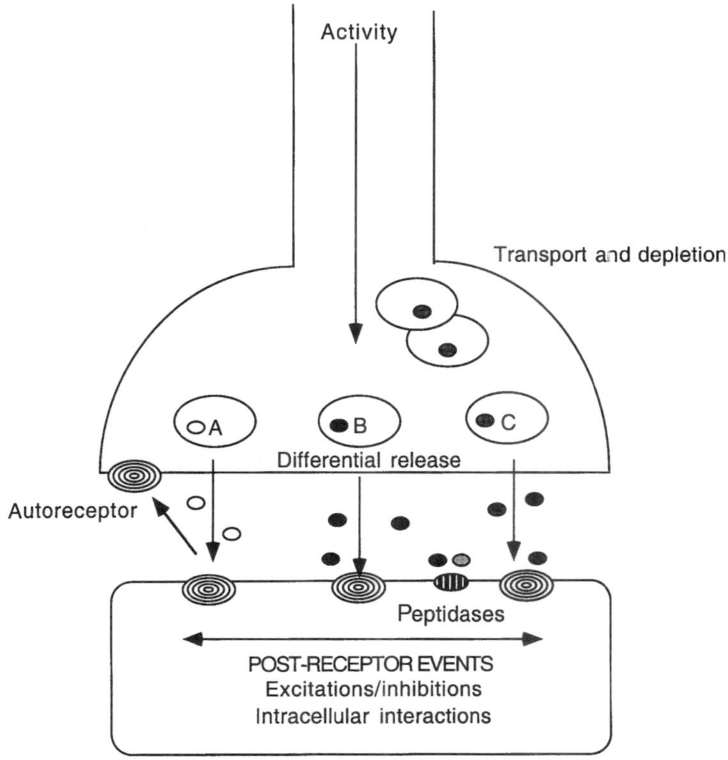

Figure 12.2 A hypothetical synapse where co-existence of peptides and 'classical transmitters' occurs. A is a 'classical transmitter' whereas B and C are peptides. The slow synthesis of peptides and the need for axonal transport may mean that in active neurons, the 'classical transmitter' may be released under all conditions, but the peptide(s) may require higher intensities of stimulation for release and be depleted if the neuron continues to fire for long periods. Competition for peptidases can lead to changes in levels of two co-released peptides. At the postsynaptic site, the receptor mechanisms of the co-existing transmitters can also produce complex changes in neuronal activity

OPIOID PEPTIDES

Despite the use of opium for thousands of years, it was only in the 1970s that the existence of opioid receptors became a reality and subsequently endogenous opioids were identified soon after as the transmitters at these receptors. To date, four receptors have been identified, the mu, delta, kappa and ORL-1 receptors. In 1992 the amino-acid sequence of the delta-opioid receptor was determined by expression cloning, and based on the expected homology to the cloned receptor, the mu and kappa receptors were also cloned. This method was extended to search for novel members of the opioid receptor gene family, and the cDNA encoding of a previously unrecognised receptor protein — the orphan opioid receptor was identified in humans, rats and mice. This new receptor, the hORL1 (human Opioid Receptor-Like 1), exhibited substantial sequence identities with opioid receptors and, once stably transfected into cells, mediated inhibition of adenylyl cyclase. The effect was abolished by diprenorphine, an opiate antagonist. However, since naloxone, the universal opioid antagonist, has low affinity for this receptor, previous functional studies that have used this, or indeed other selective antagonists, to probe opioid function have probably failed to manipulate the nociceptin/orphanin FQ system. The functional roles of nociceptin/orphanin FQ thus remain somewhat elusive and contradictory at present, yet much is known about the other opioid peptides since there are a plethora of selective agonists and antagonists at the receptors.

Mechanisms of action

The mu, delta and kappa opioid receptors are coupled to G^o and G^i proteins and the inhibitory actions of the opioids occur from the closing of calcium channels (in the case of the κ receptor) and the opening of potassium channels (for μ, δ and ORL-1). These actions result in either reductions in transmitter release or depression of neuronal excitability depending on the pre- or postsynaptic location of the receptors. Excitatory effects can also occur via indirect mechanisms such as disinhibition, which have been reported in the substantia gelatinosa and the hippocampus. Here, the activation of opioid receptors on GABA neurons results in removal of GABA-mediated inhibition and so leads to facilitation.

The four opioid receptors display *in vivo* binding preference for mu-endorphins and endomorphins, delta-enkephalins, kappa-dynorphin and ORL1-nociceptin/orphanin FQ (Table 12.2).

These peptides are not completely selective for each type of receptor since the opioid peptides show a degree of sequence homology, although modified synthetic agonists are more selective. Investigation of mu receptor-mediated controls has been hampered by the lack of an endogenous ligand for the receptor in many areas, and in particular, within the spinal cord. Very recently, two peptides (endomorphin-1 and -2) have been isolated with high affinity and selectivity for μ-opioid receptors, making it likely that they are the natural endogenous ligands for the receptor for morphine itself. Table 12.3 summarises the endogenous and synthetic agonists and the antagonists of the four opioid receptors.

Morphine acts on the mu receptor, and so do most of the clinically used opioid drugs. The detailed structure of these receptors has been described and we now have a reasonable understanding of their relative roles in physiological functions and in different pain states.

The best-understood sites of action of morphine are at spinal and brainstem/midbrain loci, producing both the wanted and unwanted effects of the opioid. The spinal actions of opioids and their mechanisms of analgesia involve (1) reduced transmitter release from nociceptive C-fibres so that spinal neurons are less excited by incoming painful messages, and (2) postsynaptic inhibitions of neurons conveying information from the spinal cord to the brain. This dual action of opioids can result in a

Table 12.2 Amino-acid sequences of endogenous opioid peptides

Nociceptin	Phe–Gly–Gly–Phe–Thr–Gly–Ala–Arg–Lys–Ser–Ala–Arg–Lys–Leu–Ala–Asn–Gln
Dynorphin A (1–17)	Tyr–Gly–Gly–Phe–Leu–Arg–Arg–Ile–Arg–Pro–Lys–Leu–Lys–Trp–Asp–Asn–Gln
Endomorphins 1 and 2	Tyr–Pro–Trp–Phe; Tyr–Pro–Phe–Phe
***Leu-* and *met*-enkephalin**	Tyr–Gly–Gly–Phe–Leu/Met

Table 12.3

	Mu (μ)	Delta (δ)	Kappa (κ)	ORL1
Endogenous ligands	β-endorphin Metorphamide Endomorphins	Methionine Enkephalin Leucine Enkephalin β-endorphin	Dynorphin A1–13 Dynorphin A1–8 Dynorphin B	Nociceptin/ orphanin FQ
Synthetic ligands	Morphine DAGO	DPDPE SNC-80 DSTBULET	U50488H Bremazocine Pentazocine	None so far
Antagonists	Naloxone β-FNA	Naltrindole Naloxone	Naloxone nor-BNI	[$Phe^1\Psi(CH_2$–NH)-Gly^2]NC (1–13)NH_2

total block of sensory inputs as they arrive in the spinal cord and is the basis for the spinal analgesic effects of these drugs. At supraspinal sites, morphine can act to alter descending pathways from the brain to the cord which involve noradrenaline and serotonin and these pathways then act to reduce spinal nociceptive activity. In addition, these sites form a link between emotions, depression and anxiety, and the level of pain and analgesia in a patient.

An intriguing area of research on opioids has been the accumulating evidence for plasticity in opioid controls. The degree of effectiveness of morphine analgesia is subject to modulation by other transmitter systems in the spinal cord and by pathological changes induced by peripheral nerve injury. Thus in neuropathic states, pain after nerve injury, morphine analgesia can be reduced (but can still be effective) and tactics other than dose-escalation to circumvent this will be briefly discussed in Chapter 21.

Finally, there is little or no clinical evidence that morphine causes psychological dependence or drug-seeking behaviour, tolerance or problematic respiratory depression in patients. These events simply do not occur when opioids are used to control pain. The reason is likely to be that the actions of morphine and the context of its use in a person in pain are neurobiologically quite different from the effects of opioids in street use. These actions of opioids are described in more detail in Chapter 23.

TACHYKININS

These are a family of peptides which include substance P, isolated in 1931 but only sequenced in 1971. This peptide has been extensively studied since it was the first major peptide to be extracted from brain but only now are useful antagonists becoming available. Two closely related peptides were then isolated from mammalian tissues and can be added to a number of other tachykinins, many of which are found in amphibians. The name tachykinins originated from the vasoactive effects of substance P but the nomenclature has been resolved into calling the three major mammalian peptides substance P, neurokinin A (NKA) and neurokinin B (NKB) with the corresponding receptors being numbered 1 to 3. The order of potencies at the three receptors as follows:

NK1 receptor: SP > NKA > NKB
NK2 receptor: NKA > NKB > SP
NK3 receptor: NKB > NKA > SP

The peptides have the following amino-acid sequences:

SP: Arg–Pro–Lys–Pro–Gln–Gln–Phe–Phe–Gly–Leu–Met
NKA: His–Lys–Thr–Asp–Ser–Phe–Val–Gly–Leu–Met
NKB: Asp–Met–His–Asp–Phe–Phe–Val–Gly–Leu–Met

A single gene gives rise to substance P but it can be produced from three different mRNAs derived from alternative splicing (α, β and γ prepro-tachykinins). Neurokinin A has the same gene and shares two of the same mRNAs as substance P whereas neurokinin B has a different gene. High levels of the tachykinins are found in many parts of the CNS including the caudate, nucleus accumbens, olfactory bulb, colliculus and spinal cord. Both the peptide itself, substance P and the NK1 receptor have been 'knocked-out' and these mice are being extensively studied in order to probe the function of this peptide. Early indications suggest that the peptide and its receptor play important roles in pain (see Chapter 21), inflammation and possibly stress, emesis, anxiety, depression and reward. This is generally backed up by studies with antagonists, although some of the early NK1 receptor antagonists lacked specificity and several blocked calcium channels.

Coupling of the receptors is very similar with all three coupling to Gq and increasing IP3/DAG and in a number of neuronal systems it has been shown that the receptors produce slow depolarising responses via the closing of potassium channels.

CHOLECYSTOKININ

Cholecystokinin (CCK) and gastrin are members of a family that share a similar C-terminal sequence (Gly–Trp–Met–Asp–Phe), the biologically active end of the molecule. CCK is the predominant form in the CNS although it also occurs in the periphery where much of the body's gastrin is found. Early problems on the localisation of CCK stemmed from the overlap of antibodies with CGRP. There is now consensus that CCK-8 is the main transmitter in the CNS although larger forms, CCK-58 and 33, and smaller versions have been located (CCK-7, 5 and 4). Interestingly, CCK-8 is sulphated at the tyrosine residue:

CCK-8: Asp–Tyr–SO_3–Met–Gly–Trp–Met–Asp–Phe

This peptide itself has no selectivity for the two CCK receptors, CCK-A and B, which have so far been established to stimulate IP3/DAG while, like substance P, can close potassium channels to increase neuronal activity. The CCK-B receptor is thought to predominate in the CNS but species differences may make this interpretation difficult. It has a wide distribution in the CNS but is also found in the gut whereas the CCK-A receptor is more restricted but is found in the hypothalamus, hippocampus and in the brainstem. There are high levels of the natural peptide, CCK-8 in cortex, hippocampus, hypothalamus, ventral tegmentum, substantia nigra, brainstem and spinal cord. CCK is one of the most abundant peptides in the brain and CCK co-exists with dopamine, substance P, 5-HT and vasopressin. Interestingly, in the dopamine areas, CCK co-exists in the mesolimbic pathways but in the nigrostriatal projections, the peptide and monoamine neurons form closely apposed but separate groups. This may be the basis for the complex interactions described for dopamine and CCK with regard to release and effects of the two transmitters—both inhibitions and enhancements of function and release/levels have been reported. Nevertheless, there is good evidence that CCK potentiates some of the motor roles of dopamine although this has yet to be demonstrated clinically.

A negative interaction between CCK, acting at the CCK-B receptor, and mu opioids has been consistently reported by electrophysiological, neurochemical and behavioural approaches. The analgesic effects of morphine are attenuated by CCK and potentiated by antagonists at the CCK-B receptor such as L-364,718 (devazepide). The mechanism is thought to involve calcium in that intracellular calcium is mobilised by CCK whereas mu receptor activation, by hyperpolarisation of the neuron, will reduce calcium entry through voltage-operated calcium channels. Thus CCK causes a physiological antagonism of the actions of morphine but this can only occur where the receptors are co-localised. This appears to be restricted to neurons in spinal cord and possibly brainstem which are associated with analgesia since antagonists potentiate only morphine analgesia, not respiratory depression. It is noteworthy that devazepide is only one of a number of recently developed non-peptide antagonists at the CCK-B receptor—agents that will allow CCK-B receptor function to be probed more fully. CCK causes satiation and inhibits feeding in animals and although this central action occurs after peripheral administration of the peptide, it is thought to happen through activation of peripheral vagal pathways to the hypothalamus. This nicely ties in with the established roles of CCK and gastrin in normal digestion.

Finally, the peptide can induce anxiety and panic in normal and anxious volunteers. Some synthetic CCK-B receptor antagonists are chemically similar to the benzodiazepine anxiolytics. Again, the clinical role of CCK manipulation in anxiety remains to be resolved.

NEUROPEPTIDE Y

Neuropeptide Y (NPY) is a large 36 amino-acid peptide found in large amounts in the brain. Although the central administration of NPY in animals causes a number of biological effects which range from control of CRF, increased food intake, anxiolysis, changes in memory and circadian rhythms, the five receptors for the peptide and the lack of any really selective tools for the Y1, 2, 4, 5 and 6 receptors (what happened to the Y3 site!) makes appraisal of this peptide as a drug target rather difficult. A selective antagonist exists only for the Y1 receptor. The marked vasoconstrictor actions of the peptide in the periphery will mean that systemic therapy will require receptor subtype selective agents.

VASOPRESSIN

Vasopressin is closely related to oxytocin and both peptides are cyclic in that they contain a disulphide bridge. Although much is known about the peripheral actions of the peptides the extent of our current knowledge of their possible CNS function is that vasopressin appears to act as a cognitive enhancer and has positive effects on learning processes in animals. Vasopressin acts on three receptors, V1a and b and a V2 receptor.

Again, the existence of non-peptide antagonists at these receptors will soon lead to a better understanding of the central roles of this peptide.

SOMATOSTATIN

Somatostatin exists in two forms, a 14 and a 28 amino-acid form called SRIF-14 and SRIF-28, respectively. Both are widely distributed in the CNS and the peptide produces inhibitory effects on neurons via G-protein-coupled opening of potassium channels. The original receptor division was twofold with a SRIF-1 and -2 receptor division but it is now clear that what are now known as sst2, 3 and 4 subtypes make up the former receptor while the sst1 and 4 receptors are the original SRIF-2 site.

Few antagonists exist at present but the distribution of the peptide with high levels in cortex, hippocampus, amyglada and spinal cord may give some clues to potential functions of the peptide.

SRIF given directly to the spinal cord is antinociceptive which would be expected from an inhibitory peptide, although some studies suggest toxic rather than physiological effects of the peptide. By contrast, the peptide appears to promote convulsions — here its role may be through disinhibition. A well-established central role in the control of growth hormone release has given rise to hopes of treatment of agromegaly and in other contexts, motor actions and increases in sleep times in animals suggest a number of roles of this peptide. There are reports of reduced brain levels of the peptide in Alzheimer's disease (Chapter 20).

NEUROTENSIN

Neurotensin is a peptide with well-established digestive functions which is also found in CNS neurons. There are two receptors (1 and 2) and a paucity of agonists apart from

Table 12.4 Potential roles of peptides

Peptide	Receptor(s)	Function	Potential indication
Opioids	Mu, delta, kappa, ORL1	Pain, anxiety, mood, reward	Chronic pain, addiction
Tachykinins	NK1–3	Inflammation, anxiolysis	Headache, anxiety
Cholecystokinin	A and B	Anxiogenesis, satiation, dopamine function, pain	Panic, eating disorders, pain, Parkinsonism, psychoses
Neuropeptide Y	Y1–6	Obesity, mood, neuronal excitability	Eating disorders, depression, epilepsy
Vasopressin	V1 and 2	Learning, memory	Amnesias
Somatostatin	Sst1–5	Analgesia	Pain
Neurotensin	Nt1 and 2	Temperature, analgesia, pain, dopamine function	Pyresis, pain, Parkinsonism, psychoses
CGRP	1 and 2	Cardiovascular, inflammation, anorexia	Headache, pain, eating disorders
Galanin	GalR1 and 2	Sensory transmission, feeding	Pain, eating disorders

the peptide itself, although selective NT1 receptor antagonists exist. The 13 amino-acid peptide has been implicated in analgesia, thermoregulation and interactions with dopamine function in the nigrostriatal and mesolimbic pathways.

CGRP

Calcitonin gene-related peptide (CGRP) is a product of the calcitonin gene with a distinct mRNA which is formed from alternative splicing in a tissue-specific manner. Thus CGRP is the main product in the CNS whereas calcitonin is found in the thyroid. The peptide is excitatory but whether there is a single receptor or two remains a point of dispute. There are large amounts of CGRP and substance P in fine sensory nerves and the two peptides are released into the periphery by antidromic stimulation where they contribute to the wheal and flare via vasodilatation through complex interactions. There is evidence that this effect of CGRP is via a potentiation of the vascular effects of SP while at the central terminals of sensory nerves the effects of substance P are enhanced by competing with it for a common peptidase so that the metabolism of SP is reduced. Hence there is interest in the potential of CGRP antagonists as therapies for inflammation and headache, although there are no useful compounds of this class at present. Other actions of CGRP have been reported such as altered food intake and thermoregulatory effects.

GALANIN

Galanin is a 29 amino-acid peptide, one amino acid longer in humans than in rats, which acts on three known receptors, GalR1–3, all of which are G-protein-linked receptors, in common with all peptides. There is a lack of any antagonist and apart from a truncated version of galanin having some GalR2 selectivity, no means of separation of the three receptors. However, the consequences of receptor activation are clear in that the GalR1 and 3 receptors are inhibitory and the GalR2 excitatory, although some mixed effects have been reported with the latter. The 1 and 3 receptors open potassium channels whereas the GalR2 receptor mobilises internal calcium, possibly via IP3 mechanisms. The distribution of the 1 and 2 receptors differs, with the former being enriched in hippocampus, spinal cord and peripheral nerves whereas GalR2 has a wider distribution. Galanin co-exists with enkephalin, NPY and substance P, modulates ACh release and is found in GABA-containing neurons in the spinal cord.

Most is known about galanin in the spinal cord and the normal almost undetectable levels of the peptide increase after nerve damage with gene induction occurring. After inflammation there is upregulation of GalR2. In normal animals spinal application of galanin has mixed effects on both spinal neurons and peripheral nerve activity and these are likely to reflect GalR1 and 2 receptors located together. Selective agents for the two receptors are needed.

CONCLUSIONS

It is clear that the peptides present as an interesting group of diverse substances. Their discovery stimulated great expectations but most of these remain to be resolved. As indicated above it seems likely that their low concentrations in the brain and their very slow turnover limits their potential for mediating high levels of activity while their slow time-course of action and apparent need to co-locate with so-called classical transmitters may preclude them from primary activity. Nonetheless they are there and, perhaps more importantly, so are their receptors. It is to be hoped that the synthesis of appropriate agonists and antagonists will make it possible to study the actions of the peptides and possibly develop appropriate therapy, even if this turns out to be a secondary line of attack.

The localisation of a particular peptide to a particular brain area and possibly associated with a particular transmitter (e.g. CCK with dopamine in mesolimbic pathways) has often prompted a prediction of function (e.g. CCK may have a role in schizophrenia). Animal studies in which the peptide has been injected into the appropriate brain area or tested on slices taken from the brain area have sometimes been taken to confirm such hypotheses. These approaches have lined up the peptides for a whole range of potential roles, some of which are listed in Table 12.4. Whether these predictions are realities will depend on the availability of chemical agents and their evaluation, not only in animals but also in humans.

REFERENCES

Darlison, MG and Richter, D (1999) Multiple genes for neuropeptides and their receptors: co-evolution and physiology. *Trends Neurosci* **22**: 81–88.

Hill, R (2000) NK1 (substance P) receptor antagonists—why are they not analgesic in humans? *Trends Pharmacol. Sci.* **21**: 244–246.

Hokfelt, T *et al.* (2000) Neuropeptides—an overview. *Neuropharmacology* **39**: 1337–1356. Also see references therein.

Uhl, GR, Childers, S and Pasternak, G (1994) An opiate receptor gene family reunion. *Trends Neurosci.* **17**: 89–93.

13 Other Transmitters and Mediators

R. A. WEBSTER

With a Section on Nitric Oxide by A. H. DICKENSON

In the preceding chapters, the synaptic pharmacology of those substances clearly established as NTs in the CNS, i.e. glutamate, GABA, ACh, NA, DA, 5-HT and certain peptides, has been discussed in some detail. There are other substances found in the CNS that could have a minor transmitter role, e.g. ATP, histamine and adrenaline, while still others that cannot claim such a property but clearly modify CNS function in some way, e.g. steroids, prostaglandins and nitric oxide. We will consider each of them in what we hope is appropriate detail.

THE PURINES, ATP AND ADENOSINE

ATP (ADENOSINE TRIPHOSPHATE)

For many years ATP has been clearly established as an important intracellular mediator of neuronal function and the provider of cellular energy. The concept that it may also be a neurotransmitter is more recent. It stems from the finding of Burnstock and his colleagues that it was the mediator of the non-adrenergic, non-cholinergic (so-called NANC) innervation of smooth muscle in the intestine and bladder (see Burnstock *et al.* 1970). Generally ATP is a co-transmitter with a wide range of other NTs and while its role may usually be secondary to them, it actually appears in some sympathetically innervated tissue to mediate the initial contraction of smooth muscle rather than the maintained tone. Structurally it consists of an adenine ring, a ribose element and a triphosphate chain (Fig. 13.1).

ATP certainly fulfils the criteria for a NT. It is mostly synthesised by mitochondrial oxidative phosphorylation using glucose taken up by the nerve terminal. Much of that ATP is, of course, required to help maintain Na^+/K^+ ATPase activity and the resting membrane potential as well as a Ca^{2+}ATPase, protein kinases and the vesicular binding and release of various NTs. But that leaves some for release as a NT. This has been shown in many peripheral tissues and organs with sympathetic and parasympathetic innervation as well as in brain slices, synaptosomes and from *in vivo* studies with microdialysis and the cortical cup. There is also evidence that in sympathetically innervated tissue some extracellular ATP originates from the activated postsynaptic cell. While most of the released ATP comes from vesicles containing other NTs, some

Neurotransmitters, Drugs and Brain Function. Edited by R. A. Webster

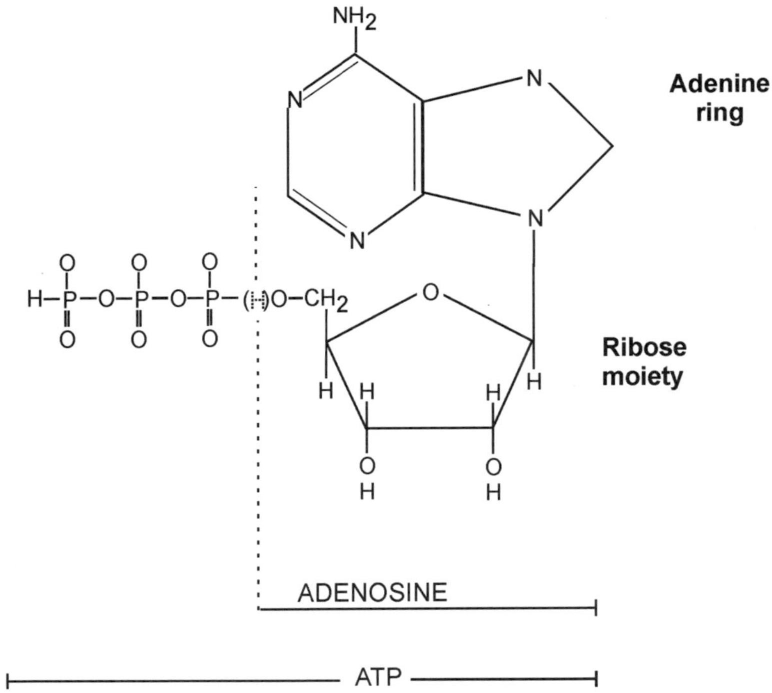

Figure 13.1 Chemical structures of, and relationship between, adenosine and adenosine 5′-triphosphate (ATP). Adenosine contains an adenine ring and ribose component. Phosphorylation of the latter's termial (C_5) hydroxy with three phosphate groups gives ATP

may be stored alone or come directly from the cytoplasm. The extracellular ATP is broken down to adenosine by ecto ATPase.

Unfortunately techniques do not exist for demonstrating purinergic nerves but purinergic receptors have been established. They are divided into two broad groups, P_1 and P_2. The former tend to be located presynaptically, are activated mainly by adenosine and have been reclassified accordingly as A_1 and A_2 (and now A_3). The latter, which respond to ATP, are postsynaptic and as with many other NTs can be divided into two families. Those linked to a fast ionotrophic effect are classified as P_{2x}, with currently six subtypes and those with slow metabotropic effects as P_{2y} with seven subtypes. It is the P_{2x} receptors that mediate the primary transmitter effects of ATP. They have been most studied and while all may be found in the CNS, P_{2x2}, P_{2x4} and P_{2x6} predominate. A schematic representation of a possible ATP (purinergic) synapse is shown in Fig. 13.2.

The role of ATP in the neural control of smooth muscle function is now, as indicated above, well established but its central actions are less clear and have only been studied closely in two areas. In slices of rat medial habenula the synaptic currents, recorded with the whole-cell patch-clamp technique that were evoked by electrical stimulation in the presence of both glutamate and GABA antagonists, were inhibited by the P_{2x} (P_{2x2} preferred) antagonist suramin and by $\alpha\beta$-me-ATP an agonist that desensitises some P_{2x} receptors but not normally the P_{2x2} form. Thus while it is difficult to characterise the precise receptor subunit involved this provides strong evidence for a neurotransmitter

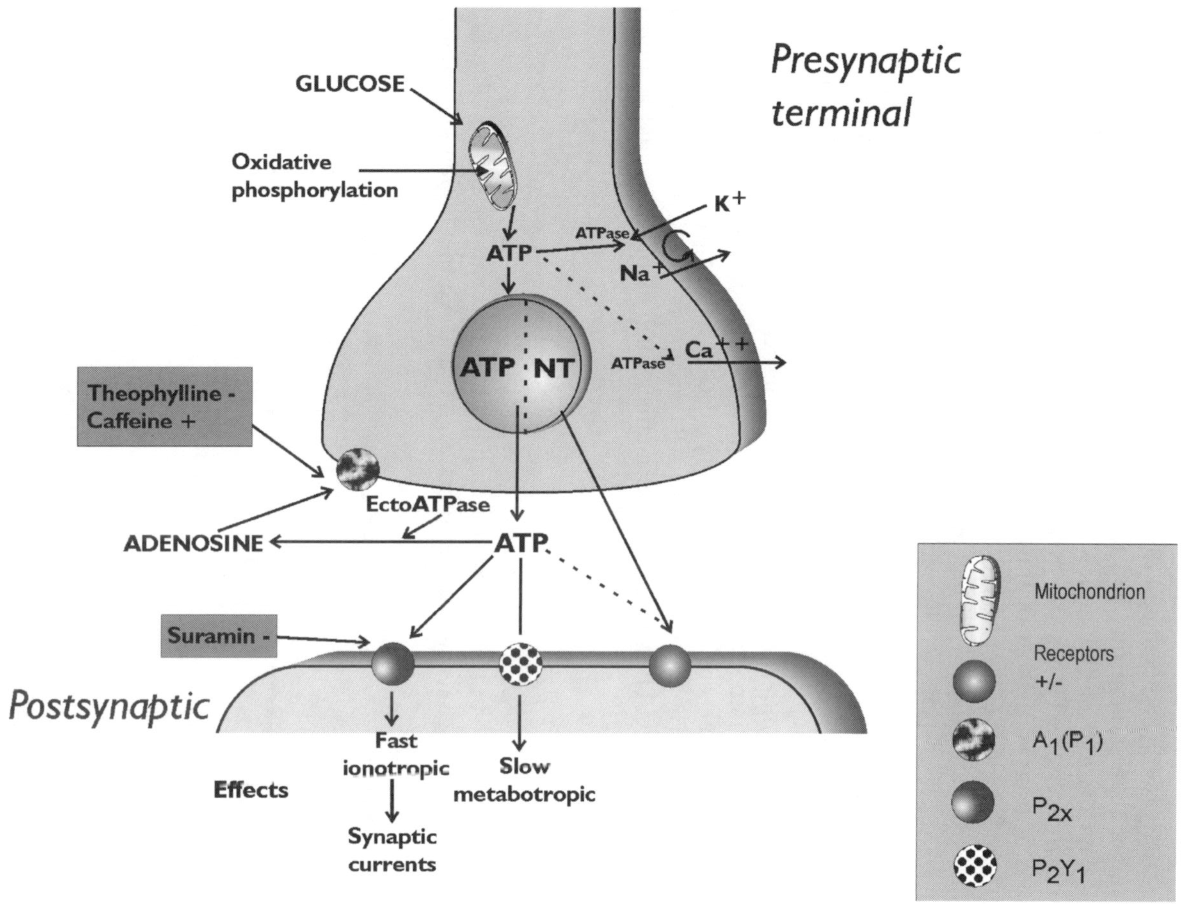

Figure 13.2 Schematic representation of a possible ATP, purinergic, synapse. The effects of ATP, synthesised intraneuronally by mitochondrial oxidative phosphorylation from glucose, on various neuronal ATPases, are shown together with its actions as a conventional neurotransmitter acting at postsynaptic P_2 and presynaptic P_1 receptors

role for ATP, although it is not known to what extent blocking P_{2x2} receptors modifies synaptic transmission when the amino acid receptors are functional. Interestingly the currents mediated by P_{2x} receptor activation are smaller and decay much more slowly than those which characterise glutamate's activation of AMPA receptors but are larger and faster than those mediated by its NMDA receptor. Thus in contrast to NMDA currents, those for ATP are less likely to be involved in the temporal integration of synaptic activity (Gibb and Halliday 1996). The P_{2x} receptor is also linked to control of calcium rather than sodium flux and its subunits have two transmembrane domains compared with the four of glutamate's AMPA and ACh's nicotinic receptor.

This above effect of ATP has also been demonstrated on neurons in lamina II of the dorsal horn in transverse slices of rat cord. Here blockade of glutamate GABA and glycine effects left a current, produced by local tissue or dorsal root stimulation which was again sensitive to the P_{2x2} antagonist suramin (Bardoni *et al.* 1997) (Fig. 13.3). Although only 5% of neurons showed this response, the expression of P_{2x} receptors there and the long established release of ATP from the peripheral terminals of dorsal root ganglia neurons and presumably therefore the central ones, have obviously raised interest in ATP being yet another NT involved in the mediation of afferent painful nociceptive stimuli (Chapter 21).

Thus the neurotransmitter role of ATP is well established in the periphary and also in sensory systems but its importance in the CNS remains to be elucidated (see Burnstock 1996). That requires the development of more specific antagonists and methods of mapping its location. The strong linkage of its P_{2x} receptors to calcium currents may also provide a role for ATP in more long-term effects such as plasticity and neuronal development and death.

ADENOSINE

This is not considered to be a neurotransmitter but it may be an important modulator of neuronal activity through its various receptors, A_1, A_2 and A_3. In addition to its ability

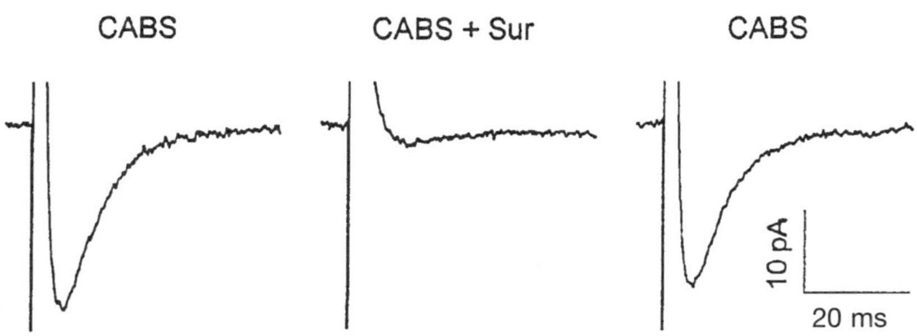

Figure 13.3 Whole-cell patch-clamp recordings of excitatory postsynaptic currents (EPSCs) from dorsal horn neurons of rat (prenatal $P_{2\text{-}13}$) spinal cord slices. The normal evoked EPSC of about 160 pA obtained by focal stimulation of nearby tissue was dramatically reduced by addition of a cocktail (CABS) of CNQX 10 μM, D-APV 50 μM, bicuculline 10 μM and strychnine 5 μM to block glutamate, $GABA_A$ and glycine receptors. The small residual EPSC shown was blocked by the ATP P_2 receptor antagonist suramin and is therefore probably mediated by released ATP. (From Bardoni *et al.* 1997 and reproduced by permission of the *Journal of Neuroscience*)

to reduce the release of a number of NTs it has a strong inhibitory effect on central neurons, enhancing neuronal after-hyperpolarisation through a Ca^{2+}-dependent change in K^+ conductance probably through A_2 receptors. The A_1 receptor appears to be negatively linked to adenylate cyclase through Gi and may mediate the presynaptic inhibition of NT release, with the A_2 acting positively through Gs.

Adenosine comes from the breakdown of ATP. This may occur either extracellularly (Fig. 13.2) or intraneuronally followed by evoked release or transport. Its basal extracellular level is 2 μM but this can increase rapidly when neuronal firing increases and can rise some twentyfold during seizures. The two enzymes responsible for its breakdown are adenosine kinase ($K_m = 2\,\mu M$) and adenosine deaminase ($K_m = 50\,\mu M$). It will be clear that as more adenosine is released during seizures, it will quickly saturate the kinase and its concentration can therefore only be controlled by deaminase. In fact deaminase but not kinase inhibitors are anticonvulsant as is adenosine and its analogues, while its antagonist theophylline is proconvulsant and a central stimulant. Adenosine has even been promoted as an endogenous 'anticonvulsant' (see Dragunow 1986). While that may not be realistic, the antiepileptic benzodiazepine drugs, in addition to their effects on GABA receptors, have been shown to increase the efflux of $[^{-3}H]$ adenosine from the rat cortex probably by blocking its uptake and adenosine is often considered to be an endogenous limiter of neuronal activity. Despite this it has also been shown to reduce fast inhibitory postsynaptic potentials (IPSPs) in the rat lateral amygdala probably by presynaptic A_1 inhibitory effect on GABA release (Henbockel and Pap 1999). Adenosine has also been considered to play a role in sleep induction (Chapter 22).

Recently much interest has been shown in the possible neuroproctive effects of adenosine but the responses are complex. Thus A_3 agonists can offer some protection given chronically before ischaemic challenge but given acutely post-challenge they can be neurotoxic (see Jacobsen 1998).

HISTAMINE

The belief that histamine (HT) has a central effect stems from the knowledge that all the classical antihistamines (H_1 receptor antagonists) used to treat allergic reaction, such as hay fever, caused marked sedation if, like mepyramine and promethazine, they can cross the blood–brain barrier, but fail to do so if, like terfenedine and cetirizine, they do not.

The major problem in establishing histamine as a transmitter in the CNS has been the difficulty in demonstrating its actual presence in neurons rather than just in the invading mast cells, in which it is concentrated and from which it is released in the periphery during allergic reactions. The development of immunohistochemical methods for the visualisation of histamine, and its synthesising enzyme histidine decarboxylase, now show there to be definite histaminergic nerves (see Tohyama *et al.* 1991). These are concentrated in the tuberomammillary nucleus in the posterior hypothalamus, not only in the rat but also in humans, and like the other monoaminergic systems (NA and 5-HT) they give off long highly branched axons which ascend in the medial forebrain bundle projecting to the cerebral cortex and hippocampus. Most histamine neurons also contain other transmitters such as GABA, substance P or enkaphalin. The whole brain concentration of histamine is relatively low (50–70 ng/g) but there is much evidence for its central action (see Schwartz *et al.* 1991).

Histamine is synthesised by decarboxylation of histidine, its amino-acid precursor, by the specific enzyme histidine decarboxylase, which like glutaminic acid decarboxylase requires pyridoxal phosphate as co-factor. Histidine is a poor substrate for the L-amino-acid decarboxylase responsible for DA and NA synthesis. The synthesis of histamine in the brain can be increased by the administration of histidine, so its decarboxylase is presumably not saturated normally, but it can be inhibited by α fluoromethylhistidine. No high-affinity neuronal uptake has been demonstrated for histamine although after initial metabolism by histamine *N*-methyl transferase to 3-methylhistamine, it is deaminated by intraneuronal MAO_B to 3-methylimidazole acetic acid (Fig. 13.4). A Ca^{2+}-dependent KCl-induced release of histamine has been demonstrated by microdialysis in the rat hypothalamus (Russell *et al.* 1990) but its overflow in some areas, such as the striatum, is neither increased by KCl nor reduced by tetradotoxin and probably comes from mast cells.

Histamine receptors were first divided into two subclasses H_1 and H_2 by Ash and Schild (1966) on the basis that the then known antihistamines did not inhibit histamine-induced gastric acid secretion. The justification for this subdivision was established some years later when Black (see Black *et al.* 1972) developed drugs, like cimetidine, that affected only the histamine stimulation of gastric acid secretion and had such a dramatic impact on the treatment of peptic ulcers. A recently developed H_2 antagonist zolantidine is the first, however, to show significant brain penetration. A further H_3 receptor has now been established. It is predominantly an autoreceptor on histamine nerves but is also found on the terminals of aminergic, cholinergic and peptide neurons. All three receptors are G-protein-coupled but little is known of the intracellular pathway linked to the H_3 receptor and unlike H_1 and H_2 receptors it still remains to be cloned. Activation of H_1 receptors stimulates IP_3 formation while the H_2 receptor is linked to activation of adenylate cyclase.

Autoradiography and receptor mRNA studies have shown H_1 receptors to be located in most of the brain areas innervated by the ascending histaminergic axons, e.g. cerebral cortex, hippocampus, limbic areas and hypothalamus. Their presence in the cerebellum is not accompanied by appropriate histaminergic innervation. Very few are found in the striatum but this region does show a high density of H_2 receptors. H_2 receptors are also found with H_1 in the cortex, hippocampus and limbic areas, but not in the hypothalamus. Although basically presynaptic the H_3 receptor is also found postsynaptically in the striatum and cerebral cortex (Pollard *et al.* 1993).

Although histamine generally inhibits neuronal firing in the cerebral cortex through H_1 receptors it causes a H_1-mediated excitation in the hypothalamus. It also appears to potentiate NMDA currents although the receptor type has not been established. In the hippocampus it has been shown to block the long-lasting hyperpolarisation (accommodation) that normally follows neuronal firing and is mediated through a Ca^{2+}-activated K^+ conduction.

From time to time it has been suggested that histamine has some role in a number of behaviours and motor activity while the established and marked sedative effect of H_1 receptor antagonists, mentioned at the start of this section, has consistently been considered to indicate a role for histamine in arousal and the sleep–waking cycle (see Chapter 22).

Histamine release in the hypothalamus is higher during the active waking than the quiescent phase of behaviour, whether this is associated with darkness (in rats) or light (rhesus monkey). The firing rate of histamine neurons is also higher during arousal

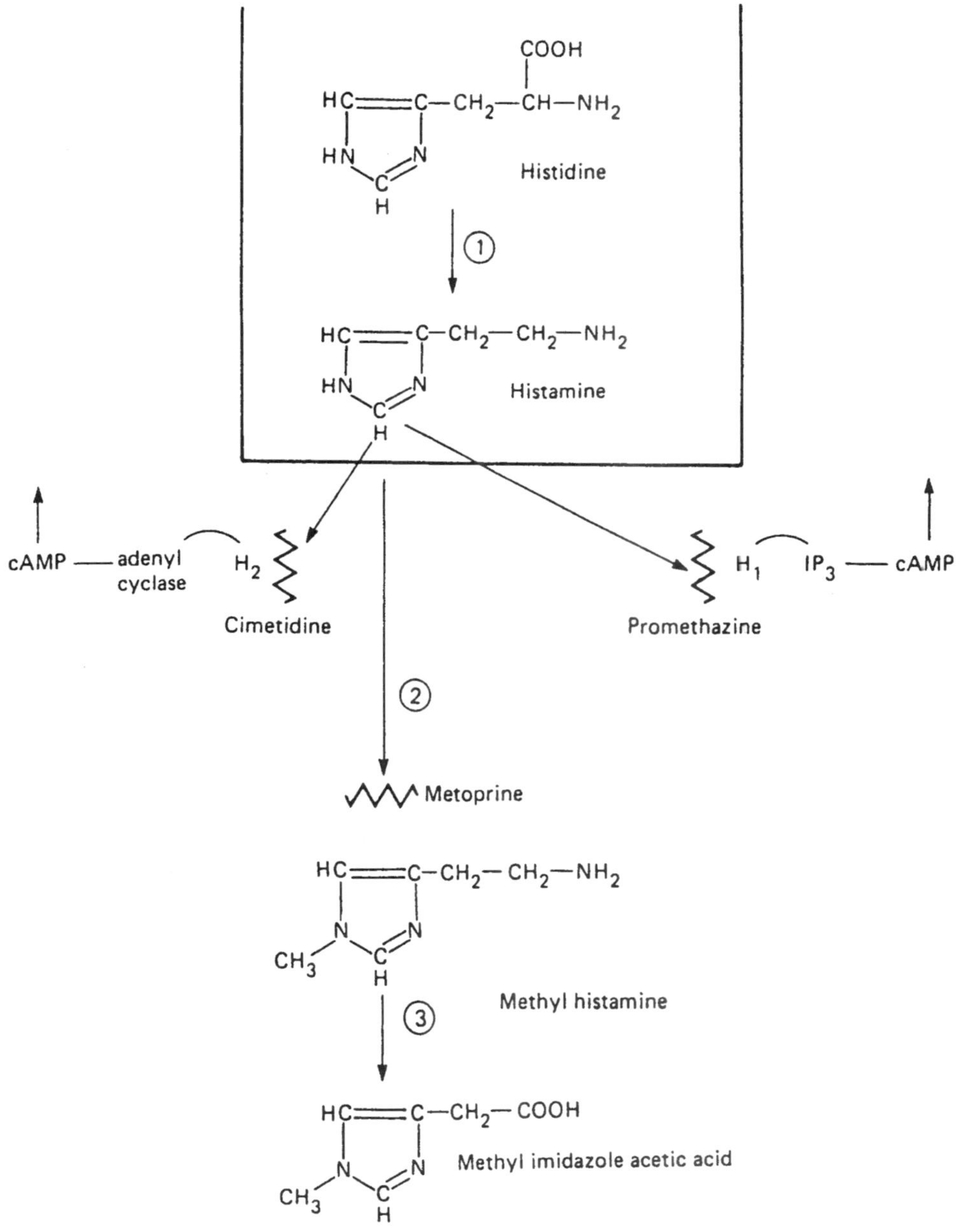

Figure 13.4 Histamine; synthesis, metabolism and receptors. Current knowledge does not justify presentation of a schematic histaminergic synapse. (1) Histidine decarboxylase; (2) histamine-*N*-methyltransferase; (3) mono amine oxidase (MAO_B)

and intraventricular histamine produced EEG arousal in animals. In the cat the H_1 antagonist mepyramine increases the slow-wave sleep pattern while direct injection into the hypothalamus of histamine itself, or an inhibitor of histamine-*N*-methyltransferase to stop histamine breakdown, produces the opposite effect, but it is still sensitive to mepyramine. Such H_1-induced waking effects have not been so clearly established in humans. In contrast to these excitatory effects elevating brain histamine levels with metoprine, an inhibitor of histamine-*N*-methyltransferase protects rodents against maximal electroshock although the specificity of the effect remains to be established. Agonists and antagonists at the H_3 autoreceptors, which should decrease and increase histamine release, have been shown to augment and reduce slow-wave sleep in rats and cats. (For a review of H_3 neuropharmacology see Leurs *et al.* 1998.)

The fact that the older H_1 antagonists, which reach the brain, are prescribed with the warning that they can affect the patient's ability to drive or operate machinery suggests that they impair pscychomotor performance. There are in fact numerous demonstrations of this using tests which require visual–motor coordination such as vigilance tasks and finger tapping. Since the slowing of such function could result from retarding information processing in the visual cortex it is interesting that the latency of components of the evoked potential, which follows presentation of a changing (reversing) black and white checkerboard pattern, is prolonged significantly in humans by the H_1 antagonist diphenhydramine, which enters the brain, but not by terfenedine which does not (Tharion, McMenemy and Rauch 1994).

There is also some evidence that histamine may be involved in food and water intake and thermoregulation (see Hough and Green 1983).

NEUROSTEROIDS

Neurosteroids differ from nearly all the other transmitters and mediators in that they are lipid-soluble and can easily cross the blood–brain barrier. Thus it is necessary to distinguish those steroids that are produced in the brain from those that find their way there from the circulation after being released from the adrenal cortex or gonads. There are many natural and synthetic steroids that have some effect on neuronal function and can be considered neuroactive but few are actually produced in the brain to act on neurons, i.e. the true neurosteroids.

Steroids which are found in the brain may be grouped as follows. The list is not exhaustive.

(a) Those formed in peripheral glands:

Corticosteroids
- Corticosterone (a glucocorticoid)
- Aldosterone (a minerolocorticoid)

Reproductive steroids
- Oestradiol (an oestrogen)
- Testosterone (an androgen)

All these steroids disappear from the brain in animals after removal of the adrenals or gonads (ovary and testis). This also applies to tetrahydrodeoxycorticosterone for although it is formed by reduction of deoxycorticosterone within the brain, its synthesis depends on that steroid coming from the blood.

(b) Those formed in both the periphery and CNS:

Progesterone (PROG) (a progestogen)

Tetrahydroprogesterone ($3\alpha5\alpha$ThPROG) or allopregnenolone, a reduced metabolite of progesterone

Pregnenolone (PREG) and its reduced (20α dihydropregnenolone) and sulphated (pregnenolone sulphate, PREGS) metabolites

The levels of these steroids are reduced, especially that of progesterone (70%) by adrenalectomy and/or gonadectomy but sufficient remains to indicate some central synthesis.

(c) Those formed within the CNS:

Dehydroepiandrosterone (DHEA) and dehydroepiandrosterone sulphate, its sulphated derivative

DHEA levels are unaffected by adrenolectomy or gonadectomy

The neurosteroids of most interest are DHEA, PREG, PREGS, PROG and 3α5αThPROG.

The chemical structures and interrelationship of the neurosteroids listed in (b) and (c) above are shown in Fig. 13.5 but the synthetic pathways are not well established. PREG and DHEA are known as 3β-hydroxy-Δ5-steroids and are also found in peripheral glands as intermediaries between cholesterol and active steroids such as PROG and testosterone. It seems that cholesterol may be the starting point of neurosteroid synthesis in the brain. Cytochrome P450scc, the specific hydroxylase needed for cleavage of the cholesterol side-chain to give PREG, has been found widely in white matter and in myelinating oligodendrocytes, but not in neurons. Enzymes are present for the conversion of PREG to PROG and both are reduced in glia and neurons. This does not occur with DHEA and very little is known of either its synthesis or metabolism.

Neurosteroids are widely and fairly evenly distributed in the brain with few noteworthy localisations but the concentration of the conjugated forms (sulphated and reduced) of PREG and DHEA can exceed that of the parent compounds. Values given by Baulieu (1997) are PREG 8.9 ng/g, with 14.2 and 9.2 for its sulphate and hydroxy metabolites, DHEA 0.24 ng/g (1.7 and 0.45) and PROG 2.2 ng/g. By contrast, although the concentration of progesterone rises some twelvefold in plasma and eight times in hippocampus of animals and humans as they pass from the follicular to luteal phase of the ovarian cycle, it increases by 300 in the cortex, suggesting a considerable variation in the ability of different brain regions to concentrate it. In considering the neurosteroids as possible NTs it should be remembered that neither specific steroid neurons nor an evoked neuronal release of steroids have been demonstrated. There are, however, receptors for them in the CNS and no shortage of actions attributed to them.

There is no doubt that steroids have behavioural effects. Given clinically in the therapy of inflammatory conditions, the glucocorticoids are considered to produce euphoria, followed after prolonged or higher dosing by depression and, of course, when we are stressed the corticocosteroids pumped out by the adrenal cortex easily enter the brain and may initiate some of the behavioural response. In women the premenstrual syndrome (irritability, depression and anxiety) is thought to be associated in some way with progesterone since not only does its concentration rise then fall during that time but in post-menopausal women the use of sequential oestrogen and progestogen hormone replacement therapy (HRT) shows that similar mood changes accompany only the addition of progestogen. More specifically, in women with epilepsy while the incidence of seizures decreases when plasma progesterone is high, it increases during the immediate premenstrual period as progesterone levels fall, rather as with withdrawing

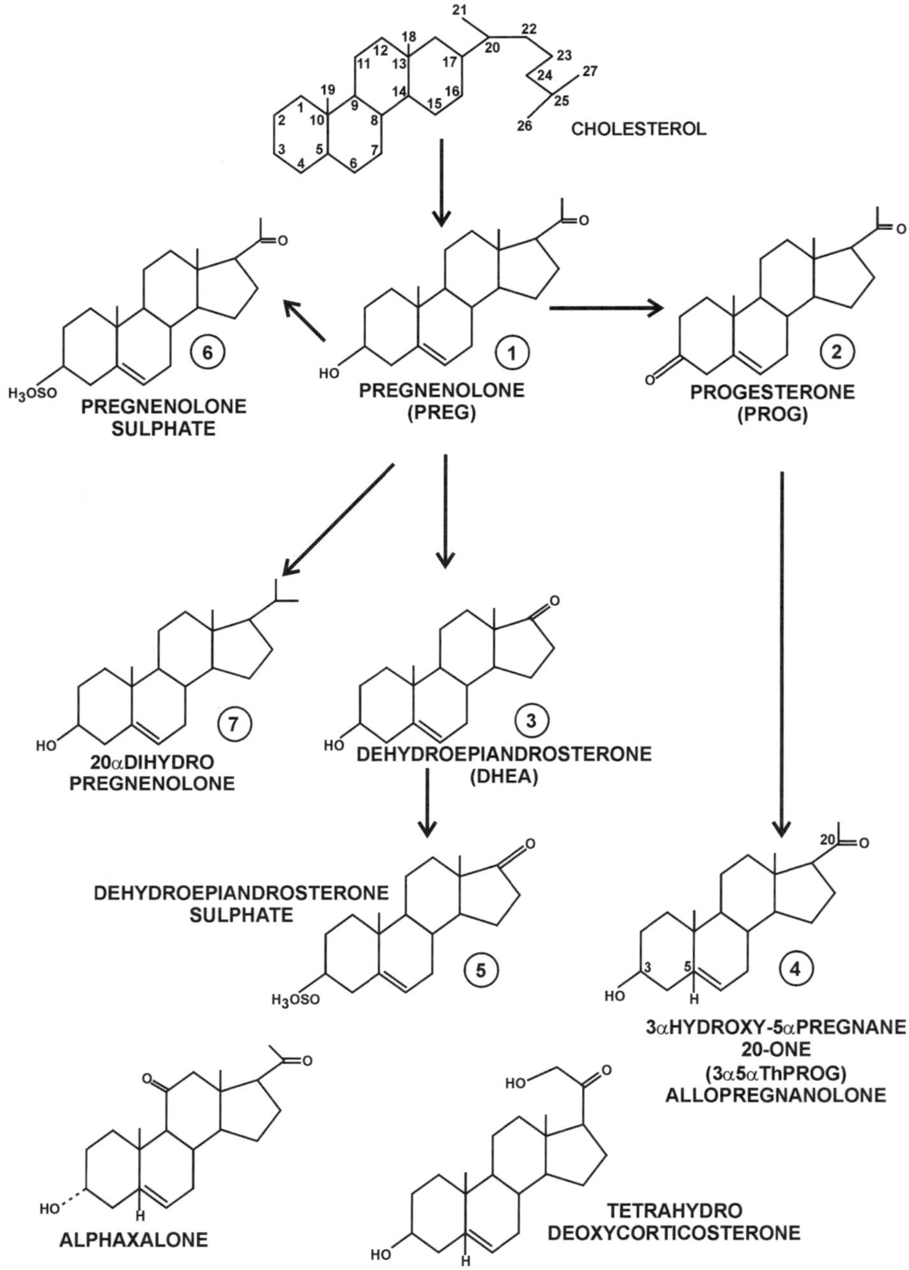

Figure 13.5 Structure and interrelationship of the major neurosteroids. Pregnenolone (PREG) (1) is synthesised from cholesterol and is then either metabolised, reduced to 20α dihydropregnenolone (7) or sulphated (6), or converted by 3β-hydroxysteroid dehydrogenase to progesterone (PROG) (2) or to dehydroepiandrosterone (DHEA) (3). The former (PROG) can then be reduced to allopregnanolone (3α5αThPROG) (4) and DHEA sulphated to dehydroepiandrosterone sulphate (5). It is PREG, DHEA and 3α5αThPROG which appear to be important centrally. The structures of alphaxalone, a steroid anaesthetic and tetrahydrodeoxy-corticosterone, which is formed in the brain from deoxycorticosterone of peripheral origin, are also shown

an antiepileptic drug. Of course, it cannot be assumed that plasma levels are reflected in the brain but in rats stressed by insertion in the Morris water maze, so that they have to swim to a safe platform (see Chapter 18), there is still some increase in the concentration of brain PROG and in particular its reduced metabolic (ThPROG), even after adrenalectomy (Purdy *et al.* 1991). The aggressiveness of castrated male mice exposed to lactating females in a cage can also be reduced by DHEA administration.

These observations, while implicating steroids in brain function and behaviour, cannot be taken as a reliable indicator of their actual effect on neuronal function. Nevertheless, some neurosteroids produce CNS depression with a rapid inhibition of neuronal excitability and one progesterone derivative, alphaxalone (3α-hydroxy-5α pregnane-11, 20 dione, see Fig. 13.5) has been used effectively as an intravenous anaesthetic in humans.

Intracellular steroid receptors, which alter gene expression, exist for corticosteroids, oestrogens and progesterone in the brain, as in the periphery but they cannot account for the relatively rapid depression of CNS function induced by some steroids. This was explained when Harrison and Simmonds (1984) discovered that alphaxalone (the steroid anaesthetic) potentiated the duration of GABA-induced currents at the $GABA_A$ receptor in slices of rat cuneate nucleus just like the barbiturates (Fig. 13.6). Of the

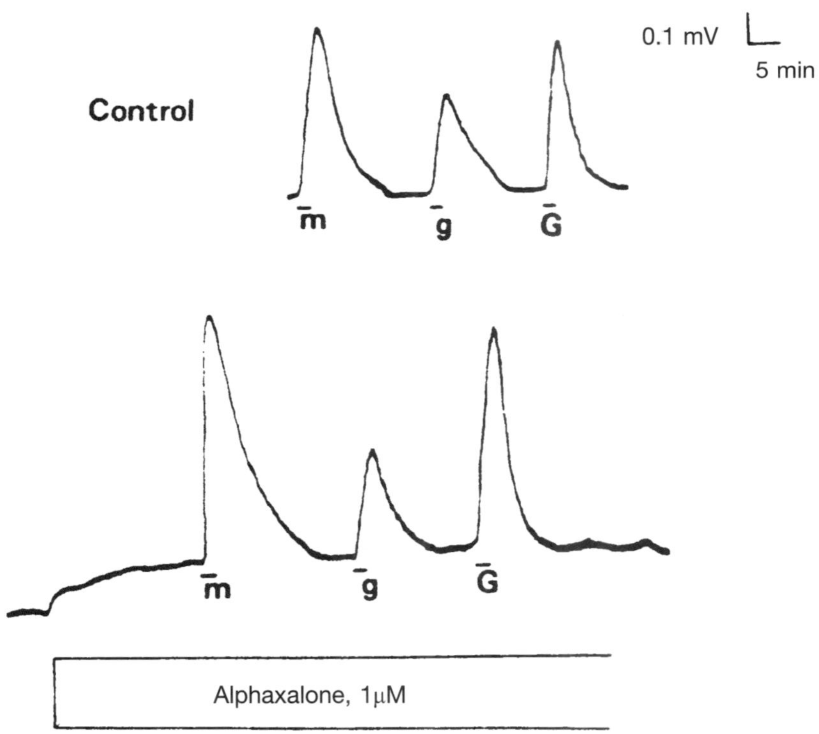

Figure 13.6 Potentiation of GABA action at the $GABA_A$ receptor by the steroid anaesthetic, alphaxalone. Depolarisations recorded extracellularly from dorsal funiculus fibres and terminals in the rat cuneate brain slice after superfusion for 2 min with muscimol 2.5 μM (m), GABA 50 μM (G) and glycine 2 mM (g). In the presence of alphaxalone (1 μM), responses to GABA and the $GABA_A$ agonist muscimol, but not those to glycine, were substantially enhanced. The effect was reversible with responses slowly returning to normal after 3 h. (From Harrison and Simmonds 1984 and reproduced by permission of Elsevier Science)

naturally occurring neurosteroids ThPROG also increases $GABA_A$ receptor currents increasing both the duration and probability of Cl^- channel opening. The sulphated metabolite of PREG is similarly active at low (nM) concentrations. These allosteric effects are still seen after maximal barbiturate potentiation and are not affected by benzodiazepine antagonists suggesting a specific and separate modulating site for the steroids (see Paul and Purdy 1992) although it has not been found. Also while their activity changes with the subunit composition of recombinent $GABA_A$ receptors no specific configuration has been established for their effectiveness but expression of α_2 with $\alpha_1 + B_1$ or $\alpha_2 + B_1$ gives a more responsive receptor than the inclusion of α_3 (Shingai, Sutherland and Barnard 1991).

Not all neurosteroids are inhibitory. The GABA potentiation seen with low concentrations of PREG sulphate changes to antagonism at higher strengths and both this compound and sulphated DHEA, which also inhibits $GABA_A$ receptors, are proconvulsant. There is also evidence that the sulphates of PREG and DHEA potentiate NMDA receptors while glucocorticoids reduce seizure threshold without affecting GABA receptors. Thus even without considering reports of effects on glycine sigma and ACh nicotinic receptors, the electrophysiology of the steroids is complex. In practice, although steriods modulate $GABA_A$ receptors at realistic nM concentrations, unphysiological μM amounts are required at other ligand-gated ion channels (see Rupprecht and Holsboer 1999).

Two unrelated steroid effects warrant some mention. The discovery that PREGS levels are significantly lower in aged than young male rats prompted an interest in its possible role in memory function. In rats, subjected to spatial memory tests in water and Y mazes, impairments in performance were mirrored by reductions in hippocampal PREGS levels (see Baulieu 1997), while aged rats with established memory impairment showed improvement after PREGS administration. Whether this depends on the known ability of PREGS to increase NMDA activity and the accepted role of that receptor in LTP maintenance and memory functions remains to be seen.

In the periphery PROG and PREG may well have an important trophic action since their production in Schwann cells has been shown to result in increased myelin synthesis in regenerating rat sciatic nerve and cultured dorsal root ganglia (see Koenig *et al.* 1995).

With so many different neurosteroids with differing and even opposing neuronal effects, much will depend on their relative concentrations at any time and any evaluation of their function must take this into consideration. Hopefully the synthesis and use of appropriate antagonists will throw more light on the physiological role of steroids in the CNS and facilitate the development and clinical use of new neuroactive steroids (see Gasior e*t al.* 1999).

ADRENALINE

The enzyme β-phenylethanolamine-*N*-methyl transferase, which is required to convert noradrenaline (NA) to adrenaline (Ad), is present in the CNS and there is histofluorometric evidence (positive staining with antibodies to that enzyme and to tyrosine hydroxylase and dopamine β-hydroxylase as well) for adrenergic cell bodies in two groups (nuclei) alongside NA neurons of the locus coeruleus (LC) but ventral and lateral (C_1) and dorsal and medial (C_2) to it. Projections go to the hypothalamus and in particular to the dorsal motor nucleus of the vagus as well as to the LC itself. Little is known of the normal function of these relatively minor pathways although stimulation of C_1 causes a release of adrenaline in the hypothalamus and a rise in blood pressure, which is blocked by mixed β-antagonists but not by β_1-antagonists administered icv (Marsden 1987). Such a β_2-mediated response would be better mediated by adrenaline rather than NA, which has little activity at such receptors (see Chapter 7).

TRACE AMINES (TRYPTAMINE, PHENYLETHYLAMINE, TYRAMINE AND OCTOPAMINE)

Decarboxylation, instead of hydroxylation, of the amino-acid precursors of DA and 5-HT results in the formation of amines that are only found in trace amounts in the CNS but have distinct effects when administered into the brain (Fig. 13.7). Since such decarboxylation can be achieved by the non-specific L-aromatic amino acid decarboxylase there is considerable potential for its occurrence, especially if there is a rise in the concentration of the appropriate precursor or some malfunction in their normal hydroxylation through rate-limiting processes. This could shunt tyrosine, tryptophan and phenylalanine through to tyramine, tryptamine and phenylethylamine rather than to the more normally formed dopa, 5-HT and tyroxine (Fig. 13.7). It is this potential for synthesis together with the known central effects of these amines when injected, that preserves an interest in them despite their very low concentrations in whole brain, i.e. phenylethylamine 1.8, *p*-tyramine 2.0, *m*-tyramine 0.3 and tryptamine 0.5 ng/g. Generally concentrations are highest in the striatum (4, 11, 0.3 and 1.5 ng/g respectively) but still very much lower than DA (10 μg/g). Unlike the catecholamines their concentration rises dramatically (50 times) after inhibition of MAO. Turnover can also be increased easily by the provision of extra substrate since decarboxylation is not rate limiting. Distinct anatomical pathways have not been identified since there is no specific enzyme involved in their synthesis that can be used for immunohistochemistry, and they are not sufficiently concentrated for ordinary histofluorescence.

TRYPTAMINE

Although tryptamine can be detected in brain there has been much debate over whether it exists separately from 5-HT or merely co-exists with it. Specific high-affinity binding sites have been demonstrated for tryptamine in rat cortex. These appear to be different from 5-HT sites but until appropriate antagonists are found it remains possible that they form a subset of the ever-increasing number of 5-HT receptors (see Chapter 9). The behavioural response in rats to tryptophan plus a MAO inhibitor (Grahame-Smith 1971) is accompanied by an elevation of brain tryptamine as well as 5-HT and is less marked if the synthesis of tryptamine is reduced by a decarboxylase inhibitor that does not have a significant effect on 5-HT levels. In fact after a MAOI, tryptamine produces a behavioural response in rats similar to that of tryptophan apart from the absence of certain features like tremor and wet-dog shake. The complexity of the situation is illustrated by studies of the effect of intra-hypothalamic injections of 5-HT and tryptamine on rat body temperature (Cox, Lee and Martin 1981). In these it was shown that 5-HT decreased temperature while tryptamine actually increased it but it was not possible to block one effect preferentially with a whole range of antagonists, despite

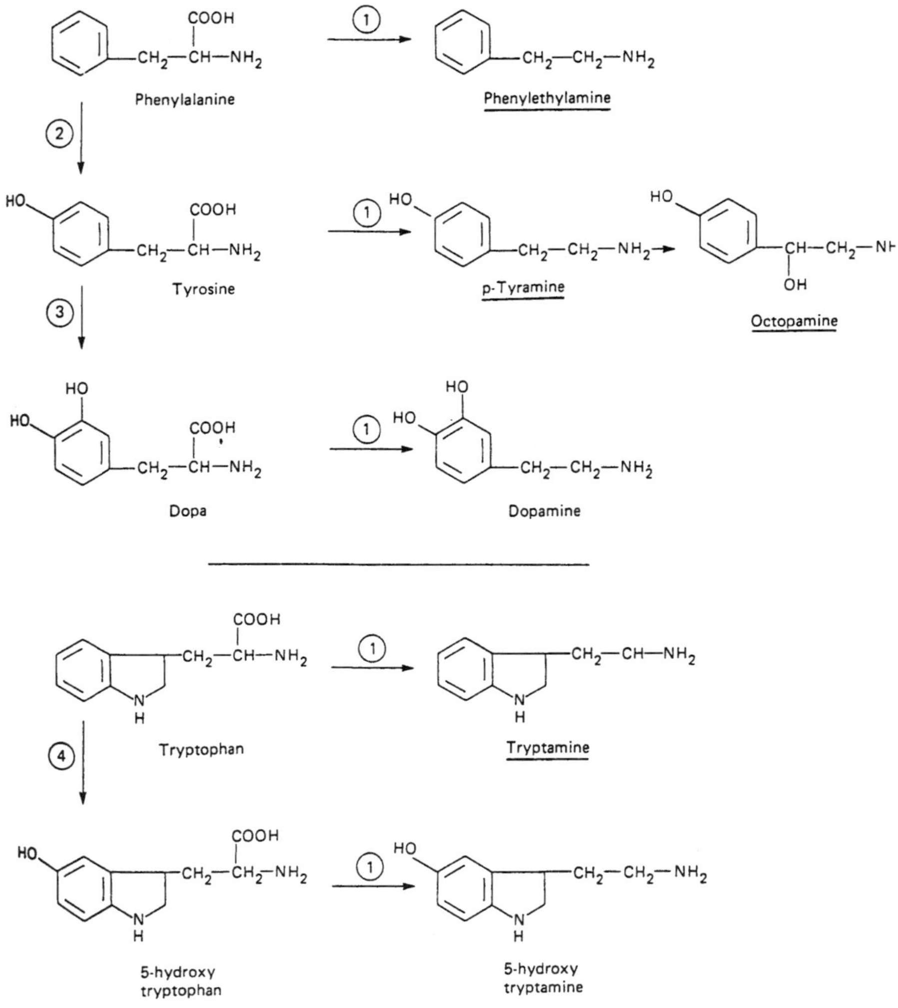

Figure 13.7 Synthesis and structure of the trace amines phenylethylamine, *p*-tyramine and tryptamine. These are all formed by decarboxylation rather than hydroxylation of the precursors of the established monoamine neurotransmitters, dopamine and 5-HT. (1) Decarboxylation by aromatic L-amino acid decarboxylase; (2) phenylaline hydroxylase; (3) tyrosine hydroxylase; (4) tryptophan hydroxylase

some differences in effectiveness. Also although it was the tryptamine and not the 5-HT response, which was abolished after destruction of 5-HT neurons with 5,7-dihydroxytryptamine and implies that tryptamine was releasing 5-HT, it was found that raphé (5-HT) neuron stimulation produces hyperthermia, like tryptamine, rather than hypothermia, like 5-HT.

These opposing effects of tryptamine and 5-HT are also seen when they are applied directly to cortical neurons by iontopheresis. Tryptamine is predominantly depressant while 5-HT is mainly excitatory. Surprisingly, the 5-HT antagonist metergoline is more effective against tryptamine and the depressant effects. When the medial Raphé nucleus is stimulated this produces inhibition of cortical neurons followed by excitation but it is the inhibition (tryptamine-like) that is blocked by metergoline. In keeping with this finding is the observation that depletion of 5-HT with pCPA reduced only the excitatory (5-HT) response while 5,7-dihydroxytryptamine, which destroys the neurons, abolished both effects (see Jones 1983). The inference from these studies and those on temperature is that some neurons in the raphé nucleus release something other than 5-HT. This might be tryptamine but if it is not, then its effects are presumably modified by tryptamine.

Possibly there is a subclass of 5-HT receptors that would be preferentially activated by tryptamine if its endogenous concentrations were ever adequate. Indeed the term 'tryptamine receptor' as first used by Gaddum to describe the effects of all indole amines may be one to which we should return.

PHENYLETHYLAMINE

The relationship of phenylethylamine to dopamine is not unlike that of tryptamine to 5-HT. Present in low concentrations in the brain there is some evidence for distinct binding sites but not for specific neurons. When injected icv it causes stereotyped behaviour similar to, but more marked than, that seen with amphetamine. These effects are blocked by neuroleptics (DA antagonists) and since phenylethylamine does not bind directly to DA receptors it is assumed to release DA, like amphetamine. This is substantiated by the fact that in rats with unilateral 6-OHDA lesions of the SN its systemic administration causes ipsilateral rotation like amphetamine (see Chapter 6). Phenylethylamine certainly increases the overflow of [^{3}H]-DA from striatal synaptosomes and slices and of endogenous DA *in vivo*, but part of this may be due to block of DA uptake. In any case such effects only occur with concentrations of 5×10^{-5} M, which are not likely to be encountered *in vivo* and it is not Ca^{2+}-dependent. Peripherally phenylethylamine causes contractions of the rat fundus just like amphetamine, tryptamine and 5-HT. These are reduced by some 5-HT antagonists, like methysergide, but not by DA antagonists. Thus some of its central effects may be mediated through a tryptamine receptor. Needless to say, the DA and amphetamine-like activity and structure of phenylethylamine, together with the facility for its synthesis in the CNS, has led to unproven suggestions of its involvement in schizophrenia. In fact there is some evidence for increased excretion of its metabolite (phenyl acetic acid) in the subgroup of paranoid schizophrenics.

TYRAMINE

p-Tyramine is produced by decarboxylation of tyrosine and is present in the CNS in higher (threefold) concentrations than *m*-tyramine, the hydroxylated derivative of phenylethylamine. In the periphery *p*-tyramine is easily hydroxylated to octopamine, which has some direct effects on α_1 adrenoceptors, unlike tyramine which functions by releasing NA. When tested on central neurons tyramine always produces the same effects as NA but they are slower and less marked, implying an indirect action. By contrast octopamine often produces the opposite effect to NA and it is probable that octopamine may have a functional role in the invertebrate CNS where it is found in higher concentrations (5 μg/g) than in the mammalian brain (0.5 ng/g). Neither tyramine nor octopamine have distinct behavioural effects, unlike phenylethylamine, and little is known of their central effects, although depressed patients have been shown to excrete less conjugated tyramine than normal subjects after challenge with tyrosine.

PROSTAGLANDINS

The main problem with any study of prostaglandins (PGs) is that although brain concentrations can exceed 0.1 μg/g, they appear to be formed on demand, rather than preformed and stored and they have very short half-lives (seconds). Also specific effective antagonists remain to be developed and PGs are widely and evenly distributed, unlike many NTs. Thus any analysis of their central effects rests heavily on either studying PG release, or their effects when applied directly (icv injection). Certainly the brain has the enzymatic ability to synthesise both prostaglandins (cycloxygenase) and leukotrienes (lypoxygenase) from arachidonic acid (AA) (see Fig. 13.8) and a number of central functions have been proposed for them (see Piomelli 1994).

When injected into the brain (often in rather large concentrations) PGE_2 but not PGF_2 is a depressant and causes sedation and catatonia. PGEs can be found in superfusates of cat cortex and their concentration is increased by direct electrical stimulation as well as by afferent nerve activation. In fact, when given intraventricularly PGE_1 and PGE_2 antagonise convulsions induced by leptazol and electroshock but whether PGs have any

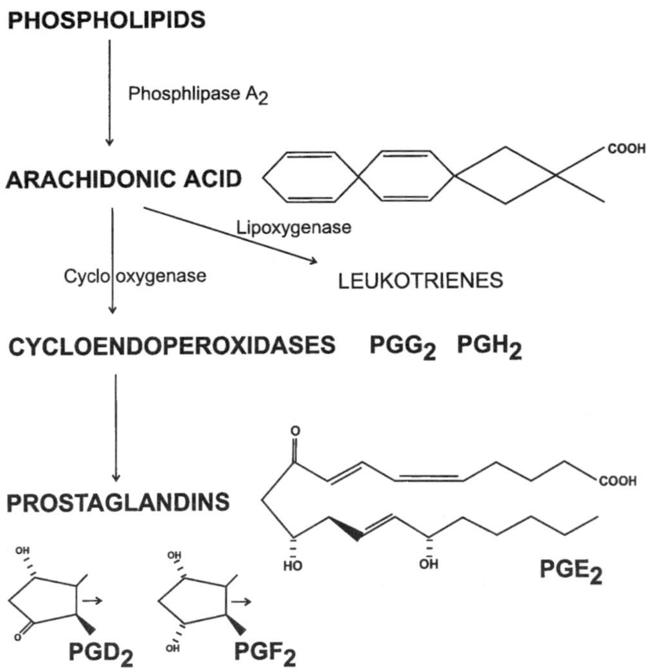

Figure 13.8 Products of arachidonic acid metabolism. The action of cyclooxygenase produces the cyclic endoperoxides PGG_2 and PGH_2 and the prostaglandins PGE_2, $PGF_{2\alpha}$ and PGD_2: lypoxygenase activity gives rise to leukotrienes. The classification D,E,F,G,H depends on the number and position of the hydroxy groups. The subscript (2) describes the number of double bonds. It is the (2) derivatives that appear to be active pharmacologically. ◀ Bonds which lie in front of the plane of the cyclopentane ring. ≡ Bonds which lie behind the plane of the ring.

role in initiating or controlling convulsive activity is uncertain. The levels of a number of PGs, especially PGD_2 and PGE_2, are reported to be significantly lowered in spontaneously convulsing gerbils and in these animals the levels of brain lypoxygenase derivatives have also been found to increase after the onset of seizures (Simmet, Seregia and Hertting 1987), although such changes could result from, rather than cause, the convulsions.

PGD_2 and PGE_2 receptors are concentrated in the preoptic region of the basal forebrain which is known to be imporant in sleep production and when injected into that area PGD_2 does induce sleep. In fact PGD_2 synthesis increases in rat cortex in the light period of a dark/light cycle when rats normally sleep and when infused into the third ventricle it induces a seemingly natural sleep at very low ($-\mu$M) concentrations.

One area of particular interest, in view of the anti-pyretic effects of cyclo-oxygenase inhibitors like aspirin, is the possible role of PGs in the control of body temperature. Thus icv injections of PGE_1 and PGE_2 elevate body temperature and PGE levels increase in CSF following pyrogen-induced fever. Unfortunately this release does not occur near the anterior hypothalamus, which is considered to control body temperature, and iontophoretically applied PGE_2 does not affect the firing of hypothalamic neurons. Also lesions of the anterior hypothalamus abolish PGE- but not pyrogen-induced fever. The situation remains to be resolved (see Wolfe 1982).

Interest in the PGs has recently reverted to their precursor arachidonic acid (AA), which seems to be able to act intracellulary as a second messenger, and also extracellularly. In this latter mode it may play a part in LTP. It is known that AA produces a long-lasting enhancement of synaptic transmission in the hippocampus that resembles LTP and in fact activation of NMDA receptors leads to the release of AA by phospholipase A_2 (see Dumuis *et al.* 1988) and inhibition of this enzyme prevents the induction of LTP. AA has also been shown to block the uptake of glutamate (see Williams and Bliss 1989) which would potentiate its effects on NMDA receptors. This would not only prolong LTP but also cause neurotoxicity.

NITRIC OXIDE

INTRODUCTION

The results of a number of studies demonstrate that the gas nitric oxide (NO) plays a functional role in the central nervous system. This all originated with the discovery that the so-called endothelium-derived relaxing factor (EDRF), found in blood vessels, and thought to be a peptide, was in fact NO. The potential roles of this freely diffusible gas have subsequently been extended to many other tissues and organs but we will concentrate on the possible neuronal roles of what is obviously a novel mediator. There are also suggestions that the closely related carbon monoxide may also have a function in the central nervous system.

Many brain and spinal cord neurons have the capacity to produce NO and experimental evidence indicates a role for this gas in neuronal transmission in animals. A major issue is that the effects of a gas are not limited to the release site and interpretation of the apparent neuronal actions of NO is complicated by the fact that some of the observed effects may be via changes in local blood flow.

Being a gas, NO can diffuse freely once produced, and so is not constrained by the usual mechanisms of release and uptake that confine most transmitters to the synapse. Likewise, the fact that it is not stored means that the criteria of presence and storage are not met by this highly labile and freely diffusible molecule. Finally, its ability to cross lipid barriers means that it is a transcellular mediator rather than a molecule that acts on a surface receptor close to its release site. Thus while it cannot be considered as a neurotransmitter, NO can still have important actions in the central nervous system.

SYNTHESIS

NO is the product of the oxidation of one of the guanidino nitrogens of the amino acid, L-arginine by the enzyme nitric oxide synthase (NOS). L-arginine is then hydroxylated and a second oxygen atom is incorporated to produce NO and citrulline (see Fig. 13.9). The production of NO occurs in many tissues. There are three isoforms of the enzyme, endothelial (eNOS), inducible (iNOS) and neuronal (nNOS). Whereas nNOS and eNOS are regulated by calcium, iNOS is not. This control stems from the production of a calcium–calmodulin complex that then binds NOS and switches on the production of NO from arginine. Then, as internal calcium levels drop, the production of NO also ceases. In many parts of the central nervous system, NMDA receptors are expressed on neurons with the capacity to produce nitric oxide. Thus, calcium influx through the NMDA receptor channel appears to trigger production of nitric oxide by activation of nNOS. To complicate matters, nNOS is also found outside the CNS in epithelial cells and skeletal muscle. Nitric oxide seems to play a much greater role in neuronal

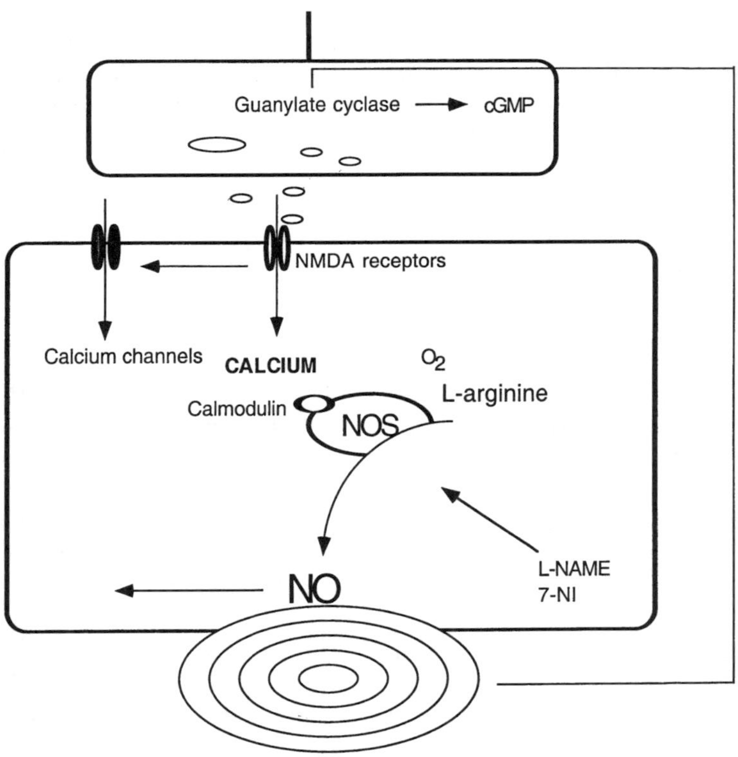

Figure 13.9 The production and actions of nitric oxide (NO). The influx of calcium through either calcium channels or NMDA receptors triggers NOS to convert L-arginine to NO. L-NAME and 7-NI inhibit this process. NO, once produced, can diffuse in a sphere and then can activate guanylate cyclase

transmission following excessive stimulation/pathology in some regions of the brain and so although the enzyme is constitutive it can clearly be unregulated.

CELLULAR ACTIONS

The main action of NO is on the enzyme-soluble guanylate cyclase. NO activates this enzyme by binding to the heme moiety and so there is an increased conversion of GTP to cGMP. This reduces intracellular calcium and this action, and also partly through activation of cGMP-dependent protein kinases, relaxes smooth muscle. The same mechanism of action occurs in neurons but NO can also inhibit other enzymes with a heme group such as cyclooxygenase and lipoxygenase. How these effects translate into what appear to be mostly excitatory effects in the CNS is unclear. Thus agents that increase NO production cause increases in neuronal excitability and vice versa. The results of a number of studies manipulating the levels of the gas show that NO plays a role as a neuronal communicator. There is, however, the problem of a lack of selective agents that modulate the production and actions of NO.

PHARMACOLOGY—INHIBITORS

Application of L-arginine and nitrates and nitrites (that donate NO) has been used to drive the system but, as always, blocking the effects of a potential mediator provides the best approach. There have been reports of a large number of putative inhibitors of NOS but there are two agents that have been widely used, L-N^G nitroarginine (L-NAME) together with the closely related L-N^G monomethylarginine (L-NMAA). These agents block NOS at the arginine site, acting as false substrates, and have no selectivity for any of the three forms of the enzyme. Thus, any study of the physiological role of NO in neurons based on the use of these compounds will be carried out in animals where the vascular effects of NO are also blocked leading to severe hypertension. This may well lead to problems of interpretation and even local application of these compounds directly within the CNS will change local blood flow.

However, more recently, a functionally selective inhibitor of nNOS has been described—7-nitroindazole (7-NI). It is puzzling that *in vitro* this compound has no selectivity for nNOS over eNOS but after systemic administration, fails to change blood pressure yet alters neuronal responses that are thought to result from production of NO. A suggested resolution of this action is that 7-NI is metabolised in the periphery but not the CNS, so that once it has crossed the blood–brain barrier, it can only act on nNOS.

FUNCTION—EXCITOTOXICITY

The proposal that NO or its reactant products mediate toxicity in the brain remains controversial in part because of the use of non-selective agents such as those listed above that block NO formation in neuronal, glial, and vascular compartments. Nevertheless, a major area of research has been into the potential role of NO in neuronal excitotoxicity. Functional deficits following cerebral ischaemia are consistently reduced by blockers of NOS and in mutant mice deficient in NOS activity, infarct volumes were significantly smaller one to three days after cerebral artery occlusion, and the neurological deficits were less than those in normal mice. Changes in blood flow or vascular anatomy did not account for these differences. By contrast, infarct size in the mutant became larger after eNOS inhibition by L-NAME administration. Hence, after middle cerebral artery occlusion neuronal NO production appears to exacerbate acute ischemic injury, whereas vascular NO protects. The data emphasise the importance of developing selective inhibitors of the neuronal isoform.

NOCICEPTION

Behavioural studies are generally unable to find a role for spinal NO in nociceptive reflexes in normal animals, whereas NO inhibitors are highly effective in blocking these same reflexes following the induction of peripheral inflammation or neuropathy. Although a complication is that NO may also play a role in peripheral vascular events during inflammation, these results do suggest that the gas is produced only under certain conditions.

The nNOS inhibitor 7-NI causes a greater inhibition of the wind-up and hyperexcitability of dorsal horn neurons (see Chapter 21) than the immediate response due to direct afferent C-fibre stimulation in normal animals. The preferential inhibition of the NMDA receptor-mediated neuronal hyperexcitability and wind-up of the neurons by 7-NI conforms to the idea that the NO generated in the spinal cord during the transmission of nociceptive information is a consequence of NMDA receptor activation. This also agrees well with a number of other observations, including electrophysiological studies in which block of NO production reduces the excitatory effects of NMDA on neurons and behavioural studies where block of NO production reduced the behavioural effects of NMDA. It seems that following the development of peripheral inflammation and consequent hyperalgesia the NMDA receptor is able to participate in spinal nociceptive reflexes providing a mechanism whereby NO is generated. Thus NOS inhibitors do block nociceptive reflexes in behavioural studies in animals with peripheral inflammation. However, once NO is generated in the spinal cord, the mechanism by which it produces its effects, such as the role played by NO in the wind-up process, has yet to be confirmed. Although NO can act in the neuron in which it is produced to increase levels of cGMP, NO can also diffuse to other neurons to produce its effects. It has been shown that activation of NMDA receptors in the cord can produce an NO-mediated release of glutamate, some of which may represent release from primary afferent terminals following the retrograde diffusion of NO. Nitric oxide can also evoke the release of CGRP and substance P from the dorsal horn of the spinal cord. An NO-evoked release of glutamate, CGRP and substance P may operate as a positive feedback system to further generate wind-up and centrally mediated hyperalgesia. Thus, the development of clinically useful neuronal NOS inhibitors could provide a novel approach to indirectly controlling NMDA receptor-mediated transmission. As with agents directly acting at the NMDA receptor–channel complex, side-effects may preclude their use.

LONG-TERM POTENTIATION

The idea of retrograde messengers such as NO has also been advanced with regard to hippocampal LTP (Chapter 20). There is a marked lack of consensus on whether NO plays a role in LTP and much discussion on why different groups find different results. The importance of the need for a diffusible messenger in the initiation of long-term changes comes from the fact that LTP is induced by activation of postsynaptic NMDA receptors yet maintained by presynaptic changes. Thus, there is a requirement for a mediator to be generated by NMDA receptor activation and then diffuse back to the presynaptic terminals. NO could fulfil this role. Unfortunately, some studies have shown that NOS inhibition blocks LTP whereas others have failed to show this.

SUMMARY AND PERSPECTIVES

NO differs from the more conventional NTs like the amino acids and monoamines in that it is not released from nerve terminals by arriving action potentials. Thus unlike them it is not a primary messenger. It could be regarded as a second messenger except that its effects appear to be mediated by the production of cGMP, itself an established second messenger. In that sense, NO is more like a G-protein. The fact that its synthesis and release from neurons, and so its actions, are dependent on and stimulated by Ca^{2+} influx, often after NMDA receptor activation, inevitably links NO to more extreme excitatory effects such as LTP, excitotoxicity, pain and possibly also epilepsy. Whether blocking its synthesis will be a more effective therapeutic approach than the use of NMDA receptor antagonists is problematic in that even if really specific NOS inhibitors are developed these effects will potentially be at least as widespread as block of NMDA receptors. Where NO inhibition may have the advantage is that it should only operate under conditions of NMDA action that are above normal and so may only affect adverse but not normal neuronal function. This should only occur in those brain areas and pathways showing that extreme level of activity. Time will tell . . .

REFERENCES AND FURTHER READING

Ash, ASF and Schild, HO (1966) Receptors mediating some actions of histamine. *Brit. J. Pharmacol.* **27**: 427–439.

Bardoni, R, Goldstein, PA, Justin Lee, C, Gee, JG and MacDermott, AB (1997) ATP P_{2x} receptors mediate fast synaptic transmission in the dorsal horn of the rat spinal cord. *J. Neurosci.* **17**: 5297–5304.

Baulieu, EE (1997) Neurosteroids: of the nervous system, by the nervous system, for the nervous system. *Recent Prog. Hormone Res.* **52**: 1–32.

Black, JW, Duncan, WAM, Durant, CJ, Ganellin, CR and Parsons, ME (1972) Definition and antagonism of histamine H_2 receptors. *Nature* **236**: 358–390.

Boulton, AA, Baker, GB, Dewhurst, NG and Sandller, M (Eds) (1984) *Neurobiology of the Trace Amines*, Humana Press, Totowa, NJ.

Bredt, DS and Synder, SH (1994) Nitric oxide: a physiologic messenger molecule. *Trends Biochem. Sci.* **63**: 175–195.

Burnstock, G (1996) Purinergic neurotransmission. *Seminars in the Neurosciences* (Ed. Burnstock, G) **8**: 171–257.

Burnstock, G, Campbell, G, Satchell, D and Smythe, A (1970) Evidence that adenosine triphosphate or a related nucleotide is the transmitter substance released by non-adrenergic inhibitory nerves in the gut. *Brit. J. Pharmacol.* **40**: 668–688.

Colasanti, M and Suzuki, H (2000) The dual personality of NO. *Trends Pharm. Sci.* **21**: 249–252.

Cox, B, Lee, TF and Martin, D (1981) Different hypothalamic receptors mediate 5-hydroxytryptamine and tryptamine induced core temperative changes in the rat. *Brit. J. Pharmacol.* **72**: 472–482.

Dragunow, M (1986) Adenosine, the brain's natural anti-convulsant. *Trends Pharmacol. Sci.* **15**: 128–129.

Dumuis, A, Sebben, M, Haynes, L, Pin, JP and Bockaert, J (1988) NMDA receptors activate the arachidonic acid cascade system in striatal neurons. *Nature* **336**: 68–70.

Edwards, FA, Gibb, AJ and Colquhoun, D (1992) ATP receptor mediated synaptic currents in the central nervous system. *Nature* **359**: 144–147.

Gasior, M, Carter, RB and Witkin, JM (1999) Neuroactive steroids: potential therapeutic use in neurological and psychiatric disorders. *Trends Pharmacol. Sci.* **20**: 107–112.

Gibb, AJ and Halliday, FC (1996) Fast purinergic transmission in the central nervous system. In *Seminars in the Neurosciences* (Ed. Burnstock, G), pp. 225–232.

Grahame-Smith, DG (1971) Studies *in vivo* on the relationship between brain tryptophan, brain 5HT synthesis and hyperactivity in rats treated with a monoamine oxidase inhibitor and L-tryptophan. *J. Neurochem.* **18**: 1053–1066.

Greene, RN and Haas, HL (1991) The electrophysiology of adenosine in the mammalian central nervous system. *Prog. Neurobiol.* **36**: 329–341.

Griffith, O and Stuehr, D (1995) NO synathases: properties and catalytic mechanisms. *Ann. Rev. Physiol.* **57**: 707–736.

Harrison, NL and Simmonds, MA (1984) Modulation of the GABA receptor complex by a steroid anaesthetic. *Brain Res.* **323**: 287–292.

Henbockel, T and Pap, HC (1999) Modulatory effects of adenosine on inhibitory postsynaptic potentials in the lateral amygdala of the rat. *Brit. J. Pharmacol.* **128**: 190–196.

Hough, LS and Green, JP (1983) Histamine and its receptors in the nervous system. In *Handbook of Neurochemistry* (Ed. Lajtha, A), Vol 6, Plenum Press, New York, pp. 187–211.

Jacobsen, KA (1998) Adenosine A_3 receptors: novel ligands and paradoxical effects. *Trends Pharmacol. Sci.* **19**: 184–191.

Jones, RSG (1983) Trace biogenic amines: a possible functional role in the CNS. *Trends Pharmacol. Sci.* **4**: 426–429.

Koenig, HL, Schumacher, M, Ferzaz, B, DeThi, A, Ressouches, A, Gennoun, R, Jung-Tester, L, Robel, P, Akwa, Y and Baulieu, EE (1995) Progesterone synthesis and myelin formation by Schwann cells. *Science* **286**: 1500–1503.

Leurs, R, Blandina, P, Tedford, C and Timmerman, H (1998) Therapeutic potential of histamine H_3 receptor agonists and antagonists. *Trends Pharmacol. Sci.* **19** 177–183.

Marsden, CA (1987) Measurements of hypothalmic adrenaline release *in vivo*—effects of drugs and electrical stimulation. *Neuropharmacol.* **22**: 823–830.

Nicholls, DG and Ward, MW (2000) Mitochondrial membrane potential and neuronal glutamate excitotoxicity: mortality and millivolts. *Trends Neurosci.* **23**: 166–174.

Paul, SM and Purdy, RH (1992) Neuroactive steroids. *FASED* **6**: 2311–2322.

Piomelli, D (1994) Eicosanoids in synaptic transmission. *Crit. Rev. Neurobiol.* **8**: 65–83.

Pollard, H, Moreau, J, Arrang, JM and Schwartz, JC (1993) A detailed autoradiographic mapping of histamine H_3 receptors in rat brain areas. *Neuroscience* **52**: 168–189.

Purdy, RH, Morrow, AL, Moore, PHJ and Paul, SM (1991) Stress-induced elevations of α amino butyric acid type A receptor-active steroids in the rat brain. *Proc. Natl. Acad. Sci. USA* **88**: 4553–4557.

Rupprecht R and Holsboer F (1999) Neuroactive steroids: mechanisms of action and neuropsychopharmacological perspectives. *Trends Neurosci.* **22**: 410–416.

Russell, WL, Henry, DP, Phebus, LA and Clemens, JA (1990) Release of histamine in rat hypothalamus and corpus striatum *in vivo*. *Brain Res.* **512**: 95–101.

Schwartz, JC, Arrang, JM, Garbary, M, Pollard, H and Ruat, M (1991) Histaminergic transmission in the mammalian brain. *Physiol. Rev.* **71**: 1–51.

Shingai, R, Sutherland, ML and Barnard, EA (1991) Effects of subunit types of the cloned $GABA_A$ receptor on the response to a neurosteroid. *Eur. J. Pharmacol.* **206**: 77–80.

Simmet, TH, Seregia, A and Hertting, G (1987) Formation of sulphidopeptide—leukotrienes in brain tissue of spontaneously convulsing gerbils. *Neuropharmacology* **26**: 107–110.

Tharion, WJ, McMenemy, DJ and Rauch, TM (1994) Antihistamine effects on the central nervous system, cognitive performance and subjective states. *Neurophsychobiol.* **29**: 97–104.

Tohyama, M, Tamiya, R, Inagok, N and Takagi, H (1991) Morphology of histaminergic neurons with histidine decarboxylase as a marker. In *Histaminergic Neurons: Morphology and Function* (Eds Watanobe, T and Wadd, H), CRC Press, Boca Raton, FL, CRC Press, pp. 107–126.

Usdin, E and Sandler, M (1976) *Trace Amines and the Brain.* Marcel Dekker, New York.

Williams, JH and Bliss, TV (1989) An *in vitro* study of the effect of lipoxygenase and cyclo-oxygenase inhibitors of arachidonic acid on the induction and maintenance of long-term potentiation in the hippocampus. *Neurosci. Lett.* **107**: 301–309.

Wolfe, CS (1982) Eicosanoids: prostaglandins, thromboxanes, leukotrienes and other derivatives of carbon-20 unsaturated fatty acids. *J. Neurochem.* **38**: 1–9.

Section C

NEUROTRANSMITTERS IN DRUG ACTION AND DISEASE STATES

In the preceding sections we have outlined and evaluated the methods by which NT function may be studied and considered the general pharmacology of the major NTs. This should enable us to consider the possible role of NTs in disease states and drug action.

The object is to determine not only whether the symptoms of a particular disorder of the CNS can be explained by the malfunction of a certain NT but whether drugs which are known to be effective in a disorder have a distinct effect on one NT system. These objectives are not unrelated since if a disorder is shown to be due to the increased activity of a particular NT then at least some drugs which are effective in its treatment could be assumed to work by decreasing the function of that NT. Similarly if a whole range of drugs are found to be effective in a certain disorder and all increase the activity of one NT then the disorder could in turn be due to a reduced function of that NT. Ideally one would hope to establish the NT malfunction that causes the disorder and then develop appropriate drugs to counter that malfunction and treat the disorder. In practice this has rarely happened, mainly because of the difficulty of establishing a true NT malfunction in humans.

14 Study and Manipulation of Neurotransmitter Function in Humans

R. A. WEBSTER

MEASUREMENTS OF NEUROTRANSMITTER FUNCTION IN HUMANS

BIOCHEMICAL APPROACHES

Before it can be established that a disorder is related to a change in the activity or use of a given NT it must be shown that one (or more) of the following measurements, made in patients with that disorder, is significantly different from the same measurements obtained in appropriately matched controls. Such measurements include (a) plasma, urine, CSF or post-mortem brain levels of the NT, or its metabolites, (b) the activity of its synthesising or degrading enzymes in post-mortem brain tissue and (c) the number or affinity of its receptors in the brain. Needless to say, any change observed must only be found in patients with that disorder and it must not be a consequence of drug therapy, diet or other identifiable factors. Even then it is still necessary to establish whether the change causes the disorder or results from it. The objective is clearly demanding and yet the methodology to realise it is often unreliable.

The value of using measurements of a NT, or its metabolites, in blood and urine as an index of its function in the CNS must be questionable. In fact their value depends to some extent on the NT being studied. ACh is rapidly metabolised synaptically and the choline quickly re-used, so there is no end-product metabolite to pass into the blood. The amino acids GABA and glutamate are not only rapidly taken up into neurons and glia but incorporated into various biochemical pathways and so are not likely to influence plasma levels. Peptides are also broken down rapidly. For these reasons much attention has been focused on the monoamines since despite uptake and re-use their metabolites can be measured in blood and urine, even if their origin is not easily established. Thus one may be interested in 5-HT within the CNS but since over 90% of the body's 5-HT is in the gut, its activity there will have much more effect on plasma levels than any alteration in its central function. The same could apply to NA except that 3:4 dihydroxyphenylglycol (MOPEG) is thought to be the main metabolite of brain NA, while 3-methoxy-4-hydroxymandelic acid (VMA) comes predominantly from peripheral NA. The restriction of DA primarily to the CNS gives it a somewhat unique position but very little metabolate reaches the plasma. If plasma concentrations are of little value then analysis of the urine is even more pointless. Because of these problems much hope was placed on the analysis of CSF but this can only be obtained by the

Neurotransmitters, Drugs and Brain Function. Edited by R. A. Webster

relatively dangerous procedure of lumbar puncture and as the composition of CSF at the base of the spine may not be the same as that in higher brain areas, yet alone reflect synaptic levels, it is rarely used. Microdialysis has been used in humans but raises ethical issues.

The value of direct studies on brain tissue itself depends to some extent on whether the disorder has a distinct neuronal lesion or just a biochemical malfunction with no clear neuronal degeneration. Post-mortem studies in Parkinsonian patients show a clear degeneration of the substantia nigra and while this was observed long before the corresponding loss in DA it has been possible to link them. On the other hand, a clear link between the long-established and diffusely distributed neurofibrillary tangles and cholinergic dysfunction in the brains of Alzheimer's patients has been more difficult to identify. Even when a distinct focus can be established, as in some forms of epilepsy, the results of studies on changes in NT levels have been equivocal.

In psychiatric disorders there is no clear neurodegeneration and the assumed biochemical fault has been even more difficult to identify. In schizophrenia the impetus to study the involvement of dopamine came from the discovery that all anti-schizophrenic drugs were dopamine antagonists but clear evidence of any malfunction of that NT, apart from a possible increase in receptor number, has not been forthcoming (see Chapter 17). Similarly with depression, the knowledge that most anti-depressant drugs increase NA (and/or 5-HT) function was a valuable lead but again direct attempts to implicate a reduced activity for either NT in humans have been disappointing (see Chapter 20).

Such difficulties prompted research workers to look for some other index of NT function in humans. These range from studies on platelets, such as abnormalities in their amine uptake and MAO activity in depressed patients, to changes in the secretion of a hormone known to be controlled by a particular NT. Thus if NA controls growth hormone release, and the secretion of the hormone is changed in depressed patients, does that confirm a role for NA in the mediation of depression?

BRAIN IMAGING

There is one experimental approach to the study of NT function that has probably advanced more in humans than in animals and that is the visualisation (imaging) of brain structures and chemicals *in situ*. Certainly the stimulus has come from clinical studies. Such neuroimaging began just over two decades ago with the use of X-ray computerised tomography (CT) which distinguishes between different brain regions through variations in their density, as measured by the differential attenuation of X-rays passed through the brain at different angles and, after elaborate computerised analysis, gives a clear image of the brain. Nuclear magnetic resonance (NMR) which utilises the signals given out by hydrogen nuclei in tissue when irradiated with radiofrequency energy provides better pictures. Both of these approaches provide anatomical detail but direct assessment of chemicals in the brain relies on positron emission tomography (PET).

This measures the distribution of a previously administered positron-emitting isotope. PET could be regarded as a form of *in vivo* autoradiography except that the radioligand is not [^{3}H] but [^{15}O], [^{13}N], [^{11}C] or [^{18}F], all of which have short half-lives (2, 10, 20 and 110 min respectively) and so the labelled ligand can only be prepared just before use. After intravenous injection the ligand can be located in the brain in a particular place within a plane (the equivalent of a slice) by positron cameras (gamma detectors) arranged around the subject's head at the appropriate level. The positron emitted from the proton of the isotope collides with an electron in the atomic orbit so that two gamma-rays are given out simultaneously at 180° to each other. Since the detectors only respond when they make two simultaneous detections, i.e. from the two emitted rays, they can record their precise origin and thus the location of the labelled ligand. The intensity of the detected emission is colour coded and reflects the concentration of ligand (Fig. 14.1). Of course, a low level of emission will be detected throughout the brain from the presence of the labelled substance and its metabolites in the blood and extracellular fluid, as well as that non-specifically located in all neuronal and glial tissue and such background activity must be distinguished from the more specific labelling.

PET has two basic uses in studying NT function.

(1) Localisation of specific NT terminals. After its injection a labelled precursor should be taken up and detected in appropriate nerve terminals (and possibly cell bodies) so that the intensity of emission reflects the density of nerve terminals and the innervation. Using this procedure it has been possible to show that very little [^{18}F] fluorodopa is concentrated in the striatum of Parkinsonian patients, compared with normals (Fig. 14.1). Whether the label remains on dopa or is transferred to dopamine will not greatly affect the result since both will label DA neurons although some will occur in noradrenergic nerve terminals.

(2) Labelling NT receptors. The injection and subsequent detection of an appropriately labelled ligand can give an indication of the density of the receptors to which it is bound. Various ligands have been used in this way to label and measure the number of DA and 5-HT receptors in schizophrenics (see Chapter 17) and the extent to which they are occupied by drugs used to treat the disorder. As with any binding study the validity of the approach depends on the specificity of the ligand for its receptor. Of course, there will always be the background activity mentioned above, but the extent of specific binding may be gauged by comparing the density of emission in any area where the NT is likely to function (e.g. striatum for DA) with that from one where it is not normally found (e.g. cerebellum). The difference between these two levels should in fact increase as unbound drug is lost (excreted). To determine the precise number of receptors and see if that varies from brain to brain (e.g. between normal and schizophrenic) is somewhat more difficult. Normally the estimation of receptor number requires a measure of specific binding at two or more ligand concentrations under equilibrium conditions (see Chapter 3), which will clearly be difficult *in vivo*, not least because the effect of different doses may be unacceptable to the patient or subject. Appropriate quantitative analysis has, to some extent, overcome this. It must also be remembered that much of the *in vivo* binding can be to presynaptic receptors and uptake sites as well as postsynaptic receptors, although drugs specific for those sites can be used to label nerve terminals.

IMPLICATIONS OF NEUROTRANSMITTER MALFUNCTION FROM DRUG STUDIES

It is perhaps not surprising, in view of all the problems associated with studying NT function directly in humans, that much attention has been focused on evaluating how drugs which are therapeutically effective in a particular disorder may modify the function of a NT. That NT may be implicated in the disorder if it can be shown that the relative therapeutic effectiveness of a range of drugs correlates significantly with some action they have on it. This requires that all the compounds must have a similar chance of reaching the CNS, or at least of providing appropriate plasma levels, and also that compounds with little or no clinical efficacy in the disorder do not modify the NT in the same way as effective drugs. Such approaches have proved useful. Thus there is a clear correlation between DA receptor antagonism, as measured by ligand binding studies and the anti-schizophrenic activity of a wide range of phenothiazine and butyrophenone derivatives (see Chapters 7 and 17). Good correlations between the analgesic potency of morphine derivatives and displacement of the labelled morphine antagonist, naloxone, helped not only to formulate the concept of opioid receptors and hence of endogenous opioids to occupy them but also the actual discovery of the enkephalins. Displacement of labelled diazepam by a wide range of other benzodiazepines, in an order in keeping with their clinical efficiency as anxiolytics, led to the realisation of endogenous benzodiazepine receptors.

Unfortunately a significant correlation between the clinical efficacy and a particular pharmacological effect of a range of drugs may give the impression that that is the only way in which that disorder may be treated. This can be counterproductive. When drugs are evaluated for therapy in a peripheral malfunction, the tendency is to try to show that they work through different mechanisms. Indeed it is considered desirable, if not essential, to develop anti-hypertensive drugs with different actions, e.g. α_1 and β antagonists, α_2 agonists, vasodilators, calcium channel blockers, etc., and then to use them in sensible combinations. Yet remarkably, when dealing with the CNS, the tendency has been to try to treat a disorder by manipulating just one NT. Consequently the assumption arose that all anti-depressant drugs must augment NA function and all neuroleptic anti-schizophrenic drugs must be DA antagonists, etc. Certainly the introduction of some drugs that are effective in depression without blocking NA uptake, which was the conceived mode of action of all the earlier tricyclic anti-depressants, not only gave a new approach to therapy but also questioned the long-held view that depression is due only to reduced NA function. There could be as many ways of controlling depression as there are of reducing blood pressure and a number of different NTs involved.

ANIMAL MODELS OF HUMAN DISORDERS

Neurotransmitter malfunction in a disorder could be studied in animals in which:

(1) the disorder has been induced by CNS lesion or changes to the animal's environment
(2) the function of a particular NT has been manipulated to see if it produces symptoms of the disorder
(3) there are spontaneous signs of the disorder or they have been produced by genetic manipulation

Specific animal models are discussed in some detail in the appropriate chapters on epilepsy, depression, schizophrenia, etc.

All such animal procedures suffer from the obvious and basic problem that laboratory animals do not behave like humans and that humans cannot reliably interpret their reactions and behaviour. Thus we know that Parkinson's disease is caused by a degeneration of the dopaminergic nigrostriatal tract but its lesion in animals does not produce any condition which resembles human Parkinsonism, except in primates, even though there are functional tests (e.g. rotational movements) which readily establish that loss of dopamine function and also respond to its augmentation (Chapter 15). By contrast, there are many ways, e.g. electrical stimulation and the administration of certain chemicals, to induce convulsions in animals and a number of effective anti-epileptic drugs have been introduced as a result of their ability to control such activity. Indeed there are some tests, as well as animals with varied spontaneous seizures, that are even predictive of particular forms of epilepsy. But then convulsions are a very basic form of activity common to most species and epileptic seizures that are characterised by behavioural rather than motor symptoms are more difficult to reproduce in animals.

It is a statement of the obvious to say that we cannot tell when a rat is anxious or depressed, assuming that they can even experience such human reactions, but they can be subjected to conditions that would make us anxious or depressed. Whether animals then react in the same way as humans is not certain but we have to assume that to be the case if the animal is to be used to detect changes in NT function that can be related to humans. This may be a more acceptable assumption if the animal model responds to drugs that are effective in humans, although it can prejudice thinking and perhaps restrict it to a consideration of just one NT. Thus since all drugs with anti-schizophrenic activity are dopamine antagonists, at least to some extent, the predictive value of any model of that condition has been evaluated by its responsiveness to DA antagonists. Consequently such tests only yield more DA antagonists—a property that can be established in the test tube. Of course, changes in other NTs may be found in such animal tests and a really genuine model of schizophrenia could generate totally different drugs. These problems are well known to experimental psychopharmacologists whose studies are becoming more sophisticated and, hopefully, more appropriate and predictive.

APPROACHES TO THE MANIPULATION OF NEUROTRANSMITTER FUNCTION IN HUMANS

Once the malfunction of a particular NT has been established in a disease state, we need to find ways by which its activity can be restored to normal. The approaches used are indicated in Fig. 14.1 and outlined below. It is assumed that no NT crosses the blood-brain barrier and so its activity must be modified indirectly.

METHODS OF INCREASING NT FUNCTION (Fig. 14.1(a))

(1) ***Increase synthesis***. This may be achieved by giving the precursor, if it crosses the blood-brain barrier. Whether this works will depend on (i) how many neurons remain to synthesise the NT, unless this can be performed extraneuronally and (ii) the availability of synthesising enzymes. Thus if synthesis is a complicated multi-stage process or is controlled by the availability of enzymes that are already reduced or working maximally in remaining neurons, this approach may prove difficult.

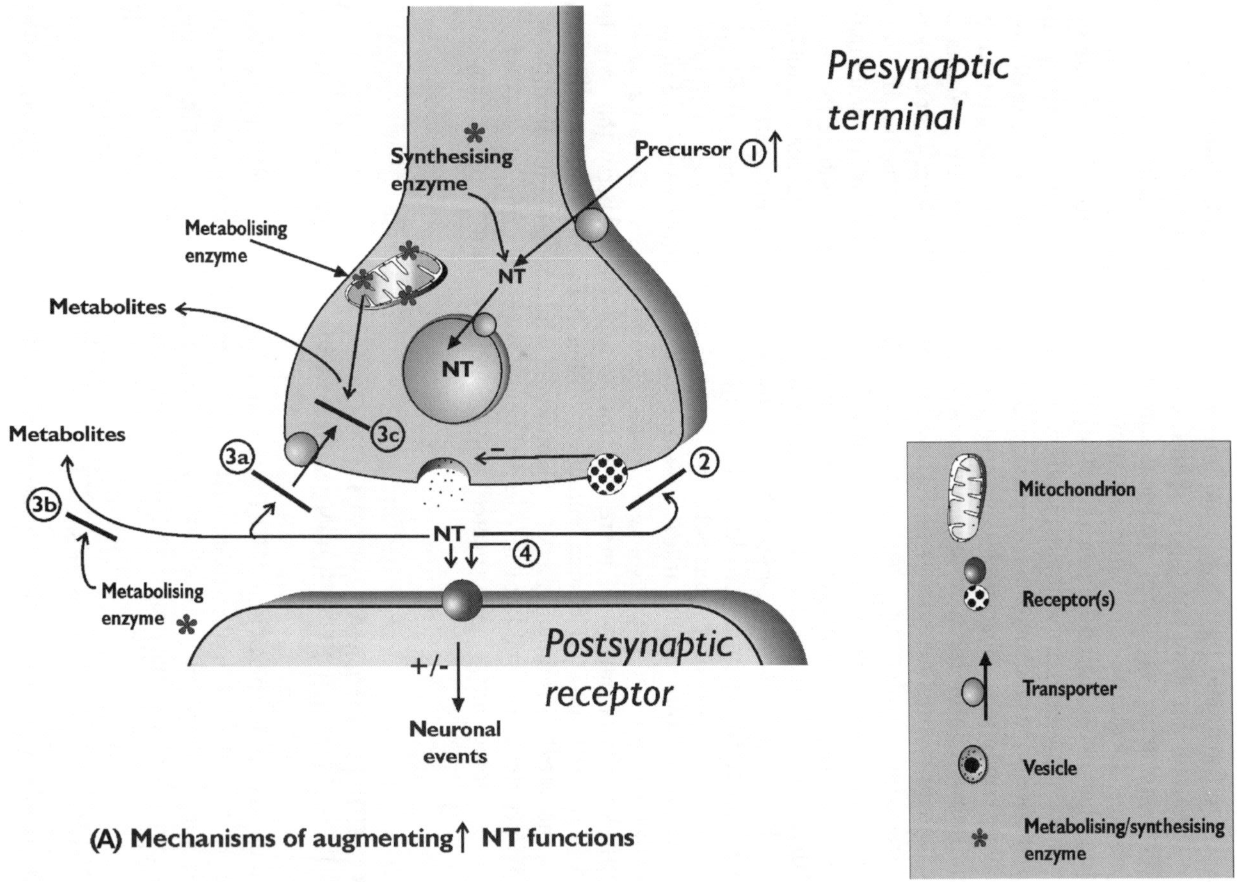

Figure 14.1 *Caption opposite*

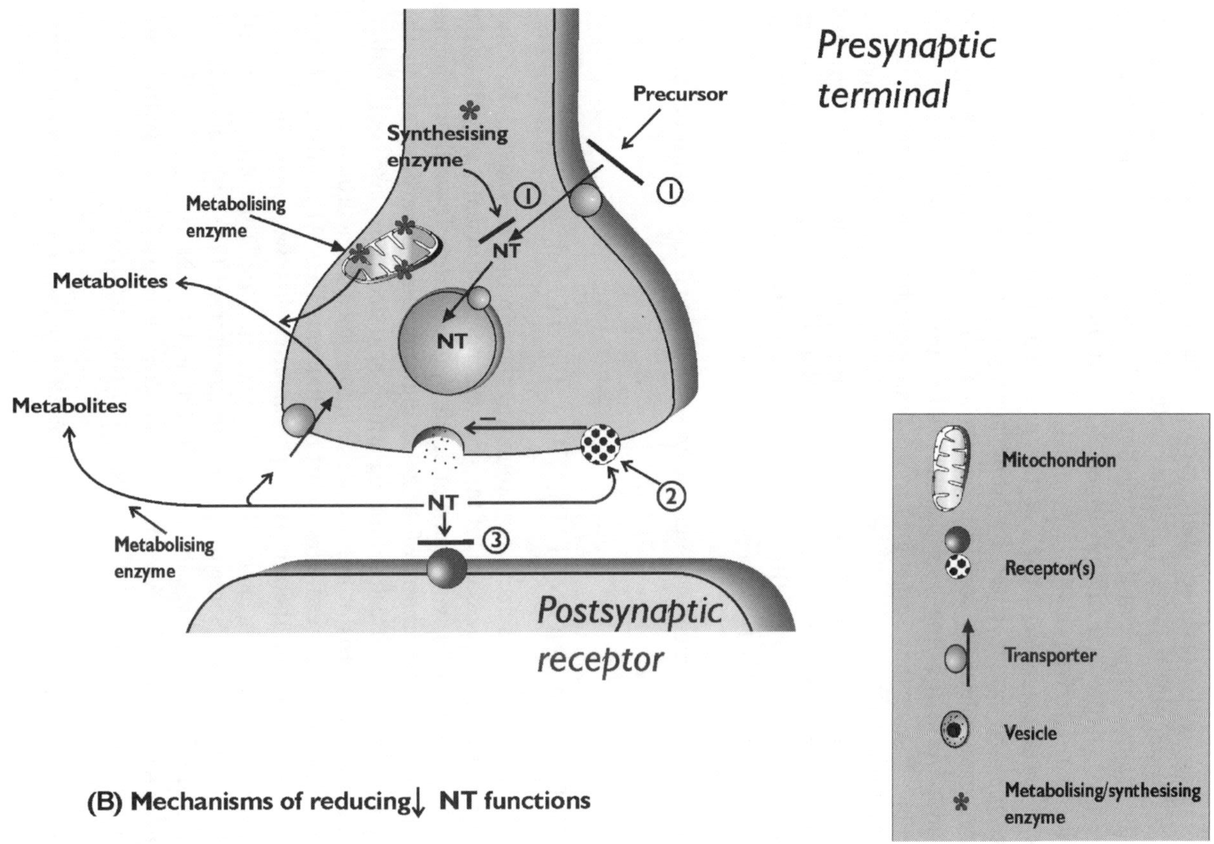

Figure 14.1 *continued* Diagrammatic representation of a neuronal synapse and the mechanisms by which the action of a neurotransmitter may be either augmented (A) or reduced (B). Augmentation of a NT (A) could involve; providing the precursor (1), increasing release by blocking autoreceptors (2), reducing destruction by blocking its reuptake (3a) or extra- (3b) or intraneuronal (3c) metabolism, or providing appropriate postsynaptic agonists (4). Reduction of NT function (B) follows blocking synthesis (1), reducing release by stimulating autoreceptor (2), or using antagonists (3). For further details see text

(2) ***Increase release***. This should follow block of any presynaptic inhibitory autoreceptors. It is not practical at present to increase the vesicular release of a particular NT.
(3) ***Reduce destruction***. This may be achieved by blocking the neuronal or glial uptake (3a) of the NT or its extra- (3b) or intraneuronal metabolism (3c). Its success depends on there still being an adequate, even if reduced, release of the NT, and the protected NT being able to work postsynaptically and not stimulate autoreceptors to reduce the synaptic release of the endogenous NT even further. If the uptake sites are outside the synapse then the protected NT may not easily gain access to the receptors located postsynaptically.
(4) ***Give an appropriate agonist***. Many of the problems associated with the above approaches may be circumvented by administering an appropriate agonist. This could be designed chemically so that it crosses the blood–brain barrier, has a long half-life, and works on the most appropriate subset of receptors, although experience has shown that sometimes more than one effect (receptor action) of the NT may be required. It would be counterproductive if the drug activated the presynaptic autoreceptors unless they happen to augment release. The synaptic action of a NT may also be increased by drugs that have an allosteric action on the receptor to increase its affinity or response to the endogenous NT, e.g. benzodiazapines at the GABA receptor.

Approaches (1)–(3) clearly depend on there being some residual neuronal function and NT release.

METHODS OF DECREASING NT FUNCTION (Fig. 14.1(b))

(1) ***Stop synthesis***. This may be achieved by inhibiting the appropriate enzyme. Its value depends on all, or at least one stage, of the synthesis being sufficiently specific to the NT involved so that only its synthesis is affected. A good example would be choline acetyltransferase in the synthesis of ACh or glutamic acid decarboxylase in the synthesis of GABA. By contrast, inhibiting amino acid decarboxylase could reduce NA, DA and 5-HT synthesis. It may be possible to reduce the neuronal uptake of a precursor if this requires a specific transport mechanism. Thus the synthesis of ACh can be reduced by blocking the uptake of precursor choline with hemicholinium.
(2) ***Reduce release***. This is most likely to be achieved by stimulating inhibitory presynaptic autoreceptors (2a). Some drugs may reduce storage (2b) and hence release, although it is unlikely that this can be targeted at just one NT.
(3) ***Give an appropriate antagonist***. As with agonists, these have the advantage that they can be designed to have a long half-life and act specifically on one type of receptor.

Currently it is not possible to increase the rate of removal (uptake) or metabolism of a NT.

RELATING NT MANIPULATION TO THE CAUSE OF THE DISORDER

To what extent the above approaches can provide successful therapy will depend on both the cause of the disorder and the manner in which the NT is used in normal neuronal function. Thus the disorder could be due to:

(1) An actual degeneration of a NT pathway or
(2) No actual degeneration but a biochemical abnormality or some circuitry failure, leading to inadequate or excessive activity of the NT.

The requirement in respect of NT function may be:

(a) That it must be released physiologically from its nerve terminals by appropriate synaptic activity in order to produce the desired effect or
(b) That it is sufficient merely to provide the NT at the synapse, without the need for it to be released physiologically.

Clearly a disorder combining (1) with (a) would mean that little improvement could be expected by manipulating the lost NT, since the nerves are no longer there to release it physiologically. The main hope then would be to try to replenish the neurons with transplants (regeneration may be possible one day) and hope they become appropriately innervated, or modify the action of some other NT which has become exaggerated (or reduced), as a result of the primary NT loss. By contrast it is easier to treat a disorder, whether characterised by neuronal degeneration ((1) above) or not (2), if it is sufficient just to provide NT (b), as appears to be the case in Parkinsonism.

Of course, the effectiveness and specificity of any of the above manipulations will depend on how widely the NT is distributed and used and whether the malfunction applies only to one area or activity. Thus trying to increase (or decrease) the activity of a NT in only one area will be difficult if it has actions elsewhere which have not been affected by the disorder. The nervous system also has remarkable adaptive powers so the synaptic loss (or increase) of a NT is generally followed by a local compensating increase (decrease) in postsynaptic receptor number. This can be a useful response initially but it will be negated by the therapeutic provision of more (less) NT. Also a change in the activity of one NT can lead to desirable compensating changes in the function of other NTs either working in conjunction with it or normally controlled by it. These will be lost by replenishment of the NT.

It must also be remembered that some NTs, like ACh, NA and 5-HT, have important peripheral as well as central roles and any attempt to modify them centrally will affect those peripheral effects as well. This may be avoided, or reduced, by utilising the blood–brain barrier. Thus if attempts made to increase the central action of a NT result in peripheral effects, these may be counteracted by using an appropriate antagonist that does not cross the blood–brain barrier. It is less easy to overcome peripheral side-effects caused by using a drug that antagonises the action of a NT, although in theory drugs that mimic or augment its action and do not cross the blood–brain barrier could be used. In fact this approach has proved valuable in treating the peripheral neuromuscular disorder of myasthenia gravis which presents as a muscle weakness caused by insufficient cholinergic activity at skeletal neuromuscular junctions. The function of ACh can be increased and the symptoms alleviated, without central side-effects, by reducing the destruction of ACh by giving the anticholinesterase drug neostigmine, which does not cross the blood–brain barrier. Of course, nothing is perfect and antimuscarinic drugs may be needed to overcome the accompanying increased peripheral parasympathomimetic effects of ACh.

Despite all these problems there has been considerable progress in the treatment of disease states through NT manipulation. Before the advent of levodopa therapy in Parkinsonism the treatment of neurological and psychiatric disorders had little scientific basis but the initial and striking success with levodopa in Parkinsonism perhaps raised false expectations. In respect of drug therapy, Parkinsonism presented with a number of advantageous features that are unlikely to be repeated in other conditions. It involves a relatively specific degeneration of one particular NT (DA) pathway, DA has a precursor (levodopa) that readily crosses the blood–brain barrier and its peripheral metabolism to DA can be stopped by decarboxylase inhibitors that themselves do not cross the blood–brain barrier. Although DA is used elsewhere in the brain, nowhere is it concentrated to the same extent as in the striatum, where the degeneration occurs, and it has few peripheral effects. Thus side-effects are relatively few. Fortunately DA also appears to be synthesised from levodopa even in the absence of DA neurons and it does not appear to have to be released synaptically in a physiological way in order to control striatal activity and reduce the symptoms of rigidity and akinesia. Even so, long-term therapy with levodopa has not been without its problems and disappointments and highlights the difficulties of replacement therapy.

15 Diseases of the Basal Ganglia

R. A. WEBSTER

Parkinson's Disease (PD)

In 1817 a general practitioner in Shoreditch, London, wrote an essay on the shaking palsy describing a neurological disorder in some of his patients which has become associated with his name. The main features are:

(1) Slowness and loss of movements known as bradykinesia or akinesia
(2) Muscle stiffness and rigidity
(3) Tremor of the limbs mainly at rest (but not in sleep)

These result in a shuffling gait, an inability to initiate even simple movements like turning, a stooped posture and micrographia (small handwriting). It is a slowly progressing degenerative disease affecting, at most, some 1% of the population above 55 years.

PATHOLOGY

It was known for most of the twentieth century that the brains of PD patients lacked the dark pigmentation characteristic of neurons in the substantia nigra. Such colouring is caused by melanin granules in their cell bodies and although its role is uncertain it is only found in humans and primates and they alone can develop the symptoms of PD. The substantia nigra (SN) is also characterised by round eosinophilic intraneuronal inclusions known as Lewy bodies, which are increased in PD. Neither neuromelonin nor Lewy bodies are confined solely to the SN but it is the neurons of the SN which degenerate in PD.

Since these neurons form the dopaminergic nigrostriatial tract (Fig. 7.1) it is not surprising that PD patients also show a loss of striatal DA. This was first detected in post-mortem studies in 1960 by Hornykiewicz and numerous studies since have shown that not only is PD associated with and presumably caused by a loss of striatal DA, but at death that loss actually reaches more than 80%. Within the striatum DA loss is greater in the putamen which has predominantly motor links with the cortex than in the caudate mucleus with its connections to cortical association areas.

Recently PET studies with 6-fluorodopa, which is taken up by DA nerve terminals in the striatum and is therefore presumably a measure of both the number of functional DA neurons in the nigrostriatal tract to it as well as its DA content, show that this is more like 50% of normal at the start of symptoms, not the 80% observed at PM (see

Neurotransmitters, Drugs and Brain Function. Edited by R. A. Webster

Fig. 15.1). In fact sequential PET measurements in selected patients as the disease progresses followed by extrapolation backwards suggests that DA loss may start some 5 years before symptoms first appear. Nevertheless it is surprising that in all PM studies of normal and PD brains the concentration of striatal DA is either normal or only 10-20% of it.

Three questions need some consideration.

(1) Is the DA loss confined to the nigrostriatal tract?
(2) Are other NTs affected?
(3) Why do the symptoms only occur when DA loss is so marked?

There is some loss (40–60%) of DA in the nucleus accumbens of the mesolimbic system in the ventral tegmentum (A10) and cortex at post-mortem but nowhere is it as marked as in the striatum. Some loss of NA, 5-HT, CCK and the enkephalins and of the markers GAD and ChAT (for GABA and ACh) have been reported in the striatum, SN and other areas but these rarely exceed 50% and could be secondary to DA loss.

The ability of the striatum to apparently function normally until it has lost much of its DA can be ascribed in part to denervation supersensitivity, the degeneration of the DA input resulting in an increase in postsynaptic DA receptors and partly to the remaining neurons producing more DA. This is supported by measurements in humans which show that the HVA:DA ratio, a measure of DA turnover, is much greater in Parkinsonism patients and by microdialysis in rats with 6-OHDA lesions of the nigrostriatal tract, when the reduction in perfusate (released) DA is very much less than that of neuronal (stored) DA.

Thus initially the nigrostriatal tract is able to compensate for the loss of neurons but eventually this fails and the symptoms of PD emerge.

ANIMAL MODELS

Since PD is caused by a relatively specific degeneration of the DA nigrostriatal tract and as there are specific toxins, for DA neurons, i.e. 6-OHDA and MPTP, it should be possible to produce appropriate experimental models. Certainly both toxins cause rotational behaviour in rats (Fig. 7.7) but no rodent shows a syndrome suggestive of PD. Tremor and akinesia can be seen, however, in primates after such toxins and these are being more widely used in experimental studies of PD and drug evaluation. Reserpine causes a depletion of all brain monoamines and produces motor defects in rats, which, even if not PD-like, do respond to DA manipulation.

BASAL GANGLIA CIRCUITRY (Fig. 15.2)

In order to understand how the symptoms of PD could arise from a loss of striatal DA and what can be done to replace it and treat PD, it is necessary to know something of basal ganglia circuitry and the role of DA in it. The scheme to be outlined should, however, be regarded as a working template rather than fully proven fact but there is much evidence for it (Fig. 15.2). Certainly the striatum, i.e. the putamen and caudate nucleus, is accepted as the main receiving area in motor circuits. Information coming to

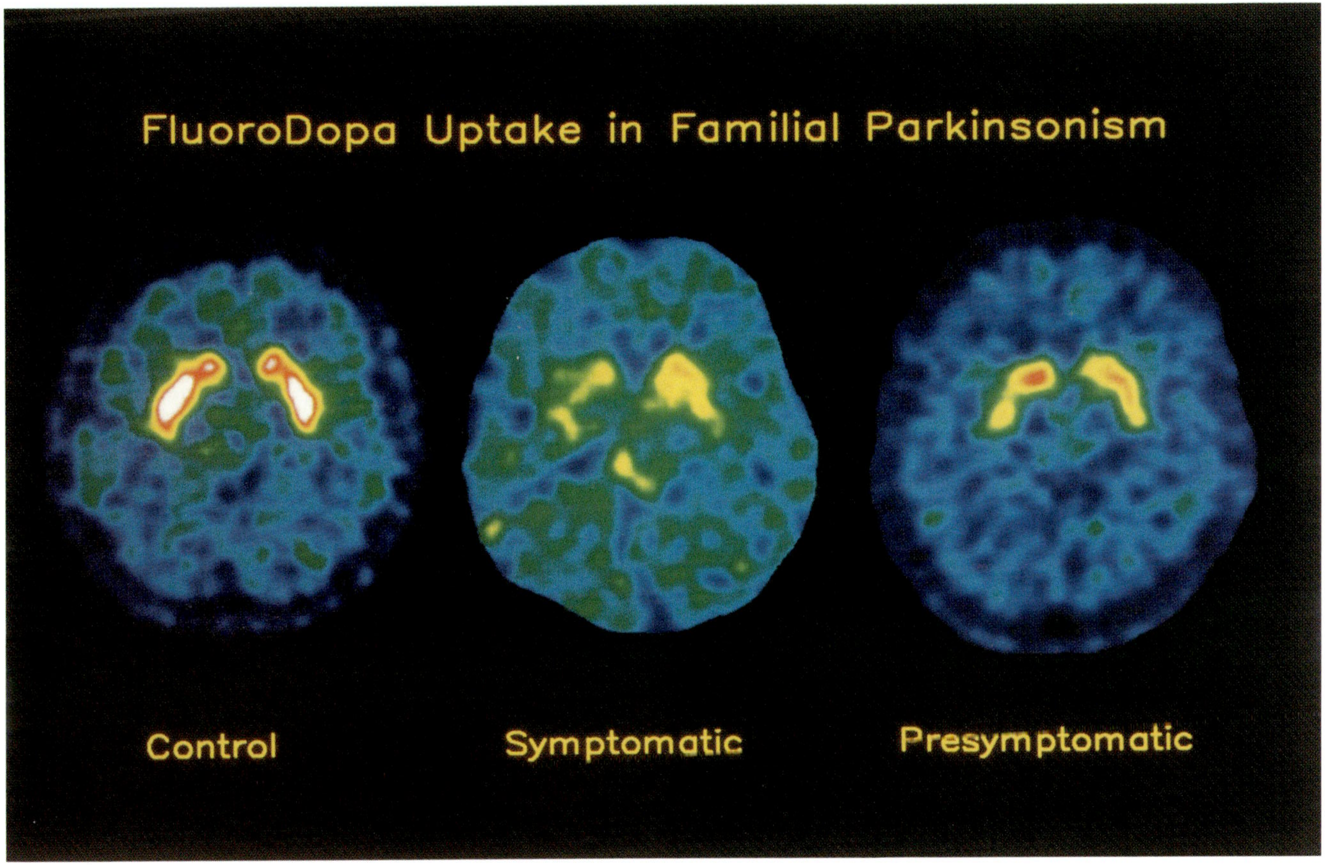

Figure 15.1 PET scans in normal and Parkinsonian patients.
A PET scan with [^{18}F] fluorodopa in a control subject shows that the striatum is heavily labelled whilst in a Parkinson patient with established symptoms there is little labelling. This patient's twin, whilst free of symptoms, also showed some loss of labelling and subsequently developed the disorder. Reproduced by kind permission of D Brooks, MRC Cyclotron Unit, Hammersmith Hospital, UK.

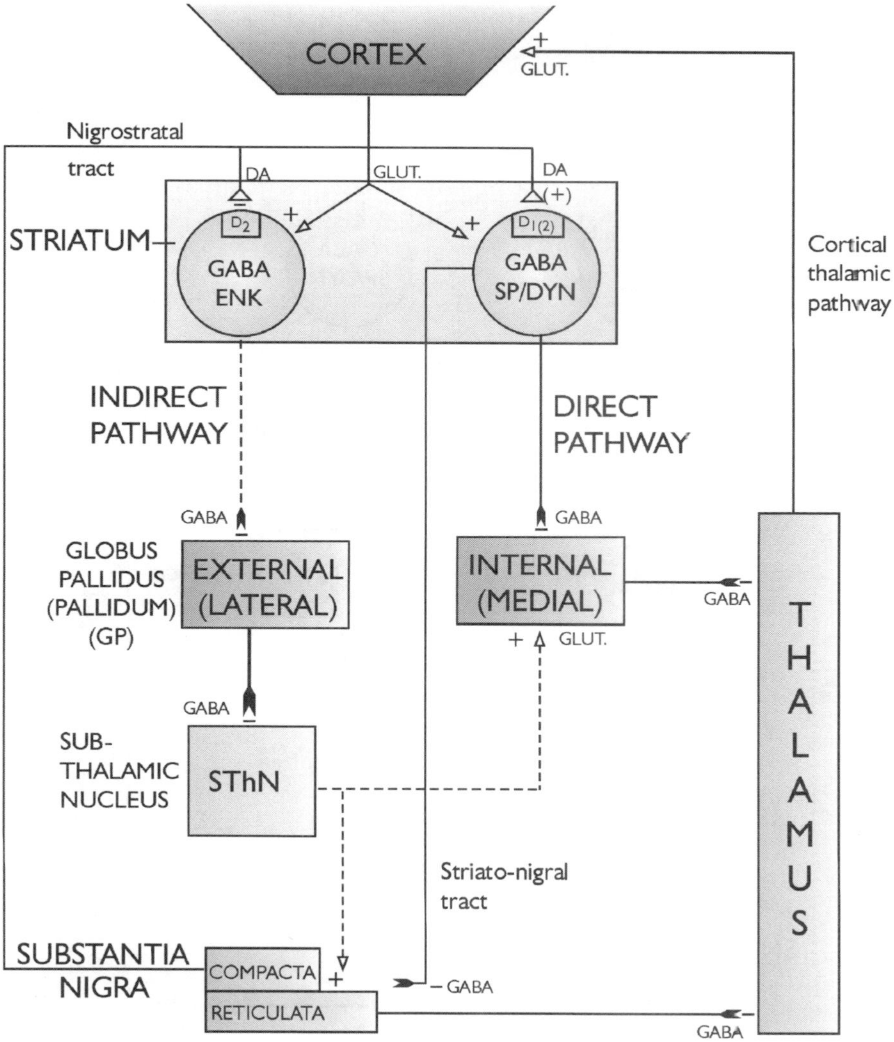

Figure 15.2(a) A schematic presentation of possible normal basal ganglia circuitry. Activity in the cortico-thalamic pathway is modulated by striatal control of the globus pallidus (pallidum) through two pathways, the indirect pathway (Ind Path) to the external pallidum/globus pallidus (GPext) and the subthalamic nucleus (SThN) and the direct pathway (Dir Path) to GPint. Scheme based on that of Chesselet and Delfs (1996) but see text. Pathway activity: ---- low; — normal; — high

it from the cortex and thalamus is processed and channelled to the pallidum (globus pallidus, GP) and to the substantia nigra reticulata.

There are two main output pathways from the striatum to the globus pallidus.

(1) The *direct pathway* (Dir Path) which makes monosynaptic contact with the internal (medial) globus pallidus (GPint) and to a lesser extent the substantia nigra reticulata (SNr).

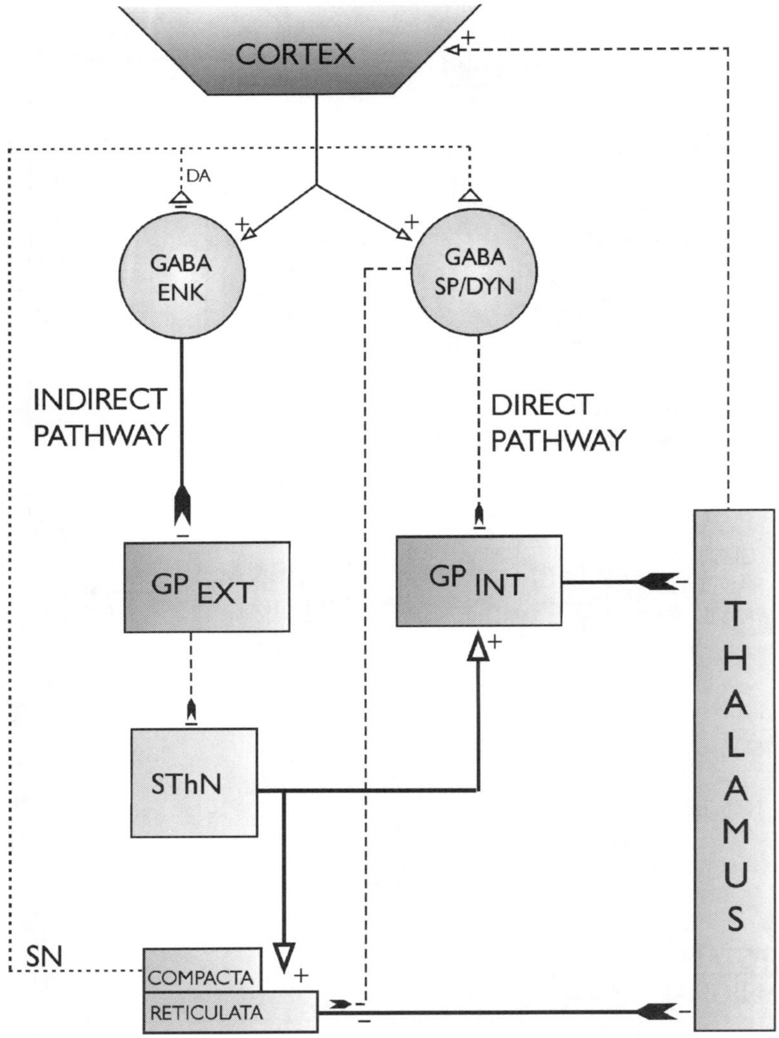

Figure 15.2(b) A schematic presentation of possible basal ganglia circuitry in Parkinson's disease. In PD there is little or no inhibitory nigrostriatal input to the striatum so the Ind Path is active and GPext is inhibited. This will then have less depressant effect on the SThN which will be free to drive the GPint (and SNr) and so reduce cortico-thalamic traffic and produce akinesia. See text for detail. Pathway activity: ---- low; — normal; **—** high

(2) The *indirect pathway* (Ind Path) which also influences GPint and SN but only after going through the external (lateral) GPext and the subthalamic nucleus (SThN).

The GPint and, to a lesser extent, the SNr modulate activity through the thalamo-cortical motor pathways (Fig. 15.2a). These outputs and both the direct and indirect pathways appear to be inhibitory.

The axons of both pathways arise from the medium spiny neurons that constitute 80% of striatal cells. These neurons release GABA but those to the Ind Path have metenkephalin as a co-transmitter and express only D_2 receptors while those to the Dir Path have dynorphin and substance P (the latter mainly to SNr) and express both DA receptors but D_1 predominates. Activation of D_2 receptors results in inhibition of the GABA/ENK neurons of the Ind Path, and probably of the Dir Path, but the D_1 effect could be excitatory on the neurons of the Dir Path as there is a reduction in substance P mRNA in the striatum after blocking its DA input.

Normally (Fig. 15.2(a)) DA inhibits the Ind Path to GPext so that this is then free to inhibit the SThN. This latter system can then no longer drive, through glutamate release, the SNr or GPint whose inhibitory outputs are reduced. The assumption is that the thalamo-cortical pathway can then function properly and movement is normal.

The role of the Dir Path is less clear. When it is active it should inhibit Gpint (and SNr) and so reduce their suppression of the thalamo-cortical pathway. If it is inhibited by DA in the striatum then the converse applies, GPint will be active and thalamo-cortical traffic will be reduced. On balance it seems, however, that DA stimulates the neurons of the Dir Path so that the GPint is inhibited and thalamo-cortical flow facilitated. Some evidence for this comes from the finding that in Huntington's Chorea when the GABA/ENK (Ind Path) neurons degenerate and the Dir Path dominates the patient suffers from dyskinesias-facilitated movement.

Whatever the precise activity of these pathways, DA obviously has a pivotal role in their control. Thus in PD (Fig. 15.2(b)) when there is little DA to inhibit the Ind Path there is more inhibition of GPext which frees the SThN to drive GPint and SNr to inhibit the thalamo-cortical link and motor activity, i.e. produce hypokinesia. The fact that lesion of GPext causes some rigidity in animals supports this. Also if the Dir Path is not driven in the absence of DA, this will also free GPint to inhibit motor activity.

So how can the abnormal pattern of striatal activity that causes akinesia be restored to normal?

THERAPY

Parkinsonism is unique among diseases of the CNS, in that it results from the known loss of a particular NT, i.e. DA, resulting from the degeneration of a particular pathway, the nigrostriatal. Dopamine also has a relatively limited distribution in the brain and few peripheral effects. It should therefore be amenable to therapy based on augmenting its function. Also since the role of DA appears to be to maintain a tonic inhibitory control on GABA output pathways from the striatum, possibly in part by an extra synaptic action (Chapter 6), it may not be necessary for it to be released physiologically from nerve terminals. Thus it may be adequate to just provide DA extracellularly.

Nevertheless it must also be expected that anything which increases DA function not only controls extrapyramidal function but also reproduces the other central effects of DA; i.e. vomiting, a reduction in prolactin secretion and some psychotic manifestations. In excess it may also cause dyskinesias. Despite these problems, the therapy of PD is one of the success stories of neurology.

It is generally assumed that DA itself cannot be used because it does not cross the blood–brain barrier although some recent mocrodialysis studies following intravenous DA indicates that this may not be so, in rats anyway. Thus PD may be treated by

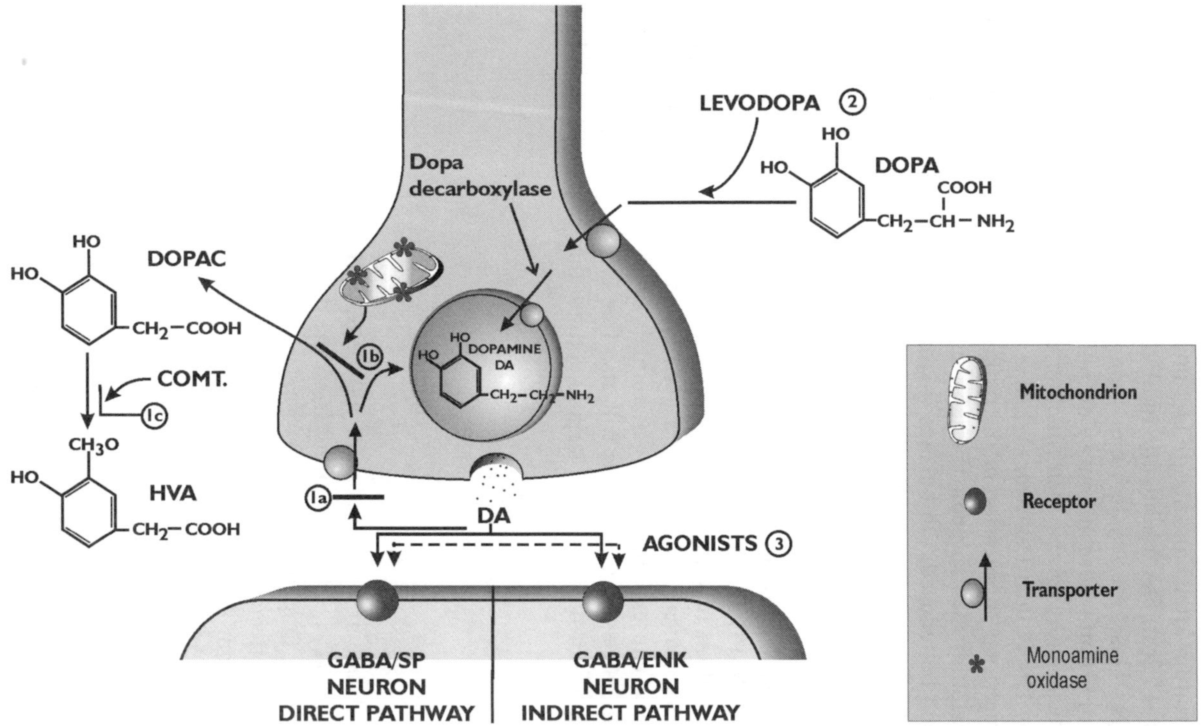

Figure 15.3 Mechanisms of augmenting dopamine function at synapses in the striatum. Synaptic DA levels may be increased by blocking its neuronal uptake (1a), inhibiting the metabolising enzymes MAO_B (1b) or COMT (1c) or by providing its precursor dopa in its levo form (2). The DA receptors may also be stimulated by appropriate D_2/D_1 agonists (3)

(A) Augmenting the action of DA (Fig. 15.3)
(B) Modifying the action of other NTs that could oppose or augment DA function
(C) Non-pharmacological approaches, e.g. grafts

Within these approaches one can:

(A) (1) Increase the action of the remaining DA by reducing its destruction
(2) Replenish DA by giving its precursor levodopa
(3) Mimic the action of DA with appropriate DA receptor agonists
(B) (4) Block the action of any NT released from the neurons of striatal output pathways that are normally inhibited by DA (Fig. 15.8)
(5) Reduce or increase the effects of any NT that could be either antagonising or augmenting the action of DA in the striatum (Fig. 15.9)
(C) (6) Transplant appropriate neural tissue into the striatum
(7) Lesion the pathways that DA no longer inhibits or stimulate opposing ones

In addition to therapy it is hoped that the cause of PD can be established and degeneration of DA neurons stopped or even reversed by pharmacological (including trophic) means or genetic manipulation.

(A) (1) INCREASING THE EFFECTIVENESS OF THE REMAINING DA

Dopamine is removed from its site of action by uptake into nerve terminals and metabolism by MAO and COMT. Drugs are available to block all these processes. The DA neuronal uptake transporter can be distinguished from that for NA and is blocked preferentially by nomifensine. Also unlike NA, which is a substrate for MAO_A, dopamine is a substrate for MAO_B for which selegiline (deprenyl) is an effective inhibitor. The efficacy of both nomifensine and selegiline might be augmented initially by supersensitivity to the remaining DA (increased receptor number) but this decreases with time and augmenting synaptic DA increases the likelihood of stimulating terminal autoreceptors and inhibiting DA release. In view of these problems and the progressive degeneration of DA neurons it is not surprising that nomifensine has little effect but selegiline does produce some improvements in the early stages of the disease and there has been much interest in the possibility that it can prevent, or at least reduce, further degeneration (see section on Aetiology).

O-methylation of DA is a secondary line of metabolism and its inhibition has little effect on the removal of DA but drugs that block this enzyme are gaining a place in prolonging the action of levodopa.

(A) (2) REPLENISHING DA: LEVODOPA

Early use

Irrespective of whether or not DA can cross the blood-brain barrier it will certainly be destroyed after oral administration by MAO and COMT in the gut and liver before achieving an adequate plasma concentration. Levodopa, by contrast, is not a good substrate for MAO, although metabolised by COMT (Fig. 15.4) and is transported across the gut and blood–brain barrier.

Early attempts to treat PD with dopa failed because the doses used were too small. This arose partly from the fear that it would be converted to NA as well as DA and so

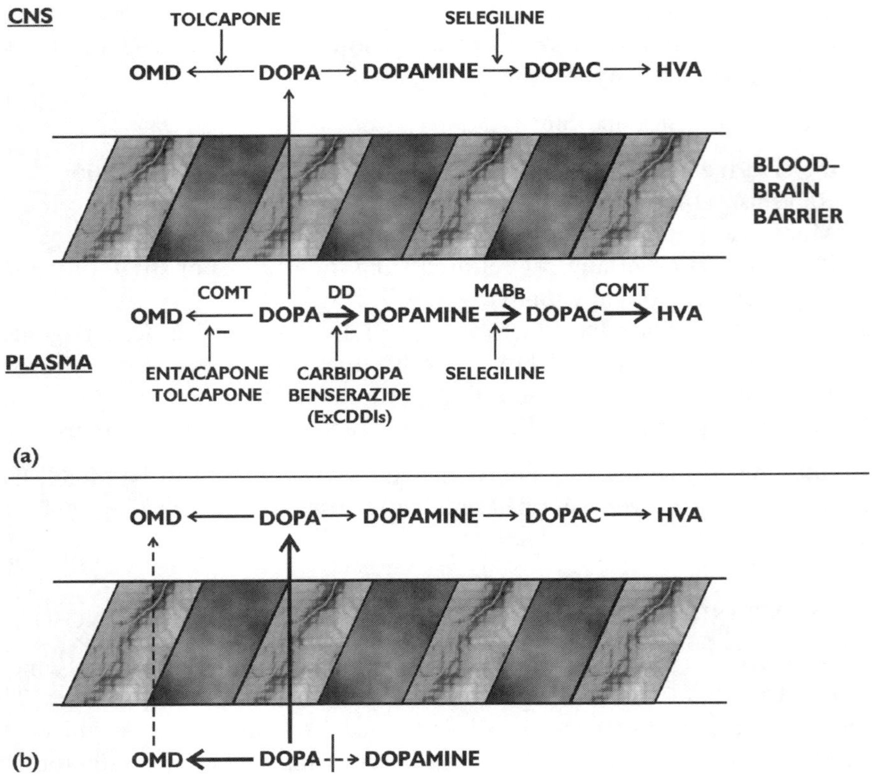

Figure 15.4 The central and peripheral metabolism of levodopa and its modification by drugs. (a) **Levodopa** alone. After oral administration alone most dopa is rapidly decarboxylated to DA in the gut and blood with some *o*-methylated (COMT) to *o*-methyl/dopa (OMD). Only a small amount (3%) enters the CNS to be converted to DA. (b) **After an extracerebral dopa** decarboxylase inhibitor. Blocking just the peripheral dopa decarboxylase (DD) with inhibitors like carbidopa and benserazide, that cannot enter the CNS (extra cerebral dopa decarboxylase inhibitors, ExCDDIs), stops the conversion of levodopa to DA peripherally, so that more enters the CNS or is *o*-methylated peripherally to OMD.

The deamination of DA to DOPAC can be prevented by MAO_B inhibitors such as selegiline while COMT inhibitors stop its further *o*-methylation to HVA and the conversion of dopa to OMD. COMT inhibitors can act just peripherally (entacapone) or in the CNS as well (tolcapone). DD—dopa decarboxylase; MAO—monoamine oxidase; COMT—catechol-*o*-methyl transferase

raise BP. In fact DA synthesis is favoured, since dopa decarboxylase is widely distributed and never saturated, but further synthesis to NA is limited by the restriction of dopamine *B*-hydroxylase to vesicles in NA nerve terminals.

Fortunately Cotzias and his colleagues (1967), after finding that large doses of levodopa (in excess of 3 g) were effective in South American miners suffering from manganese poisoning and showing symptoms akin to PD, tried them successfully in Parkinsonism patients. Subsequently lower doses (0.5–1.0 mg) were tried universally and found to be effective. It is generally accepted that the improvement is very good in 35% of patients, good in 30% and moderate in 30% with some (<5%) not really responding. Indeed its effect is so dramatic that the validity of PD diagnosis in the non-responders is questioned.

It is not the object of this text to cover the detailed pharmacology and use of drugs but levodopa must be an exception. Its use in PD illustrates the problems that still have to be overcome even after the cause of a disease of the CNS has been established and a treatment devised.

Mode of action

Levodopa itself has virtually no affinity for DA receptors and if its conversion to DA is stopped by inhibiting its decarboxylation in the brain then it has no behavioural (DA-like) effects in animals. More importantly, it loses its efficacy in primates with experimental (MPTP) Parkinsonism. Thus it must be converted to DA. Unfortunately most of the DA nerves that would normally do that in the striatum have degenerated and the remainder are already working overtime. Nevertheless it is known from experimental studies, after virtually complete destruction of the nigrostriatal tract, that systemic dopa can still increase striatal DA. Presumably conversion must take place in other neurons or as dopa crosses the blood-brain barrier. Whichever is correct, dopa will increase DA not only in the striatum but elsewhere in the brain and so side-effects occur such as vomiting (60% of patients), dyskinesia (80%), some psychoses (25%) and a reduction in prolactin secretion. Although some phasic hypertension may be seen, the dominant cardiovascular effects are cardiac arrythmias and hypotension (50%) probably through reduced sympathetic activity either due to DA displacing NA in peripheral sympathetic nerve terminals or a reduction in central sympathetic outflow.

Unfortunately levodopa (only the levo form of dopa is active) has a very short plasma half-life (½*t*) of 1½–2 hours. Also it is estimated that only 30% of an oral dose reaches the circulation and less than 10% of that gets into the CNS. How, then, can the efficacy of levodopa be improved?

Adjuncts

Decarboxylase inhibitors

A glance at Fig. 15.4 will show that levodopa is metabolised primarily by dopa decarboxylase to DA and by COMT to 3-methoxy tyrosine, but usually referred to as OMD (*o*-methyldopa).

Blocking the conversion to DA would appear stupid unless this could be restricted to the periphery. More dopa would then be preserved for entry into the brain, where it could be decarboxylated to DA as usual. Drugs like carbidopa and benserazide do precisely that and are used successfully with levodopa. They are known as extracerebral dopa decarboxylase inhibitors (ExCDDIs). Carbidopa (α-methyldopa hydrazine) is structurally similar to dopa but its hydrazine group ($NHNH_2$) reduces lipid solubility and CNS penetration (Fig. 15.4).

ExCDDIs certainly improve the efficacy and duration of action of levodopa so that it can be given in a smaller dose (e.g. 25%) and generally in a 4:1 ratio, levodopa:ExCDDI. As might be expected, some DA side-effects such as dyskinesia and psychoses are worse, but hypotension is less (no peripheral effects of DA) and vomiting is actually much reduced or abolished. This is because the chemoreceptor trigger zone of the vomiting centre while in the brain is on the blood side of the blood–brain barrier and will not be stimulated since no DA is formed peripherally (Fig. 15.5). That an

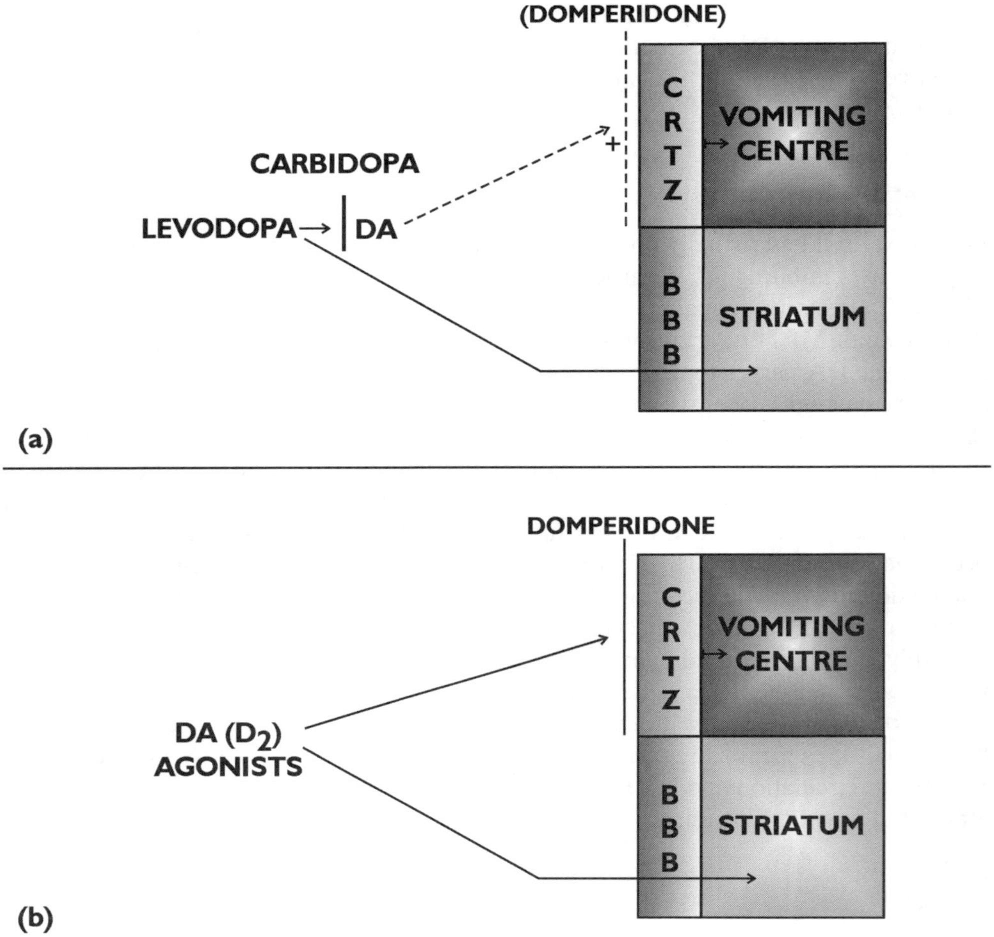

Figure 15.5 Counteracting the emetic effect of (a) levodopa and (b) a dopamine agonist in the therapy of PD.

DA produces vomiting by acting on the chemo receptor trigger zone (CRTZ) of the vomiting centre (VC) outside, on the blood side, of the blood–brain barrier. When levodopa is given with an extracerebral dopa decarboxylase inhibitor like carbidopa it is not converted to DA peripherally and so there is no stimulation of the CRTZ. The emetic effect of a DA (D_2) agonist can be prevented by a D_2 antagonist like domperidone which acts only peripherally. This could also prevent the emetic effect of any DA formed peripherally from levodopa. Since neither carbidopa nor domperidone enter the CNS they do not modify the central effect of either levodopa or a DA agonist

ExCDDI does reduce the peripheral metabolism of dopa in humans and increase the amount entering the brain in animals is shown in Fig. 15.6.

COMT inhibitors

Levodopa is a better substrate for COMT than MAO and when given with an ExCDDI most of it is *o*-methylated to OMD (Fig. 15.4). Recently COMT inhibitors have been developed which act either just peripherally (entacopone) or centrally as well

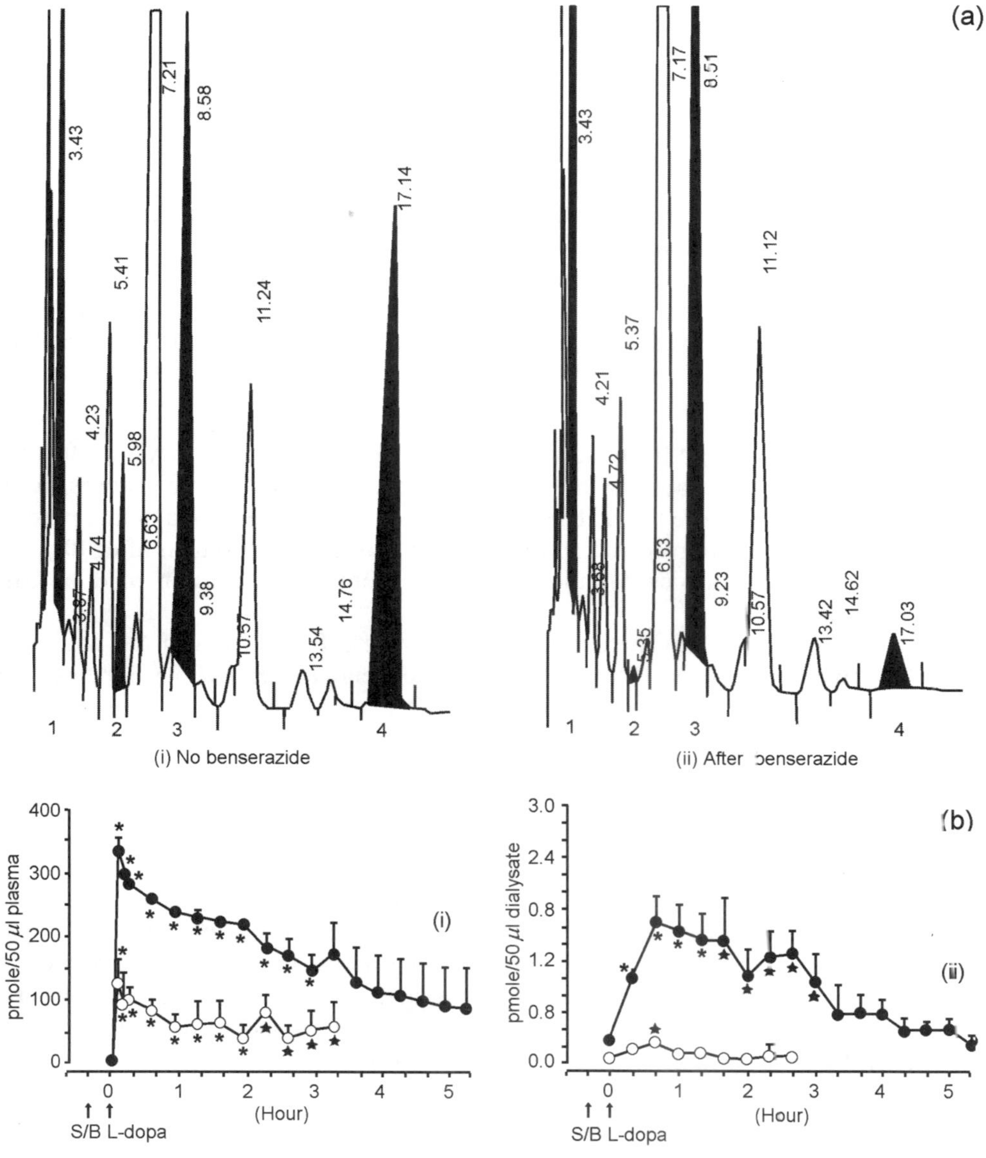

Figure 15.6 The effect of the extra cerebral dopa decarboxylose inhibitor, benzerazide, on the metabolism of levodopa. (a) HPLC chromatogram of human plasma pooled after being taken at varying times over 4 h after levodopa given orally (250 mg) either alone (i) or (ii) following benserazide (50 mg) to a PD patient. Peaks 1 = dopa, 2 = DOPAC, 3 = OMD, 4 = HVA. These peaks have been filled in to highlight them. Dopamine is not seen in either trace due to its rapid metabolism to DOPAC and HVA. Peaks for both these metabolites are seen in (i) but they are much reduced in (ii) indicating that in the presence of benserazide very little DA had been formed peripherally. The OMD peak is much greater after benserazide, which again indicates that dopa has been *o*-methylated (by COMT) rather than decarboxylated. (Unpublished data, Bovingdon, M and Webster, RA). (b) Time-course of dopa levels in microdialysates from the striatum of an anaesthetised rat after 15 mg/kg levodopa given intravenously 20 min after saline (open circles) or 5 mg/kg benserazide (filled circles). After benserazide dopa CSF levels rose eight times. (Unpublished figure, Wan Ya Chang and Webster, RA)

(tolcopone). Both have been tried clinically and shown to prolong the plasma half-life and effect of dopa (+ExCDDI).

Long-term effects

After some 5 years of treatment, most patients show

(1) Abnormal involuntary movements (AIMs), manifest mainly as dyskinesias at the peak plasma level of dopa.
(2) End-of-dose akinesia, i.e. a quicker return to the symptoms of PD.
(3) The ON–OFF effect in which patients experience abrupt swings, perhaps in a matter of minutes, between the extreme of AIMs (1) and akinesia (2). A patient may be walking fairly well but then become suddenly akinetic and fixed before quickly moving again.

These effects could result from the progression of the disease but as they are a feature of levodopa therapy a change in the central response to levodopa or changes in its peripheral kinetics are more likely. The latter does not occur since the maximum plasma concentration, the time to reach it and the plasma half-life are still similar after 10 years of treatment to those achieved initially, although continuous infusion of dopa can smooth out the swings.

As the disease progresses there will presumably be a further loss in the ability of striatal neurons to synthesise DA whether from endogenous or exogenous dopa but the occurrence of dyskinesias suggests that enough DA can somehow still be formed. In view of the complexity of the striatal output pathways and the critical role of DA in controlling the balance between them it is perhaps not surprising that sudden swings in motor function can occur especially with a drug like levodopa that affects both outputs (see below).

Attention has been given to the possibility that some of the above motor effects may arise from a metabolite of levodopa. The prime suspect is OMD which has a half-life of some 20 hours and reaches plasma concentrations three- to fourfold those of dopa. Suggestions that it may compete with dopa for entry across the blood–brain barrier or act as a partial agonist (effective antagonist) have not been substantiated experimentally although it does reduce DA release from rat striatal slices. Also if free radical production through deamination of DA is neurotoxic (see below) then this would be increased by levodopa.

Despite all these problems, levodopa improves the life of the PD patient, effectively slows progession of the disease and prolongs life.

(A) (3) DA AGONISTS

Since there is no reduction in striatal postsynaptic DA receptors in PD then drugs acting directly on them should be effective and in theory present some advantages over levodopa. They do not have to be converted to DA, they can be designed to be long-acting, cross the blood–brain barrier and act on specific DA receptors. In practice they have been somewhat disappointing.

Despite the fact that in quantitative studies of motor performance they often appear to produce more benefit than levodopa and are less likely to cause dyskinesia and ON–OFF fluctations, patients tend to prefer levodopa. Possibly unless DA function is

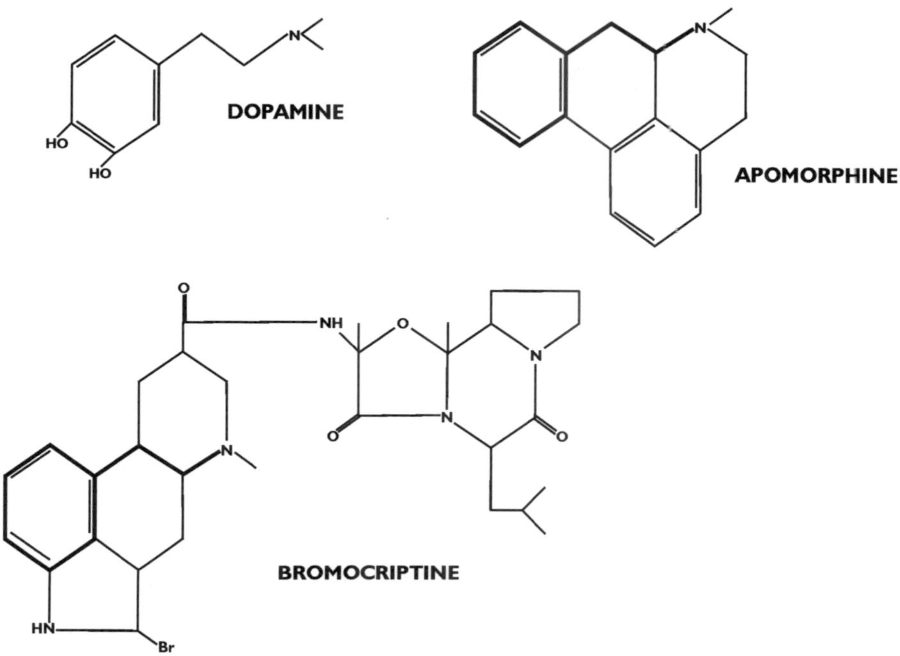

Figure 15.7 Chemical structure of some DA agonists. The structure of DA is shown for comparison and its configuration emphasised in the agonist structure where appropriate

pushed towards the extreme of causing dyskinesia, as with levodopa, there is inadequate release from akinesia. Dyskinesia seems preferable to akinesia.

With most DA agonists there are the other expected signs of increased DA activity such as hallucinations, psychosis and hypotension which can be worse than with levodopa. Fortunately vomiting can be countered by giving the DA antagonist domperidone. This does not cross the blood–brain barrier and so counteracts only the peripheral (chemoreceptor trigger zone) effect of the DA agonist (Fig. 15.5).

Bromocriptine was the first DA agonist used followed by pergolide and lisuride and more recently the very long-acting cabergoline. They are all erot derivatives (Fig. 15.7) and act predominantly on D_2 receptors. Other non-ergolines under test include ropinirole and pramipexole. Apomorphine is a D_2 and D_1 receptor agonist which was tested in PD long before any other agonist but abandoned because of vomiting. Recently with anti-emetic (domperidone) back-up it has been shown to be effective and even reduce the ON–OFF fluctuations of levodopa. Unfortunately it is ineffective orally.

DA agonists have rarely been used alone or before levodopa although there is interest in whether this latter approach would avoid the early use of levodopa and so reduce the time over which it was given and delay the appearance of its side-effects. Generally monotherapy, especially long term, with DA agonists is not considered adequate. This may be because there is some evidence that they still require the presence of endogenous DA to be fully effective. In fact studies in marmosets with experimental (MPTP-induced) Parkinsonism show that after inhibition of central dopa decarboxylase and thus the synthesis of striatal DA, specific D_1 and D_2 agonists were inactive alone and

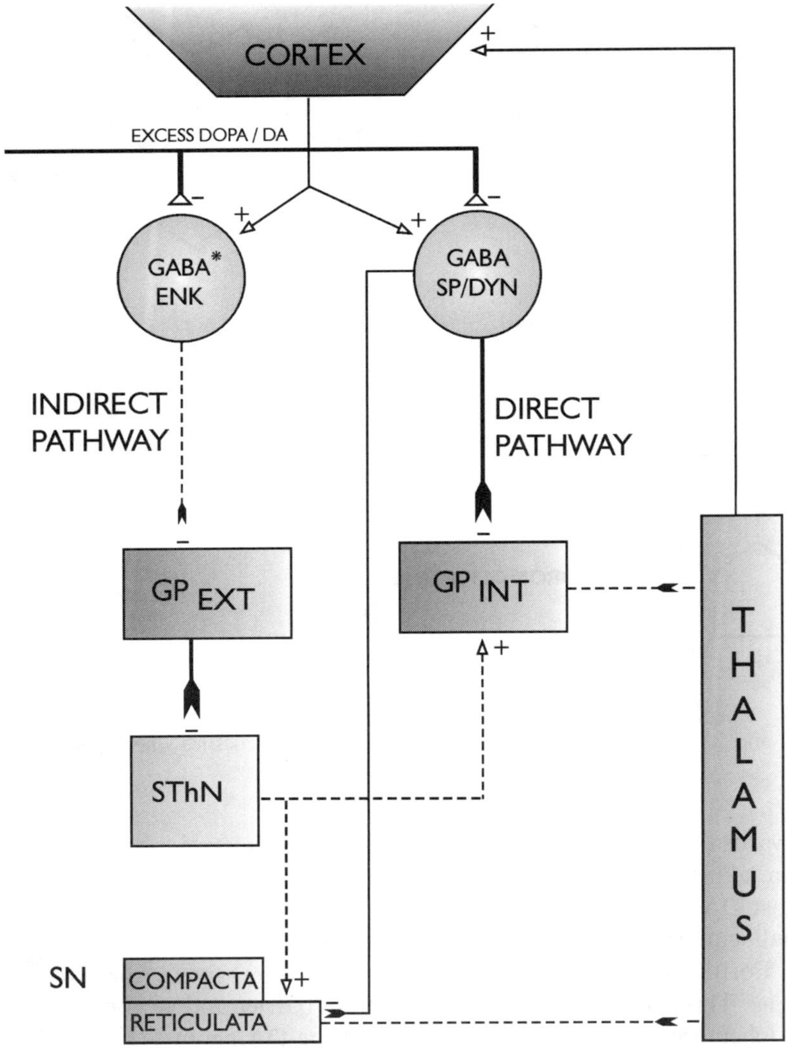

Figure 15.8 A schematic presentation of possible basal ganglia circuitry after excess dopa. High DA levels formed from excess levodopa could so depress the Ind Path that there is little inhibition of GPext which is then able to control the SThN leaving it unable to drive GPint or SNr. Cortico thalamic traffic will then be facilitated and dyskinesias could develop. This may possibly be augmented by DA driving the Dir Path and further inhibiting GPint. See text for detail. *Neurons lost in Huntington's Chorea. Pathway activity: ---- low; — normal; — high

less effective in combination unless given in high doses that could be inappropriate clinically (Treseder, Jackson and Jenner 2000).

DOPA, DA AGONISTS AND STRIATAL FUNCTION (Fig. 15.8)

The production of DA from levodopa should restore the normal striatal picture from that proposed for PD by inhibiting striatal output pathways (Fig. 15.8). It may probably also reduce glutamate release from the excitatory cortical input to the striatum which drives the output pathways. However, too much dopa (DA) could swing the balance in favour of the Dir Path and so facilitate thalamo-cortical activity to produce dyskinesias (see Fig. 15.8). The possible importance of the D_1 effects of levodopa is substantiated by the finding that after treating rats with levodopa and carbidopa for four weeks it was the decrease in substance P mRNA expression on neurons of the D_1-controlled Dir Path rather than the increase in ENK mRNA expression of neurons on the D_2-controlled Ind Path, induced by 6-OHDA leisons, that was reversed (Jolkkonen, Jenner and Marsden 1995).

Since D_2 (but not D_1) receptors are expressed on neurons of the Ind Path, then D_2 agonists will have the same effect on this pathway as levodopa and overcome the hypokinesia. Their inability to activate D_1 receptors could mean, however, that while they are less likely to cause dyskinesias, for the reasons given above, their ability to dampen the GPint may also not be sufficient to give the required facilitation of motor function. Conversely, the absence of D_1 receptors on the Ind Path explains why their agonists cannot influence it and so appear unable to reduce hypokinesia.

Although D_1 agonists alone are considered to have little value in the treatment of PD, the knowledge that the mixed agonist apomorphine (and indeed levodopa) appears to be more effective than a D_2 agonist alone and as experimental evidence indicates that the full DA behavioural effect can only be achieved by stimulating both D_2 and D_1 receptors (Chapter 7, Fig. 7.9), the obsession with D_2 stimulation in PD is perhaps surprising. There are as many D_1 as D_2 receptors in the striatum and it is unlikely that they are all redundant. Unfortunately few specific full D_1 agonists have been available for evaluation until recently (see Hagan *et al.* 1997). Some show promise in both animal models and humans, although the reported absence of dyskinesias is perhaps surprising in view of the considered role of D_1 receptors in their initiation (see above). Nevertheless, treatment with specific D_1 and D_2 agonists in controlled combinations could be useful.

The efficacy of DA agonists, even if not total, does show that striatal function can be reinstated to some extent by merely flooding it with the equivalent of DA and that this does not have to be released physiologically.

Summary: DA augmentation

Clearly there are a number of ways of treating PD based on the concept of augmenting DA but clinical advice is not the object of this text. There is much discussion and no little disagreement on the subject. Views are conditioned by the knowledge that the disorder is progressive, requiring long-term therapy and tempered by the cost of some agonists. Perhaps the consensus now is to start therapy as late as possible, keep it to the minimum and only increase dose or add drugs as is absolutely necessary. Hardly any patient avoids polypharmacy but the order of prescription is probably to augment existing DA with MAOI, then either replenish with levodopa or use DA agonist. There is a developing consensus that since levodopa so frequently causes motor complications (e.g. dyskinesias) it is better to keep that in hand and start with a D_2 agonist. In fact a recent multicentre 5-year trial of ropinirole compared with levodopa showed it to have similar efficacy to levodopa but producing fewer dyskinesias. To these approaches must be added adjuncts such as ExCDDIs, antiemetics, antimuscarinics and possibly amantidine.

Amantidine

This is an anti-viral agent that has weak levodopa-like effects but its mode of action is not really known. Since the most likely effect is considered to be the release of DA it is not surprising that its value is limited when most DA neurons have been destroyed.

Co-transmitters

Although CCK is known to co-exist with DA in nigrostriatal nerve terminals its precise role is not yet sufficiently understood to be manipulated to advantage.

(B) (4) MODIFYING STRIATAL OUTPUT

Knowledge of the striatal output pathways could provide another approach to therapy either by inhibiting the effects of GABA, at different points along the indirect or direct pathways, or those of its co-transmitters, enkephalin, dynorphin and substance P. Since GABA is, of course, widely distributed and its antagonism is primarily proconvulsant manipulating its function specifically in the basal ganglia is not a current option, unless molecular biology establishes a distinct subset of receptors there and drugs can be found to block them. Much the same might be said of the peptides but some recent research requires consideration.

There is evidence from some experimental studies that metENK can decrease GABA release in the GPext while dynorphin reduces GLUT release in GPint. The former effect would reduce the inhibition of GPext neurons by the Ind Path (just as DA would in the striatum) leaving them with greater control of the SThN and hence reduced stimulation of GPint. Dynorphin inhibition of glutamate release within GPint would have the same effect (Fig. 15.9). Since the increased output of this nucleus is believed to cause akinesia these processes could be of benefit in PD.

Preliminary data indicate that in the reserpinised rat or MPTP marmoset, the enkephalin agonist (SNC80) reduces PD-like symptoms without causing increased activity, i.e. no trend to dyskinesias. Enadoline, a dynorphin-like kappa opioid agonist also has similar effects in the same models. Whether it would be similar in humans remains to be seen.

Despite the fact that neither delta nor kappa agonists caused hyperkinetic (dyskinesia-like) activity in the above studies, antagonism of these receptors with naloxone can apparently diminish such activity induced in animals by long-term dosage with levodopa and has been shown to work in preliminary human studies (Henry and Brotchie 1996). Of course, levodopa-produced DA might be expected to inhibit striatal GABA/ENK output to GPext sufficiently to ensure that very little met ENK was actually released to be antagonised, although dynorphin release from the Dir Path could be maintained so that blocking its inhibitory effects on glutamate release would result in decreased output from GPint and a shift away from dyskinesia.

Clearly many more data are needed on the release of these peptides and their function in GP before their possible role in PD can be properly evaluated but they illustrate an interesting alternative approach to therapy.

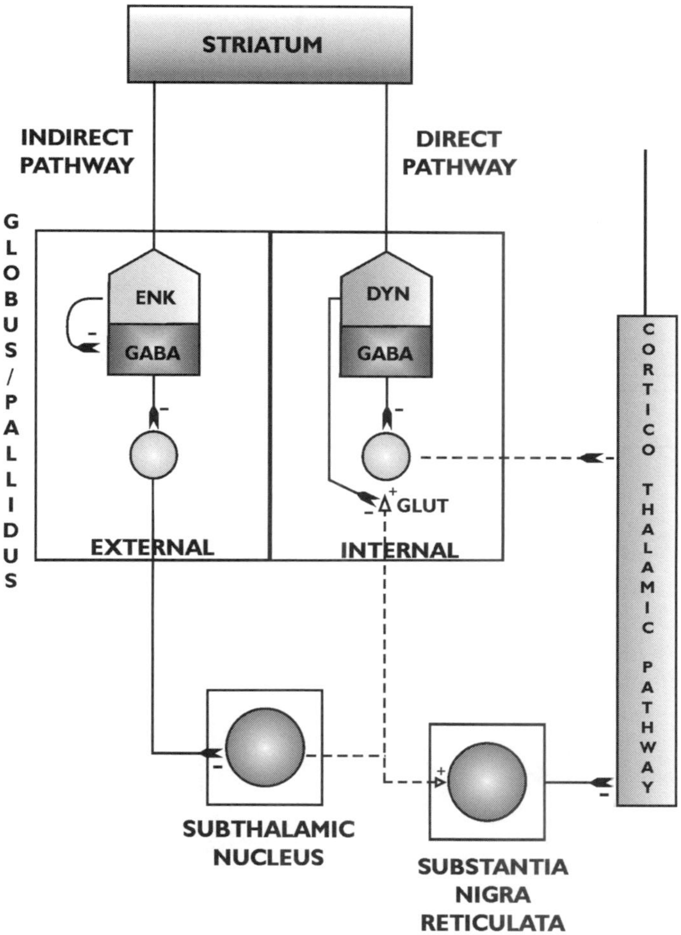

Figure 15.9 Peptide modulation of striatal input to the globus pollidus. Enkephalin released from axon terminals of neurons of the indirect pathway (see Fig. 15.2 for details) is thought to inhibit GABA release from the same terminals so that feedback (auto) inhibition is reduced. This will free the neurons to inhibit the subthalamic nucleus (SThN) and its drive to GPint and SNr which in turn will have less inhibitory effect on cortico-thalamic traffic and possibly reduce akinesia. Dynorphin released from terminals of neurons of the direct pathway may also reduce glutamate release and excitation in the internal globus pallidus and further depress its inhibition of the cortico-thalamic pathway. High concentrations of these peptides may, however, result in dyskinesias. (See Henry and Brotchie 1996 and Maneuf *et al.* 1995)

(B) (5) MANIPULATING OTHER STRIATAL NTs

Acetylcholine

Antimuscarinic drugs such as atropine have been used to modest effect in the treatment of PD for more than a century attenuating tremor and rigidity but with little effect on akinesia. Currently benzhexol and benztropine are sometimes added to levodopa therapy but peripheral effects such as dry mouth, blurred vision and constipation are unpleasant. They are also often used to counteract neuroleptic-induced extrapyramidal effects.

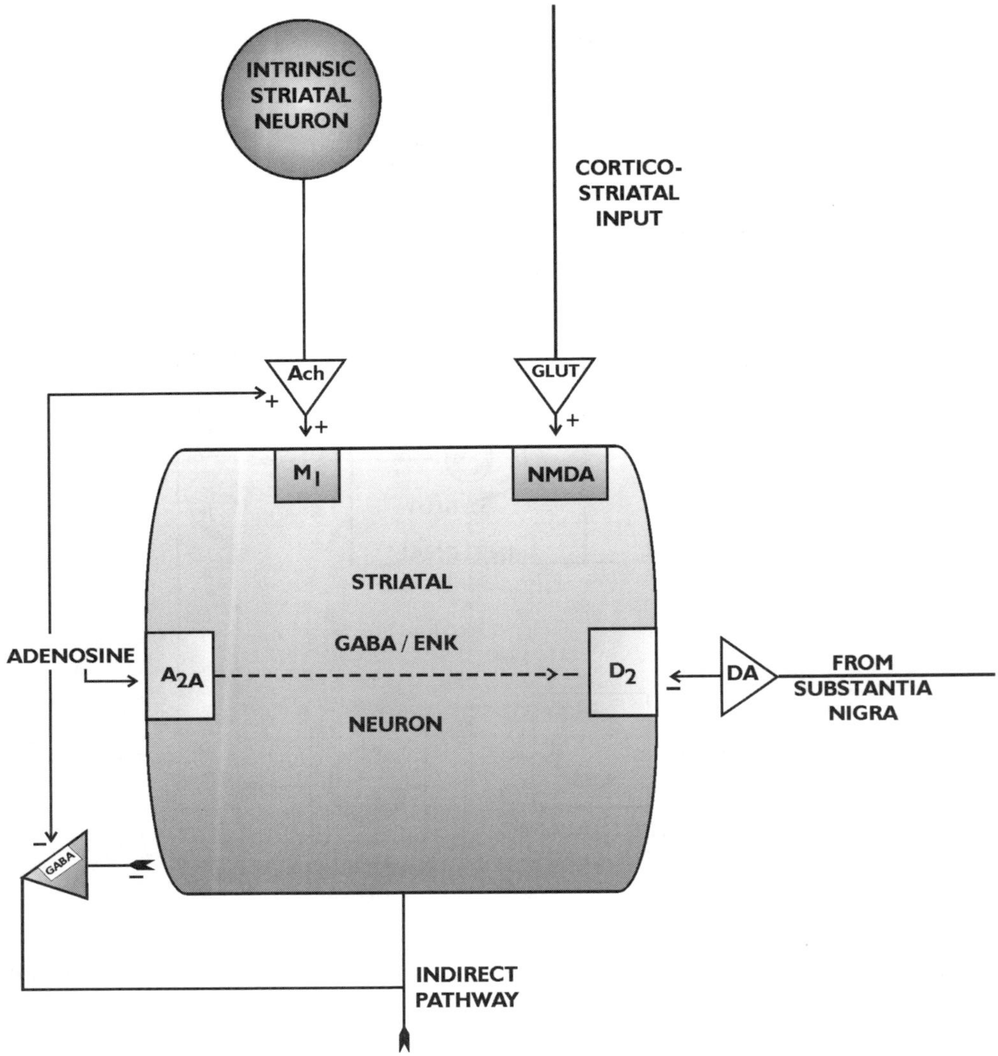

Figure 15.10 Neurotransmitter interactions on the striatal GABA (ENK) neurons of the indirect pathway to globus pallidus. Stratal GABA (ENK) neurons are normally inhibited by both DA, released from nigrostriatal nerve terminals and GABA from their recurrent collaterals. Excitation is mediated by ACh released from intrinsic interneurons and glutamate from cortico-striatal afferents. To compensate for the absence of DA-mediated inhibition in PD the excitation could be reduced by antagonising the actions of ACh (a) at M_1 receptors (antimuscarinics), glutamate (NMDA antagonists) (b) or possibly adenosine. Through its A_{2A} receptor adenosine (c) appears to counter D_2 receptor activity and increase ACh and reduce GABA release, all of which would increase neuron excitability. (See Ferre *et al* 1997 and Richardson *et al* 1997)

ACh is released from the large non-spiny striatal interneurons (Fig. 15.10) which only represent some 5% of total striatal neuron number. Since ACh is excitatory and DA inhibitory on striatal neurons, various schemes have been proposed to balance their antagonistic action but the role of ACh in striatal function (and PD) appears to be relatively minor.

Dopamine inhibits cholinergic neuron firing and ACh release in the striatum predominantly through D_2 receptors. Released ACh probably stimulates the GABA/ENK neurons through M_1 receptors opposing the inhibitory action of DA on them. Clearly in the absence of DA more ACh will be released and its excitatory effect on the GABA/ENK neurons will not be counteracted. Since these neurons and the Ind Path are mainly associated with the akinesia of PD it is perhaps surprising that antimuscarinics have little effect on this symptom.

The excitatory muscarinic receptors on GABA/ENK neurons are M_1 but those on the GABA/SP neurons are probably M_4 and inhibitory. A study of more specific M_1 and M_4 antagonists in PD therapy may be appropriate.

Excitatory amino acids

It would be surprising if these were not implicated in PD. Although most striatal neurons release GABA they are driven by cortical and thalamic inputs releasing glutamate. So in the absence of DA to inhibit them in PD the antagonism of glutamate is an alternative possible approach, providing it can be restricted to the striatum. Certainly intrastriatal (and pallidal) injections of NMDA and AMPA receptor antagonists alleviate motor symptoms in rodent and primate models of PD. The fact that striatal NMDA receptors belong to the subgroup NMDA-2RB, which have a high affinity for antagonists of the glycine and polyamine site (Chapter 10), provides an opportunity for some selectivity of action. The non-competitive antagonist at the NMDA receptor polyamine site, ifenprodil, can in fact reduce PD-like symptoms when injected either intrastriatally or even intravenously in 6-OHDA bilaterally lesioned marmosets, but proved ineffective systemically in humans. Other drugs acting on NMDA receptors may be better.

Adenosine

Binding and mRNA measurements show A_{2A} receptors on the GABA/ENK neurons with D_2 receptors and A_1 receptors on the GABA/SP neurons which express mainly D_1 receptors (Fig. 15.9). The A_{2A} receptors have been most studied.

Activation of the A_{2A} receptor has been shown to increase ACh and reduce GABA release from striatal synaptosomes. Both effects could increase the activity of the striatal GABA neurons, the latter by reducing their inhibition through GABA released by recurrent collaterals. There are also studies which show that the A_{2A} receptor activation is associated with both reduced D_2 receptors binding and DA-induced behaviour. Thus blocking these effects with an A_{2A} antagonist should augment DA inhibition of GABA/ENK neurons and A_{2A} antagonists have been shown to reverse DA antagonist-induced increases in proenkephalin mRNA in those neurons.

The A_{2A} antagonist KF1787 has also been found to improve the motor impairment in marmosets after MPTP while the non-selective adenosine antagonist theophylline augmented levodopa effects in Parkinsonian patients. Whether these responses reflect a specific effect in the striatum is unclear but in order to be effective, these drugs would require ongoing adenosine activity which, it must be remembered, is mainly depressant on neurons. The role of the A_1 receptor, more concentrated on GABA/SP neurons and linked to D_1 receptors, is even less clear although there is evidence that these two adenosine receptors have reciprocal effects. Whether A_1 activation would result in reduced activity of the Dir Path and possibly alleviate levodopa-induced dyskinesias remains to be seen, but it does counteract D_1-driven GABA release in that pathway.

Summary of therapy

Apart from dopamine many NTs such as glutamate, GABA, various peptides, adenosine and ACh are all involved in striatal function but their wide distribution in the CNS makes it difficult to restrict any manipulation of their activity to the striatum after systemic drug administration.

By contrast, the unique loss of DA in PD and its relative restriction to the striatum makes it more amenable to manipulation and augmenting dopamine function is currently the only realistic and effective therapy for PD. Nevertheless, increasing knowledge of basal ganglia circuitry and modifying other NTs involved could lead to some improvement in overall therapy while DA-based therapy itself could be improved. The swings in response to levodopa might be avoided by using DA agonists that are not simply the most potent and specific for D_2 receptors, as this may overcompensate for DA loss, while if levodopa-induced dyskinesias depend on D_1 receptor stimulation then using levodopa with a D_1 antagonist or partial agonist might overcome them.

(C) (6) TRANSPLANTS AND CELL REPLACEMENT

Perhaps all the problems of drug therapy would be avoided if neural tissue capable of synthesising DA could be inserted into the striatum. Foetal mesencephalic tissue has been implanted in the striatum of PD patients and it survives sufficiently for axons to extend, branch and innervate neurons. Experimental studies show DA is formed and in patients PET scans show increased fluorodopa uptake, some function is restored and the dose of levodopa can be reduced. Although DA release cannot be measured directly in patients, there is in fact indirect evidence for it from studies in one transplant patient (Piccini *et al.* 1999). These showed that 10 years after a graft into one putamen the number of D_2 receptors there as measured by PET scans of the binding of the specific D_2 antagonist (^{11}C) raclopride was normal but upregulated (by 44%) in the non-grafted putamen. Since postsynaptic DA receptors are known to increase in number if DA release is reduced this was taken to indicate that DA release was still reduced on the untreated side but restored to normal on the graft side. Also amphetamine, which releases DA to compete with and reduce raclopride binding, did this more effectively on the grafted side—another indication of greater DA release.

Unfortunately transplants require 6–7 foetal brains to obtain enough transplantable material for one patient, which itself raises ethical considerations, and as the tissue cannot diffuse its influence is restricted, even with multiple injection sites, and only a fraction (approx. 20%) of the neurons survive. Also without knowledge of the cause of PD the transplant could meet the same fate as the original neurons. The concept, however, demands perseverance and a number of variants are being tried.

Some ethical and practical concerns may be overcome by the use of porcine rather than human foetal cells and their potential is on trial. Certainly xenotransplants can survive in the human brain partly because it does not show the same immunoreactivity as the rest of the body but recipients will still require some immunosuppressant drugs. Attempts are also being made, with some success, to expand mesencephalic dopamine neurons *in vitro* by the use of nerve growth factors, and so produce large numbers for transplant.

Non-neuronal transplants such as adrenal chromaffin cells have been tried but do not survive although some L-dopa-producing cell lines (e.g. PC12) or glomus cells of the carotid body do produce DA *in vivo* and may provide the equivalent of a continuous infusion of dopa (and DA) directly into the brain. Expression of tyrosine hydroxylase to promote dopa and DA synthesis in striatal cells by direct gene transfer *in vivo* or in cultures for subsequent transplanting, may also be possible. (See Dunnett and Bjorklund 1999 for a review of these approaches.)

(C) (7) LESIONS

The therapeutic strategies outlined above are aimed primarily at blocking the activity of the striatal indirect output pathway so that the SThN drive to GP and its output inhibition of thalamo-cortical transmission is reduced (Fig. 15.2). The same effect could be achieved quite specifically and permanently by lesioning the SThN or GP. Both have been tried in limited numbers and shown to have some success. Surprisingly, stimulation of SThN and GP through chronically implanted electrodes is also effective but since this required high-frequency stimulation (100 Hz) it is possible that this is blocking rather than initiating impulse flow and is like a temporary lesion.

AETIOLOGY AND PREVENTION

If the symptoms of PD arise when nigra cell loss results in a particular depletion of striatal DA (e.g. 50% or more) and, as is generally assumed, there is a gradual loss of nigra cells during ageing then we should all develop PD if we live long enough. Fortunately this is not the case as many people can reach 90 or 100 years without developing PD. In fact, PM studies show that in normal subjects nigra DA cell loss proceeds at 4–5% per 10 years but in PD sufferers it occurs at almost ten times this level (Fearnley and Lees 1991).

Thus either the gradual loss of nigral cells and striatal DA is accelerated for some reason in certain people, so that these markers fall to below 50% of normal around 55–60 years, or some people experience a specific event (or events) during life which acutely reduces DA concentration. This could be to a level which is not enough to produce PD at the time but ensures that when a natural ageing loss of DA is superimposed on it the critical low level will be reached and PD emerge before natural death. The first possibility is likely to have a genetic basis but although examples of familial PD are rare there is typically an increased incidence (2–14) of the disease in the family of a PD patient and initial PET studies show a much higher (53%) loss of DA neuron labelling in the monozygotic than the dizygotic twin of a PD sufferer even if the disorder is not clinically apparent.

While a number of gene markers have been identified in different families there is no consistent mutation although parkin on chromosome 6 and α synuclein on 4 have aroused most interest. Mutations of the gene encoding the latter, such as threonine replacing alanine on amino acid 53 (A53T) or phenylalanine for alanine on 30 (A30P) have certainly been established in particular families with inherited PD. In fact ablation of the gene encoding α synuclein has been shown to produce locomotor defects in mice and surprisingly in the fruitfly *Drosophila melanogaster*. By expressing normal human α synuclein in all the nerve cells of *Drosophila*, Feany and Bender (2000) found no neuronal abnormalities but with wild-type α synuclein or the mutants A53T and A30P they observed premature and specific death of dopaminergic neurons. Additionally some neurons showed intracellular aggregates that resembled Lewy bodies and were composed of the α synuclein filaments seen in the human counterpart. Of course, flies cannot be said to develop PD but unlike normal ones, the transengic fly found it more difficult to climb the sides of a vertical vial.

The fact that some schizophrenics show PD symptoms when given DA antagonists has been considered to indicate that they already have a reduced DA function and are asymptomatic potential PD patients but the high incidence of PD side-effects after neuroleptics and its occurrence in young people (20–30 years) argues against this. A viral infection can lead to PD as evidenced by its high incidence (50%) in survivors of an outbreak of encephalitic lethargica in Europe around 1920. Toxins can also be inducers.

In 1982 there was a small outbreak of PD among Californian heroin addicts taking what was thought to be a methadone substitute, but due to a mistake in synthesis turned out to be a piperidine derivative MPTP (1-methyl-4-phenyl-1,2,3,6-tetra hydropyridine). By any route, even cutaneous or inspired, this causes a specific degeneration of nigral DA neurons in humans and primates but not in rodents, which may indicate some link with melanin (not found in rodents). MPTP itself is not the active factor but requires deamination by mitochondrial MAO_B to a charged pyridium MPP^+ which is taken up specifically by DA neurons. MAO_B inhibitors such as selegiline prevent MPTP-induced PD in primates. The production of MPP^+ generates free radicals as does the oxidation of DA itself.

Free radicals and peroxides are highly reactive substances and can damage DNA, membrane lipid and cell protein and initiate lipid peroxidation to destroy all membranes. Hydrogen peroxide (H_2O_2) can actually be produced by the oxidation of DA, under the influence of MAO_B and is potentially toxic to SN neurons (Fig. 15.11). Normally such H_2O_2 would be detoxified by glutathione but glutathione activity is low in brain so that H_2O_2 can accumulate. While a reduction in glutathione itself is not sufficient to destroy nigral cells, since its direct inhibition alone does not have that effect, the rise in H_2O_2 coupled with its conversion to toxic radicals could do so. This process is also favoured by the high levels of free iron in the substantia nigra which are augmented in PD patients. Iron is normally bound in the body by ferritin but as this is low in the brain the iron will increase and facilitate the production of free radicals. Thus the SN, sitting as it does with high DA levels, ample MAO for converting it to H_2O_2, little chance to detoxify it but plenty of iron for free radical production, is ready to self-destruct. Whether this is enhanced by dopa therapy and the provision of more DA is uncertain but it has been shown that systemic L-dopa does undergo auto-oxidation in rat striatum to a semiquinone (Serra *et al.* 2000). This process is inhibited by antioxidants and enhanced by manganese and, of course, miners of this element are known sometimes to develop Parkinsonism-like symptoms and as indicated above, were the first patients to be shown to respond to L-dopa therapy. Whether antioxidants should be given with L-dopa may bear investigation although when one such agent, tocopherol, was tested alone, i.e. on otherwise untreated PD patients, it failed to retard the development of symptoms.

The reliance of free radical and MPP^+ production on MAO_B activity stimulated considerable interest in the possibility that blocking this enzyme could prevent the

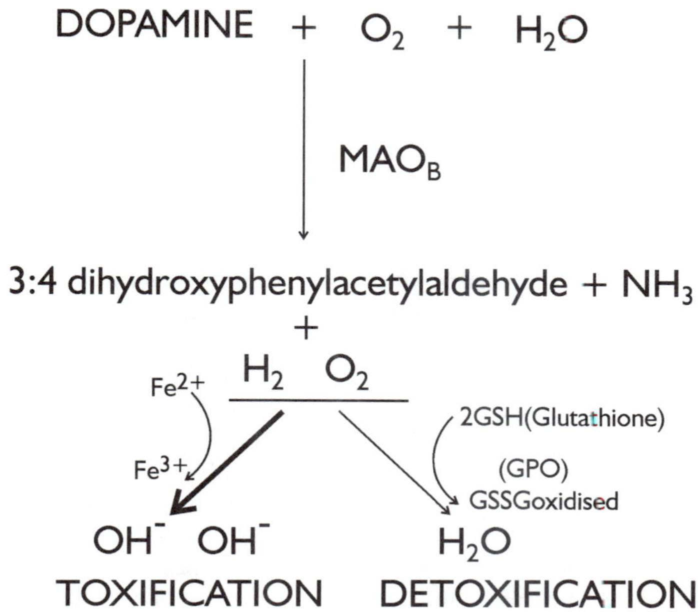

Figure 15.11 Possible scheme for the formation of free radicals from the metabolism of dopamine. Normally hydrogen peroxide formed from the deamination of DA is detoxified to H_2O along with the production of oxidised glutathione (GSSG) from its reduced form (GSH), by glutathione peroxidase. This reaction is restricted in the brain, however, because of low levels of the peroxidase. By contrast the formation of the reactive OH-radical (toxification) is enhanced in the substantia nigra because of its high levels of active iron and the low concentration of transferin to bind it. This potential toxic process could be enhanced by extra DA formed from levodopa in the therapy of PD (see Olanow 1993 and Olanow *et al.* 1998)

progression of PD. Trials with selegiline (deprenyl) on hundreds of new untreated patients initially suggested that this was the case since almost twice as many patients in the control group needed to receive levodopa in the first year compared with those on the MAO_B inhibitor. Unfortunately analysis after discontinuing the inhibitor showed no difference in the subsequent development of PD, implying that the inhibitor was just relieving the symptoms (i.e. levodopa-like effects). Nevertheless interest in the possible preventive effects of MAO_B inhibition persists. Certainly the slow progression of PD and neuronal loss means that if the cause can be established and countered, it might be possible to stop disease development especially if treatment could start in the very early or preclinical stages. In fact neurotrophic factors such as glial cell line or brain-derived neurotrophic factors (GDNF or BDNF) have shown some beneficial effect on 6-OHDA or MPTP-depleted DA function when injected intracerebrally in animals.

Nitric oxide has also been implicated in PD. Thus animals with MPTP-induced Parkinsonism not only show extensive gliosis in the substantia nigra (like humans) in which the glial cells produce NO, but Liberatore and colleagues have found that in iNOS (inducible nitric oxide synthase) knock-out mice the toxicity of MPTP is halved. Since NO releases iron from ferritin and produces toxic peroxinitrate in the presence of superoxide radicals it could accelerate, even if it does not initiate, dopaminergic cell death (see Hirsch and Hunot 2000 for further details).

Huntington's Chorea

This disease takes its name from George Huntington who observed and studied it in families in New York at the end of the nineteenth century. It is an inherited autosomal dominant disease with the child of an affected parent having a 50% chance of inheriting the gene and then unavoidably suffering from the disease. It is characterised by choreas (dyskinesias) which start in the extremities (fingers) but spread to the face, limbs and whole body even though they disappear during sleep. Akinesia develops as in PD and like that disease HC is initially a disorder of the basal ganglia but starts earlier (30–45 years). It is more progressive, invariably resulting in death within 20 years, as motor impairment makes any function difficult and emotional disturbances and dementia develop. Fortunately it affects only 0.01% of the population and hopefully prenatal diagnosis will become available and its occurrence further reduced.

Early neuron loss occurs in the striatum and especially of the GABA/ENK neurons which project to the GPext (Fig. 15.2) and referred to as the indirect pathway in the discussion on PD. This would leave the GABA/DYN neurons and the direct pathway dominant, which appears to be the requirement for dyskinesia (Fig. 15.8). Thus, unlike PD, there is no loss of neuronal input to the striatum or of DA levels but a marked reduction in striatal GABA as well as its co-transmitting peptides, especially metenkephalin but also dynorphin and substance P. It is not possible to replace adequately the lost GABA but the early dyskinesias can be reduced by using DA antagonists which would primarily block the D_2 inhibitory effect on the remaining GABA/ENK neurons and so help to restore balance. Knowledge of basal ganglia circuitry (Fig. 15.2) suggests, however, that the antagonists probably need to have some effect on the D_1 receptors and GABA neurons of the direct pathway as well. Whether adenosine A_1 agonists or opiate antagonists (see above) could usefully reduce the activity of those neurons remains to be evaluated. Unfortunately any process reducing dyskinesias could encourage akinesia. This is the opposite problem to that in PD but a reminder of the pivotal role of the striatum and DA in movement control (see Fig. 7.8).

Since HC is such a progressive disorder with clear neuronal loss, it is not surprising that NT manipulation has been of little value in therapy and that there is no drug treatment of any significance. More hope rests on a genetic approach and the mutated gene has in fact been identified and cloned but its precise role remains uncertain. For details of its structure, possible actions and appropriate models see Reddy, Williams and Tagle (1999).

ADDENDUM

The first proper double blind trial of embryonic implants in 40 PD patients (20 undergoing just surgery without any implant), has shown no improvement in patients over 60 years but some clinical benefit (fewer symptoms between levodopa dosing) in those below that age. Unfortunately some of these responders eventually developed dyskinesias, a sign of too much dopamine, and further implants were halted until the technique has been re-evaluated, see Freed, CR *et al. N Engl J Med* 2001, **344**: 710–719.

REFERENCES

Calne, DW (1997) Treatment of Parkinson's disease. *New Engl. J. Med.* **329**: 1021–1027.

Chesselet, M-F and Delfs, JM (1996) Basal ganglia and movement disorders. *Trends Neurosci.* **19**: 417–422.

Chiara, G, Morelli, M and Consolo, S (1994) Modulatory function of neurotransmitters in the striatum: ACh/dopamine/NMDA interactions. *Trends Neurosci.* **17**: 228–233.

Cotzias, GC, von Woert, MH and Schiffer, LM (1967) Aromatic amino acids and modification of Parkinsonism. *New Engl. J. Med.* **276**: 374–379.

Dunnett, SB and Bjorklund, A (1999) Prospects for new restorative and neuroprotective treatments in Parkinson's disease. *Nature* **399** (Neurological disorders supplement): A32–A39.

Ehringer, H and Hornykiewicz, O (1960) Verteilung von Noradrenalin and Dopamin im Gehirn des Menschen und ihr Verholten bei Erkrankungen des Extrapyramidalen systems. *Klinische Wochenschrift* **38**: 1236–1239.

Feany, MB and Bender, WW (2000) A *Drosophila* model of Parkinson's disease. *Nature* **404**: 394–398.

Fearnley, J and Lees, AJ (1991) Parkinson's disease: neuropathology. *Brain* **114**: 2283–2301.

Ferre, S, Fredholm, BB, Morelli, M, Popoli, P and Fuxe, K (1997) Adenosine–dopamine receptor–receptor interaction as an integrative mechanism in the basal ganglia. *Trends Neurosci.* **20**: 482–487.

Goetz, CG and Diederich, NJ (1996) There is a renaissance of interest in pallidotomy for Parkinson's disease. *Nat. Med.* **2**: 510–514.

Hagan, JJ, Middlemas, DN, Sharpe, PC and Poste, GH (1997) Parkinson's disease: prospects for improved therapy. *Trends Pharmacol. Sci.* **18**: 156–163.

Henry, B and Brotchie, JM (1996) Potential of opioid antagonists in the treatment of levodopa, induced dyskinesias in Parkinson's disease. *Drugs and Ageing* **9**: 149–158.

Hirsch, EC and Hunot, S (2000) Nitric oxide, glial cells and neuronal degeneration in Parkinsonism. *Trends Pharmacol. Sci.* **21**: 163–165.

Jolkkonen, J, Jenner, P and Marsden, CD (1995) L-Dopa reverses altered gene expression of substance P but not enkephalin in the caudate-putamen of common marmosets treated with MPTP. *Brain Res. Mol. Brain Res.* **32**: 297–307.

Kawaguchi, Y, Wilson, CJ, Augood, ST and Emson, PC (1995) Striatal interneurones: chemical physiological and morphological characterization. *Trends Neurosci.* **18**: 527–535.

Maneuf, YP, Mitchell, IJ, Crossman, AR, Woodruff, GN and Brotchis, JM (1995) Functional implications of Kappa opioid receptor mediated modulation of glutamate transmission in the output regions of the basal ganglia in rodent and primate models. *Brain Res.* **683**: 102–108.

Mercuri, NB, Bonci, A and Bernardi, G (1997) Electrophysiological pharmacology of the autoreceptor mediated responses of dopaminergic cells to antiparkinsonian drugs. *Trends Pharmacol. Sci.* **18**: 232–235.

Olanow, CW (1993) A radical hypothesis for neurodegeneration. *Trends Neurosci.* **16**: 434–444.

Olanow, CW, Jenner, P and Beal, ME (1998) Cell death and neuroprotection in Parkinson's disease. *Ann. Neurol. (Suppl.)* **44**: S1–S196.

Piccini, P, Brooks, DJ, Bjorklund, A, Gunn, RN, Grasby, PM, Ornella, R, Brundin, P, Hagell, P, Rehncrona, S, Widner, H and Lindvall, O (1999) Dopamine release from nigral transplants visualised *in vivo* in a Parkinsonian patient. *Nat. Neurosci.* **2**: 1137–1140.

Quinn, N (1995) Drug treatment of Parkinson's disease. *Brit. Med. J.* **310**: 575–580.

Reddy, HP, Williams, M and Tagle, DA (1999) Recent advances in understanding the pathogenesis of Huntington's disease. *Trends Neurosci.* **22**: 248–255.

Richardson, PJ, Kase, H and Jenner, PG (1997) Adenosine A_{2A} receptor antagonists as new agents for the treatment of Parkinson's disease. *Trends Pharmacol. Sci.* **18**: 338–344.

Serra, PA, Esposita, G, Enrico, P, Mura, M, Migheli, R, Delogu, MR, Miele, M, Desole, MS, Grella, G and Miele, E (2000) Manganese increases L-dopa auto-oxidation in the striatum of the freely moving rat: potential implications to L-dopa long-term therapy of Parkinson's disease. *Brit. J. Pharmacol.* **130**: 937–945.

Treseder, SA, Jackson, M and Jenner, P (2000) The effects of central aromatic amino acid (dopa) decarbosylase inhibition on the motor actions of L-dopa and dopamine agonists in MPTP-treated primates. *Brit. J. Pharmacol.* **129**: 1355–1366.

16 The Epilepsies

R. A. WEBSTER

DEFINITION

Epilepsy was defined by Hughlings Jackson as 'An episodic disorder of the nervous system arising from the excessive synchronous and sustained discharge of a group of neurons'. It is a considered and apt definition that highlights important aspects of the disorder that are relevant to our understanding and treatment of it. The fact that it is episodic means that attacks, in whichever form they arise, can occur frequently within minutes or hours of each other or at intervals of weeks, months or years. Such unpredictability complicates treatment. Epilepsy is neither a degenerating nor generally a worsening disorder but therapy needs to be maintained to avoid the possibility, however remote, of a seizure with all its potential personal and social problems.

That an episode arises and spreads from the synchronous as well as excessive discharge of a group of neurons (focus) means that not only must those neurons be in some way predisposed to so discharging but they can also recruit neurons that are otherwise normal. How that discharge manifests itself, i.e. which type of epilepsy occurs, will depend not only on where the abnormal focal neurons are located but also to what extent the activity they initiate can and does spread through the brain. There are consequently a number of different forms of epilepsy, i.e. the epilepsies.

CLASSIFICATION

Epileptic seizures are classified broadly as (a) partial or (b) general:

(a) **Partial seizures or epilepsy (PE)**. As the name implies, these begin and generally remain localised. They may be simple or complex with the symptoms dependent on the cortical area affected. The former may just involve involuntary contractions of a group of muscles or a single limb (Jacksonian motor epilepsy) or abnormal but localised sensory disturbances (Jacksonian sensory epilepsy). They rarely last more than a couple of minutes and consciousness is not impaired.

Complex partial seizures manifest themselves as bizarre behaviours which are also known as psychomotor or temporal lobe epilepsy, since a lesion (focus) is often found in that brain area. Repetitive and apparently purposeful movements vary from simple hand clenching or rubbing to more bizarre hand movements and walking. These can last a few minutes, often disrupt other ongoing activity or speech and the patient has no subsequent memory of them. Complex seizures may develop from simple ones.

Neurotransmitters, Drugs and Brain Function. Edited by R. A. Webster

(b) **Generalised seizures**. These involve more, or even the whole, of the brain including the reticular system so that consciousness is lost, although in some instances (absence seizures) this is more a loss of awareness rather than any collapse. The two main forms are:

(1) Grand mal (GM) or tonic–clonic seizures (TCS). This is probably what everyone recognises as 'epilepsy'. It starts with a tonic spasm of all musculature and rigid extension of the body, a temporary cessation of respiration, generally salivation and often defecation and micturition. After about one minute this gives way to violent synchronous clonic jerking movements (convulsions) which may continue for a few minutes. The patient may remain unconscious for a longer period before recovering. In some cases the tonic and occasionally the clonic phrase can exist alone.

(2) Petit mal (PM) or absence seizures (AS). These are less dramatic and generally occur in children. They entail a brief and abrupt loss of awareness (consciousness) in which the patient suddenly ceases ongoing activity or speech and stares vacantly for a few seconds before recovering equally quickly. Motor disturbances are rare apart from blinking of the eyes and the patient has no recollection of the event.

In addition to the above main categories seizures can be just myoclonic, isolated clonic jerks, or atonic, loss of postural control with just head drooping or the patient actually falling.

Epileptic seizures affect 0.5% of the population, are more common in the young and, except for partial seizures, often decrease with age. Convulsions associated with metabolic disturbances are not considered to be epileptic.

Many seizures are associated with distinctive EEG patterns (Fig. 16.1). Perhaps the most striking is the 3 per second spike wave activity seen in most leads (cortical areas) in absence seizures, which can be invoked by hyperventilation. Otherwise distinctive EEG patterns are usually only found during an actual seizure, with burst spiking seen alongside clonus in TCS and abnormal discharges with the behavioural patterns of partial epilepsy and in particular that originating in the temporal lobe.

ANIMAL MODELS OF EPILEPSY

These are normally based on the use of either electrical stimulation or chemical convulsants. When applied generally, i.e. an electric shock to the whole brain or convulsants injected systemically, the resulting convulsions are indicative of generalised seizures. If they are applied locally to specific brain areas, the same approaches induce activity indicative of partial seizures. Also some animals can be bred in which seizures either occur spontaneously or can be induced easily by appropriate sensory stimulation.

MODELS OF GENERALISED SEIZURES

(1) Electric shock

In the maximal electric shock (MES) test a supramaximal stimulus is applied bilaterally through corneal or auricular electrodes to induce tonic hind limb extension in rats or

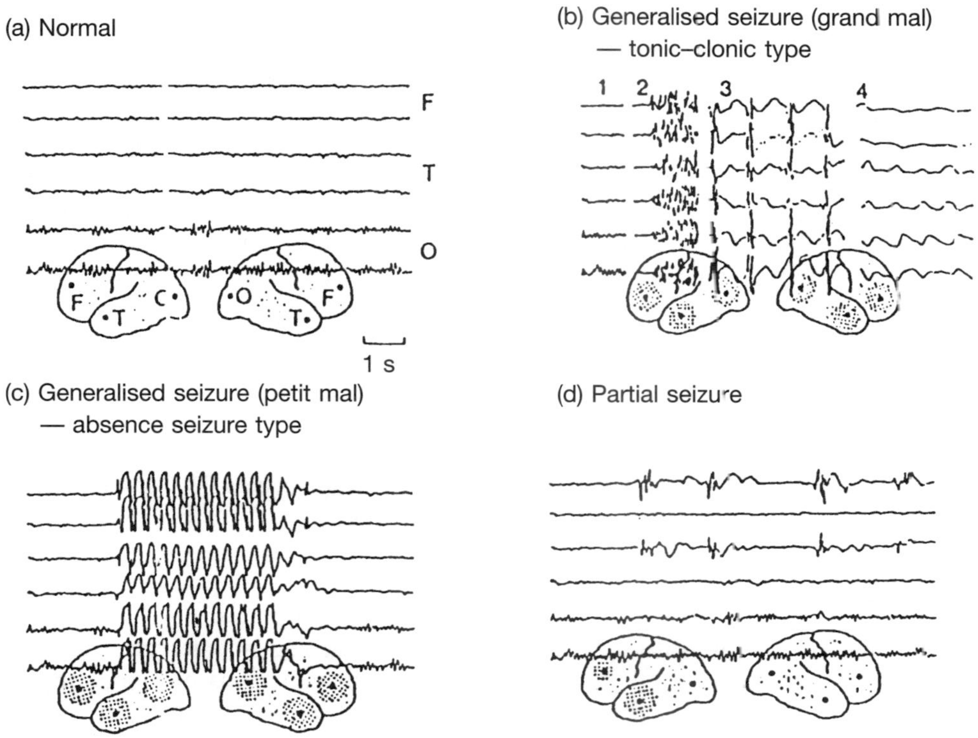

Figure 16.1 EEG patterns in human epilepsies. Electroencephalograms are shown for, a normal subject (a), those suffering from general tonic–clonic seizures (grand mal, (b)), generalised absence seizures (petit mal, (c)), and partial seizures (d). Spikes are seen in both right- and left-sided leads from all three cortical areas, frontal (F), temporal (T) and occipital (C) in the generalised seizures (b, c) but only in the occipital leads in the example of partial seizures (d). In the EEG recorded during the generalised tonic–clonic seizure the normal tracing (1) is followed by the onset of the tonic phase (2), the clonic (convulsive) phase 3 and post-convulsive coma 4. The typical $3\,s^{-1}$ spike and wave discharge of petit mal (c) may be seen during routine recording or induced by procedures such as hyperventilation but the spiking of grand mal and partial epilepsy is only seen during seizures. (Reproduced from Eliesson *et al* (1978), *Neurological Pathophysiology*, 2nd edn, Oxford University Press, New York with permission)

mice. Anticonvulsant activity is determined by measuring the dose of drug required to protect 50% of the stimulated animals (ED or PD_{50}) and is predictive of efficacy in TCS.

(2) Chemical convulsants

A number of different chemicals have been used including GABA antagonists such as bicuculline or picrotoxin. Strychnine convulsions have no predictive value since they arise through antagonism of spinal (glycine-mediated) rather than cortical inhibition. The most commonly used agent is pentylenetetrazol (PTZ), also called leptazol. Anticonvulsant activity is again assessed as the dose required to protect 50% of animals, usually mice, against the clonic seizures induced by a dose of PTZ that would otherwise produce them in almost every mouse injected, the so-called CD_{97} (convulsive dose in

97% of animals). The absence of a marked tonic component to the seizure may be significant since the ability to protect animals against PTZ convulsions is predictive of a drug's potential efficacy in absence seizures rather than TCS, despite the fact that PTZ causes convulsions.

The anticonvulsant activity of a drug may also be evaluated by measuring its ability to raise the convulsive threshold, i.e. the amount of applied current or infused PTZ required to just evoke a seizure. Comparison of the efficacy of drugs in the threshold and maximal seizure tests may distinguish between their abilities to raise seizure threshold or reduce seizure spread and development.

MODELS OF PARTIAL SEIZURES

(3) Focal

Partial seizure activity can be induced by the localised application of chemicals such as cobalt or alumina to the cortex or the injection of chemicals such as PTZ or kainic acid directly into particular brain areas like the hippocampus.

(4) Kindling

If a subconvulsive stimulus is applied, generally in rats, at regular intervals, e.g. daily for some two weeks to a specific brain area, especially the amygdala or hippocampus, then eventually full localised (partial) or secondary generalised seizures develop. A similar effect can be obtained by the repeated localised injection of subconvulsive doses of some convulsants. The ability of a drug to reduce the kindled seizure itself may be indicative of value in partial seizure but if it slows the actual development of kindling that may indicate some ability to retard epileptogenesis.

SPONTANEOUSLY EPILEPTIC (GENETIC) ANIMALS

Various animals show spontaneous epilepsy or seizures that can be readily induced by sensory stimulation (see Jobe *et al.* 1991). Tottering mice display seizures that resemble absence attacks behaviourally, in their EEG pattern and response to drugs. DBA/2 mice show reflex seizures to audiogenic stimuli while photically-induced seizures can be obtained in the Senegalese baboon, *Papiopapio*, which are similar to generalised tonic–clonic epilepsy.

PREDICTIVE VALUE

It has become clear that drugs which are effective in protecting mice against PTZ are effective in absence seizures while those able to control the tonic response to maximal electroshock are effective in tonic–clonic seizure. Some drugs are effective in only one test and clinical condition whilst a few are active in both (Table 16.1). Experimental focal seizures are indicative of partial seizures.

It could be argued that an antiepileptic drug should really stop the development of epilepsy, i.e. epileptogenesis, and not merely control seizures which would make them just anticonvulsant. If the development of kindling reflects the process of epileptogenesis then drugs effective against its progression should stop the development of

Table 16.1 Comparison of the experimental and clinical activities of established antiepileptic drugs

	Activity against convulsions induced by		Effectiveness clinically in	
	Electroshock	Pentylene tetrazol	Clonic–tonic seizures	Absence seizures
Phenytoin	++	−	+	−
Carbamazepine	++	−	+	−
Phenobarbitone	+(+)	+	(+)	−
Na valproate	+	+	+	+
Clonazepam	(+)	++	(+)	+
Ethosuximide	−	+	−	+

Notes:
The data for the experimental studies gives a semi-quantitative guide to relative activities based on ED_{50} values, i.e. ++ = active, + = some effect, − = not active at non-toxic doses. Clinical comparisons are not related to recommended doses but simply indicate whether a drug is effective (+) or not (−). Generally, drugs that are to be used clinically to control tonic–clonic seizures control electroshock but not pentylenetetrazol-induced convulsions in rats and mice, whilst the converse applies to drugs effective in absence seizures. Na valproate is effective in both experimental models and is used in both clinical conditions, although in all cases higher doses have to be used than for any other drug.

human seizures. Phenytoin and carbamazepine do not stop the development of kindling, although acutely they reduce the fully kindled seizure, and in studies of post-traumatic epilepsy following brain damage in humans (car accidents) these drugs stop the appearance of seizures in the first week or so but do not control epileptogenesis, since seizures can develop subsequently in those patients after therapy has stopped. Generally drugs that increase GABA function or block NMDA receptors retard kindling.

CAUSE AND PATHOLOGY

With such a diversity of seizures it would be surprising if a common cause of epilepsy had been found or even existed, although it is conceivable that a focus might arise in the same way wherever it was found. The actual symptoms would then be determined simply by the location of the focus and their extent, partial or general, by how easily or widely the influence of the focal neurons spread. Other factors might then control that spread and could vary from one region to another depending on local neuronal circuitry and NT utilisation. The fact that different drugs with different mechanisms of action are effective in different epilepsies may support that view.

There is, however, no clear neuropathology. Epilepsy may be secondary to focal lesions such as congenital malformations, infarcts, tumours, cysts or inflections but fortunately many patients with these problems do not develop epilepsy. Again epileptic seizures may occur in those suffering from Huntington's Chorea or Alzheimer's disease. Brain damage such as neuronal loss and glial proliferation may in fact be seen in epileptics but these changes may be secondary to, rather than the cause of, epilepsy. They probably reflect the consequences of intense neuronal activation since in patients dying in status epilepticus they appear to be of recent origin and can be induced in animals by systemic or locally administered convulsant (see Meldrum and Corsellis 1984).

Extensive brain damage or lesions are certainly not essential for convulsions. These merely require appropriate conditions. Everyone is capable of having a convulsion, indeed their induction has been a common treatment for depression. The convulsive threshold of an epileptic, or more precisely that of some of their neurons, is just lower than normal.

There is no known genetic basis for most of the common epilepsies apart from juvenile myoclonic epilepsy and childhood absence epilepsy which are dependent on inheritance of two or more susceptible genes, although genetic factors might more generally determine predisposition. Single distinct mutant genes have been established, however, in three rare forms of epilepsy (less than 1% of total), namely generalised epilepsy with febrile seizures, benign familial neonatal convulsions and autosomal dominant epilepsy (see McNamara 1999). These each encode a part of some voltage-gated ion channel which are believed to be respectively the β subunit of a Na^+ channel (SCNIB), novel K^+ channels and the α subunit of cholinergic nicotinic receptors (CHRNA4). All could lead to increased neuronal excitability and in fact co-expression in oocytes of the Na channel α subunit with the β subunit found in febrile convulsions produces a channel that inactivates more slowly than when it is expressed with normal β subunits.

DEVELOPMENT OF AN EPILEPTIC SEIZURE

A seizure is accompanied by a burst of spikes in the EEG. Between these so-called *ictal* phases are solitary EEG *interictal* spikes. Each of them represents the field potential associated with a burst of action potentials in a group of neurons within the epileptic focus (Fig. 16.2).

Focal neurons when activated show an abnormal excitatory postsynaptic potential (EPSP) called the paroxysmal depolarising shift (PDS) (Fig. 16.2) which leads to an abnormal burst of action potentials at frequencies up to $200\,s^{-1}$. While such neurons can be found anywhere they are more common in the CA3 region of the hippocampus and layer 4 of the cerebral cortex. Neurons showing this burst firing are also called Group I, pacemaker or epileptic neurons and their activation always results in a burst discharge and not a single impulse. Thus they could have a persisting abnormality in membrane or ion channel excitability. What we need to know is not only how such neurons arise but how their influence can spread to affect neighbouring neurons to produce the interictal spike and, more importantly, how this can sometimes, and at immensely variable intervals, develop into a full ictal discharge and seizure (Fig. 16.3) (see Prince 1992; Prince and Connors 1986). There is obviously a graduation in excitability from the group I focal neurons through group II neurons, found either in or immediately adjacent to the focus that may fire normally but can be recruited to display abnormal firing, and their surrounding normal neurons. The gradual progression of an epileptic EEG during the infusion of PTZ in a rat is shown in Fig. 16.4.

The development of a focus is likely to be determined by one or more of the following:

(1) changes in the intrinsic properties and excitability of the focal neuron
(2) a reduction in normal GABA-mediated inhibitory controls
(3) an increase in excitatory coupling between neurons

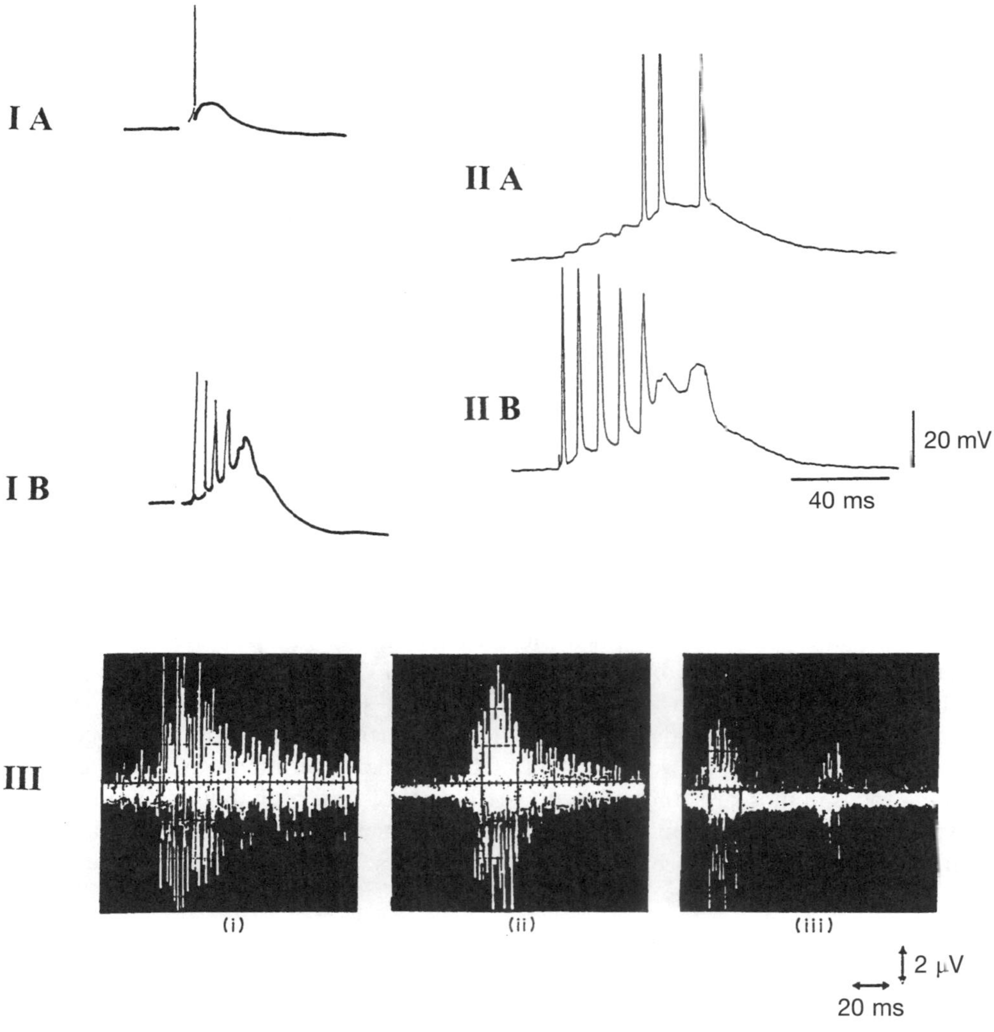

Figure 16.2 Electrophysiological events in the development of an epileptic focus and EEG interictal spike. Intracellular recordings generally show that afferent stimulation of a normal cortical neuron produces one action potential superimposed on a small depolarisation (approx. 10 mV), the excitatory postsynaptic potential of the form drawn in (IA). In a focal type-I epileptic neuron, as found in the CA3 region, the same stimulus can produce a much larger depolarisation, the paroxysmal depolarising shift (PDS) and a burst of spikes (IB). Other neurons must then be recruited and this is shown to be possible in the intracellular recording from two monosynaptically connected CA3 neurons in the hippocampal slice preparation in which each action potential in the presynaptic neuron (IIA) elicits an excitatory potential in the postsynaptic cell which eventually shows a burst of potentials (IIB). Once a number of neurons are recruited there is an almost synchronous discharge of cortical neurons which give rise to an EEG interictal spike. This can be seen from the extracellular recording made with a glass-coated tungsten microelectrode in the cortex of an anaesthetised rat after topical application of the GABA antagonist bicuculline (III). The burst shown in (i) gives rise to a large EEG spike while the other discharges (ii and iii) correspond to medium and small EEG spikes respectively. (II reproduced from Wong *et al.* 1986 and III from Neuropharmacology **30**: Zia-Gharib and Webster 1991 with permission from Elsevier Science.) See also Fig. 2.14

ORIGIN OF FOCAL NEURONS (A in Fig. 16.3)

(i) Properties of focal neurons

Focal neurons must either possess inherent abnormal electrophysiological characteristics or develop them as a result of morphological changes induced in them or around them following some event. There is little evidence of any abnormality in the intrinsic electrophysiological properties of individual neurons studied in brain slices from human focal cortical or hippocampal tissue, although the possibility of some unidentified genetic change in the characteristics of certain ion channels remains possible. By contrast, in electrically kindled rats, NMDA receptors on dentate gyrus granule cells show some plasticity, which at the channel level is manifest by prolonged bursts, clusters and increased agonist potency. Although these changes persist through the kindled state and must therefore be transferred to new receptors, the molecular basis is not known (see Mody 1998). Brain damage can, however, modify neuron function and so possibly make some of them hyperexcitable and focal.

(ii) Reduced inhibition

It has been known for many years that inhibitory interneurons in the spinal cord are very vulnerable and easily destroyed by a reduction in blood supply and that in their absence motoneurons become much more excitable. So it is possible that localised ischemia or hypoxia in the brain could equally well cause a selective loss of GABA inhibitory interneurons and increased excitability of some pyramidal cells. Certainly there is morphological evidence for the loss of such interneurons from occlusion experiments in rodents, as well as a loss of GABA nerve terminals around a cortical alumina focus in monkeys and reduced GABA uptake, and probably therefore GABA nerve terminals, during brain dialysis in epileptic patients. Despite these findings, any neuronal loss reported in human epilepsy appears to be confined to the larger pyramidal neurons, and these do not release GABA.

(iii) Increased excitation

It is equally well known that if a neuron dies, or is destroyed, then any other neuron, which had been innervated by it, gradually becomes supersensitive to the NT it released. In the case of degenerating pyramidal cells this would be glutamate, the excitatory NT. Not surprisingly, undercutting the cortex in animals to produce a deafferentation of some of its neurons not only renders them more likely to show epileptic-like discharges but neurons in hippocampal slices from kindled rats and human focal cortex show supersensitivity to the excitatory amino acids. Such supersensitivity could make some neurons so easily activated that they become ‘epileptic’.

The rate of development of such experimentally induced supersensitivity following denervation or hypoxia is similar to that seen in animals with focal (alumina) lesions but quicker than epileptogenesis following focal pathology (injuries) in humans. Also it must be remembered that although neurons may become supersensitive to glutamate this will no longer be released synaptically from the afferent terminals of the degenerating neurons although its release from others could produce inappropriate, disorganised and extended activation. Indeed there are some morphological changes that would support this.

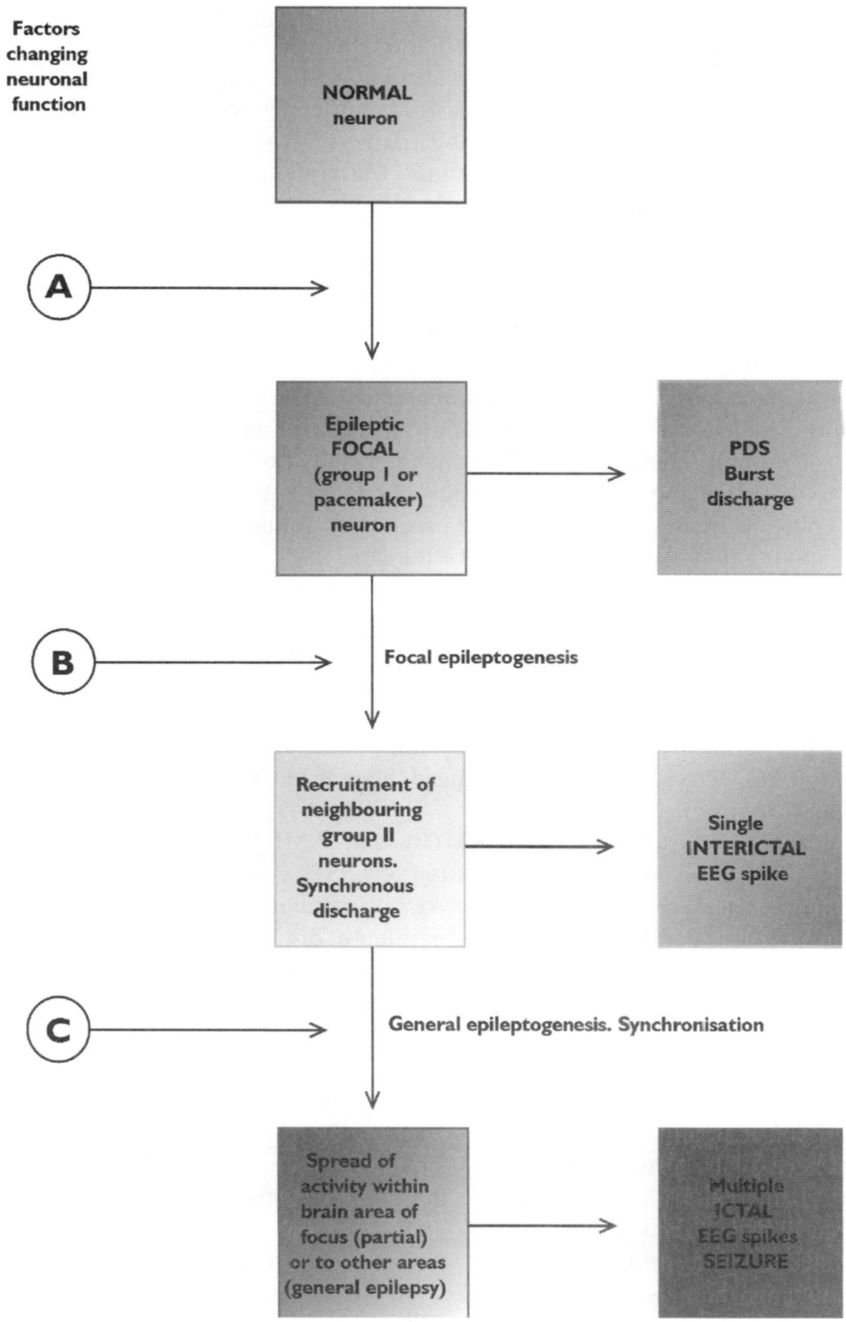

Figure 16.3 Changes in neuronal function required for the development of epileptic seizures. The factors that may control or induce the changes in neuronal function that turn a normal neuron into a focal one (A) recruit other neurons (focal epileptogenesis) to produce an interictal EEG spike (B) and ensure the spread of activity (general epileptogenesis) to full ictal activity (C) are discussed in the text. They include alterations to various ion channels, especially those for Na^+, a reduction in local inhibitory activity or an increase in local excitatory drive. The electrophysiological counterparts of some of the events involved are shown in Fig. 16.2

The dendrites of neurons adjacent to those which degenerate also show extensive growth and sprouting which could facilitate abnormal and disorganised synaptic transmission and cause hyperactivity. It is also known that the dendrites of cells around an alumina focus in monkeys, as well as in human epileptic brain, lose their spinous processes, which might contribute to the paroxysmal discharge by facilitating the spread of depolarisation to the neuron soma. Certainly an increase in the number of Na^+ channels on the dendrites of spinal motoneurons, which would facilitate the occurrence of reactive dendritic Na^+ spikes, has been seen after axotomy.

ORIGIN OF INTERICTAL AND ICTAL SPIKES (B and C in Fig. 16.3)

There are many studies on the induction and spread of spiking in animals both *in vivo* and in isolated brain slices, generally initiated by the use of GABA antagonists or removal of Mg^{2+} ions (*in vitro*). Unfortunately since neither of these events is likely to occur in or around a human epileptic focus the results do not tell us much about how focal activity arises and spreads in humans. This needs to be achieved by the use of human epileptic tissue even though the procedures found to control experimentally induced spiking may well be applicable to humans.

There have been a number of observations which show increased excitation and/or reduced inhibition in slices prepared from human epileptic brain tissue. Thus burst discharges can be evoked with stimuli that would not do so in normal animal tissue and these can be blocked by NMDA receptor antagonists. The inhibitory postsynaptic currents (IPSCs) in hippocampal dentate granule cells in slices prepared from temporal lobe epileptic tissue are in fact reduced by stimulation that activates NMDA currents (Isokawa 1996), which are more prolonged than usual and show changes in slope conductance.

It is perhaps not surprising that NMDA and AMPA receptor mechanisms are important in epileptogenesis. The summation of EPSPs through activation of recurrent polysynaptic excitatory pathways is necessary to mediate the large depolarisation of neurons in and around a focus and the intense discharge and extracellular field potentials of the interactal EEG spike, although these may only occur if counteracting inhibition is reduced. There is in fact some evidence of morphological changes in human epileptic hippocampal tissue that would facilitate such excitatory circuits with aberrant networks of collaterals from axons of individual mossy fibre neurons ramifying through to the CA3 and other regions (Isokawa *et al.* 1993). Also the increase in extracellular K^+ following increasing neuronal activity may itself reinforce the activity by directly depolarising nerve terminals and neurons. High extracellular K^+ would also counteract K^+ efflux and so initiate a prolonged low depolarisation that would facilitate repetitive firing.

From this survey it is clear that just as normal neuronal function requires appropriately balanced inhibitory and excitatory controls so the generation of interictal spikes depends on disturbances in both. Clearly activity cannot spread without the activation of excitatory circuits, in which NMDA receptors play an important role, but it will be much facilitated by reduced inhibition (Masukawa *et al.* 1989). These observations may help to explain the establishment of a focus and the development of the interictal spike, but why activity can only spread to seizure proportions, at certain times, is less clear. It will, however, again require overactivity of excitatory circuits inadequately controlled by inhibitory processes. Since these controls are mediated by

NTs it is now appropriate to consider what evidence there is for a malfunction of NT activity in epilepsy, particularly in those responsible for primary excitation and inhibition, i.e. the amino acids. Before doing so the epileptogenesis of absence seizures (petit mal) justifies separate consideration.

ORIGIN OF ABSENCE SEIZURES

There is much evidence that absence seizures originate in the thalamus probably due to some malfunction of neuronal Ca^{2+} channels. The sudden synchronous bilateral nature of the slow-wave discharge (SWD) in the EEG which typifies this condition was justifiably considered by Jasper (see Jasper and Drooglewer-Fortuyn 1997) to require a subcortical focus and he was able to reproduce them in anaesthetised cats by 3 Hz stimulation of the intralamina thalamus, which in conscious animals also produced absence-like behavioural symptoms such as staring and unresponsiveness. Also in rats with genetic absence epilepsy (GAER) such symptoms are not only accompanied by a synchronous 7–9 Hz SWD but this coincides with high-amplitude discharges in the lateral part of the thalamus, the lesion of which inhibits SWDs.

Within the thalamus the reticular nucleus, which contains predominantly GABA neurons, sends axons to all the other thalamic muclei and although it does not appear to directly drive any thalamic projection to the cortex it receives collaterals from both thalamo-cortical and cortico-thalamic pathways and is well positioned to influence cortico-thalamic activity. If its neurons are stimulated while slightly hyperpolarised they show repetitive burst discharges in rat brain slices followed by a marked after-hyperpolarisation, i.e. oscillatory activity (Avanzini *et al.* 1992). Pharmacological studies *in vivo* in the genetically prone rat show that this depends on the activity of certain Ca^{2+} and Ca^{2+}-activated K^{+} conductances and that blocking Ca^{2+} channels just in the reticular nucleus reduces the cortical SWDs. In fact cloning studies in mutant mice strains with features of absence epilepsy show defects in the subunit structure of these channels (Fletcher *et al.* 1996), although why such an effect on channels that have a very widespread distribution should manifest itself in rhythmic activity only in thalamic neurons is uncertain. It may, however, depend on a particular inhibitory control and hyperpolarisation induced locally by GABA, which certainly invokes rhythmic activity when applied to firing neurons and potentiates SWDs in GAERs. In fact this response is probably mediated by $GABA_B$ rather than $GABA_A$ receptors since not only does baclofen ($GABA_B$ agonist) have a similar effect to GABA but when GABA is applied to thalamic neurons it produces a bicuculline-insensitive long-lasting but slight hyperpolarisation which is followed by a low-threshold calcium potential (LTCP) and spike. This T-type Ca^{2+} channel is common in GAERs and larger than normal in thalamic GABA neurons.

NEUROTRANSMITTERS IN EPILEPTIC ACTIVITY

Changes in NT levels and function have been

(1) Looked for in

- (a) human epileptic tissue
- (b) animals in which convulsions have been induced experimentally

(c) animals with spontaneous (genetically disposed) epilepsy

(2) Induced in animals to see how they modify convulsive threshold and intensity

These approaches will be considered in respect of the different NTs although most interest has centred on the amino acids not only because of their possible involvement in the pathology, as already emphasised, but because increased neuronal activity in epilepsy must reflect, even if it is not initiated by, augmented glutamate and/or reduced GABA function.

AMINO ACID MEASUREMENTS

Human studies

Reduced GABA uptake during microdialysis has been mentioned and there are reports of reduced levels of GABA in the CSF of chronic epileptics and of its synthesising enzyme glutamic acid decarboxylase (GAD) in some samples of temporal lobe tissue removed during surgery to alleviate focal seizures. Other reports find no change in GAD but an increase in $GABA_A$ receptors.

Animal studies

In addition to the loss of GAD staining (i.e. GABA) neurons and inhibitory symmetrical synapses around an alumina focus in primates (see above), studies with a chronically implanted cortical cup over a cobalt lesion (focus) in rats show an increased release of glutamate that is associated with spiking (Dodd and Bradford 1976).

Numerous acute experiments with cortical cups show that systemic convulsants increase the release of ACh but rarely that of glutamate. Even the marked convulsant EEG seen after PTZ infusion in the rat (Fig. 16.4) is not accompanied by any rise in glutamate release. This may not mean that it does not occur but that the avid uptake mechanism for glutamate ensures that levels do not rise above basal, unless the stimulation is very extreme. This may explain why perfusates of the lateral ventricle, obtained during kindled seizures induced by the stimulation of the amygdala, showed elevated glutamate levels, but only after very intense neuronal disharges. Basal GABA levels are often too low to even detect in such studies.

If kindling is regarded as a model of the development of epilepsy (epileptogenesis) then following changes in NT function, after or through its development, may be of more value than merely monitoring release during convulsions. Unfortunately results have been inconclusive. Kindling induced by the intraventricular injection of folic acid in rats produced significant increases in cortical glutamate and aspartate, but only the latter correlated directly with increased spiking. With kindling induced by electrical stimulation of the frontal cortex the only change observed alongside the increase in after-discharge was a reduction in glutamine, although this could reflect its utilisation in providing the extra glutamate required for spiking and epileptic activity.

Animals with spontaneous epilepsy

These have yielded few data apart from reports of reduced GABA and taurine in the CSF of baboons with spontaneous seizures.

AMINO ACIDS, MANIPULATION

GABA

Experimentally all GABA antagonists induce convulsions. These include the genuine receptor antagonist bicuculline, which competes with GABA for its recognition site on the $GABA_A$ receptor and picrotoxin, which binds to a different site more closely related to the chloride ion channel.

Reducing the availability of GABA by blocking the synthesising enzyme GAD also promotes convulsions. This may be achieved by substrate competition (e.g. 3-mercapto propionic acid), irreversible inhibition (e.g. allylglycine) or reducing the action or availability of its co-factor pyridoxal phosphate (e.g. various hydrazides such as semi-carbazide). In fact pyridoxal phosphate deficiency has been shown to be the cause of convulsions in children.

Clearly since a reduction in GABA function causes convulsions, then augmenting its function should provide an anticonvulsant action. This may be achieved in a number of ways as listed in Table 16.2 and indicated in Fig. 16.6. For more detail see Chapter 9.

Agonists and prodrugs

$GABA_A$ receptor agonists like muscimol and (Fig. 16.7) are active against PTZ in mice and amygdala kindling in rats but ineffective in the photosensitive baboon and in fact produce rhythmic spike and wave discharges in the EEG despite poor brain penetration. These discharges have also been seen in the few humans on which the drugs have been tested unsuccessfully.

The reason for this disappointing response is uncertain but may be due to desensitisation of the $GABA_A$ receptors, or the actual inhibition of GABA inhibitory neurons through somatic autoreceptors which could disrupt the precise timing of physiological inhibition. Activation of the $GABA_B$ receptor with baclofen has no

Table 16.2 Drug augmentation of GABA function

A	B
1 GABA receptor, agonists	$GABA_A$ — muscimol $GABA_B$ — baclofen
2 Gabamimetics	Progabide
Prodrugs	Gabapentin
3 GABA-t inhibitors	Ethanolamine-*o*-sulphate (EOS) Na^+ valproate γ vinyl-GABA (vigabatrin)
4 Uptake inhibitors	
Neuronal	DABA.ACHC
Glial	Nipecotic acid-tiagabine
5 Allosteric enhancement	Benzodiazepines
6 Chloride channel openers	Barbiturates

Notes:
Mechanisms are listed under A and examples of drugs that utilise them under B. All compounds that increase the action of endogenous GABA (1–5) augment neuronal inhibition and have an anticonvulsant action. Drugs that act directly on GABA receptors have not so far proved effective. Barbiturates do not really augment GABA function; they do not act on GABA receptors or modify its destruction, but can open Cl^- channels and so increase neuronal inhibition and thus the action of GABA.

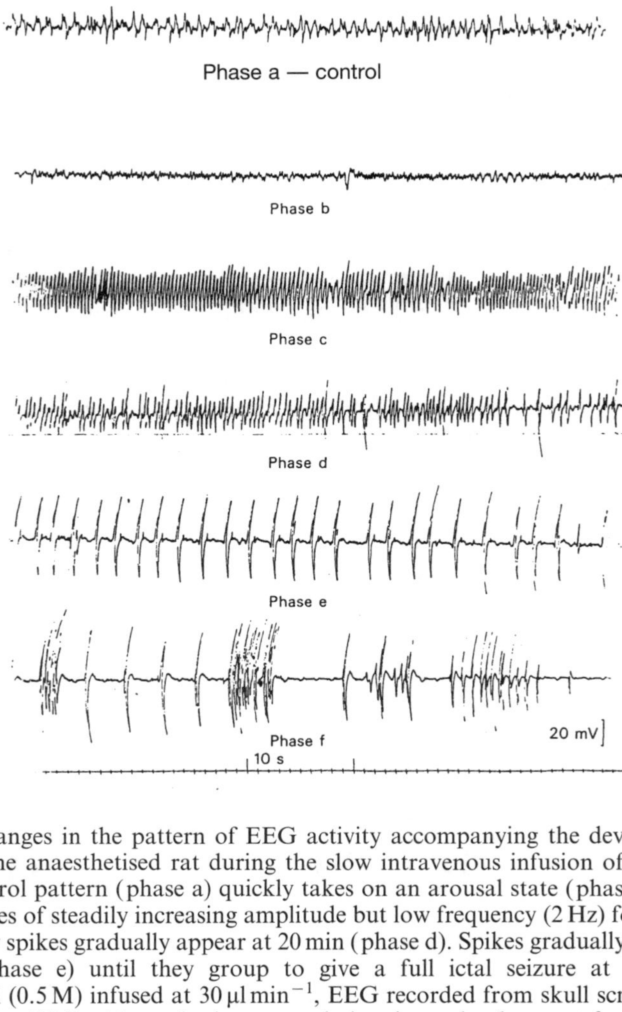

Figure 16.4 Changes in the pattern of EEG activity accompanying the development of a full ictal seizure in the anaesthetised rat during the slow intravenous infusion of pentylenetetrazol. The normal control pattern (phase a) quickly takes on an arousal state (phase b, 2–5 min). This gives way to waves of steadily increasing amplitude but low frequency (2 Hz) for 8–18 min (phase c) on which a few spikes gradually appear at 20 min (phase d). Spikes gradually predominate after some 26 min (phase e) until they group to give a full ictal seizure at 30 min (phase f). Pentylenetetrazol (0.5 M) infused at 30 μl min^{-1}, EEG recorded from skull screw electrodes over the parietal cortex. While this study does not mimic seizure development from a specific focus, since PTZ given systemically can act throughout the brain, it illustrates how cortical activity can become synchronised even without a primary focus. (Reproduced with permission of Macmillan Press Limited from Kent and Webster 1983)

general anticonvulsant effect even though it reduces reflex epilepsy in photosensitive baboons and spiking in hippocampal slices. That GABA function is important, however, in the control of epileptogenic activity is illustrated in Fig. 16.5 which shows that spiking induced in the cortex of the anaesthetised rat by leptazol occurs more readily if GABA function is reduced by the local application of its antagonist bicuculline but retarded if GABA itself is applied.

GABA-t inhibitors

GABA transaminase is a mitochondrial enzyme which, like GAD, requires pyridoxal phosphate as co-factor. It is present in both neurons and glia and while secondary to

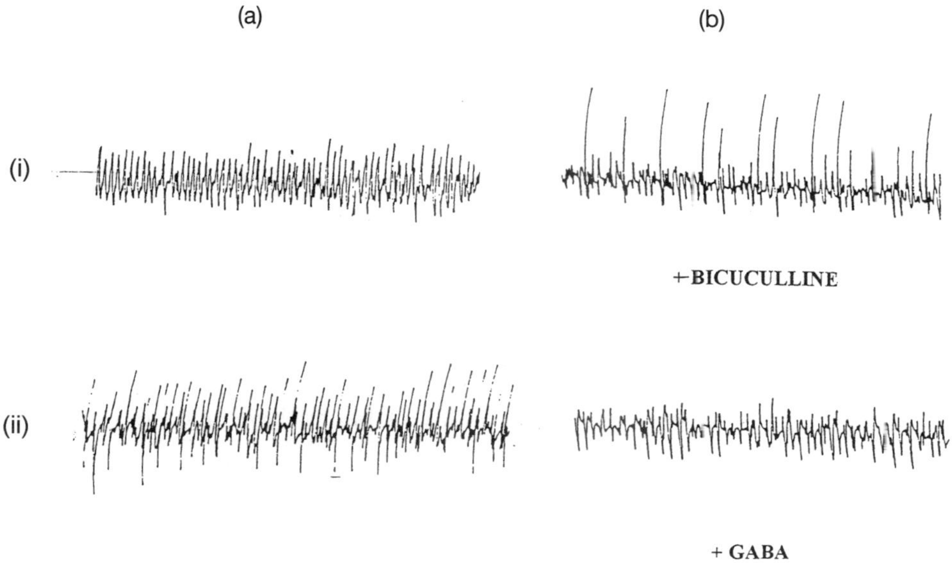

Figure 16.5 The importance of GABA in controlling the development of EEG epileptic spiking. The EEG records shown were taken from the anaesthetised rat during the infusion of pentylenetetrazol (PTZ). They were obtained from screw electrodes (a) in the skull over one parietal cortex and from electrodes within a cortical cup (b) on the other exposed parietal cortex which was superfused with artificial CSF to which drugs could be added. Thus while the whole cortex received PTZ only that area adjacent to the cup could be influenced by the drugs. Under control conditions the developing epileptogenic EEG was identical in both recordings. Records from the screw electrodes (a) showed the expected progressive change from wave-like (i) to spiking (ii) similar to phases c and d in Fig. 16.4. When the cortex under the cup electrodes (b) was exposed to the GABA antagonist bicuccilline the EEG had already developed spiking (bi) while that from the screw electrodes (ai) still remained wave-like. By contrast, when GABA was in the cup the EEG within it developed more slowly with wave-like activity (bii) persisting when spiking had already developed in the record from the screw electrodes (aii). Clearly GABA retards the development of spiking. (Unpublished figure but see Kent and Webster 1986 for detail and drug concentrations)

uptake in the degradation of GABA a number of inhibitors have proved effective experimentally and some clinically. Ethanolamine-O-sulphate was one of the first tested. It produces a large (fortyfold) and sustained increase in brain GABA accompanied by a reduction in seizures induced by maximal electroshock. Gabaculine and aminooxyacetic acid are similar but are ineffective in man whereas γ-vinyl GABA (vigabatrin) has proved useful clinically. The use of this and sodium valporate is considered later.

Uptake inhibitors

GABA is removed from the synapse by a high-affinity sodium and chloride-dependent uptake into GABA neurons and surrounding glia. Blocking this process potentiates the inhibitory action of GABA applied directly to neurons *in vivo* and *in vitro*. Some inhibitors show specificity for glia and others for neuronal uptake, although since recent molecular cloning has revealed four distinct GABA transporters (Chapter 9) this simple classification may require modification. Probably because of structural similarities to GABA, few of these compounds show brain penetration but tiagabin, a lipophilic form of nipecotic acid, has been tried successfully in refractory epilepsy.

Receptor modulators

Benzodiazepines bind to a specific site on the GABA chloride ionophore, which differs from that for GABA itself, but when occupied augments the binding and action of GABA to increase the frequency of opening of chloride ion channels. Thus they augment GABA inhibition. Many of them are potent anticonvulsants, especially when tested against PTZ and retard the development of kindling. Unfortunately their clinical value is limited by the development of tolerance.

Barbiturates also potentiate the action of GABA but as they can do this by directly increasing the duration of opening of the chloride ion channel, independently of the GABA or benzadiazepine receptor sites, they cannot strictly be considered to augment GABA. Some such as phenobarbetone are, however, of proven clinical value.

Glutamate

NMDA receptor antagonists such as AP5 and AP7 were first shown to be anticonvulsant following introcerebroventricular injection into DBA/2 mice susceptable to audiogenic seizures. In addition, they offer protection to PTZ, reduce the after-discharge in amygdala kindled rats and can actually retard the development of kindling. Although AP7 has some effect in photosensitive baboons, systemically active compounds have proved difficult to synthesise. Recently felbamate, an antagonist at the glycine-sensitive site on the NMDA receptor, has shown systemic anticonvulsant activity and clinical efficacy.

Inhibition of glutamate release was thought to be the mode of action of lamotrigine. It reduces MES and kindling and also glutamate (and to a lesser extent GABA) release induced in brain slices by veratridine, which opens sodium channels. But it now seems likely that the actual block of sodium channels is its primary action (see later).

The epileptic discharges induced in hippocampal slices by tetanic stimulation has been shown to be accompanied by reduced GABA-mediated IPSPs (Stelzer, Slater and Bruggencate 1987). Since AP7 not only reduced the discharges but also restored the response to GABA some linkage between NMDA and $GABA_A$ receptors seems probable. In fact the interaction between glutamate and GABA probably means that both of them and possibly their different receptors may need to be manipulated appropriately to control convulsive activity. This has been shown in fact experimentally when bicuculline was infused intravenously for short periods in the rat to give a burst of epileptic-like spiking in the EEG. Superfusion of the cortex using the cup technique with the glutamate AMPA antagonist CNQX or the $GABA_B$ agonist baclofen reduced the actual number (initiation) of spikes but not their amplitude, while NMDA antagonists (AP7) and the $GABA_A$ agonist muscimal reduced the size (development and spread of excitation) and not the number of spikes (Zia-Gharib and Webster 1991). Clearly more than one aspect of amino acid function may need to be controlled.

Other NTs have been implicated in the aetiology of epilepsy but direct evidence is lacking. They will be considered briefly.

ACETYLCHOLINE (ACh)

Cholinergic agonists, e.g. carbachol, applied to the rat cortex cause focal spiking and even seizures which can also be induced by large doses of CNS-penetrating anti-cholinesterases such as physostigmine (reversible inhibitor) or di-isopropylfluoro-phosphate (irreversible). Many studies have also shown that cortical ACh release increases in proportion to EEG activity during the administration of a wide range of convulsants. Nevertheless while cholinergic-induced seizures can be suppressed by antimuscarinic drugs they have no effect against any epilepsy in humans and ACh release presumably reflects rather than directly causes cortical activity.

MONOAMINES

The widespread and diverging nature of ascending monoamine pathways to the cortex suggest that NA and 5-HT are more likely to have a secondary modifying rather than a primary effect on the initiation of epileptic activity. In reality this is the case and their secondary role is even a minor one. Generally a reduction in monoamine function facilitates experimentally induced seizures (see Meldrum 1989) while increasing it reduces seizure susceptibility. The variability of the procedures used and results obtained do not justify more detailed analysis here.

Some mention should perhaps be made of dopamine, considering its role in the control of motor function. It is perhaps not surprising that DA agonists like apo-morphine block the myoclonus induced in photosensitive baboons and audiogenic seizures in DBA/2 mice while neuroleptics (DA antagonists) may have a weak procon-vulsant effect in humans. Also in rats with absence seizures dopa, apomorphine and D_1 agonists reduce facial clonus and spike and wave discharges, while the D_1 antagonist SCH 23390 increases them. Nevertheless, there is no evidence of a significant role for DA (or NA and 5-HT) in human epilepsies.

ADENOSINE

A number of studies have shown that adenosine inhibits neuronal firing both *in vitro* and *in vivo* and is itself released during intense neuronal activity. It can protect against PTZ seizures in rodents while the antagonist theophylline is proconvulsant. No clear picture of its role in human epilepsy has emerged.

APPROACHES TO THE CONTROL OF EPILEPTIC ACTIVITY

Irrespective of the cause of epilepsy, the spread of seizure activity will be attenuated by either decreasing the excitation or increasing the inhibition of neurons. This may be achieved in a number of ways, either *directly* by

(a) blocking excitatory voltage-gated Na^+ (or possibly Ca^{2+}) channels (1)
(b) increasing the opening of inhibitory Cl^- channels (2)

or *indirectly* by

(c) reducing the release of the excitatory NT, glutamate (3) or its action at NMDA receptors (4)

(d) increasing the availability (and release) of the inhibitory NT, GABA by blocking its reuptake (5) or metabolism (6) or activating the GABA receptor either directly (7) or through the benzodiazepine receptor (8).

These effects, to which the above numbers (1)–(8) refer, are shown in Fig. 16.6.

How the drugs currently available for the treatment of epilepsy may utilise these mechanisms will now be considered.

ANTIEPILEPTIC DRUGS (AEDs)

There is no shortage of AEDs (Fig. 16.7) but it is not appropriate to consider them in detail in this text other than to see how their mechanisms of action comply with and illustrate those proposed above (Fig. 16.6) for the control of epileptic seizures (see Meldrum 1996; Upton 1994). The decision on which drug to use depends not only on their proven efficacy in a particular type of epilepsy (some drugs are inactive in certain forms) but also what side-effects they have — many are sedative — how they interact with other drugs and how often they need to be taken. Compliance is a problem over a long period if dosing is required more than once a day. It is probably acceptable in reality, if not scientifically, to divide the drugs into old-established AEDs and new AEDs. Only the latter have been developed chemically to modify the known synaptic function of the amino acids.

OLD AEDs

Phenobarbitone was the first AED and was introduced in 1912. It was largely replaced in 1932 by phenytoin for the management of tonic–clonic seizures and partial and secondary epilepsy. Carbamazepine followed, then ethosuximide for absence seizures and valproic acid. These remained, apart from the introduction of the benzodiazepines, the mainstay of therapy until the last decade. They were introduced solely on their ability to control experimentally induced seizures. Their mechanisms of action were unknown and no thought was given to the possibility of NT modification and in fact subsequent research has shown that with the exception of the benzodiazepines none of them work primarily through NT manipulation. They act directly on neuronal excitability.

Phenytoin and carbamazepine

An effective AED might control seizures and not be too sedative, by stopping a neuron from firing excessively without affecting its ability to respond normally. This is how phenytoin is believed to work. Studies in cultured spinal cord neurons (Macdonald and McLean 1986) have shown that concentrations of phenytoin equivalent to those occurring clinically do not affect the resting membrane potential or the shape of a single-action potential but reduce the rapid discharge induced by depolarising the neuron, while leaving the first action potential intact (Fig. 16.8). It is believed to block voltage-dependent sodium channels (not those mediating the synaptic currents) after their activation, i.e. when they become inactivated, and so maintains them in that inactivated state and unresponsive. These effects, which are also shown by carbamazepine, would explain their effectiveness experimentally against maximal electroshock-induced

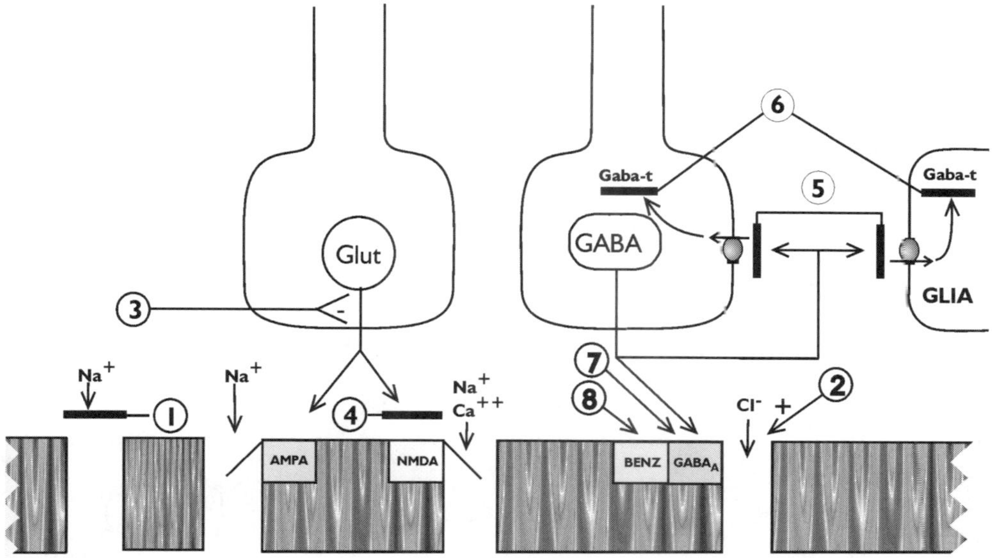

Figure 16.6 Possible sites of action of antiepileptic drugs. Antiepileptic drugs either directly affect ion channels to reduce Na^+ (1) or increase Cl^-(2) influx, depress glutamate release (3) or its action through NMDA receptors (4), or potentiate the effect of GABA by reducing its destruction by uptake (5) or metabolism by GABA transaminase (6), acting directly on $GABA_A$ receptors (7) or potentiating that effect of GABA through an action on benzodiazepine receptors that allosterically alter the $GABA_A$ site (8). Currently there are no clinically useful drugs that act as glutamate receptor antagonists

seizures and clinically in focal and generalised epilepsy. Also, since they act only on the inactivated channel, they will not affect normal neuronal function, which is why in the experimental study, the first action potential remains unaltered. Neither compound is of any value against absence seizures and may exacerbate them. They have no clear effect on NT function although there is some evidence that *in vivo* they may potentiate GABA-induced chloride currents.

Ethosuximide

This is really only effective against absence seizures. Experimentally it has no effect on the voltage-gated sodium channels affected by phenytoin but has been reported to suppress the transient T-type calcium currents in the thalamic neurons which are the origin of the 2–3 Hz spike and wave discharge characteristic of this form of epilepsy (see Mody 1998 for detail). Since these discharges are thought to arise from oscillations in excitability induced by changes in the T-type calcium current (see section above on the origin of absence seizures), this would obviously be a neat explanation of its efficacy in that condition. Unfortunately some workers have not been able to repeat this finding at clinically equivalent concentrations and consider ethosuximide to reduce a special persistent Na^+ channel and a Ca^{2+}-activated K^+ channel.

Barbiturates and benzodiazepines (B & Bs)

As outlined above (see also Chapter 9), these drugs have been found to influence the Cl^- channel of the $GABA_A$ receptor. Phenobarbitone acts directly to prolong its

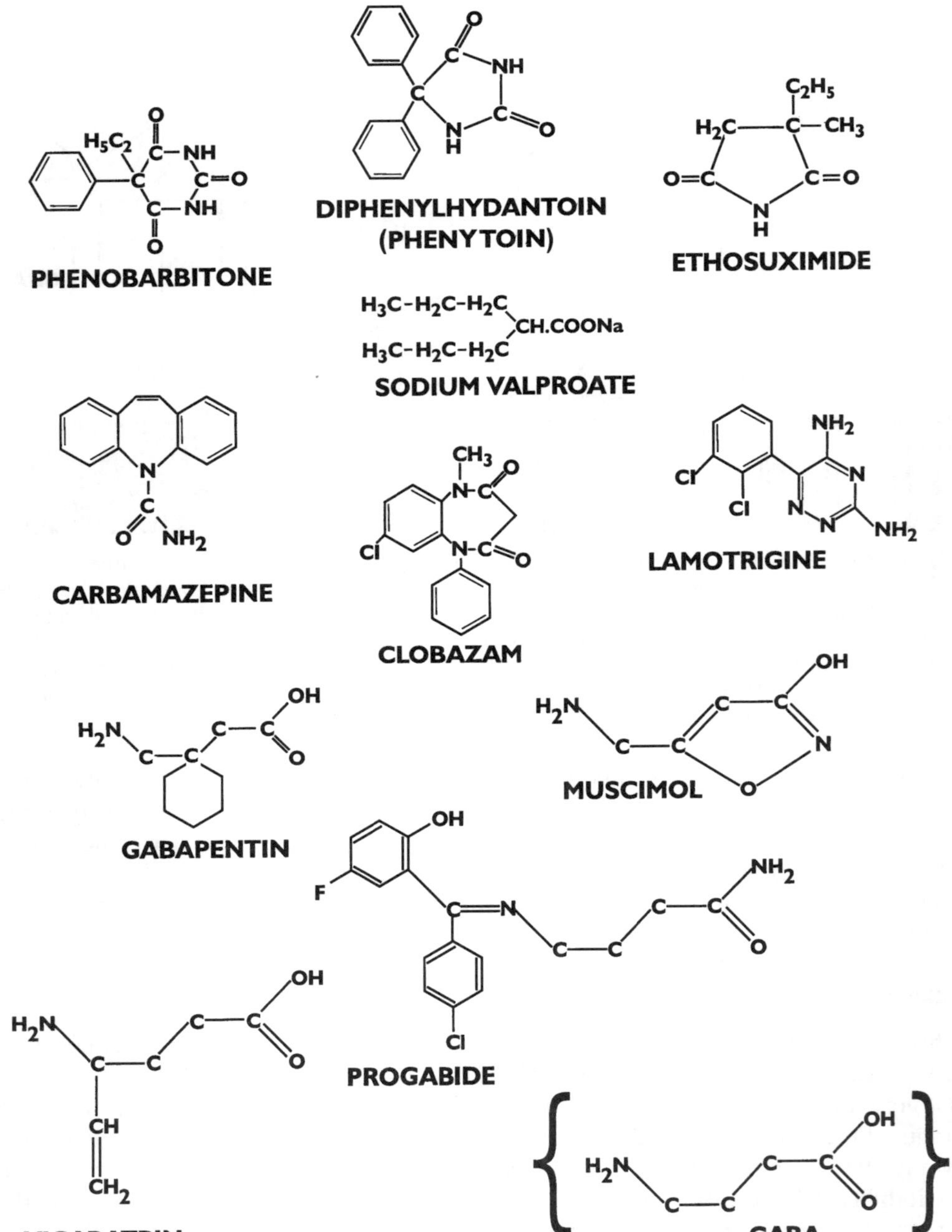

Figure 16.7 The structure of some established antiepileptic drugs (AEDs) and some newer ones. Note that while the structures of phenytoin and ethosuximide are similar and also close to that of phenobarbitone, they are effective in different forms of epilepsy. Vigabatrin, progabide and gabapentin are clearly related to GABA. Muscimol is a $GABA_A$ agonist but is not an effective antiepileptic drug

Table 16.3 Possible mechanisms of action and features of some antiepileptic drugs

	Use	Mode of action	Comments (half-life, hours)
Established drugs			
HYDANTOINS			
Diphenylhdantoin (phenytoin)	GM (PE)	1	Widely used. Hyperplasia of gums. Anti-folate. Teratogenic. Ineffective in PM (20–80)
DIBENZAPINES			
Carbamazepine	GM FE TLE	1	Improves mood. Related to tricyclic antidepressants. Drug of choice in FE (10–20)
SUCCINIMIDES			
Ethosuximide	PM (AS)	1	Drug of choice for PM, with Na valproate (20–60)
BARBITURATES			
Phenobarbitone	GM/FE	2	Sedative. Withdrawal fits. Little used (50–100)
Primidone	GM/PE		Works partly by conversion to phenobarbitone in body
BENZODIAZEPINES			
Diazepam	SE	8	Given intravenously in SE (<100)
Clonazepam	ME SE		Diazepam largely replaced by clonazepam
Clobazam	PM		Adjunct to other anti-epileptics. Partly as an anxiolytic
SHORT-CHAIN FATTY ACIDS			
Sodium valproate	GM PM ME	6 also 1 (and 2)	Inhibition of GABA metabolism too slow to explain initial anti-convulsant effect. Increasing use in ME, PM, GM (5–15)
Newer drugs			
Lamotrigine	PE GM (AS)	4 (1)	Fewer side effects (24)
Gabapentin	PE GM	?	Excreted unchanged
Vigabatrin	PE (GM)	6	Exacerbates AS (PE)

Notes:
The numbers (1–8) refer to their sites of action as shown in Fig. 16.6. All compounds may produce some overt signs of CNS depression, e.g. ataxia, sedation, dizziness.

opening time (mechanism 2 in Fig. 16.8), while the benzodiazepines modify the $GABA_A$ receptor allosterically and increase the likelihood (frequency) of Cl^- channel opening. The benzodiazepines are particularly effective against experimentally induced PTZ seizures.

Phenobarbitone may be as effective as phenytoin and carbamazepine in partial and generalised tonic–clonic seizures but its other central effects such as sedation, depression, listlessness and cognitive impairment mar its usefulness.

Clonazepam, a typical 1:4 benzodiazepine, is effective in absence seizures, myoclonic jerks and tonic–clonic seizures and given intravenously it attenuates status epilepticus. It is less sedative than phenobarbitone but tolerance develops and its withdrawal, as

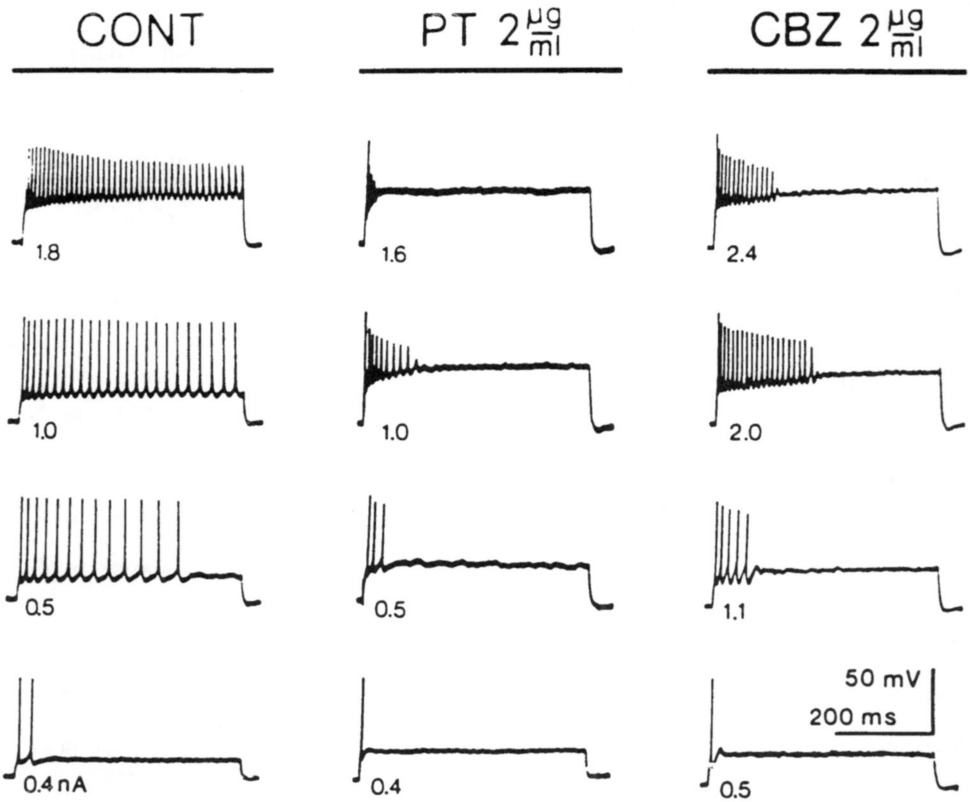

Figure 16.8 Cellular action of phenytoin and carbamazepine. Each column shows the response of a spinal cord neuron in culture to four increasing directly applied current pulses (amplitude in nA given at start of each sweep. Under control conditions (CONT)) the progressive depolarisations (bottom to top of each column of traces) induce increasing sustained discharges, whereas in the presence of phenytoin (PTZ) and carbamazepine (CBZ) firing cannot be maintained although the initial action potential remains. (Reproduced from MacDonald and McLean 1986.) These drugs are thought to bind to Na^+ channels after they have been active (opened) and maintain them in the inactivated state. Thus they do not affect the initial response but stop neurons from maintaining the abnormal sustained discharge that would be characteristic of epileptic activity. Resting membrane potentials (E_m) are shown at the bottom of each column and amplitude (mV) and time (ms) at the bottom right

with phenobarbitone, can precipitate seizures. Although still used in refractory myoclonic epilepsy, when its depressant effect on the spinal cord may be significant, clonazepam, like phenobarbitone, is rarely used now, but the more recently introduced 1:5 benzodiazepine clobazam is quite often used as an adjunct (not in the United States). While there is some belief and evidence that clonazepam and clobazam are more effective than other benzodiazepines as anticonvulsants nothing is known specifically about their modes of action that supports this view. The reported inhibitory effects of B & Bs on a calcium-sensitive NT release in synaptosomes is difficult to evaluate in terms of their *in vivo* anticonvulsant activity.

Valproic acid (sodium valproate)

Introduced initially for absence seizures, this drug is now known to be effective in and used to treat tonic–clonic seizures and most types of epilepsy. It was found to inhibit GABA transaminase and so elevate GABA concentrations and inhibition. This is achieved, however, over a slower time-course than its anti-seizure effect, especially experimentally, which is now thought to be due to its phenytoin-like, use-dependent block of sodium channels. Since, unlike phenytoin, the full effect of valproate takes some weeks to develop, its slower effect on GABA metabolism and activity should not be ignored.

NEW AEDs

Most of these have been used mainly as add-on therapy although some are now being used alone.

Lamotrigine

One unwanted side-effect of phenytoin is its anti-folate activity. A programme of synthetic chemistry to manipulate the structure of the anti-folate compound pyrimethium to try to replace that property with anticonvulsant activity resulted in the synthesis of lamotrigine. It proved to be an effective AED in partial and generalised epilepsy but experience has found it also to be of value in absence seizures.

Experimentally it was shown to reduce the release of glutamate and to a lesser extent GABA, induced in small brain slices by veratridine, a sodium ion channel opener. It now appears that its primary effect is prolonging the inactivation of sodium channels in a use-dependent manner much like phenytoin, although in a recent study of intracellularly recorded activity of striatal neurons in the rat corticostriatal slice preparation some differences emerged. While both drugs reduced experimentally induced repetitive firing, phenytoin was more effective against those induced by direct current activation of the neurons and also inhibited the EPSPs induced by the direct application of glutamate. By contrast, lamotrigine had little effect on the glutamate response but was more active against those induced by corticostriatal tract stimulation, suggesting that part of lamotrigine's action may still reside presynaptically in reducing glutamate release (Calabresi *et al.* 1999).

Vigabatrin (γ vinyl GABA)

This drug is chemically related to GABA, is an irreversible inhibiter of GABA transaminase and appears to produce its antiepileptic effect through that mechanism. Not only does it increase brain GABA levels in animals it also elevates them up to threefold in human CSF and in the occipital cortex of normal and epileptic patients as shown by nuclear magnetic resonance spectroscopy. An interesting decrease in glutamate may be secondary to the rise in GABA. It is effective in partial and secondary generalised epilepsy, but since its mode of action requires the regeneration of new enzyme (GABA-t) its effect far outlasts its plasma life. A worrying intramyelinic oedema in rat nerves has fortunately not been seen in humans or primates.

Tiagabine

Drugs that block the neuronal and in particular the glial uptake of GABA, like diaminobutyric acid and nipecotic acid respectively, proved effective anticonvulsants experimentally but had to be administered directly into the ventricals (intra-cerebroventrically). Attaching nipecotic acid to a lipophilic component to increase brain penetration resulted in tiagabine. Surprisingly, it appears to act preferentially on the GABA transporter GAT_1 which, although found on astrocytes, is more associated with nerve terminals. Microdialysis in rats shows it increases extracellular GABA and prolongs the post-excitatory hyperpolarisation of neurons. It has proved effective in partial and secondary generalised epilepsy but prolonged post- and possibly presynaptic actions of the increased GABA could present problems.

Gabapentin

This drug, which is a cyclohexone analogue of GABA, was synthesised in the hope that it would be an agonist for GABA receptors which could cross the blood–brain barrier. Its efficacy in drug-resistant partial and secondary generalised epilepsy means that it certainly must enter the brain but it does not bind to GABA receptors. Despite this, it appears to increase GABA brain levels in epileptic patients and weak potentiation of GAD and inhibition of GABA-t have been described. It does not appear to affect sodium or calcium channels even though experimentally chronic dosing blocks repetitive neuronal firing. Specific binding sites have been shown for it on neuronal membranes which appear to be a leucine transporter, but their significance is not clear.

OTHER NEW AEDs

The last few years has seen an explosion in AEDs. Some of those mentioned above may fall by the wayside and others appear. At the time of writing, we could include felbamate, zonisamide oxcarbazepine and topiramate. They all appear to have a phenytoin-like action on sodium channels, although topiramate appears to also potentiate the action of GABA on $GABA_A$ receptors like the benzodiazepines but through a different site.

SUMMARY

It will be apparent that all the possible mechanisms of action for anticonvulsant drugs outlined above (Fig. 16.6) have not been realised by those drugs currently available. The efficacy of glutamate NMDA antagonists is still restricted to experimental studies. No clinically useful drug has been developed and its synthesis will depend not only on finding a compound capable of entering the brain but also on the realisation of the hope that focal NMDA receptors may prove to be different from others. It may then be possible to target them specifically and avoid widespread depression. Lamotrigine does reduce the release of glutamate but this may be secondary to the blockade of sodium channels.

No directly acting $GABA_A$ receptor agonists have been found and it is likely that they would be too depressant (widespread in action) unless focal GABA, like NMDA, receptors have undergone some changes to become specifically targetable. Drugs that decrease the destruction of GABA such as GABA-t inhibitors (vigabratrim) and uptake blockers (tiagabine) have, however, been developed.

Despite all these approaches, drugs acting directly on neuronal ions channels are still the most effective AEDS.

Whether one drug with one mechanism of action will ever be adequate in the therapy of epilepsy is uncertain. Even drugs which apparently have a similar mechanism of action on sodium channels, such as phenytoin, carbamazepine, valproic acid and lamotrigine have different uses as only the latter two are effective in absence seizures. This could reflect some action additional to that on sodium channels (e.g. GABA-t inhibition for valproate) or an effect on a particular type of sodium channel that is different by virtue of some change in its α subunits. In fact the additional clinical effect of some new AEDs (e.g. vigabatrin and tiagabine) in patients not properly controlled by old AEDs like phenytoin could indicate the need for increased GABA function as well as sodium channel block for proper seizure control. The obvious complexity of NT and ion channel interactions in the control of neuronal function may well mean that the proper control of seizures may require the appropriate manipulation of more than one NT and one neuronal function.

Newer AEDs do have some advantages in that they tend to have fewer effects on the metabolism of each other or other drugs. By contrast, phenobarbitone is one of the most potent inducers of the microsomal enzyme system (cytochrone P_{450}) responsible for the metabolism of drugs. Phenytoin and carbamazepine have a similar but less marked effect while valproate inhibits the system.

One thing is certain. All the new AEDs are much more expensive than the older ones and one might therefore question the justification of their use. The reason is that the older ones have limited efficacy and not-inconsiderable toxicity. Indeed even with polytherapy the seizures are not always adequately controlled. So are there other approaches?

OTHER TREATMENTS

Surgery

If there is a clear established focus then maybe the best treatment is to remove it. This is, of course, both difficult and expensive but its use is expanding with about 500 operations per year in the UK. It is only considered in cases of partial (not general) epilepsy when conventional drug therapy has failed and a clear focus can be established. The advent of sophisticated assessments, such as MIR, long-term EEG telemetry, in-depth electrode recording and PET studies of blood flow and diazepan binding has now made this possible. Most commonly part of the anterior temperal lobe is removed, 70% of patients become seizure-free and neurological (mainly visual) and psychiatric problems are surprisingly few (5–10%).

Gliosis

This is not really a treatment but there is a view that glial cells can protect against seizures since the enzyme systems they possess (e.g. Na–K^+ATPase and carbonic anhydrase) facilitate the regulation of ion movements and reduce the spread of seizures. Certainly ageing, a fatty diet, and phenytoin itself increase glial cell count while decreasing seizure susceptibility. In fact inhibition of carbonic anhydrase and the production of bicarbonate was one of the first treatments for epilepsy and a recent discovery that under certain circumstances intracellular bicarbonate can depolarise neurons has created a fresh interest in it.

REFERENCES

Avanzini, G (1992) Generalised epileptogenesis. In *Epileptogenic and Excitotoxic Mechanisms* (Eds Avanzini, G, Fariello, R, Heinemann, U and Mutani, R), John Libbey, London pp. 29–40.

Avanzini, G, de Curtis, M, Marescaux, C, Panzica, F, Spriefico, R and Vergnes, M (1992) Role of the thalamic reticular nucleus in the generation of rhythmic thalamo-cortical activities subserving spike and waves. *J. Neural Trans.* **35** (Suppl): 85–95.

Calabresi, P, Centonze, D, Marfia, GA, Pisani, A and Bernardi, G (1999) An *in vitro* electrophysiological study on the effects of phenytoin, lamotrigine and gabapentia on striatal neurons. *Brit. J. Pharmacol.* **126**: 689–696.

Dodd, PR and Bradford, HF (1976) Release of amino acids from the mature cobalt induced epileptogenic focus. *Brain Res.* **111**: 377–388.

Fletcher, CF, Lutz, CM, O'Sullivan, TN, Shaughnessy Jr, JD, Hawkes, R, Frankel, WN, Copeland, NG and Jenkins, NA (1996) Absence epilepsy in tottering mutant mice is associated with calcium channel defects. *Cell* **87**: 607–617.

Isokawa, M (1996) Decrement of $GABA_A$ receptor mediated inhibitory postysynaptic currents in dentate granule cells in epileptic hippocampus. *J. Neurophysiol.* **75**: 1901–1908.

Isokawa, M, Levesque, MF, Babb, TL and Engel Jr, J (1993) Single mossy fibre axonal systems of human dentate granule cells studied in hippo-campal slices from patients with temporal lobe epilepsy. *J. Neurosci.* **13**: 1511–1522.

Isokawa, M, Levesque, M, Fried, I and Engel, J Jr (1997) Glutamate currents in morphologically identified human dentate granule cells in temporal lobe epilepsy. *J. Neurophysiol.* **77**: 3355–3369.

Jasper, HH and Drooglewer-Fortuyn, J (1997) Experimental studies of the functional anatomy of petit mal epilepsy. *Ass. Res. Nerv. Ment. Dis. Proc.* **26**: 272–298.

Jobe, PC, Mishra, PK, Ludwig, N and Dailey, JW (1991) Scope and contribution of genetic models to an understanding of the epilepsies. *Crit. Rev. Neurobiol.* **6**: 183–215.

Kent, AP and Webster, RA (1983) The evaluation of different types of anti-convulsant drug activity against leptazol-induced epileptogenic activity in the anaesthetised rat. *Brit. J. Pharmac.* **78**: 553–559.

Kent, AP and Webster, RA (1986) The role of GABA and excitatory amino acids in the development of the leptazol-induced epileptogenic EEG. *Neuropharmacology* **25**: 1023–1030.

McNamara JO (1999) Emerging insights into the genesis of epilepsy. *Nature* **399**: Suppl A15–22.

Macdonald, RL and McLean, MJ (1986) Anticonvulsant drugs: mechanisms of action. *Adv. Neurol.* **44**: 713–736.

Masukawa, LM, Higashima, M, Kim, JH and Spencer, DD (1989) Epileptiform discharges evoked in hippocampal brain slices from epileptic patients. *Brain Res.* **493**: 168–174.

Meldrum, BS (1989) The epilepsies. In *Neurotransmitters, Drugs and Disease* (Eds Webster, RA and Jordan, CC), Blackwell, pp. 301–334.

Meldrum, BS (1996) Update on the mechanism of action of antiepileptic drugs. *Epilepsia* **37** (Suppl. 6): 4–11.

Meldrum, BS and Corsellis, JAN (1984) Epilepsy. In *Greenfield's Neuropathology*, 4th edn (Eds Blackwood, W and Corsellis, JAN), Edward Arnold, London, pp. 921–950.

Mody, I (1998) Ion channels in epilepsy. *Int. Rev. Neurobiol.* **42**: 199–226.

Prince, DA (1992) Basic mechanisms of focal epileptogenesis. In *Epileptogenic and Excitotoxic Mechanisms* (Eds Avanzini, G, Fariello, R, Heinemann, U and Mutani, R), John Libbey, London, pp. 17–27.

Prince, DA and Connors, BW (1986) Mechanisms of interictal epileptogenesis. *Adv. Neurol.* **44**: 275–299.

Stelzer, A, Slater, NT and Bruggencate, G (1987) Activation of NMDA receptors blocks GABA ergic innervation in an *in vitro* model of epilepsy. *Nature* **326**: 698–701.

Upton, N (1994) Mechanisms of action of new anti-epileptic drugs. *Trends Pharmacol. Sci.* **15**: 456–462.

Wong, RKS, Traub, RD and Miles R (1986) *Cellular Basis of Neuronal Syncronyic Epilepsy* **44**: 583–592.

Zia-Gharib, F and Webster, RA (1991) Effects of compounds modulating amino acid transmission on the development and control of bicuculline-induced epileptogenic spiking in the rat. *Neuropharmacology* **30**: 995–1009.

17 Schizophrenia

R. A. WEBSTER

SYMPTOMS AND GENERAL ASPECTS

Schizophrenia may be defined as the progressive disintegration of an individual's personality and of the relationship between that person and the world in general. The main symptoms are:

(1) *Auditory hallucinations.* Voices repeating the person's thoughts and commenting on their actions.
(2) *Thought disorders.* The patient believes that other people can read and control their thoughts.
(3) *Physical catatonia.* An ability to maintain exaggerated and often bizarre postures for long periods.
(4) *Emotional problems.* Withdrawal, diminished emotions and response, reduced speech.

The first three characteristics are considered to be the 'positive' symptoms of the disorder. The fourth are described as 'negative' symptoms although they can be divided into true negative symptoms, i.e. diminished emotions and speech and reactive ones, i.e. social apathy and withdrawal brought on by the positive symptoms. Schizophrenics do not have a split personality. Normally their reaction to the positive symptoms is to withdraw quietly but occasionally they will react violently to the voices they hear and shout at them.

There are a number of drugs that reduce the positive symptoms and in so doing can make the patient less withdrawn. Consequently they appear to produce some beneficial effect on the negative symptoms. The good responders with positive symptoms have been categorised as type I schizophrenics and those with true negative symptoms and poor response to drugs as type II (Crow 1985). Approximately 1% of the population may develop schizophrenia during life and generally it appears in late adolescence or early adulthood (18–30 years). A general assessment of treatment is that some 25% recover fully and an almost equal number not at all, with many of them requiring long-term hospitalisation. The remaining half have fluctuating episodes often requiring chronic therapy.

AETIOLOGY

Schizophrenia is not a neurodegenerative disease but there is some general neuropathology. There is also evidence for a genetic influence. In monozygotic twins with

Neurotransmitters, Drugs and Brain Function. Edited by R. A. Webster

identical genes, if one twin develops schizophrenia there is a 50% chance that the other will, even it seems if they live apart but as the offspring of both have an equal chance of showing symptoms the tendency can lie dormant. Certainly the siblings of a schizophrenic show an increased risk of developing the disorder.

Although there is no specific lesion nearly all schizophrenic brains show some pathology such as reduced neuron number and brain size or some minor lesions, particularly in amygdala hippocampus and prefrontal cortex, where PET studies also show reduced blood flow. Surprisingly such changes are often more pronounced on the left side. There is also evidence of increased ventricular size, especially in those with true negative symptoms. Glyosis is not apparent, lesions are not ongoing and many could have arisen at birth.

THERAPY — OUTLINE

In 1952 reserpine, an alkaloid extract from the Indian snakewort plant, *Rauwolfia serpentina*, which had been used in that country to treat 'madness', was first tried in schizophrenia. The beneficial impact on patients and the hospital wards was dramatic, as was that a year later of chlorpromazine, a phenothiazine derivative and haloperidol, a butyrophenone. These latter two drugs and closely related derivatives remained the mainstay of therapy for almost 40 years.

Chlorpromazine had been shown to produce a tranquil state in animals and since it had a similar effect in humans it became known as a major tranquiliser but the term is rarely used today. Sometimes the drugs used to treat schizophrenia are called antipsychotics but more commonly neuroleptics. Leptic means to activate (take hold of) and in animals these compounds produce a state of maintained motor tone known as catalepsy. This is an extrapyramidal effect and in schizophrenics the neuroleptics can cause a number of extrapyramidal side-effects (EPSs) including Parkinsonism. The new term 'neuroleptic' is unsatisfactory as a description of clinically useful drugs. It really describes a condition (catalepsy) seen in animals and is more indicative of a compound's ability to produce EPSs than to treat schizophrenia. 'Antipsychotic' is more descriptive but could imply a more general efficacy in psychoses than is the case. It would seem more appropriate to call a drug that is used to treat schizophrenia an 'antischizophrenic' just as we use the terms 'antidepressant' or 'antiepileptic' irrespective of how the drug works. Despite such personal reservations, the term 'neuroleptic' will be used in this text.

The ability of neuroleptics to produce EPSs immediately suggests that they reduce or antagonise dopamine (DA) function and this is supported by a number of other observations (Table 17.1).

Table 17.1

CNS effect	Known change in DA function	Neuroleptic effect	Presumed change in DA function
Parkinsonism	Reduced	Induction of Parkinsonism	Reduced
Elevated plasma prolactin	Reduced	Elevated proloctin	Reduced
Vomiting	Increased	Anti-emetic	Reduced
Hallucinations	Increased	Decrease hallucinations in schizophrenics	Reduced?

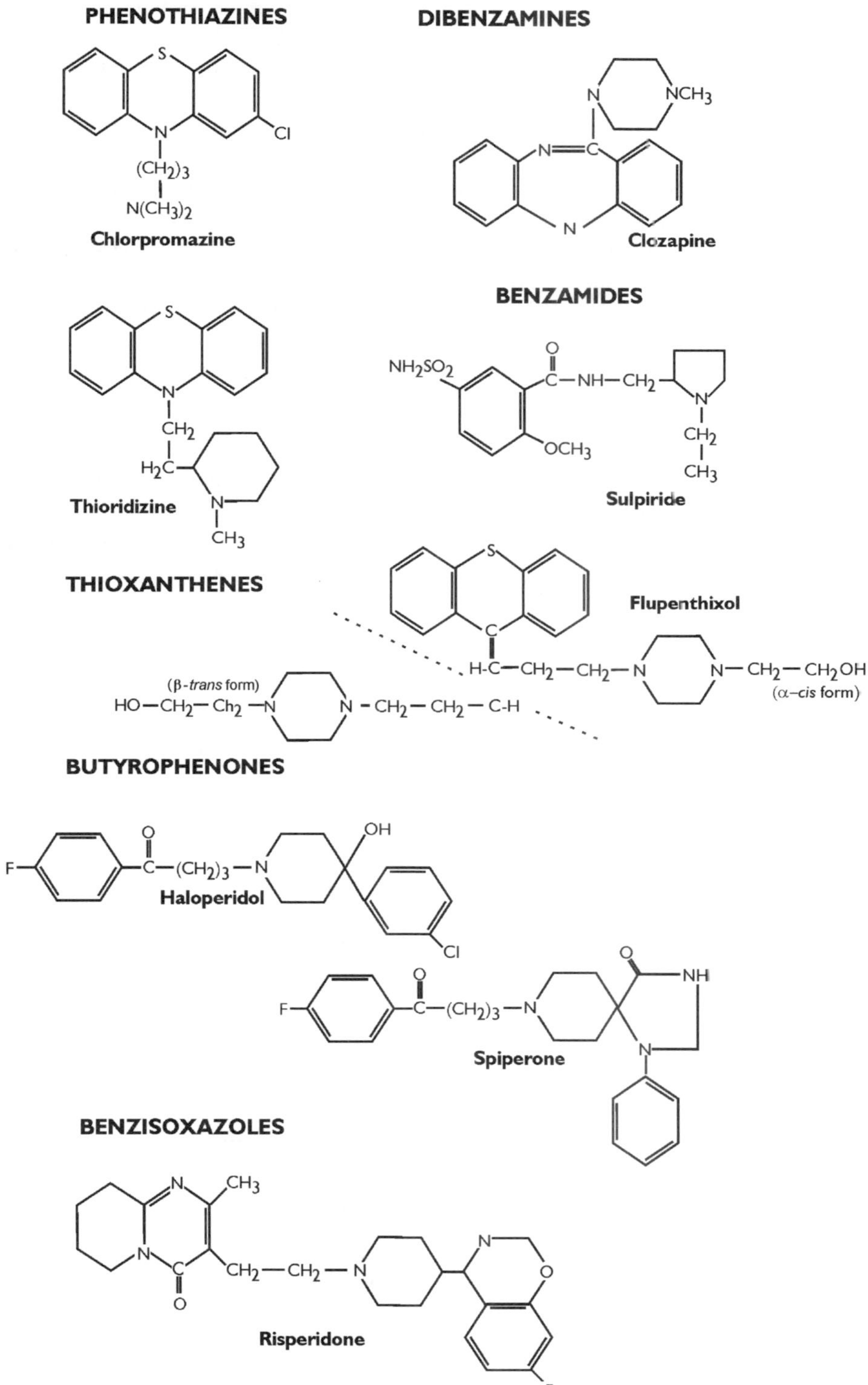

Figure 17.1 The structures of some neuroleptic drugs

In addition, amphetamine causes hallucinations in humans similar to those in schizophrenia and in rats it induces stereotyped behaviour (rearing, grooming and sniffing). This is dependent on the release of DA and blocked by the neuroleptics, which are DA antagonists.

There is now a whole range of neuroleptics (Fig. 17.1) but their ability to block the D_1-receptor-mediated stimulation of adenylate cyclase does not correlate with clinical potency. By contrast, their potency in displacing DA or more commonly an appropriately (^3H) labelled ligand, such as haloperidol, from D_2 binding sites on striatal membranes shows a surprisingly good correlation with clinical efficacy (Fig. 17.2). Most effective neuroleptics are indeed dopamine D_2-receptor antagonists. The importance of DA antagonism generally is underlined by the finding that while the thioxanthene flupenthixol exists in two forms α (*cis*) and β (*trans*) only the former is effective in schizophrenia and it is a hundred times more potent as a DA antagonist than the β form.

This raises two obvious questions:

(1) If effective neuroleptics are DA antagonists, is there any evidence for increased DA function in schizophrenia?
(2) How can blocking DA-mediated activity overcome the symptoms of schizophrenia and which DA pathways are involved?

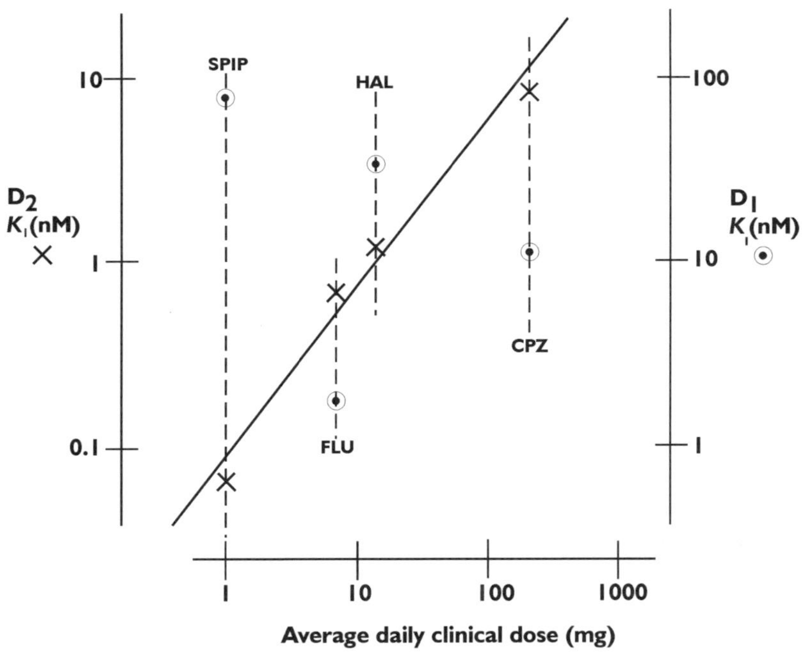

Figure 17.2 Comparison between the clinical dose of some neuroleptic drugs in the therapy of schizophrenia and their affinity for D_1 receptors, measured indirectly by inhibition (ki) of dopamine stimulated adenylate cyclase ((•) right-hand ordinate) or D_2 receptors indicated as displacement of haloperidol binding ((×) left-hand ordinate). Data are given for only four selected compounds but many more neuroleptics fall on the regression line between clinical dosing and D_2 antagonism (see Seeman 1980, 1992). The clinical doses used are based on those generally prescribed while K_1 (nM) values are averaged from a number of published figures. SPIP—spiperone, FLU—fluphenazine, HAL—haloperidol, CPZ—chlorpromazine

IS DA FUNCTION INCREASED IN SCHIZOPHRENICS?

There is no evidence of a general overactivity in DA function in schizophrenic patients. Plasma prolactin is not reduced, so the DA inhibitory control of its release is normal; there is no recorded increase in DA turnover as CSF and plasma levels of its major metabolite HVA are normal; and dyskinesias, which would reflect increased DA activity, are rare. PM studies have shown no consistent increases in DA brain levels, although some reports show an increase in the left amygdala, or in the activity of enzymes involved in its synthesis (tyrosine hydroxylase) or metabolism (MAO). For a review of the neurochemistry see Reynolds (1995).

Many post-mortem measurements have been made of the number of D_2 receptors in the striatum of schizophrenics, even though the striatum is unlikely to be the seat of schizophrenic symptoms. These invariably showed an increase above normal but this was not always significant if studied in patients who had not been on neuroleptic therapy. Neuroleptics alone would, by virtue of being DA antagonists, produce the equivalent of denervation supersensitivity and automatically increase DA receptor number. PET studies on newly diagnosed untreated patients were disappointingly inconclusive, possibly due to the lack of specificity of the ligands used. Generally it is felt that there might be a slight increase in striatal D_2 receptors in schizophrenia which is independent of neuroleptic treatment. If this was so and DA release remained normal, then increased DA function would follow. It should be borne in mind, however, that an increase in receptor number is normally the response to a defect in NT release (transmission).

Possibly increased DA function is not the actual cause of schizophrenia and its symptoms are just mediated by normally functioning DA systems that appear overactive because of the loss of some counteracting function or other NT(s). To date there is no evidence to fully implicate any other NT but there is growing interest in 5-HT and glutamate (see below).

BLOCKADE OF DA PATHWAYS IN SCHIZOPHRENIA

There are three main ascending DA pathways in the brain (Fig. 7.2).

(1) The nigrostriatal from substantia nigra (SN), the A9 nucleus
(2) The mesolimbic from the ventral tegmentum (VTA, A10) to the nucleus accumbens, olfactory tubercule, amygdala and pyriform cortex
(3) The mesocortical also from the VTA (A10) but to the prefrontal cortex (PFC)

Sometimes (2) and (3) are grouped together and called the mesocorticolimbic pathway. It is not clear which pathway is responsible for which schizophrenic symptom.

The VTA (A10) neurons innervating the cortex certainly show features that distinguish them from those in A9. They have a faster basal discharge rate (10 Hz, cf. 3 Hz in A9), a higher turnover of DA, fewer autoreceptors and are less easily inhibited by DA agonists (Bannon and Roth 1983; Farde *et al.* 1989).

There is no doubt that the nigrostriatal pathway is concerned with motor function and blocking DA transmission in it with most neuroleptics would certainly produce signs of Parkinsonism (see Chapter 15). The nucleus accumbens (and some other subcortical regions) are generally assumed to be concerned with psychotic effects although its core is also regarded as part of the basal ganglia. In rats it is involved in motor function since the locomotor activity caused by low doses of amphetamine is abolished by 6-OHDA lesions of the nucleus, and DA antagonist injections into it. By contrast stereotypy induced by high doses of amphetamine is dependent on the striatum.

The prefrontal cortex (PFC) and in particular the dorsal lateral part (DLPFC) appear to be particularly important in schizophrenia (Kerwin 1992). Lesions there are known to produce functional defects in humans reminiscent of many of the negative symptoms of schizophrenia, such as attention and cognitive defects and withdrawal. Despite this, no specific pathology is seen in the DLPFC in schizophrenics although there is some atrophy and neuronal loss which are normally old and could be congenital. That being so, it is necessary to explain why the symptoms become apparent only in adolescence.

Weinberger (1987) points out that myelination is not complete in DLPFC until around the age of 20 and that lesions of that area in young monkeys do not seem to affect their behaviour immediately but do impede their ability to perform delayed response tasks in later life. Early adulthood is also apparently a time of maximal DA activity in the brain as evidenced by its concentration, turnover and receptor number. Thus it is possible (Weinberger 1987) that the full effect of DLPFC lesions will only manifest itself as behavioural defects when the DA system is fully functional. Lesions within the DLPFC would obviously make it difficult for DA to function properly in that region and this could initiate negative symptoms. How this would account for the positive mesolimbic symptoms is less clear but no area of the brain works in isolation and the prefrontal cortex (PFC) has intricate relationships with the basal ganglia projecting to, and receiving inputs from, them. In fact, 6-OHDA-induced lesions in the PFC in rats, which destroy DA cortical afferents (Pycock, Kerwin and Carter 1980), somehow result in increased subcortical DA function, as evidenced by increased HVA levels, receptor number and motor response to apomorphine and amphetamine. Whether this results from reduced DA inhibition in PFC is uncertain but stimulation of, or local injections of glutamate into, the ventromedial prefrontal and ventral anterior cingulate cortices have been found to increase A10 neuron firing and DA release in the nucleus accumbens.

Although there is no evidence that the DA afferents to DLPFC are damaged in schizophrenics, if the cortical pathology does reduce the ability of DA to function there, this would be equivalent to deafferentation and, as in the experimental studies, lead to increased subcortical mesolimbic activity and positive symptoms (Fig. 17.3). Unfortunately there is no good evidence that the nucleus accumbens is more active in schizophrenics or is even the origin of positive symptoms (but see 'Animal models'). Nevertheless it is a useful working hypothesis.

A DA antagonist could certainly counter the increased mesolimbic activity and the positive symptoms. On the other hand, they would not be expected to reduce negative symptoms if these arise through an already inadequate DA influence. This fits with clinical experience because most of the neuroleptics are ineffective in treating negative symptoms. In fact if the negative symptoms do result from loss of the actual cortical neurons, rather than input to them, they will be difficult to reverse and much will depend on the precise role of DA in the DLPFC (see later).

ANIMAL MODELS

There are few experimental models. Even if appropriate lesions could be produced it will always be difficult to tell if an animal is experiencing hallucinations. Neuroleptics

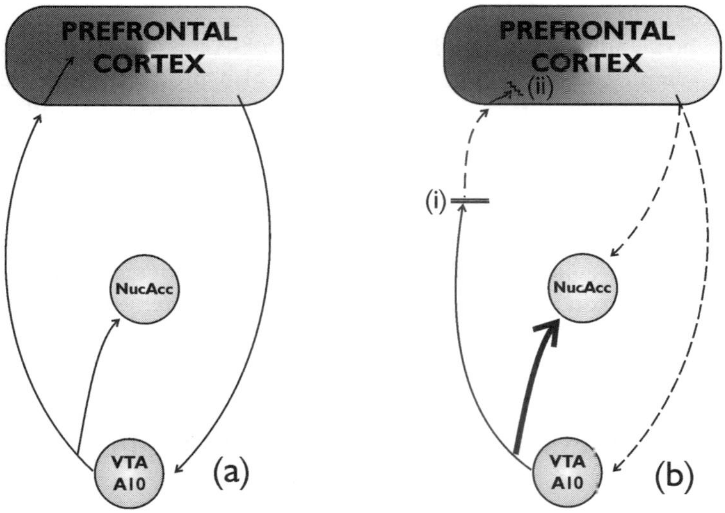

Figure 17.3 Effect of lesions in the prefrontal cortex on the activity of the dopamine mesolimbic pathway. DA neurons from the ventral tegmentum (A10 nucleus) not only innervate the prefrontal cortex (PFC) and limbic areas and in particular the nucleus accumbens (NucAcc) but are influenced by descending projections to them from the prefrontal cortex. Under normal circumstances the system appears to be balanced (a). Experimental lesions of the DA pathway to prefrontal cortex in rats (i) disturbs the balance and appears to increase the activity of VTA neurons and DA input to the limbic system (Pycock, Kerwin and Carter 1980), possibly through a change in cortico-limbic or cortico-VTA activity (b). While there is no evidence for such a lesion in the schizophrenic brain the general pathology found in the PFC could reduce the influence of DA there (ii) and so be equivalent to deafferentation. This could then result in a change in cortical influence on the limbic system and VTA resulting in increased activity in the DA mesolimbic pathway. Such increased mesolimbic activity is thought to mediate the positive symptoms of schizophrenia (see Weinberger 1987)

block DA-induced stereotypy or locomotor activity but this merely reflects their DA antagonism and restricts the discovery of new neuroleptics to those which antagonise DA. They are little better than *in vitro* binding studies.

More recently animal models based on the startle response have been developed which do in fact reflect some of the behavioural changes seen in schizophrenia (Geyer *et al.* 1990). It is believed that schizophrenics cannot adequately process (filter) incoming sensory information, become inundated with it and show cognitive impairment. The startle reflex is a motor response to sensory input (sensorimotor reflex) which is common to both animals and humans. The whole body reaction of rats to a sound or tactile (air-puff) stimulus can be monitored in a special chamber (stabilimeter) while in humans eyelid movements or electromyograms from the facial muscles can be monitored. In both species, if a smaller subthreshold stimulus (the pre-pulse) is presented some time (100–1000 ms) before the actual startle inducing stimulus (pulse) is given, then the response to the standard pulse is inhibited. Such pre-pulse inhibition (PPI) is attenuated in schizophrenics and in rats in which DA activity has been supplemented by giving apomorphine or amphetamine; although in neither case is the response to the actual pulse altered. This DA-increased inhibition of PPI in rats is counteracted by neuroleptics as is the attenuation of PPI in schizophrenics.

Based just on these results, PPI could simply be just another index of DA function but phencyclidine, an NMDA rather than DA receptor antagonist, which exacerbates schizophrenic symptoms and induces such symptoms when abused, also inhibits PPI (Bashki, Swerdlow and Geyer 1994). In contrast with DA augmentation, which initiates only positive symptoms in humans, phencyclidine also produces negative symptoms and its inhibition of PPI in rats is not affected by DA antagonists. Thus PPI may be one model that is not solely dependent on DA, although phencyclidine does enhance DA release in the mesolimbic system.

The injection of 6-OHDA into the rat nucleus accumbens produces the expected proliferation of DA receptors and resulting supersensitivity so that doses of apomorphine lower than normal produce a significant attenuation of PPI. This is not seen after the production of supersensitivity by toxin lesions of the substantia nigra and prefrontal cortex. The effects of amphetamine were also mainly modified by accumbens lesions. Thus as DA agonists primarily augment the positive symptoms such findings link these with the accumbens.

OTHER NEUROTRANSMITTERS IN SCHIZOPHRENIA

Neurotransmitters other than dopamine have been implicated in the aetiology of schizophrenia. But, as with DA, most of the evidence for their possible involvement has come from finding that their activity is modified by neuroleptic drugs (see later section on atypical neuroloptics), rather than from any evidence of their malfunction in schizophrenic patients. Nevertheless 5-HT and glutamate justify some consideration although interest again stems from drug observations, namely that LSD which is a 5-HT_{2A} receptor agonist and phencyclidine a glutamate NMDA antagonist can both cause hallucinations and schizophrenic-like symptoms.

Consistent changes in 5-HT levels or receptor number have not been reported in schizophrenic brain but a possible genetic link between 5-HT and schizophrenia comes from the finding of allelic variations in genes encoding 5-HT receptors and in particular polymorphism of the 5-HT_{2A} receptor gene in schizophrenics. There is also an indicator from animal studies which shows that habituation to the startle response is slowed by the hallucinogenic drug LSD and that this slowing is blocked by 5-HT_2 antagonists. Interestingly, schizophrenic patients, apart from showing attenuation of pre-pulse inhibition, also show a much slower habituation in response to a repeated startle stimulus which might reflect an inability to show selective attention through not being able to dismiss a repeated stimulus. DA does not appear to modify such habituation.

While there are some reports of increased NMDA and non-NMDA receptor number in various cortical regions of schizophrenics including the prefrontal cortex, there are also indications of impaired glutamate innervation, such as reduction in its neuronal uptake sites (Ishimaru, Kurumaji and Toru 1994). Also it has been found that levels of the mRNA for the NRI subunit of the NMDA receptor in the hippocampus and its D-aspartate binding sites in the temporal cortex are both reduced more on the left than right side in schizophrenic brain. This is another indication of greater malfunction on the left side of the brain and the possibility that some schizophrenic symptoms arise from an imbalance between cross-cortical activity.

NEUROLEPTICS (ANTISCHIZOPHRENIC DRUGS)

There is no shortage of these. The established ones belong to four main chemical groups (Fig. 17.1), the phenothiazines, thioxanthenes, butyrophenones and dibenzazepines.

They are all DA antagonists acting predominantly at D_2 receptors. As a result, they reduce the positive symptoms of schizophrenia but as their potency in this respect increases in line with their affinity for D_2 receptors, so does their tendency to produce extrapyramidal side-effects. They block DA inhibition of prolactin release and the resulting raised plasma levels can lead to amenorrhoea (reduced gonadal function) and galactorrhoea (lactation) in both sexes. They have little beneficial effect on negative symptoms. Neuroleptics showing this pattern of activity are called *typical neuroleptics*. Outside of schizophrenia many of them are used as anti-emetics (not motion sickness) counteracting the effects of DA on the chemoreceptor trigger zone of the vomiting centre. Some compounds, e.g. thioridizine and clozapine produce few extrapyramidal side-effects and are therefore known as *atypical* neuroleptics*. Even when effective the anti-schizophrenic action of all neuroleptics takes 2–3 weeks to develop and only clozapine has any appreciable effect on the negative symptoms.

Before trying to determine how the neuroleptics may reduce some of the symptoms of schizophrenia through modifying NT function it is necessary to consider:

(1) What are the effects of DA antagonism on the function of DA neurons themselves?
(2) Why do neuroleptics take 2–3 weeks to work?
(3) What forms of extrapyramidal side-effects occur and how they might arise?

NEUOLEPTICS AND THE FUNCTION OF DA NEURONS

The consequences of DA antagonism on DA neuron activity are shown diagrammatically in Fig. 17.4. Acutely neuroleptics increase the firing of DA neurons and the release of DA. This is because DA antagonists:

(1) Block the inhibitory feedback effect of released DA on terminal autoreceptors.
(2) Block the action of DA on similar inhibitory receptors on the DA neuron cell body itself.
(3) Block postsynaptic DA receptors on neurons inhibited by released DA which can initiate positive feedback to the DA neurons.

The receptor mediating all three effects appears to be the D_2 (or D_3) receptor.

Thus initially neuroleptics may increase DA turnover and possibly even its action depending on the degree of postsynaptic block. This may also change as the block induces compensating increases in receptor number. Clearly it is a rather fluid and somewhat uncertain situation.

LATENCY OF NEUROLEPTIC EFFECT

There is no reason why DA receptor block should not occur as soon as the antagonist reaches the brain. Antagonism of DA agonist-induced behavioural or electrophysiological

---

*Reference has been made already to the shortcomings of the term 'neuroleptic'. We now have a situation in which the drugs that are most useful in schizophrenia are regarded as atypical. While the term was introduced to cover those neuroleptics that do not cause EPSs, it has become synonymous with clozapine which has additional advantages over other neuroleptics (e.g. reduces negative symptoms, see text). Thus it is not always clear what is meant or covered by atypical. Hopefully this distinction between the neuroleptics will become unnecessary as better compounds are developed and the older ones become obsolete.

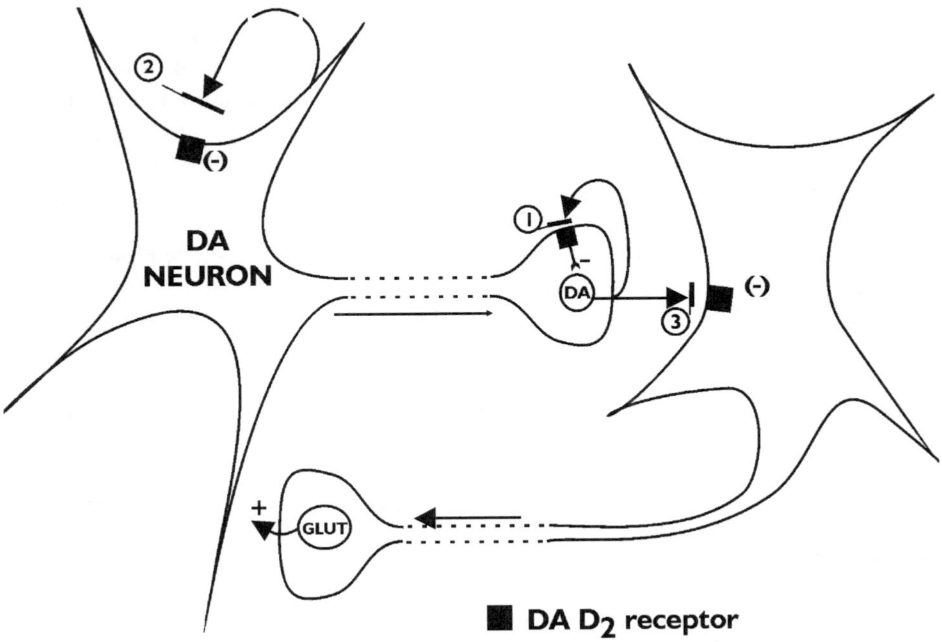

Figure 17.4 The effect of neuroleptics on the activity of DA neurons. Although neuroleptics (DA antagonists) are used primarily to inhibit the postsynaptic effects of released DA they also increase the activity of the DA neuron itself since they (1) inhibit the effect of synaptic DA on nerve terminal autoreceptors and so increase DA release; (2) block inhibitory DA autoreceptors on the soma of the DA neuron so that they cannot be stimulated by endogenous DA, possibly released from the neuron's own dendrites; and (3) facilitate feedback excitation to the DA neuron from those neurons normally inhibited by distally released DA. All the DA receptors involved are D_2 (or possibly D_3). —Blocked by D_2 antagonists (neuroleptics)

effects is immediate and the elevated plasma prolactin level plateaus in humans within a few days. So why is the antipsychotic effect so slow?

One possibility is that even with a potentially effective drug, the necessary readjustments in the neuronal circuitry to reverse or compensate for the disorder-induced malfunction just requires time. Another hinges on the degree of polarisation of A10 and A9 neurons.

These neurons are usually not very active but DA antagonists increase their excitability through the mechanisms outlined above so that their firing rates rise, the pattern of discharge changes from single- to multiple-spike burst discharges and the proportion of neurons firing increases. These changes are also aided by the fact that the excitatory inputs to A9 and A10 neurons normally promote a dendritic release of DA which through inhibitory soma D_2 autoreceptors will automatically counteract the excitation (Fig. 17.5). Clearly when these autoreceptors are blocked by acute neuroleptic administration in rats they cannot be activated by released DA, and the neurons fire much more frequently.

It was found, however, that if neuroleptic administration was continued for two weeks then neuronal firing stopped. Also while the neurons could not be made to fire by the excitatory NT glutamate, the inhibitory NT GABA activated them by reducing the

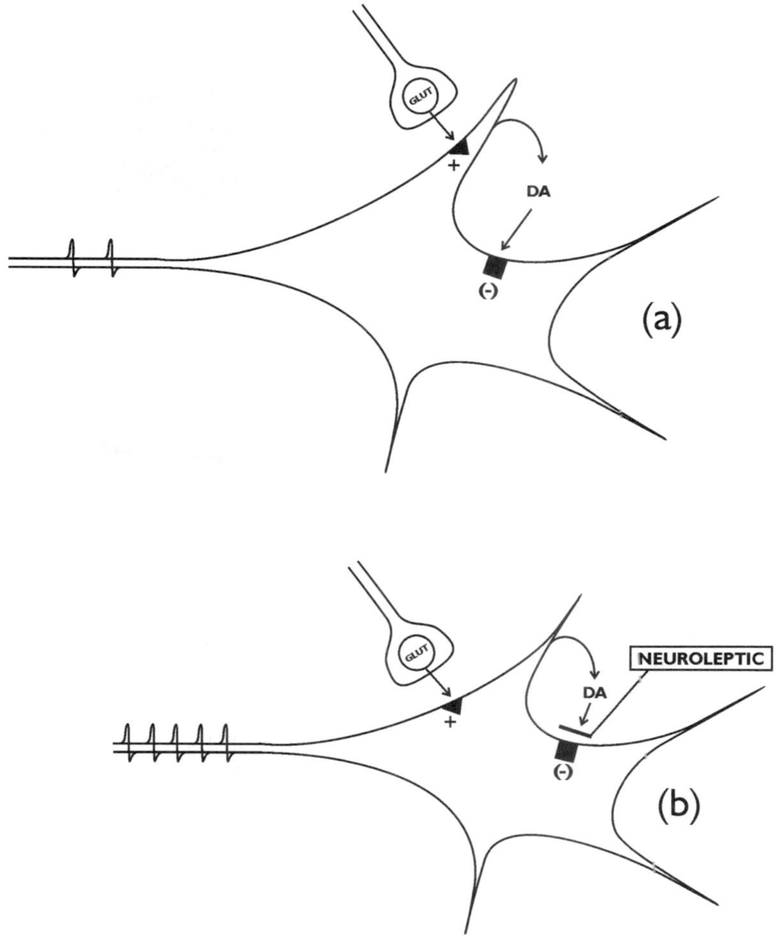

Figure 17.5 Possible scheme for the initiation of depolarisation block of DA neurons. In (a) the excitatory effect of glutamate released on to the DA neuron from the afferent input is counteracted by the inhibitory effect of DA, presumed to be released from dendrites, acting on D_2 autoreceptors. In the absence of such inhibition due to the presence of a typical neuroleptic (b) the neuron will fire more frequently and eventually become depolarised. Atypical neuroleptics, like clozapine, will be less likely to produce the depolarisation of A9 neurons because they are generally weaker D_2 antagonists and so will reduce the DA inhibition much less allowing it to counteract the excitatory input. Additionally some of them have antimuscarinic activity and will block the excitatory effect of ACh released from intrinsic neurons (see Fig. 17.7)

depolarisation (Grace *et al.* 1997). Thus it appears that due to their continuous intense activity the neurons eventually become permanently depolarised (confirmed by intracellular recording) and inactive (Fig. 17.6). This would obviously reduce output from the DA nuclei A9, A10 and have just the same effect as antagonising the postsynaptic action of DA released from their axons' terminals in the striatum, nucleus accumbens and cortex, etc.

Two features require some comment. The induction of depolarisation block in DA neurons needs afferent input to the nuclei, since prior lesion of the striatum and nucleus

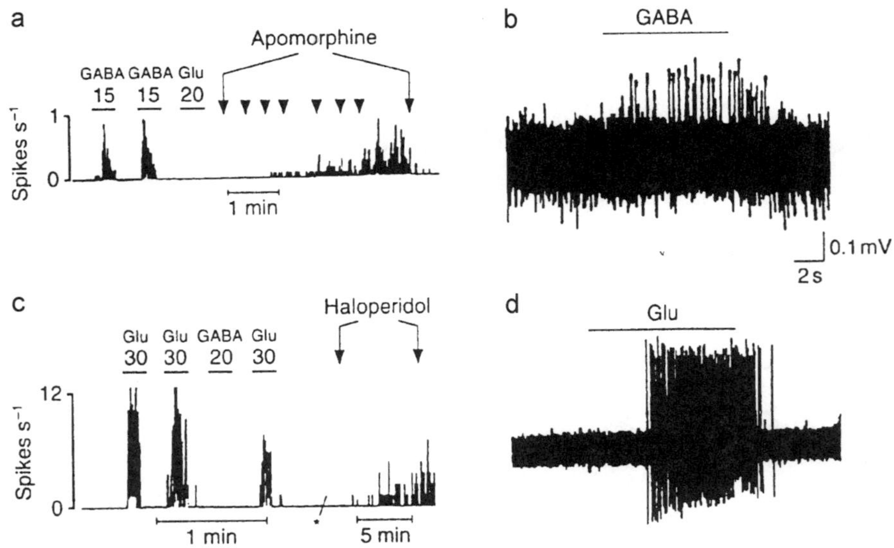

Figure 17.6 Response of DA neurons in the substantia nigra of normal rats and those having received chronic neuroleptic (haloperidol) treatment. (a) The firing rate histogram (spikes s^{-1} monitored at 10 s intervals) recorded extracellularly from a presumed DA neuron after three weeks dosing with haloperidol (0.5 mg kg^{-1} day^{-1}). Iontophoretic application of GABA the inhibitory NT, at the currents indicated (nA) actually induced firing which is shown as an oscilloscope sweep in (b). The excitatory NT glutamate failed to stimulate the neuron but apomorphine, which normally has an inhibitory effect through D_2 receptors, increased firing when given intraventricularly in increasing amounts from 2–5 to 160 μg kg^{-1}. These effects are consistent with the neurons being overstimulated and depolarised as a result of chronic neuroleptic dosing and so requiring to be hyperpolarised (inhibited) in order to become active. In a neuron from an untreated rat (c), GABA produces the expected inhibition and glutamate the rapid excitation, shown as an oscilloscope sweep in (d). The DA antagonist haloperidol (0.1 mg kg^{-1}) systemically also induced firing by blocking DA inhibition. (Reproduced from Grace *et al.* 1997 with permission from Elsevier Science)

accumbens prevent its development in A9 and A10 respectively and an established block can be reversed by such lesions. This again emphasises the importance of feedback loops in DA neuron function and schizophrenia as discussed above.

Second, although typical neuroleptics produce depolarisation block of both A9 and A10 neurons, the atypical neuroleptics only induce it in A10 neurons (Chiodi and Bunney 1983). So after an atypical neuroleptic the A9 neurons of the nigrostriatal tract remain functional, which would explain why EPSs are not seen. Another difference is seen with the expression of an immediate-early gene, *c-fos*, and although its functional significance is not clear, typical neuroleptics induce its protein production in both the striatum and nucleus accumbens while the atypicals only achieve it in the accumbens.

The slow time-course of depolarisation block not only offers an explanation for the latency of action of neuroleptic drugs but its occurrence may explain how they actually reduce DA function. Whether it explains their antischizophrenic effect is less certain since it is not possible to determine if such depolarisation occurs in patients on neuroleptic drugs. Certainly if this is how neuroleptics work it cannot be claimed that they have returned brain function to normal.

THE EXTRAPYRAMIDAL SIDE-EFFECTS (EPSs) OF NEUROLEPTIC DRUGS

These take three basic forms

(1) Acute dyskinesias
(2) Parkinsonian-like symptoms, e.g. akinesia
(3) Tardive dyskinesias

Dyskinesias are thought to be due to increased DA function, which would not be an obvious effect of a DA antagonist but the early acute ones could reflect the increase in DA neuron firing and release produced by such drugs, in the manner described above, overcoming the postsynaptic DA receptor block achieved in the striatum.

It is not surprising that a DA antagonist (especially those acting primarily on D_2 receptors) should produce the symptoms of Parkinsonism, a disorder caused by inadequate DA function (see Chapter 15), nor that its intensity or rate of onset over some weeks or months should increase with D_2 antagonistic potency. Tolerance to this adverse effect can develop without affecting antipsychotic activity but the speed with which Parkinsonism resolves after stopping therapy may be from 3 to 12 months and can persist indefinitely in some cases.

The late (tardive) dyskinesias, which mainly involve facial muscles, can take months or years to develop. They occur in 20–40% of patients, may not cease after stopping the drug and in fact can get worse, or even start then. Since they can be reduced temporarily by increasing neuroleptic dose it would appear that they do really result from DA overactivity and that the antagonism is not adequate. Certainly many experimental studies show that long-term neuroleptic dosing causes a compensatory increase in DA receptor number which would predispose to dyskinesias. Against this view are the findings that the increase in receptor number may precede dyskinesias by many weeks, receptor number but not dyskinesias routinely decline after drug withdrawal and while all patients should develop increased receptor number only some show dyskinesias. The dyskinesias are also more common in schizophrenics with clear negative symptoms and most brain damage and, since they have been seen in some untreated schizophrenics, could be a latent feature brought out by neuroleptics. Of course if the A9 neurons have been depolarised by the neuroleptics (see above) it is difficult to see how they can become so active unless the depolarisation also wears off.

ATYPICAL NEUROLEPTICS

Typical neuroleptics reduce the positive symptoms of schizophrenia at the expense of producing EPSs but the so-called atypical neuroleptics have less tendency to cause EPSs. With most of them, e.g. thioridizine, that is the extent of their atypicality but a few others, such as clozapine (and to a lesser extent risperidone and olanzapine) also reduce negative symptoms. Clozapine can even be effective in patients refractory to other neuroleptics. It is clearly a special drug, so special in fact that although it was once withdrawn because it causes agranulocytosis in some patients (2%), it has been reintroduced, alongside careful blood monitoring, for refractory cases. It will be given special consideration below.

MODE OF ACTION

There is certainly evidence that whereas typical neuroleptics are equally active in mesolimbic/cortical areas as well as the striatum, the atypical drugs are much less effective in the latter. This has been shown by (1) increased DA turnover through DOPAC and HVA production *in vitro*, (2) augmented DA and DOPAC release by microdialysis *in vivo* and (3) increased *c-fos*-like expression.

How the atypical neuroleptics achieve this differential effect is less clear but they could achieve some control of schizophrenia without producing EPSs by:

(1) Acting primarily on a particular subset of DA receptors
(2) Antagonising (or augmenting) some other NT(s) instead of, or in addition to, DA
(3) Having a particular but appropriate profile of DA and other NT (antagonistic) effects

These possibilities will be considered in turn.

Significance of different DA receptors

So far we have generally just alluded to the neuroleptics as DA receptor antagonists. The reader will know that there are five such receptors (Chapter 7). Clearly, if the DA released at the terminals of one dopaminergic tract acted on a subset of DA receptors that were different from those found postsynaptically at other tracts then some specificity of antagonist action might be achieved. Unfortunately there is no evidence that different pathways innervate different DA receptor populations and as with the use of agonists in PD, the D_2 receptor is dominant. Specific D_1 antagonists have no antischizophrenic effect and antischizophrenic efficacy increases with neuroleptic affinity (potency) at D_2 receptors — as unfortunately does the tendency to produce EPSs. Thus there is no great advantage in producing more potent D_2 antagonists, other than that less drug needs to be incorporated into long-term release depot preparations.

PET studies show that at effective therapeutic plasma concentrations most neuroleptics occupy some 80% of brain D_2 receptors (in the striatum at least) and this is therefore considered to be a requirement for efficacy (Pilowsky, Costa and Eli 1992; Farde 1996). If that is so then clozapine, which occupies only 20–40% of the D_2 receptors at a therapeutic concentration, must have some other action which accounts for its therapeutic effectiveness.

Its activity at D_1 receptors has been put forward as a possibility and although it has a relatively higher affinity for D_1 than D_2 receptors, compared with typical neuroleptics, it is still a weak antagonist at both and in the absence of evidence for D_1 (or D_5) receptor involvement in schizophrenia the significance of any D_1 antagonism is unclear.

K_1 (nM) values for clozapine at D_2 and D_1 receptors are 56 and 141 compared with 0.5 and 27 for haloperidol giving D_1/D_2 ratios of 2.5 and 54 for the two drugs. A relatively strong block of D_1 compared with D_2 receptors may not be the answer for schizophrenia but it could reduce the tendency to produce dyskinesias, if this depends on D_1 receptor activation (see Fig. 17.2).

Among the D_2 family of receptors (D_2, D_3 and D_4) the D_2 receptor itself seems to be the most important. At a therapeutic concentration, most neuroleptics, except clozapine (and risperidone), should, according to *in vitro* binding studies, be occupying 50–70% of brain D_2 receptors. The picture is similar for D_3 receptors but only clozapine (and risperidone and olanzapine) occupy more than 50% of D_4 receptors at a therapeutic dose.

This relative selectivity of clozapine for D_4 receptors with their restricted location, even if it is in small numbers, to the prefrontal cortex has stimulated much interest in their involvement in schizophrenia and the control of negative symptoms. There has been one report (Seeman, Guan and Van Tol 1993), refuted by others, of a sixfold increase in D_4 receptors in schizophrenic brain. Unfortunately the measurements were made in striatum rather than cortex and depended on the difference in the binding of a D_2, D_3, D_4 antagonist nemonopride compared with that of a D_2 and D_3 antagonist raclopride. D_4 occupancy was thus inferred rather than established by a specific D_4 antagonist. When such a selective D_4 antagonist, L-745,870, became available and was tested in 38 schizophrenics it proved ineffective at what were considered to be doses sufficient to occupy 50% of the D_4 receptors (Bristow *et al.* 1997). It has not been used apparently to assess D_4 receptor number in schizophrenic brain.

There are few specific drugs for D_3 receptors but D_3 knock-out mice show no behavioural defects. Thus the significance of any DA receptor other than the D_2 still remains to be established (see Seeman and Van Tol 1994; Sokoloff and Schwartz 1995; Strange 1994).

Involvement of other NTs

Acetylcholine

Neuroleptic-induced Parkinsonism (but not tardive dyskinesias) can be reduced by antimuscarinic drugs. It is generally assumed that neuroleptic antagonism of DA-mediated inhibition in the striatum leaves the excitatory muscarinic action of ACh unchecked (Fig. 15.9) so that blocking it will restore normality. The atypical neuroleptics thioridizine and clozapine both have potent inherent antimuscarinic activity with PA_2 values (7–8) similar to that for atropine and more than a hundredfold that of a typical neuroleptic-like haloperidol. Thus each compound has the ability to nullify its own antidopamine effect in the striatum and stop Parkinsonian symptoms developing (Fig. 17.7) without affecting DA antagonism, and possible antischizophrenic effects elsewhere. There is no evidence that antimuscarinic activity has any effect on schizophrenia and thioridizine has no more effect on negative symptoms than typical neuroleptics. Of course, since clozapine is also a weaker D_2 antagonist than thioridizine this automatically reduces its ability to produce EPSs anyway.

5-HT

Some neuroleptics, including clozapine, are potent 5-HT-receptor antagonists and the possible role of 5-HT in the action of neuroleptics and the development of schizophrenia has recently generated much interest (Busatto and Kerwin 1997). This has centred primarily on 5-HT_{2A} receptors found in the limbic cortex, which are linked to neuronal excitation and believed to mediate the hallucinogenic effects of drugs such as lysergic acid diethylamide (LSD).

Generally most atypical neuroleptics have higher affinity for 5-HT_2 than D_2 receptors while typical ones retain a preference for the D_2 receptor. There is, however, no infallible separation since chlorpromazine (typical neuroleptic) is more active at 5-HT_{2A}

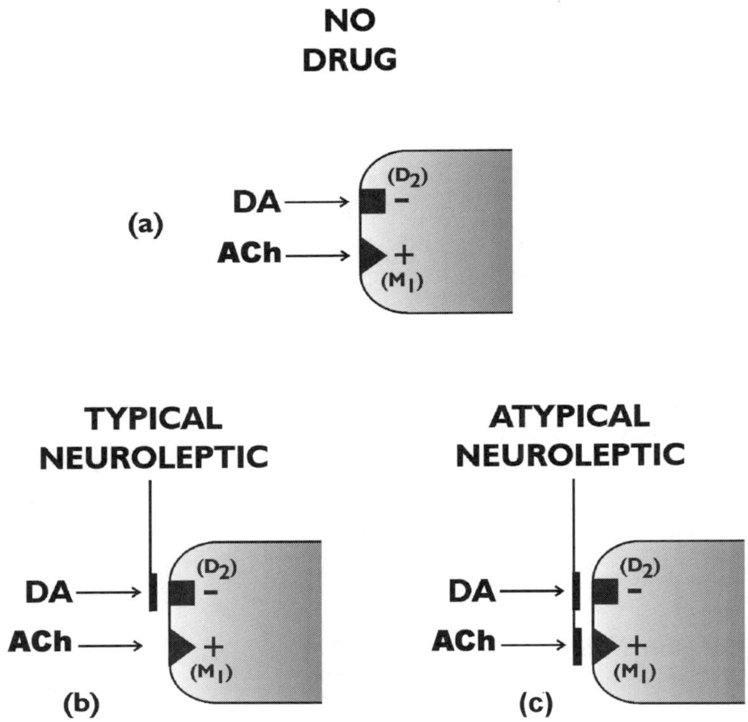

Figure 17.7 Possible mechanism by which atypical neuroleptics with antimuscarinic activity produce few EPSs. Normally the inhibitory effects of DA released from nigrostriatal afferents on to striatal neuron D_2 receptors is believed to balance the excitatory effect of ACh from intrinsic neurons acting on muscarinic (M2) receptors (a). Typical neuroleptics block the inhibitory effect of DA which leaves unopposed the excitatory effect of ACh (b) leading to the augmented activity of the striatal neurons and EPSs (see Fig. 15.2). An atypical neuroleptic with intrinsic antimuscarinic activity reduces this possibility by counteracting the excitatory effects of released ACh as well as the inhibitory effects of DA (c). Thus the control of striatal neurons remains balanced

than D_2 receptors and remoxidpride (atypical) more active at D_2 than 5-HT_{2A}. Also differences in the values for the dissociation constants between experimental studies (see later) make comparisons between D_2 and 5-HT_{2A} potency somewhat difficult.

No neuroleptic has purely 5-HT_{2A} antagonist activity and a pure 5-HT_{2A} antagonist drug may not have neuroleptic activity. Risperidone, an atypical neuroleptic with some benefit against negative symptoms, is the most potent of all neuroleptics at 5-HT_{2A} receptors (K_1: 0.2 nM). Some *in vitro* measurements show it to have up to 25 times more affinity for 5-HT_{2A} than D_2 receptors and PET studies indicate that at therapeutic doses it displaces a 5-HT_2 ligand in preference to a D_2 one. Clozapine is also claimed to occupy over 80% of 5-HT_2 and less than half this number of D_2 receptors at clinical doses. Neuroleptics with 5-HT_2 antagonist activity not only produce fewer EPSs but 5-HT_2 antagonists reduce neuroleptic-induced EPSs.

Fibres from 5-HT neurons in the raphé nucleus innervate and yet, despite the observed 5-HT_{2A} receptor link with neuronal excitation, appear to inhibit DA neurons in the SN (A9). Thus antagonism of 5-HT released onto them would increase their firing and so reduce the likelihood of EPSs, although how 5-HT_{2A} antagonists can overcome the established motor side-effects of another neuroleptic is less clear if that compound has already caused a depolarising block of the neurons.

The mechanism by which 5-HT_2 antagonism could ameliorate schizophrenic symptoms and what effect 5-HT has on mesolimbic and mesocortical pathways through A10 neurons is even less certain. It is more likely that 5-HT's action occurs postsynaptically in the limbic system or PFC. The probability that neuroleptics benefit from a particular balance of DA and 5-HT_{2A} antagonism is developed later.

The 5-HT_3 receptor is found appropriately in mesocortical areas and while behavioural studies with their antagonists in rodents showed potential antipsychotic activity, they have proved ineffective in patients. 5-HT_{1A} agonists may be more useful. They have been found to increase the extracellular concentration of DA in the frontal cortex of rats but diminish apomorphine-induced stereotypy (striatal effect). So they could be of some benefit, especially against negative symptoms, without causing EPSPs (see Chapter 9).

Noradrenaline

Many of the neuroleptics are α-adrenoceptor antagonists. Some, like chlorpromazine, block α_1 postsynaptic receptors while clozapine (and risperidone) are as potent at α_2 as D_2 receptors. There is no evidence that either of these actions could influence striatal or mesolimbic function but NA is considered important for function of the prefrontal cortex and any increase in its release, achieved by blocking α_2-mediated autoinhibition, might contribute to a reduction in negative symptoms and provide a further plus for clozapine (see Nutt *et al.* 1997). Centrally, however, most α_2-receptors are found post-synaptically and their function, and the effect of blocking them, is uncertain.

Glutamate

Although there is no evidence that any of the neuroleptics have any significant effect on glutamate receptors, it will be of no surprise to learn that clozapine, but not pure D_2 or 5-HT_2 antagonists nor any typical neuroleptic, can overcome phencyclidine disruption of PPI in animals. Interestingly, the efficacy of clozapine (but not risperidone or olanzopine) is increased by the antiepileptic drug lamotrigine that has inhibition of glutamate release as one of its actions (see Chapter 16). Also glycine (and serine) have been shown to improve the negative symptoms by what is assumed (but not proven) to be a potentiation of NMDA receptor activity, but they can make positive symptoms worse.

Profile of NT antagonism in neuroleptic action

In deciphering the role of the different NTs, or more precisely their antagonists, in the antischizophrenic action of neuroleptic drugs it must be remembered that published binding data and calculated dissociation constants vary considerably, which, of course, affects correlation coefficients made with clinical activity. Factors to bear in mind are:

(1) *In vitro* binding studies use different cell lines or membrane preparations and generally only yield the apparent dissociation constants for a number of antagonists obtained by comparative displacement of one labelled ligand. Unfortunately few

such ligands are specific for the receptor being analysed, i.e. they bind to other receptors to differing extents as do the displacing compounds. Reported values for clozapine's binding affinity vary from 84–388 nM depending on the D_2 ligand being displaced. Real dissociation constants can be obtained from direct measurements of the binding of the neuroleptic alone in labelled form but because neuroleptics also bind to more than one receptor, the preparation must express only the receptor being studied.

(2) PET studies have almost always centred on measurements of binding and DA receptor number in the striatum rather than other DA-innervated areas of more significance in schizophrenia. Also in PM measurements of receptor number it is invariably the striatum which is used, because of its high density of DA receptors.

(3) Functional activity (clinical effect, catalepsy in animals, etc.) is invariably correlated with plasma concentrations whereas the brain levels of many neuroleptics, which are very lipophilic compounds, could be much higher. Some clinicians also believe that many newer compounds achieve atypical status compared with older ones because they are used at minimal dosage while older ones are prescribed at established levels which may be unnecessarily high.

Despite these problems it remains necessary to attempt some explanation in terms of differential NT antagonism, of why clozapine is so effective (see Reynolds 1997) in that it causes fewer EPSs, reduces negative symptoms and is effective in some patients refractory to other drugs. Considering these benefits in turn:

(1) *Reduced EPSs*. This may be achieved with clozapine because it is a:

  (a) Relatively weak D_2 antagonist. The one thing that is reasonably certain about the neuroleptics is that irrespective of the role of D_2 antagonism in controlling schizophrenia the more potent the D_2 antagonist, the more likely are EPSs. Just as Parkinsonian symptoms only occur in PD patients when 50–80% of the DA innervation to the striatum is lost (Chapter 15) so neuroleptic-induced Parkinsonism only follows blockage of some 80% of D_2 receptors. Clozapine only achieves about half of this at therapeutic doses and its weak binding may allow DA to override its antagonism at appropriate times in the striatum. Thus clozapine has little potential for inducing EPSs and what it has could be reduced by its other activities.

  (b) Potent antimuscarinic. ACh excitation counteracts DA inhibition in the striatum.

  (c) Strong 5-HT_2 antagonist. Compounds with this property appear to reduce EPSs.

  (d) Relatively strong D_1 antagonist. This may not stop PD symptoms but could reduce dyskinesias (Fig. 15.8).

  As a result of these features clozapine is likely to have little effect on A9 (SN) neurons and does not cause their depolarisation in chronic dosing.

(2) *Negative symptoms*. These may be reduced because either clozapine antagonises appropriate receptors in the prefrontal cortex or it does not act as an antagonist there. This apparently stupid statement is prompted by the lack of knowledge of what is required to reduce negative symptoms. D_4 and D_1 receptors are found in the prefrontal cortex and only clozapine among current neuroleptics is more active at both of these than the D_2 receptor. Thus on this basis it is well placed to block DA's

influence but if negative symptoms follow an impairment of DA input (see above) further blockage is undesirable. In fact clozapine would have to augment DA function and based on the knowledge that D_1 receptor activation appears to be required for optimal cognitive performance it has been suggested that neuroleptics should optimise activation of D_1 receptors in addition to blocking D_2 receptors (Lidow, Williams and Goldman-Rakic 1998). Little is known of the effect of DA or its agonists on cortical neurons, although most studies show it to be inhibitory. Even less is known about clozapine's action on neuronal firing but in one study on the prefrontal cortex of anaesthetised rats it was found to mimic the action of the DA agonist apomorphine, an effect blocked by haloperidol (Dalley and Webster 1993). A number of microdialysis studies have also shown that it is the only neuroleptic to increase DA efflux in the prefrontal cortex although most of them have that effect in the striatum. So perhaps clozapine can in some way increase DA transmission in PFC, even if that is achieved through initially antagonising an effect of DA or another NT. Recently risperidone has also been shown to increase both 5-HT and DA release in the rat prefrontal cortex (Knoble and Weinberger 1997) but possibly through α_2 and 5-HT receptor antagonism. In view of the strong antimuscarinic activity of clozapine it is interesting that cholinergic overactivity has been reported to induce behaviour in animals that was thought to reflect negative symptoms.

(3) *Refractory cases respond to clozapine*. If D_2 antagonism is considered necessary, or at least desirable, for counteracting positive symptoms it is surprising that a relatively weak D_2 antagonist like clozapine should not only be so effective but also prove successful in patients who have not responded to other neuroleptics more potent at D_2 receptors.

Certainly clozapine can avoid EPSs by only blocking a fraction of D_2 receptors but that seems insufficient on its own to make clozapine so effective in schizophrenia. That is probably achieved by a unique combination of other blocking actions, at D_1, D_4, 5-HT_2, α_2 and possibly other receptors (see Fig. 17.8). It may simply be that clozapine is so effective because it is so 'dirty', a view held for many years about the first neuroleptic chlorpromazine. Indeed it is unlikely that the varied symptoms of such a complicated disorder could be rectified by manipulating just one NT.

Unfortunately although much is known about the pathways and receptors involved in extrapyramidal activity and the mechanism of the EPSs that follow neuroleptic therapy and even the possible origin of negative symptoms in the prefrontal cortex, the precise site of origin and NT involvement in the overriding positive symptoms is less clear. Until that is corrected, permutations of NT antagonisms are likely to multiply with the neuroleptics.

PERSPECTIVE

What is obvious from all the experimental evidence is that it is easier to unravel the cause of the undesirable than it is to explain the desirable effects of neuroleptic drugs. EPSs occur because such drugs all have some D_2 antagonist activity and so reduce DA transmission in the striatum. The degree to which they achieve this and whether they are typical or atypical depends on their affinity for D_2 striatal receptors, since EPSs

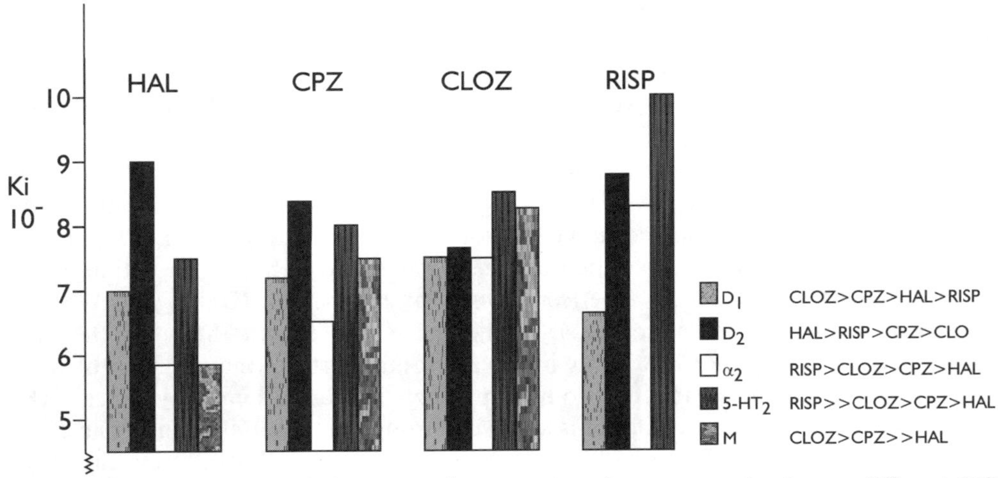

Figure 17.8 Comparison of the antagonist potencies of some neuroleptics on different NT receptors. Data are shown for haloperidol (HAL), chlorpromazine (CPZ), clozapine (CLOZ) and risperidone (RISP) acting on dopamine D_1 and D_2, 5-HT_2 (S_2), alpha (α_2) adrenoceptors and cholinergic muscarinic receptors (M). The height of each column shows an average of the dissociation constants obtained from a number of publications (see Seeman 1990). The values, which can vary some fiftyfold, are expressed as the negative logarithms (i.e. $9 = 10^{-9}$ M,lnM) so that the higher the column, the more potent the compound. The order of potency of the four compounds at each receptor is shown alongside

diminish with low D_2 affinity and their ability to block ACh muscarinic or 5-HT_2 or other receptors. Trying to translate from *in vitro* binding studies to an explanation of antipsychotic effectiveness is also made more difficult by the fact that they do not readily distinguish between agonist and antagonist activity. More functional studies of neuroleptic activity in different brain areas is required.

Measuring the expression of the early-immediate gene *c-fos* supports the striatal role of neuroleptics in the induction of EPSs because although all neuroleptics induce such expression in both the nucleus accumbens and striatum, the atypical neuroleptics do so more in the accumbens while clozapine, but not risperidone, also achieve it in the prefrontal cortex (Robertson, Matsumura and Fibiger 1994). How this arises is uncertain but since risperidone is a more potent 5-HT_2 antagonist than clozapine, it cannot be through that mechanism.

Establishing the possible site of action of a drug *in vivo* first and then trying to unravel what it actually does at the cellular or molecular level is an alternative approach to the analysis of drug action. In this respect much was, and is, hoped of PET (SPECT) studies in humans and non-human primates. Of course, these tell us primarily where drugs are not located and therefore certainly do not act. Locating their labelled form in particular brain regions does, however, indicate where they may act, although a high concentration in one area does not automatically make that the drug's primary site of action. Nevertheless, this approach does help to clarify the origin of EPSs since although both typical and atypical drugs appear to bind to limbic and cortical areas to a similar extent it is only the typical ones that show high striatal levels.

On this evidence one can confidently equate EPS with neuroleptic DA receptor (D_2) antagonism in the striatum and possibly a reduction in the positive symptoms of schizophrenia through similar action in the limbic system (nucleus accumbens).

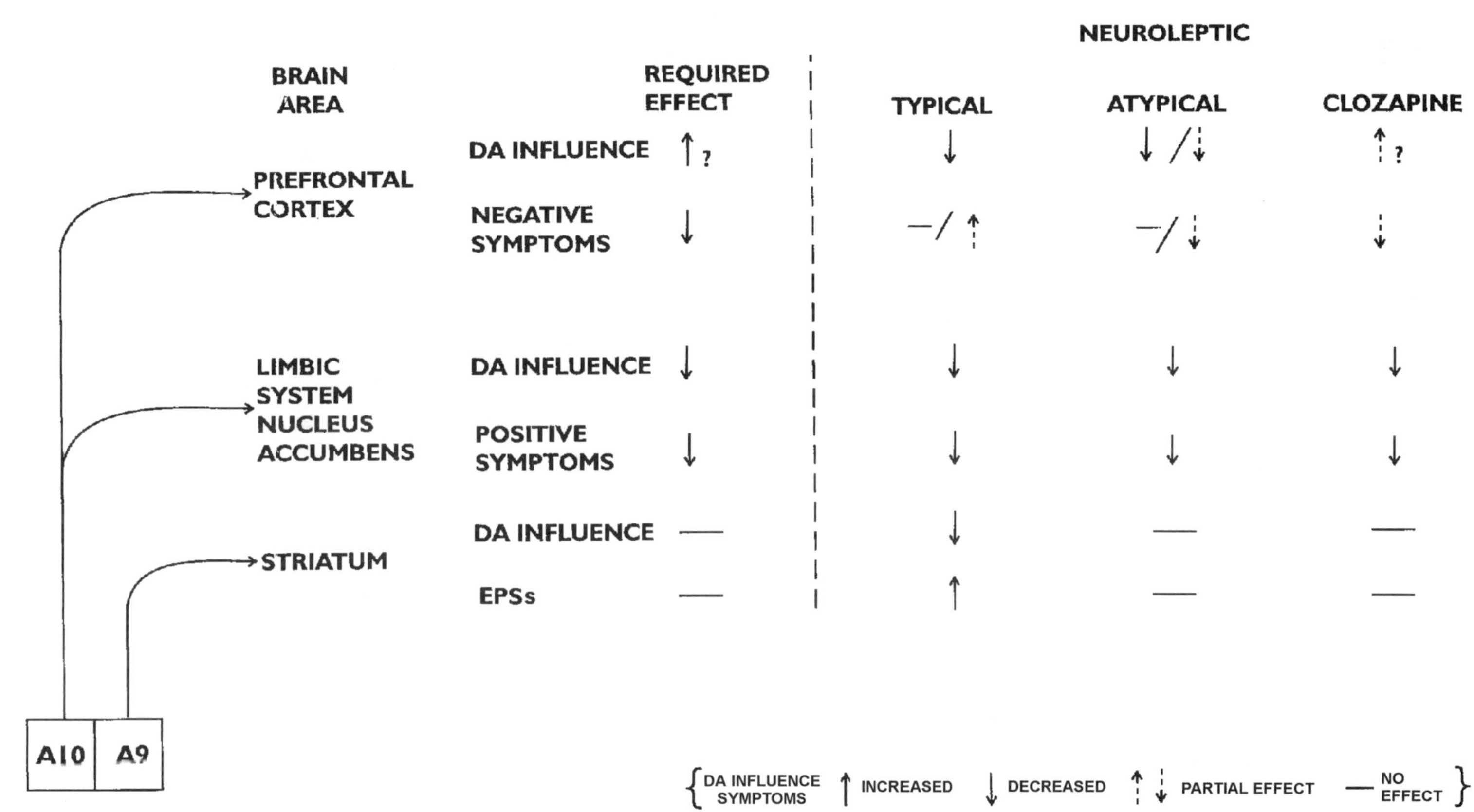

Figure 17.9 Schematic representation of the proposed activity profile of an ideal neuroleptic. The figure shows DA pathways to the prefrontal cortex, mesolimbic nucleus accumbens and striatum; the effects required for an ideal drug on the DA influence and symptoms there and to what extent they are met by most typical and atypical neuroleptics and by clozapine. Note that while all atypical neuroleptics induce few extrapyramidal side-effects (EPSs) few of them, apart from clozapine, have much beneficial effect in overcoming negative symptoms of schizophrenia

Whether the amelioration of negative symptoms results from an action in the cortex and, in particular, the prefrontal cortex requires further study. The fact that clozapine, the atypical drug that is currently most effective in this respect, has actions there which are not shown by other compounds is encouraging even though the precise mechanism by which it works remains to be elucidated.

It appears that an ideal neuroleptic may need to reduce DA activity in the mesolimbic system (nucleus accumbens) to counter the positive symptoms of schizophrenia, increase it in the prefrontal cortex to overcome negative symptoms and have little or possibly no effect on it in the striatum so EPSs do not arise (Fig. 17.9). No wonder we still await the ideal drug.

REFERENCES

Bannon, ML and Roth, RH (1983) Pharmacology of mesocortical dopamine neurons. *Pharmacol. Rev.* **35**: 53–68.

Bashki, VP, Swerdlow, NR and Geyer, MA (1994) Clozapine antagonises phencyclidine-induced deficits in sensorimotor gating of the startle response. *J. Pharmacol. Exp. Ther.* **271**: 787–794.

Bristow, LJ, Kramer, MS, Kulagowski, J, Patel, S, Ragan, CI and Seabrook, GR (1997) Schizophrenia and L745,870, a novel dopamine D_4 receptor antagonist. *Trends Pharmacol. Sci.* **18**: 186–188.

Busatto, GF and Kerwin, RW (1997) Perspectives on the role of serotonergic mechanisms in the pharmacology of schizophrenia. *J. Psychopharm.* **11**: 3–12.

Chiodi, LA and Bunney, BS (1983) Typical and atypical neuroleptics: differential effects of chronic administration on the activity of A9 and A10 midbrain dopamine neurones. *J. Neurosci.* **3**: 1607–1619.

Crow, TJ (1985) The two syndrome concept: origins and current status. *Schizophr. Bull.* **11**: 471–486.

Dalley, JW and Webster, RA (1993) Dopamine-like effects of clozapine on spontaneously active neurons in the rat prefrontal cortex. *J. Psychopharm.* **7**: A7.

Farde, L (1996) The advantage of using positron emission tomography in drug research. *Trends Neurosci.* **19**: 211–214.

Farde, L, Wiesel, FA, Nordstrom, AL and Sedvall, G (1989) D_1 and D_2 dopamine receptor occupancy during treatment with conventional and atypical neuroleptics. *Psychopharmacology* **99** (Suppl): 528–531.

Geyer, MA, Swerdlow, NR, Mansbach, RS and Braff, DL (1990) Startle response models of sensorimotor gating and habituation deficits in schizophrenia. *Brain Res. Bull.* **25**: 485–498.

Grace, AA, Bunney, BS, Moore, H and Todd, CL (1997) Dopamine-cell depolarization block as a model for the therapeutic actions of antipsychotic drugs. *Trends Neurosci.* **20**: 31–37.

Ishimaru, M, Kurumaji, A and Toru, M (1994) Increases in strychnine-insensitive glycine binding sites in cerebral cortex of chronic schizophrenics: evidence for glutamate hypothesis. *Biol. Psychiat.* **35**: 84–95.

Kerwin, R (1992) A history of frontal and temporal lobe aspects of the neuropharmacology of schizophrenia. *J. Psychopharm.* **6**: 230–240.

Knoble, MB and Weinberger, DR (1997) Dopamine, the prefrontal cortex and schizophrenia. *J. Psychopharm.* **21**: 123–131.

Lidow, MS, Williams, GV and Goldman-Rakic, PS (1998) The cerebral cortex: a case for a common site of action of antipsychotics. *Trends Pharmacol. Sci.* **19**: 136–140.

Nutt, DJ, Lalies, MD, Lione, LA and Hudson, AL (1997) Noradrenergic mechanisms in the prefrontal cortex. *J. Psychopharm.* **11**: 163–168.

Pilowsky, LS, Costa, DC and Eli, PJ (1992) Clozapine single photon emission tomography and the D_2 dopamine receptor blockade hypothesis of schizophrenia. *Lancet* **340**: 199–202.

Pycock, CJ, Kerwin, RW and Carter, CJ (1980) Effect of lesion of cortical dopamine terminals on subcortical dopamine receptors in rats. *Nature* **286**: 74–77.

Reynolds, GP (1995) Neurotransmitter systems in schizophrenia. *Int. Rev. Neurobiol.* **38**: 305–359.
Reynolds, GP (1997) What is an atypical antipsychotic? *J. Psychopharm.* **11**: 195–199.
Robertson, GS, Mastumura, H and Fibiger, HC (1994) Induction patterns of *Fos*-like immunoreactivity in the forebrain as predictors of atypical antipsychotic activity. *J. Pharmacol. Exp. Ther.* **271**: 1058–1061 (see also *Neuropsychopharmacology* **14**: 105–110).
Seeman, P (1980) Brain dopamine receptors. *Pharmacol. Rev.* **32**: 229–313.
Seeman, P (1990) Atypical neuroleptics: role of multiple receptors, endogenous dopamine and receptor linkage. *Act. Psychiatr. Scand.* **82** (Suppl. 358): 14–20.
Seeman, P (1992) Dopamine receptor sequences: therapeutic levels of neuroleptics occupy D_2 receptors, clozapine occupies D_4. *Neuropsychopharmacology* **7**: 261–284.
Seeman, P, Guan, HC and Van Tol, HHM (1993) Dopamine D_4 receptors elevated in schizophrenia. *Nature* **365**: 441–445.
Seeman, P and Van Tol, HHM (1994) Dopamine receptor pharmacology. *Trends Pharmacol. Sci.* **15**: 61–72.
Sokoloff, P and Schwartz, JC (1995) Novel dopamine receptors half a decade later. *Trends Pharmacol. Sci.* **16**: 270–275.
Strange, PG (1994) Dopamine D4 receptors: curiouser and curiouser. *Trends Pharmacol. Sci.* **15**: 317–319.
Weinberger, DR (1987) Implications of normal brain development for the pathogenesis of schizophrenia. *Arch. Gen. Psychiat.* **44**: 660–669.

18 Alzheimer's Disease (AzD)

R. A. WEBSTER

Everybody suffers some intellectual and memory impairment with age. If it becomes very marked or occurs earlier in life (40+) it is known as dementia. Although it may be caused by alcoholism, cardiovascular disease such as multiple infarcts, and is often seen in the later stages of Parkinsonism, the most common cause is a neurodegenerative one, namely, Alzheimer's disease (AzD). In fact this is the primary and sole cause in over half the cases of dementia and is a contributory cause in a further quarter and the younger the patient, the more likely is the dementia to be of the Alzheimer type.

Alzheimer's disease generally presents itself as a relatively isolated failure of memory for recent events, particularly for names. Speech problems, disorientation with respect to time and place follow along with depression that can be interrupted by aggression. All aspects of higher brain function are then affected, memory loss becomes virtually total and movements very slow. Eventually the patient becomes almost totally incapacitated, doubly incontinent and bed-bound in which terrible state they may survive for 1–2 years. It is not surprising that its appearance is devastating not only to the patient but more particularly to family and friends. It can last from 3 to 20 years but 7 to 10 years is more common and while it may start in one as young as 20 it usually waits until well after 40. Some 10% or more of the population over 65 may suffer from it, a figure that more than doubles beyond 80 years. Also as life expectancy increases and the population becomes more aged the actual incidence will increase. The cost is thus becoming immense. In the United Kingdom alone, the annual cost to the health and social services of caring for people with AzD is estimated at over £2.0 billion (a hundred times more in the United States) but the total cost to society could be double that.

Despite its characteristic symptoms and even after the exclusion of other established causes, AzD can only be reliably diagnosed by neuropathology and microscopic examination of the brain. Indeed that is how it came by its name. In 1907, a German physician, Alois Alzheimer, described two distinct post-mortem changes in the brain of a woman patient who had died with an unusual mental illness. These were the now characteristically accepted markers of the disease, namely senile plaques and neurofibrillary tangles (Fig. 18.1).

PATHOLOGY

Microscopically the brains of AzD patients often show neuronal loss and some atrophy, much as in Down Syndrome, as well as widened sulci and narrowed gyri. Since,

Neurotransmitters, Drugs and Brain Function. Edited by R. A. Webster

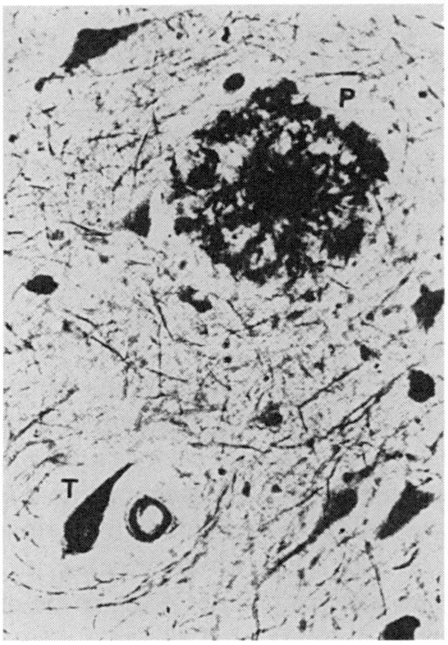

Figure 18.1 Typical tangle (T) and plaque (P) as visualised by silver impregnation in the cerebral cortex of a case of Alzheimer's disease. The extracellular plaque (10–50 μm diameter) consists of a central core of amyloid surrounded by glial processes and a number of neurites in a ring formation. The intracellular cytoplasmic tangle is composed of helical filaments in a paired format. (Reproduced with permission of Academic Press from Wischik and Crowther 1986)

however, such changes are not uncommon in elderly people (75+) these features can only really be considered indicators of the disease when found in younger patients. This does not apply to the plaques and tangles.

SENILE PLAQUES

Appropriate silver staining and immunohistochemical localisation of β-amyloid show these to be extracellular lesions which in their typical neurite form are roughly spherical in shape (10–50 μm diameter) with a central core of amyloid surrounded by glial processes and a rim of neurites. The amyloid can sometimes exist alone (compact plaque), when the neurites no longer react to silver staining or in a diffuse state (primary plaque) before neurites have formed. It is unclear whether the development of neuritic from diffuse plaques causes neurofibrillary pathology and neuronal dysfunction or results from those processes. Plaques are, however, indices of neuronal death, generally of large pyramidal cells. They are found mostly in the cerebral cortex, especially the hippocampus and frontal temporal area, and while most common in AzD brain they also occur in Down Syndrome and in pugilistic (brain damage) dementia and can even be found sparsely in the normal ageing brain.

NEUROFIBRILLARY TANGLES

These are intraneuronal cytoplasmic lesions found predominantly in large pyramidal cells, again, mostly within the hippocampus and frontal temperal cortex, and while they can be seen in some other conditions, e.g. post-encephalatic Parkinsonism and Down Syndrome, they are generally considered to be more specific to AzD than the plaques. The tangles are composed of tau® protein, which normally promotes polymerisation of the microtubules that maintain cell structure, but for some reason has become hyperphosphorylated and deposited as helical filaments in a characteristic entwined paired format which disrupts neuron function. Hirano bodies, which are intraneuronal eosinophile inclusions, are also seen in AzD.

FORMATION OF β-AMYLOID AND ITS EFFECTS

Most cases of AzD show cerebrovascular amyloid deposits and the amyloid protein of senile plaques is the same as that found in blood vessels. It is referred to as β-amyloid protein and is part of a 695, 751 or 770 amino-acid amyloid precursor protein APP, which is a transmembrane protein and although its precise function is not clear, it is widely distributed and APP knock-out mice show reduced motor function. Normally so-called short 40 amino-acid-soluble derivatives of APP are produced by proteolytic cleavage of APP within the β (A4) amino-acid sequence but APP can also be cleaved

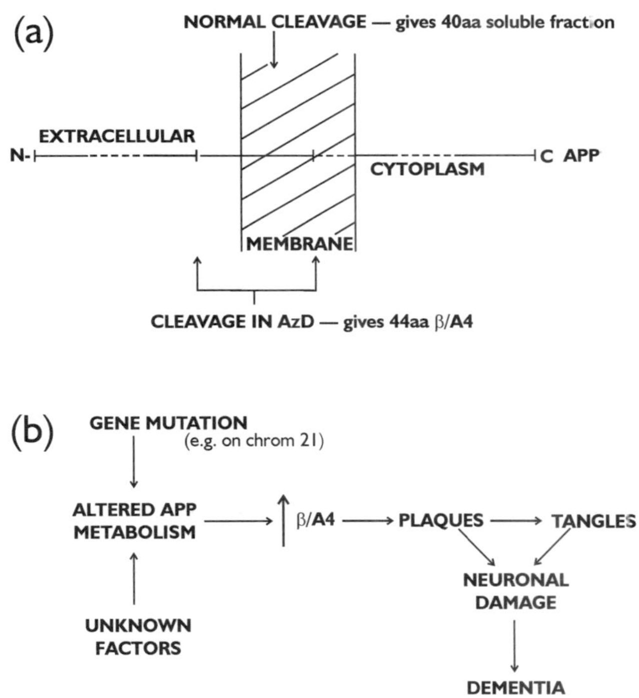

Figure 18.2 Production of senile plaque β/A4 amyloid protein. Amyloid β4 protein (β/A4) is part of a 695, 751 or 770 amino-acid amyloid precursor protein APP. This is a transmembrane protein which is normally cleared within the β/A4 amino acid sequence to give short 40 amino-acid soluble derivatives. It seems that under some circumstances as in Alzheimer's disease, APP is cleared either side of the β/A4 sequence to release the 42/43 amino acid β/A4 which aggregates into the amyloid fibrils of a senile plaque (a). (See also Fig. 18.5.) Some factors, e.g. gene mutation, must stimulate this abnormal clearage leading to the deposition of β/A4 amyloid protein as plaques and tangles and the death of neurons (b)

either side of the β sequence to liberate the longer 42 (or 43) amino-acid-insoluble β-amyloid protein (Fig. 18.2). One possibility is that in AzD this process is excessive and the insoluble amyloid β protein (Aβ) aggregates to form the amyloid fibrils and core of the senile plaques. The protein may also stimulate the phosphorylation of tau and the production of neurofibrillary tangles. How it kills neurons is unclear. Suggestions include the production of free radicals, sensitisation to glutamate and increased Ca^{2+} influx. The last has been shown in *in vitro* studies but these tend to use concentrations in excess of those found in the brain and often with shorter and soluble synthetic forms of Aβ. Certainly the direct injection of β-amyloid or neurotic plaques into rat brain does not appear to kill neurons but continuous infusion of Aβ (1–40) into the cerebral ventricles of rats does lead to impairment of learning and memory (Nitta *et al.* 1997).

In fact no consistent correlation has been found between the appearance, distribution and number of amyloid plaques and either neuronal loss or the degree of dementia, although the latter correlates with the number of neurofibrillary tangles, which tend to precede plaques in appearance by some years. Also cortical amyloid deposits can be found in non-demented elderly patients. Thus the basic question appears to be: does the disease process, whatever that is, cause the development of AzD as well as the production of β-amyloid or is there production of β-amyloid, which then causes AzD? Consensus supports the latter but it is not proven.

AETIOLOGY

If β-amyloid deposition is responsible for AzD, it is important to know what causes its production. Since AzD is most common in the elderly and as β-amyloid is found in the normal aged brain, it is likely that AzD depends on some predisposing factor that increases amyloid production and which may strike early in life but is more likely to become apparent during ageing. There is strong evidence for a genetic component. Mutations of the APP gene on chromosome 21, the apolipoprotein E (ApoE) gene on chromosome 19, the presenilin 1 (Ps1) gene on chromosome 14 and presenilin 2 (Ps2) gene on chromosome 1 have all been implicated in AzD.

GENETIC MUTATION

A number of family mutations of the APP gene on chromosome 21 have been found, generally in early-onset AzD patients in different countries, all of which lead to increased β-amyloid production. Also chromosome 21 is abnormally trisomic in Down Syndrome and most Down sufferers develop AzD if they reach 40 years. In transgenic mice, expressing familial AzD mutations of APP, the overexpression of APP is accompanied by increased amyloid deposition but whether this is due to the mutation or overexpression of APP is uncertain. Also not all the animals show memory loss and that tends to precede the amyloid disposition.

One of the most likely risk factors for AzD is the patient's genotype for apolipoprotein E (ApoE), which is believed normally to be involved in neuronal repair and growth, but is also found in plaques and tangles. Three distinct forms of ApoE, E2, E3 and E4 are encoded on chromosome 19 but it is the ApoE, E4 allele that occurs at a much higher frequency in late-onset AzD patients (50%) compared with controls (16%) and binds to and possibly increases the formation of β-amyloid. Many early-onset cases of familial AzD are associated with mutations in the genes for PS1 and PS2 on chromosomes 14 and 1 respectively. The precise physiological role of these 463 and 448 amino-acid transmembrane proteins is unclear but plasma and brain tissue from patients with PS mutations contain above-normal levels of the β-amyloid protein as do transgenic mice expressing PS mutations and cells transfected with mutant PS.

Thus all the above genetic mutations can lead to increased amyloid deposition and possibly AzD (see Smith 1998). Unfortunately familial AzD represents only the minority of cases and so other causes need to be considered.

HEAD INJURIES

It has been estimated that up to 15% of head injuries may lead to AzD with dementia being common among boxers (*dementia pugilistica*). Certainly such trauma is associated with diffuse amyloid deposits (not plaques) and a number of neurofibrillary tangles apparently identical to those in AzD.

ALUMINIUM

Reported positive associations between AzD and a high aluminium level in drinking water promoted that element as a risk factor for, or cause of, AzD. Since then aluminium in silicate form has been found in plaques and tangles and shown to impair the axonal transport of neurofilament. However, the occurrence of high brain levels of aluminium, either through environmental exposure or dialysis encephalopathy, is not associated with a greater incidence of AzD and the neurofibrillary changes it produces appear different from those of AzD. Currently while aluminium is accepted to be neurotoxic, it is thought to be a more likely cause of neurological impairments than AzD.

INFLAMMATION

The finding that patients treated with non-steroidal anti-inflammatory drugs (NSAIDs) like asprin were less likely to develop AzD stimulated the suggestion that AzD may have an inflammatory component and indeed NSAIDs have been shown to have a protective effect against AzD. It remains to be seen whether this is a true anti-inflammatory effect or whether the NSAIDs are protecting by reducing free radical production. Smoking appears to offer an accepted protection against AzD.

SUMMARY

Even if there is a link between the presence of tangles and plaques and the emergence of AzD, it is by no means certain how those markers could be responsible for all the symptoms. They do not seem to be sufficiently numerous or widely spread to disrupt brain function to the extent that eventually occurs in AzD, although their preferential location in the hippocampus and the known association of that area with memory processing could explain the loss of that faculty.

Since therapy for AzD, like that for the other major neurodegenerative disorder Parkinsonism, could depend on establishing to what extent its pathology is associated with the loss of neurotransmitter function, it is important to consider NT changes in AzD.

NEUROTRANSMITTER CHANGES IN AzD

The NT most consistently implicated in AzD is ACh.

ACETYLCHOLINE

It is 20 years since a 50% reduction was noted in the level of choline acetyltransferase (ChAT), the enzyme responsible for ACh synthesis, in the frontal cortex and hippocampus of AzD patients (Bowen *et al.* 1976). This has since been confirmed by others (see Perry 1986). ACh itself was not easily measured at that time but a reduced synthesis of ACh from ^{14}C glucose was observed in brain tissue from AzD patients. There is in fact a significant correlation between the reduction in ChAT and both the increased number of plaques and tangles at death and the severity of mental impairment six months before death (Perry *et al.* 1978). ACh loss is not global, no change being found in the striatum or some parts of the cortex. Recently reduced ACh levels have been reported in CSF obtained by lumbar puncture, though it is surprising that it survived degradation (Tohgi *et al.* 1996).

Since ACh is mostly synthesised in nerve terminals, the reduction in cortical ChAT must reflect a loss of cholinergic nerve terminals and as there are few cholinergic neurons in the cortex, these must be the endings of axons that come from cholinergic neurons in the subcortical nucleus basalis (Fig. 6.7). In fact there is a dramatic loss (<70%) of such neurons in AzD, especially in younger patients, although there is some evidence that the loss of cortical ChAT is greater than the cell loss and that degeneration starts in the cortical terminals and proceeds retrogradely to the cell bodies. Plaques and tangles are also found in the nucleus basalis but lesion of it does not induce their formation in the cortex and their cortical location does not just coincide with cholinergic innervation.

No overall reduction in cholinergic muscarinic receptors was found but recent studies with relatively specific ligands show a loss of presynaptic M_2 receptors, in keeping with the loss of terminals, but no reduction in postsynaptic M_1 receptors. Some acetylcholinesterase is found in plaques.

ACh AND β-AMYLOID

Low concentrations of solubilised β-albumin inhibit ACh release in slices from rat hippocampus and cortex areas which show degeneration in AzD, but not in slices from the striatum which is unaffected. While not totally specific to ACh, since some inhibition of NA and DA and potentiation of glutamate release have been reported, this effect is achieved at concentrations of Aβ below those generally neurotoxic. Since β-amyloid can inhibit choline uptake it is also possible (see Auld, Kar and Quiron 1998) that in order to obtain sufficient choline for ACh synthesis and the continued function of cholinergic neurons, a breakdown of membrane phosphatidyl choline is required leading to cell death (so-called autocannibalism). β-amyloid can also reduce the secondary effects of M_1 receptor activation such as GTPase activity and IP_3 production. To what extent these events can occur *in vivo*, let alone with insoluble β-amyloid, which forms the plaques, is not clear but soluble β-amyloid itself is also increased significantly in AzD brain and when infused into the ventricles of rats reduces ChAT activity.

MONOAMINES

Of course, cholinergic neurons are not the only ones with axon terminals in the cortex and if their degeneration does originate in the cortex then other afferants and their neurons could also be affected. This contention is supported by reported reductions in the number of NA neurons in the locus coeruleus, and 5-HT neurons in dorsal raphé but these are less marked (approximately 50%) than the loss of cholinergic neurons. Accompanying reductions in cortical NA and 5-HT are also seen but are again lower than those for ChAT but 5-HT_2 receptors are reduced (43%).

SOMATOSTATIN

Among a number of peptides studied it is only the reduction of somatostatin in the temporal, parietal and frontal cortices that correlates with the severity of dementia in AzD, although corticotrophin-releasing factor is lower. Reductions in somatostatin do not generally parallel those of ChAT, its concentration being almost normal in the hippocampus and nucleus basalis, where ChAT levels are lowest and there is no evidence that it is localised in cholinergic neurons.

GLUTAMATE

Despite the loss of cortical pyramidal cells no reduction in glutamate levels has been found generally in AzD, except in parts of the hippocampus where the density of glutamate nerve terminals is very high. Here the NT pool could form a sufficiently major part of the total tissue content that any reduction in that measure would indeed reflect a loss in NT glutamate. A reduction in the sodium-dependent D-[^{3}H] aspartate binding, which is presumed to label glutamate nerve terminals, has been shown for some (e.g. temporal) but not all areas of the cortex or the hippocampus despite the widespread loss of neurons. This picture is also complicated by the binding of aspartate to glial cells that multiply to occupy lost neuronal space. Although there are some reports of a reduction in the number of glutamate NMDA receptors in the hippocampus, this has not been found generally in the cortex.

Some of the symptoms of AzD are similar to those seen in patients with cortico-cortical disconnection, i.e. a loss of cortical association fibres running from one part of the cortex to another. These include difficulties in recognising known objects through sensory inputs such as touch or smell (*agnosia*), producing or understanding spoken or written words (*aphasia*), and initiating or performing familiar movements (*apraxia*). All these impairments show not only a loss of memory but also an inability to link (associate) the activity of different cortical functions and areas. Since the fibres that normally link the areas arise from glutamate-releasing pyramidal cells, their degeneration would implicate some loss of glutamate.

ANIMAL TESTS OF MEMORY FUNCTION

Since these test an animal's ability to initially perform and then repeat a simple behavioural task, the animal can be said to have learnt some function, which means they have remembered or memorised certain actions. They are tests of memory only in so far as memory is an essential part of learning which may be defined as the process by which an experience is somehow incorporated into the brain so that it can be retrieved. Animal tests are, of course, very basic but human memory can be much more complex since we can memorise occurences, events and impressions that do not actually require active accquisition or learning. Also animal tests, whether induced by drugs or used to test for a drug effect, can be influenced by the drug's possible modification of attention, arousal and motor function. Length of memory can be evaluated by varying the time between the initial learning and the subsequent tests.

The tests generally involve some form of maze but the simplest is the passive avoidance test. In this the animal learns that in a certain environment it will be punished with an electric shock for some particular action, like stepping onto a special part of the floor of the test chamber. The test of memory is how long the rat avoids (remains passive to) making the movement that will initiate the shock. Of course, drugs that reduce the animal's anxiety also modify the response. Using a maze in its simplest T shape, the animal is placed at the base of the vertical arm and a food reward at the end of one of the horizontal arms. Clearly the animal has to learn which arm contains the reward. Memory is assessed by the time taken for a food-deprived animal to reach the reward and the number of false arm entries. This simple system can be made more complex by introducing many more arms and branches but the principle is the same.

In a radial maze a number of arms of equal length radiate from a central point, where the animal is placed. Initially food is placed at the end of each arm and the rat is expected to learn that fact by exploring and entering each arm. The test of memory is to see whether on re-exposure to the maze the rat remembers only to enter an arm not previously visited and so still containing food.

The Morris (1984) water maze is a large circular glass tank (1.5 m diameter) filled with opaque (e.g. dye-treated) water to a depth of some 50 cm. A small platform, large enough to take the rat, is placed in the water with its top about 1.5 cm below the surface, so it cannot be seen. When placed in the water the rat finds and escapes to the platform, the position of which is apparently learnt by reference to peripheral visual markers. Memory is demonstrated by the rat's ability to swim to the platform when put back into the water at various points and measured by the time or the length of path taken to do so (Fig. 18.3).

NEUROTRANSMITTERS IN MEMORY PROCESSING AND APPROPRIATE SYNAPTIC FUNCTION

Since AzD is characterised by an impairment of memory, which is a normal brain function, then a consideration of which NTs and brain circuitry are implicated in the laying down and retrieval of memory may provide an indication of not only which NTs we should expect to be affected in AzD but also which need to be manipulated to therapeutic advantage. Again, most evidence points to ACh and glutamate.

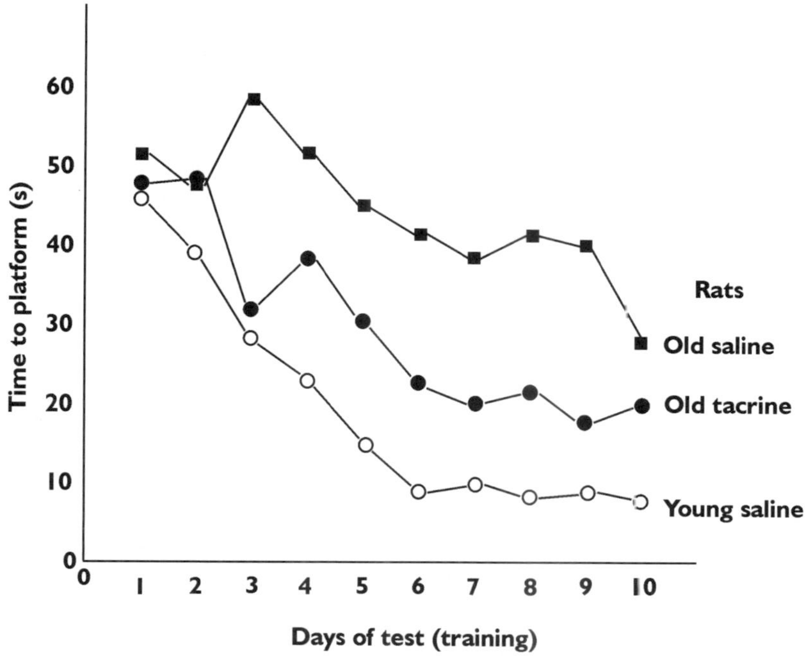

Figure 18.3 Acquisition of the water maze task as an index of learning and memory. In this example rats were trained daily to find a platform just submerged below water in a circular (150 cm) glass tank painted black and the time taken to reach it recorded. Young (4-month) saline-treated rats (o) quickly learnt and from day 6 consistently swam to the platform in less than 10 s. By contrast, aged (22-month) rats (■) took significantly longer to acquire the task and by day 10 were still taking about 30 s. The anticholinesterase drug tacrine (30 mg kg^{-1} daily.ip) given to such aged rats (•) significantly reduced the time to acquire (6 days) and perform the task (20 s). Adapted from Yavich *et al.* (1996)

ACETYLCHOLINE

ACh is implicated in memory for two main reasons:

(1) It has been known for many years that antimuscarinic drugs like hyoscine, which enter the brain, cause amnesia when used clinically, e.g. pre-operatively, to reduce bronchial secretions. In experimental studies in both humans and animals they disrupt both the acquisition and the performance of learned behaviour. Anticholinestrase drugs have the opposite effect. It is by no means certain, however, that the memory defects induced by antimuscarinics are identical to those seen in AzD.
(2) Exocytoxic lesions of the nucleus basalis with excitatory amino acids such as quisqualic and ibotenic reduce cortical ChAT activity and impair memory performance in animals. Unfortunately although quisqualate is more effective than ibotenic in reducing ChAT, it has less effect on memory (acquisition of passive avoidance), suggesting some additional effect of ibotenic acid not concerned with the ACh. Nevertheless, the memory defect induced by ibotenic acid is similar to that obtained with hyoscine and can be reversed, as studies in the rat water maze test show, by implanting fibroblasts with high ChAT activity in the cortex to secrete ACh. Anticholinesterase inhibition and foetal brain grafts containing cholinergic

neurons have also been shown to partially reverse the effects of lesions of the nucleus basalis.

So if ACh is involved in memory function, what does it do? Any attempt to answer that question has to follow some consideration of how memory is thought to be processed. Many neuroscientists believe that memory is achieved by changes in the strength of synaptic connections (activation) between neurons and that increases in such synaptic activity somehow reinforce the pattern of neuronal activity during the memorising of an event so that it can be more easily restored later. One form of such plasticity is long-term potentiation (LTP), which has been mostly studied in the hippocampus where, as in other areas associated with memory, there is the appropriate complex synaptic morphology.

That the hippocampus is important for memory is generally accepted. This is not because it is a site of major degeneration in AzD, that finding can only be used to account for the memory loss if memory is known to be dependent on the hippocampus, but because lesions of that region are known to impair memory. Case reports in the medical literature are rightly mistrusted but few people have felt inclined to disregard the evidence presented by one 27-year-old male mechanic who underwent bilateral hippocampal removal for intractable epilepsy in Montreal in 1953. While that condition was improved the operation has not been repeated because memory loss was almost total, so while he appeared to behave reasonably normally (and still does), he cannot remember where he lives, what he has just eaten or the person he met a few minutes previously.

Long-term potentiation can be defined as the increased effectiveness (potentiation) of synaptic transmission which may last for hours (possibly days) and is triggered experimentally by a brief burst of high-frequency stimulation of presynaptic inputs so that the response to any following input is much greater than normal. It was first demonstrated *in vivo* (Bliss and Loma 1973) but much studied *in vitro*. There is considerable debate as to whether the potentiation is of pre- or postsynaptic origin, or both, and while neurons can discharge spontaneously at an appropriate tetanic frequency (e.g. 200 Hz) it is not known how this may arise in normal neurophysiological processing. So to what extent LTP is essential to the memory process is unclear but there is no disputing the fact that despite all the evidence for the involvement of ACh in memory, antimuscarinic drugs do not affect LTP. ACh does, however, have the ability to partially depolarise neurons by reducing K^+ efflux (Chapter 6) and so make them more likely to fire repetitively to an incoming impulse. On the other hand, LTP is blocked by glutamate NMDA antagonists.

GLUTAMATE

As outlined above, changes in glutamate levels and function in AzD are much less clear-cut than for ACh, despite the fact that the lost pyramidal cells presumably use glutamate as a NT. On the other hand, glutamate, unlike ACh, does appear to be essential for LTP and if that is important in the memory process then so is glutamate. LTP is increased by NMDA agonists as well as being blocked by NMDA antagonists, which also decrease learning in animals. Such drugs have not been risked in patients but phencyclidine, which has been used as an anaesthetic (and drug of abuse), is known to cause amnesia and has been found to directly block NMDA receptor channels.

OTHER NTs

Just as there is less degeneration of monoamine than cholinergic neurons in AzD, so they have less influence on memory function. Generally in both animals and humans, increases in NA activity (α_1 agonists, α_2 antagonists) improve cognition although both positive and negative effects have been reported with α_2 agonists and cardiovascular effects cannot be ruled out in all these studies. 5-HT_3 antagonists such as zacopride and ondansetron have been shown to produce some improvement in cognitive performance in animal and human studies. Removal of the posterior pituitary in rats shortens retention of a conditioned avoidance response, an effect which can be overcome by the administration of vasopressin. Variable but generally weak positive effects on cognition have been seen with this peptide in humans. Opioids tend to impair and their antagonists improve memory in animals (see McGough, Introlni-Collison and Castellano 1993).

THERAPY

Therapy should be aimed at either

(a) The manipulation of NTs known to be affected by the neurodegeneration or
(b) The attenuation and possible reversal of any cause of AzD such as amyloid production and deposition

It must be made clear from the outset, however, that there is currently no way of stopping the progression of (i.e. curing) the disease and that the most that can be achieved at present is some restoration of memory function in the early stages.

MANIPULATION OF NEUROTRANSMITTERS

AUGMENTING CHOLINERGIC FUNCTION

Since ACh appears to be important in memory processing and as its concentration is significantly reduced in appropriate brain areas in AzD then augmenting its action should at least improve memory function. ACh activity may be increased by

(1) Enhancing its synthesis (and presumed release) through giving the precursor choline
(2) Stopping its degradation by cholinesterase with anticholinesterase drugs
(3) Reproducing its action with appropriate agonists—(a) muscarinic, (b) nicotinic

Approaches (1) and (2) depend on sufficient cholinergic function remaining to make its supplementation feasible, while all three methods suffer from the fact that not only does ACh have other central effects (e.g. in striatum), but also numerous peripheral parasympathomimetic ones, such as increased bronchial and gastric secretion or reduced heart rate.

Increased synthesis

This requires the provision of the precursor choline which is often given as lecithin (phosphatidyl choline), a natural source of choline found in many foods such as eggs and fish. Large doses (9–10 g) have to be given, probably to overcome the body's natural ability to restrict plasma choline levels, and the fact that only a very small percentage is converted to ACh. Brain penetration is modest but uptake into cholinergic nerve terminals is through a sodium-dependent high-affinity system that is normally adequately supplied and possibly saturated with choline. In any case, ACh can only be synthesised from choline in cholinergic nerve terminals, many of which will have degenerated, and just increasing the activity of those remaining may not be adequate. Whether choline could reverse the choline leakage and resulting autocannibalism (see above) of cholinergic neurons is not known.

Reduced degradation

ACh is metabolised extraneuronally by the enzyme acetylcholinesterase, to reform precursor choline and acetate. Blocking its activity with various anticholinesterases has been widely investigated and some improvement in memory noted. Such studies have invariably used reversible inhibition because of the toxicity associated with long-term irreversible inhibition of the enzyme. Physostigmine was the pilot drug. It is known to improve memory in animals and some small effects have been seen in humans (reduces number of mistakes in word-recall tests rather than number of words recalled), but it really needs to be given intravenously and has a very short half-life (30 min).

The limited effectiveness of physostigmine did, however, encourage the development of longer-acting orally effective anticholinesterases such as tacrine (tetrahydroaminoacrydine), velnacrine and donepezil.

Clinical evaluation of anticholinesterases and other drugs in AzD

The newer anticholinesterases have all been subject to large and often multicentred trials. These take various forms but generally include an initial assessment of disease severity over a few weeks while on placebo alone, then a drug-dose evaluation before the chosen drug dose(s) is compared directly with placebo for some weeks in two groups. Confirmation of any drug effect is usually obtained by finishing with all patients on placebo. Although performed double-blind generally, only patients that respond in the early evaluation period enter the final drug trial and those with severe AzD are excluded altogether. Results from a simpler Phase III drug study showing some efficacy for donepezil are shown in Fig. 18.4 (Rogers *et al.* 1998).

The evaluation of drug effectiveness in AzD is not without its difficulties. There is a need to record changes in both cognitive function and general performance. Two primary measures are the Alzheimer's Disease Assessment Scale of cognition (ADAS-cog) and the Clinician's Global Impression of Change (CGIC). The former measures such things as memory, language, orientation, reason and praxis, on a 0–70 scale range. The higher the score, the more severe the condition, and as most patients normally decline at the acquisition rate of 5–10 extra points a year, any reduction of 4 or more points is considered a drug effect. The CGIC scale, as its name implies, is a more global measure of patient function not only in cognition but also in general behaviour and daily living obtained by the clinician interviewing both patients and carers. On a 7-point scale, improvement is represented by 1, worsening by 7 and no change by 4. Other measures include the patient's own evaluation of quality of life (QoL) noting their general feelings and ability to eat, sleep and relate to others. Generally improvements in

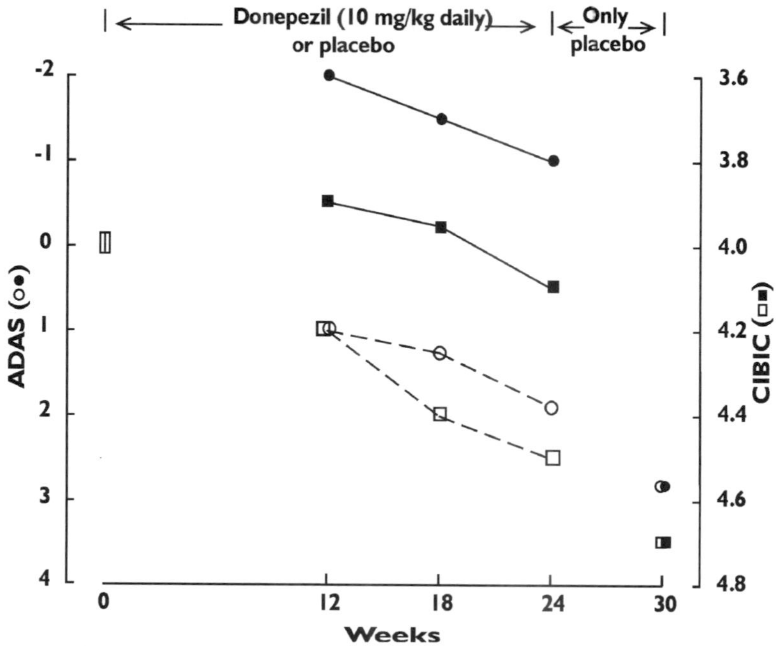

Figure 18.4 Clinical assessment of the efficacy of the anticholinesterase drug, donepezil, in Alzheimer's disease. Summary redrawing of some of the results of a large double-blind placebo-controlled trial by Rogers *et al.* (1998) © Lippincott Williams & Wilkins in which the effect of donepezil ($10\,mg\,day^{-1}$) was tested on men and women over 50 years with uncomplicated Alzheimer's disease ($N = 157$) compared with placebo ($n = 162$), using a number of measures including the Alzheimer's Disease Association Assessment Scale of cognition (ADAS-cog) and a more global assessment (CIBIC plus) equivalent to the Clinician's Interview-based Impression of Change scale based on clinical, patient and family assessment of cognition and behaviour. The results show that donezepil has had a significant ($p = 0.009$–0.0001) beneficial effect by both the ADAS (●—●) and CIBIC (■–■) assessments, when compared with placebo (○–○ and □–□) from 12th to 24th week. Two further features, characteristic of such therapy, are also apparent; (i) the drug has a greater effect on cognition (ADAS) than on overall state of health (CIBIC) and (ii) it does not retard the progress of the disorder (no difference between drug and placebo groups 6 weeks after cessation of drug). ABAS scores range from 0 to 70 (min–max symptoms) with patients normally deteriorating at a rate of 7–11 extra points per year so that any reduction in that rate constitutes an improvement. The CIBIC scale scores 1–7, with 1 = marked improvement, 4 = no change and 7 = worsening

cognition (ADAS-cog scores) are more easily achieved than in the overall quality of life (CGIC) (see Fig. 18.4) and is a useful reminder that AzD is not just a loss of memory.

The initial enthusiasm for tacrine and velnacrine, which are the anticholinesterases most studied clinically, has been tempered by the fact that not all patients respond. Most show the peripheral parasympathomimetic effects of cholinesterase inhibition, e.g. dyspepsia and diarrhoea, as well as nausea and vomiting, and about half of the patients develop hepatotoxicity with elevated levels of plasma alanine transaminase. While some peripheral effects can be attenuated with antimuscarinics that do not enter the brain, these add further side-effects and the drop-out rate from such trials is high (<75%) in most long-term studies. Donepezil appears to show less hepatotoxicity but its long-term value remains to be determined.

Generally, anticholinesterases produce some improvement in cognitive function and the quality of life in the early stages of AzD but that needs to be balanced against side-effects.

Some of the cognitive improvements with tacrine, which is chemically related to amidopyridine, may be due to blockage of K^+ channels.

Use of agonists

Muscarinic

Since most postynaptic cholinergic receptors in the brain are muscarinic and as they do not appear to be reduced in AzD, despite some degeneration of pyramidal neurons, the use of muscarinic agonists could be worth while. Early studies with bethanecol, arecoline and oxotremorine, mixed M_1 and M_2 agonists, showed little benefit and newer drugs have not been much better. Peripheral cholinergic effects are a problem and central infusion, which has been tried with bethanecol to no great effect, is hardly a practical proposition. There is, however, a realisation that more appropriate drug or drug combinations could be developed and tried. Thus, theoretically anyway, the requirement is for a specific M_1 agonist that readily crosses the blood–brain barrier. Avoiding M_2 receptor stimulation will also mean no activation of presynaptic auto-receptors and counterproductive inhibition of ACh release, and fewer peripheral effects. These latter could also be avoided with an M_1 antagonist that does not cross the blood–brain barrier. Even then successful therapy may be negated by a requirement for ACh to be released physiologically from appropriate neurons.

Nicotinic

Despite the paucity of nicotinic receptors in the brain there is considerable evidence that AzD is less common among smokers. Whether this is due to the action of inhaled nicotine is uncertain, but nicotine is known to stimulate presynaptic receptors on cholinergic nerve terminals which, unlike the muscarinic ones, result in increased ACh release.

MODULATING GLUTAMATE FUNCTION

If long-term potentiation (LTP) is important in memory function and as it can be blocked by glutamate NMDA antagonists (see above) then increasing NMDA activity is of putative value in AzD. In reality this presents a problem because overstimulation of the receptor could not only increase neuronal function up to convulsive level but even cause neurotoxicity. Briefly, NMDA applied to rat cortex causes retrograde degeneration of cholinergic neurons in nucleus basalis while NMDA antagonists prevent anoxic destruction of cultured hippocampal neurons and brain damage caused by cerebral vascular occlusion in rodents. The ischaemia the latter produces causes such an excessive neuronal discharge and release of glutamate that the intense activation of NMDA receptors produces a prolonged neuronal depolarisation, Ca^{2+} entry and cell death. Possibly a weak partial NMDA agonist, or a drug acting at one of the NMDA receptor subsites (see Chapter 10) like that for glycine, may be of some value. Whether glutamate-induced neuronal death, which is enhanced by β-amyloid, plays any part in the actiology of AzD is uncertain but controlling glutamate activity so that it can be increased enough to facilitate memory processes without undue excitation of neurons will be difficult.

GABA

Although there is no neuropathological evidence to implicate GABA in AzD it is known that agonists at the benzodiazepine receptor site not only augment GABA function but also cause amnesia. So it is possible that an inverse agonist, or perhaps even an antagonist, for the benzodiazepine receptor could have the opposite effect and improve memory. In humans, one antagonist, the β carboline derivative ZK93426, showed some improvement in learning and memory tests. It also improves acquisition in animal-learning tests and counteracts the impairment caused by scopolamine, as does the β-carboline inverse agonist DMCM. The fear of inducing anxiety or even convulsions with inverse benzodiazepine agonists has prompted the evaluation of partial inverse agonists (see Abe, Takeyama and Yoshimura 1998).

OTHER NTs

There have been few attempts to manipulate the monoamines in AzD and those using selegiline, the MOA_B inhibitor, have shown little effect although the $5\text{-}HT_3$ antagonist, ondansetron, may give a slight improvement.

Despite the clear loss of somatostatin in AzD a synthetic analogue L-363586 had no beneficial effect on memory loss.

ATTENUATION OF DEGENERATION

Even if NT manipulation had provided an effective therapy in AzD it would still be important to stop the progression of degeneration and the disease process itself. The failure of the NT approach makes it even more vital.

β-AMYLOID

While the precise cause of AzD remains unknown, the evidence implicating β-amyloid is such as to justify attempts to reduce its involvement. The activity of β-amyloid might be reduced by:

(a) stopping its production by reducing the phosphorylation and proteolysis of APP
(b) increasing its breakdown
(c) counteracting its toxic effects through plaque formation

APP is normally cleaved within the $A\beta$ sequence by an unidentified protease, so-called α-secretase, so that most of the extracellular APP is released in a soluble form into the extracellular fluid (see Checler 1995). When β-amyloid is formed another protease (β) splits APP so that the complete $A\beta$ sequence persists at the extracellular end of the remaining membrane and intracellular APP chain. This is then cleaved by anaother protease (γ-secretase) to release the β-amyloid (Fig. 18.5). Potentiation of α- or blockage of β- and γ-secretase could reduce the production of $A\beta$ which becomes insoluble and is precipitated (see Hardy 1997).

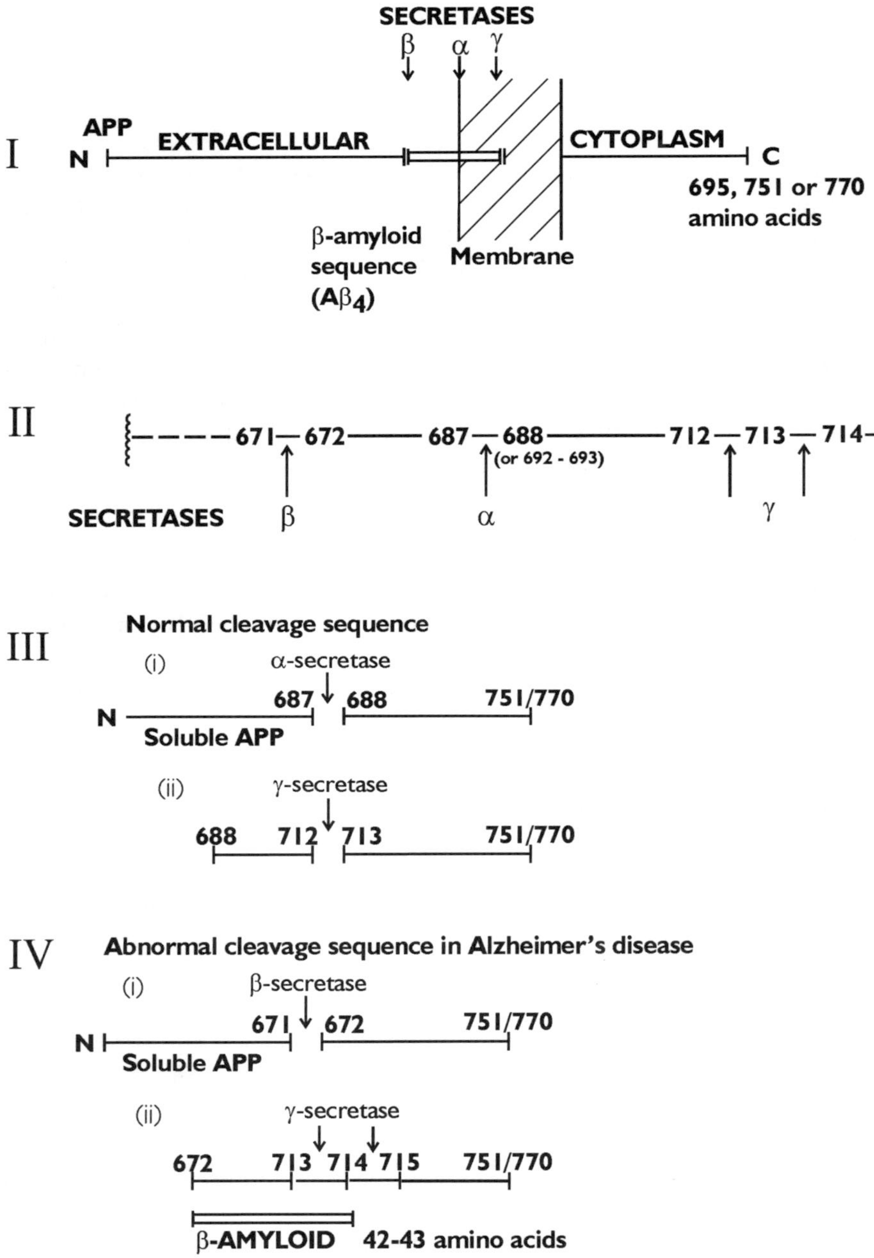

Figure 18.5 Schematic representation of possible cleavage sites of APP by α, β and γ-secretase and the production of β-amyloid protein. (I) This shows the disposition of APP molecules in 695, 751 and 770 amino-acid chain lengths. Much of it is extracellular. The β-amyloid ($A\beta_4$) sequence is partly extracellular and partly in the membrane. (II) An enlargement of the β-amyloid sequence. (III) Normal cleavage of APP by α-secretase occurs in the centre of $A\beta_4$ sequence to release the extracellular APP while the remaining membrane and intracellular chain is broken down by γ-secretase to give two short proteins that are quickly broken down. (IV) In Alzheimer's disease β rather than α-secretase activity splits off the extracellular APP to leave the full $A\beta_4$ sequence remaining attached to the residual membrane and intracellular chain. 42/43 amino acid β-amyloid sequence is then split off by γ-secretase activity

These must be worthwhile objectives and the recent identification by a number of research groups (see Skovronsky and Lee 2000 for description and details) of β-secretase as the membrane-bound aspartyl protease (BACE), β-site APP cleaving enzyme, paves the way for developing possible chemical inhibitors of its activity for experimental and clinical evaluation, although that remains for the future.

NEUROTROPHIC FACTORS

Whether or not the production of β-amyloid can be curtailed, it would be desirable to either replace the damaged neurons or encourage the remaining functional ones to ramify further and exhibit more influence. The former, which requires tissue or cell line grafts, is currently not feasible and barely investigated experimentally but there is much interest in the possible use of neurotrophic proteins (neurotrophins) that encourage neuronal growth and differentiation.

A number of these have been isolated and identified but the first to be discovered (see Levi-Montalcini 1987), and the most studied, is nerve growth factor (NGF) which, despite its name, is not universally effective on all neurons. In the periphery it is mainly released in tissues containing sympathetic nerves that take it up and transport it retrogradely to the cell body where it acts. In the brain, however, it has more influence on cholinergic than noradrenergic or other neurons so that NGF protein and MRNA expression is highest in cholinergic innervated areas of the brain such as the hippocampus and cortex while its binding sites (receptors) are mainly in subcortical regions with cholinergic neurons like the nucleus basalis. In fact injection of NGF into the latter's projection areas like the hippocampus and cortex result in its uptake and transport back to the nucleus. So it may be assumed that normally the cortically produced NGF is transported back to cholinergic subcortical neurons where it exerts its trophic action. Certainly NGF increases ChAT production when added to cultured cholinergic neurons and its intraventricular infusion in rats and primates prevents the loss of ChAT activity in and degeneration of, cholinergic neurons caused by transection of the septal hippocampal cholinergic pathway, or ibotenic acid injection into the nucleus basalis. Intraventricular NGF has also been shown to improve learning and memory in aged rats and those with lesions to cholinergic pathways. So if NGF is so important for the growth and function of the cholinergic neurons, that appear so vulnerable in AzD, can they be restored and AzD controlled by administering NGF? Before that question can be answered some practical problems have to be overcome, namely how to obtain and administer it.

If immune reactions are to be avoided then recombinant human factor should be used and that cannot be produced in large quantities. In any case, it is a large protein that will have to be injected directly into the brain. Even if these problems can be overcome the spread and intensity of any NGF effect has to be restricted so that excessive neuritic growth and inappropriate increases in synaptic connections do not occur.

In addition to these problems there is no evidence of reduced NGF in AzD although levels and receptor number are lower in the nucleus basalis. In fact the levels of NGF were found to be increased in the cortex and hippocampus (Scott *et al.* 1995) and while this could just be due to fewer cholinergic fibres to transport it away from the cortex it does suggest its synthesis is normal and possibly even increased. At least it throws doubt on the value of augmenting NGF as a therapy for AzD.

Nevertheless NGF from mouse mandibular gland has been infused into the right lateral ventricle of two patients (67 and 57 years) for three months at a rate of 75 μg/h. The younger showed no change in memory performance; the older some improvement after one month, which ceased after the infusion was stopped. Both patients had various reversible side-effects such as back pain and weight loss.

OTHER DRUG THERAPY

In the face of the failure of rational approaches in the treatment of AzD it is perhaps not surprising that there have been many less rational ones. These include the use of vasodilators and nootropics. The former, such as hydergine, a mixture of ergot alkaloids, are intended to increase cerebral blood flow and neuronal metabolism despite some reduction in blood pressure, while the latter, like piracetam, are metabolic stimulants that increase cerebral metabolism and ATP production. Neither are of proven value in AzD.

Although there is no evidence that the neuronal degeneration of AzD results, as in cardiovascular ischaemia, from the excitotoxicity of increased intracellular Ca^{2+}, some calcium channel blockers have been tried in AzD. They have had little effect but surprisingly a pyrrolidone derivative nefiracetam, which opens L-type voltage-sensitive calcium channels (VSCCs) reduces both scopolamine- and β-amyloid-induced impairments of learning and memory in rats (Yamada *et al.* 1999). This effect can be overcome by VSCC antagonists, but nefiracetam has not been tried in humans.

SUMMARY

Clearly there is a long way to go in the treatment of AzD particularly as even the most active of the generally ineffective therapies discussed above have only addressed the early symptoms of memory loss. In such a progressive and debilitating disease that is only a small step. Finding a NT malfunction is obviously a long way from providing an effective treatment, let alone a cure. That is more likely to come from attempts to reduce neuronal degeneration (see Selkoe 1999).

REFERENCES AND FURTHER READING

Abe, K, Takeyama, C and Yoshimura, K (1998) Effect of S-8510, a novel benzodiazepine inverse agonist, on basal forebrain lesion-induced dysfunction in rats. *Eur. J. Pharmacol.* **347**: 145–152.

Allen, NRP and Burns, A (1995) The treatment of Alzheimer's disease. *J. Psychopharm.* **9**: 43–49.

Auld, DS, Kar, S and Quiron, R (1998) β-amyloid peptides as direct cholinergic neuromodulations: a missing link? *Trends Neurosci.* **21**: 43–49.

Bliss, TVP and Loma, T (1973) Long-lasting potentiation of synaptic transmission in the dentate area of the anaesthetised rabbit following stimulation of the perforant path. *J. Physiol.* **232**: 331–356.

Bowen, DM, Smith, CB, White, P and Davison, AN (1976) Neurotransmitter-related enzymes and indices of hypoxia in senile dementia and other abiotrophies. *Brain* **99**: 454–466.

Checler, F (1995) Processing of the β-amyloid precursor protein and its regulation in Alzheimer's disease. *J. Neurochem.* **65**: 1431–1444.

Davis, RE, Emmerling, MR, Jaen, JC, Moos, WH and Spiegel, K (1993) Therapeutic intervention in dementia. *Crit. Rev. Neurobiol.* **7**: 41–83.

Hardy, J (1997) Amyloid, the presenilins and Alzheimer's disease. *Trends Neurosci.* **20**: 154–159.

Levi-Montalcini, R (1987) The nerve growth factor 35 years later *Science* **237**: 1154–1162.

Maragos, WF, Greenamyre, JT, Penny Jr, JB and Young, AB (1987) Glutamate dysfunction in Alzheimer's disease: an hypothesis. *Trends Neurosci.* **10**: 65–68.

McGough, JL, Introlni-Collison, IB and Castellano, C (1993) Involvement of opioid peptides in learning and memory. In *Opioids II. Handbook of Experimental Pharmacology*, Springer-Verlag, New York, pp. 429–448.

Morris, RGM (1984) Development of a water-maze procedure for studying spatial learning in the rat. *J. Neurosci. Meth.* **11**: 47–60.

Neve, RL and Robakis, NK (1998) Amyloid disease: a re-examination of the amyloid hypothesis. *Trends Neurosci.* **21**: 15–19.

Nitta, A, Fukuta, T, Hasegawa, T and Nabeshima, T (1997) Continuous infusion of β-amyloid proteins into cerebral ventricle induces learning impairment and neuronal and morphological degeneration. *Jpn J. Pharmacol.* **73**: 51–57.

Perry, EK (1986) The cholinergic hypothesis — ten years on. *Brit. Med. Bull.* **42**: 63–69.

Perry, EK, Tomlinson, BE, Blessed, G, Bergmann, K, Gibson, PH and Perry, RH (1978) Correlation of cholinergic abnormalities with senile plaques and mental test scores in senile dementia. *Brit. Med. J.* **2**: 1457–1459.

Rogaeu, EI, Sherrington, R, Rogaevo, EA, Levesque, G, Ikeda, M, Liang, Y, Chl, H, Lin, C, Holman, K, Tsuda, T, Mar, L, Sorbi, S, Nacmias, R, Placentini, S, Amaduco, L, Chumakov, L, Cohen, D, Lannfelt, L, Fraser, PE, Rommens, JM and St George-Hyslop, PH (1995) Familial Alzheimer's disease in kindreds with missense mutations in a gene on chromosome I related to Alzheimer's disease type 3 gene. *Nature* **376**: 775–778.

Rogers, SL, Farlow, MR, Doody, RS, Mohs, R, Friedhoff, LT and the Donepezil Study Group (1998) A 24-week, double blind, placebo-controlled trial of donepezil in patients with Alzheimer's disease. *Neurology* **50**: 136–145.

Scott, SA, Mufson, EJ, Weingartner, JA, Skou, KA and Crutcher, KA (1995) Nerve growth factor in Alzheimer's disease: Increased levels throughout the brain coupled with decline in nucleus basalis. *J. Neurosci.* **15**: 6213–6221.

Selkoe, DJ (1999) Translating cell biology into therapeutic advances in Alzheimer's disease. *Nature* **399** (Suppl. 6738): A23–A31.

Skovronsky, DM and Lee, VM-Y (2000) β-secretase revealed: starting gate for race to novel therapies for Alzheimer's disease. *Trends Pharmacol. Sci.* **21**: 161–163.

Smith, MA (1998) Alzheimer's disease. *Int. Rev. Neurobiol.* **42**: 1–44.

Tohgi, H, Abe, T, Kimura, M, Saheki, M and Takahashi, S (1996) Cerebrospinal fluid acetylcholine and choline in vascular dementia of Binswanger and multiple small infarct types as compared with Alzheimer-type dementia. *J. Neural Trans.* **103**: 1211–1220.

Wischik, CM and Crowther, RA (1986) Subunit structure of the Alzheimer tangle. *Brit. Med. Bull.* **42**: 51–56.

Yamada, K, Tanaka, T, Mamiya, T, Shiotani, T, Kameyama, T and Nabeshima, T (1999) Improvement by nefiracetam of β-amyloid — (1-42) — induced learning and memory impairments in rats. *Brit. J. Pharmacol.* **126**: 235–244.

Yavich, L, Sirvio, J, Haapalinna, A, Puumala, T, Koivisto, E, Heinonen, E and Reikkinen, PJ (1996) The systemic administration of tacrine or selegiline facilitate spatial learning in aged Fisher 344 rats. *J. Neural Trans.* **103**: 619–626.

19 Anxiety

S. C. STANFORD

INTRODUCTION

Emotional states that would now be classified as anxiety were recognised as long ago as the classical Greek period but have undergone many phases of medical classification since then. Nowadays, 'anxiety' is a term used loosely to cover the clusters of physiological and emotional changes shared by several disorders in which anxiety is a major component (Table 19.1). The current diagnostic criteria for these disorders are defined in the *Diagnostic and Statistical Manual* of the American Psychiatric Association (4th edition: 'DSM–IV' (1994)). The extent to which they share a common neurobiological basis is far from clear but it is evident that different anxiety disorders do not all respond to the same drug treatments. In particular, whereas generalised anxiety disorder (GAD) is treated preferentially with so-called 'anti-anxiety' agents (e.g. the benzodiazepines), these compounds are regarded as relatively ineffective in relieving panic disorder (except, possibly, at high doses) and they are of no benefit at all in treating phobias. In fact, it is antidepressant drugs, especially the selective serotonin reuptake inhibitors (SSRIs), that are turning out to be the most effective treatments for some anxiety disorders and their use has undoubtedly been encouraged by fears that prolonged treatment with benzodiazepines might induce a dependence syndrome. The pharmacology of antidepressants is described in Chapter 20.

The beneficial effects of antidepressants in anxiety are often interpreted as support for a neurobiological link between anxiety and depression. Also, because anxiety often progresses to depression and because these disorders can co-exist in the same patients, it has even been suggested that they might be different manifestations of a single problem (Tyrer 1989). Whether or not this is the case, it is clear that key features of all anxiety disorders, especially the sympathetic arousal, resemble the response to aversive stimuli ('stress') and this overlap has strongly influenced research into the neurobiology of anxiety. However, whereas anxiety drives people to seek medical help, the response to stress is a normal physiological event. A distinctive feature of anxiety, therefore, is that it can be regarded as an inappropriate stress response that is chronic or intermittent, for which the stimulus is either not obvious (as in GAD) or irrational (as in the phobias), or provokes a prolonged emotional disturbance (as in post-traumatic stress disorder).

There are two main approaches to research into anxiety. The first is to establish experimental models of anxiety in animals and humans in order to discover its neurobiological basis. The second is to investigate the actions of anti-anxiety drugs in the brain in the hope that this will give some clues to the cause(s) of anxiety. This chapter will discuss evidence from both these lines of research.

Neurotransmitters, Drugs and Brain Function. Edited by R. A. Webster

Table 19.1 Anxiety disorders

The diagnostic criteria for, and general features of, disorders in which anxiety is a prominent component are described in detail in the *Diagnostic and Statistical Manual* of the American Psychiatric Association (DSM–IV, 4th edition (1994)) and are regarded as either 'phobias' or 'anxiety states':

PHOBIC DISORDERS: profound fear of, and avoidance of, a dreaded object or situation.

Agoraphobia:	Fear of places or situations from which escape is difficult—can occur with or without a history of panic disorder
Social phobia:	Fear of social or performance situations
Specific phobia:	Fear of a specific object or situation

ANXIETY STATES

Panic disorder (with or without agoraphobia):	Recurrent unexpected panic attacks
Generalised anxiety disorder:	At least 6 months of persistent and excessive anxiety and worry
Obsessive compulsive disorder:	Obsessions (which cause anxiety) and or compulsions (which serve to neutralise anxiety)
Post-traumatic stress disorder (PTSD):	Re-experiencing of traumatic event outside normal human experience with increased arousal and avoidance of stimuli associated with the trauma
Acute stress disorder:	Symptoms similar to PTSD but occur within 1 month of the traumatic event
Anxiety disorder due to a general medical condition:	e.g. Disorders of thyroid function, cardiovascular system, respiratory system, head injury, etc.
Substance-induced anxiety disorder:	e.g. Caffeine, cocaine, alcohol
Anxiety disorder not otherwise specified:	Prominent symptoms of anxiety that do not fit any of the above categories

SYMPTOMS AND SIGNS OF ANXIETY
(Modified from Nutt 1990)

Mood:	Apprehension, worry, difficulty in concentration, irritability, insomnia
Cognitions:	Fear of (for example): death, ineffectiveness, failure, humiliation, mental illness
Somatic:	Cardiovascular (tachycardia, palpitations), sweating, respiration, GIT, muscle tension, tremor, muscle aches or soreness, nausea, exaggerated startle reflex, increased urinary frequency
Behaviour:	Hypervigilance, nail-biting, scratching

ANIMAL MODELS OF ANXIETY

All preclinical animal models of anxiety involve exposing animals (usually rats or mice) to environmental stimuli that disrupt their normal pattern of behaviour (Table 19.2). Obviously, it can never be confirmed that animals are actually experiencing the equivalent of human anxiety and so the validity of all preclinical models rests largely on confirming that the change in behaviour is prevented by drugs that have established anti-anxiety effects in humans.

An influential theory proposed by Gray (1987) suggests that environmental stimuli that induce anxiety in both rats and humans fall into three major groups, all of which present 'threats' to the individual. One of these is 'novelty' in which the subject's innate

Table 19.2 Environmental stimuli that induce changes in behaviour which are prevented by anti-anxiety drugs

'*Ethological models*'
Elevated plus-maze
Social interaction test
Light/dark shuttle-box
Isolation-induced ultrasonic vocalisation (rodent pups)
Models dependent on conditioned cues
Fear-potentiated startle reflex (conditioned fear)
Four-plate test (conditioned fear)
Geller–Seifter test ('signals' of conflict)
Vogel conflict test ('signals' of conflict)
Frustrative non-reward ('signals' of non-reward)

tendency to explore ('approach') novel stimuli is opposed by a tendency to avoid them; it is the conflict between approach and avoidance that gives rise to anxiety. The two other forms of anxiety-inducing ('anxiogenic') stimuli are those that are normally regarded as neutral but, as a result of innate factors (genetic programming) or the subject's previous experiences (associative learning), are interpreted as a signal for a stimulus that the subject would normally avoid. The signal can either warn that behaviour which is reinforced by reward will also be punished (e.g. by a footshock to rats; 'conflict') or warn that the reward will not materialise ('non-reward').

Testing the effects of drugs on animals' behavioural response to novel environmental stimuli offers the major advantage that it relies on evaluating changes in their innate behaviour (so-called 'ethological' models). Contrasting with this, testing the effects of drugs on animals' behavioural response to an environmental 'signal' requires extensive prior training (see below). The important point about this approach is that it is the 'signal' or 'threat' of the aversive event that triggers anxiety, rather than the aversive event itself. By analogy, environmental conditions that are associated with a threat of attack will provoke anxiety in humans whereas an actual attack triggers an (appropriate) acute stress response that recruits the 'fight or flight' (stress) reaction. In the following sections, specific behavioural models used to study anxiety and the effects of anti-anxiety drugs are described.

EVALUATING DRUG EFFECTS ON INNATE BEHAVIOUR (ETHOLOGICAL MODELS)

Most of these models evaluate the effects of drugs on the behaviour of animals when they are exposed to a novel environment. Novelty normally reduces animals' exploratory activity but established anti-anxiety drugs consistently increase exploration of, and approaches to, the novel stimulus and reduce the neophobic ('avoidance') reaction. There are several examples of tests based on this principle (Table 19.2) but two that are widely used are the 'plus-maze' and the 'social interaction' tests.

The plus-maze

This consists of a raised platform with four narrow arms, two of which have walls ('closed arms') and two which do not ('open arms') (Fig. 19.1). When placed on the

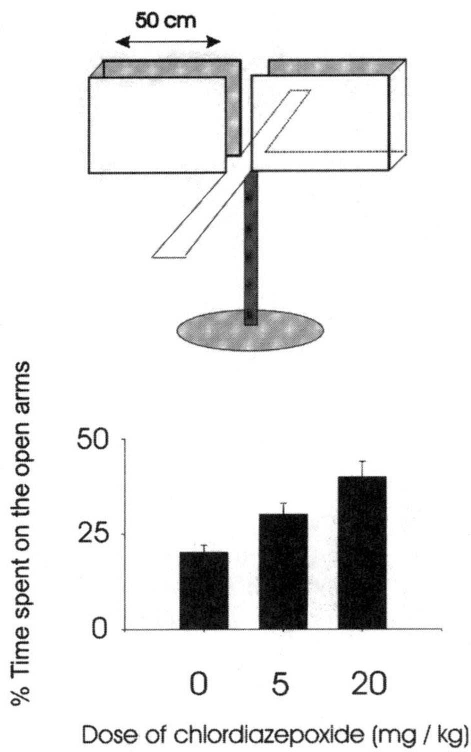

Figure 19.1 The elevated plus-maze. (*Top*) The apparatus is arranged with two open arms, two closed arms and a central zone, raised above the ground. Animals are placed in the central zone (usually facing an open arm) and their movements scored for: number of entries to the open and closed arms and the percentage time spent in the open arms. (*Bottom*) Chronic administration (5 days) of the anti-anxiety drug, chlordiazepoxide, increases the percentage time spent on the open arms to approximately 50% of the total. (Figure kindly provided by S. E. File)

apparatus for the first time, animals explore all zones of the maze but spend most time (approximately 75%) in, and make most entries to, the closed arms. Pretreatment with an anti-anxiety drug increases exploration of the open arms so that approximately equal times are spent on the open and closed arms of the maze. When interpreting results from this test, it is important to establish that any drug effects are independent of non-specific effects on the animals' overall locomotor activity (i.e. their ability to make appropriate movements), particularly since many anti-anxiety drugs are highly sedative when given acutely. Detailed insight into some of the many assumptions and refinements of the use of the plus-maze is to be found in Rodgers and Dalvi (1997).

Social interaction test

In this test, it is the interaction (sniffing, grooming, etc.) between two rats in a test arena that is scored. Social interaction is dependent on the familiarity of the animals with the test arena (social interaction is reduced in an unfamiliar arena) and the intensity of illumination (social interaction is reduced in bright light). The reduction in social interaction under aversive conditions (unfamiliar arena and bright light) is prevented by pretreatment with anti-anxiety drugs (File and Hyde 1979). However, it is again important to establish that any drug effects are directed specifically at the behavioural response to the test environment, rather than overall locomotor activity.

MODELS THAT REQUIRE CONDITIONING

There are several models that depend on monitoring changes in animals' behaviour when they are exposed to conditioned threatening cues. One of these, the fear-potentiated startle reflex, rests on the development of an exaggerated startle on presentation of the conditioned cue. Although this response is prevented by anti-anxiety drugs, there is considerable debate over whether 'fear' is the same as 'anxiety'.

A model that fits better with Gray's criteria is the Geller–Seifter ('conflict') test. This is named after the two scientists who developed it and is still often used to screen putative anti-anxiety drugs (Geller, Kulak and Seifter 1962). Briefly, animals are trained to associate the pressing of a lever with a food reward ('operant' or 'instrumental conditioning'). After reaching a stable response on the lever, the rats are then trained to realise that when a (normally) neutral stimulus is presented, such as a buzzer or a light, they will experience a mild footshock, *as well as* receive the reward, when they press on the lever. This invokes a classic approach/avoidance conflict and animals invariably respond on the lever less frequently when the conditioned cue is presented (the 'punished phase'). An important distinction is that their response on the lever in the absence of the signal (the 'unpunished' phase) is unaffected. Anti-anxiety drugs abolish the inhibition of responding during the punished phase but do not affect unpunished responding (Fig. 19.2). Obviously, a drug that increases punished responding could be increasing animals' overall activity (as with amphetamine) but this can be excluded if it has no effect on lever responses during the unpunished phase. A drug-induced reduction in the discomfort caused by the footshock (as is achieved with analgesics) or amnesia (i.e. under the influence of the test drug the animal forgets that the cue warns of a footshock) must be ruled out also.

There are many variations of this model, a commonly used example being the Vogel licking (conflict) test. This evaluates the effects of drugs on the punished phase of drinking from a water spout (Vogel *et al.* 1980) which has the advantage that the animals do not have to be trained to initiate the behavioural response (drinking). However, the increase in baseline fluid intake induced by some anti-anxiety drugs, in the absence of any anxiogenic stimuli, can be a confounding factor.

INDUCING ANXIETY IN HUMANS

One advantage of studying humans is that it is possible to confirm that a given experimental intervention does actually induce anxiety in the subject. A disadvantage is that any research into the neurobiological changes that underlie the subjects' psychological status is limited to the analysis of accessible tissue samples, such as plasma or urine. Such measurements will, at best, be indirect indications of what is happening in the brain. As a result, research of anxiety in humans has concentrated on drugs with a known pharmacological target (usually a neurotransmitter receptor) and has compared their effects in anxious patients and normal subjects. Some treatments that induce or

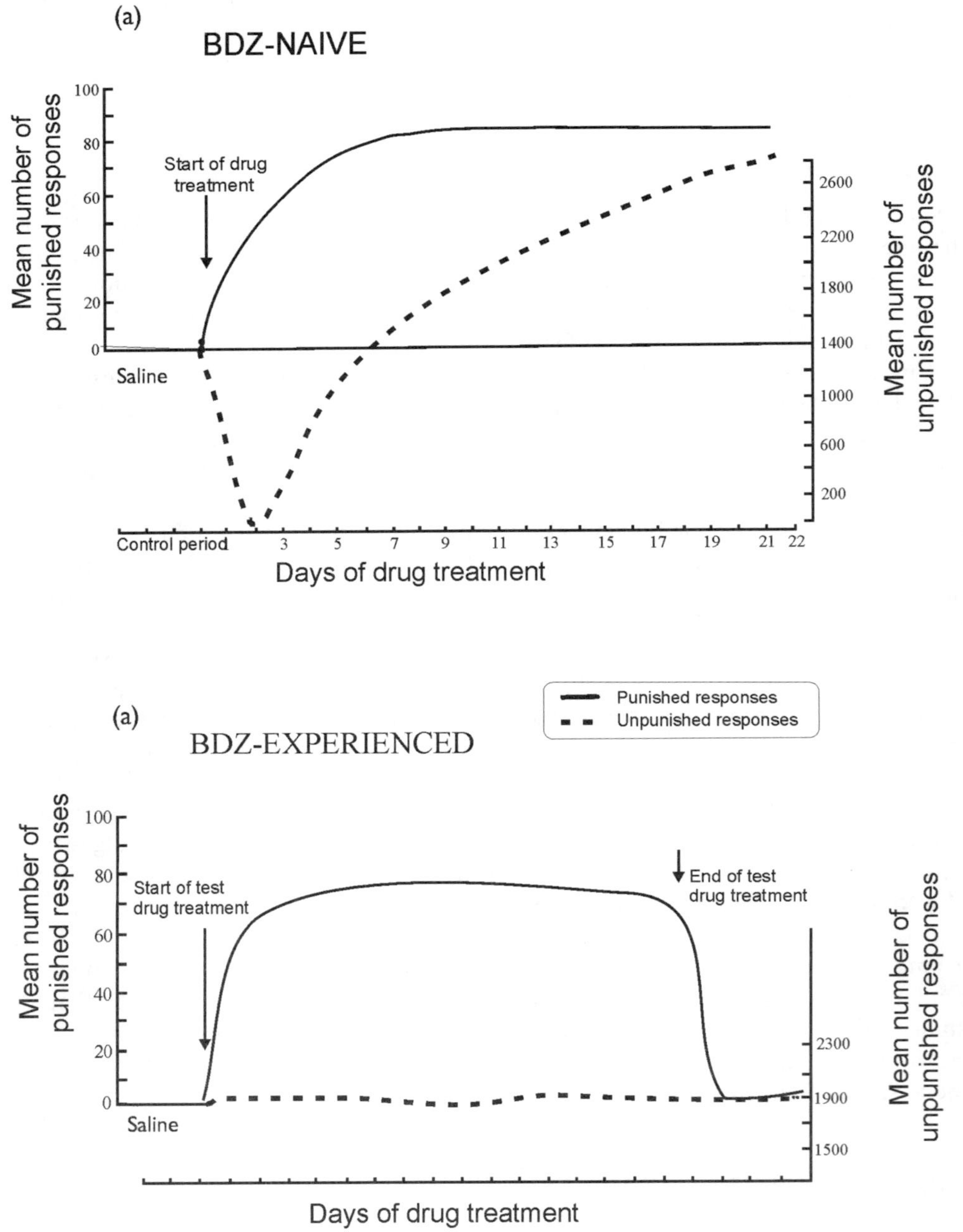

Figure 19.2 (a) Average curves of punished and unpunished responses of rats, never previously treated with a benzodiazepine (BDZ), during several days of administration of a test BDZ. The apparent delay in the increase in punished responses is due to the reduction in all responses (including unpunished ones) at the start of drug administration. The progressive recovery of unpunished responses reflects the development of tolerance to the sedative effects of the test compound. (b) Average curves of punished and unpunished responses of rats, previously treated with a BDZ, during administration of a test BDZ. Note the immediate increase in punished responses and the lack of a decline in unpunished responses, indicating pre-existing tolerance to the sedative effects of the test compound. (Based on Margules and Stein 1968)

Table 19.3 Substances that induce panic attacks in humans

Sodium lactate (mechanism unresolved)
CO_2 (inhalation) (mechanism unresolved)
Caffeine (adenosine, A_2 receptor antagonism?)
Yohimbine (α_2-adrenoceptor antagonism?)
m-Chlorophenylpiperazine (mCPP, 5-$HT_{2A/2C}$ receptor agonism?)
CCK-4 (CCK_B receptor agonism?)
FG7142 (benzodiazepine receptor inverse agonism?)

exacerbate anxiety in humans are listed in Table 19.3 and their presumed neurobiological targets have formed the bases of theories to explain the cause(s) of this disorder. A full appraisal of this topic is beyond the scope of this chapter but the links between drugs that affect central monoamine transmission and anxiety are discussed in later sections. Details of findings from research in humans can be found in Ballenger (1990) and Coupland, Glue and Nutt (1992).

DRUG TREATMENTS FOR ANXIETY

The oldest anti-anxiety agent is undoubtedly alcohol and it is certain that this drug is still routinely self-administered for this purpose. Towards the end of the eighteenth century, bromide salts were used to relieve conditions akin to anxiety despite the risk of a characteristic toxic delirium, known as 'bromism'. Alternative treatments, such as paraldehyde and chloral hydrate, were also widely used but these too had adverse effects; the former can cause psychosis but the latter is still used as a sedative and anaesthetic agent.

By the turn of the century, barbiturates (e.g. pentobarbitone, Fig. 19.3) were gradually replacing these treatments. Early reports (one as early as 1903) described a 'toxic' behavioural reaction to barbiturates that was attributed to a form of poisoning. It was not until the 1930s that it was recognised that this adverse behavioural effect of barbiturates in fact represented a drug-withdrawal syndrome (Seevers and Tatum 1931). This, together with the overt sedation caused by barbiturates, their narrow therapeutic index and their lethal toxicity in overdose, motivated the search for non-sedative anti-anxiety agents. One compound to carry such claims was meprobamate ('Miltown'; Fig. 19.3), synthesised in the 1950s. However, the initial enthusiasm over the use of this compound as a treatment for anxiety rapidly abated because it too proved to be a potent sedative and, of even more concern, it induced dependence and was widely abused.

This background set the scene for the arrival of the benzodiazepines. The first of these was chlordiazepoxide ('Librium') launched in 1960, followed by diazepam ('Valium'; Fig. 19.3). Like their predecessors, but with greater justification, these drugs were claimed to relieve anxiety at non-sedative doses (see Sternbach, Randall and Gustafson 1964). However, the benzodiazepines are members of the sedative/hypnotic group of anti-anxiety drugs, which also include alcohol, meprobamate and the barbiturates. This means that the liability of all these compounds to induce sedation, or even hypnosis (sleep), is largely a question of dose (Fig. 19.4) although it is offset by the rapid development of tolerance to their sedative effects. Like their predecessors, the

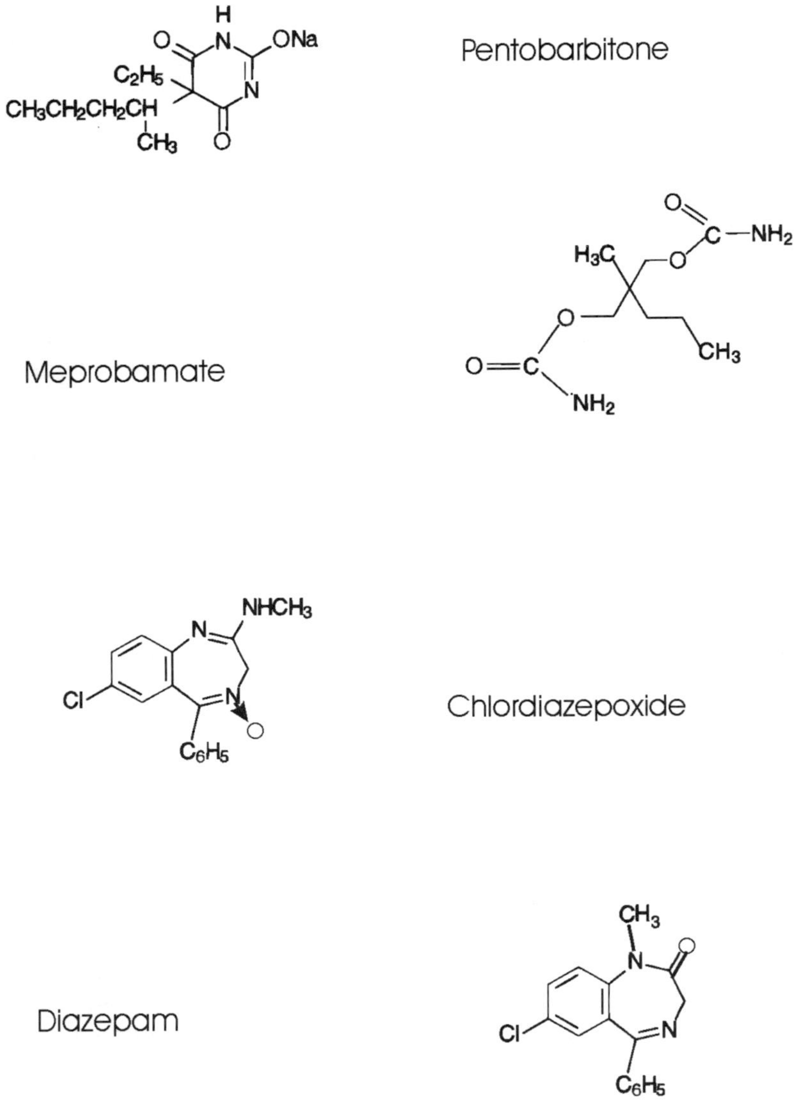

Figure 19.3 The chemical structure of some leading sedative/hypnotic anti-anxiety agents

actions of benzodiazepines are not restricted to relief of anxiety: they also induce ataxia, muscle relaxation (used in relief of muscle spasm), anterograde amnesia (used to relieve dental phobia) and increase seizure threshold (used to treat some forms of epilepsy).

BENZODIAZEPINES AND BENZODIAZEPINE RECEPTORS

MOLECULAR TARGETS FOR THE GABA$_A$ RECEPTOR

The first clues to the mechanism of action of benzodiazepines came from landmark experiments (Squires and Braestrup 1977; Moehler and Okada 1977) which showed that

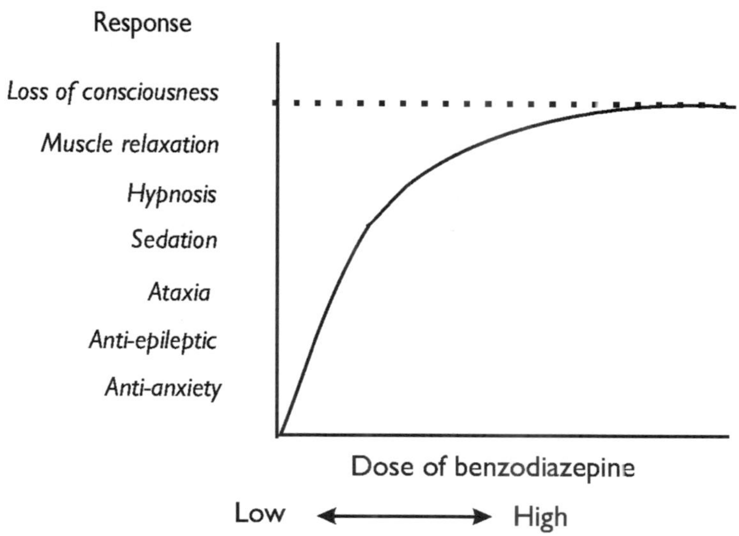

Figure 19.4 The activity spectrum of the benzodiazepines. Motor impairment and CNS depression increases with drug dose. (Based on data for chlordiazepoxide (Sternbach, Randall and Gustafson 1964))

[^{3}H]diazepam binds to a specific site in the brain. Studies of solubilised receptors confirmed that this binding site was a component of the $GABA_A$ receptor which incorporates a Cl^- channel. GABA did not compete with [^{3}H]benzodiazepine for binding to this receptor and so it was clear that their binding domains were not the same. It was soon realised that there is an allosteric interaction between them such that binding of [^{3}H]benzodiazepines is increased by GABA (Fig. 19.5). This is thought to be due to an interaction between the GABA recognition site on a β-subunit of the $GABA_A$ receptor and the benzodiazepine recognition site on an α-subunit (see Chapter 11 and Doble and Martin 1996). The overall effect of benzodiazepines is to augment the increase in Cl^- conductance caused by GABA and thereby potentiate its inhibitory actions; they achieve this by increasing the probability (and as a consequence, the frequency) of Cl^- channel opening. This action is thought to play a crucial role in the anti-anxiety effects of these drugs. The progressive increase in CNS depression, as drug dose is increased, is attributed to an increase in receptor occupancy, although the extent to which binding in different brain regions contributes to these different actions is not known.

Barbiturates bind non-competitively to yet another, functionally distinct, domain on the receptor which is thought to be directly associated with the Cl^- channel itself. Although there is an allosteric interaction with GABA, as with the benzodiazepines, barbiturates also directly increase Cl^- conductance by increasing the duration of channel opening. Thus, by contrast with the benzodiazepines, the barbiturates activate this receptor even in the absence of GABA. This explains why antagonists of the GABA binding site, such as bicuculline, block the actions of benzodiazepines, but not those of the barbiturates, and probably accounts for the greater toxicity of the barbiturates in overdose. In contrast, so-called 'cage convulsants' (e.g. picrotoxinin, pentylenetetrazol and *t*-butylbicycloorthobenzoate ('TBOB')) are thought to bind directly to another site on the Cl^- channel and to *reduce* Cl^- conductance. In recent years, the range of

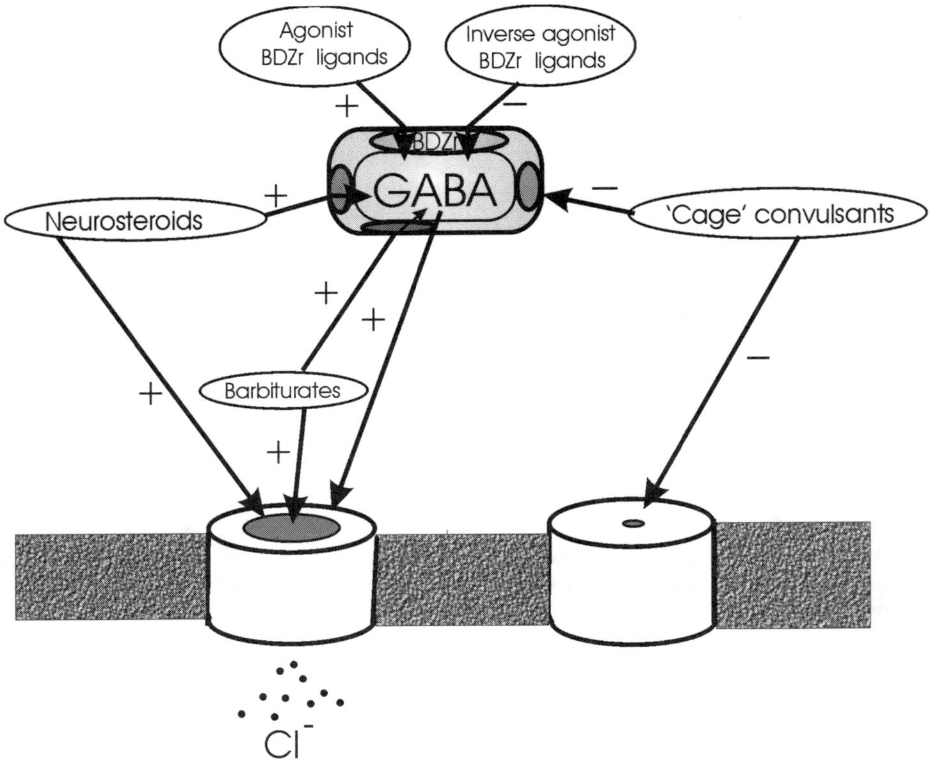

Figure 19.5 A schematic diagram of the $GABA_A$ receptor. Binding of GABA to its domain on the receptor opens a Cl^- channel. This action of GABA is augmented (+) by agonist benzodiazepines that bind to their own domain (BDZr) on the $GABA_A$ receptor and trigger an allosteric interaction with the GABA binding site. Other compounds that act in this way include the neurosteroids (e.g. allopregnanolone) and barbiturates but these compounds also bind to the Cl^- channel directly and, at high concentrations, increase Cl^- conductance in the absence of GABA. Some compounds, such as BDZr inverse agonists, have the opposite effect: i.e. they bind to the BDZr site but reduce Cl^- channel opening through a negative allosteric interaction (−) with the GABA binding site. The so-called 'cage convulsants' (e.g. picrotoxinin, pentylenetetrazol and *t*-butylbicycloorthobenzoate ('TBOB'), which also reduce Cl^- conductance, are thought to bind directly to a site on the Cl^- channel

compounds shown to have binding domains on the $GABA_A$ receptor has steadily increased and it is thought that there could be even more.

BENZODIAZEPINE RECEPTOR SUBTYPES

The discovery of the benzodiazepine receptor was quickly followed by their subdivision as evidenced by (see also Chapter 11):

(1) The biphasic dissociation of [^{3}H]flunitrazepam binding (Chiu, Dryden and Rosenberg 1982).
(2) Competition binding studies showing that when using compounds like β-CCE (ethyl-β-carboline-3-carboxylate), which bind to the benzodiazepine receptor, the displacement curve for [^{3}H]flunitrazepam was shallow in the hippocampus and

cortex. In contrast, in the cerebellum, the curve was steep with a Hill coefficient of 1 (Duggan and Stephenson 1988).

(3) Photoaffinity labelling confirming that there was more than one type of benzodiazepine receptor, rather than multiple sites on the same receptor. In this technique [^{3}H]flunitrazepam binding is carried out under ultra-violet light which renders most of the ligand binding irreversible. Purification of the radioligand/receptor complex revealed a 51 kDa protein in the cerebellum (now known to be the α1 subunit of the 'BZ$_1$ receptor') but quite different protein(s) in the hippocampus (now known to be the α2 and α3 subunits of the 'BZ$_2$ receptor') (Sieghart and Karobath 1980).

The discovery of these receptor subtypes kindled the hope that it would be possible to develop subtype-selective drugs with specific clinical actions: i.e. that it would at last be feasible to produce an anti-anxiety agent that was genuinely non-sedative. It is now known that the GABA$_A$ receptor comprises different combinations of subunits, probably in a pentameric complex (see Chapter 11), and so it might even be possible to develop drugs that target different subunits, thereby increasing their functional specificity. One such compound has already been developed, the imidazopyridine zolpidem (Fig. 19.6), which was initially regarded as a BZ$_1$ receptor ligand but is now classified as a high-affinity ligand for the α1 subunit of GABA$_A$ receptors (in contrast to α2 and α3 subunits which produce the so-called BZ$_2$ receptor). Notwithstanding this selectivity, zolpidem turns out to be a potent hypnotic agent. Zopiclone is another hypnotic, available in the clinic; this drug displaces benzodiazepines from the GABA$_A$ receptor but lacks subunit selectivity and does not seem to target precisely the same binding domain as the benzodiazepines.

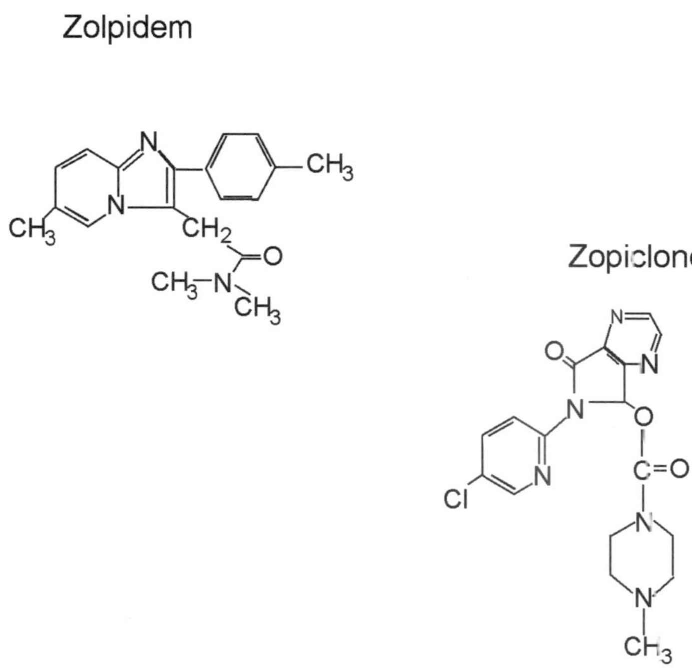

Figure 19.6 The chemical structure of the imidazopyridine and benzodiazepine (BZ$_1$) receptor ligand, zolpidem, and the cyclopyrrolone, zopiclone

Whether simple augmentation of $GABA_A$ receptor function accounts for the anti-anxiety effects of these compounds remains equivocal. If this was the case then other agents that augment GABAergic transmission such as inhibitors of GABA uptake (e.g. vigabatrin) or metabolism (e.g. tiagabine) should also have anti-anxiety effects. Indeed, there are reports of their anti-anxiety effects in patients receiving these treatments for relief of epilepsy. There is also some supporting evidence from preclinical studies but the behavioural effects of these drugs in animal models are less robust than are those of the benzodiazepines. It remains to be seen whether this is because they are just less effective anti-anxiety agents than the benzodiazepines or whether existing preclinical models show a bias that detects preferentially the anti-anxiety effects of benzodiazepines.

'PERIPHERAL' BENZODIAZEPINE RECEPTORS

In the 1980s, a further binding site for benzodiazepines was identified and, because it was first discovered in the rat adrenal gland, the term 'peripheral benzodiazepine receptor' was coined. This is regrettably confusing because this receptor has now been found in the brain also (Awad and Gavish 1987). These benzodiazepine receptors differ from those described above in a number of important respects, not least because they do not affect, nor are they affected by, GABA binding. They also have their own specific ligands: the isoquinolone, PK 11195, and the benzodiazepine, Ro 4864, neither of which binds to the $GABA_A$ receptor. Moreover, the benzodiazepine, clonazepam, which is a high-affinity, partial agonist ligand for the benzodiazepine domain on the $GABA_A$ receptor, does not bind to the 'peripheral' receptor.

Another difference is that these peripheral benzodiazepine receptors are located mainly intraneuronally, on mitochondrial outer membranes, rather than on the plasma membrane. In the brain, they are associated with glial cells but in the periphery they are found in a range of tissues, including mast cells and platelets. Their function is still a matter of intense debate but one possibility is that they regulate cholesterol uptake and, secondary to this, the synthesis of neurosteroids (Do Rego *et al.* 1998). Since neurosteroids also have a binding domain on the $GABA_A$ receptor, there might be some indirect functional coupling between these two types of receptors after all. To some extent, the possibility of such an interaction is supported by evidence that the density of peripheral benzodiazepine receptors differs in inbred strains of rats which are distinguished by their behavioural reactivity ('fearfulness') to novel stimuli (Drugan *et al.* 1987). Other possible, albeit controversial, functions of these receptors are reviewed in Doble and Martin (1996).

OTHER LIGANDS FOR THE BENZODIAZEPINE RECEPTOR

Among the early indications that the benzodiazepines were not the only compounds to bind to the benzodiazepine receptor were findings that emerged from a search for an endogenous ligand for this receptor site. This effort produced a non-benzodiazepine ligand, ethyl-β-carboline-3-carboxylate (β-CCE), and this pointed the way to a whole family of compounds that are high-affinity ligands for this receptor. However, not all turned out to share the properties of the protypical benzodiazepines (anti-anxiety, anticonvulsant, etc.). Some, including β-CCE itself, had the opposite effects in animals: i.e. they *induced* anxiety and *reduced* seizure threshold and some, such as 3-carbomethoxy-4-ethyl-6,7-dimethoxy-β-carboline (DMCM), caused overt seizures.

These new recruits to the activity spectrum were named 'inverse agonists' and subsequent studies confirmed that they reduce the affinity of GABA for its binding site on the $GABA_A$ receptor and attenuate the $GABA_A$ receptor-mediated increase in Cl^- conductance (Fig. 19.5).

THE BENZODIAZEPINE RECEPTOR AGONIST/INVERSE AGONIST SPECTRUM

The rich portfolio of compounds that bind to the benzodiazepine receptor includes many compounds which, despite not being benzodiazepines, share the properties of the prototypical benzodiazepines, chlordiazepoxide and diazepam. However, all these groups of compounds, including the benzodiazepines themselves, span the activity spectrum: from full inverse agonist to full agonist. In between these extremes are compounds which have either partial agonist or partial inverse agonist activity and some are antagonists (Fig. 19.7). This spectrum of actions reflects the overall effects of these drugs on native receptors and is usually assessed in whole animals. However, the synthesis of receptors comprising different combinations of subunits has shown that the activity of these drugs depends greatly on subunit composition. For instance, $GABA_A$ receptors have been characterised to which diazepam does not bind at all (see Chapter 11).

The first antagonist to be developed was the (imidazo)benzodiazepine, flumazenil. This compound blocks the actions of both agonists and inverse agonists *in vitro*. It will

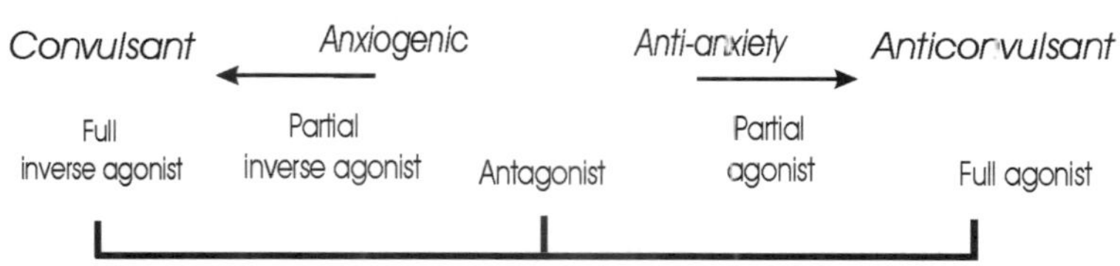

	Full inverse agonist	Partial inverse agonist	Antagonist	Partial agonist	Full agonist
Benzodiazepines	Ro19-4603	Sarmazenil	Flumazenil	Bretazenil	Diazepam
β-carbolines	DMCM	FG7142	ZK93426	Abecarnil	ZK94323
Cyclopyrrolones				Rp60503	Zopiclone
Imidazopyrimidines	Ru34000			Ru32698	
Imidazopyridines		S-8510		Alpidem	Zolpidem
Phenylquinolines				Pk8165	
Pyrazoloquinolinones			CGS8216	CGS9895	
Triazolopyridazines				Cl218872	

Figure 19.7 The activity spectrum for different generic groups of compounds that bind to the 'benzodiazepine' domain on the $GABA_A$ receptor

also block the effects of these agents *in vivo*, but there is some argument over whether it is a true antagonist: i.e. whether it really lacks any intrinsic activity (as would be required of an antagonist) or whether it is merely a weak partial agonist. The subunit composition of the $GABA_A$ receptor could be one confounding factor in resolving this question. For instance, flumazenil has been reported to *augment* the action of GABA at cloned receptors comprising α_4 β_2 γ_2 subunits. Apparent effects of the antagonist *in vivo* could also depend on whether there is any tonic activation of the benzodiazepine receptor by an endogenous ligand. Flumazenil is available in the clinic for intravenous infusion to reverse benzodiazepine-induced sedation (e.g. in the post-anaesthetic context) or coma (after overdose). However, because it has a half-life of only 1 h in humans, it is only of realistic benefit in reversing the actions of agonist benzodiazepines with a short half-life, such as midazolam.

The potential benefits of benzodiazepine partial agonists are as non-sedative, anti-anxiety agents. Because of their low efficacy, it was predicted that a partial agonist should not induce sedation even if their receptor occupancy exceeds that normally required for an anti-anxiety effect when using a full agonist. One such compound, bretazenil, has been developed but failed to reach the clinic because it displayed some sedative activity and, more problematic, there were end-of-dose rebound effects that were undoubtedly exacerbated by its short half-life. Currently, the partial agonist, abecarnil (a β-carboline), is undergoing clinical trials. For the current status of the development of partial agonists and other promising benzodiazepine receptor ligands see Cheetham and Heal (2000).

Even benzodiazepine inverse agonists might yet find some useful applications such as in the relief of cognitive deficits (which are increased by benzodiazepine full agonists) (Abe, Takeyama and Yoshimura 1998). With the rapidly expanding understanding of different combinations of subunits that comprise the $GABA_A$ receptor, it is hoped to develop compounds that target specific subunit combinations and improve cognitive function in dementia but which lack any proconvulsant or anxiogenic actions.

THE QUEST FOR THE ENDOGENOUS BENZODIAZEPINE RECEPTOR LIGAND ('ENDOZEPINES')

The discovery of the opioid receptor, followed by isolation of endogenous opioids, provided the impetus for a search for an endogenous ligand for the established benzodiazepine receptor. Although many candidates have emerged (De Robertis *et al.* 1988; Table 19.4), most are present in the CNS at concentrations far too low for them to be feasible endogenous modulators of $GABA_A$ receptor function. However, three candidates have been given prominent attention, albeit for different reasons, and are worthy of mention.

Table 19.4 Putative endogenous ligands for the benzodiazepine binding domain on the $GABA_A$ receptor

β-carbolines (β-CCB)
Desmethyldiazepam
'Endozepines' (unknown chemical structure)
Peptides (nepenthin, octodecaneuropeptide)
Purines (inosine, hypoxanthine, guanosine, nicotinamide)
Thromboxane A_2

The first, β-CCE, was the product of an arduous attempt to isolate an endogenous ligand from human urine. Although subsequently found to be an artefact of the extraction process, this compound turned out to be a ligand for the benzodiazepine receptor, nonetheless, and was the first inverse agonist to be identified. The anxiogenic effects in humans of its more stable congener, FG 7142, are described graphically in a report by Dorow *et al.* (1983). β-Carbolines are realistic candidates for an endogenous ligand because they can be synthesised in the brain (Han and Dryhurst 1996) but, although other members of this group of compounds have at various times been suggested to fulfil the role of an endogenous ligand, none has been confirmed as such.

Another, more recent, candidate is an endogenous propeptide, 'diazepam binding inhibitor' (also known as Acyl-CoA Binding Protein (DBI/ACBP)), which yields 'octodecaneuropeptide' ('ODN') and 'triakontatetraneuropeptide' ('TTN') (Costa and Guidotti 1991). Both these peptides are neuroactive and ODN turns out to have inverse agonist activity at $GABA_A$ receptors both *in vivo* and *in vitro* and to have marked effects on behaviour (e.g. Reddy and Kulkarni 1998). However, there is scepticism as to whether the brain can manufacture sufficient peptide to regulate the ubiquitous $GABA_A$ receptor on a moment-to-moment basis. Currently, the binding of TTN to the peripheral benzodiazepine site, and its effect on neurosteroid synthesis, is attracting greater interest (Do Rego *et al.* 1998).

Finally, the presence in human post-mortem brain tissue of the active metabolite of diazepam, desmethyldiazepam, raised some curiosity and frank alarm (Sangameswaran *et al.* 1986). At the time of its discovery in the brain it was thought that there was no enzyme system capable of producing such halogenated compounds and that its presence in the brain reflected dietary intake from an environment contaminated by overuse of its parent compound. However, its discovery in stored brain tissue which had been obtained before the synthesis of the benzodiazepines allayed these fears. It is now thought possible that some benzodiazepines, including desmethyldiazepam, occur naturally and that they are taken in as part of a normal diet (Table 19.5).

Although, by analogy with the opioids, one would expect there to be an endogenous ligand for the widely distributed benzodiazepine receptor, its existence remains uncertain and we must be alert to the possibility that any such ligand(s) could have either agonist or inverse agonist activity.

Table 19.5 Examples of plants containing ligands for benzodiazepine receptors

Plant source	Active agent(s)
Valeriana officinalis	Hydroxypinoresinol (a lignan)
Hypericum perforatum L. Hypericaceae (St John's Wort)	Unknown
Matricaria recutita L.	5,7,4′-trihydroxyflavone (apigenin)
Passiflor coeruleus L.	Chrysin
Wheat grain	Diazepam, desmethyldiazepam, lormetazepam
Potato	Diazepam, desmethyldiazepam, lormetazepam
Karmelitter Geist	Amentoflavon

ENDOGENOUS LIGANDS AND BENZODIAZEPINE RECEPTORS: AN EXPLANATION FOR THE CAUSE OF ANXIETY

The undisputed efficacy of benzodiazepines in relief of anxiety led to the question of whether this disorder could arise from abnormal concentrations in the brain of an endogenous ligand or a malfunction of the benzodiazepine/GABA receptor system. An important study, aimed at distinguishing between these possibilities, has been carried out in humans (Nutt *et al.* 1990) and was based on the premise that anxiety could be caused by either:

(1) Inadequate activity of an endogenous ligand which is a benzodiazepine receptor agonist and suppresses anxiety. In this case, the administration of the antagonist, flumazenil, should induce anxiety in normal subjects and exacerbate anxiety in anxious patients.
(2) Excessive activity of an endogenous ligand which is a benzodiazepine receptor inverse agonist and induces anxiety. In this case, the administration of flumazenil should relieve anxiety in anxious patients and have no, or sedative, effects in healthy subjects.
(3) Dysfunction of the $GABA_A$ receptor complex such that the effects of all benzodiazepine receptor ligands are shifted in the direction of inverse agonism. In this case, flumazenil (which normally has zero efficacy) should induce anxiety in anxious patients but have no effects in healthy subjects because they have normal receptors.

To distinguish between these possibilities, flumazenil was administered to panic patients and control subjects. The results of the experiment were consistent with the third possibility: flumazenil induced panic attacks in 8 of 10 patients but not in control subjects (Fig. 19.8). Unfortunately, the change(s) in the benzodiazepine receptor or its coupling to the rest of the $GABA_A$ receptor are unknown, as are the stimuli that could explain this functional change. Recent studies suggest that the binding of $[^{11}C]$flumazenil is abnormally low in panic patients (Malizia *et al.* 1998), although this finding does not relate in any obvious way to the '$GABA_A$ receptor shift' hypothesis. However, this is the only tested theory so far to connect panic anxiety directly with a disorder of the $GABA_A$ receptor. The receptor shift theory could also explain why benzodiazepines are ineffective in treating panic disorder but, because these drugs do effectively relieve generalised anxiety, it seems that the theory might explain the origin of the former, but not the latter disorder, and that they have different causes.

MONOAMINES IN ANXIETY

NORADRENALINE

The first suggestion that abnormal noradrenergic transmission was linked with anxiety came from Redmond's laboratory in the 1970s when he drew attention to the similarities in the symptoms and signs of anxiety with those of the acute stress response (Redmond and Huang 1979). He went on to stimulate the locus coeruleus of (chair-restrained) monkeys and showed that this caused behavioural changes, some of which resembled a cluster of behaviours displayed by the animals when under threat. This work led to the proposal that anxiety was due to (or exacerbated by) excessive

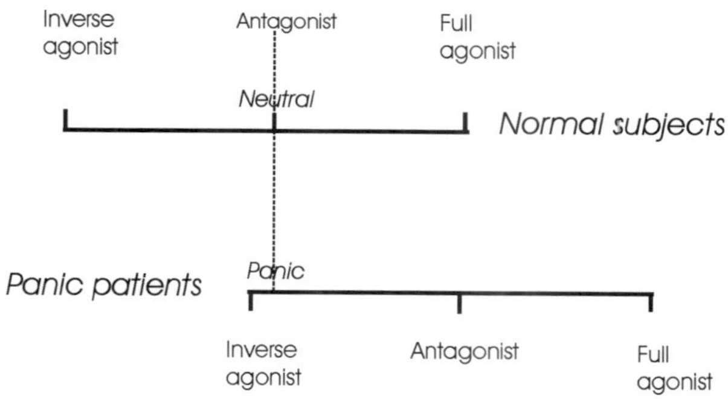

Figure 19.8 A schematic representation of the $GABA_A$ receptor shift hypothesis. This proposes that patients with panic disorder have dysfunctional $GABA_A$ receptors such that the actions of drugs that behave as antagonists in normal subjects are expressed as inverse agonism in panic patients. It is unlikely that this theory extends to generalised anxiety disorder (GAD), for which benzodiazepine agonists are highly effective treatments, but it could explain why these drugs are relatively ineffective at treating panic disorder. (Based on Nutt *et al.* 1990)

noradrenergic transmission in the brain. This hypothesis has survived, largely unchallenged, for over 20 years.

Redmond's explanation of anxiety has been underpinned by investigations of changes in noradrenaline release in humans given the anxiogenic drug, yohimbine. This is an α_2-adrenoceptor antagonist that increases the firing rate of, and release of noradrenaline from, noradrenergic neurons by blockade of presynaptic α_2-adrenoceptors on the neuronal cell bodies and terminals, respectively. Increases in noradrenaline release, inferred from measurement of the noradrenaline metabolite, 3-methoxy, 4-hydroxyphenylglycol (MHPG), in plasma, have shown that the noradrenergic response in panic patients who experience a panic attack with yohimbine is greater than that in either panic patients who do not express this response or in normal patients (Bremner *et al.* 1996). Unfortunately, the noradrenergic response to yohimbine is not exaggerated in patients with GAD, suggesting that the aetiology of this form of anxiety could differ from that of panic disorder. Nevertheless, the α_2-adrenoceptor agonist, clonidine, which has the opposite effect to yohimbine on noradrenergic neurons, is sometimes used to relieve anxiety, especially that associated with alcohol and opiate withdrawal. However, it is not a viable long-term treatment for anxiety because of its effects on the cardiovascular system.

One complication with the above concept is that, in some brain regions, the majority of α_2-adrenoceptors are postsynaptic and so a *reduction* in α_2-adrenoceptor-mediated noradrenergic transmission, after treatment with yohimbine, cannot be ruled out as a causal factor for the anxiety induced by this drug. Another problem is that yohimbine is also a 5-HT_{1A} receptor agonist, a 5-HT_{1D} partial agonist and a 5-HT_{2B} antagonist; all these receptors could be involved in anxiety (see below). Furthermore, a finding that argues against excessive noradrenergic transmission as a cause of anxiety is that stimulating the locus coeruleus in humans induces a pleasant sensation, rather than anxiety (Libet and Gleason 1994) and that not all anxiogenic challenges increase plasma MHPG (Silverstone *et al.* 1994).

Measurements of noradrenaline release in animals have not helped to resolve this confusion. Microdialysis studies *in vivo* have confirmed that anxiogenic doses of yohimbine do increase the extracellular concentration of noradrenaline in the frontal cortex of rats but an anxiogenic dose of the benzodiazepine inverse agonist, FG 7142, does not. Moreover, anxiogenic doses of these two drugs have opposite effects on the rats' response to a novel environment: the increase in noradrenaline efflux caused by this aversive stimulus is *reduced* by yohimbine but *increased* by FG 7142 (Mason, Heal and Stanford 1998). A further complication is that, despite its anxiogenic effects, yohimbine actually *augments* the anti-anxiety effects of benzodiazepines in the 'conflict' test (Söderpalm, Blomqvist and Söderpalm 1995) while a more selective α_2-adrenoceptor antagonist, idazoxan, has anti-anxiety, rather than anxiogenic, effects in animal models of conflict (La Marca and Dunn 1994).

These observations question the role of noradrenaline as an initiator of anxiety as does the finding that the anti-anxiety drug, buspirone (see Chapter 9), increases the concentration of noradrenaline in the extracellular fluid in the frontal cortex of freely-moving rats (Done and Sharp 1994). Whether this is because buspirone is metabolised to 1-(2-pyrimidinyl)-piperazine (1-PP), which is an α_2-adrenoceptor antagonist, is uncertain. Unfortunately, no studies have investigated the effects of chronic administration of this drug on noradrenergic transmission; this could be important because, unlike benzodiazepines, buspirone is effective therapeutically only after several weeks of treatment.

The finding that infusion of the β-adrenoceptor agonist, isoprenaline, has an anxiogenic effect in humans implicates this receptor subytype also but little (if any) isoprenaline crosses the blood–brain barrier and so any anxiogenic effects are likely to be an indirect consequence of the autonomic arousal it will cause (i.e. increased heart rate, reduced salivation, etc.). Of course, this alone does not rule out a role for these receptors in the psychological component of anxiety. It has long been claimed that peripheral responses can serve as 'interoceptive cues' and cause secondary (anxiogenic) changes in the brain. This is the 'James–Lange hypothesis' which broadly suggests that we experience anxiety because of an increase in heart rate and a dry mouth, rather than the other way round. Blocking such β-adrenoceptor-mediated effects in anxiety would also explain the 'anti-anxiety' effects of their antagonists (e.g. propranolol) because β-adrenoceptors are postsynaptic in the CNS and so their antagonists would blunt noradrenergic transmission. However, one complication is that propranolol, like many other β-adrenoceptor antagonists, is also a 5-HT_{1A} receptor antagonist which could contribute to its anti-anxiety effects (see below). Another is that β-adrenoceptor antagonists are only of clear benefit in prevention of 'situational' anxiety ('competition nerves') which should really be regarded as a normal stress response rather than an anxiety disorder. Indeed, subjects claim that, whereas these drugs relieve the peripheral manifestations of anxiety, they have no appreciable effects on its psychological component. This is supported by evidence that these drugs are of little, if any, long-term benefit in GAD (Nutt 1990) and that they can even exacerbate this condition.

If excessive noradrenergic transmission is a causal factor in anxiety, then it would be predicted that a lesion of central noradrenergic neurons would have an anti-anxiety effect in behavioural models of this condition. Unfortunately, the behavioural effects of such lesions are notoriously inconsistent and there are many reports of negative findings (e.g. Salmon, Tsaltas and Gray 1989). One study has even shown that a lesion of central noradrenergic neurons, induced by the selective neurotoxin, DSP-4, abolishes the anti-anxiety effects of tricyclic antidepressants and MAO inhibitors, but not those of the benzodiazepine, alprazolam, or the barbiturate, phenobarbitone (Fontana,

McMiller and Commissaris 1999). This suggests that the central noradrenergic system is actually needed to express the anti-anxiety effects of some drugs, but not others.

Finally, many early studies suggested that benzodiazepines attenuate the increase in turnover of noradrenaline in the brain caused by stressful stimuli such as footshock and restraint (Taylor and Laverty 1969; reviewed by Stanford 1995). It has also been reported that they prevent the phasic increase in firing rate of neurons in the locus coeruleus caused by such stimuli (Rasmussen and Jacobs 1986). These actions would certainly support Redmond's theory. However, recent microdialysis studies suggest that the actions of these drugs might not be so straightforward. Anti-anxiety doses of the benzodiazepine, diazepam, reduced spontaneous efflux of noradrenaline but had no effect on the noradrenergic stress response on exposure to a novel environment (Dalley, Mason and Stanford 1996). This points to an important limitation of many studies in this area, namely that stimuli used to investigate the neurochemical effects of test anti-anxiety drugs are usually crude and involve somatosensory stress, often involving physical discomfort. This approach disregards the criteria, defined by Gray (1987; discussed above) for the types of environmental stimuli that trigger anxiety in rodent models or humans. In fact, surprisingly few studies have investigated the effects of *anxiogenic* stimuli on neurochemical changes in the brain. This is probably because techniques of sufficient sensitivity to detect the neurochemical changes provoked by these procedures have been developed only recently.

In one such study, using *in vivo* microdialysis, exposure to an aversive novel environment (a brightly lit, novel arena) increased the concentration of extracellular noradrenaline (suggestive of increased noradrenaline release) in both the rat frontal cortex and the hypothalamus. However, if animals were trained to associate the sound of a tone (which becomes a conditioned cue) with imminent transfer to the aversive environment, a different pattern of noradrenaline responses ensued. After a series of such conditioning trials, the sound of the tone alone increased the concentration of extracellular noradrenaline in the rat frontal cortex, but not the hypothalamus (McQuade and Stanford 2000). This suggests that the noradrenergic innervation of these two brain areas might have different roles in the response to conditioned and unconditioned aversive environmental stimuli. They also suggest that noradrenergic neurons innervating the frontal cortex are recruited in the response to anxiogenic environmental signals (of the type described by Gray) whereas those projecting to both brain regions could have a role in coordinating or triggering the flight/fight response to unconditioned stimuli. Clearly, different components of the central noradrenergic system could have different roles in anxiety, a possibility that is considered in more detail later.

5-HYDROXYTRYPTAMINE

It has been known for many years that aversive stimuli increase serotonergic transmission (reviewed by Chaouloff 1993) and so it was inevitable that exaggerated serotonergic transmission in a hypothetical brain 'punishment system' became linked with anxiety. Unfortunately, much of the evidence for this idea was gleaned from unreliable measures of changes in 5-HT concentration in rodent brain tissue post-mortem after experience of moderately severe forms of stress *in vivo*. Nevertheless, this concept was encouraged by reports that a reduction in 5-HT transmission, following administration of either the 5-HT synthesis inhibitor, *p*-chloroamphetamine (*p*CPA), or the selective neurotoxin, 5,7-dihydroxytryptamine (5,7-DHT), usually increased responding in the punished (anxiety) phase of a 'conflict' test.

Equally inevitably, many confounding factors have come to light which undermine this simplistic explanation of anxiety. For instance, 5-HT release in the brain suppresses food and fluid intake and so its depletion will distort results from any procedure that relies on changes in ingestive behaviour (such as conflict tests). In contrast, benzodiazepines increase food intake. Ethological models avoid this problem but offer others instead: e.g. both *p*CPA and 5,7-DHT neurotoxic lesions of 5-HT neurons in the dorsal Raphé nucleus increase social interaction and yet lesions of neurons in the median Raphé nucleus decrease it. Moreover, *p*CPA exaggerates the flight/fight response induced by stimulation of the periaquaductal grey (PAG), an action that is hard to reconcile with the anti-anxiety effect of this drug in the social interaction test. Studies using *in vivo* microdialysis have complicated the picture even more: e.g. the concentration of extracellular 5-HT is increased in the forebrain of rats tested in the plus-maze and yet isolation rearing, which increases animals' behavioural reactivity to the plus-maze, abolishes the increase in extracellular 5-HT (Marsden *et al.* 1993). Another complication is that different serotonergic nuclei project to different brain areas (reviewed by Jacobs and Azmitia 1992) and it is now thought that different components of the brain 5-HT system have different roles in behaviour (see below). Finally, the wide variety of responses to 5-HT mediated by its many receptor subtypes is problematic (see Barnes and Sharp 1999), especially when coupled with the dubious selectivity of most test drugs.

In recent years, attention has been directed to the azapirones such as gepirone, ipsapirone and, in particular, buspirone since this is the only one which is available clinically as an anti-anxiety agent. It was developed as a neuroleptic, because it is a dopaminergic D_2 receptor antagonist, but turned out to be more effective as a treatment for anxiety. In preclinical models, buspirone generally produces behavioural changes that are consistent with an 'anti-anxiety' effect but this has been much harder to demonstrate than for the benzodiazepines. These anti-anxiety effects of buspirone were thought to rest on its activation of inhibitory 5-HT_{1A} autoreceptors on serotonergic cell bodies in the Raphé nuclei (Dourish, Hutson and Curzon 1986). The ensuing reduction in the firing rate of serotonergic neurons and 5-HT release in the terminal field fitted well with the idea that a hyperresponsive serotonergic system was a causal factor in anxiety. Evidence that both *p*CPA and 5,7-DHT lesions abolish the anti-anxiety effects of this drug in conflict tests (Eison *et al.* 1986) was deemed to support this scheme, although it is hard to understand why such a lesion should abolish the effects of a drug that is supposed to work through inhibition of 5-HT release.

One problem with the idea that the anti-anxiety effects of buspirone are mediated by its activation of 5-HT_{1A} autoreceptors is that it disregards the postsynaptic 5-HT_{1A} receptors which are found in many key areas of the limbic system (e.g. frontal cortex, hippocampus and hypothalamus). Buspirone would be expected to activate these receptors, regardless of any inhibition of neuronal firing rate and 5-HT release. A solution to this problem was offered by the suggestion that buspirone is a full agonist at presynaptic receptors but only a partial agonist at postsynaptic sites. An alternative suggestion was that there is a greater receptor reserve on cell bodies than postsynaptically. In either case, a reduction of neuronal firing mediated by buspirone's activation of inhibitory autoreceptors would be accompanied by minimal postsynaptic effects.

More recently, evidence favouring an anti-anxiety effect mediated by activation of postsynaptic 5-HT_{1A} receptors has emerged. However, it must be remembered that, since buspirone is a partial agonist, its postsynaptic effects will depend on the degree of tonic activation of the target receptor(s). Indeed, this could account for the many conflicting reports in the literature. It would certainly explain why a neurotoxic lesion of serotonergic neurons has no effect on the anti-anxiety effects of ipsapirone in the conflict test (Przegalinski, Chojnacka-Wojcik and Filip 1992) whereas buspirone prevents the anxiogenic effects of 5-hydroxytryptophan, which increases 5-HT transmission, in rats placed in a novel environment. Antagonists of 5-HT_{1A} receptors (e.g. WAY100635) have now been developed as well and, over a limited range of doses, they too have anti-anxiety effects in preclinical models (Cao and Rodgers 1997). Unfortunately, it is still unclear whether this is due to a presynaptic action (i.e. an increase in 5-HT release and transmission) or postsynaptic action (i.e. a reduction in 5-HT_{1A}-mediated transmission) (see Beckett and Marsden 1997).

Overall, the extent to which postsynaptic receptors contribute to the actions of 5-HT_{1A} receptor agonists seems to depend on the behavioural test used, whether the drugs are administered systemically or locally and, if the latter, into which brain region (see Handley 1995). Moreover, an important limitation of much of this work is that buspirone is effective in humans only after prolonged administration and yet most experimental studies have investigated its behavioural sequelae only after acute drug administration. The outcome of the few chronic studies that have been attempted seems to differ across different models, with buspirone being ineffective in the plus-maze but effective in conflict tests (see Handley 1995). An understanding of the effects of chronic treatment with buspirone is a notable gap in the field because it is possible that its anti-anxiety effects rest on long-latency changes in receptor populations that culminate in a shift in the balance of 5-HT responses from one set of receptors to another.

Unfortunately, research in humans has not helped to resolve these difficulties. There is some evidence which implicates increased 5-HT transmission as a causal factor in anxiety: For instance, some patients experienced anxiety when given the anti-obesity agent, fenfluramine (now withdrawn), which increases release of 5-HT in the brain. Also, *m*-1-(3-chlorophenyl) piperazine (*m*CPP), which is a 5-HT_{2C} partial agonist (and has some, albeit even lowers efficacy, at 5-HT_{2A} and 5-HT_{1B} receptors) reliably induces panic in patients. Finally, unlike other antidepressants, trazodone is ineffective in treatment of panic disorder, probably because it is metabolised to *m*CPP which is anxiogenic. On the other hand, there is no consistent evidence that administration of the 5-HT precursor, 5-hydroxytryptophan, causes anxiety in humans and selective serotonin reuptake inhibitors (SSRIs), which are thought to increase 5-HT transmission after chronic administration, do not cause anxiety disorders. Indeed they are now used routinely to treat them. In short, evidence for either an excess or a deficit in serotonergic transmission as a causal factor in anxiety in humans is equally (un)convincing (Bell and Nutt 1998).

Currently, hopes for compounds with greater clinical efficacy and faster onset of action than buspirone rest on the development of selective ligands for 5-HT receptors. So far, antagonists of 5-$HT_{2A/C}$ (e.g. ritanserin), 5-HT_3 (e.g. ondansetron) and 5-HT_4 (e.g. zacopride) receptors have all been explored but their anti-anxiety effects are, at best, equivocal. Full appraisals of the role of 5-HT systems in anxiety and the actions of anti-anxiety drugs are to be found in Handley (1995), Barnes and Sharp (1999) and Olivier, van Wijngaarden and Soudijn (2000).

INTEGRATED THEORIES OF ANXIETY

The evidence outlined so far does little to explain how monoamines or anti-anxiety drugs might influence anxiety states. To achieve this, an integrated view of the relevant brain systems is required, together with an appreciation of how their function is regulated.

One scheme focuses on the roles of the septum and hippocampus. Detailed justification of this theory is beyond the scope of this chapter but can be found in Gray (1987). Briefly, the 'septohippocampal system' is thought to form part of a neuronal network that functions as a 'comparator', i.e. it compares anticipated and actual stimuli. It is envisaged that, when the comparator detects a mismatch between events that are suggested by 'signals' and prevailing stimuli (as in novelty, conflict or frustrative non-reward), a 'behavioural inhibition system' is activated by the septohippocampal system (Fig. 19.9). This system arrests ongoing behaviour and increases vigilance, as is evident in animal models of anxiety (e.g. suppression of rewarded responses or exploration). Ascending noradrenergic and serotonergic inputs are thought to activate this behavioural inhibition system, with these two monoamines playing complementary roles. Moreover, there is extensive evidence that anti-anxiety drugs prevent activation of the behavioural inhibition system by blunting monoaminergic transmission in the hippocampus.

An additional ('defence') system was proposed as early as the 1960s which mediates the flight and fight response. This comprises the amygdala, hypothalamus and central grey in the midbrain. It is generally agreed that the periaquaductal grey area (PAG) of the central grey is responsible for eliciting the flight/fight response which incorporates autonomic changes and analgesia as well as the locomotor response. Gray (1987) proposes that the central grey is normally inhibited by the (ventromedial) hypothalamus and that the influence of the hypothalamus is governed in opposing ways by the behavioural inhibition system and the amygdala. Whereas the former augments hypothalamic inhibition of the flight/fight response, the latter inhibits it, thereby releasing the flight/fight response (Fig. 19.9).

A more recent hypothesis, which incorporates many features of Gray's hypothesis, has concentrated on the central serotonergic system and proposes that different serotonergic pathways underlie GAD and panic (Fig. 19.10). This theory, like that described above, focuses on the amygdala as part of the neuronal 'defence' system and highlights evidence for its key role in the response to conditioned fear. The amygdala is thought to be a major target for conditioned sensory inputs and to organise the conditioned fear response (LeDoux and Muller 1997); this is effected by its connections to the hypothalamus and PAG. Different zones of the PAG seem to evoke different components of this response: whereas stimulation of the dorsal PAG (dPAG) evokes 'explosive running', the ventral PAG (vPAG) is responsible for 'freezing', both of which are common features of a panic attack.

Serotonergic neurons, originating in the dorsal Raphé nucleus (DRN), innervate both the amygdala and the PAG. In the former region, they are thought to augment active avoidance of aversive signals by exaggerating the amygdalar response to conditioned aversive stimuli (Deakin and Graeff 1991; Graeff *et al.* 1996). Excessive serotonergic activity of neurons originating in the DRN is proposed to underlie anticipatory (or 'learned') anxiety which is regarded as akin to GAD. This response could be modulated, at the level of both the DRN and the amygdala, by neuronal inputs from both the frontal

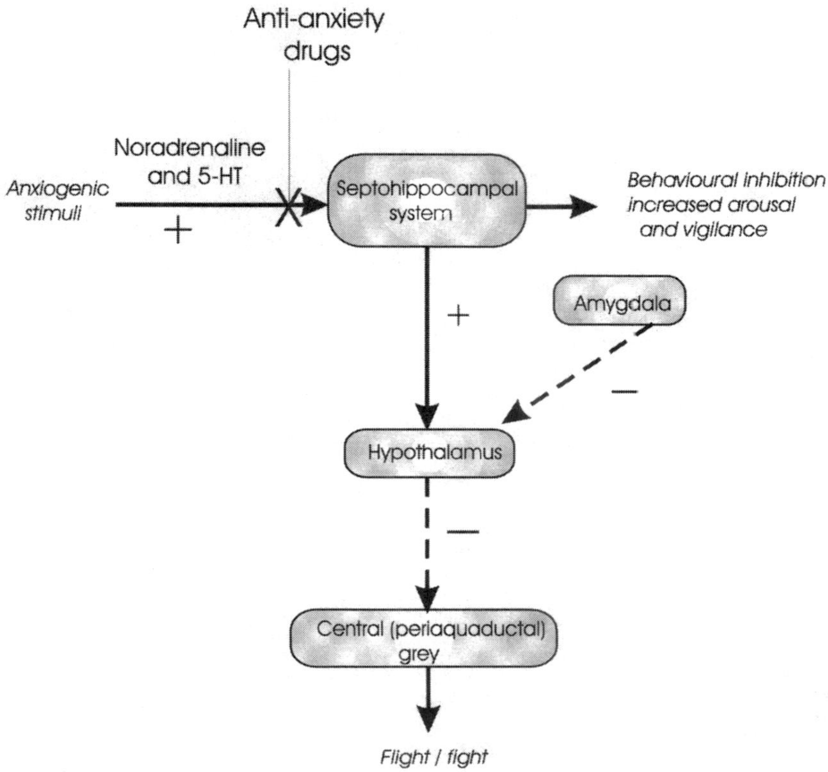

Figure 19.9 A schematic representation of key elements of Gray's explanation for anxiety and the flight/fight response. Noradrenergic and serotonergic inputs to the septohippocampal component of a neuronal 'comparator' activate a behavioural inhibition system which suppresses ongoing behaviour and increases vigilance ('anxiety'). Inputs from the behavioural inhibition system also augment the activity of the (ventromedial) hypothalamus which suppresses the flight/fight response generated in the periaquaductal grey. In contrast, the amygdala inhibits hypothalamic activity and releases the flight/fight response. Anti-anxiety drugs are thought to inhibit monoaminergic activation of the behavioural inhibition system

cortex, which is thought to process the perception of sensory information, and the hippocampus, which processes contextual (environmental) cues.

Deakin and Graeff (1991) further propose the existence of a pathway, again arising in the DRN, which inhibits activation of the PAG. It is suggested that a reduction of serotonergic transmission in this area releases the flight/fight response. Under normal conditions, activity in this system is governed by higher centres in the forebrain (the cortex and hippocampus) so that, when interpretation of prevailing stimuli deems it appropriate, the flight/fight response is suppressed. A deficit in serotonergic inhibition of the PAG is thought to be the origin of panic. There are several ramifications of this interesting theory. For instance, during low arousal states, a decline in the activity of forebrain serotonergic systems would diminish the inhibition of the PAG. This would ensure that threatening stimuli would evoke a protective escape response by default until cortical systems switch off the PAG response, if appropriate, as arousal increases (Handley 1995). It could also explain why patients often report that they are woken up during the night by their panic attacks.

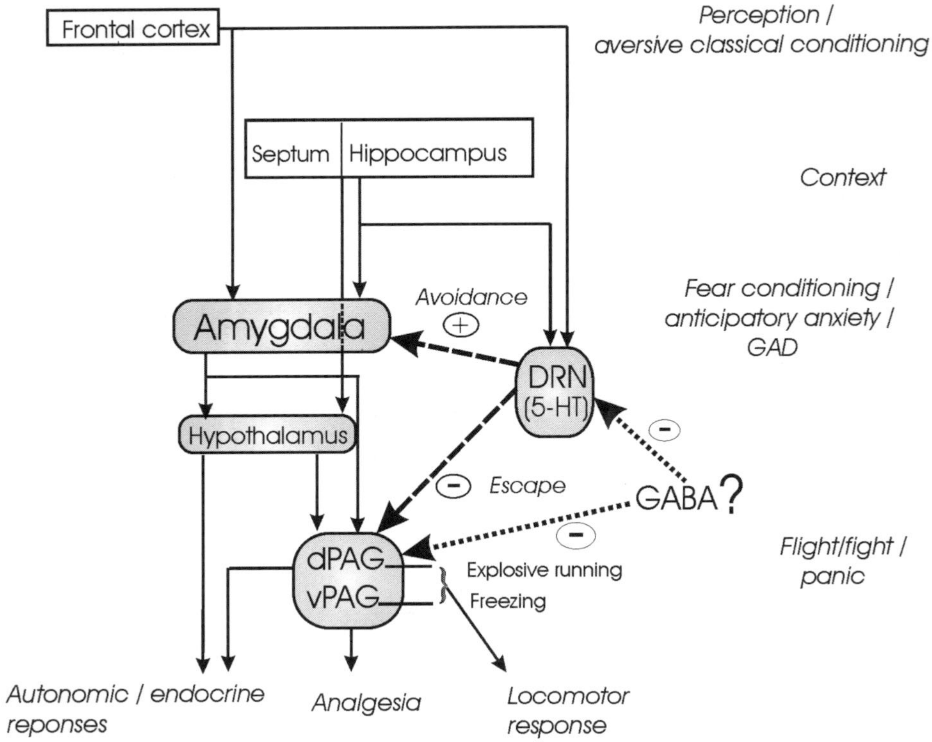

Figure 19.10 The influence of 5-HT pathways (represented by the dashed lines) projecting from the dorsal Raphé nucleus (DRN) on brain regions comprising the 'brain defence system'. This system comprises the amygdala, hypothalamus and periaquaductal grey (PAG) and coordinates behavioural and neuroendocrine responses to conditioned and unconditioned aversive stimuli. Activity within the defence system is governed by higher centres, such as the frontal cortex and hippocampus. Serotonergic neurons projecting from the dorsal Raphé nucleus are proposed to activate the amygdala (+) thereby promoting the response to conditioned aversive stimuli (anxiety). Projections from this nucleus to the dorsal and ventral periaquaductal grey (dPAG and vPAG) are thought to suppress (−) the flight/fight response to aversive stimuli. A deficit in serotonergic transmission to this brain region is thought to underlie panic. Possible targets for anti-anxiety drugs, acting via the $GABA_A$ receptor, are indicated by the dotted arrows. See text for further details

There is a good deal of evidence that postsynaptic $5\text{-HT}_{2A/2C}$ receptors mediate the actions of 5-HT in both the amygdala and the PAG (Deakin, Graeff and Guimaraes 1992). Thus local infusion of $5\text{-HT}_{2A/2C}$ antagonists into the amygdala has an anti-conflict effect in animals while their systemic administration might have (albeit controversially) anti-anxiety effects in humans. In contrast, these drugs promote the flight/fight response to aversive stimuli. This leads to the prediction that drugs that relieve anxiety, through inhibition of 5-HT transmission in the amygdala, will exacerbate panic by inhibiting the restraining influence of 5-HT in the PAG. In fact, this has been offered as an explanation for the panic attacks experienced by some patients given buspirone. It could also explain the increase in panic attacks in the early stages of treatment with antidepressants. These drugs first decrease the firing rate of serotonergic neurons and the terminal release of 5-HT; recovery of neuronal firing and increased release of 5-HT, like the relief of panic by these drugs, requires prolonged treatment (see Chapter 20).

Obviously, any explanation of anxiety must account for the actions of benzodiazepines. Gray's theory suggests that they inhibit monoaminergic inputs to the septohippocampal system and switch off behavioural inhibition. A related suggestion is that, whereas the behavioural inhibition system is located in the medial septum/dorsal hippocampus, there is also a 'safety system' in the lateral septum/ventral hippocampus. According to this scheme, benzodiazepines might activate this latter system and generate spurious safety signals (see Handley 1995). Alternatively, Deakin and Graeff's theory suggests that benzodiazepines could directly inhibit activity generated in the PAG. However, they could also inhibit the activity of serotonergic neurons in the DRN and suppress the amygdala response to conditioned fear stimuli. In this case, suppression of the serotonergic inhibitory inputs to the PAG might also be anticipated, an action that could explain why benzodiazepines are ineffective in treating panic disorder.

The finding that noradrenergic neurons innervating the frontal cortex, but not those projecting to the hypothalamus, respond to conditioned environmental cues (McQuade and Stanford 1999) suggests that there could be a similar subdivision of function in this monoamine system as well. However, activation of 5-HT receptors modulates release of noradrenaline (Stanford 1999) and vice versa (Gobert *et al.* 1997). The function of both these neuronal systems is influenced by GABAergic systems. Clearly, any theory for anxiety must eventually take account of evidence that serotonergic and noradrenergic systems do not operate independently.

PEPTIDES

Many peptidergic and amino acid receptors have been suggested as potential targets for anti-anxiety drugs. The outcome of preclinical and clinical investigations (together with further references) are detailed in Jackson and Nutt (1996). So far, no compounds have emerged as clear candidates for the clinic, not least because pharmacokinetic considerations and adverse effects in the periphery are common confounding factors. How any of the peptides could actually influence the neuronal networks considered to be involved in anxiety, as outlined above, is uncertain but the targets which hold greatest promise for the future are:

(1) *The angiotensin system*: The possibility that drugs targeting this system might be efficacious in anxiety first came to light from self-reports of patients who were being treated with inhibitors of angiotensin-converting enzyme ('ACE' inhibitors) for their hypertension. Since then, these agents (e.g. captopril) have been found to mimic the effects of benzodiazepines in a range of preclinical models. Experiments with selective angiotensin receptor antagonists, e.g. the AT_1 receptor antagonist, losartin, have so far confirmed that these agents are effective in many preclinical models whereas AT_2 receptor antagonists are not.

(2) *Cholecystokinin receptor ligands*: The actions of receptor-selective ligands have been explored following reports that peptides targeting CCK receptors can induce anxiety in humans. CCK_B receptor agonists have anxiogenic effects in animal models whereas antagonists have an anti-anxiety effect, including the prevention of behavioural changes resulting from ethanol withdrawal (Wilson, Watson and Little

1998). In humans, CCK_B receptor ligands can induce a panic attack, especially in panic patients, whereas antagonists (e.g. CL988) have the opposite effect. Nevertheless, the results from clinical trials of this compound are not promising, mainly due to low bioavailability and unacceptable side-effects.

(3) *Neurokinin receptors*: NK_2 receptor agonists (e.g. GR64349) have an anxiogenic profile in animal models while the antagonists (GR100679) have an anti-anxiety effect. However, NK_1 receptor antagonists have also been reported to have anti-anxiety activity in the social interaction test (File 1997).

(4) *Neuropeptide Y receptors*: The NPY_1 receptor agonist, (leu^{31}, Pro^{34})neuropeptide Y has anticonflict activity. Although NPY_2 receptor ligands are generally thought to lack anti-anxiety effects, there is some evidence that they are active in preclinical models (Kask, Rago and Harro 1998). The picture is complicated by findings that NPY itself can have anti-anxiety or anxiogenic effects, depending on dose (Nakajima *et al.* 1998).

(5) *Corticotropin-releasing factor (CRF) receptors*: CRF is well known for its role in the stress response and for its interactions with monoamine systems in the brain. As might be expected, this hormone has anxiogenic activity in preclinical models. The CRF receptor antagonist, α-helical CRF^{9-14}, prevents the actions of anxiogenic drug treatments (such as ethanol withdrawal) but seems to be inactive in preclinical tests when given alone. Interestingly, it also seems to prevent the anxiogenic effects of NPY_1 antagonists suggesting some functional interactions between these two peptide systems (Kask, Rago and Harro 1997).

(6) *NMDA receptors*: The non-competitive NMDA receptor antagonist, dizocilpine, and competitive antagonists such as AP-5, have anti-anxiety activity in preclinical models. However, the psychotropic effects of other NMDA receptor antagonists, such as the potent hallucinogen, phencyclidine, warn against these compounds being realistic targets for future drug development. Antagonists of the glycine site on the NMDA receptor (e.g. HA-966) also have anti-anxiety effects in preclinical models and are more promising.

(7) *Adenosine receptors*: The possibility that adenosine receptors will be a useful target is suggested by the marked anxiogenic effects of the adenosine receptor antagonist, caffeine. Whether it is this action of caffeine (which has many molecular targets in the brain) that explains its anxiogenic actions is not at all certain and, so far, selective adenosine receptor agonists have not yielded promising results.

CONCLUSIONS

Without doubt, the benzodiazepines are the most successful of the anti-anxiety agents in respect of their safety and tolerability and so one might ask why there is a need to search for better agents at all. One problem is that, while they are highly efficacious in treating GAD, the benzodiazepines are not without their drawbacks, particularly in respect of concern about their potential for dependency and their clear liability for abuse. Another is the need to develop better treatments for other manifestations of anxiety. Novel agents, targeting peptidergic systems, might provide solutions to both these problems. It is only through the combined efforts of all the approaches outlined in this chapter that we are likely to identify the cause(s) of anxiety and develop the ideal treatment.

REFERENCES

Abe, K, Takeyama, C and Yoshimura, K (1998) Effects of S-8510, a novel benzodiazepine receptor partial inverse agonist, on basal forebrain lesioning-induced dysfunction in rats. *Eur. J. Pharmacol.* **347**: 145–152.

Awad, M and Gavish, M (1987) Binding of [^{3}H]Ro 5-4864 and [^{3}H]PK 11195 to cerebral cortex and peripheral tissues of various species: species differences and heterogeneity in peripheral benzodiazepine binding sites. *J. Neurochem.* **49**: 1407–1414.

Ballenger, JC (1990) *Neurobiology of Panic Disorder*, Wiley-Liss, New York.

Barnes, NM and Sharp, T (1999) A review of central 5-HT receptors and their function. *Neuropharmacology* **38**: 1083–1152.

Beckett, S and Marsden, CA (1997) The effect of central and systemic injection of the 5-HT_{1A} receptor agonist 8-OHDPAT and the 5-HT_{1A} receptor antagonist WAY100635 on periaquaductal grey-induced defence behaviour. *J. Psychopharmacol.* **11**: 35–40.

Bell, CJ and Nutt, DJ (1998) Serotonin and panic. *Brit. J. Psychiatry* **172**: 465–471.

Bremner, JD, Krystal, JH, Southwick, SM and Charney, DS (1996) Noradrenergic mechanisms in stress and anxiety. II. Clinical studies. *Synapse* **23**: 39–51.

Cao, BJ and Rodgers, RJ (1997) Influence of 5-HT_{1A} receptor antagonism on plus-maze behaviour in mice. II WAY 100635, SDZ 216-525 and NAN-190. *Pharmacol. Biochem. Behav.* **58**: 593–603.

Chaouloff, F (1993) Physiopharmacological interactions between stress hormones and central serotonergic systems. *Brain Res.* **18**: 1–32.

Cheetham, SC and Heal, DJ (2000) Antidepressant and anxiolytic drugs. In *Biological Psychiatry* (Eds Bittar, EE and Bittar, N), JAI Press, Stanford, CT, pp. 551–567.

Chiu, TH, Dryden, DM and Rosenberg, HC (1982) Kinetics of [^{3}H]-labelled flunitrazepam binding to membrane-bound benzodiazepine receptors. *Mol. Pharmacol.* **21**: 57–65.

Costa, E and Guidotti, A (1991) Diazepam binding inhibitor (DBI): a peptide with multiple biological actions. *Life Sci.* **49**: 325–344.

Coupland, N, Glue, P and Nutt, DJ (1992) Challenge tests: assessment of the noradrenergic and GABA systems in depression and anxiety disorders. *Molec. Aspects Med.* **13**: 221–247.

Dalley, JW, Mason, K and Stanford, SC (1996) Increased levels of extracellular noradrenaline in the frontal cortex of rats exposed to naturalistic environmental stimuli: modulation by acute systemic administration of diazepam or buspirone. *Psychopharmacology* **127**: 47–54.

Deakin, JFW and Graeff, FG (1991) 5-HT and mechanisms of defence. *J. Psychopharmacol.* **5**: 305–315.

Deakin, JFW, Graeff, FG and Guimaraes, FS (1992) 5-HT receptor subtypes and the modulation of aversion. In *Central Serotonin Receptors and Psychotropic Drugs* (Eds Marsden, CA and Heal, DJ), Blackwell Scientific Publications, Oxford, pp. 147–174.

De Robertis, E, Peña, C, Paladini, AC and Medina, JH (1988) New developments on the search for endogenous ligand(s) of central benzodiazepine receptors. *Neurochem. Int.* **13**: 1–11.

Doble, A and Martin, IL (1996) *The $GABA_A$/Benzodiazepine Receptor as a Target for Psychoactive Drugs*, RG Landes Co., Austin, TX.

Done, CJ and Sharp, T (1994) Biochemical evidence for the regulation of central noradrenergic activity by 5-HT_{1A} and 5-HT_2 receptors: Microdialysis studies in the awake and anaesthetised rat. *Neuropharmacology* **33**: 411–421.

Do Rego, JL, Mensah-Nyagan, AG, Feuilloley, M, Ferrara, P, Pelletier, G and Vaudry, H (1998) The endozepine triakontatetraneuropeptide diazepam-binding inhibitor [17–50] stimulates neurosteroid biosynthesis in the frog hypothalamus. *Neuroscience* **83**: 555–570.

Dorow, R, Horowski, R, Paschelke, G, Amin, M and Braestrup, C (1983) Severe anxiety induced by FG7142, a β-carboline ligand for benzodiazepine receptors. *Lancet* **ii**: 98–99.

Dourish, CT, Hutson, PH and Curzon, G (1986) Putative anxiolytics 8-OHDPAT buspirone and TVXQ 7821 are agonists at 5-HT_{1A} autoreceptors in the raphé nuclei. *Trends Pharmacol. Sci.* **7**: 212–214.

Drugan, RC, Basile, AS, Crawley, JN, Paul, SM and Skolnick, P (1987) 'Peripheral' benzodiazepine binding sites in the Maudsley Reactive rat: selective decrease confined to peripheral tissues. *Brain Res. Bull.* **18**: 143–145.

DSM–IV: *Diagnostic and Statistical Manual of Mental Disorders*, 4th edition (1994), American Psychiatric Association, Washington, DC, pp. 393–444.

Duggan, MJ and Stephenson, FA (1988) Benzodiazepine binding site heterogeneity in the purified $GABA_A$ receptor. *Eur. J. Pharmacol.* **154**: 293–298.

Eison, AS, Eison, MS, Stanley, M and Riblet, LA (1986) Serotonergic mechanisms in the behavioural effects of buspirone and gepirone. *Pharmac. Biochem. Behav.* **24**: 701–707.

File, SE (1997) Anxiolytic action of a neurokinin1 receptor antagonist in the social interaction test. *Pharmacol. Biochem. Behav.* **58**: 747–752.

File, SE and Hyde, JR (1979) A test of anxiety that distinguishes between the actions of benzodiazepines and those of other minor tranquilisers and of stimulants. *Pharmacol. Biochem. Behav.* **11**: 65–69.

Fontana, DJ, McMiller, LV and Commissaris, RL (1999) Depletion of brain norepinephrine: differential influence on anxiolytic treatment effects. *Psychopharmacology* **143**: 197–208.

Geller, I, Kulak, JT and Seifter, J (1962) The effects of chlordiazepoxide and chlorpromazine on a punished discrimination. *Psychopharmacologia* **3**: 374–385.

Gobert, A, Rivet, JM, Cistarelli, L, Melon, C and Millan, MJ (1997) Alpha2-adrenergic receptor blockade markedly potentiates duloxetine- and fluoxetine-induced increases in noradrenaline, dopamine and serotonin levels in the frontal cortex of freely moving rats. *J. Neurochem.* **69**: 2616–2619.

Graeff, FG, Guimarães, FS, De Andrade, TGC and Deakin, JFW (1996) Role of 5-HT in stress, anxiety and depression. *Pharmacol. Biochem. Behav.* **54**: 129–141.

Gray, JA (1987) *The Psychology of Fear and Stress*, 2nd edition, Cambridge University Press, Cambridge.

Han, QP and Dryhurst, G (1996) Influence of glutathione on the oxidation of 1-methyl-6-hydroxy-1,2,3,4-tetrahydro-beta-carboline; chemistry of potential relevance to the addictive and neurodegenerative consequences of ethanol use. *J. Med. Chem.* **39**: 1494–1508.

Handley, SL (1995) 5-Hydroxytryptamine pathways in anxiety and its treatment. *Pharmac. Ther.* **66**: 103–148.

Jackson, HC and Nutt, DJ (1996) Anxiety and panic disorders. In *Neurotherapeutics, Emerging Strategies* (Eds Pullan, LM and Patel, J), Humana Press Inc., Totowawa, NJ, pp. 85–131.

Jacobs, BL and Azmitia, EC (1992) Structure and function of the brain serotonin system. *Pharmacol. Rev.* **72**: 165–229.

Kask, A, Rago, L and Harro, J (1997) Alpha-helical CRF(9-41) prevents the anxiogenic-like effect of NPY Y1 receptor antagonist BIBP3226 in rats. *Neuroreport* **8**: 3645–3647.

Kask, A, Rago, L and Harro, J (1998) Anxiolytic-like effect of neuropeptide Y (NPY) and NPY13-36 microinjected into vicinity of locus coeruleus in rats. *Brain Res.* **788**: 345–348.

La Marca, S and Dunn, RW (1994) The α_2-antagonists idazoxan and rauwolscine but not yohimbine or piperoxan are anxiolytic in the Vogel lick-shock conflict paradigm following intravenous administration. *Life Sci.* **54**: 179–184.

LeDoux, JE and Muller, J (1997) Emotional memory and psychopathology. *Phil. Trans. Roy. Soc. Lond. B* **352**: 1719–1726.

Libet, B and Gleason, CA (1994) The human locus coeruleus and anxiogenesis. *Brain Res.* **634**: 178–180.

Malizia, AL, Cunningham, VJ, Bell, CJ, Liddle, PF, Jones, T and Nutt, DJ (1998) Decreased brain GABA(A)-benzodiazepine receptor binding in panic disorder: preliminary results from a quantitative PET study. *Arch. Gen. Psychiatry* **55**: 715–720.

Margules, DL and Stein, L (1968) Increase of 'antianxiety' activity and tolerance of behavioural depression during chronic administration of oxazepam. *Psychopharmacologia* **13**: 74–80.

Marsden, CA, Bickerdike, M, Cadogen, A-K, Wright, I, Rex, A and Fink, H (1993) Serotonergic mechanisms and animal models of anxiety. In *Anxiety, Neurobiology, Clinic and Therapeutic Perspectives* (Eds Hamon, M, Oilat, H and Thiebot, M-H), John Libbey Eurotext Ltd, 232, pp. 75–82.

Mason, K, Heal, DJ and Stanford, SC (1998) The anxiogenic agents, yohimbine and FG 7142 disrupt the noradrenergic response to novelty. *Pharmacol. Biochem. Behav.* **60**: 321–327.

McQuade, R and Stanford, SC (2000) A microdialysis study of the noradrenergic response in rat frontal cortex and hypothalamus to a conditioned cue for aversive, naturalistic environmental stimuli. *Psychopharmacol.* **148**: 201–208.

Moehler, H and Okada, T (1977) Properties of [3H]diazepam binding to benzodiazepine receptors in rat cerebral cortex. *Life Sci.* **20**: 2101–2110.

Nakajima, M, Unui, A, Asakawa, A, Momose, K, Ueno, N, Teranishi, A, Baba, S and Kasuga, M (1998) Neuropeptide Y produces anxiety via Y2-type receptors. *Peptides* **19**: 359–363.

Nutt, DJ (1990) The pharmacology of human anxiety. *Pharmacol. Ther.* **47**: 233–266.

Nutt, DJ, Glue, P, Lawson, C and Wilson, S (1990) Flumazenil provocation of panic attacks. Evidence for altered benzodiazepine receptor sensitivity in panic disorder. *Arch. Gen. Psychiatry* **47**: 917–925.

Olivier, B, van Wijngaarden, I and Soudijn, W (2000) 5-HT_3-receptor antagonists and anxiety: a preclinical and clinical view. *Eur. Neuropsychopharmacol.* **10**: 77–95.

Przegalinski, E, Chojnacka-Wojcik, E and Filip, M (1992) Stimulation of postsynaptic 5-HT_{1A} receptors is responsible for the anticonflict effect of ipsapirone in rats. *J. Pharm. Pharmac.* **44**: 780–782.

Rasmussen, K and Jacobs, BL (1986) Single unit activity of locus coeruleus neurons in the freely moving cat II. Conditioning and pharmacological studies. *Brain Res.* **371**: 335–344.

Reddy, DS and Kulkarni, SK (1998) The role of GABA-A and mitochondrial diazepam-binding inhibitor receptors on the effects of neurosteroids on food intake in mice. *Psychopharmacology* **137**: 391–400.

Redmond, DE and Huang, YH (1979) New evidence for a locus coeruleus-norepinephrine connection with anxiety. *Life Sci.* **25**: 2149–2162.

Rodgers, RJ and Dalvi, A (1997) Anxiety, defence and the elevated plus-maze. *Neurosci. Biobehav. Rev.* **21**: 801–810.

Salmon, P, Tsaltas, E and Gray, JE (1989) Disinhibition by propranolol and chlordiazepoxide of nonrewarded lever-pressing in the rat is unaffected by dorsal noradrenergic bundle lesion. *Neuropharmacology* **28**: 207–210.

Sangameswaran, L, Fales, HM, Friedrich, P and De-Blas, AL (1986) Purification of a benzodiazepine from bovine brain and detection of benzodiazepine-like immunoreactivity in human brain. *Proc. Natl. Acad. Sci. USA* **83**: 9236–9240.

Seevers, MH and Tatum, AL (1931) Chronic experimental barbital poisoning. *J. Pharmacol. Exp. Ther.* **42**: 217–231.

Sieghart, W and Karobath, M (1980) Molecular heterogenity of benzodiazepine receptors. *Nature* **286**: 285–287.

Silverstone, PH, Rue, JE, Franklin, M, Hallis, K, Camplin, G, Laver, D and Cowen, PJ (1994) The effects of administration of *m*CPP on psychological, cognitive, cardiovascular, hormonal and MHPG measurements in human volunteers. *Int. Clin. Psychopharmacol.* **9**: 173–178,

Söderpalm, B, Blomqvist, O and Söderpalm, B (1995) The yohimbine-induced anticonflict effect in the rat, part I. Involvement of noradrenergic, serotonergic and endozepinergic(?) mechanisms. *J. Neural Trans.* **100**: 175–189.

Squires, RF and Braestrup, C (1977) Benzodiazepine receptors in rat brain. *Nature* **266**: 732–734.

Stanford, SC (1995) Central noradrenergic neurones and stress. *Pharmac. Ther.* **68**: 297–342.

Stanford, SC (1999) SSRI-induced changes in catecholaminergic transmission. In Selective Serotonin Reuptake Inhibitors (SSRIs): Past, Present and Future, Ron Landes Bioscience, Austin, TX, pp. 148–170.

Sternbach, LH, Randall, LO and Gustafson, SR (1964) 1,4-Benzodiazepines (chlordiazepoxide and related compounds). In Medicinal Chemistry, Volume 1, Psychopharmacological Agents (Ed. Gordon, M), Academic Press, New York & London, pp. 137–224.

Taylor, KM and Laverty, R (1969) The effect of chlordiazepoxide, diazepam and nitrazepam on catecholamine metabolism in regions of the rat brain. *Eur. J. Pharmacol.* **8**: 296–301.

Tyrer, P (1989) *The Psychopharmacology of Anxiety*, Oxford Medical Publications, Oxford.

Vogel, RA, Frye, GD, Wilson, JH, Kuhn, CM, Koepke, KM, Mailman, RB, Mueller, RA and Breese, GR (1980) Attenuation of the effects of punishment by ethanol: comparisons with chlordiazepoxide. *Psychopharmacology* **71**: 123–129.

Wilson, J, Watson, WP and Little, HJ (1998) CCK(B) antagonists protect against anxiety-related behaviour produce by ethanol withdrawal, measured using the elevated plus-maze. *Psychopharmacology* **137**: 121–131.

20 Depression

S. C. STANFORD

INTRODUCTION

Depression and mania are both 'affective disorders' but their symptoms and treatments are quite distinct. Mania is expressed as heightened mood, exaggerated sense of self-worth, irritability, aggression, delusions and hallucinations. In stark contrast, the most obvious disturbance in depression is melancholia that often co-exists with behavioural and somatic changes (Table 20.1). Some individuals experience dramatic mood swings between depression and mania. This is known as '*bipolar disorder*' which, like mania itself, is treated with lithium salts or neuroleptics.

The focus of this chapter is '*unipolar depression*' in which patients suffer long bouts of depression interspersed with periods of remission. Unipolar depression is regarded as either '*reactive*' or '*endogenous*': whereas the former is apparently triggered by a recent life event involving 'loss' (e.g. bereavement or redundancy), the latter has no obvious origin, although it is often associated with disease (e.g. viral infection, diabetes, neoplasia), drug use/misuse or chronic pain.

A major difficulty in unravelling the neurobiological basis of depression is that it is not a simple, unitary disorder. Also, whereas about 33% of patients recover spontaneously, about the same proportion do not respond to any treatment and only about 60–70% of patients who do respond show any improvement with the first drug of choice. So far, there is no explanation for these different prognoses. There is an urgent need to find a solution to these problems, though, because this is a debilitating disorder that affects about 1 in 10 women and 1 in 50 men at some stage in their life. In over half of these cases, the depression recurs and about 15% of depressives commit suicide. A particular concern is that, in the UK, about 1 in every 200 depressed patients attempts suicide by overdosing on their drug treatment (Henry and Rivas 1999).

These statistics alone stress the urgency of finding more effective and safer treatments for depression but this requires a better understanding of its underlying neurobiology and the mechanism(s) of action of existing drug treatments. This chapter outlines some of the progress that has been made so far.

LINKING DEPRESSION AND CENTRAL MONOAMINES

A link between the central monoamines, 5-hydroxytryptamine (5-HT) and noradrenaline, and depression was forged some 40 years ago and arose from clinical experience with the drugs, reserpine and iproniazid. At that time, reserpine was used as an

Neurotransmitters, Drugs and Brain Function. Edited by R. A. Webster

Table 20.1 Symptoms and signs of depression

Psychological	Depressed mood
	Anhedonia
	Low self-esteem
	Guilt
	Cognitive deficits
	Poor concentration
	Feeling of 'hopelessness'
	Suicidal thoughts
	Hypochondria
Somatic	Unintentional weight change
	Fatigue
	(Appetite and sleep disturbance)
	Aches and pains
Behavioural	Psychomotor agitation or retardation
	Self-neglect
	Sleep disturbance

antihypertensive and neuroleptic agent but a major problem was that it induced suicidal depression in some (but, interestingly, only about 5–10%) patients. From studies in rats it emerged that reserpine consistently caused behavioural changes, such as motor inactivity and sedation (the 'reserpine syndrome'), which resemble certain features of depression in humans. These behaviours, and consequently those in humans, were attributed to the depletion of neuronal vesicular stores of monoamines and the reduction in monoamine transmission caused by this drug.

In contrast, iproniazid, introduced in 1951 for treatment of tuberculosis, induced euphoria and was described as a 'psychic energiser'. In fact, these patients, when given iproniazid, could become quite disruptive and this action was regarded as an undesirable side-effect! However, its beneficial effects in depression were soon recognised and it was regarded as the first effective antidepressant drug. Studies of peripheral sympathetic neurons, later extended to noradrenergic neurons in the brain, showed that iproniazid irreversibly inhibits the catalytic enzyme, monoamine oxidase (MAO). Because only cytoplasmic monoamines are accessible to MAO, inhibition of this enzyme first increases the concentration of the pool of soluble transmitter but this leads to a secondary increase in the stores of vesicle-bound transmitter: i.e. the pool available for impulse-evoked release (Fillenz and Stanford 1981).

Iproniazid also prevents the 'reserpine syndrome' in rats. Reserpine blocks vesicular uptake of monoamines which, as a consequence, leak from the storage vesicles into the cytosol. Although these monoamines would normally be metabolised by MAO, they are conserved when a MAO inhibitor (MAOI) is present, and so co-administration of reserpine and a MAOI leads to accumulation of monoamines in the neuronal cytosol. It is now known that, when the concentration of cytoplasmic monoamines is increased in this way, they are exported to the synapse on membrane-bound monoamine transporters. The ensuing increase in monoamine transmission, despite the depletion of the vesicular pool, presumably accounts for the effects of iproniazid on the behaviour of reserpine-pretreated rats.

In 1958, another agent, imipramine, was discovered by chance to have beneficial effects in depression. This compound is not a MAOI and its actions were first described as 'a complete riddle'. Axelrod's group in Washington (Hertting, Axelrod and Whitby

1961) later found that imipramine blocked neuronal reuptake of noradrenaline from the synapse. This blockade prolongs and augments the actions of released transmitter and was assumed to explain imipramine's antidepressant effects.

Drawing all this evidence together, Schildkraut (1965) concluded that depression was caused by a functional deficit of noradrenergic transmission in the brain. He also thought that the rebound depression and fatigue, which are experienced after the arousing effects of amphetamine have worn off, were due to depletion of neuronal stores of noradrenaline. However, Schildkraut made a clear distinction between the effects of antidepressants and the arousal induced by amphetamine, describing the latter as 'stimulation' and 'excitement'. To this day, there is controversy over whether or not amphetamine has a beneficial effect in depression.

Later theories continued to focus on the monoamines. One suggested that there is a malfunction of neurons which release 5-HT (Coppen 1967). Another proposed that a deficit in both noradrenergic and serotonergic transmission is to blame (Maas 1975). Others have argued that an imbalance in the functional output of these two systems is the key factor (Ricci and Wellman 1990). However, they all share a common theme: that disruption of some aspect of monoaminergic transmission in the brain is a causal factor in depression. It is remarkable that, although this theory is often challenged, it has not yet been replaced by a validated alternative and, to this day, central noradrenergic and/or serotonergic systems are primary targets for all established antidepressant drugs.

THE NEUROBIOLOGY OF DEPRESSION

Attempts to find the cause(s) of depression have adopted two main approaches. One is to look for the neurobiological basis of depression in human subjects and animal models of this condition. The second is to investigate the pharmacology of established antidepressant agents to see whether they consistently augment some, and ideally the same, neurobiological targets in the brain.

HUMAN STUDIES

The objective of these studies is to find a neurochemical marker for depression. For obvious reasons, the majority has looked for changes that might affect monoamine function and so the following sections concentrate on these neurotransmitters. (Evidence suggesting that a dysfunction of the gluocorticoid hormonal system could be involved is discussed later.) Most techniques compare depressed and non-depressed (control) subjects and measure:

(1) The concentration of monoamines and their metabolites in accessible tissue samples (e.g. blood and urine).
(2) Indices of neurotransmitter function on lymphocytes (e.g. β-adrenoceptor binding) or platelets (e.g. α_2-adrenoceptor binding, 5-HT uptake).
(3) Monoamine concentrations or receptor binding in brain tissue post-mortem.
(4) Changes in neuroendocrine responses to challenges with drugs that target specific monoamine receptors (e.g. the α_2-adrenoceptor agonist, clonidine, stimulates growth hormone secretion while β-adrenoceptor activation stimulates melatonin secretion from the pineal gland).

Table 20.2 Neurochemical markers for depression

Marker	Tissue	Usual finding in depression	
NA or MHPG concentrations	Post-mortem brain, urine, CSF, plasma	NCC	
Tyrosine hydroxylase immunoreactivity	Post-mortem brain	NCC	
α_1-adrenoceptor binding	Post-mortem brain	↓	Some brain areas
α_2-adrenoceptor binding	Post-mortem brain	↑	High-affinity site in some brain areas
	Platelets	NCC	
β_1-adrenoceptor binding	Post-mortem brain	↓	Certain cortical areas of suicide victims
β_2-adrenoceptor binding	Lymphocytes	↓	
—cAMP response		↓	
5-HT/5-HIAA concentration	Post-mortem brain,	NCC	
	CSF, urine	NCC	
5-HT uptake	Platelets	↓V_{max}	
[^{3}H]imipramine binding	Platelets	NCC	
5-HT_1 receptor binding	Post-mortem brain	↓	Hippocampus
5-HT_{1A} receptor binding	Post-mortem brain	NCC	
5-HT_{1D} receptor binding	Post-mortem brain	↑	Globus pallidus
5-HT_2 receptor binding	Platelets	↑	Suicides
	Post-mortem brain	NCC	
D_1-/D_2-receptor binding	Post-mortem brain	No change	Drug-free suicides
GABA receptor binding	Post-mortem brain	NCC	
NMDA receptor	Post-mortem brain	↓	Affinity of glutamate site ligands and allosteric coupling to high-affinity glycine sites (frontal cortex of suicides)
Cortisol concentration	Plasma	↑	
Dexamethasone response		↓	

Notes:
Measurements in depressed patients compared with normal subjects, euthymic controls or patients suffering from an unrelated psychiatric disorder. NCC: No consistent change. The changes indicated are based on the most frequently published findings.

Unfortunately, such measurements are fraught with difficulties. For instance, it is not at all certain that neurochemical changes in the plasma or urine give any reliable indication of what is happening in the brain. Measurements in post-mortem brain tissue do not have this problem but the unavoidable delay in collecting tissue samples introduces another. Confirmation of the diagnostic status of the subjects is often difficult (especially retrospectively) and any drug treatments they had taken could distort the results. Another confounding factor is the possibility that any neurochemical changes are expressed only while the patient is experiencing a depressive phase ('state') rather than persisting during remission or recovery ('trait').

So far, evidence for abnormal peripheral (Elliott 1992) or central (Horton 1992) monoamine function in depression is equivocal, and no consistent biochemical markers have emerged to provide a firm link between the two (Table 20.2). One widely cited finding is that subjects who have attempted violent suicide form a neurochemically distinct group because the concentration of the 5-HT metabolite, 5-HIAA, in their CSF is lower than normal, suggesting that a deficit in 5-HT release is associated with suicide

(Åsberg, Traskman and Thoren 1976). However, this abnormality is now believed to be associated with a deficit in control of behavioural impulsivity, rather than depression.

Evidence for a link between monoaminergic transmission and the therapeutic effects of antidepressant agents is more convincing. Depletion of noradrenaline stores (achieved by administration of the noradrenaline synthesis inhibitor, α-methyl-*p*-tyrosine) causes a resurgence of depression in patients who are in remission following treatment with antidepressants that selectively target noradrenergic neurons. However, patients who respond to antidepressants that act primarily on serotonergic neurons are unaffected (Delgado *et al.* 1993). Conversely, a deficiency of tryptophan in the diet, which depletes 5-HT stores in the brain, reinstates depression in patients who have responded to drugs presumed to increase serotonergic transmission, but not those acting on noradrenergic neurons (Salomon *et al.* 1993). It seems that the therapeutic effects of different antidepressants could well rest on augmenting particular components of central monoamine transmission, whether or not depression itself is explained by a deficit in the functional output of these neurons.

ANIMAL STUDIES

The use of animal models for depression has two main objectives. One is to provide a behavioural model that can be used to screen potential antidepressant treatments. For this, the behaviour does not have to be an animal analogue of depression: all that is needed is for it to be consistently prevented by established antidepressant agents (i.e. no false negatives) but not by drugs which have no antidepressant effect in humans (i.e. no false positives).

A second objective is to produce behavioural changes in animals that are analogous to depression so that the model can be used to discover its neurobiological cause(s). This is a far more demanding problem and its success rests on satisfying at least three criteria (see Willner 1984): *face validity* (i.e. the behaviour looks like depression), *construct validity* (i.e. the causes and consequences of the behavioural change are the same as in depression) and *predictive validity* (i.e. the behaviour is reliably prevented only by drugs which have antidepressant effects in humans).

Procedures that have been suggested as models of depression and used to look for neurochemical changes that parallel the onset of the behavioural change, as well as to test how antidepressants affect the behaviour, are listed in Table 20.3. Those that have been used most, either as a drug screen or in research into the neurobiology of depression, are as follows.

Table 20.3 Procedures that have been used as animal models for depression or as a preclinical screen for novel antidepressant drugs

Procedures that have been widely studied:
Forced swim test ('behavioural despair')
Inescapable shock ('learned helplessness')
Olfactory bulbectomy
Novel (or controversial) models of depression that await further validation:
Chronic isolation housing
Chronic mild stress
Raphé lesion-induced muricidal behaviour
Tail suspension test

Olfactory bulbectomy

The olfactory bulb has multiple connections with limbic areas of the brain and receives afferent neurons from both the Raphé nuclei and the locus coeruleus. Bilateral removal of the olfactory bulbs in rats induces many behavioural, neuroendocrine and immunological changes including: hyperphagia, decreased rapid eye movement (REM) sleep, hyperactivity and deficits in the acquisition of conditioning behaviour (passive avoidance) (Van Riezen and Leonard 1990). These changes echo many of the problems experienced by depressives (disruption of appetite and sleep patterns and cognitive deficits). Also, as in many depressed patients, the concentration of plasma corticosteroids is increased in rats after olfactory bulbectomy (see later). This pattern of changes suggests that bulbectomy disrupts links between limbic areas of the brain and the hypothalamus. Its validity as a model of depression is supported by findings that all physiological and behavioural changes resulting from bulbectomy, that have been reported so far, are normalised by antidepressants from different generic groups. Only tranylcypromine, a MAOI, is ineffective in this model and this could be because it is metabolised to the CNS stimulant, methamphetamine, which is also ineffective.

Learned helplessness

This model was developed after pioneering experiments carried out in the USA by Overmier and Seligman (1967) who reported profound behavioural changes in dogs after their exposure to inescapable, uncontrollable stress (footshock). Subsequent work has concentrated on rats and mice, which show a similar behavioural response. This is expressed as appetite and sleep disturbance, general passivity and, on re-exposure of subjects to the stress, a failure to attempt to escape ('escape deficits'), even when this is feasible.

This behavioural syndrome, rather emotively called '*learned helplessness*', is widely believed to share many features of depression, not least because both culminate in psychomotor retardation and both are linked with experience of uncontrollable, unpredictable stress. Whether or not learned helplessness really is an analogue of depression remains controversial (Maier 1993). Nevertheless, escape deficits in rats are prevented by pretreatment with antidepressants from different generic groups. Other psychotropic agents, such as CNS stimulants and neuroleptics, are generally ineffective.

One of the earliest and most consistent findings with this model was a marked depletion of noradrenaline stores in certain brain regions, particularly the cortex, hippocampus and hypothalamus, of mice that have been exposed to inescapable shock. Moreover, when noradrenaline stores are depleted by other means (e.g. by administration of reserpine or α-methyl-*p*-tyrosine) the development of escape deficits is accelerated. One school of thought proposes that exhaustion of releasable noradrenaline in the neurons projecting to these brain regions underlies learned helplessness (Anisman and Zacharko 1991). Others highlight the depletion of noradrenaline stores in the locus coeruleus and suggest that a reduction in the release of noradrenaline in this nucleus diminishes the α_2-autoreceptor-mediated feedback inhibition of neuronal firing and that the resulting neuronal hyperactivity explains learned helplessness (Weiss *et al.* 1981). These two theories obviously differ in respect of whether it is an increase or a decrease in noradrenergic transmission in the terminal field that could account for depression. However, they are both consistent with Schildkraut's theory in that depression could arise from a deficit in noradrenaline release, be it in either the terminal field or the cell body region.

One problem with both these theories is that disruption of noradrenergic transmission by selective adrenoceptor antagonists has little impact on the development of escape deficits. However, such antagonists do prevent the reversal of learned helplessness by antidepressants (reviewed by Stanford 1995). Also, it would be most unlikely that a deficit in only one neurotransmitter system fully accounts for learned helplessness. Indeed, there is plenty of evidence for a role for 5-HT in learned helplessness: for instance, this behaviour is reversed by microinjection of 5-HT into the prefrontal cortex (Davis *et al.* 1999). Finally, it is clear that opioid, GABAergic and cholinergic systems (among others) are all linked with this behavioural deficit and even dihydropyridine antagonists of Ca^{2+} channels prevent its development.

In short, the widespread neurochemical disruption during learned helplessness suggests that antidepressant drugs could prevent this syndrome by targeting any of several different neurotransmitter systems.

The swim test

Another behavioural model focuses on the immobile, floating posture (sometimes called '*behavioural despair*') which develops in rodents during a brief swim when they are put into a pool of water from which they cannot escape. Many antidepressants delay the onset and duration of immobility and this action has been widely adopted as a preclinical screen for novel compounds (Porsolt *et al.* 1979). However, drugs that selectively inhibit reuptake of 5-HT (SSRIs, see below), and which are highly effective antidepressants, are generally ineffective at diminishing immobility in the swim test, a finding that somewhat undermines its validity as a model of depression. Also, the many false positives (e.g. anticholinergics, amphetamine, sodium valproate) emphasise why it is important to distinguish drug effects on the emotional impact of this test from their non-specific effects on animals' motor activity.

SUMMARY

All these animal models express behavioural deficits that are paralleled by some abnormality in noradrenaline and/or 5-HT function but it is unlikely that the monoamines are the only neurotransmitters to influence these complex behaviours. Nevertheless, the behavioural deficits all respond, with varying degrees of specificity, to established antidepressants and central monoamines appear to have a crucial role in the therapeutic effects of these drugs. For a more detailed review of this subject see Stanford (1995).

NEUROCHEMISTRY OF ANTIDEPRESSANTS

The monoamine hypothesis predicts that drugs which increase the concentration of noradrenaline and/or 5-HT in the synapse should relieve depression. This could be achieved in two ways, as illustrated in Figure 20.1:

(1) The first is by preventing their intraneuronal destruction, thereby making more transmitter available for release (e.g. the MAO inhibitors). Some antidepressants

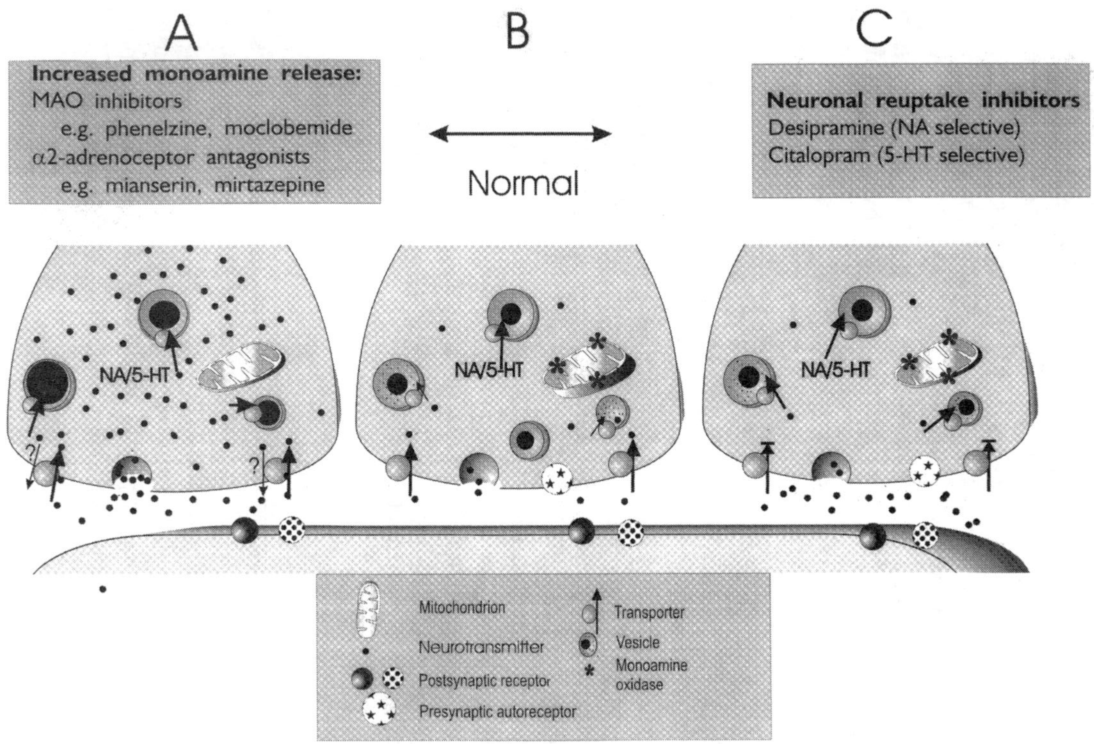

Figure 20.1 Schematic diagram illustrating how antidepressants increase the concentration of extraneuronal neurotransmitter (noradrenaline and/or 5-HT). In the absence of drug (b), monoamine oxidase on the outer membrane of mitochondria metabolises cytoplasmic neurotransmitter and limits its concentration. Also, transmitter released by exocytosis is sequestered from the extracellular space by the membrane-bound transporters which limit the concentration of extraneuronal transmitter. In the presence of a MAO inhibitor (a), the concentration of cytoplasmic transmitter increases, causing a secondary increase in the vesicular pool of transmitter (illustrated by the increase in the size of the vesicle core). As a consequence, exocytotic release of transmitter is increased. Blocking the inhibitory presynaptic autoreceptors would also increase transmitter release, as shown by the absence of this receptor in the figure. In the presence of a neuronal reuptake inhibitor (c), the membrane-bound transporter is inactivated and the clearance of transmitter from the synapse is diminished

are α_2-adrenceptor antagonists and the ensuing increase in monoamine release is thought to account for their antidepressant effects (e.g. mianserin).

(2) A second way of increasing synaptic concentrations of noradrenaline and 5-HT is to block their neuronal reuptake. Several groups of compounds act in this way and can be classified according to their relative selectivity for the noradrenaline and 5-HT transporters.

As described in the introduction, the first generation of antidepressant drugs comprised the MAO inhibitors and the reuptake blockers (e.g. imipramine) which became known as the tricyclic antidepressants (TCAs). The following section starts with a discussion of these two groups of compounds. Subsequent research concentrated on developing drugs that prevent the reuptake of either noradrenaline or 5-HT, like imipramine, but which lack its side-effects. Some drugs even combine reuptake inhibition with actions which increase transmitter release. Examples of all these types of compounds are given in Table 20.4. Whereas the newer antidepressants are a great improvement in terms of safety and tolerability, imipramine still remains the benchmark for efficacy. Full appraisals of antidepressants that are already in the clinic and those that are currently under development, together with their likely clinical and commercial impact, are to be found in Cheetham and Heal (2000) and Heal and Cheetham (1999).

MAO INHIBITORS (MAOIs)

With the exception of tranylcypromine (a phenylcycloalkylamine), the first MAOIs (e.g. iproniazid, isoniazid, phenelzine, isocarboxazid) were derivatives of hydrazine (originally used as a rocket fuel) (Fig. 20.2). All are irreversible inhibitors of the enzyme and restoration of MAO activity requires the synthesis of new enzyme.

As described above, because MAO is bound to mitochondrial outer membranes, MAOIs first increase the concentration of monoamines in the neuronal cytosol, followed by a secondary increase in the vesicle-bound transmitter. The enlarged vesicular pool will increase exocytotic release of transmitter, while an increase in cytoplasmic monoamines will both reduce carrier-mediated removal of transmitter from the synapse (because the favourable concentration gradient is reduced) and could even lead to net export of transmitter by the membrane transporter. That MAOIs increase the concentration of extracellular monoamines has been confirmed using intracranial microdialysis (Ferrer and Artigas 1994).

The main problems with early, irreversible MAOIs were adverse interactions with other drugs (notably sympathomimetics, such as ephedrine, phenylpropanolamine and tricyclic antidepressants) and the infamous '*cheese reaction*'. The cheese reaction is a consequence of accumulation of the dietary and trace amine, tyramine, in noradrenergic neurons when MAO is inhibited. Tyramine, which is found in cheese and certain other foods (particularly fermented food products and dried meats), is normally metabolised by MAO in the gut wall and liver and so little ever reaches the systemic circulation. MAOIs, by inactivating this enzymic shield, enable tyramine to reach the bloodstream and eventually to be taken up by the monoamine transporters on serotonergic and noradrenergic neurons. Like amphetamine, tyramine reduces the pH gradient across the vesicle membrane which, in turn, causes the vesicular transporter to fail. Transmitter that leaks out of the vesicles into the neuronal cytosol cannot be metabolised because

Table 20.4 Main groups of antidepressant drugs affecting monoamine uptake, release or receptors

Group	Specific mechanism	Examples	Additional notable actions
Inhibition of momoamine uptake			
Tricyclics	Preferential inhibition of noradrenaline *in vivo* (except clomipramine)	*Imipramine* *Clomipramine* *Desipramine* *Dothiepin*	Potent antagonists of: M-receptors, α_1-adrenoceptors and H_1-receptors. Some are antagonists of 5-HT_2 receptors
'NARI'	Inhibition of noradrenaline reuptake	*Maprotiline* *Viloxazine* *Reboxetine*	H_1- and α_1-antagonism
'SSRI'	Inhibition of 5-HT reuptake	*Citalopram* *Fluoxetine* *Fluvoxamine* *Paroxetine* *Sertraline*	Various
SNRIs	Inhibition of noradrenaline and 5-HT reuptake	*Venlafaxine* *Milnacipran*	
Increased monoamine release			
MAOIs	Irreversible, non-selective inhibition of MAO (causes secondary increase in monoamine release)	*Phenelzine* *Pargyline* *Isocarboxazid* *Tranylcypromine*	
RIMA	Reversible, selective inhibition of MAO_A (causes a secondary increase in noradrenaline and 5-HT release)	*Moclobemide* *Pirlindole*	
Tetracyclic	α_2-adrenoceptor antagonist	*Mianserin* *Mirtazepine*	H_1-antagonist, 5-HT_1, 5-HT_2 and 5-HT_3 antagonist and some α_1-antagonism (especially mianserin)
	α_2-adrenoceptor antagonist and some 5-HT uptake inhibition	*Trazodone* *Nefazodone*	5-HT_{1A} and 5-HT_2 antagonist and some α_1- (especially trazodone) and H_1 antagonism
Mechanism unknown			
'Atypical'		*Iprindole*	

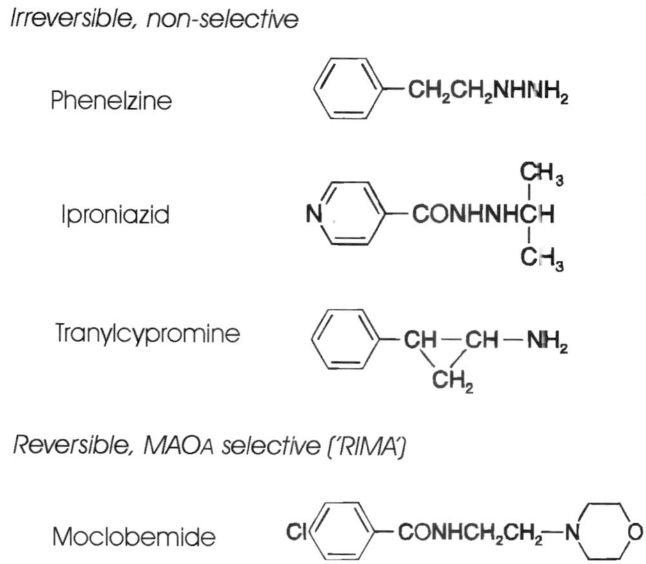

Figure 20.2 The chemical structure of some well-known MAO inhibitors. Most of these drugs irreversibly inhibit both MAO_A and MAO_B but reversible inhibitors (RIMAs), such as moclobemide, inhibit MAO_A only

MAO has been inhibited. As a result, transmitter accumulates in the cytoplasm and is exported into the synapse via the membrane-bound transporter. The ensuing (impulse-independent) sympathetic arousal can be disastrous, culminating in a hypertensive crisis and stroke. Although this process is a pharmacological curiosity and certainly contributed to the demise of MAOIs, it is possibly overrated (Tyrer 1979): it has been estimated that the number of deaths associated with the use of the MAOI, tranylcypromine, amounts to only 1 per 14 000 patient years. However, this sequence of events echoes exactly the acute actions of methylenedioxymethamphetamine (MDMA, 'Ecstasy') and undoubtedly accounts for some of the deaths attributed to this drug.

The discovery that MAO has two isoenzymes with different distributions, substrate specificity and inhibitor sensitivity has helped to rehabilitate the MAOIs to some extent. These isoenzymes are the products of different genes on the X-chromosome and share about 70% sequence homology. Whereas noradrenaline and 5-HT are metabolised preferentially by MAO_A, tyramine and dopamine can be metabolised by either isoenzyme. Selective inhibitors of MAO_A (e.g. moclobemide; Da Prada *et al.* 1989) should therefore be safe and effective antidepressants whereas the selective MAO_B inhibitor, selegiline, should not have any appreciable antidepressant activity (Table 20.5).

Both these predictions are borne out by clinical experience despite the snag that only MAO_B is found in serotonergic neurons (Saura *et al.* 1996). So far, there is no explanation for this anomaly. However, the lack of a tyramine-induced pressor effect with moclobemide probably owes more to the fact that it acts as a reversible inhibitor of $\underline{M}AO_{\underline{A}}$ (RIMA) than to its isoenzyme selectivity. Its reversible inhibition of MAO_A means that, should tyramine ever accumulate in the periphery, it will displace

Table 20.5 Irreversible MAO inhibitors and RIMAs

	MAO_A	Non-selective	MAO_B
Substrates	Noradrenaline 5-Hydroxytryptamine	 Tyramine Dopamine	β-Phenylethylamine Benzylamine
Inhibitors			
Irreversible	Clorgyline	Iproniazid Isocarboxazid Phenelzine Tranylcypromine	Selegiline Pargyline
Reversible ('RIMA')	Befloxaton Brofaromine Moclobemide Pirlindole Toloxatone		

RIMA: Reversible Inhibitor of Monoamine $Oxidase_A$.

moclobemide from the enzyme, thereby ensuring that it is metabolised before reaching sympathetic neurons (Benedetti *et al.* 1983).

TRICYCLIC ANTIDEPRESSANTS (TCAs)

The tricyclic antidepressants (TCAs) derive their name from their three-ringed molecular structure (Fig. 20.3) and emerged, in 1958, from a search for better neuroleptics than chlopromazine among the phenothiazines. The prototype, imipramine, turned out to be ineffective in treating the positive symptoms experienced by schizophrenics but it did relieve their depression (negative symptoms). In fact, imipramine is still the standard agent against which novel antidepressants are compared in clinical trials.

All TCAs are either secondary- or tertiary-amines of a dibenzazepine nucleus (Fig. 20.3), and they all inhibit neuronal reuptake of noradrenaline and/or 5-HT but are much less potent as dopamine reuptake blockers. A common claim is that secondary amines (e.g. desipramine) are preferential inhibitors of noradrenaline uptake whereas the tertiary derivatives (e.g. imipramine, doxepin and amitryptyline) preferentially inhibit 5-HT uptake. However, when Richelson and Pfenning (1984) actually compared the effects of a wide range of antidepressants on the synaptosomal uptake of [^{3}H]monoamines *in vitro*, and compared their K_is, instead of merely ranking IC_{50}s collected from different studies, they found that tertiary- and secondary-substituted compounds were equipotent inhibitors of [^{3}H]noradrenaline uptake. Moreover, all the TCAs turned out to be more potent inhibitors of [^{3}H]noradrenaline than of [^{3}H]5-HT uptake. Tertiary amines are even less convincing inhibitors of 5-HT reuptake *in vivo*, because any such action is diminished by their metabolism to secondary amines (e.g. imipramine to desipramine; amitriptyline to nortriptyline). Only clomipramine retains any appreciable 5-HT uptake blocking activity *in vivo* with (an unimpressive) five-fold selectivity for 5-HT versus noradrenaline.

Set against this background is the finding that the inhibition of [^{3}H]noradrenaline uptake by the neuroleptic, chlorpromazine, is even greater than that of imipramine and yet chlorpromazine has no apparent antidepressant effects. This serves as a testimony

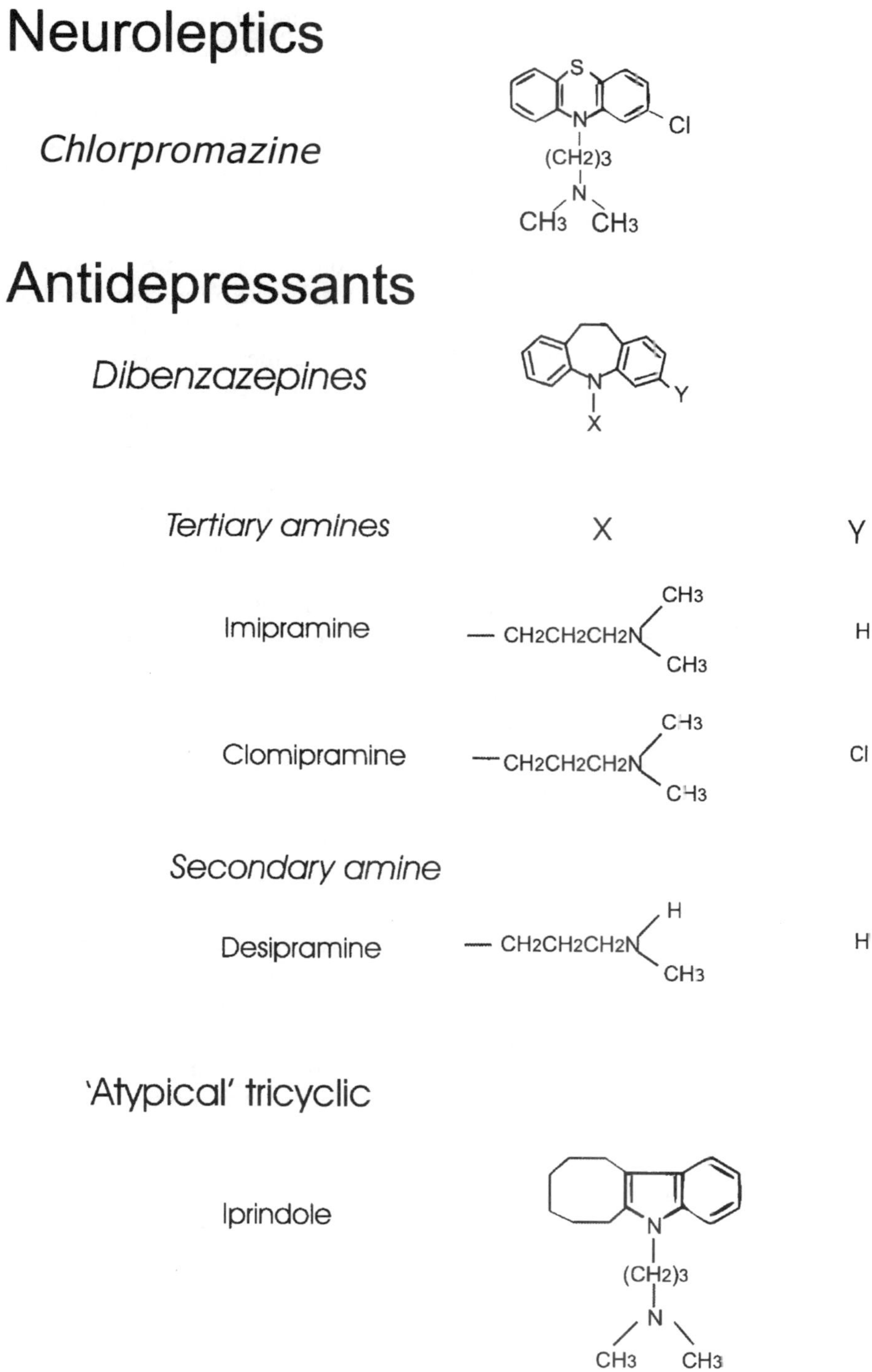

Figure 20.3 The basic structure of tricyclic antidepressants with some well-known examples

to the complex pharmacology of chlorpromazine and also as a warning that studies *in vitro* can be poor predictors of drug effects on patients' mood and behaviour.

The major drawback of the TCAs is their adverse side-effects. These are explained by their high affinity for histamine H_1- and α_1-adrenoceptors and all five of the muscarinic (M-) receptor subtypes. They consequently induce sedation (possibly through H_1-receptor antagonism), anticholinergic effects, such as dry mouth and blurred vision (M-receptor antagonism), orthostatic hypotension and dizziness (α_1-adrenoceptor antagonism). In fact, the tertiary TCA, doxepin, is considerably more potent as an H_1-receptor antagonist (K_B: 56 pM) than the standard H_1-receptor antagonist, mepyramine (K_B: 1 nM). The combination of inhibition of noradrenaline uptake and anticholinergic (antivagal) effects accounts for the pronounced cardiotoxicity of the older TCAs. This is of particular concern when treating the elderly, especially if patients have a history of cardiac problems, and, together with hyperpyrexia and convulsions, these effects explain the toxicity of TCAs in overdose. Other side-effects include loss of libido and stimulation of appetite which leads to weight gain. Little is known about the physiological bases of these actions which, although not life-threatening, are important because they undermine patient compliance.

The adverse side-effects of the TCAs, coupled with their toxicity in overdose, provoked a search for compounds which retained their monoamine uptake blocking activity but which lacked the side-effects arising from interactions with H_1, α_1-adrenoceptors and muscarinic receptors. One of the first compounds to emerge from this effort was iprindole, which has an indole nucleus (Fig. 20.3). This turned out to be an interesting compound because it has no apparent effects on monoamine uptake and is not a MAO inhibitor. This, together with its relatively minor antimuscarinic effects, led to it commonly being described as an 'atypical' antidepressant. Mechanisms that could underlie its therapeutic actions have still not been identified but, in any case, this drug has now been withdrawn in the UK.

α_2-ADRENOCEPTOR ANTAGONISTS

Another compound to emerge from the refinement of the TCAs is the tetracyclic agent, mianserin (Fig. 20.4). Like iprindole, this drug lacks antimuscarinic activity and, since it also has no significant effects on monoamine reuptake or MAO activity, it was regarded as another 'atypical' agent. However, it is now known to act as an α_2-adrenoceptor antagonist, an action that will increase the release of noradrenaline through blockade of autoreceptors on the cell bodies and terminals of noradrenergic neurons (see Chapter 8). In the sense that this process will increase synaptic concentrations of noradrenaline, and is thought to explain (or contribute to) its therapeutic effects, mianserin is not at all 'atypical'. Of course, it is likely that this drug will also block postsynaptic α_2-adrenoceptors, unless it specifically targets a different subtype of this receptor family, but this evidently does not prevent its therapeutic effects.

Mianserin will also increase 5-HT release through inhibition of α_2-heteroceptors on serotonergic neurons. Whether this contributes to its antidepressant actions is uncertain because it is a potent antagonist of 5-$HT_{2A/2B/2C}$ receptors and because, like other antidepressants (and despite being an antagonist), its chronic administration leads to downregulation of these receptors. However, this action of mianserin might well limit or reduce any co-existing anxiety and insomnia. A recent addition to this class of

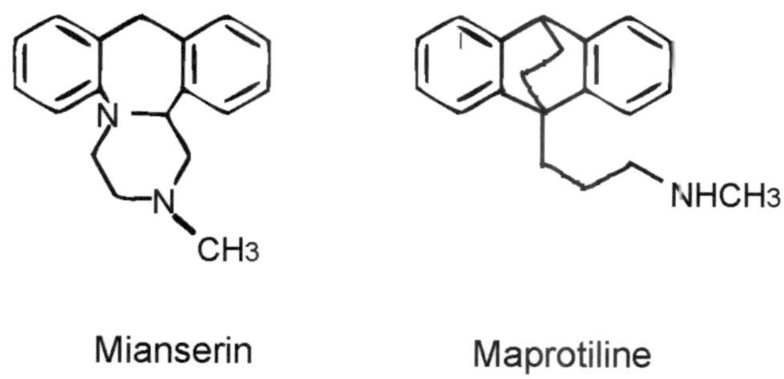

Figure 20.4 The chemical structure of mianserin and maprotiline

antidepressants is mirtazepine, which is an analogue of mianserin but with fewer pronounced side-effects.

SELECTIVE NORADRENALINE REUPTAKE INHIBITORS (NARIs)

Once it was realised that the adverse effects of the TCAs were due to their interactions with transmitter receptors, rather than their inhibition of noradrenaline reuptake, one objective in the development of novel drug treatments was to produce a 'clean' and selective noradrenaline reuptake inhibitor. The first of these was maprotiline, a bridged tricyclic agent (Fig. 20.4) which has a four hundred and fifty-fold selectivity for inhibition of noradrenaline versus 5-HT uptake *in vitro*. Although it has little antimuscarinic activity, its antidepressant activity is compromised because it is highly sedative, probably because of its appreciable H_1-receptor antagonism, and it is also an α_1-adrenoceptor antagonist. Viloxazine, an oxazine derivative of propranolol, is a bicyclic agent which similarly inhibits noradrenaline uptake more than that of 5-HT (hundred-fold selectivity *in vitro*) but which has little anticholinergic or antihistaminic activity. The latest NARI to be recruited in the clinic (1997) is another bicyclic antidepressant, reboxetine (Dostert, Benedetti and Poggesi 1997).

SELECTIVE SEROTONIN REUPTAKE INHIBITORS (SSRIs)

An alternative strategy was to develop drugs that are selective inhibitors of 5-HT reuptake but which, because they are chemically unrelated to the TCAs, would be unlikely to share their side-effects. This direction of research was prompted by the finding, in the late 1960s, that imipramine inhibited 5-HT reuptake, as well as that of noradrenaline, and was reinforced by the evidence that the TCA, clomipramine, was a preferential 5-HT reuptake inhibitor. The first selective serotonin reuptake inhibitor, zimelidine, was tested in the clinic in 1971 but, although it proved to be an effective antidepressant, it was subsequently withdrawn because it could apparently induce the serious neurological disorder, Guillain-Barré syndrome. Nevertheless, other SSRIs quickly followed and five agents are currently available in the UK: fluvoxamine, fluoxetine, paroxetine, sertraline and citalopram (Fig. 20.5).

The SSRIs are all chemically unrelated but their benefits and adverse effects are broadly similar. Their efficacy in depression is not superior to that of the TCAs but their side-effects (nausea, agitation, akathisia and sexual dysfunction), although sometimes problematic, are not life-threatening. They are also considerably safer

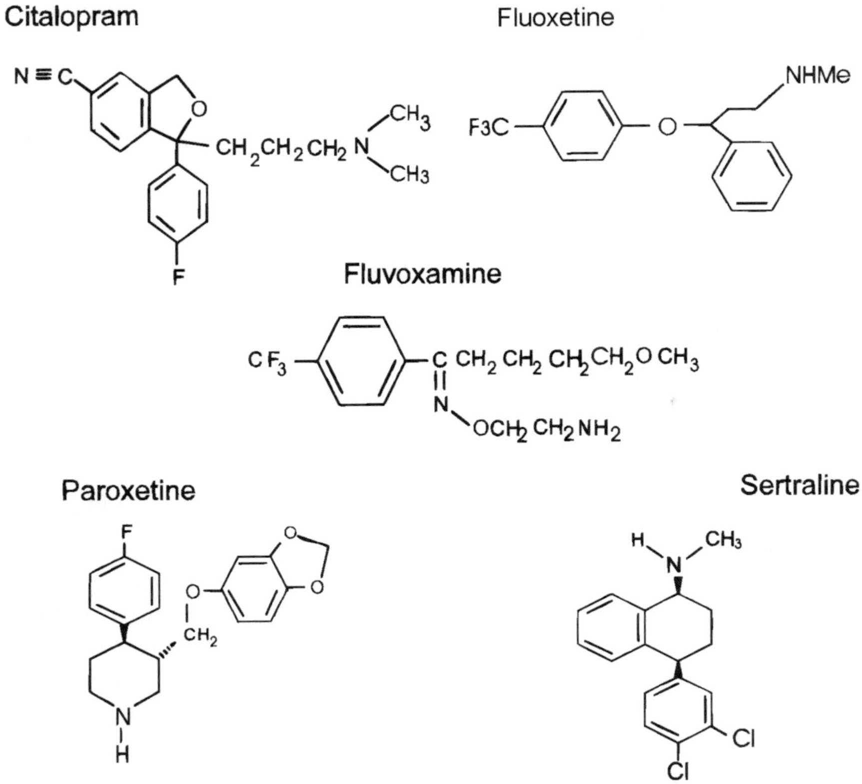

Figure 20.5 The chemical structure of the selective serotonin reuptake inhibitors (SSRIs)

than the TCAs in overdose, but the excessive activation of serotonergic systems can culminate in the 5-HT syndrome, a life-threatening delirium. All SSRIs have other clinical applications, such as in treatment of bulimia, anxiety disorders (e.g. obsessive compulsive disorder, panic disorder, social phobia) and Seasonal Affective disorder.

Paroxetine is the most potent inhibitor of 5-HT reuptake but, in terms of distinguishing one compound from another, their preferential selectivity for inhibition of 5-HT rather than noradrenaline reuptake is the key criterion. Citalopram is by far the most selective *in vitro* (1500–3000-fold) and fluoxetine, the most frequently prescribed SSRI in the UK, is the least selective of all these agents (see Stanford 1999). In fact, it is worth questioning whether fluoxetine is a true SSRI at all.

The most reliable estimates of the selectivity of fluoxetine for inhibition of 5-HT, versus noradrenaline reuptake, put this at twenty-fold, with a high K_i for noradrenaline uptake of between 1 and 10 μM. However, its active metabolite, norfluoxetine, is an even more effective inhibitor of noradrenaline uptake (K_i: 0.1 μM). After chronic administration, the concentration of fluoxetine in the plasma of patients is between 0.5 and 1.5 μM and is thought to be even higher in the brain. Thus, even accounting for pharmacokinetic factors, such as protein binding, the brain concentrations of fluoxetine and norfluoxetine could well be high enough to inhibit noradrenaline reuptake. Similarly, the plasma concentration of citalopram (285 nM) after chronic administration of the recommended therapeutic dose is about a hundred times greater than its K_i for inhibition of 5-HT uptake (1–10 nM), and its corresponding brain concentration is ten-fold greater still. With a K_i for inhibition of noradrenaline uptake of 4 μM, even this drug, the most selective of all the SSRIs, could still express this inhibition in patients.

It is perhaps not surprising that, even after taking into account pharmacokinetic differences between these drugs, the therapeutic doses of the SSRIs do not parallel their K_i for inhibition of 5-HT reuptake. For instance, citalopram is about a thousand times more selective than fluoxetine for inhibition of 5-HT uptake, and yet their clinically effective doses are similar. In short, not only is their selectivity for the 5-HT transporter *in vitro* a poor predictor of their efficacy *in vivo* but it has to be questioned whether any of these compounds actually work by blocking 5-HT uptake alone.

Of course, it has to be borne in mind that there are functional interactions between serotonergic and noradrenergic neurons in the CNS. Indeed, intracerebral microdialysis studies in rats have confirmed that, with the exception of fluvoxamine, all SSRIs increase the concentration of extracellular noradrenaline whether they are given systemically, or by local intracranial infusion. Such an increase could result from activation of 5-HT heteroceptors on noradrenergic neurons. There is plenty of evidence that activation of 5-HT_2, and possibly 5-HT_3 receptors, in the terminal field increases noradrenaline release. There is also evidence that activation of presynaptic 5-HT_{1A}, and possibly 5-HT_2, receptors, increases the activity of noradrenergic neurons in the locus coeruleus. The complex interactions between serotonergic and noradrenergic neurons that could mediate SSRI-induced changes in noradrenaline release are discussed in more detail in Stanford (1999).

Because the SSRIs are derived from different chemical groups, their receptor interactions vary from compound to compound but, apart from paroxetine, none of them shows any appreciable binding to muscarinic receptors, a prime objective of their development. However, compared with other SSRIs, fluoxetine binds with moderately high affinity to human 5-HT_{2A} (K_i: 280 nM) and 5-HT_{2C} receptors (K_i: 55 nM); sertraline is a relatively potent ligand for α_1-adrenoceptors, α_2-adrenoceptors and D_1 receptors and citalopram shows appreciable binding to 5-HT_{1A}, α_1-adrenoceptors and H_1 receptors (Table 20.6; Stanford 1996). The extent to which any of these receptor interactions affects the efficacy of these compounds is not known.

SEROTONIN AND NORADRENALINE REUPTAKE INHIBITORS (SNRIs)

Over the last ten years or so, the emphasis on selectivity of action has waned. This is because the relative importance of blocking noradrenaline and 5-HT reuptake remains uncertain and it is possible that it could be beneficial to block both. Some drugs that act in this way have already been developed. It is hoped that this approach might increase the response rate of patients who are resistant to more selective drug treatments and even reduce the therapeutic lag that dogs their predecessors. As yet, there is not enough information on these compounds to know whether or not this has turned out to be the case.

One of these compounds, venlafaxine (licensed in the UK in 1996), is regarded as an inhibitor of both 5-HT and noradrenaline reuptake but this is based on its actions *in vitro*. At low doses *in vivo*, it is a more potent inhibitor of 5-HT (K_i: 39 nM) than noradrenaline reuptake (K_i: 210 nM). Moreover, its active metabolite, *O*-demethylvenlafaxine, is a weaker inhibitor of NA reuptake, and has a longer half-life, than its parent compound. However, at high doses, venlafaxine inhibits reuptake of both these monoamines but has negligible activity at muscarinic, H_1-receptors or α_1-adrenoceptors and

Table 20.6 Rank order of affinity for receptor binding of the SSRIs

Receptor	K_d or K_i (nM)		
	< 1 μM	1–10 μM	> 10 μM
α_1	sert	zimel > cital. > fluox ⩾ parox > fluvox	
α_2	zimel	sert	fluox > fluvox ⩾ cital > parox
5-HT_{1A} (rat)			fluox > cital > fluvox > sert
5-HT_{2A} (rat)		fluox	sert ⩾ cital > fluvox
5-HT_{2C} (pig)		fluox	sert > cital ≫ fluvox
DA_2			sert > fluox > parox
Muscarinic	parox > sert	fluox > cital	zimel > fluvox
H_1	cital	zimel > fluox	parox ⩾ sert ≫ fluvox
σ_1	fluvox > sert > fluox > cital	parox	
σ_2		sert = cital > fluvox	fluox > parox

Notes:
cital = citalopram; fluox = fluoxetine; fluvox = fluvoxamine; parox = paroxetine; sert = sertraline; zimel = zimelidine.
Sequences derived from within study K_ds or K_is.

so lacks the problematic side-effects of the TCAs. Milnacipran is another agent in this group, and has only a two-fold preference for noradrenaline versus 5-HT reuptake inhibition. Finally, another SNRI, sibutramine, has been found to induce weight loss, for reasons that are not fully understood, and it is licensed for use as an anti-obesity agent rather than as an antidepressant.

SEROTONIN REUPTAKE AND RECEPTOR INHIBITORS

A final group of antidepressants targets both uptake and release of monoamines. These are triazolopyridine derivatives and include trazodone and the more recent addition, nefazodone. Trazodone is a weak inhibitor of 5-HT uptake but shows appreciable binding to 5-HT_{1A} receptors, α_1-adrenoceptors and H_1-receptors and so shares some of the disadvantages of the TCAs. It is also a 5-$HT_{2A/2C}$ receptor antagonist and an α_2-adrenoceptor antagonist, an action that is thought to contribute to its antidepressant actions. A related compound that has recently been introduced into the clinic is nefazodone. This is another weak 5-HT reuptake inhibitor with 5-HT_{2A} antagonist effects but it also inhibits uptake of noradrenaline to some extent. It has a lower affinity for the receptors that are responsible for the unwanted side-effects of trazodone, in particular α_1-adrenoceptors and muscarinic receptors.

Both these compounds have several metabolites and one of these, albeit constituting only 1% of the total, is *m*-chlorophenylpiperazine (*m*CPP). This is a 5-HT_{2C}-receptor agonist/5-HT_{2A} antagonist and has been suggested to contribute to the antidepressant effects of these compounds. In fact, 5-HT_{2C} receptor agonists are currently being explored as potential antidepressants. This is interesting because *m*CPP induces anxiety in humans (Rotzinger *et al.* 1999) and trazodone is contraindicated in the treatment of patients experiencing depression with panic attacks. The enzyme responsible for this metabolic product, CYP2D6, shows genetic polymorphism and so it is possible that the accumulation of *m*CPP is more problematic in some individuals than others.

Ultimately, agonist drugs that directly activate monoamine receptors would appear to be a logical development in this field. Unfortunately, the peripheral side-effects of such compounds could well limit their acceptability even if we were to discover what subset of receptors to target.

NEUROBIOLOGICAL CHANGES INDUCED BY CHRONIC ADMINISTRATION OF ANTIDEPRESSANTS

The actions of all the compounds described so far seem to underpin the monoamine hypothesis. Yet an outstanding problem in treating depression is that the therapeutic response is both slow and progressive: a significant improvement usually takes at least 2–3 weeks and sometimes much longer. Obviously, if we are to explain the therapeutic effects of antidepressants, we must search for long-term neurochemical changes that occur after their prolonged administration.

NORADRENERGIC TARGETS

The first indication that some neurochemical changes developed only after prolonged treatment with antidepressants came from landmark experiments carried out by Vetulani and Sulser in the mid-1970s (Vetulani *et al.* 1976). They found that repeated, but not a single, administration to rats of any of the antidepressants which were available at that time (i.e. MAOIs, TCAs, iprindole and even simulated electroconvulsive therapy) attenuated the increase in cAMP in the cerebral cortex induced by β-adrenoceptor agonists. They suggested that antidepressants desensitised β-adrenoceptors by uncoupling the receptor from what is now recognised as the Gs-protein so that it can no longer synthesise the β-adrenoceptor second messenger, cAMP. Shortly afterwards, it was found that this desensitisation was usually paralleled by downregulation of β_1- (but not β_2-) adrenoceptors. This action is even shared by repeated electroconvulsive shock (Stanford and Nutt 1982) but not by drugs that are ineffective in relieving depression (e.g. neuroleptics).

A logical conclusion from this work was that depression is caused by hyperresponsive β-adrenoceptors. At first, this might seem to undermine Schildkraut's suggestion that depression is caused by a deficit in noradrenergic transmission. However, proliferation of receptors is the normal response to a deficit in transmitter release and so the opposite change, downregulation of β-adrenoceptors by antidepressants, would follow an increase in the concentration of synaptic noradrenaline. This would be consistent with both their proposed mechanism of action and the monoamine theory for depression.

Nonetheless, there are many reasons to be confident that β-adrenoceptor desensitisation does not explain the therapeutic effects of antidepressants. First, with the development of more selective ligands for use in radioligand binding studies, it became evident that β-adrenoceptor downregulation can occur after only 2–3 days of drug treatment (Heal *et al.* 1989). Second, maprotiline, most of the SSRIs, and even some of the newer TCAs have no effect on β-adrenoceptor binding or function. Third, and the greatest problem of all, citalopram *increases* the β-adrenoceptor-mediated cAMP response without changing receptor density. Evidently, we must look elsewhere to find an explanation for the neurobiology of depression and its treatment.

SEROTONERGIC TARGETS

There is a good deal of evidence that the therapeutic effects of antidepressants could involve adaptive changes in 5-HT_{1A} receptors. Postsynaptic 5-HT_{1A} receptor responses became implicated because the hyperpolarisation of hippocampal CA3 pyramidal neurons that follows ionophoretic administration of 5-HT was found to be increased after chronic treatment with most (but not all) antidepressants (Chaput, de Montigny and Blier 1991). Others suggested that antidepressants attenuate postsynaptic 5-HT_{1A} responses because the hypothermia, evoked by their activation, is diminished by antidepressants (Martin *et al.* 1992).

More recently, a series of studies using microdialysis *in vivo* has suggested that long-latency changes in presynaptic 5-HT_{1A} receptors could underlie the therapeutic lag in antidepressant treatment. In these experiments, a single dose of either a SSRI (e.g. fluoxetine or paroxetine), or a MAOI (e.g. tranylcypromine) increased the concentration of extracellular 5-HT in the dorsal Raphé nucleus but not in the brain areas to which these neurons project (e.g. the frontal cortex or striatum; see Hervás *et al.* 1999). The suggested explanation for this regional difference was that the accumulation of extracellular 5-HT in the Raphé nuclei, caused by the SSRIs blocking its reuptake, activates somatodendritic 5-HT_{1A} receptors and so inhibits the firing of serotonergic neurons. This results in reduced impulse flow to their terminals so that extracellular 5-HT does not increase there despite blockade of its reuptake (Fig. 20.6). Obviously, if

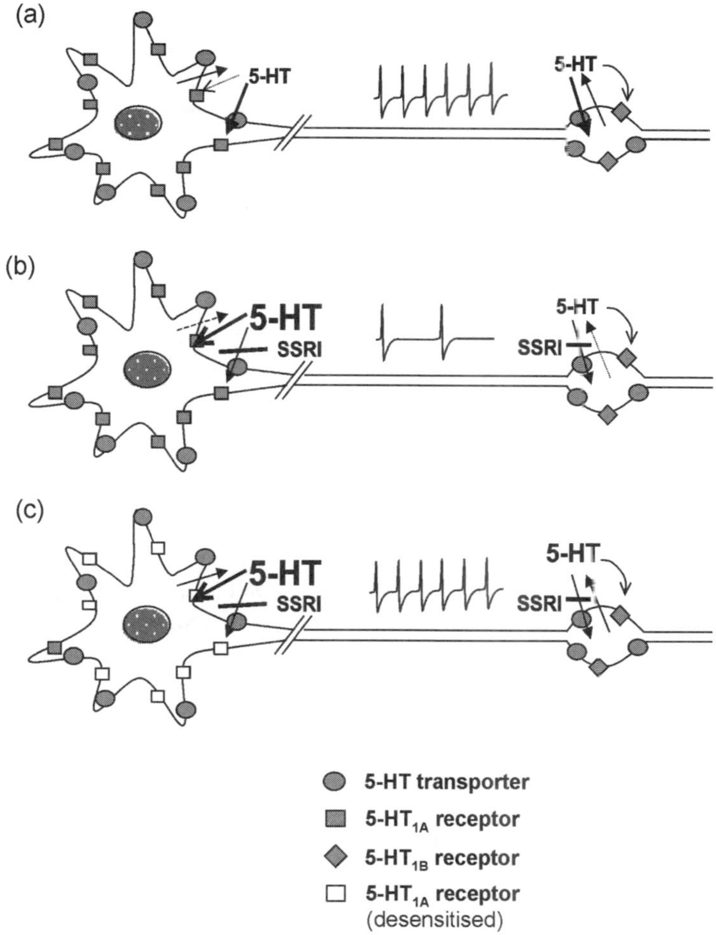

Figure 20.6 Schematic representation of the effects of 5-HT reuptake inhibitors on serotonergic neurons. (a) 5-HT is released at the somatodendritic level and by proximal segments of serotonergic axons within the Raphé nuclei and taken up by the 5-HT transporter. In these conditions there is little tonic activation of somatodendritic 5-HT_{1A} autoreceptors. At nerve terminals 5-HT_{1B} receptors control the 5-HT synthesis and release in a local manner. (b) The blockade of the 5-HT transporter at the level of the Raphé nuclei elevates the concentration of extraneuronal 5-HT to an extent that activates somatodendritic autoreceptors (5-HT_{1A}). This leads to neuronal hyperpolarisation, reduction of the discharge rate and reduction of 5-HT release by forebrain terminals. (c) The exposure to an enhanced extracellular 5-HT concentration produced by continuous treatment with SSRIs desensitises Raphé 5-HT_{1A} autoreceptors. The reduced 5-HT_{1A} function enables serotonergic neurons to recover cell firing and terminal release. Under these conditions, the SSRI-induced blockade of the 5-HT transporter in forebrain nerve terminals results in extracellular 5-HT increases larger than those observed after a single treatment with SSRIs. (Figure and legend taken from Hervás *et al.* 1999 with permission)

this is correct, then blocking 5-HT_{1A} receptors in the Raphé nuclei should prevent these changes. This was confirmed by the finding that SSRIs did increase the concentration of extracellular 5-HT in the cortex and failed to reduce neuronal firing rate if the 5-HT_{1A} receptor antagonist, WAY 100635, was co-administered, either systemically or by infusion directly into the dorsal Raphé nucleus.

More importantly for this discussion is the finding that chronic administration of an antidepressant produces a similar increase in the concentration of extracellular 5-HT in the terminal field together with recovery of neuronal firing. Presumably this is because the prolonged elevation of extracellular 5-HT around the neurons in the Raphé causes progressive desensitisation of the somatodendritic 5-HT_{1A} receptors. At this point, inhibition of their firing does not occur and so more 5-HT is released in the cortex (see Hervás *et al.* 1999).

If long-latency 5-HT_{1A} receptor downregulation explains the antidepressant therapeutic lag, then 5-HT_{1A} receptor antagonists might reduce the delay in treatment response. This prediction has been tested in the clinic using combined treatment with paroxetine and the mixed β-adrenoceptor/5-HT_{1A} antagonist, pindolol and the majority of studies report a successful outcome (see Hervás *et al.* 1999). However, it remains uncertain whether this effect of pindolol is due to its actions at presynaptic 5-HT_{1A} receptors. If, as suggested earlier, postsynaptic 5-HT_{1A} receptors are involved in the therapeutic effects of antidepressants, then co-administration of a 5-HT_{1A} receptor antagonist of this receptor might well diminish any antidepressant effect. Pindolol is said to avoid this problem by its selective antagonism of presynaptic, but not postsynaptic, 5-HT_{1A} receptors, but this is controversial.

A related strategy would be to inactivate the $5\text{-HT}_{1B/1D}$ autoreceptors which are found on serotonergic nerve terminals and so prevent feedback inhibition of 5-HT release in the terminal field. These drugs would not prevent the impact of indirect activation of 5-HT_{1A} receptors, and the reduced neuronal firing, by SSRIs (described above), but they would augment 5-HT release in the terminal field once the presynaptic 5-HT_{1A} receptors have desensitised. Selective $5\text{-HT}_{1B/1D}$ antagonists have been developed only recently but will doubtless soon be tested in humans.

OTHER TRANSMITTER SYSTEMS

The extensive literature on long-latency changes in neurotransmitter receptors following chronic administration of antidepressants reflects the intense effort that has been invested in the search for the cause of their therapeutic lag. Indeed, apart from developing compounds that help patients who currently do not respond to any existing treatment, the most pressing problem in this field is to reduce the delay in treatment response. Yet, despite the numerous investigations of the effects of antidepressants on a wide range of transmitter receptors, few consistent findings have emerged. Results tend to vary not only from laboratory to laboratory and between different brain regions but they also vary with the species and compound tested. The most promising changes are summarised in Table 20.7 but, so far, these do not fit into a scheme that explains either depression or its reversal by antidepressants.

Obviously one limitation of all this work is that the drug effects have been tested in 'normal' animals. So far, the neurochemical changes induced by long-term drug treatment have not been tested in combination with procedures such as learned helplessness, but it cannot be assumed that they will be the same as those in normal (non-depressed) subjects.

THE HPA AXIS AND DEPRESSION

There are clear links between depression and disruption of the neuroendocrine system. Thyroid and gonadal hormone secretion are both abnormal in depression but most

Table 20.7 Neurochemical changes generally found after chronic administration of antidepressant drugs or repeated electroconvulsive shock

Neurochemical marker	Usual finding	Exceptions
Tyrosine hydroxylase expression (locus coeruleus)	↓	
Turnover rate (noradrenaline, dopamine and 5-HT)	↓	
α_1-adrenoceptor density (cortex)	↑	
α_2-adrenoceptor density (cortex)	↓	Inconsistent effects with SSRIs. Some studies find reduction in locus coeruleus, only
β-adrenoceptor binding (cortex)	↓	β_1-adrenoceptors, only. Generally no change with SSRIs
β-adrenoceptor-mediated cAMP response	↓	Inconsistent changes with SSRIs
5-HT_{1A} receptor density (cerebral cortex)	↓	
5-HT_2 receptor binding (rat cortex)	↓	Increased with simulated ECT. No change with SSRIs. Species differences in change
DA_1 receptor-mediated responses	↓	
DA_2 receptor-mediated responses	↑	
NMDA receptor: affinity of glycine for strychnine-insensitive site	↓	

attention has been devoted to the hypothalamic–pituitary–adrenocortical (HPA) axis (Mussleman and Nemeroff 1993). This is a complex system with many interlinked feedback and feedforward controls. However, a key role is thought to be served by corticotropin-releasing factor (CRF) which is released from neurons in the paraventricular nucleus (PVN) in the hypothalamus. From here, CRF is carried to the anterior pituitary where it triggers release of adrenocortical hormone (ACTH) into the systemic circulation. In turn, ACTH promotes release of glucocortioid hormones from the adrenal cortex. Not all CRF release is directed at the HPA system: extra-hypothalamic CRF is found in many limbic areas, including the locus coeruleus and Raphé nuclei (Fig. 20.7).

Normally, circulating glucocorticoids (of which cortisol is the most prominent in humans) cause feedback inhibition of ACTH release so that cortisol secretion is, to some extent, self-limiting. However, many patients suffering from major depression have an increased concentration of plasma cortisol but reduced ACTH secretion. The latter abnormality seems to be partly due to a reduction in the number of CRF receptors in the pituitary, although it is thought that decreased ACTH secretion could provoke the adrenal hyperplasia which is common in depression. This would result in excessive secretion of cortisol and contribute to the inhibition of ACTH release (Musselman and Nemeroff 1993).

Also, a high proportion of depressed patients do not show the reduction in cortisol secretion which is seen when normal subjects are challenged with the synthetic glucocorticoid, dexamethasone, that normally decreases further release through feedback

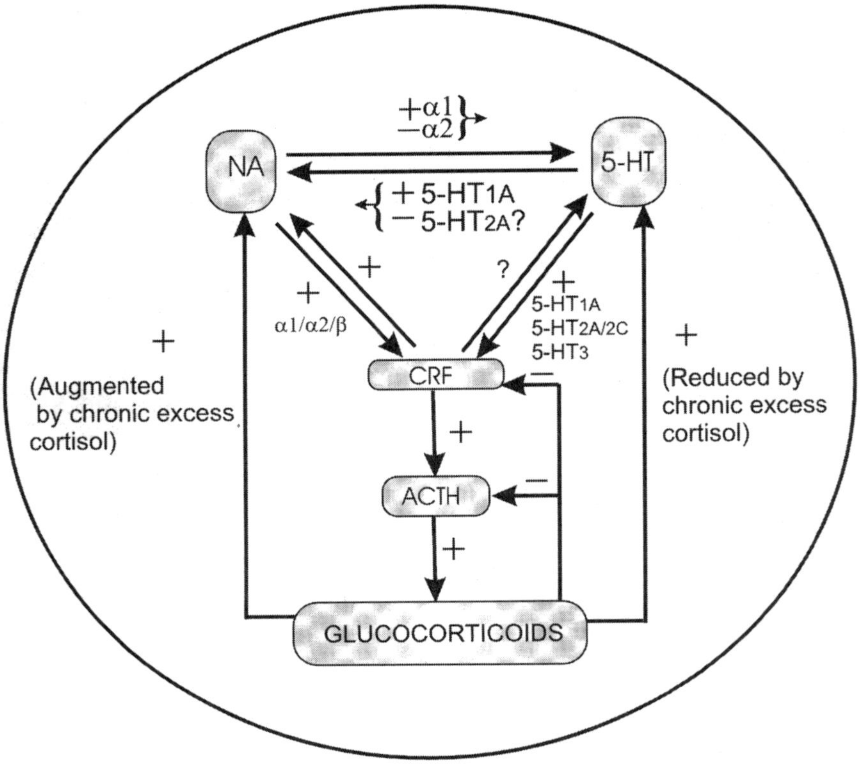

Figure 20.7 Possible interactions between hormones of the HPA axis, extrahypothalamic CRF and the central monoamines, noradrenaline and 5-hydroxytryptamine. See text for a full explanation

inhibition. This is the '*dexamethasone suppression test*' (Carroll, Curtis and Mendels 1976). For most patients whose depression is relieved by antidepressants or electroconvulsive therapy, both the raised concentration of circulating cortisol and the negative 'dexamethasone suppression' response are resolved. This suggests that depression is associated with a defect in the regulation of glucocorticoid secretion and the locus of this disorder could be glucocorticoid receptors in the hippocampus. Evidence that CRF secretion is increased in depressives supports the idea that these receptors, which depress CRF secretion, are hypofunctional in depression (Ritchie and Nemeroff 1991). Also, transgenic mice which are deficient in glucocorticoid receptors exhibit many features of depression; these extend to disruption of feeding and cognitive deficits as well as abnormal HPA function. Antidepressant treatment causes a long-latency increase in hippocampal glucocorticoid receptor binding and the concentration of mRNA for these receptors. Since this happens even in cultured fibroblasts it is thought to involve an action at the level of the genome. Increased glucocorticoid receptor function is thought to restore the feedback effects of cortisol on neurons that regulate CRF secretion (Barden, Reul and Holsboer 1995).

In addition to these actions, glucocorticoids modify the function of several transmitters and/or their receptors, notably GABA, acetylcholine, noradrenaline, 5-HT and sigma (σ)-receptors for which the endogenous ligand is unknown. Conversely, the paraventricular nucleus in the hypothalamus receives noradrenergic and serotonergic inputs from the brainstem, both of which can activate CRF release, so there are complex reciprocal interactions between CRF and monoamine function in the brain. Electrophysiological studies of the locus coeruleus suggest that antidepressants might influence these interactions because chronic administration of antidepressants, from different generic groups, blocks the activation of noradrenergic neurons induced by CRF. However, different antidepressants seem to achieve this through different mechanisms. Some appear to block CRF release (e.g. mianserin) while others show physiological antagonism (e.g. sertraline) (Curtis and Valentino 1994).

AN OVERVIEW OF THE NEUROCHEMISTRY OF DEPRESSION

Separate lines of research have implicated either the noradrenergic, serotonergic or the HPA axis in depression, and there is more evidence, not covered here, that other neuroendocrine systems are involved as well. Yet, all this effort has so far failed to identify disruption of any single transmitter or hormone system as the sole culprit. This points to disruption of the *interactions* between these different systems as the cause of the problem.

Concentrating on the systems highlighted in this chapter, there is plenty of evidence for mutual interactions between the noradrenergic and serotonergic neuronal systems and the HPA hormones: inappropriate release or dysfunctional receptors at any point in these interactions could easily disrupt the balance between these different factors. (Figure 20.7 incorporates some of the interactions that have been characterised so far, but there are doubtless many others.) For example, it is clear that either hyper-responsive α_2-adrenoceptors or hyporesponsive α_1-adrenoceptors could diminish the release of 5-HT evoked by noradrenaline. Conversely, hyporesponsive 5-HT_{1A} receptors and possibly hyperresponsive 5-HT_2 receptors would diminish noradrenaline release from neurons in the locus coeruleus. A disorder of the HPA axis could affect the monoamines in two ways: either directly through effects of CRF on monoamine release or through its effects on glucocorticoid secretion. For instance, whereas CRF can modulate the release of these monoamines directly, 5-HT release is increased by cortisol, but this is blunted by prolonged excessive cortisol secretion (such as can occur in depression). Also, α_2-adrenoceptors, which normally limit release of noradrenaline, are desensitised after chronic exposure to excess cortisol. From this perspective, any single neurochemical factor could have far-reaching effects on all these (and other) neurohumoral systems and could lead to the mood and behavioural changes that culminate in depression. In other words, whereas the expression of an abnormal neurochemical response would be linked with one transmitter system, the problem could lie in another. If this is so, the prospects for finding either a marker for, or a definitive cause of, depression are gloomy, if not misguided.

However, experience proves that depression can be reversed by drugs that augment serotonergic and noradrenergic transmission (and reinstated by a deficit in the synthesis of these monoamines). These, then seem to be crucial targets that ultimately determine mood. This would explain why, despite numerous neurochemical options for the causes of depression, all antidepressants developed so far (and even those discovered by chance) target these neuronal systems. Whatever the cause of depression, therefore, its relief seems to rest on appropriate secretion of these monoamines. This would be entirely consistent with their pervading influence throughout the limbic system (regarded as the governor of mood and behaviour), a characteristic not found so far for any other neurotransmitter.

What remains to be seen is whether it will be possible to accelerate the neurochemical readjustments triggered by antidepressant drugs that target these systems, so as to reduce the latency in their therapeutic effects. We also need to discover why some patients do not respond to any drug treatment. Is this because existing antidepressants simply fail to initiate the appropriate combination of changes in monoaminergic transmission in these patients or do they have a disorder that affects neuronal systems that function in parallel with (or override) the monoamines? If the former is the case, then drugs with combinations of actions which modify monoaminergic transmission in ways that differ from those of established antidepressants might prove to be effective. In the latter case, a new approach to development of antidepressant drugs, targeting completely different transmitter systems, is needed. Obviously, there is a pressing need for future research that will distinguish between these possibilities. Overall, it is 'depressing' to realise that, despite all the advances in our understanding of central neurotransmitters and brain function, the drug treatment of depression has not advanced significantly since the discovery of imipramine nearly half a century ago.

REFERENCES

Anisman, H and Zacharko, RM (1991) Multiple neurochemical and behavioural consequences of stressors: implications for depression. In *Psychopharmacology of Anxiolytics and Antidepressants* (Ed. File, SE), Pergamon Press, New York, pp. 57–82.

Åsberg, M, Traskman, L and Thoren, P (1976) 5-HIAA in the cerebrospinal fluid: a biological chemical suicide predictor. *Arch. Gen. Psychiatry* **33**: 1193–1197.

Barden, N, Reul, JMHM and Holsboer, F (1995) Do antidepressants stabilize mood through actions on the hypothalamic–pituitary–adrenocortical system? *Trends Neurosci.* **18**: 6–10.

Benedetti, M, Boucher, T, Carlsson, A and Fowler, CJ (1983) Intestinal metabolism of tyramine by both forms of monoamine oxidase in the rat. *Biochem. Pharmacol.* **32**: 47–52.

Carlsson, A (1999) The discovery of the SSRIs: a milestone in neuropsychopharmacology and rational drug design. In *Selective Serotonin Reuptake Inhibitors (SSRIs): Past, Present and Future* (Ed. Stanford, SC), RG Landes Co., Austin, TX, pp. 1–7.

Carroll, BJ, Curtis, GC and Mendels, J (1976) Neuroendocrine regulation in depression, II: discrimination of depressed from non-depressed subjects. *Arch. Gen. Psychiatry* **33**: 1051–1058.

Chaput, Y, de Montigny, C and Blier, P (1991) Presynaptic and postsynaptic modifications of the serotonin system by long-term administration of antidepressant treatments. An *in vivo* electrophysiologic study in the rat. *Neuropsychopharmacology* **5**: 219–229.

Cheetham, SC and Heal, DJ (2000) Antidepressant and anxiolytic drugs. In *Biological Psychiatry* (Eds Bittar, EE and Bittar, N), JAI Press, Stanford, CT, pp. 551–567.

Coppen, A (1967) The biochemistry of affective disorders. *Brit. J. Psychiatry* **113**: 1237–1264.

Curtis, A and Valentino, RJ (1994) Corticotropin-releasing factor neurotransmission in locus coeruleus: a possible site of antidepressant action. *Brain Res. Bull.* **35**: 581–587.

Da Prada, M, Kettler, R, Keller, HH, Burkard, WP, Muggli-Maniglio, D and Haefely, WE (1989) Neurochemical profile of moclobemide, a short-acting and reversible inhibitor of monoamine oxidase type A. *J. Pharmacol. Exp. Ther.* **248**: 400–414.

Davis, LL, Yonkers, KA, Trivedi, M, Kramwer, GL and Petty, F (1999) The mechanism of action of SSRIs, a new hypothesis. In *Selective Serotonin Reuptake Inhibitors (SSRIs): Past, Present and Future* (Ed. Stanford, SC), RG Landes Co., Austen, TX, pp. 171–185.

Delgado, PL, Miller, HL, Salomon, RM, Licinio, J, Heninger, GR, Gelenberg, AJ and Charney, DS (1993) Monoamines and the mechanism of antidepressant action: effects of catecholamine depletion on mood of patients treated with antidepressants. *Psychopharmacol. Bull.* **29**: 389–396.

Dostert, P, Benedetti, MS and Poggesi, I (1997) Review of the pharmacokinetics and metabolism of reboxetine, a selective noradrenaline reuptake inhibitor. *Eur. Neuropsychopharmacol.* **7** (Suppl. 1): S23–S35.

Elliott, JM (1992) Peripheral markers in anxiety and depression. *Molec. Aspects Med.* **13**: 173–190.

Ferrer, A and Artigas, F (1994) Effects of single and chronic treatment with tranylcypromine on extracellular serotonin in rat brain. *Eur. J. Pharmacol.* **263**: 227–234.

Fillenz, M and Stanford, SC (1981) Vesicular noradrenaline stores in peripheral nerves of the rat and their modification by tranylcypromine. *Brit. J. Pharmacol.* **73**: 401–404.

Heal, DJ, Butyler, SA, Hurst, EM and Buckett, WR (1989) Antidepressant treatments, including sibutramine hydrochloride and electroconvulsive shock, decrease beta1- but not beta2-adrenoceptors in rat cortex. *J. Neurochem.* **53**: 1019–1025.

Heal, DJ and Cheetham, SC (1999) SSRIs: Where now, where next? In *Selective Serotonin Reuptake Inhibitors (SSRIs): Past, Present and Future* (Ed. Stanford, SC), RG Landes Co., Austen, TX, pp. 187–218.

Henry, JA and Rivas, CA (1999) SSRIs and suicide. In *Selective Serotonin Reuptake Inhibitors (SSRIs): Past, Present and Future* (Ed. Stanford, SC), RG Landes Co., Austin, TX, pp. 95–109.

Hertting, G, Axelrod, J and Whitby, LG (1961) Effect of drugs on the uptake and metabolism of H^3-norepinephrine. *J. Pharmacol. Exp. Ther.* **134**: 146–153.

Hervás, I, Raurich, A, Romero, L, Cortés, R and Artigas, F (1999) SSRI-induced functional changes in serotonergic neurones. In *Selective Serotonin Reuptake Inhibitors (SSRIs): Past, Present and Future* (Ed. Stanford, SC), RG Landes Co., Austin, TX, pp. 127–145.

Horton, RW (1992) The neurochemistry of depression: evidence derived from studies of post-mortem brain tissue. *Molec. Aspects Med.* **13**: 191–203.

Maas, JW (1975) Biogenic amines and depression. Biochemical and pharmacological separation of two types of depression. *Arch. Gen. Psychiatry* **32**: 1357–1361.

Maier, SF (1993) Learned helplessness, relationships with fear and anxiety. In *Stress: from Synapse to Syndrome* (Eds Stanford, SC and Salmon, P), Academic Press, London and New York, pp. 207–243.

Martin, KF, Phillips, I, Hearson, M, Prow, MR and Heal, DJ (1992) Characterization of 8-OH-DPAT-induced hypothermia in mice as a 5-HT1A autoreceptor response and its evaluation as a model to selectively identify antidepressants. *Brit. J. Pharmacol.* **107**: 15–21.

Musselman, DL and Nemeroff, CB (1993) Neuroendocrinology of depression. *Clin. Neurosci.* **1**: 115–121.

Overmier, JB and Seligman, ME (1967) Effects of inescapable shock upon subsequent escape and avoidance responding. *J. Comp. Physiol. Psychol.* **63**: 28–33.

Porsolt, RD, Bertin, A, Blavet, A, Deniel, M and Jalfre, M (1979) Immobility induced by forced swimming in rats: effects of agents which modify central catecholamine and serotonergic activity. *Eur. J. Pharmacol.* **57**: 201–210.

Ricci, LC and Wellman, MM (1990) Monoamines: biochemical markers of suicide. *J. Clin. Psychol.* **46**: 106–116.

Richelson, E and Pfenning, M (1984) Blockade by antidepressants and related compounds of biogenic amine uptake into rat brain synaptosomes: most antidepressants selectively block noradrenaline uptake. *Eur. J. Pharmacol.* **104**: 277–286.

Ritchie, JC and Nemeroff, CB (1991) Stress, the hypothalamic–pituitary–adrenal axis and depression. In *Stress, Neuropeptides and Systemic Disease* (Eds McCubbin, JA, Kaufmann, PG and Nemeroff, CB), Academic Press, London, pp. 181–197.

Rotzinger, S, Bourin, M, Akimoto, Y, Coutts, RT and Baker, GB (1999) Metabolism of some 'second' and 'fourth' generation antidepressants: iprindole, viloxazine, buproprion, mianserin, maprotiline, trazodone, nefazodone and venlafaxine. *Cell. Molec. Neurobiol.* **19**: 427–442.

Salomon, RM, Miller, HL, Delgado, PL and Charney, D (1993) The use of tryptophan depletion to evaluate central serotonin function in depression and other neuropsychiatric disorders. *Int. J. Clin. Psychopharmacol.* **8**: 41–46.

Saura, J, Bleuel, Z, Ulrich, J, Mendelowitsch, A, Chen, K, Shih, JC, Malherbe, P, Da Prada, M and Richards, JG (1996) Molecular neuroanatomy of human monoamine oxidases A and B revealed by quantitative enzyme radioautoradiography and *in situ* hybridization histochemistry. *Neurosci.* **70**: 755–774.

Schildkraut, JJ (1965) The catecholamine hypothesis of affective disorders: a review of supporting evidence. *Am. J. Psychiat.* **122**: 509–522.

Stanford, SC (1995) Central noradrenergic neurones and stress. *Pharmac. Ther.* **68**: 297–342.

Stanford, SC (1996) Prozac: panacea or puzzle? *Trends Pharmacol. Sci.* **17**: 150–154.

Stanford, SC (1999) SSRI-induced changes in catecholaminergic transmission. In *Selective Serotonin Reuptake Inhibitors (SSRIs): Past, Present and Future* (Ed. Stanford, SC), RG Landes Co., Austin, TX, pp. 147–170.

Stanford, SC and Nutt, DJ (1982) Comparison of the effects of repeated electroconvulsive shock on α_2- and β-adrenoceptors in different regions of the rat brain. *Neurosci.* **7**: 1753–1757.

Tyrer, P (1979) Clinical use of monoamine oxidase inhibitors. In *Psychopharmacology of Affective Disorders* (Eds Paykel, ES and Coppen, A), Oxford University Press, Oxford, pp. 159–178.

Van Riezen, H and Leonard, BE (1990) Effects of psychotropic drugs on the behaviour and neurochemistry of olfactory bulbectomised rats. *Pharmac. Ther.* **47**: 21–34.

Vetulani, J, Stawarz, RJ, Dingell, JV and Sulser, F (1976) A possible common mechanism of action of antidepressant treatments. *Naunyn-Schmiedeberg's Arch. Pharmacol.* **293**: 109–114.

Weiss, JM, Goodman, PA, Lostito, BG, Corrigan, S, Charry, JM and Bailey, WH (1981) Behavioral depression produced by an uncontrollable stressor: relationship to norepinephrine, dopamine and serotonin levels in various regions of rat brain. *Brain Res. Rev.* **3**: 167–205.

Willner, P (1984) The validity of animal models of depression. *Psychopharmacology* **83**: 1–16.

21 Pain and Analgesia

A. H. DICKENSON

INTRODUCTION

The mechanisms of pain and the ability to control pain may vary in different pain states. This is of particular importance in consideration of a rational basis for the treatment of both inflammatory and neuropathic pain where the damage to tissue and nerve leads to alterations in both the peripheral and central mechanisms of pain signalling. In respect of existing drug therapies, this plasticity, the ability of the system to change in the face of a particular pain syndrome, explains the effectiveness of NSAIDs in inflammatory conditions and yet is also responsible for some of the limitations in the effectiveness of opioids in neuropathic pain.

For many years, the neurobiological basis for understanding the causes and improving the treatment of pain states remained somewhat unclear. Fortunately, the development of a number of animal models of inflammation and nerve injury, produced by manipulation of either peripheral tissue or nerves, has greatly aided our understanding of the mechanisms of pain and realised many examples of this plasticity. Over the past two decades our knowledge of the pharmacology of pain and analgesia has made enormous strides so that whereas 25 years ago we had a rudimentary idea that morphine worked somewhere in the nervous system we can now recite the complete amino-acid sequence of the four opioid receptors. In parallel with advances in the opioid pharmacology a bewildering list of interacting mediators, transmitters and receptors, some peripheral, some central and some located at both sites, has been established as parts of the initiation and conception of pain.

In recent years, further progress has been made in our understanding of both acute and chronic pain mechanisms that can be largely attributed to advancements in molecular biology and genomic techniques, as well as the use of animals. This has fundamentally altered our understanding of the pathophysiology of pain mechanisms, allowed us to explore new targets for pain relief and has led to the hope of development of novel analgesics. Unfortunately, despite this progress, the management of pain remains a major clinical concern and is still inadequate in many cases and a significant problem even to this day. Not only does it bring undesirable sensations, it can often impair the quality of living for many if not effectively treated.

In broad terms, pain can be divided into two categories, acute and chronic, which differ in their aetiology, mechanisms and pathophysiology. Acute pain and its associated responses are provoked by noxious stimulation and/or disease, or by abnormal function of muscle or viscerae which does not involve actual tissue damage. Although acute pain conditions may last for a length of time if not treated effectively, many cases of acute

Neurotransmitters, Drugs and Brain Function. Edited by R. A. Webster

pain often resolve within days or weeks. In contrast, chronic pain can persist for a long period of time (3 months is usually considered as the transition point from acute to chronic) and results from damage and/or pathology in peripheral tissues or viscerae, or from dysfunction or lesions to the nervous system, either peripheral or central. Pain after tissue damage can be considered as inflammatory pain whereas nerve damage is termed neuropathic pain.

In fact, pain provides a model system for examining how the CNS deals with sensory inputs from both the external and internal areas of the body. Not surprisingly, the intensity, duration and origins of the pain will all have a bearing on the mechanisms underlying the final perception of the patient. Many clinicians, noting the advances made in the basic sciences and our understanding of the pathology and physiology of pain, appear a little disillusioned by the lack of new therapies, and indeed, any magic bullet. First, drug development rests in the hands of the pharmaceutical companies— yet more and more of them are becoming involved in analgesic research, a hopeful change. Second, animal studies really only shed light on efficacy—side-effects may confound the clinical utility of agents. Finally, given the plethora of mediators of pain and analgesia the chance of a single drug being effective in all pain states is unlikely. Combination therapy is possibly the best approach, in that targeting more than one mechanism or site may be fruitful.

PERIPHERAL EVENTS IN THE INITIATION OF PAIN

SENSORY RECEPTORS

Pain is initiated as activation of peripheral sensory fibres by injury or an insult to tissue but is perceived as a sensation through central responses. Thus, it may be relieved either by reducing its initiation by drugs acting peripherally or by drugs acting centrally to reduce the transmission and effects of nociceptive messages sent to the spinal cord and brain. Knowing the mediators involved in both the initiation and transmission of nociceptive impulses provides targets for drug therapy and pain control.

The first stage in the transmission of acute pain involves activation of specialised sensory receptors, the nociceptors, on peripheral C-fibres. These receptors include mechano-, chemo- and thermoreceptors. The terminology of 'receptor' for transmission of somatosensory information can incorporate the type of nerve fibre they are activating, the proposed transduction mechanisms, as well as the form of the adequate stimulus which activates them. Generally, the nerve fibres which respond to non-painful, low-threshold stimulation are the Aβ-fibres and their associated endings. By contrast, Aδ-fibres can be nociceptive or non-nociceptive while nociceptors associated with C-fibres are often termed polymodal since they can respond to a variety of adequate stimuli. The transduction mechanism associated with the free endings of these latter fibres has still to be ascertained. Some C-fibres can, however, also convey low-threshold information while some Aδ-fibres have also been shown to behave as polymodal receptors in their own right with Aδ-mechanoreceptors behaving like C-polymodal afferents after sensitisation.

Primary afferent fibres mediating painful inputs

The somatosensory primary afferent fibre, which conveys sensory information to the spinal cord, can be classified into several classes, according to the transduction

Table 21.1 Classification of somatosensory primary afferent fibres innervating the skin

Primary afferent fibre type	Mean diameter (μm)	Myelination	Mean conduction velocity (m/s)
Aβ	6–12	Myelinated	25–70
Aδ	6–5	Thin myelinated	10–30
C	0.2–1.5	None	<2.5

properties of the individual nerve fibre. The properties of each afferent fibre are summarised in Table 21.1 and their termination sites in the spinal cord are shown in Fig. 21.1.

The afferent fibres differ in their conduction velocity and degree of myelination, and can be distinguished by their diameter. The large diameter Aβ-fibres are myelinated by Schwann cells and hence have a fast conduction velocity. This group of nerve fibres innervates receptors in the dermis and is involved in the transmission of low-threshold, non-noxious information, such as touch. The Aδ-fibre is less densely myelinated and conveys both non-noxious and noxious sensory information. The unmyelinated C-fibre conveys high-threshold noxious inputs and has the slowest conduction velocity of all three fibre types.

Aβ-fibres

The large diameter Aβ-afferent fibre enters the dorsal horn of the spinal cord through the medial division of the dorsal root. It then descends through the medial region of lamina I or II, or alternatively, curves around the medial (central) edge of the dorsal horn down to the ventral horn. On reaching deeper laminae, laminae IV and V, the Aβ-fibres ascend back up into laminae III and IV where they repeatedly subdivide and form a characteristic termination pattern. The densest arborisation appears to occur in lamina III. Axons originating from specialised muscle stretch receptors have collaterals that pass ventrally to make monosynaptic connections with neurons of laminae V, VI and VII. Some also extend to laminae VIII and IX of the ventral horn where they synapse directly onto motor neurons and form the basis of monosynaptic reflexes.

Aδ-fibres

The termination pattern exhibited by Aδ-fibres is entirely different from that of large Aβ-fibres. Aδ-fibres travel extensively in Lissauer's tract, overlying the dorsal horn and their terminals form a plexus at the surface of the spinal cord Aδ-fibres from high-threshold mechanoreceptors distributed to laminae I, II outer and V. Projections also appear to terminate on the contralateral side, in lamina V. Aδ-fibre innervations from deep tissues (muscles and joint) have been shown to terminate exclusively in lamina I, or in laminae IV and V.

C-fibres

Extensive studies have investigated the organisation and termination patterns of C-fibres, employing various techniques including Golgi staining, degeneration techniques and HRP transport. Unmyelinated C-fibres enter the spinal cord through the lateral part of the dorsal white matter, including Lissauer's tract. Studies have shown that unmyelinated primary afferents terminate in the superficial dorsal horn, although there is conflicting evidence as to whether the terminations are restricted to lamina II or whether it includes both laminae I and II. Current evidence suggests that lamina II is the main termination area for cutaneous primary afferent C-fibres while that for Aδ-fibres is in lamina I.

TISSUE DAMAGE AND CHEMICAL MEDIATORS

These polymodal receptors, on C-fibres, can be selectively activated by noxious thermal and mechanical stimuli. In the case of the former modality, we now suspect that a recently characterised receptor-channel (vanilloid receptor 1, VR1) that responds to capsaicin, the extract of hot peppers, may also be responsible for the generation of action potentials after application of heat. Although the endogenous ligand for this receptor is unclear, it may be anandamide, the cannabinoid. The peripheral terminals of small-diameter neurons, especially in conditions of tissue damage like inflammation, are excited by a number of endogenous chemical mediators. These can be released from local non-neuronal cells, the afferent fibres themselves, and from products triggered by activation of the body's defence mechanisms. These chemical mediators then interact to sensitise nociceptors so that afferent activity to a given stimulus is increased. This is known as primary hyperalgesia.

Some of the most important components in inflammation are the products of arachidonic acid metabolism. Arachidonic acid, a component of cell membranes, is liberated by phospholipase A2 and subsequently metabolised by two main pathways which are controlled by two different enzymes, cyclo-oxygenase (COX) and lipoxygenase. This metabolism gives rise to a large number of eicosanoids (leukotrienes, thromboxanes, prostacyclins and prostaglandins) (see Chapter 13). These chemicals do not normally activate nociceptors directly but, by contrast, reduce the C-fibre threshold and so sensitise them to other mediators and stimuli. Thus the value of both steroids and the non-steroidal anti-inflammatory (NSAIDs) drugs in pain after tissue damage is based on their ability to block the conversion of arachidonic acid to these mediators. It should be emphasised that these drugs can only prevent further conversion and will not change the effects of eicosanoids that have already been produced. The short half-life of these mediators makes this fact less important than it would be if the mediators had long-lasting effects. Importantly, a second inducible form of COX, COX-2, has been described. COX occurs in two isoforms, 1 and 2. The former is a constitutive enzyme, always present so that its inhibition affects not only inflammation but also other actions of the products leading to gastric and renal side-effects. By contrast, COX-2 is induced in the periphery by tissue damage and a new generation of selective COX-2 inhibitors have improved therapeutic profiles over existing non-selective drugs. Several novel agents with actions on this latter enzyme are effective in inflammatory pain. Interestingly, COX-2 is normally present in the brain and spinal cord and so may be responsible for some of the central analgesic effects of NSAIDs.

Bradykinin is another chemical with important peripheral actions but, as yet, cannot be manipulated in any direct way by drugs. It is a product of plasma kininogens that find their way to C-fibre endings following plasma extravasation in response to tissue injury. Bradykinin receptors have been characterised and here again, there are two forms. The B_1-receptor is constitutively expressed less than the B_2-receptor, but in chronic inflammation, it is upregulated. Pain may arise via the activation of the B_2-receptor, which is abundant in most tissues and which can activate C-polymodal receptors. The response to bradykinin can be enhanced by prostaglandins, heat and serotonin, indicating the extent of interactions between these peripheral pain mediators.

Hydrogen ions accumulate in tissue damaged by inflammation and ischaemia and so pH is lowered. These protons may activate nociceptors directly via their own family of ion channels as well as sensitising them to mechanical stimulation. Acid-sensing ion channels (ASICS) are a family of sodium channels that are activated by protons — of special interest is one type found only in small dorsal root ganglion neurons that possibly are responsible for activation of nociceptors. Although the transduction of mechanical stimuli is poorly understood, ASICs are closely related to channels that respond to stretch.

VASCULAR DAMAGE, HEADACHE AND MIGRAINE

Serotonin, 5-hydroxytryptamine (5-HT), is released from a number of non-neuronal cells such as platelets and mast cells and can produce an excitation of nociceptive afferents via the activation of its large number of receptors, e.g. 5-HT_{1A}, 5-HT_2 and 5-HT_3 as well as sensitising nociceptors, especially to bradykinin. The key role, but not the mechanisms of action, of 5-HT in the pain associated with migraine and other headaches is well established but little is known about the actions of this mediator in other non-cranial pains. The aura of neurological symptoms and/or signs in migraine is thought to be caused by a vascular or a neuronal mechanism, or a combination of the two. One theory suggests that changes in the vasculature are responsible for causing migraine whereas a second theory proposes that the vascular changes only mediate the pain and symptoms of migraine. A third theory suggests the primary abnormality is neuronal and originates within the brain itself.

The original hypothesis was that vasoconstriction of intracranial vessels leads to a reduced blood flow, which results in cerebral hypoxia. If the arterioles are constricted sufficiently to cause a reduction in regional cerebral blood flow (rCBF), the brain tissue is hypoperfused, which can cause neurological deficits thought to be responsible for the 'aura'. Wolff, who proposed this idea, stated that following the vasoconstriction of the cranial vessels, vasodilatation of these vessels occurred which gave rise to the pain (via the stretching of nerve endings in the vascular walls), and which also resulted in a change in regional cerebral blood flow. There are some weaknesses in the theory that the primary problem is within the vasculature. As the progression of the symptoms does not respect vascular territories it is unlikely to be primarily due to spasm within the vasculature. The blood flow changes are more consistent with a primary neuronal event causing secondary vascular changes. Another factor that makes the theory of a primary vascular abnormality untenable is that the headache may begin while cortical blood flow is still reduced.

A related idea is that peripheral nerves are the source of the problem and then cause the associated vascular changes via release of 5-HT and other inflammatory mediators. The observation that the changes in the vasculature do not follow vascular anatomy has led to a new theory, that of 'spreading depression'. Here, the primary abnormality is within the brain itself, a spreading decrease in electrical activity, that moves at a rate of 2–3 mm/min from the site of origin across the cortex. This transient wavefront suppresses both evoked and spontaneous neuronal activity. In spreading depression, the depolarisation is limited to one hemisphere, and there is a refractory period for further spreading depression of up to 3 min. Any decrease in neuronal firing leads to an increase in metabolism which would result in a decrease in rCBF, via autoregulation.

Sumatriptan is an agonist at 5-HT_{1B} and 5-HT_{1D} receptors. It has three distinct pharmacological actions.

Stimulation of the presynaptic inhibitory 5-HT_{1D} receptors on trigeminal Aδ-fibres inhibits the release of calcitonin gene related peptide (CGRP) which on release forms peripheral ends of sensory fibres via the antidromic axon reflex, causes vasodilatation. Sumatriptan therefore inhibits dural vasodilatation. 5-HT_{1D} receptors on trigeminal C-fibres are also activated by the drug, inhibiting the release of substance P (SP) and neurokinin A (NKA) and therefore blocking neurogenic inflammation and dural plasma extravasation.

Direct attenuation of the excitability of neurons in the trigeminal nuclei, as 5-HT_{1B}/5-HT_{1D} receptors on pain transmission neurons in the trigeminal nucleus caudalis and in the upper cervical cord, are activated. Stimulation of these receptors is caused by second-generation triptans that cross the blood–brain barrier such as zolmitriptan, naratriptan, rizatriptan and eletriptan.

Direct vasoconstriction is mediated by the stimulation of vascular 5-HT_{1B} receptors. These receptors are also found systemically, so coronary arteries also undergo vasoconstriction. Sumatriptan constricts cerebral arteries, but if the vasculature is normal, this does not affect rCBF.

Mast cells, as well as releasing 5-HT, can also release histamine which causes vasodilatation, oedema and itch and ATP and adenosine are also involved in inflammatory conditions. Substance P and CGRP released from the peripheral terminals of primary afferents (via axon reflex) can also cause the mast cells to degranulate and release 5-HT. The peptides cause a number of effects including vasodilatation, plasma extravasation and mast cell degranulation and ATP can result in a direct nociceptor activation via activation of P2X receptors. Other factors such as Nerve Growth Factor (NGF) and cytokines are also important at the peripheral level and resultant changes in the phenotype of the sensory neurons have been shown to be one of the resultant effects. Thus, NGF is upregulated in the area of tissue damage and then binds to its high-affinity receptor, the trkA receptor, one of the tyrosine kinase family. NGF and the receptor are then internalised and transported to the cell body in the dorsal root ganglion. Here, there is a resultant change in gene expression so that the gene for pre-pro tachykinins is turned on. Thus tissue damage causes complex changes in the transduction of painful stimuli. Figure 21.1 shows some of the mediators at the peripheral level with their receptors.

NERVE DAMAGE

Neuropathic pain states are thought to be generated in the peripheral sensory neurons by events within the nerve itself and so are independent of peripheral nociceptor activation. Damage to peripheral nerves can be caused by a number of pathological, metabolic and viral causes. According to the terminology guide of the International

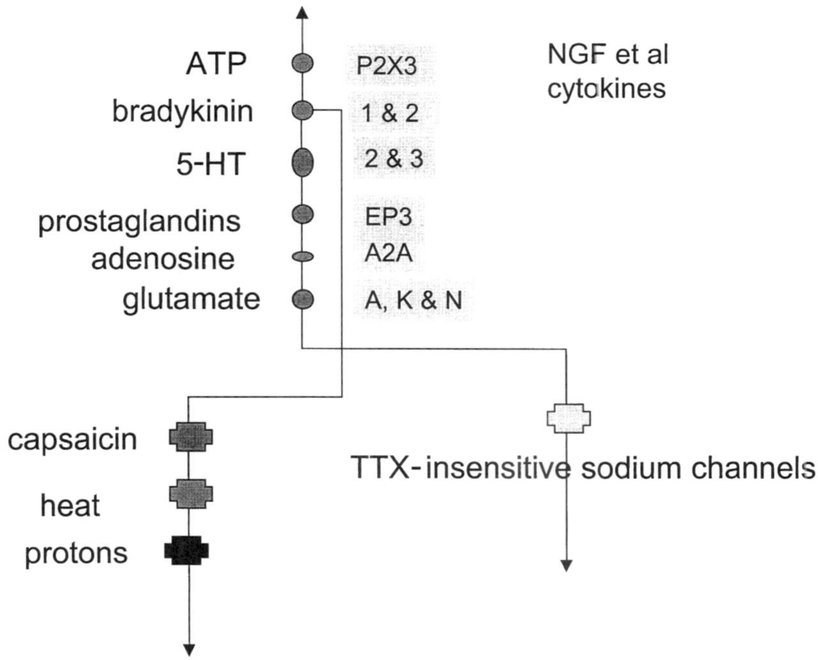

Figure 21.1 Some of the mediators of pain at the peripheral level with their receptors. Note that with regard to 5-HT, the cranial mechanisms have been omitted for clarity

Association of Pain, neuropathic pain is defined as '*pain initiated or caused by a primary lesion, dysfunction in the nervous system*'. Neuropathy can be divided broadly into peripheral and central neuropathic pain, depending on whether the primary lesion or dysfunction is situated in the peripheral or central nervous system. In the periphery, neuropathic pain can result from disease or inflammatory states that affect peripheral nerves (e.g. diabetes mellitus, herpes zoster, HIV) or alternatively due to neuroma formation (amputation, nerve transection), nerve compression (e.g. tumours, entrapment) or other injuries (e.g. nerve crush, trauma). Central pain syndromes, on the other hand, result from alterations in different regions of the brain or the spinal cord. Examples include tumour or trauma affecting particular CNS structures (e.g. brainstem and thalamus) or spinal cord injury. Both the symptoms and origins of neuropathic pain are extremely diverse. Due to this variability, neuropathic pain syndromes are often difficult to treat. Some of the clinical symptoms associated with this condition include spontaneous pain, tactile allodynia (touch-evoked pain), hyperalgesia (enhanced responses to a painful stimulus) and sensory deficits.

Neuropathy elicits a number of changes in nerves, in terms of activity, properties and transmitter content. The recent advent of a number of animal models of neuropathic pain states has facilitated understanding of the peripheral mechanisms involved. Damaged nerves may start to generate ongoing ectopic activity due to the accumulation and clustering of sodium channels around the damaged axons and there is also evidence that mechanoreceptors become highly sensitive to applied stimuli. This aberrant activity can then start to spread rapidly to the cell body in the dorsal root ganglia. Nerve fibres can start to cross-excite each other and the same occurs in the cell bodies.

In addition to changes within the nerve, sympathetic afferents become able to activate sensory afferents via as yet poorly characterised α-adrenoceptors. These interactions between adjacent sensory and autonomic nerve axons and between ganglion cells result in excitation spreading between different nerve fibres. These peripheral ectopic impulses can cause spontaneous pain and prime the spinal cord to exhibit enhanced evoked responses to stimuli, which themselves have greater effects due to increased sensitivity of the peripheral nerves.

This peripheral activity may be a rational basis for the use of systemic local anaesthetics in neuropathic states since ectopic activity in damaged nerves has been shown to be highly sensitive to systemic sodium channel blockers. This too is probably part of the basis for the analgesic effects of established effective anti-convulsants that block sodium channels such as carbamazepine, although central actions are important and may even predominate. The precise actions of excitability blockers therefore remains hazy as does any clear basis for the effectiveness of antidepressants and other adrenergic agents in the treatment of neuropathic pain as both central and peripheral actions, including sympathetic effects are possible.

It has been clearly shown recently that C-fibres can generate action potentials via unique sodium channels with very low TTX sensitivity that are different from those found in other tissues. These channels may become important targets for drugs in neuropathic and other pains since a systemic agent with selectivity for those channels would only block C-fibre activity. However, a complex regulation of these channels after nerve injury makes appraisal of their place in the control of this type of pain difficult—the TTX-resistant channels translocate from the cell bodies of the injured nerves to the site of injury and yet are upregulated in adjacent ganglia. Furthermore, TTX-sensitive channels also upregulate and a novel channel is induced.

CENTRAL EVENTS IN THE TRANSMISSION OF PAIN

SENSORY TRANSMISSION IN THE SPINAL CORD

Morphology of the spinal cord dorsal horn

The spinal cord is arranged in such a way that primary afferents originating from different regions of the body display specific somatotopic organisations upon entry into the cord. Hence in any given segment, there is a definite laterality (ipsilateral/contralateral) and a three-dimensional organisation (rostrocaudal, mediolateral, dorsoventral) of the afferent terminations.

The spinal cord is classically divided into white and grey matter (Fig. 21.2). The grey matter can be organised into ten different laminae, which run continuously along the entire length of the spinal cord. Within a given section of a spinal cord, each lamina can be seen as a layer of functionally distinct cells. Laminae I to VI comprise the dorsal horn, laminae VII to IX the ventral horn, and lamina X is the substantia grisea centralis which surrounds the central canal.

Lamina I

Lamina I forms the outer layer of the dorsal horn and contains the large marginal cells of Waldeyer and plays an important role in nociception since it is the layer in which

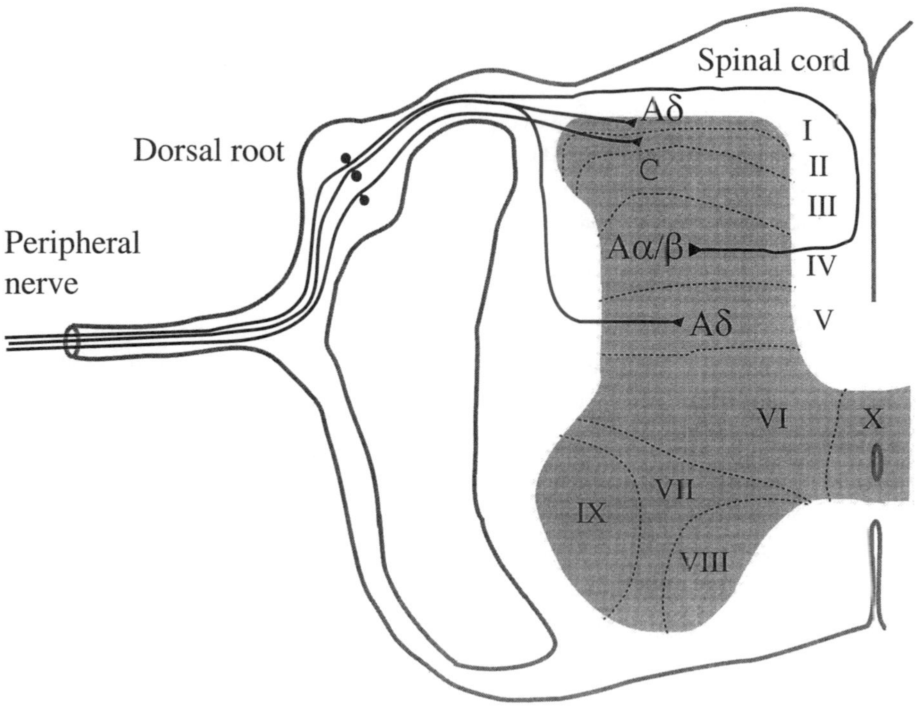

Figure 21.2 The anatomical organisation of the spinal cord, showing the grey and white matter with the laminae terminal zones of the different afferent fibre types

some nociceptive afferents terminate. It contains a number of cell types, including nociceptive-specific neurons, which received Aδ- and C-fibre input, neurons which only respond to innocuous thermal stimuli and wide dynamic range (WDR) neurons. Many marginal cells appear to be projection neurons, which contribute to the lateral spinocervical (SCT), spinoreticular and spinothalamic tracts (STT). The projections also extend to the periaquaductal grey (PAG), parabrachial nucleus and the nucleus submedius.

Lamina II

Lamina II is also known as the substantia gelatinosa (SG) and can be divided into two layers, the outer layer (IIo) and the inner layer (IIi). This layer is densely packed with small neurons and lacks myelinated axons. Neurons with cell bodies in IIi receive inputs from low-threshold mechanoreceptive primary afferents, while those in IIo respond to inputs from high-threshold and thermoreceptive afferents. The intrinsic cells which comprise the SG are predominantly stalk and islet cells. Stalk cells are found located in lamina IIo, particularly on the border of lamina I, and most of their axons have ramifications in lamina I although some also project to deeper layers. These cells are thought to predominantly relay excitatory transmission. Islet cells, on the other hand, are located in IIi and have been demonstrated to contain the inhibitory neurotransmitters, γ-aminobutyric acid (GABA), glycine and enkephalins in their dendrites. Hence these cells have been proposed to be inhibitory interneurons.

Lamina III

The cell bodies in lamina III are generally larger and less densely packed than those in the substantia gelatinosa. The main cell type of lamina III includes projection cells, which contribute to the SCT and postsynaptic dorsal column (PSDC). The dendrites of SCT cells are confined to lamina III and do not reach laminae I and IIo. However, those of PSDC are not flattened in the mediolateral plane and extend to laminae I and II, thus forming monosynaptic connections with small primary afferent fibres.

Laminae IV to VI

Lamina IV is composed of heterogeneous sized cells and is less densely packed than lamina III due to the number of nerve axons passing in this layer. At least three types of neurons have been identified in lamina IV, based on different dendritic projection patterns and these include SCT and PSDC cells. Another cell type has been described which has a dendritic pattern similar to SCT and PSDC, but with local axon terminations. Somas of STT cells are also found in lamina IV.

The cells comprising lamina V are more diverse than those of lamina IV and their dendrites extend vertically toward the superficial layers. Cell bodies in lamina V contribute to three projection pathways, the SCT, PSDC and STT. However, the STT cells appear to be predominant in this lamina. Lamina V plays an important role in nociception since it receives both Aδ- and C-fibre inputs. Some cells in lamina V also respond to cutaneous low- and high-threshold mechanical stimuli and receive nociceptive inputs from the viscerae. Many of these neurons also project onto mononeurons and so act as interneurons in the polysynaptic withdrawal reflex to noxious stimuli.

Lamina VI forms the base of the dorsal horn and can be found only in certain levels of the spinal cord, the cervical and lumbar regions. Few data have been reported on the cell composition of lamina VI. Cells of lamina VI are small compared to those of lamina V and some axons appear to contribute to the STT and SCT pathways.

NEUROTRANSMITTERS AND DRUGS

Nociceptive sensory information arriving from primary afferent fibres enters via the dorsal horn and on entering the spinal cord undergoes considerable convergence and modulation. The spinal cord is an important site at which the various incoming nociceptive signalling systems undergo convergence and modulation and is under ongoing control by peripheral inputs, interneurons and descending controls. One consequence of this modulation is that the relationship between stimulus and response to pain is not always straightforward. The response of output cells could be greatly altered via the interaction of various neurotransmitter systems in the spinal cord, all of which are subject to plasticity and alterations during pathological conditions.

The arrival of action potentials in the dorsal horn of the spinal cord, carrying the sensory information either from nociceptors in inflammation or generated both from nociceptors and intrinsically after nerve damage, produces a complex response to pain. Densely packed neurons, containing most of the channels, transmitters and receptors found anywhere in the CNS, are present in the zones where the C-fibres terminate and while excitatory mechanisms are of importance, the role of controlling inhibitory transmitter systems is perhaps paramount.

Since glutamate is the main excitatory neurotransmitter in the CNS it is not unexpected to find that the vast majority of primary afferents synapsing in the dorsal horn of the spinal cord, regardless of whether they are small or large diameter, utilise this transmitter. It has an excitatory effect on a number of receptors found on both postsynaptic spinal neurons, leading to a depolarisation via three distinct receptor subclasses, the α-amino-3-hydroxy 5-methyl-4-isoxazelopropionic acid (AMPA) receptor, the N-methyl-D-aspartate (NMDA) receptors and the G-protein-linked metabotropic family of receptors. In addition, presynaptic kainate receptors for glutamate have been described in the spinal cord. Most is known about the first two receptors, the AMPA and NMDA receptors, named after chemical analogues of glutamate with selective actions on these sites (see Chapter 11).

Glutamate is released in response to both acute and more persistent noxious stimuli and it is fast AMPA-receptor activation that is responsible for setting the initial baseline level of activity in responses to both noxious inputs and tactile stimuli. However, if a repetitive and high-frequency stimulation of C-fibres occurs there is then an amplification and prolongation of the response of spinal dorsal horn neurons, so-called wind-up (Fig. 21.3). This enhanced activity results from the activation of the NMDA-receptor. If there are only acute or low-frequency noxious or tactile inputs to the spinal cord the activation of the NMDA-receptor is not possible. The reason is that under normal physiological conditions the ion channel of this receptor is blocked by the normal levels of Mg^{2+} found in nervous tissues. This unique Mg^{2+} plug of the channel requires a repeated depolarisation of the membrane to be removed and allows the NMDA receptor-channel to be activated. Here it is likely that the co-release of the peptides such as substance P and CGRP that are found in C-fibres with glutamate is responsible for a prolonged slow depolarisation of the neurons and subsequent removal of the block. Not only do AMPA receptor antagonists have no effect on wind-up but the brief depolarisation produced by this receptor would not be expected to produce any prolonged removal of the block, unlike the long-lasting slow (several seconds) activations caused by peptides. The lack of peptides in large Aβ afferent fibres explains the lack of wind-up after low-threshold stimuli. This NMDA receptor activation has been clearly shown to play a key role in the hyperalgesia and enhancement of pain signalling seen in more persistent pain states including inflammation and neuropathic conditions.

There are a number of antagonists at the multiple regulatory sites found on the NMDA receptor and its channel, including the licensed drugs, ketamine, a potent channel blocker, and the weaker agents, dextromethorphan and memantine. These drugs have been shown to be antinociceptive in a number of animal models of inflammation and nerve damage and there are also data from volunteer and clinical studies to support this. Overall, these studies indicate that it is likely that aberrant peripheral activity is amplified and enhanced by NMDA-receptor-mediated spinal mechanisms in tissue damage and neuropathic pain and that the receptor is critical for both the induction and maintenance of the pain. Thus, therapy after the initiating damage can still be effective. Although there is much good clinical evidence for the effectiveness of agents acting as antagonists at the NMDA-receptor complex, especially ketamine, and although some individual patients get good pain relief in nerve injury situations, the majority cannot achieve complete pain control. This is partly because adequate dosing is prevented by the narrow therapeutic window of the existing drugs.

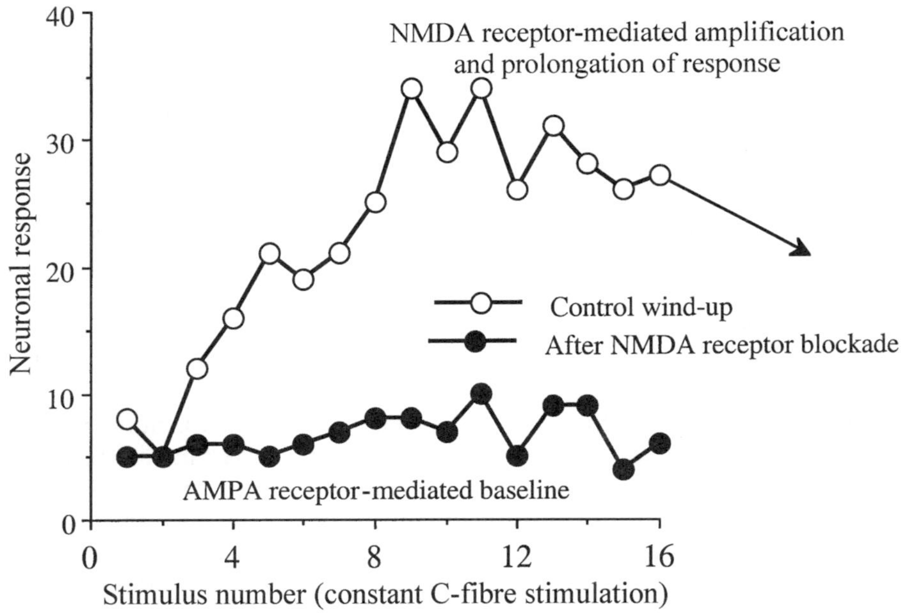

Figure 21.3 Wind-up in a dorsal horn neuron. Note the increased response to a constant peripheral stimulus as the NMDA receptor is activated. (Unpublished data)

Ultimately, the broad use of glutamate receptor/channel antagonists in the treatment of pain will therefore depend on strategies that increase their therapeutic window over existing drugs. These may include drugs acting on subtypes of the receptor (NR2B receptor antagonists are analgesic but side-effects have not been fully evaluated), drugs with different use-dependent block of the channel or more practically, use of low-dose NMDA blockers in combination with another agent.

As neurons become more active, then ion channels, other than sodium channels, open in their membranes. There are a number of voltage-operated calcium channels (see Chapter 3) that are critical for both transmitter release and neuronal excitability. Successful results in animals with agents that block neuronal voltage-sensitive calcium channels would also suggest that there is an increase in central neuronal excitability after both inflammation and nerve damage. N-type channels, blocked by ω-conotoxin, a marine snail toxin, have been shown to play a key role in behavioural allodynia and the neuronal responses to low- and high-threshold natural stimuli after nerve damage, and in the C-fibre-evoked central hyperexcitability that follows inflammation. Blockers of this channel (SNX-111 or ω-conotoxin) are considerably more effective after nerve injury (spinal nerve ligation) and since the channel is voltage operated then these results again suggest increased excitability of the spinal cord after injury. Less is known about P-type channels but ω-agatoxin GVIA, a selective blocker, is effective against persistent inflammatory inputs through central spinal actions. Unfortunately, since calcium channels are extensively distributed in all excitable tissue it is necessary to give blockers used for analgesia by the spinal route.

Gabepentin is an antiepileptic drug that has analgesic activity in neuropathic pain states from varying origins. Two recent randomised controlled trials of gabapentin in patients, one group with postherpetic neuralgia and another with diabetic neuropathy, concluded that gabapentin was effective in the treatment of these pain states. It has also been reported that gabapentin is effective in pain due to peripheral nerve injury and central lesions, with particular effectiveness on paroxysmal pain and allodynia. How gabapentin works is not clearly established but it is thought the drug may interact with calcium channels in that it becomes attached to the so-called gabapentin-binding protein, itself associated with a subunit of the calcium channel. This action would fit with the evidence that N-type calcium channel blockers are more effective in reducing behavioural and electrophysiological responses to sensory stimuli after both nerve injury and tissue damage, conditions where it appears that N-type calcium channels are upregulated.

The influx of calcium through activation of the NMDA channel and also voltage-operated calcium channels may be a mechanism through which further profound changes in nociceptive processing occur. Rises in internal calcium in neurons is a key means by which genes can be activated. The protooncogene markers *c-fos* and *c-jun* can be observed in dorsal horn neurons only minutes after the application of noxious stimulation, either mechanical or thermal or from tissue damage. The one functional piece of evidence at present for the consequences of gene induction is the increase in the mRNA and dynorphin production in some dorsal horn cells, although the physiological consequences of this are unknown.

A comparatively new putative nociceptive transmitter is the gas nitric oxide (NO), and many studies have provided much indirect evidence for a spinal role of this gas during prolonged nociceptive events. NO therefore appears to have a role during prolonged chronic pain states which have been associated with NMDA-receptor activation. It has been proposed that NMDA-receptor activation and the associated Ca^{2+} influx results in the generation of NO by activation of the enzyme, nitric oxide synthase (NOS). The NOS antagonist, L-NAME, abolishes hyperalgesia in neuropathic animals, reduces pain-related behaviour after inflammation and blockers of the production of NO prevent wind-up. One proposed action of NO is as a retrograde transmitter feeding back from spinal neurons onto presynaptic sites to further increase transmitter release from C-fibres. The synthesis of inhibitors of the neuronal version of NOS which lack hypertensive effects yet are antinociceptive suggests possible therapeutic uses of NOS inhibitors.

This positive feedback may also be due to the spinal generation of prostanoids, following both NMDA- and substance P-induced activation of neurons. It is now recognised that in addition to the well-documented production of prostaglandins in peripheral tissues there can be central neuronal synthesis, again with calcium being the trigger. It is not yet known how important this central action is to the analgesic effects of systemic NSAIDs but, as mentioned earlier, COX-2 is constitutive in the spinal cord and further upregulated by peripheral inflammation.

There are important inhibitory systems built into the control of events following C-fibre stimulation. Thus, during peripheral noxious stimulation, spinal mechanism, driven by NMDA-receptor-mediated activity, can become active to damp down further neuronal responses, the purine, adenosine (see Chapter 13), appears to be involved in this type of control and has been reported to be effective in humans with neuropathic pain. It is thought that the depolarisations caused by activation of the NMDA receptor increase the metabolic demand on neurons and so ATP utilisation increases. ATP then is metabolised to adenosene and the purine then acts on its inhibitory A_1 receptor in the spinal cord to reduce further neuronal activity—a negative feedback. Thus there are potential indirect targets for the control of NMDA events. These transmitter systems are summarised in Fig. 21.4.

CENTRAL INHIBITORY SYSTEMS

GABA

γ-Amino butyric acid (GABA) has been firmly established as the major inhibitory neurotransmitter in the central nervous system. The extensive distribution and influence of GABAergic terminals suggests the nervous system operates under considerable restraint, with GABA acting as a tonic controller of excitation. This is also true for the spinal cord where GABA is concentrated in interneurons of the superficial dorsal horn. About one-third of neurons in the superficial spinal cord, the main site of termination of Aδ- and C-fibre afferents, contain GABA. In addition, there is evidence that GABA can co-exist with either glycine, galanin, enkephalin or neuropeptide Y in separate populations of neurons. GABAergic terminals contact more Aδ-fibre terminals than C-fibre terminals, and in support of this anatomical data, the benzodiazepine (Bz), midazolam, has weak depressive effects on C-fibre-evoked responses, but marked effects on Aδ-fibre-evoked responses. In addition, the $GABA_A$ antagonist bicuculline facilitates C-fibre-evoked activity less than the profound potentiation of Aδ-fibre-evoked responses. Both presynaptic and postsynaptic $GABA_A$ receptor-mediated mechanisms are documented in the spinal cord. Several studies have demonstrated Bzs to be analgesic, whereas others have found no antinociceptive properties. In addition, there are contradictory reports of Bzs both potentiating and antagonising morphine analgesia. This diversity of results, however, is the product of many different experimental protocols, models of nociception and routes of administration. In addition, the sedative and myorelaxant effects of these compounds must be considered and these will always limit the usefulness of GABAergic agonists.

In the spinal cord GABA can also activate the G-protein-linked $GABA_B$ receptor, also found pre- and postsynaptically. Baclofen modulation of nociceptive transmission is seen under inflammatory conditions in animals but in humans the drug appears to lack any analgesic effect.

OPIATES

Opiate receptors

Almost all clinically used opioid drugs act on the mu opioid receptor, the receptor for morphine, and they can be highly effective analgesics in many patients unless the pain is due to nerve damage where some patients have inadequate control. The assessment of the analgesic effectiveness of opioids in both animals and in patients is complicated by the fact that the type of neuropathy and the extent, duration and intensity of the symptoms will vary. There is no real consensus from clinical studies on the efficacy of morphine in neuropathic pain states. Dose escalation with morphine was shown to produce good analgesia in one study and others have reported that, in general, morphine could be effective in a group of patients with neuropathy. Another study concluded that opioids were entirely ineffective and finally, opioid analgesia was less in

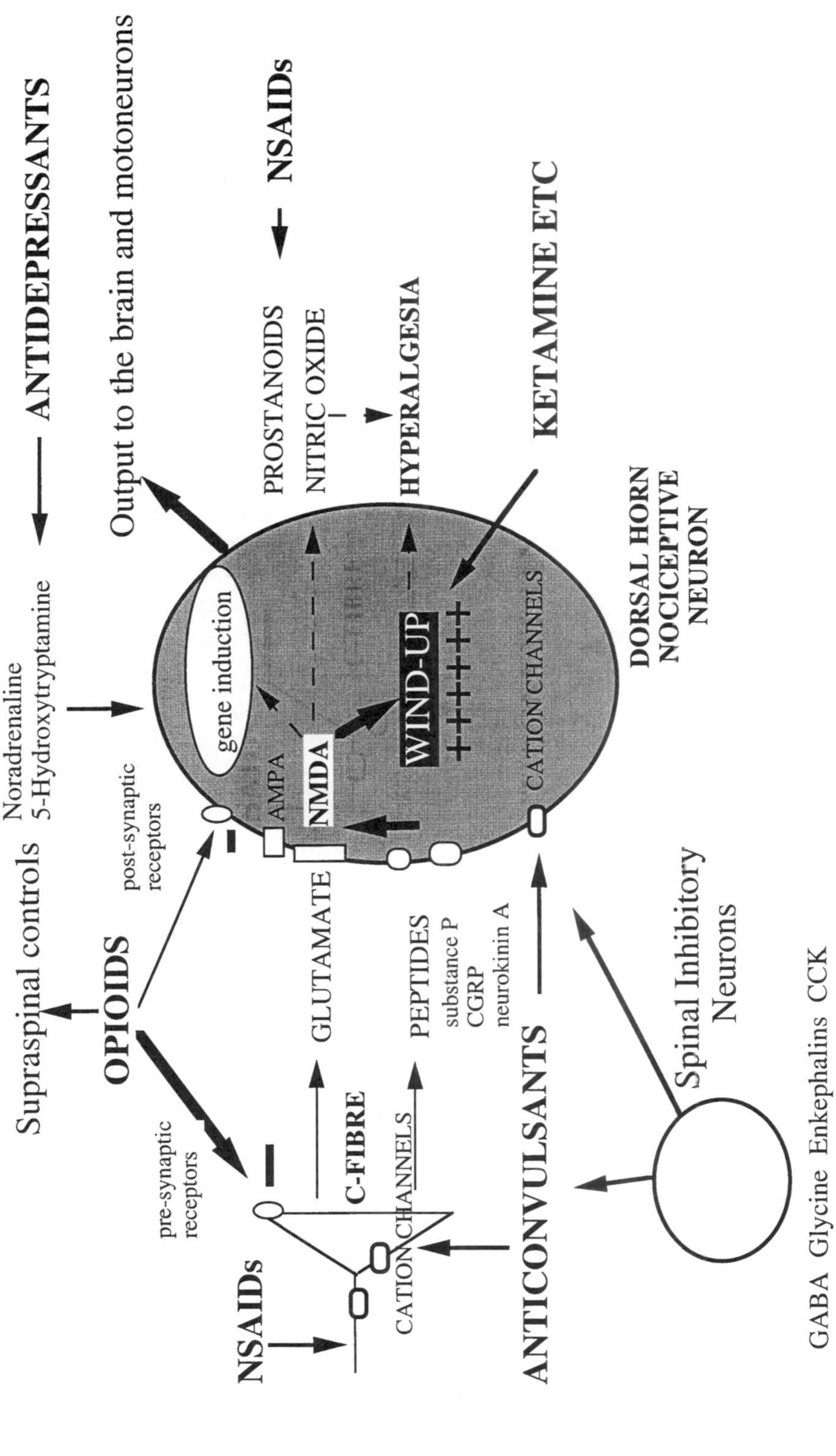

Figure 21.4 An overview of the pharmacological systems involved in the transmission and modulation of pain together with some drugs that act on these systems

neuropathic pain patients as compared to a group with nociceptive pain. Resolution of this problem has important implications yet a similar series of discrepant results can be found in the animal literature.

Following the description and then isolation of opioid receptors, there were three known receptors for the opioids, the mu, delta and kappa opioid receptors, but a novel fourth receptor, the orphan receptor, has been characterised very recently. This newly discovered opioid receptor-like (ORL-1) receptor appears to be linked to an inhibitory receptor despite the endogenous agonist having been named nociceptin (orphanin FQ). The receptor system does not appear to be anything like the traditional opioids. Overall, this peptide produces spinal analgesia but may well function as an 'anti-opioid' at supraspinal sites and cause excitation of C-fibres in the periphery. The central effects of nociceptin include a low abuse potential compared to morphine, and so provide an opportunity for the development of alternative analgesics to morphine. However, sufficiently selective tools for the receptor are lacking; the peptide itself is the only agonist available at present, and the putative antagonist appears to be at best a partial agonist. The apparently paradoxical site-dependent antinociceptive/hyperalgesic effects of this peptide are yet to be resolved.

The actions of all clinically used opiates can now be explained in terms of their acting as agonists at one of the four opiate receptors found in the brain, spinal cord and peripheral nervous system. All four receptors are inhibitory (Table 21.2).

The opioid receptors are for the endogenous opioids, peptide transmitters, β-endorphin, endomorphins, enkephalins, dynorphins and nociceptin. Thus all the problems of drugs based on peptides need to be overcome in order for the roles of these

Table 21.2 The four opioid receptors with transmitters and drugs acting on the various receptors together with the effector mechanisms and the effects of receptor activation for each receptor

Receptor	Mu	Delta	Kappa	ORL-1
Endogenous opioid	β-endorphin Endomorphins	Enkephalins	Dynorphins	Nociceptin
Synthetic agonist	Morphine Codeine Fentanyl Pethidine	DSTBULET DPDPE	U50488H Pentazocine Oxycodone?	–
Antagonists	Naloxone Beta FNA	Naloxone Naltrindole	Naloxone Not BNI	Not naloxone
Effector mechanism	G-protein opens K^+ channel	G-protein opens K^+ channel	G-protein closes Ca^{2+} channel	G-protein opens K^+ channel
Effects	Hyperpolarisation of neurons, inhibition of neurotransmitter release			
	Analgesia Relief of anxiety Euphoria Nausea Constipation Cough suppression Dependence	Similar to mu but less marked	Analgesia Aversion Diuresis	Analgesia Hyperalgesia

receptors to be elucidated. The use of morphine and naloxone, non-peptides with mu selectivity has been responsible for the wealth of knowledge about the mu receptor but much less is known about the delta and ORL-1 receptors. Kappa opioids have weak actions in many animal studies and also cause aversive effects—clinical studies with these drugs have been discontinued. Side-effects are due to the peripheral and central receptors whereas the analgesic effects are due to the interaction of opioid with central receptors. The degree of analgesia can be limited by the side-effects.

These issues make appraisal of different opioid receptors as a target in the development of opioid analgesics lacking the side-effects of mu-receptor-selective agonists such as morphine rather difficult. Progress has been limited in terms of new synthetic opioids that act on the delta receptor, partly due to the peptide nature of the endogenous opioid transmitters but also poor selectivity of drugs between the mu, delta and kappa receptors. The kappa receptor, where synthetic drugs have been produced, does not appear to be a viable analgesic target at present due to central and peripheral side-effects but delta receptor-selective compounds appear to have limited analgesic effects in primate behavioural studies.

There is little new with regard to the mu receptor, the main target for opioid drugs. The receptor is remarkably similar in structure and function in all species studied so animal studies will be good predictors for clinical applications. Although there have been suggestions of subtypes of the receptor, the cloned mu receptors have all been identical.

SPINAL OPIATE ANALGESIA

Opioids act in the brain and within the dorsal horn of the spinal cord, where their actions are better understood. The actions of opioids important for analgesia and their side-effects involve pre- and postsynaptic effects: (1) reduced transmitter release from nerve terminals so that neurons are less excited by excitatory transmitters, and (2) direct inhibitions of neuronal firing so that the information flow from the neuron is reduced but also inhibitions of inhibitory neurons leading to disinhibition. This dual action of opioids can result in a total block of sensory inputs as they arrive in the spinal cord (Fig. 21.5). Thus any new drug would have to equal this dual action in controlling both transmitter release and neuronal firing.

C-fibre stimulation will release a number of transmitters in the spinal cord including substance P, CGRP, glutamate and aspartate. By actions on their receptors the peptides produce slow depolarising responses of dorsal horn neurons which in concert with the fast AMPA and delayed NMDA receptor-mediated depolarisations produced by the excitatory amino-acids activate ascending, local and motoneurons to cause both the sensation of pain and the withdrawal reflex to the stimulus (see Fig. 21.4). There is good reason to believe that the spinal processing of pain is highly plastic and can be altered in different pain states.

The opiate receptors in the spinal cord are predominantly of the mu and delta type and are found in the C-fibre terminal zone (the substantia gelatinosa) in the superficial dorsal horn. Considerable numbers of ORL-1 receptors are also found in this area. Up to 75% of the opiate receptors are found presynaptically on the C-fibre terminals and when activated inhibit neurotransmitter release. The opening of potassium channels will reduce calcium flux in the terminal and so there will be a resultant decrease in

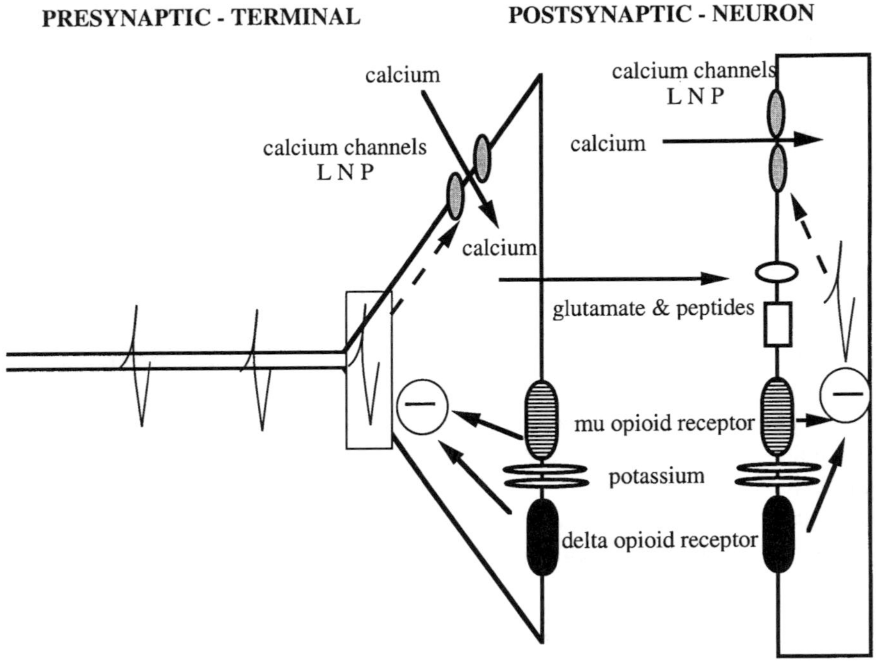

Figure 21.5 Mechanisms of opioid analgesia at the spinal level. Action potentials in nociceptive afferent fibres invade the terminal and by opening calcium channels (L, N and P-type) cause the release of glutamate and peptides that further transmit pain subsequent to activation of their postsynaptic receptors. Presynaptic opioid receptor activation (mu- and delta-mediated effects have been most clearly shown) opens potassium channels which hyperpolarise the terminal, so reducing transmitter release and inhibiting the postsynaptic neuron

release of all the transmitters in the C-fibres. The remaining postsynaptic receptors appear to hyperpolarise the dendrites of projection neurons and interneurons and disinhibit inhibitory interneurons; the net result is further inhibition of the C-fibre-induced activity. This spinal action of opiates can be targeted by using the intrathecal or epidural routes of administration which have an advantage over systemic application of avoiding the side-effects mediated by opiate receptors in the brain and periphery.

Complete C-fibre inhibitions can be produced under normal conditions but opiates do not always produce a complete analgesia in some clinical situations, especially when the pain arises from nerve damage. Reasons for this are suspected to be excessive NMDA-mediated activity which is hard to inhibit and the mobilisation of cholecystokinin in the spinal cord which can act as a physiological antagonist of opiate actions. The idea that pre-emptive analgesia aids post-operative pain relief by preventing the pain-induced activation of these systems is becoming popular.

SUPRASPINAL OPIATE ANALGESIA

There are other important sites of opiate actions located in the 5-HT and noradrenergic nuclei of the brainstem and midbrain including the raphé nuclei, the periaquaductal grey matter and the locus coeruleus. Opiate receptors in these zones (mu, delta and kappa) when activated alter the level of activity in descending pathways from these zones to the spinal cord. The mechanisms of action of opioids at supraspinal levels are still poorly understood. One idea is that descending controls filter sensory messages at the spinal level allowing a pain message to be extracted from the incoming barrage. Supraspinal morphine is thought to reduce these controls so blurring the perception of pain. The second theory is that morphine turns on descending controls which simply inhibit spinal pain transmission. The relative roles of the 5-HT receptors in the spinal cord are yet unknown but the spinal target for the noradrenaline released from descending pathways is alpha-2 receptors which have similar actions and distribution to the opiate receptors. Sedation and hypotension with alpha-2 agonists presently limit their use as analgesics.

Opioid actions at a number of other supraspinal sites (thalamic levels, the amygdala and the sensory cortex) are likely to be of relevance to analgesia.

SIDE-EFFECTS OF OPIATES

CENTRAL

The large numbers of opioid receptors in areas of the brainstem such as the solitary tract and adjacent areas are probably related to respiratory effects of opiates, cough suppression and nausea and vomiting. Opiates acting in the brainstem reduce the sensitivity of the respiratory centres to pCO_2 and this is the most common cause of death from overdose with street use of opiates.

Opiates activate the chemoreceptor trigger zone in the medulla (by disinhibition) to cause nausea and vomiting, and cough suppression also occurs because of the inhibitory effects of opiates on the brainstem nuclei in the cough reflex pathway. Dextromethorphan is the non-opiate isomer of the opiate levorphanol and is an effective cough suppressant.

Sites in the monoamine nuclei such as the well-demonstrated actions of opioids on noradrenergic transmission in the locus coeruleus and enhancing dopamine-release in the ventral tegmental area (again via disinhibition) are likely to be associated with reward processes and so relate to dependence. Psychological dependence does not appear to occur in the presence of pain. Thus, although a patient prescribed morphine over a prolonged period of time will show signs of physical dependence, requiring slow reductions in dose at the end of treatment to avoid withdrawal, drug-seeking behaviour in these patients is very rare. However, with street use, psychological dependence on opioids is rapid to develop and overwhelming. Thus, it would appear that pain prevents psychological dependence. The reason for this is unclear but it could result from the fact that pain is aversive, in that the stimulus produces not only a sensation of pain but also an unpleasant psychological effect. Perhaps this latter characteristic of pain switches off the reward systems in the cortex. Thus opioids in the presence of pain cannot trigger reinforcement.

The relative extent of the unwanted effects caused by selective agonists at the different opioid receptors is of great importance in determining if non-mu opioids will have better spectra of actions as compared to morphine. However, there are good indications that the kappa and delta receptor agonists cause less respiratory depression than mu opioids. A lack of dependence is also seen with kappa agonists but is accompanied by aversive or non-rewarding effects that limit the usefulness of these agents in humans. The endogenous enkephalins are rapidly degraded. Kelatorphan, an inhibitor of the peptidases which degrade the enkephalins, was thought be a novel route to analgesia by prolonging the duration of their actions. This protection of the enkephalins by the peptidase inhibitors has no dependence liability but as yet no peptidase inhibitor selective for the opioid peptides has been reported in humans.

PERIPHERAL

There are a number of side-effects of opiates that are due to their actions on opiate receptors outside the central nervous system. Opiates constrict the pupils by acting on the oculomotor nucleus and cause constipation by activating a maintained contraction of the smooth muscle of the gut which reduces motility. This diminished propulsion coupled with opiates reducing secretion in the gut underlie the anti-diarrhoeal effect. Opiates contract sphincters throughout the gastrointestinal tract. Although these effects are predominantly peripheral in origin there are central contributions as well. Morphine can also release histamine from mast cells and this can produce irritation and bronchospasm in extreme cases. Opiates have minimal cardiovascular effects at therapeutic doses.

OPIATE AGONISTS

All clinically used opiates have the same pharmacology since they all act on the mu receptor with the exception of the kappa agonist, pentazocine. Opiates are used to relieve moderate to severe pain whatever the cause (accidents, post-operative pain, cancer, etc.) and are used pre-, intra- and post-operatively. The mu opiates differ only in potency and pharmacokinetics. Examples are:

- Codeine: a weak opiate which is orally effective and is used for mild pains.
- Methadone: long duration and orally effective, thereby useful in weaning off heroin.
- Fentanyl: highly potent but with a short duration of action, used for short analgesia in surgical settings.
- Heroin (diacetylmorphine): a highly lipophilic drug but has very weak or no affinity for opiate receptors. It penetrates the brain rapidly whereupon it is metabolised to morphine which then binds to the mu receptor.
- Tramadol: a weak opioid that also blocks the reuptake of NA and 5-HT — these combined actions synergise to give a good analgesia that lacks some of the typical opioid side-effects.

OPIATE ANTAGONISTS

There are now selective antagonists for all three opiate receptors (see Table 21.2) but with the exception of naloxone they are experimental tools for probing the functional roles of the opiate receptors. Naloxone is a potent competitive antagonist at all three receptors with highest affinity for the mu receptor. It will rapidly reverse all opiate actions but has a short half-life compared to morphine itself. It is used in cases of overdose, usually to reverse the respiratory depression but with the cost of also reversing the analgesia.

INTERACTIONS WITH OTHER NEUROTRANSMITTERS

Some opioids, such as methadone and ketobemidone, have been reported to bind additionally to NMDA receptors and so may be different in their pharmacological profile. However, it is very unclear that this has any bearing on their effects in patients, especially in cases where morphine effectiveness is reduced, such as in neuropathic pain. In terms of changes in opioid systems relevant to the control of pain after nerve injury, nerve damage can lead to a loss of opioid receptors such as the marked reduction in spinal opioid receptor number seen after nerve section. Although this may be an explanation of the poor effectiveness of opioids in post-amputation pains, less severe nerve damage, where opioids can also lack effectiveness, only slightly alters opioid receptor number. However, the levels of the non-opioid peptide, cholecystokinin (CCK), can determine the potency of morphine and the peptide may, in turn, be upregulated after nerve damage. Activation of the CCK_B receptor mobilises internal calcium whereas opioid receptors hyperpolarise — these actions of CCK thereby physiologically antagonise those of opioids. Antagonists at the CCK_B receptor have been predicted to enhance or restore morphine analgesia after nerve injury but none have been tested in patients as yet.

As discussed earlier, the changes that occur in the periphery and spinal cord after nerve damage can result in overexcitability of spinal neurons so that a hypersensitive state is induced. The N-methyl-D-aspartate (NMDA) receptor is a major candidate in the generation of hyperalgesic states in neuropathic and tissue damage pain states. Quite simply, if neuronal excitability is dramatically increased then opioid controls may be insufficiently efficacious unless doses are increased sufficiently to increase the degree of inhibition required to balance the level of excitation. Here, the combination of a low dose of opioid, increasing inhibition, with a drug that blocks excitation such as ketamine may result in synergistic or additive effects that result in the desired degree of analgesia without adverse side-effects. Other combinations could include the use of anti-convulsants with opioids.

In common with neuropathy, NMDA receptor activation occurs after inflammation but here opioid actions are enhanced since CCK levels decrease. Thus, this augmented opioid actions may counter the increased excitability without the need for large increase in doses of opioid.

BEHAVIOUR AND PAIN

Finally, as outlined above, descending monoamine systems, originating in the midbrain and brainstem that act through the spinal release of noradrenaline and 5-HT, modulate the spinal transmission of pain. $Alpha_2$ adrenoceptors appear to be important in this role but it is unlikely that behavioural effects such as sedation can be separated from the analgesia. Since both noradrenaline and 5-HT are key transmitters in the control of mood and anxiety and yet also participate in the control of sensory events that lead to pain we can start to see links between state of mind and the level of pain experienced. This may be just one early step in the understanding of some of the chemistry of the psychological aspects of pain. Independently of their effects on mood, antidepressants increase activity in these descending control systems and are used as analgesics in neuropathic pain states.

Individual differences in levels of pain, in the transition from acute to chronic pain, in susceptibility to neuropathic pain after nerve damage and in analgesic effectiveness may have a genetic basis. There is marked variability in animal genetic strains in terms of the sequelae of tissue and nerve damage and even in their responses to morphine. Given the huge range of human phenotypes, this may indicate important individual differences in susceptibility to pain and analgesia but we have no way of monitoring this possibility.

REFERENCES

Dickenson, AH (1995) Spinal cord pharmacology of pain. *Brit. J. Anaesthesia* **75**: 192–200.

Dray, A, Urban, L and Dickenson, AH (1994) Pharmacology of chronic pain. *Trends in Pharm. Sci.* **15**: 190–197.

Goadsby, P (2000) The pharmacology of headache. *Progress in Neurobiology* **62**: 509–525.

McMahon, SB, Lewin, GR and Wall, PD (1993) Central excitability triggered by noxious inputs. *Current Opinion in Neurobiol.* **3**: 602–610.

Pain series (1999) *The Lancet* **353**: 1607–2237.

Yaksh, TL (1999) Spinal systems and pain processing. *Trends in Pharm. Sci.* **20**: 329–337.

Section D

NEUROTRANSMITTERS AND BEHAVIOUR

22 Sleep and Waking

R. A. WEBSTER AND S. C. STANFORD

INTRODUCTION

There have been many references in this book to the role of neurotransmitters in the control of CNS excitability. It is therefore appropriate, but possibly foolhardy, to see if the two natural extremes of that excitability, namely sleep and waking, can be explained in terms of neurotransmitter activity. Of course, these states are not constant: our sleep can be deep or light and, even when we are awake, our attention and vigilance fluctuate, as the reading of these pages will no doubt demonstrate. Also, the fact that we sleep does not mean that our neurotransmitters are inactive: this would imply that sleep is a totally passive state, whereas all the evidence suggests that it is an actively induced process, subject to refined physiological control.

In order to explain the physiological characteristics of the sleep–waking cycle, as well as how this might be controlled by different neurotransmitters and modified by drugs, we need to know which areas and pathways in the brain are vital to the induction and maintenance of this rhythmic behaviour. Essentially, these brain systems can be resolved into two interacting networks. One is responsible for the basic circadian rhythm and ensures that our sleeping and waking periods normally occur at regular intervals. A second system fine-tunes this process and ultimately determines our precise functional status on the sleep–waking continuum.

THE NEURAL BASIS OF CIRCADIAN RHYTHMS

It is most probable that sleep and waking stem from an inherent cycle of neuronal activity that can be influenced dramatically by changes in sensory stimulation. This is demonstrable not only in humans and laboratory animals, but also in invertebrates. Thus, while we cannot be sure that other animals sleep in the same way that we do, they do show a circadian cycle of motor activity. In some (nocturnal) species, such as the rat, this activity is actually highest during darkness. Even aplysia, the sea hare, has such a rhythm but this is more like that of humans in being maximally active during daylight (diurnal).

These rhythms seem to be innately programmed although they can be adjusted. For instance, in a normal environment, the sleep–waking cycle of humans is obviously synchronised ('entrained') with the (24-h) dark–light cycle whereas it assumes a period of around 25–27 h in a (time-free) environment where there are no diurnal cues. Interestingly, when humans are in a time-free environment, the change in the rhythm of

Neurotransmitters, Drugs and Brain Function. Edited by R. A. Webster

body temperature does not follow the change in the sleep–waking cycle. Generally, it becomes shorter (to as little as 20 h), rather than longer, which suggests that these cycles are regulated in different ways. Entrainment has also been shown in aplysia which, after exposure to a normal dark–light cycle, retains a cyclic pattern of activity for a number of days even if subjected to continuous light.

At its most fundamental level, the circadian cycle rests on the influence of so-called 'clock genes'. These genes have been studied most extensively in insects but they have also been found in humans. Their protein products enter the cell nucleus and regulate their own transcription. This feedback process is linked to exposure to light and so it is not surprising that visual inputs are important for maintenance of circadian rhythms. However, it is not the reception of specific visual information, transmitted in the optic nerve to the lateral geniculate nucleus (LGN) and visual cortex (i.e. visual discrimination), that is responsible for the rhythm but the more simple, almost subconscious, reception of light.

The fibres conveying this sensation arise in the retina but diverge from the optic nerve and travel in the retinohypothalamic tract (RHT) to innervate the suprachiasmatic nucleus (SCN), a small nucleus which is found in the anterior hypothalamus above the optic chiasma (Fig. 22.3). Destruction of the RHT leads to 'free-running' rhythmic behaviour and so this pathway seems vital for coupling the circadian rhythm to the light cycle. A deficit in information carried in this pathway could help to explain why the blind often suffer from disrupted sleep patterns. Another prominent input to the SCN comes from the intergeniculate leaflet (in the lateral geniculate nucleus (LGN) complex) via the geniculohypothalamic tract (GHT) and, whereas the retinohypothalamic pathway seems to be essential for light-entrainment of the circadian rhythm, the LGN seems to be influenced by rhythmic variations in non-photic inputs such as changes in motor activity. Of course, the LGN is obviously influenced too by visual inputs and, together with the GHT projection to the SCN, can be regarded as an indirect retino-hypothalamic pathway which appears to be inhibitory on SCN neurons. A neuronal input to the SCN from 5-HT neurons in the median Raphé nucleus is another possible route for setting the circadian clock (entrainment) by non-photic stimuli (Fig. 22.1).

Destruction of the SCN, the target of all these pathways, abolishes the synchronised circadian rhythms in locomotor and autonomic function which clearly points to this nucleus as a crucial centre for the control of cyclic function. However, there seems to be some topographical organisation of the neurons in the SCN in respect of their function and the transmitters they release. Whereas those in the dorsomedial zone of this nucleus (or nuclei, since it is paired) contain arginine vasopressin (AVP) or angiotensin II and GABA, neurons in the ventrolateral zone contain vasoactive intestinal peptide (VIP), gastrin-releasing peptide (GRP) and GABA. It is these latter neurons which form the core of the nucleus and show rhythmic pacemaker function. In fact, when maintained in culture, they even display a metabolic rhythm which has the same phase as that of SCN neurons *in vivo*. Unfortunately the presence of GABA in these neurons means that they must be inhibitory and so could not directly stimulate any brain function when activated, e.g. by light inputs, although they could dampen melatonin secretion (see below).

Neurons within the SCN innervate those hypothalamic areas which have a crucial role in the regulation of the reproductive cycle, mood and sleep/arousal, as well as regions such as the basal forebrain and the thalamus which help to determine the state of arousal. They also project to the pineal gland to govern the synthesis and release of

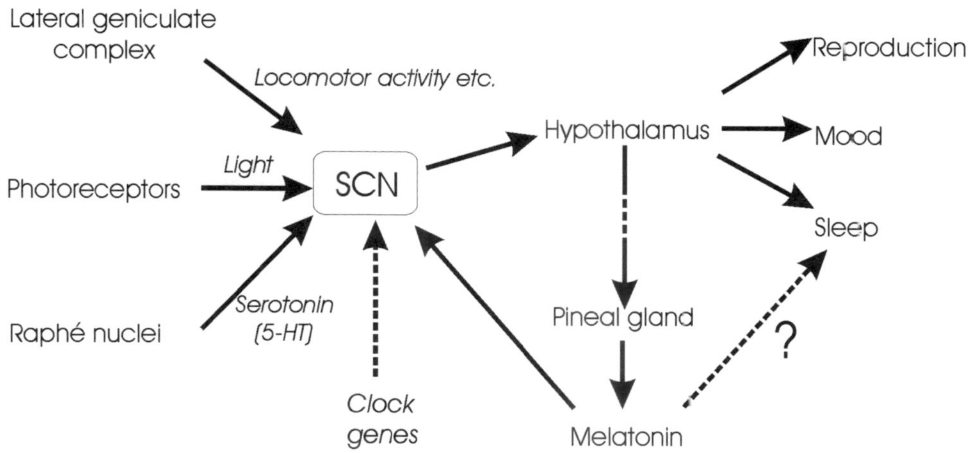

Figure 22.1 Pathways projecting to and from the suprachiasmatic nucleus (SCN). Inputs from photoreceptors in the retina help to 'reset' the circadian clock in response to changes in the light cycle. Other inputs derive from the lateral geniculate complex and the serotonergic, Raphé nuclei and help to reset the SCN in response to non-photic stimuli. Neurons in the SCN project to the hypothalamus, which has a key role in the regulation of the reproductive cycle, mood and the sleep–waking cycle. These neurons also project to the pineal gland which shows rhythmic changes in the rate of synthesis and release of the hormone, melatonin

the hormone, melatonin, which is another factor involved in the control of the 24-h cycle.

MELATONIN

Some lower vertebrates (e.g. frogs and lizards) have what is commonly described as a 'third eye': the pineal gland. This is found in the dorsal cranium and is linked to the diencephalon. Despite its trivial name, the pineal gland does not contribute to discriminative vision and its role is merely to detect changes in light intensity so that, in animals with a clear photoperiod, it couples physiological rhythms with the length of the day–light cycle. In mammals, the pineal is not exteriorised but it persists as a brain appendage for the secretion of the hormone, melatonin. (*N*-acetyl 5-methoxytryptamine) which, like 5-hydroxytryptamine (5-HT), is an indole derivative (Fig. 22.2). Melatonin is not a normal metabolite of neuronal 5-HT but it is synthesised from that amine in the pineal gland by the enzyme, 5-hydroxytryptamine *N*-acetyltransferase. This is a rate-limiting process that shows a circadian rhythm with maximal activity occurring during darkness. The product of this reaction, *N*-acetyl 5-hydroxytryptamine, is methylated, to form melatonin, by the enzyme hydroxyindole-*O*-methyltransferase.

The rate of melatonin synthesis is controlled primarily by the release of noradrenaline from sympathetic fibres originating in the superior cervical ganglion. The activity of these neurons and, consequently, the synthesis and release of melatonin, follows a circadian rhythm such that sympathetic input and melatonin synthesis are both increased in the dark. This coupling with the light cycle certainly involves the SCN since destruction of this nucleus greatly reduces the fluctuations in melatonin production. Moreover, retrograde transneural tracing has shown that there is a neuronal pathway

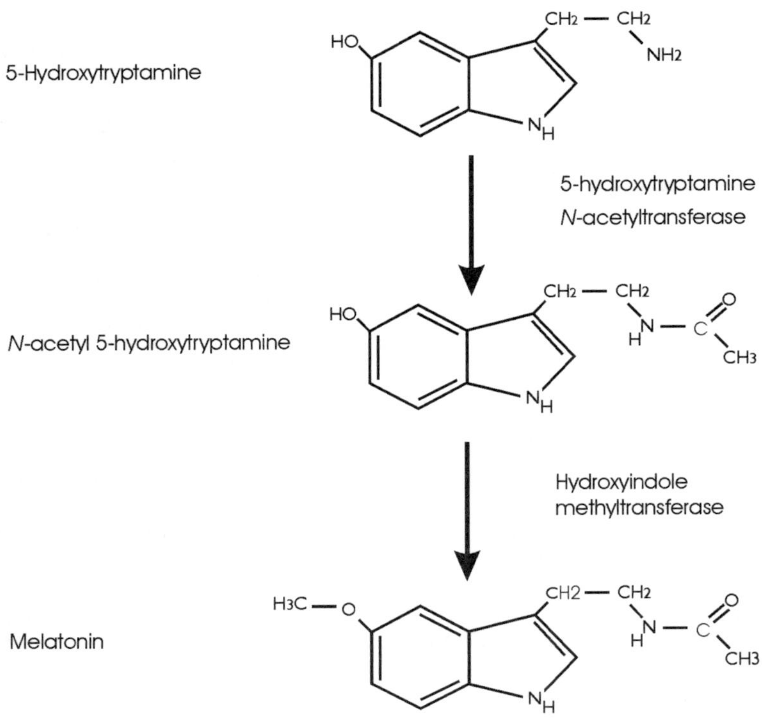

Figure 22.2 The biosynthetic pathway for melatonin

that connects the SCN with sympathetic innervation of the pineal via the paraventricular nucleus of the hypothalamus.

The effect of noradrenaline on melatonin synthesis appears to be mediated through β-adrenoceptors, using cyclic AMP as their second messenger, although studies on cultured pinealocytes suggest that this process is potentiated by activation of α_1-adrenoceptors (see Hagan and Oakley 1995). However, there is evidence that melatonin synthesis in the pineal is also regulated by dopamine and 5-HT. Finally, some melatonin is synthesised in the retina where the rate-limiting enzyme is tryptophan hydroxylase; this process is rhythmic, even in cultured retinal cells, and it seems to adjust to shifts in the light–dark cycle.

The precise role of melatonin in sleep and waking is uncertain but it seems to act as a 'go-between' for the light and biological cycles and evidence suggests that it has a reciprocal relationship with the SCN (Fig. 22.3). Its actions are mediated by (ML_1) receptors which are found predominantly in the SCN as well as thalamic nuclei and the anterior pituitary. These are G protein-coupled receptors, with seven transmembrane domains, that inhibit adenylyl cyclase. Their activation by melatonin, or an ML_1 agonist such as 2-iodomelatonin, restores the impaired circadian cycle in aged rats.

In humans, poor sleep correlates with low plasma melatonin and can be improved by melatonin administration. This therapeutic approach has been tried especially in individuals whose sleep rhythms are disrupted by shift-work, blindness or 'jet-lag' but its benefits are as yet unconfirmed and, in any case, the mechanisms by which it might reset sleep patterns are unclear. Of course, it must be remembered that other body

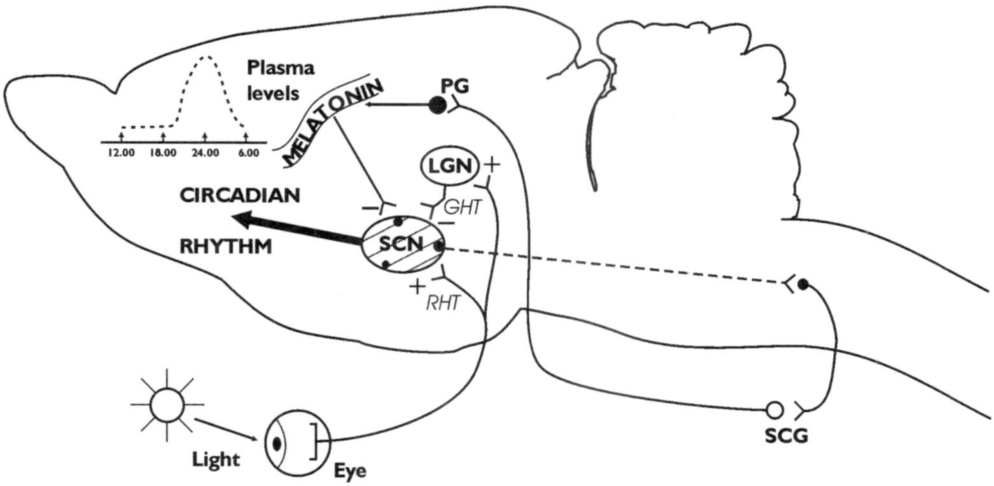

Figure 22.3 Possible links in the induction of circadian rhythm between daylight, the suprachiasmatic nucleus and melatonin release from the pineal gland. Some fibres in the optic nerve, projecting from the eye to the lateral geniculate nucleus (LGN) in the thalamus, innervate the suprachiasmatic nucleus (SCN) in the anterior hypothalamus, via the retinohypothalamic tract (RHT). Others project to the SCN from the LGN in the geniculohypothalamic tract (GHT). The release of melatonin into the circulation from the pineal gland (PG) is maximal at night and appears to be controlled partly by noradrenaline released from sympathetic nerves originating in the superior cervical ganglion (SCG). Melatonin receptors are found in the SCN, the removal of which dampens melatonin secretion

functions show a circadian rhythm, some of which, such as corticosteroid production (high in morning) and body temperature (low during sleep) could all influence the state of arousal. However, the night-time peak for melatonin secretion normally coincides with the trough for body temperature, and these two events could well be linked. Nevertheless, whether melatonin affects sleep itself, rather than merely the entrainment of the sleep rhythm, is controversial (for a detailed review of this topic see Arendt *et al.* 1999).

SLEEP

Defining sleep is not at all straightforward but its general features comprise (see Hendricks, Sehgal and Pack 2000) (1) a sterotypical, species-specific posture; (2) an absence of voluntary movements; (3) elevated threshold for arousing stimuli; (4) reversibility on stimulation of the individual (or organism). The following sections outline what is known about how these changes come about and how they are regulated.

THE ELECTROENCEPHALOGRAM (EEG)

Probably the most important breakthrough in sleep research came in the mid-1930s when it was discovered that the profile of the electroencephalogram (EEG) changed markedly during the sleep–waking cycle (Fig. 22.4). To this day, the EEG is a major

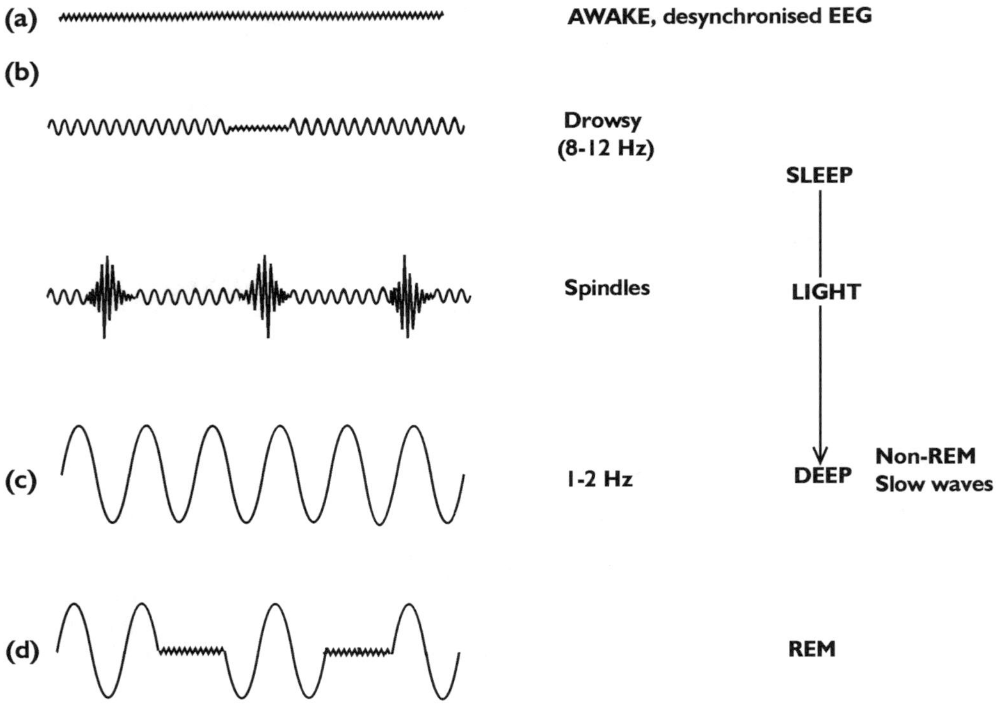

Figure 22.4 Idealised EEG-like patterns in sleep and waking. When we are awake and aroused the EEG is desynchronised (a). As we become drowsy and pass into sleep the EEG waves become more synchronised with 8–12 Hz alpha waves (b), sleep spindles then appear (c) before the EEG becomes even more synchronised with slow (about 1–2 Hz) high-voltage waves characteristic of deep slow-wave sleep (SWS). About every 90 min this pattern is disrupted and the EEG becomes more like that in arousal (d) except that the subject remains asleep. This phase of sleep is also characterised by rolling, rapid eye movements, the so-called REM sleep. SWS is consequently also known as non-REM sleep. These tracings have been drawn to show the main features of the different EEG phases of sleep and as such are much simpler than those that are actually recorded

focus of sleep research but is usually complemented by measurements of muscle tone (the electromyogram, EMG) and eye movements (the electro-occulogram, EOG) which also show marked changes during the sleep cycle.

When we are aroused and awake, the EEG is random (desynchronised) with multiple high-frequency (of at least 15 Hz), low-amplitude γ(gamma)-wave forms. As we become drowsy and close our eyes, the EEG becomes more synchronised and a clear rhythm emerges (stage 1 sleep): this is α(alpha)-rhythm which has a frequency of 8–12 Hz. At the onset of sleep (stage 2), θ(theta)-waves (4–7 Hz) are evident but these are disrupted to some extent by the intermittent appearance of waves, known as K-complexes and 'sleep spindles'. The former are single spikes whereas the latter are short trains of pulses (12–14 Hz). Progressing still further into the sleep state (as assessed by the EMG and EOG), the EEG becomes even more synchronised so that slower (about 1–2 Hz) and larger waves become more prominent. These are the δ(delta)-waves which are associated with stage 4 (deep) sleep, often called 'slow-wave sleep' (SWS). At the same time as all these changes are developing, the threshold for arousal by sensory stimuli increases.

It was not until much later (1953) that another phase of the sleep cycle was discovered. At about 90 min after the onset of sleep, the EEG becomes desynchronised and, in fact, it bears a strong resemblance to that seen in stage 1, apart from the appearance of so-called 'PGO-waves' (see below). Also, rapid eye movements, resembling those while reading in the awake state, are evident on the EOG: this is REM (rapid eye movement) or 'paradoxical' sleep. However, in adults, other physiological changes that occur during REM sleep are quite different from those of stage 1. In particular, there is a flaccid paralysis of the limb muscles together with a loss of fine control of body temperature and other homeostatic mechanisms. It is often maintained that dreaming is restricted to these periods of REM sleep, which occur some three or four times during the night, each lasting about 30 min. However, it is now thought that dreams also occur during SWS but that these are more logical and more consistent with normal life events than are those occurring during REM sleep.

This sleep pattern, seen in adults, takes some time to develop and appears in infants only around 6 months to one year after birth. Instead, as new parents will testify, young babies have a sleep cycle that lasts only around 3–6 h. Further striking differences are that babies' REM sleep accounts for as much as half the sleep cycle (compared with only a quarter in the adult) and is accompanied by increased motor activity with spasmodic movements of the limbs and facial muscles, rather than the muscle atonia seen in adults. In fact, the adult sleep cycle can take up to 20 years to stabilise and its pattern changes again in the elderly who show a reduction in the duration of SWS, an increase in the proportion of REM sleep, and increased daytime 'napping'.

The functions of these different phases of sleep are not at all clear but chronic sleep deprivation does eventually lead to death. It seems to be the slow-wave component of sleep (SWS) that is vital and it is thought to serve a restorative purpose. This would be consistent with its greater occurrence during the early stages of the sleep cycle when hormone secretion supports anabolic metabolism. If subjects are wakened every time they enter a period of REM sleep (evidenced by the EEG) there appears to be no overt harmful effect on their behaviour. In fact, REM sleep deprivation has even been used, with some claims of success, as a treatment for minor depression. However, there is an unproven belief that REM sleep is important for memory consolidation.

ORIGIN OF THE EEG

It appears that the voltage waves recorded in the EEG represent the summation of synaptic potentials in the apical dendrites of pyramidal cells in the cortex. These cells generate sufficient extracellular current for it to reach, and be recorded from, the cranium and scalp. Although these waves originate from the cortex rather than the SCN, the distinctive REM and non-REM phases of sleep still remain after destruction of the SCN but they then occur randomly over the 24-h cycle. This is a further indication that the SCN is at least partly responsible for setting the overall circadian rhythm of the sleep cycle.

The more synchronised the activity of the cortical neurons, the greater the summation of currents and the larger and slower the EEG wave, as in the sleep pattern (Fig. 22.4). While there are some dissociations between EEG pattern and behavioural states, the EEG offers one way of determining experimentally the pathways (and neurotransmitters) that control arousal and sleep, and can be regarded as an important objective measurement of the cortical correlates of sleep and waking.

The slow (deep sleep) δ-waves probably originate in the cortex because they survive separation from, or lesions of, the thalamus. However, the rhythm and appearance of spindles in earlier phases of the sleep cycle do depend on links with the thalamus (see Steriade 1999). Unlike stimulation of the specific sensory relay nuclei in the thalamus, which only affects neurons in the appropriate sensory areas of the cortex, the non-specific nuclei can produce responses throughout the cortex and may not only control, but also generate, cortical activity. Certainly, *in vitro* studies show that neurons of the non-specific reticular thalamic nucleus (NspRTN) can fire spontaneously at about 8–12 Hz (equivalent to EEG α-rhythm) or lower, and that low-frequency stimulation of this area can induce sleep.

Maintenance of these frequencies relies on the degree of depolarisation of the thalamic neurons (Jahnsen and Llinas 1985) and this, in turn, depends on the nature and intensity of their afferent inputs. The NspRTN and other thalamic nuclei receive reciprocal inputs from the cortex and it is possible that it is the ensuing oscillations in neuronal activity in this circuit between the cortex and thalamus that give rise to the sleep spindle waves in stages 2–4. In fact, it has been suggested that the stronger and clearer these oscillations become, the more likely it is that there will be loss of consciousness.

Apart from neuronal inputs originating in the cortex, thalamic afferents (see Fig. 22.5) come from:

(1) *Collaterals from neurons of neighbouring specific thalamo-cortical relay nuclei.* Because these neurons are themselves activated by sensory inputs transmitted along the spinothalamic tract, this provides one way in which sensory stimuli can influence cortical activity generally, as well as specifically.
(2) *Ascending inputs from the brainstem ascending reticular activating system (ARAS).* As described below, these seem to be particularly important and probably disrupt the thalamo-cortical synchrony.

SLEEP AND WAKING CENTRES

One of the first experiments to investigate the brain mechanisms that might be involved in regulation of sleep and waking showed that after transection of the brain of cats, so that the cerebrum was separated from the brainstem, the animal displayed continuous sleep. Conversely, transection that separated the entire brain, including the brainstem, from the spinal cord (at the level of Cl) caused continuous arousal. Jouvet (1974) extended this work by showing that a lesion at a specific site in the pons abolished REM sleep, together with the associated muscle atonia and EEG changes, but did not affect SWS. All this work suggested the existence, not only of 'sleep' and 'waking' centres in the brain, but also that a separate brain area was responsible for REM sleep. Later studies confirmed the existence of these brain centres in that stimulation of the anterior hypothalamus, at a frequency similar to that of the sleep spindles in the EEG, induced sleep whereas stimulation of a zone of the brainstem, that came to be known as the ascending reticular activating system (ARAS), induced arousal (Moruzzi and Mayoun 1949).

The generally accepted view is that the stimulatory drive for the ARAS comes from collaterals of the classical ascending sensory pathways. Indeed, this is another way in which sensory stimuli can affect our state of arousal (Fig. 22.5). The diffuse activating

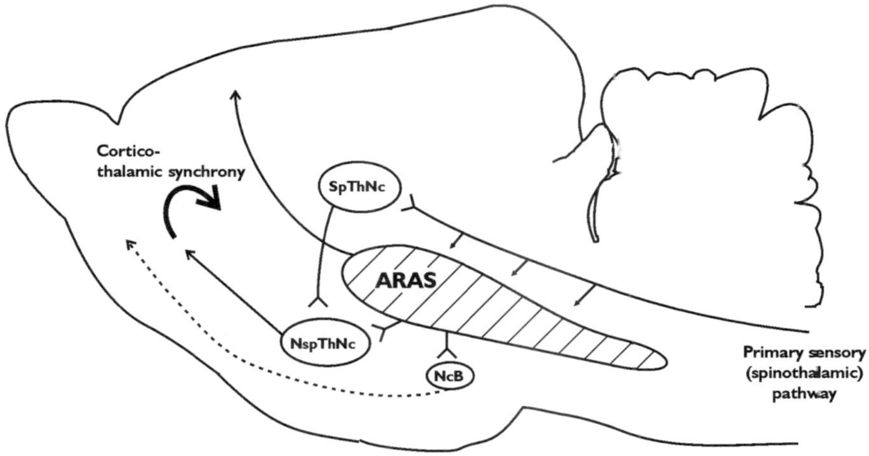

Figure 22.5 Pathways involved in cortico-thalamic synchrony and EEG arousal. The ascending reticular activating system (ARAS) extends from the cephalic medulla through the pons and mid-brain to the thalamus (see Moruzzi and Mayoun 1949). It is activated by impulses in collaterals of the spinothalamic sensory pathway running to specific thalamic nuclei (SpThNc) and in turn activates much of the cortex, partly through the non-specific thalamic nuclei (NspThNc), which also receive inputs from SpThNc and also via the nucleus basalis (NcB). Its stimulation is followed by EEG arousal. It is probable that reciprocal links between cortical areas and the thalamus, particularly NspThN, lead to slow-wave (8 Hz) cortical EEG synchrony and, in the absence of appropriate sensory input and ARAS activity, a sleep state

system ensures that all sensory stimuli, whatever their strength or modality, contribute collectively to cortical arousal. This is possible because part of any sensory input is diverted to the ARAS and so prevents the cortex from reverting to its basic slow-wave oscillating rhythm. Thus, not only will the sensory cortex be more responsive to any primary sensory input it receives, but its activation keeps us alert. In this respect, the ARAS can be considered to contribute to our circadian rhythm by helping to ensure that we have an active cortex and so stay awake when we have adequate stimulation. Nevertheless, humans deprived of diurnal cues (such as when they are confined in an insulated, 'time-free' chamber) still show a sleep–waking cycle, although it progressively adopts a longer time period.

In addition to the excitatory drive, there are also inhibitory neurons from the anterior hypothalamus which provide one route for suppressing activity in the ARAS. Another inhibitory influence comes from the spinal cord. Together, these links could help to ensure smooth progression from one state of arousal to another. Also, during REM sleep, pontine–geniculate–occipital (PGO) waves travel to the cerebral cortex and spinal cord and it is this wave of activity, passing through intermediate brain regions, that is thought to blunt sensory and motor function.

It is important to emphasise that a lesion of the reticular system disrupts a number of afferent inputs to the cortex. Particularly important in this respect are the monoaminergic (especially noradrenaline, 5-HT and histamine) and cholinergic pathways. When the ascending inputs from these neurons are destroyed, sleep is passive and not at all like natural sleep which, as detailed above, has distinct phases and depends on brainstem influences on cortical function. How these different neurotransmitters might influence sleep and arousal will be considered next.

NEUROTRANSMITTER SYSTEMS

Based on the above account of the neuronal pathways thought to be responsible for the basic sleep–wake cycle, the neurotransmitters that are most likely to be involved in the cycle are those which:

(1) Are released either in the cortex or the non-specific thalamic nuclei.
(2) Augment, or more probably, break up thalamic-cortico synchrony and its tendency to promote slow-wave EEG activity and non-REM sleep. Whether this results in full arousal, or merely a temporary disruption of sleep to give REM periods without full awaking, will depend on the balance of inputs and the overall state of cortical activity.

Some of these inputs come from cholinergic, histaminergic, noradrenergic and 5-HT neurons. These neurons innervate the cortex more than the thalamus and their possible roles will be considered in the following sections. This material draws on studies designed to show: which neurotransmitters are associated with those brain structures concerned with sleep and waking; how their function may change during the cycle; to what extent pharmacological manipulation of their activity influences the cycle; and how drugs which modify our state of arousal affect neurotransmitters.

ACETYLCHOLINE

Studies of several animal species, ranging from rats to sheep, have shown that the release of acetylcholine (ACh) into cortical cups (see Chapter 4 and 6) is increased in proportion to cortical (EEG) activity, being maximal during convulsions and lowest under deep anaesthesia. These findings are consistent with evidence that cortical arousal (EEG desynchronisation) is increased by injection of ACh into the carotid artery of animals, or by direct stimulation of the ascending reticular system (ARAS), and that both these actions are blocked by the muscarinic receptor antagonist, atropine. It has even been shown in humans that REM sleep is induced by intravenous infusion of centrally-acting cholinomimetic agents, such as arecoline or physostigmine (an acetylcholinesterase inhibitor), and, again, the effects of these treatments are inhibited by atropine. Yet antimuscarinic drugs do not have any marked sedative effects on behavioural arousal. This could mean that sedation requires recruitment of the 'sleep' system, as well as blockade of arousal.

As outlined previously (Chapter 6), cholinergic neurons are located in two broad groups of nuclei, both of which are linked to the ARAS and thalamus (Fig. 22.6). One group lies rostrally in the basal forebrain, within the nucleus basalis, medial septum and diagonal band. This system is more active during the waking state than during sleep and blocking its effects could well explain how antimuscarinic drugs inhibit EEG desynchronisation. The nucleus basalis, which sends diffuse projections to the cortex and hippocampus, has also been linked with memory function (Chapter 18).

The second cluster of neurons lies more caudally, near the pons, in the pedunculo-pontine (PPT) and laterodorsal tegmental (LDT) nuclei (see Fig. 22.6) and could be regarded as part of the ARAS (see McCormick 1992). It innervates the non-specific thalamic nuclei as well as some more specific ones like the lateral geniculate nucleus (visual pathway), the pontine reticular formation and occipital cortex. Because long

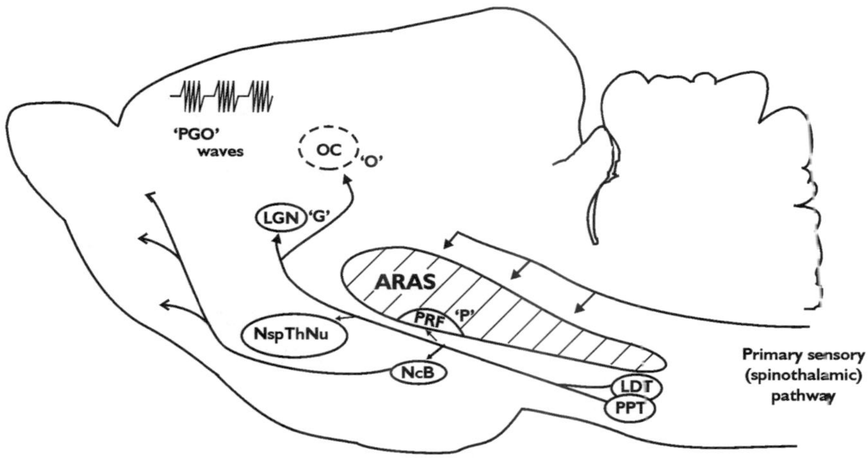

Figure 22.6 Cholinergic influences on sleep and arousal. Cholinergic neurons are found primarily either rostral to the ascending reticular activating system (ARAS) in the nucleus basalis (NcB) caudally in the pedunculo pontine tegmentum (PPT) nucleus. The former, which innervate much of the cortex, receive inputs from the ARAS and appear to be partly responsible for maintaining the EEG and behavioural arousal. The latter innervate non-specific (NspThNu) and specific (SpThN) thalamic nuclei, including the lateral geniculate nucleus as well as the pontine reticular formation (PRF) and occipital cortex (OC). The high-voltage pontine–geniculo–occipital (PGO) waves they initiate in all three areas are characteristic of REM sleep, which is reduced by their destruction

bursts of high-voltage waves occur in all these three terminal areas during REM sleep, forming the pontine–geniculo–occipital (PGO) waves described above, they could derive from the PPT (see Hobson 1992).

In fact, there is a good deal of evidence to support this suggestion. First, more than half the neurons in the PPT fire rhythmically only when PGO waves are evident and their firing starts immediately before the PGO waves appear. Second, in cats, REM sleep is augmented by direct injection of either carbachol, or more selective muscarinic agonists, or the anticholinesterase, neostigmine, into the pontine reticular formation (one of the projection sites for PPT). Third, REM sleep is abolished by lesion of the PPT nucleus but, interestingly, not by lesion of the LDT.

Overall, there are compelling reasons to believe that cholinergic pathways not only play a part in arousal but also contribute to the induction of the 'arousal-like' features of REM sleep.

HISTAMINE

Although histamine has mixed excitatory and inhibitory effects on central neurons, those antihistamines (H_1-receptor antagonists) that enter the brain produce sedation; this indicates that the predominant overall effect of histamine is excitatory. The preferred explanation for this rests on evidence that histaminergic neurons in the posterior hypothalamus are active in waking and silent in deep SWS and REM sleep.

The histamine neurons in the tuberomammillary nucleus, in the posterior hypothalamus, project to the cortex and thalamus and receive an afferent input from

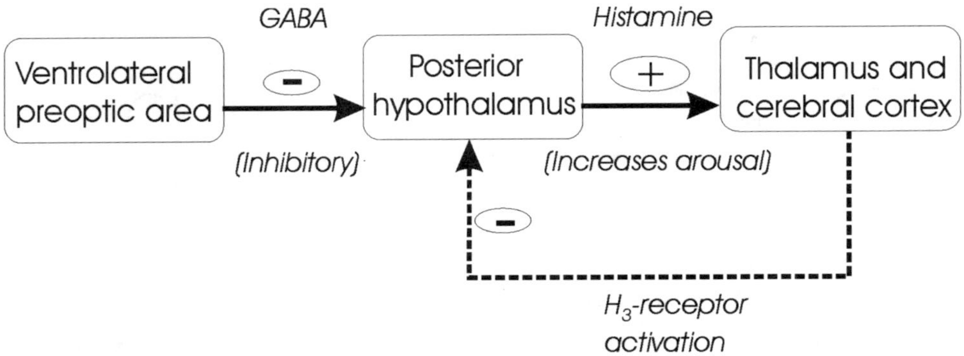

Figure 22.7 Histamine influences on sleep and arousal. The activity of histamine-releasing neurons increases with arousal and diminishes during sleep. Both H_1 antagonists and agonists of H_3-autoreceptors depress release of histamine and reduce arousal (see text for details)

GABAergic neurons in the ventrolateral preoptic area (VLPO). Since the VLPO is more active in SWS sleep, this phase of the sleep cycle could depend in part on GABAergic inhibition of histamine-releasing neurons that project to the cortex (Fig. 22.7). What activates the VLPO is not clear, however. There also seems to be some feedback control of histamine release because H_3-receptor agonists, that activate the autoreceptors on histamine-releasing neurons and reduce release of this transmitter, augment SWS while H_3-receptor antagonists have the opposite effect. Finally, other effects of histamine that could contribute to increased arousal are increasing the activity of excitatory cholinergic neurons in the basal forebrain and inhibition of neurons in the hypothalamic preoptic area which promote sleep.

A much higher profile has recently been claimed for histamine in the control of circadian rhythm (see Jacobs, Yamatodani and Timmerman 2000). When injected intracerebroventricularly in rats it appears to alter locomotor and drinking rhythms in a somewhat complex manner depending on when it is given in the light–dark cycle, being most active when the animals are in constant darkness. Some of the effects can also be mimicked by increasing the amount of endogenous histamine released with the H_3 autoreceptor antagonist thioperamine. Certainly histamine has both excitatory (H_1) and inhibitory (H_2) effects on SCN neuron firing and autoradiography has revealed the presence there of H_1 receptors. Since glutamate and 5-HT have been shown to increase histamine release in the SCN and GABA to inhibit it, the above authors consider histamine to be the final mediator of their effects. Whether this is so remains to be seen for, despite the sedative effects of some H_1 antagonists, rhythm changes have not been reported with their long-term clinical use.

NORADRENALINE

Although some studies show that noradrenaline inhibits neuronal firing it is generally considered to increase behavioural activity and arousal. This impression is borne out to the extent that CNS stimulants, like amphetamine, increase release of noradrenaline and produce behavioural and EEG arousal, while reserpine, which reduces noradrenaline storage and hence release, causes psychomotor retardation. It is also supported by evidence that the firing rate of neurons projecting from the locus coeruleus is greater during waking (1–2 Hz) than during SWS (0.2–0.5 Hz) and is increased even more as behaviour progresses from vegetative or consummatory activities (e.g. grooming or feeding) to vigilance. Furthermore, stimulation of the locus coeruleus in cats causes EEG desynchronisation and increases arousal, while a neurotoxic lesion of these neurons leads to EEG synchrony, increases SWS and reduces REM sleep. In fact, some ('REM-off') cells in the locus coeruleus stop firing altogether during REM sleep. Because a reduction in the activity of noradrenergic neurons precedes the onset of sleep, this change in activity is thought to have a permissive role in sleep induction.

How all these actions of noradrenaline are manifest is not clear and, unfortunately, most experiments in this area have been carried out on anaesthetised animals which, arguably, are not ideal for investigating mechanisms underlying arousal! One of the few investigations to have been carried out in unanaesthetised rats has shown that infusions of noradrenaline into the nucleus basalis of the medial septum increases waking (and the γ-wave activity of the waking phase), but reduces the γ-waves of SWS.

These changes, which are thought to be mediated by activation of β-adrenoceptors, suggest that noradrenaline increases cholinergic influences on arousal, in the nucleus basalis, at least (Cape and Jones 1998). However, a fairly common side-effect of β-adrenoceptor antagonists, used clinically to relieve hypertension, is sleep disturbance which is expressed as nightmares, insomnia and increased waking. Clearly, these drugs must have additional actions either in other brain centres, or non-selective effects on other (possibly 5-HT_{1A}) receptors that have quite different effects on arousal. It has even been suggested that β-blockers disrupt sleep patterns by inhibiting melatonin synthesis and release, but this is controversial.

In contrast, α_2-adrenoceptor agonists are well-known for their sedative effects. Since their activation of presynaptic α_2-autoreceptors will reduce noradrenergic transmission, by depressing the firing of neurons in the locus coeruleus and release of noradrenaline from their terminals, this action is entirely consistent with the proposal that increased noradrenergic transmission increases arousal. Although this presynaptic action of α_2-agonists would explain their sedative effects it must be borne in mind that many α_2-adrenoceptors in the brain are in fact postsynaptic. Their role (if any) in sedation is unclear but it must be inferred that, if they make any contribution to sedation, then either a specific brain region or a specific α_2-adrenoceptor subtype is involved. Another possible confounding factor is that many α_2-adrenoceptor ligands have an imidazoline structure (see Chapter 8) and the recently discovered imidazoline receptors are also thought to influence the sleep cycle and arousal. Even less is known about the role of α_1-adrenoceptors on arousal partly because most drugs acting at these receptors do not readily cross the blood–brain barrier.

The role of noradrenergic neurons from the locus coeruleus on behaviour during the waking phase is rather controversial. It is doubtful that noradrenaline release is actually required for waking because animals with more than a 90% lesion of these neurons are still capable of staying awake, although they are rather subdued. Nevertheless, the single-unit activity of these neurons is increased by sensory stimuli ranging from those that cause physical discomfort (e.g. tailpinch) to environmental stimuli (e.g. tones and light flashes), especially those that provoke orientation to the stimulus (e.g. approach of the experimenter). The evoked neuronal response typically shows a brief (phasic) burst of activity followed by a quiescent period of post-stimulus inhibition but this response, along with behavioural arousal, habituates on successive presentation of the stimulus.

Such findings have led to suggestions that neurons in the locus coeruleus complex serve as a central 'alarm' system while others have argued that their increased neuronal firing during the waking period mediates changes in 'selective attention'. It has even been suggested that the tonic activity of these neurons could determine overall arousal, whereas the more transient, phasic, response determines 'attentiveness'. In fact, these neurons could serve all these purposes, thereby helping to protect the individual from threatening stimuli as well as directing attention to interesting, or salient environmental features (see also Chapter 8).

Few studies have investigated the role in behaviour of noradrenergic neurons originating in the nuclei of the lateral tegmental area (see Chapter 8). However, what little evidence there is suggests that they respond primarily to unconditioned environmental stimuli but are capable of adaptive changes in their activity on repeated presentation of the stimulus. Because noradrenergic neurons arising in the lateral tegmental nuclei have numerous reciprocal connections with other brainstem nuclei involved in homeostasis (e.g. regulating blood pressure and heart rate), it is likely that they make an important contribution to the adjustments in the activity of the peripheral autonomic system during the various states of sleep and waking (see Goldstein 1995).

DOPAMINE

The role of this neurotransmitter in the sleep–waking cycle has not received as much attention as that devoted to noradrenaline and interpretation of existing evidence is not straightforward. On the one hand, the firing rate of neurons projecting from the dopaminergic neurons in the ventral tegmental area does not vary across the sleep–waking cycle and, in any case, the dopaminergic innervation of the cortex is much more restricted than that of noradrenaline or 5-HT. On the other hand, drugs that modify dopaminergic transmission do affect arousal albeit in complex ways (see Gottesmann 1999).

Low doses of the dopamine agonist, apomorphine, induce SWS and, in humans, dopamine agonists can induce somnolence which is a problem when treating Parkinson's disease. This action is thought to be due to activation of presynaptic D_2-autoreceptors and some antagonists of this receptor increase waking state and reduce both non-REM and REM sleep. That a reduction in firing of dopaminergic neurons is associated with reduced arousal is consistent with evidence that local infusion of GABA into the dopaminergic ventral tegmental area also reduces waking. However, others have suggested that activation of postsynaptic D_2-receptors in the dorsal striatum is responsible.

By contrast, high doses of dopamine agonists increase arousal and cortical desynchronisation, possibly by activating postsynaptic D_2-receptors. Indeed, local infusion of dopamine into the nucleus accumbens increases waking, an effect blocked by the D_2-receptor antagonist, haloperidol. Such an action is consistent with the general improvement in sleep (especially sleep continuity) in patients treated with neuroleptics, such as haloperidol and clozapine, which share D_2-receptor antagonism as a common target. However, the various changes seen in the different phases of the EEG seem to depend on the actual compound tested.

5-HYDROXYTRYPTAMINE

This neurotransmitter presents something of a paradox in respect of its role in sleep and waking behaviour, although its importance to both is undoubted. Early experiments suggested that an increase in 5-HT transmission actually helps to induce sleep (see Jouvet 1974). Thus *p*CPA, which blocks the synthesis of 5-HT, causes insomnia in cats and reduces SWS; this insomnia is reversed by giving the 5-HT precursor, 5-hydroxytryptophan (5-HTP), which bypasses the *p*CPA block. Also, a lesion of the dorsal Raphé nucleus (DRN) produces insomnia, the degree of which is proportional to the loss of 5-HT neurons and the decrease of 5-HT turnover in their projection areas. Despite such lesions, sleep patterns return to normal after some days and, if they are made in new-born rats, sleep patterns normalise after a few weeks, suggesting that they are not solely dependent on 5-HT.

In contrast to this evidence that 5-HT activity decreases arousal, antidepressants are generally thought to increase serotonergic transmission while the central depressant, reserpine, reduces it, although it must be remembered that both these treatments affect central noradrenergic transmission as well. Nevertheless, direct stimulation of Raphé neurons, or systemic administration of a 5-HT precursor, actually increases waking. This suggests that 5-HT has either an excitatory influence on behaviour and/or an inhibitory effect on sleep. This view is supported by electrophysiological recordings of the activity (firing frequency) of neurons in the cat DRN. Insofar as it can be certain that it is serotonergic neurons that are being monitored in this nucleus, these studies have shown that, during quiet waking, their activity is about 2–3 spikes/s, but that this rate decreases progressively and becomes less regular as sleep progresses to SWS. In fact, these neurons become virtually totally quiescent during REM sleep and this reduction in activity is probably effected by GABAergic inputs to the DRN. For a review of all this evidence, see Jacobs and Azmitia (1992).

Assigning a particular role for changes in 5-HT transmission in sleep is confounded by the existence of 5-HT neurons in several distinct Raphé nuclei (Fig. 22.8; see also Chapter 9). These project to different regions of the brain but the differences in their functional influences are, as yet, poorly understood. Most studies have in fact investigated the DRN, which innervates forebrain areas, but it does seem that other serotonergic nuclei in the medulla show a similar pattern of responses. Thus, neurons in the 'inferior' Raphé nuclei (the Raphé magnus (NRM), the nucleus Raphé obscurus (NRO) and the nucleus Raphé pallidus (NRP)) (see Fig. 22.8) which project to the lower brainstem and spinal cord, all show a reduced discharge during SWS when compared with that in the awake subject. However, their firing rate is generally higher than in the DRN. Moreover, unlike DRN neurons, those in the NRO and NRP continue to fire, albeit at a reduced frequency, during REM sleep. The implications of these differences in the regulation of the sleep cycle are unclear.

The role of 5-HT transmission in waking behaviour is even less clear. The tonic activity of DRN neurons during 'active' waking is certainly greater than during 'quiet' waking but it is not increased further by arousing or threatening stimuli. However, environmental stimuli that provoke behavioural orientation induce a marked phasic increase in serotonergic neuronal activity (see Chapter 9) suggesting that they do have some role in the response to stimuli requiring attention.

A link between 5-HT release and increased waking is supported by evidence from *in vivo* microdialysis of cats and rats. This has confirmed that the extracellular concentration of 5-HT in all brain regions studied to date is lower during both SWS and REM sleep than in the awake state (see Portas, Bjorvatn and Ursin 2000). Interestingly, if behaviour is maintained at a constant level, the activity of 5-HT neurons does not show circadian variation although 5-HT turnover in the brain areas to which they project

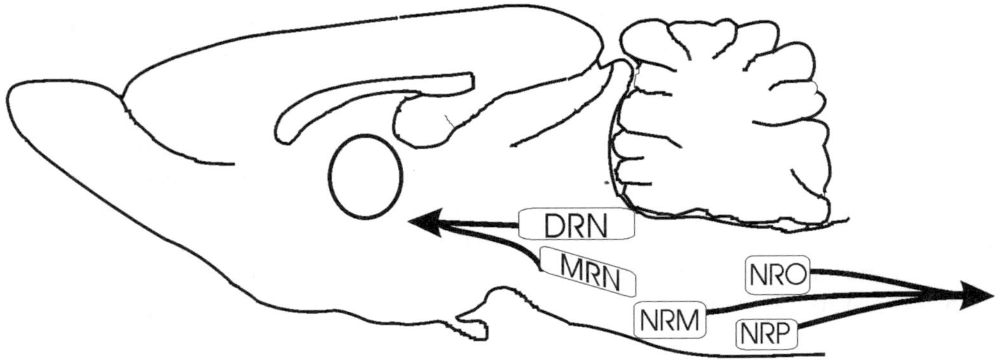

Figure 22.8 The distribution of brainstem Raphé nuclei. Neurons that release 5-HT are clustered in two groups of nuclei in the pons and upper brainstem. The 'superior' group, which projects to forebrain areas, includes the dorsal Raphé nucleus (DRN) and the median Raphé nucleus (MRN). The 'inferior' group projects to the medulla and spinal cord and includes the nucleus Raphé pallidus (NRP), the nucleus Raphé obscurus (NRO) and the nucleus Raphé magnus (NRM)

does show such a rhythm. The reasons for this apparent dissociation between firing rate and transmitter release are not clear but it does suggest that neuronal firing rate is not necessarily a reliable indicator of transmitter release in the terminal field.

One specific theory for the role of 5-HT in arousal suggests that serotonergic transmission serves to coordinate target cell responses by adjusting their excitability to match the subjects' general level of arousal. In so doing, they are responsible for gating motor output and coordinating this with homeostatic and sensory function (Jacobs and Azmitia 1992; Jacobs and Fornal 1999). This would be consistent with evidence that, like the noradrenergic system, increases in the firing rate of neurons in the DRN precede an increase in arousal. The frequency of discharge would code the state of arousal and prime target cells for forthcoming changes in the response to sensory inputs.

Apart from the problem of trying to associate the effects of 5-HT with specific nuclei, there is also no clear picture of which 5-HT receptors mediate any of these changes in sleep and waking. This is not least because of the large number of receptor subtypes, the limited receptor selectivity of most test drugs, species differences in the response, as well as time- and dose-related differences in the response to any given agent. 5-HT is also known to affect noradrenaline and dopamine release in the brain (see Stanford 1999) and such interactions undoubtedly explain some of the inconsistencies between the early findings and recent studies of the role of these different 5-HT neurons in sleep.

Nevertheless, it is evident that activation of many different receptor subtypes affect the sleep–waking cycle. For instance, recent evidence suggests that activation of 5-HT_{1A}, 5-HT_{1B}, $5\text{-HT}_{2A/C}$ and 5-HT_7 receptors in the SCN all affect circadian rhythms. Activation of 5-HT_{1B} (presynaptic) receptors in the retinohypothalamic tract is thought to attenuate 5-HT release and so blunt light inputs to the SCN and reduce its photic regulation. In contrast, postsynaptic 5-HT_7 receptors, 5-HT_{2C}, and possibly postsynaptic 5-HT_{1A} receptors, are thought to have an important role in photic entrainment and to mediate phase-shifts in circadian rhythms (reviewed by Barnes and Sharp 1999). In addition to these effects on circadian rhythms, it is clear that 5-HT receptors affect sleep more directly. A detailed review of this subject is to be found in Portas, Bjorvatn and Ursin (2000) but key findings are summarised here.

Table 22.1 Effects of activation of 5-HT receptors on sleep–waking cycle

Receptor	REM	Waking	Location	Effect on 5-HT transmission
5-HT_{1A}	↑	↓	Presynaptic	↓
5-HT_{1A}	↓	↓	Postsynaptic	↑
5-HT_{1B}	↓	↑	Not known	?
5-$HT_{2A/2C}$	↓	↑	?Postsynaptic	↑
5-HT_3		↑	Postsynaptic	↑

The actions of 5-HT_{1A} receptor agonists in rats depend on their route of administration (Bjorvatn and Ursin 1998). When they are given systemically they cause a transient increase in waking time and a reduction in SWS and REM sleep which is followed by a delayed increase in SWS. This latter response is possibly mediated by activation of inhibitory postsynaptic 5-HT_{1A} receptors in the nucleus basalis (Table 22.1). Certainly, local infusion of 5-HT_{1A} agonists into this area increases SWS. Another contributory factor is suggested by the reduction in waking and increase in SWS following intrathecal infusion of 8-OH-DPAT. This is thought to reflect inhibition of primary sensory afferents, by activation of presynaptic 5-HT_{1A} receptors, an action which would be conducive with induction of sleep. However, infusion of low concentrations of the 5-HT_{1A} agonist, 8-OH-DPAT, into the DRN to activate autoreceptors induces a type of REM sleep which is explained by a reduction in the firing rate of 5-HT neurons. In turn, this is presumed to result in disinhibition of mesopontine cholinergic neurons in the PPT and LTD nuclei which are responsible for REM sleep. Such a scheme is supported by evidence that local infusion of a 5-HT_{1A} agonist into these areas reduces REM sleep, presumably by inhibition of mesopontine cholinergic neurons by postsynaptic 5-HT_{1A} receptors

Administration of 5-HT_{1B} receptor agonists increases waking time and reduces REM sleep. This is consistent with recent evidence gathered from 5-HT_{1B}-receptor knock-out mice which exhibit more REM sleep and less SWS than the wild-type. Moreover, 5-HT_{1B} agonists reduce, while antagonists increase, REM sleep in the wild-type mouse, but neither type of compound has any effect in the knock-outs (Boutrel *et al.* 1999). Unfortunately, it is not known whether these actions are mediated by presynaptic, postsynaptic or heteroceptors and therefore whether 5-HT activity is increased or decreased. It is also not helped by the limited selectivity of test agents.

5-$HT_{2A/2C}$ agonists increase waking and reduce SWS and REM sleep in humans and rats, possibly through an action in the thalamus. Conversely, blockade of 5-HT_{2A} receptors, e.g. by ritanserin, increases SWS, an action that might contribute to the beneficial effects of antidepressants that share this action. However, these findings are confounded by evidence that activation of 5-HT_{2C} receptors increases SWS.

Infusion of 5-HT_3 receptor agonists into the nucleus accumbens increases waking and reduces SWS, although REM sleep is unchanged. These effects of 5-HT_3 receptor activation are prevented by co-administration of a D_2-receptor antagonist. This is consistent with evidence that activation of 5-HT_3 receptors can increase dopamine release and points to functional interactions between these two groups of neurons that affect the sleep–waking cycle. Such interactions will certainly confound any attempts to define the specific role of 5-HT in the regulation of sleep and arousal.

Overall, 5-HT transmission seems to increase during waking and to decline in sleep although it may only reach its minimal level, in some neurons anyway, during REM sleep. Whether its role is simply to prime target cells to enable an increase in the motor activity associated with waking, as has been suggested, remains to be seen.

ADENOSINE

It is perhaps not surprising that, since adenosine has been presented as an endogenous inhibitor of neuronal function with its antagonists, like theophylline, being stimulants (see Chapter 13), it should have been implicated in sleep induction.

In fact EEG studies have shown that administration of an adenosine A_1 agonist increases SWS in humans and induces it in sleep-deprived rats while adenosine also inhibits the important cortical activating brainstem cholinergic neurons. Of more physiological significance is the finding from microdialysis in rats that the extracellular concentration of adenosine progressively increases in the hippocampus, reaching a maximum at the end of the animal's active (lights off) period. After that, it falls sharply within an hour as the animal enters the quiet (lights on) sleepy period (see Huston *et al.* 1996). Of course, the hippocampus is not generally associated with sleep patterns and whether these studies establish adenosine as a potential sleep inducer, or merely as an 'activity-restrictor' that facilitates sleep, is unclear.

AMINO ACIDS

Since most excitatory transmission is mediated by glutamate this must be involved in the sleep–waking cycle. It certainly mediates the input of the retinohypothalamic tract to the SCN, apart from afferent inputs more generally to the ARAS, etc. So far, specific *in vivo* manipulation of the direct glutamate input to the SCN has not been possible.

The fact that SCN neurons contain GABA, and that this appears to be the neurotransmitter released by the geniculohypothalamic tract onto the SCN, clearly puts it in a prime position for regulation of sleep rhythms. However, its precise role is unclear, not least because it can act as an excitatory, as well as an inhibitory, neurotransmitter in this nucleus and that these varied responses appear to follow a circadian rhythm (see Chapter 11). Again, specific manipulation of this pathway is difficult although GABA enhancement generally (e.g. by benzodiazepines) is, of course, sedative (see later section on drug-induced sleep).

SLEEP FACTORS

In classical times, sleep was thought to be induced by sleep factors (vapours) emanating from food in the stomach. To this day, and despite the encyclopedic evidence that neurotransmitters have discrete effects on sleep and arousal, the idea still lingers that there are sleep-inducing ('somnogenic') factors. These are thought to have a pervading influence on sleep throughout the brain, although the stomach is no longer regarded as their source! This view was strongly encouraged by experiments, carried out in the early twentieth century, by Pieron in Paris, who showed that the CSF of sleep-deprived dogs contained a substance that had a somnogenic effect when infused into non-sleep-deprived animals. Since then, many candidate sleep substances have emerged, some of which are more convincing than others.

The first serious attempts to identify and characterise an endogenous somnogenic agent was carried out by Pappenheimer and colleagues (see Pappenheimer 1983) who found that transferring samples of CSF from sleep-deprived goats into normal rabbits increased the latter's REM sleep. Chemical extraction from thousands of rabbit brains and many gallons of human urine yielded a sleep factor and established it as a muramyl peptide. Unfortunately muramyl peptides are not synthesised by mammalian cells but are components of bacterial cell walls. Apart from the obvious possibility of mere contamination, it is not clear how the substance turned up in the CSF and brain tissue. Despite this setback, and some scepticism about whether somnogenic peptides exist at all, research still continues in this area and many candidates have been suggested. These include well-known peptides such as prolactin, CCK-8, VIP and somatostatin as well as some novel ones such as δ-sleep-inducing peptide. (For a full review of this subject, see Garcia-Garcia and Drucker-Colin 1999.)

Another line of research has produced convincing evidence that the pro-inflammatory cytokines, interleukin IL-1β and TNFα modify the sleep cycle: these agents generally increase non-REM sleep and suppress REM sleep. IL-6 also reduces REM sleep and SWS in the first half of the sleep cycle but subsequently increases SWS. However, all these responses vary with dose, test species and even time of day. These factors are produced by T-cell lymphocytes but their receptors are associated with neurons, astrocytes, microglia and endothelial cells. Because these agents induce nitric oxide synthase, and there is some evidence that nitric oxide increases waking, possibly through modulation of ACh release in the medial pontine reticular formation, there is no need for them to cross the blood–brain barrier (although there is evidence that they do). Nevertheless, how these factors actually cause changes in the sleep cycle is as yet unclear. An indirect effect via changes in the rate of prostaglandin synthesis (see below) is one possibility but others include modulation of 5-HT$_{2A}$-mediated serotonergic transmission and suppression of glutamatergic neuronal activity through an adenosine-dependent process.

Prostaglandins, in particular PG_{D2}, have also been shown to act as sleep-promoting substances. PG_{D2} is synthesised in the arachnoid membrane and choroid plexus and its receptors are prevalent in the basal forebrain. Moreover, its concentration in the CSF shows a circadian rhythm and increases during sleep deprivation. It is not yet known how PG_{D2} influences sleep but when it is infused locally, it changes the firing rate of neurons in the preoptic and basal forebrain areas in ways suggesting that it promotes sleep. Like the interleukins, and TNFα, mechanisms proposed to explain these actions include modification of monoaminergic or adenosinergic transmission.

Finally, the endogenous fatty acid amide, oleamide, has somnogenic effects. This compound is chemically related to the endogenous ligand for cannabinoid receptors, anandamide. Although oleamide has even been reported to augment anandamide binding to cannabinoid (CB$_1$) receptors, it is still not known whether this action is relevant to its somnogenic effects. Oleamide has been shown to potentiate (benzodiazepine-sensitive) GABA$_A$ receptor responses through a mechanism that seems to involve the γ-subunit. However, modifications of 5-HT$_2$, muscarinic, metabotropic glutamate and NMDA receptor function have all been suggested as possible mechanisms.

DRUG-INDUCED SLEEP

Setting aside the general anaesthetics, which do not directly modify the function of any particular neurotransmitter, all the drugs that are used to induce sleep, i.e. the 'hypnotics', augment the function of GABA and so directly depress neuronal function and probably facilitate cortico-thalamic synchrony. Most of them are benzodiazepines and even those that are not, like zopiclone (a cyclopyrrolone) and zolpidem (an imidazopyridine), act on the benzodiazepine receptor (see Chapter 19). Many benzodiazepines have a long half-life (20+ h) and a similar spectrum of activity, being both anxiolytic and sedative, and unless these effects are actually required during the day after the hypnotic action (as would occur with nitrazepam and flurazepam) it is important to use those benzodiazepines with a short half-life: e.g. temazepam, lorazepam

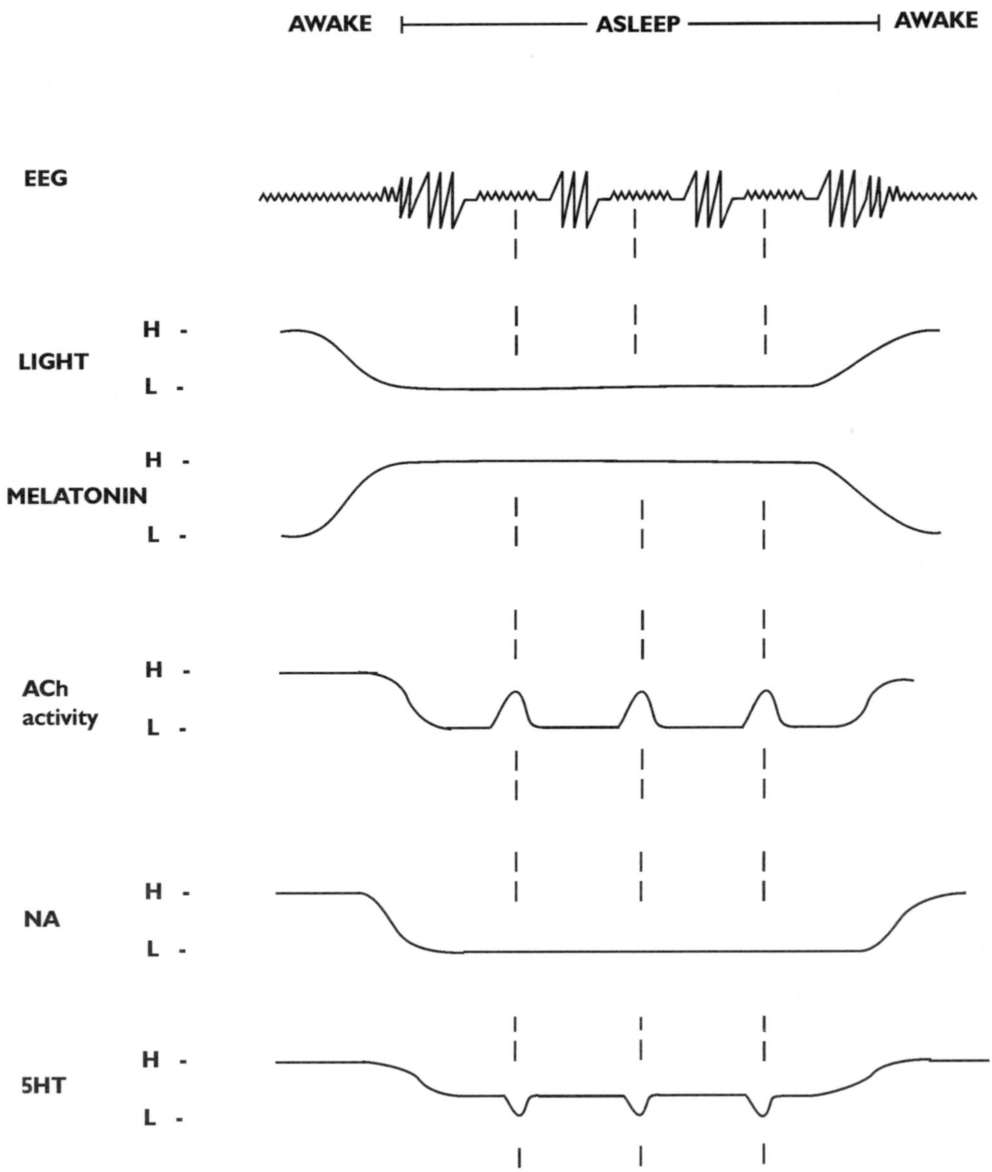

Figure 22.9 Summary of the influence of varying factors on sleep and waking. The EEG is shown diagramatically in the typical arousal (awake) state and in both non-REM (slow wave) and REM sleep. Appropriate activity levels, high or low, are shown for the different factors such as light input, melatonin secretion or ACh, NA, and 5-HT function in the different phases

and lormetazepam ($T^{1/2}$ s = 6–10 h). All hypnotics appear to increase SWS at the expense of REM sleep and this has been suggested as a cause of irritability and possibly even the cognitive deficits claimed to be associated with use of these drugs

That hypnotic drugs do not produce a natural sleep should not be surprising in view of the fact that they merely augment GABA and depress neuronal function, when sleep is clearly a very complex phenomenon involving the integrated activity of a number of neurotransmitters. To what extent it might be possible to induce sleep by simultaneously blocking the action of ACh, noradrenaline, 5-HT and histamine is not known. It would be an interesting experiment but the peripheral and other central effects are too numerous and dangerous to contemplate its trial.

SUMMARY

Sleep appears to rely on synchrony in cortico-thalamic reciprocal pathways in which GABA plays an important part such that sleep can be enhanced by augmenting GABA function. Probably such synchrony is the state to which the nervous system and our bodies return unless it can be disrupted as a result of stimulation by appropriate afferent inputs. Some of this information may come from the SCN, which is activated simply by the reception of light, but other diffuse, ascending inputs from the reticular activating system certainly cause EEG arousal also. In turn, activation of this system depends on normal sensory inputs to the body since it receives collaterals from classical sensory axons projecting to specific thalamic nuclei. Part of this activating effect appears to be mediated by cholinergic neurons of the nucleus basalis but other ascending projections from brainstem nuclei utilising acetylcholine, noradrenaline and histamine also disrupt the cortical synchrony thereby causing EEG desynchronisation and behavioural arousal.

To what extent we depend on these neuronal projections to stay awake is uncertain but it seems clear that noradrenaline, histamine, 5-HT and cholinergic ones all become less active, or quiescent, when sleep starts except that, periodically, certain cholinergic neurons in the prepontine nucleus discharge and initiate REM sleep while the neurons releasing the classical monoamines generally cease firing (Fig. 22.9). The role of other chemicals in sleep induction is even less clear, although melatonin release is certainly increased during sleep.

This, of course, is only an outline of what has been observed during sleep and waking. It is evident that, over the last 50 years, a great deal has been learned about the sleep pattern and its underlying neurobiology. Indeed, it is now clear that the induction and regulation of sleep and arousal involves the concerted influences of a wide range of neurotransmitters and, possibly, non-neuronal factors. However, we still cannot explain how sleep occurs or how these neurotransmitter systems are actually activated, inhibited or coordinated to control our sleep and waking. Also, we still do not know why sleep is necessary at all. The challenge of these questions still remains.

REFERENCES AND FURTHER READING

Arendt, J, Middleton, B, Stone, B and Skene, D (1999) Complex effects of melatonin: Evidence for photoperiodic responses in humans? *Sleep* **22**: 625–635.

Barnes, NM and Sharp, T (1999) A review of central 5-HT receptors and their function. *Neuropharmacology* **38**: 1083–1152.

Berridge, CW (1998) Modulation of forebrain electroencephalographic action and behavioral state by locus coeruleus–noradrenergic system: involvement of the medial septal area. *Adv. in Pharmacol.* **42**: 744–748.

Bjorvatn, B and Ursin, R (1998) Changes in sleep and wakefulness following 5-HT1A ligands given systemically and locally in different brain regions. *Rev. Neurosci.* **9**: 265–273.

Boutrel, B, Franc, B, Hen, R, Hamon, M and Adrien, J (1999) Key role of 5-HT1B receptors in the regulation of paradoxical sleep as evidenced by 5-HT1B knock-out mice. *J. Neurosci.* **19**: 3204–3212.

Cape EG and Jones, BE (1998) Differential modulation of high-frequency gamma-electroencephalogram activity and sleep-wake state by noradrenaline and serotonin microinjections into the region of cholinergic basalis neurons. *J. Neurosci.* **18**: 2653–2666.

Garcia-Garcia, F and Drucker-Colin, R (1999) Endogenous and exogenous factors on sleep-wake cycle regulation. *Prog. Neurobiol.* **58**: 297–314.

Goldstein, DS (1995) *Stress, Catecholamines and Cardiovascular Disease*, Oxford University Press, Oxford.

Gottesmann, C (1999) Neurophysiological support of consciousness during waking and sleep. *Prog. Neurobiol.* **59**: 469–508.

Hagan, RM and Oakley, NR (1995) Melatonin comes of age. *Trends Pharmacol. Sci.* **16**: 81–83.

Hendricks, JC, Sehgal, A and Pack, AI (2000) The need for a simple animal model to understand sleep. *Prog. Neurobiol.* **61**: 339–351.

Hobson, JA (1992) Sleep and dreaming: induction and mediation of REM sleep by cholinergic mechanism. *Curr. Opin. Neurobiol.* **2**: 759–763.

Huston, JP, Haas, HL, Boix, F, Pfister, M, Decking, U, Schrader, J and Schwarting, RKW (1996) Extracellular adenosine levels in neostriatum and hippocampus during rest and activity periods of rats. *Neuroscience* **73**: 99–107.

Jacobs, BL and Azmitia, EC (1992) Structure and function of the brain serotonin system. *Physiol. Rev.* **72**: 165–229.

Jacobs, BL and Fornal, CA (1999) Activity of serotonergic neurons in behaving animals. *Neuropsychopharmacology* **21**: 9S–15S.

Jacobs, EH, Yamatodani, A and Timmerman, H (2000) Is histamine the final neurotransmitter in the entrainment of circadian rhythms? *TIPS* **21**: 293–298.

Jahnsen, H and Llinas, R (1985) Ionic basis for the electroresponsiveness and oscillatory properties of guinea-pig thalamic neurons *in vitro*. *J. Physiol.* **349**: 227–247.

Jouvet, M (1974) The role of monaminergic neurons in the regulation and function of sleep. In *Basic Sleep Mechanisms* (Eds Petre Quadens, O and Schlag, JD), Academic Press, New York, pp. 207–236.

McCormick, DA (1989) Cholinergic and noradrenergic modulation of thalamocortical processing. *Trends Neurosci.* **12**: 215–220.

McCormick, DA (1992) Neurotransmitter actions in the thalamus and cerbral cortex and their role in neuromodulation of thalamocortical activity. *Prog. Neurobiol.* **39**: 3778–3788.

Moruzzi, G and Mayoun, HW (1949) Brainstem reticular formation and activation of the EEG. *EEG Clin. Neurophysiol.* **1**: 455–473.

Pappenheimer, JR (1983) Induction of sleep by muramyl peptides. *J. Physiol.* **336**: 1–11.

Portas, CM, Bjorvatn, B and Ursin, R (2000) Serotonin and the sleep–wake cycle: special emphasis on microdialysis studies. *Prog. Neurobiol.* **60**: 13–35.

Stanford, SC (1999) SSRI-induced changes in catecholaminergic transmission. In *Selective Serotonin Reuptake Inhibitors: Past, Present and Future* (Ed Stanford, SC), RG Landes Bioscience, Austin, TX, pp. 147–170.

Steriade, M (1992) Basic mechanisms of sleep generation. *Neurol.* **42**: 9–18.

Steriade, M (1999) Coherent oscillations and short-term plasticity in corticothalamic networks. *Trends Neurosci.* **22**: 337–344.

Steriade, M, McCormick, DA and Sejnowski, TJ (1993) Thalamocortical oscillations in the sleeping and aroused brain. *Science* **262**: 679–685.

23 Drug Dependence and Abuse

A. H. DICKENSON

OVERVIEW

There are pharmacological, social and legal issues to consider in this context of drug dependence and, of course, drug users and abusers vary enormously. A chronic alcoholic is very different from a weekend user of cannabis and any consideration of the topic has to consider legal and social issues as well as pharmacological effects of the drugs.

The key points can be outlined as follows:

- The use of a drug produces both physical and psychological effects on an individual.
- Even though the drug is used for non-medical reasons, there will still be sought-after and unwanted effects.
- The desired effects may be pleasurable or the drug may simply be used to escape from the world — here the sought-after effect is oblivion.
- The continued use of a drug can lead to dependence on it so that the effects of abstinence or withdrawal from it will reinforce further use.
- Dependence can be compounded by tolerance, so that more of a drug is needed to produce the desired effect.
- The individual may then withdraw from society into a drug subculture.
- Possession of a non-medical drug is illegal.

The cost of the drug may well lead on to further problems with the law as commonly, burglary, prostitution and dealing in the drug become necessary to finance the habit. In fact, it has been estimated that a large part of crime results from the need for drugs.

A recent survey by the Institute for the Study of Drug Dependence reveals that while drug use has been steadily on the increase since the 1960s, particularly in the late 1980s and early 1990s, latest figures from across the UK suggest that in some areas like south-east England, drug use associated with the dance culture may be 'levelling out'. A perennial issue is the unreliability of the data on drug use. Although drugs associated with the dance culture are primarily the amphetamine type, including Ecstasy, it is now felt that heroin use has been spreading involving younger people from 14 years old onwards. Although heroin use remains a minority activity, there is probably more heroin available than ever before at low prices and in smaller, more affordable quantities.

Neurotransmitters, Drugs and Brain Function. Edited by R. A. Webster

The number of individuals having ever taken an illicit drug is estimated currently at around 28% of the UK adult population, with around a quarter of 16–29-year-olds having taken a drug within the last 12 months. Not only are young people coming into contact with drugs at a younger age than before but a wider range of drugs are available, including those currently not controlled under the Misuse of Drugs Act such as amyl nitrite and ketamine.

Government-sponsored campaigns and media attention can have varying effects on drug consumption. 'Scare campaigns' may dissuade some from using drugs but there are suggestions that non-selective claims that all drugs cause harm is leading young people to ignore entirely all aspects of the campaigns, relying on their own experiences and those of their peers. There appears to have been a positive impact of campaigns on the misuse of solvents as the most recent figures show only a slight increase in the number of deaths which reached an all-time low in 1994.

The media has paid much attention to fuel the fears of a crack 'epidemic' but these have not been realised although the drug, once very rare in the UK has found a level of consumption in the drug-using community where it causes substantial problems for users and their families.

Overall, the rates of HIV among drug injectors are steadily declining and great credit must be paid to the damage-limitation strategies (needle exchanges, free condoms, etc.) that have been highly successful in keeping the incidence of HIV and AIDS in the drug-misusing community to levels far lower than in other European countries. However, rates of hepatitis C among this same group remain high.

The most reliable numerical data available probably comes from the number of people registered seeking help for their drug habit. Here the number becoming addicted to the notifiable drugs (mainly opiates) continues to rise steadily and is now about 30 000. Research suggests that now perhaps one in three come forward, possibly as a result of more common prescribing of methadone as a replacement therapy and general concern about drugs and HIV/AIDS.

Another guide to the extent of the problem is the number of drug seizures by the authorities. In 1998, this increased by 8% to 14 000 with the largest increases in heroin and cocaine (20–30%) although 76% of the total seizures are still cannabis. The number of drug-related prosecutions was just under 130 000 in that year of which 90% were for possession, and the majority of the cases dealt with were cannabis. This is despite many police forces giving warnings and cautions for low-level possession of cannabis rather than proceeding with prosecution.

DEFINITIONS AND DRUG CLASSIFICATION

Drug dependence has been defined as a state, psychological and/or physical, resulting from an interaction between a drug and an organism characterised by a compulsion to take the drug on a continuous or periodic basis to experience its psychic effects and/or avoid the discomfort of its absence. Figure 23.1 shows the interactions between a drug and an individual. This definition of dependence (WHO) covers all forms of drug dependence which may be psychological or physical or combinations of both, accompanied or not by tolerance to the drug. Because of these complexities drug dependence is classified somewhat on the basis of the effects produced or nature of the dependence-producing compound. The major groups to be considered are:

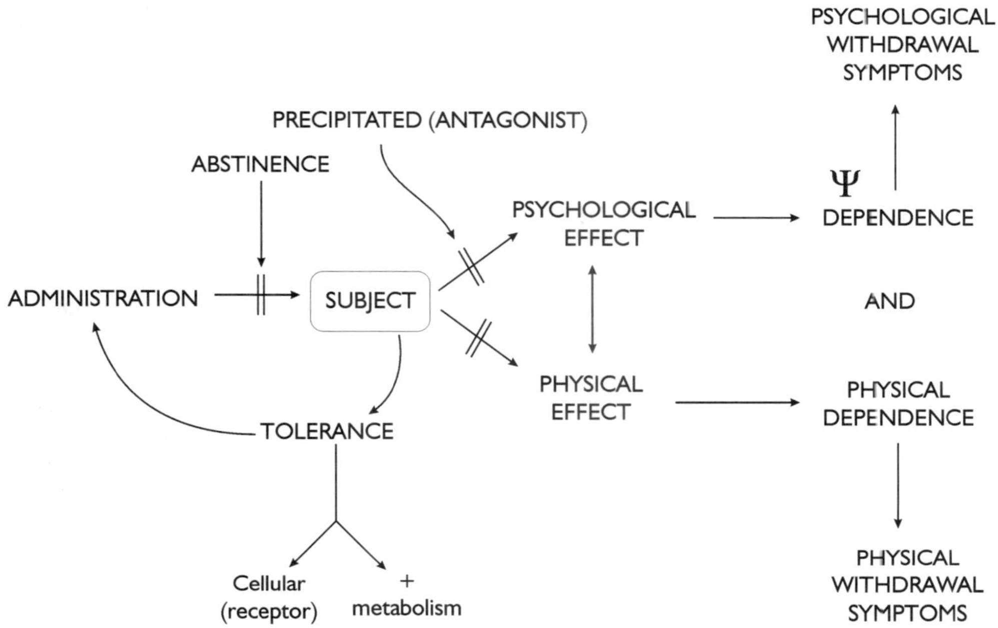

Figure 23.1 The diagram shows the stages that may occur after administration of a non-medical drug. The details of each of the potential consequences are given in the text but note that whereas all drugs must have some psychological effect and so may trigger psychological (Ψ) dependence, physical dependence is only obvious with depressant drugs

Depressants: alcohol, barbiturates
Opioids: heroin, morphine, methadone
Stimulants: amphetamines, cocaine, Ecstasy
Psychedelics: LSD, mescaline, 'magic mushrooms'
Inhalants: glue, solvents
Cannabis:

Legal definitions of drugs
Class A: cocaine, heroin, LSD, methadone, mescaline, morphine, opium, Ecstasy plus all class B when taken in injectable form
Class B: amphetamine, cannabis, codeine, barbiturates
Class C: benzodiazepines

Schedules:

1	Drugs with no therapeutic use (cannabis, LSD) and so are not prescribed
2&3	Drugs with medical use—heroin and morphine for pain relief, amphetamine for narcolepsy and cocaine These can be possessed by doctors, pharmacists and nurses and prescribed
4	Benzodiazepines which can be possessed by those above and also others as long as they are a medicinal product (use for epilepsy, anxiety and sleeping pills)
5	Cough medicines, etc. with small amounts of opioids which are non-prescribed, and can be bought over the counter

Acute administration of a drug of abuse will produce acute effects related to that drug. The psychological effects must somehow reinforce the administration of the drug. On repeated use *tolerance* may develop leading to an increase in the dose of drug required to produce the required effect. *Psychological dependence* occurs and is defined as 'a condition in which a drug produces a feeling of satisfaction and a psychic drive that requires administration of the drug to produce pleasure or avoid discomfort'. Psychological dependence varies from mild to strong depending on the drug used. *Physical dependence* is 'an adaptive state that manifests itself by intense physical disturbance when the drug is discontinued'. Physical dependence is not produced by all drugs of abuse and is most pronounced after use of depressant drugs such as alcohol or heroin. If a drug usage is halted *withdrawal* or *abstinence* occurs, the symptoms of which can be psychological (i.e. cravings, discomfort, etc.) and/or physical on the basis of whether physical and psychological dependence are present. To avoid withdrawal symptoms drug administration is continued and a cycle is set up (see Table 23.1).

DRUGS USED

OPIOIDS

General

Heroin (medical name diamorphine) is one of a group of drugs called 'opiates or opioids' which are derived from the sap of the ripe opium poppy. Opium is the dried milk of the opium poppy. It contains morphine and codeine, both effective and widely used analgesics, along with heroin which can be made from morphine and in its pure form is a white powder. The main sources of street heroin for the UK are the Golden Crescent countries of South-west Asia, mainly Afghanistan, Iran and Pakistan. Today street heroin usually comes as an off-white or brown powder whereas for medical use it is usually tablets or an injectable liquid. A number of synthetic opioids are also manufactured for medical use and all have similar effects. Methadone, a drug which is often prescribed as a substitute drug in the treatment of heroin addiction, is a weaker but long-lasting orally effective opioid and is usually prescribed as a syrup.

Table 23.1 Effects of various classes of drugs of abuse

Drug	Psychological dependence	Physical dependence	Physical withdrawal symptoms	Tolerance
Depressants				
Alcohol	Mild–strong	Marked	Intense	Irregular
Opiates	Strong	Marked	Severe	Marked
Stimulants				
Cocaine	Strong	None	None	None
Amphetamines	Mild–strong	Low	Mild	Marked
Hallucinogens				
LSD	Variable	None	None	Marked
Cannabis	Mild–strong	None	None	Weak

Heroin can be smoked ('chasing the dragon'), sniffed or prepared for injection. Opioids prescribed for medical use may be used for non-medical reasons, especially by heroin users who cannot otherwise get hold of heroin.

The sudden influx of smokable heroin in the 1980s caused a dramatic increase in use, because it was no longer necessary to inject the drug in order to obtain its effects. Despite new initiatives to try to reduce heroin use it has continued to increase and there is concern about the wider availability and use of cheap heroin among young people, particularly in deprived areas.

There are many debates about the best way of tackling heroin use. The UK government participates with other countries in attempting to cut off the supply of heroin, although given that the source of the drug can shift rapidly it is not clear how effective this approach is. Likewise, removal of dealers from the street appears to simply allow others to move in to supply the constant demand. There is also a debate in the UK about the substitute drug, methadone, which is similar to heroin, except the user does not get the same 'high' as with heroin. The idea is to gradually reduce the dose of methadone until the person is able to come off drugs without suffering withdrawal symptoms. The problem is that many users seem to quickly go back on heroin so that some doctors prescribe methadone on a maintenance basis, not reducing the dosage until the person feels ready to give up, a process that can be lengthy. One school of thought would claim that this approach simply keeps people dependent on a different drug. The opposite view is that methadone keeps people away from the dangerous street market in heroin, with the associated risks of crime and overdose. Unfortunately, many users obtain methadone legally and then sell it to buy street heroin.

Recent years have also seen the development of needle exchange schemes whereby users of injectable heroin can receive clean equipment rather than sharing needles to minimise the threat of hepatitis and HIV. These schemes seem to be very effective although it has been claimed that they encourage injecting. Another issue is heroin-related crime, especially theft, burglary and forgery, as a dependent user will need about £50 a day to pay for the drug. The cost to the community is unknown but some police forces have estimated that up to one-third of crime relates to drug use.

Legal

Heroin and other opiates are controlled under the Misuse of Drugs Act making it illegal to possess them or to supply them to other people without a prescription. Heroin is treated as a Class A drug where the maximum penalties are 7 years' imprisonment and a fine for possession and life imprisonment and a fine for supply.

Effects

Heroin and other opiates are drugs that depress the nervous system. The desired effect on the street is not the analgesia that is the reason for their medical use but the feeling of warmth, reduction in anxiety and detachment. The effects of both smoked and obviously injected heroin are rapid and then last several hours but this varies with how much is taken and the route of application. The feeling produced by the drug has been described as 'being wrapped in cottonwool'. Mental anguish is removed and hence the use of the drug as a means of escape from social and other pressures becomes clearer. Although initial use can result in a feeling of nausea and even actual vomiting these unpleasant reactions are subject to tolerance. With high doses of heroin a marked sedation takes over and people can fall asleep. Excessive doses can lead to severe sedation and vomiting — the combination can be lethal. This is further compounded by the fact that opioids can depress the respiratory centre and with non-medical use the most common form of death is from respiratory failure.

With regular use tolerance develops as does psychological and physical dependence. The withdrawal syndrome leads to unpleasant flu-like symptoms which may include aches, tremor, sweating, chills and muscular spasms. These fade after about a week but can be a major deterrent to giving up the drug. While many people do successfully give up long-term heroin use, coming off and staying off heroin can be very difficult. Fatal overdoses can happen, especially when users take their initial dose after a break during which tolerance has faded, or when opiate use is combined with use of other drugs such as alcohol, tranquillisers or other opiates. Many regular heroin users will use other opiates or depressant drugs when they cannot get hold of heroin.

If is often difficult to know exactly what is being taken because the purity of street heroin varies and it is often mixed with adulterants. Injecting increases these risks and also puts users at risk of a range of infections including hepatitis and HIV if injecting equipment is shared. Regular injectors may suffer a wide range of health problems including chronic constipation, damaged veins, heart and lung disorders.

Mode of action

Opiates are part of the depressant group of drugs. Alcohol is thought to produce its effects by a general depressant action on most neural systems and although this is true there is some evidence that the drug exerts preferential effects on some pharmacological systems and thereby acts to block NMDA receptors and Ca^{2+} channels while enhancing $GABA_A$-mediated inhibitions.

Barbiturates are rare these days but these depressants produce their pharmacological effects by increasing the duration of Cl^- channel opening associated with $GABA_A$ receptors (see Chapter 11).

Opiates produce more discreet inhibitory effects since they bind to and activate inhibitory opioid receptors which, due to their restricted distribution, cause less widespread effects than those of the barbiturates and alcohol. Activation of the opioid receptors leads to a decrease in release of other neurotransmitters (glutamate, NA, DA, 5-HT, ACh, many peptides, etc.) and direct hyperpolarisation of cells by opening of K^+ channels and decreasing Ca^{2+} channel activity via predominant actions on the mu opiate receptor (see Chapter 12).

LSD

General

Lysergic acid diethylamide (LSD) is an hallucinogenic drug that is made from ergot, a fungus found growing wild on rye and other grasses. It is a white powder, but as a street drug, it is a liquid absorbed into paper sheets. The sheets are cut into tiny squares like postage stamps or transfers and often have pictures or designs on. LSD is also sometimes dropped onto sugar cubes or formed into tablets or small capsules. Only tiny amounts are needed to have an effect and the strength of LSD can vary greatly. It is usually taken orally.

Although often thought of as a drug that was popular in the 1960s and 1970s, LSD is still used by many and in a national UK survey published in 1997 around 10% of those aged 16–29 said they had tried it at least once, with 50% of those saying that they had tried it during the year preceding the survey. One survey of club-goers listed LSD as their fourth favourite drug after cannabis, Ecstasy and amphetamines.

LSD was first discovered in 1938 by a medicinal chemist while working on the synthesis of drugs that might have use in disorders of the CNS. The chemist, Albert Hofmann, was the first to take the drug in 1943 when he, supposedly inadvertently, took it on a Friday evening—he then reported the dreamlike state caused by the drug with a vivid description of the visual changes and other perceptual effects of the drug. In the next decade, psychiatrists used LSD in some patients with a variety of mental problems and the drug was also tried unsuccessfully by the US military as a 'truth drug' and as a possible chemical weapon. In the early 1960s LSD was used for pleasure and among hippie groups LSD taking was seen as a religious experience and a way of getting in touch with the self, other people and the environment. In 1966 its use was made illegal in the UK.

Legal

LSD is a class A drug under the Misuse of Drugs Act. It is not available for medical use and is illegal to possess or supply. Maximum penalties are 7 years' imprisonment and a fine for possession and life imprisonment and a fine for supply.

Effects/risks

The strength of LSD preparations varies but one dose will usually result in a mild hallucinogenic experience and two or three doses in a full-blown 'trip'. This begins about 30 min after taking LSD and can last up to 12 h. The effects vary greatly depending on dose level, how the user feels and the situation they are in. Users often report visual effects such as intensified colours, distorted shapes and sizes and movement in stationary objects. Distortion of sound and changes in the sense of time and place are also common. Tactile changes can also occur. The general view is that the block or interference with 5-HT mechanisms removes an inhibitory control which in turn leads to an increase in sensory inputs into the brain and spinal cord. The increased sensory barrage is then misinterpreted by the brain leading to the perceptual changes. The word 'hallucination' is not strictly true since it is a distortion of reality that is perceived, not the perception of something that is not there.

Emotional reactions vary greatly. Some people claim they become more aware of themselves and other people and describe LSD trips as being similar to a religious or spiritual experience. Feelings of being separated from the body are also common. Unpleasant or frightening experiences are more likely if the user is already anxious or takes the drug when depressed—this can lead to paranoia. The time-course of the effects of LSD is prolonged and, of course, once LSD is taken there is no going back until it wears off, so a bad trip can be very disturbing. If users become anxious there is no antidote but they can often be talked down and reassured by others.

There is no evidence of LSD use leading to physical dependence or fatal overdose, although people have died through accidents occurring under the influence of the drug since their sensory perception is altered. Some LSD users experience 'flashbacks' when the 'trip' is re-experienced some time afterwards. This has also been seen in animal studies where the changes in the firing of 5-HT neurons recurs after the effects of the drug have worn off. Flashbacks tend to be short-lived.

OTHER HALLUCINOGENS

A number of mushrooms, liberty cap (psilocybe), psilocybin, fly agaric, *Amantia muscaria* and the peyote cactus contain hallucinogenic agents. They are usually eaten raw but can be dried out and stored or cooked into food or made into a tea and drunk. The effects are highly variable and whereas 20–30 liberty caps would be required to give a full dose, just one fly agaric mushroom would produce similar actions. Some recent local surveys in the UK have found between 12% and 15% of 16-year-olds claiming to have used these at least once.

Vast numbers of hallucinogenic plants and fungi were used by ancient tribes and civilisations usually as a means of entering the spiritual world. Fly agaric mushrooms were used by medicine men or 'shamans' of North-east Asia and Siberia whereas other species were sacred to the Aztecs of Mexico at the time of the Spanish invasion in the sixteenth century. The use of mushrooms and other hallucinogenic plants is less common in European history, although witches used hallucinogenic plants from the potato family, especially deadly nightshade and henbane, which contains a number of cholinergic antagonists.

Legal

The law on mushrooms is complex. It is not illegal to pick, possess or use liberty caps in their raw state. However, the Misuse of Drugs Act controls psilocybin and psilocin, the active ingredients in liberty cap mushrooms when they are 'separated' from the mushroom or where the mushrooms are prepared for use. This means that drying out the mushrooms and storing them for later use or making them into a tea or cooking with them can be an offence. The law is sill unclear but preparing mushrooms for use, rather than eating them raw, has led to a small number of prosecutions. It would seem, however, that fly agaric mushrooms which, although hallucinogenic, do not contain either psilocybin or psilocin, are not illegal under the MDA even if they are prepared.

Effects/risks

The effects of liberty caps are similar to a mild dose of LSD and can vary greatly depending on the mood, situation and expectation of the user. Effects come on after about half an hour and last up to 9 h, depending on how many are taken. Users often laugh a lot and feel more confident. Some people find that they feel sick, or indeed vomit and high doses result in a mild to moderate trip with visual and sound distortions. A bad trip like that with LSD can include feelings of anxiety and paranoia and, again, flashbacks can be experienced some time later. Possibly, the greatest risk is picking the wrong type of mushroom and being poisoned.

Like LSD, tolerance develops very rapidly so the next day it might take twice as many liberty caps to repeat the experience and so most users only use mushrooms occasionally. Physical dependence and withdrawal symptoms do not result from regular use though some people may become psychologically dependent and feel a desire to use on a regular basis. At present there is no evidence of serious health damage from long-term use.

Fly agaric use is more likely to result in unpleasant effects, including nausea and vomiting, stiffness of joints and lack of coordination. High doses (anything more than one fly agaric mushroom) may result in intense disorientation and even possibly convulsions.

Mode of action of hallucinogens

LSD and the other hallucinogenic drugs are thought to work by interfering with the 'filter' exerted by 5-HT neuronal systems on incoming sensory activity. How this is done is unclear but LSD can inhibit the activity of 5-HT neurons by activation of the somatic 5-HT_1 autoreceptor. By acting as an agonist at this site LSD slows the firing of 5-HT neurons in the dorsal raphé and so leads to a reduced postsynaptic action of the monoamine. There is a very good correlation between the potency of a range of hallucinogenic drugs and their ability to reduce 5-HT neuronal activity. As mentioned earlier, this reduction in activity may allow increased sensory inputs to enter the CNS. It may then be that the brain misinterprets the sensory information leading to the perceptual errors and also the different senses may converge and are then confused. The CNS, faced with a marked increase in incoming sensory information, fails to cope. However, LSD and the other hallucinogens probably also have postsynaptic actions on 5-HT receptors which contribute to their actions.

CANNABIS

General

Cannabis or marijuana comes with many acronyms such as bhang, black, blast, blow, blunts, Bob Hope, bush, dope, draw, ganga, grass, hash, hashish, hemp, herb, marijuana, pot, puff, skunk, smoke, spliff, wacky backy, weed, etc. Some of the names are based on the country of origin such as Afghan, Colombian, homegrown, Lebanese, Moroccan, Pakistani, etc. Its main source is *Cannabis sativa*, a bushy plant that grows in many parts of the world and is also cultivated in the UK. The main active ingredients in cannabis are the tetrahydrocannabinols (THC). These are the chemicals that cause the main effects on the brain and although the most prevalent is the Δ^9 THC, there are many others that add to the effect. Different forms of cannabis come from different parts of the plant and have different strengths. 'Hashish' or 'hash' is the commonest form found in the UK, a resin scraped or rubbed from the dried plant and then pressed into brown/black blocks. It is mostly imported from Morocco, Pakistan, the Lebanon and Afghanistan. 'Grass', 'bush' and 'ganga' or 'marijuana' is imported from Africa, South America, Thailand and the West Indies and is the chopped, dried leaves of the plant. It is also cultivated in the UK, sometimes on a large scale to sell but sometimes by individuals in their homes or greenhouses for their own use. In general, the herbal form or marijuana is usually not as strong as the resin form although particularly strong herbal forms such as 'skunk' have recently been cultivated in Holland.

In the UK the drug is usually smoked rolled into a cigarette or joint, often with tobacco. The herbal form is sometime made into a cigarette without using tobacco or it can also be smoked in a pipe, brewed into a tea or cooked into cakes. Of course, the fibre of the cannabis plant is non-psychoactive and hemp has a long history, being used to make rope, mats, clothing, cooking oil, fuel and varnishes.

Cannabis is the most widely used illegal drug in the UK and easily the illegal drug most likely to have been tried by young people. Probably over 5 million people have used it at least once and many people are regular users. It is not surprising that cannabis is the most-seized drug and that the large majority of court cases involve this drug.

The debate about 'legalising' or 'decriminalising' the drug for personal or medical use has become a topical issue. This step was taken in Holland and since it did not appear to lead to more use of drugs 'decriminalisation' has been supported by the civil liberties movement in that the adoption of a more liberal approach to possession of the drug for personal use is unlikely to lead to problems. Others are very much against the idea on both health and moral grounds but the former view has been taken by many police forces who now no longer prosecute those found with small amounts of the drug. There are many issues to debate, few of which have been discussed in detail in the UK. Currently, there is discussion of the medical aspects of the pharmacology of cannabis. There are suggestions, based on anecdotes, animal studies or pressure group opinions, that the drug can be useful to treat glaucoma, in the control of the muscle spasms that are one of the symptoms of multiple sclerosis and for appetite stimulation in cases of chemo- and radiotherapy. The status of cannabis is such that doctors cannot prescribe smokable cannabis to their patients, although synthetic THC preparations (nabilone) are available for nausea. There is growing pressure on the British government to change the law so that the required controlled clinical studies on the potential effects of cannabis can be carried out.

Legal

Cannabis is controlled under the Misuse of Drugs Act. It is illegal to grow, possess or supply to another person. A particular restriction on cannabis (and opium) is the offence of allowing your house (or any other premises you have responsibility for) to be used for growing cannabis or smoking it. Under the Misuse of Drugs Act cannabis is a class B drug. The maximum penalty for supply is 14 years' imprisonment plus a fine whereas the maximum penalty for possessing it for personal use is 5 years' imprisonment plus a fine. These maximum penalties are only rarely imposed except where there is very large-scale supplying or trafficking. Most prison sentences for cannabis possession and small-scale supply are less than one year. Fines for possession are generally between £20 and £100.

Effects/risks

Smoking cannabis causes a number of physical effects including increased heart rate, decreased blood pressure, bloodshot eyes, increased appetite and mild dizziness. The effects are rapid in onset and start within a few minutes and may last several hours depending on how much is taken. When eaten the effects are slower in onset but then longer in duration. Eating cannabis may mean a large dose is taken at once, making it difficult to avoid any unpleasant reactions.

Cannabis has a mild sedative effect, not unexpected with the receptors for this drug being inhibitory. The experience can vary greatly depending on the user's mood and also expectation. It is said that a person learns the effects of the drug. Many people find that when they first use cannabis nothing much happens. Generally cannabis makes people relax and they may become giggly and very talkative or alternatively quieter and subdued. The former effects may be due to disinhibitory actions of the cannabinoids. Users often report that they become more aware of music and colours and that time seems to stand still. While under the influence of cannabis, short-term memory may be impaired but this goes as the effects of the drug wear off. Accidents are more likely especially if people drive or operate machinery while on the drug since judgement and motor coordination are reduced and a mild ataxia ensues.

Some people find that cannabis makes them anxious and paranoid, both inexperienced users or people who are anxious or those who consume strong varieties or high doses of cannabis. Very heavy use by people who already have a predisposition to mental health problems may lead to very distressing experiences.

There is no conclusive evidence that moderate, long-term use of cannabis causes lasting damage to physical or mental health. However, it is probable that frequent inhalation of cannabis smoke over a period of years will contribute towards bronchitis and other respiratory disorders and possible cancers of the lung and parts of the digestive system. Risks are greater if cannabis is smoked with tobacco.

There is no physical dependence associated with cannabis use. Regular users who stop smoking do not suffer withdrawal symptoms in the same way as with drugs like alcohol or the opioids. Even so, regular users can become psychologically dependent and come to rely on using cannabis, either as an aid to relaxation or as a social prop. Someone who uses cannabis excessively may appear apathetic, lack energy and motivation and perform poorly at their work or education. This state may carry on for weeks after stopping use of the drug. However, such a condition seems rare, is similar to what would be expected from someone who drinks too much or regularly uses other depressant drugs and it is likely that the effects of cannabis suit someone who is amotivational rather than the drug leading to a particular syndrome.

If is often claimed that cannabis is a 'gateway' drug in that its use leads to use of drugs like heroin or cocaine. Although the bulk of heroin and cocaine users have used cannabis the vast majority of people who have used cannabis have never used heroin or cocaine.

Mode of action

Cannabis is now known to interact with specific receptors, the CB1, a slow G-protein-linked inhibitory receptor being the main type in the CNS. The pharmacology of cannabis commenced with the finding that the active principle in cannabis was Δ9 tetrahydrocannabinol (THC). Early studies showed a number of inhibitory actions of the cannabinoids that, although akin to opioid actions in some sites, were not reversed by naloxone. A breakthrough came when a high-affinity ligand, CP-55490, was synthesised. Radiolabelling of this ligand allowed the demonstration of binding sites in the CNS for this ligand. In 1992, the central CB1 receptor was cloned and found to be a member of the G-protein-linked superfamily of receptors. This receptor has now been found to be linked to the opening of potassium channels, the closing of calcium channels and the inhibition of adenyl cyclase. The aminoacid sequence of the receptor has been deduced from human brain and the rat and human variants show 97% homology. Later in 1993, a second receptor, the CB2 receptor, with 40% identity to the CB1 receptor was reported. This receptor is found mainly on cells in the immune system. Alternate splicing of the CB1 receptor gives rise to a shorter version of the receptor but this version is very poorly expressed and may not serve any function. In a similar way to events following the description of opioid receptors, the next stage after the discovery of the receptors was the search for the endogenous ligands. Now, anandamide, arachidonylethanolamide (named after an Indian god of bliss) and arachidonyl-glycerol (2AG) have been proposed as endogenous ligands. They are certainly present in the CNS and act as agonists and there is much research on their synpatic actions. Anandamide has about threefold higher affinity for the CB1 receptor. It is formed by a calcium-dependent hydrolysis of a membrane phospholipid and release is stimulated by depolarising agents. It is subject to a selective uptake mechanism into neurons and glia and following reuptake anandamide is hydrolysed to arachidonic acid and ethanolamine. 2AG is found in the brain at hundredhold higher levels than anandamide and is hydrolysed more rapidly. Regarding the receptors, detailed mapping of their distribution has been done using autoradiography, *in situ* hybridisation and immunohistochemistry. There are high levels of CB1 receptors in the hippocampus and they probably underlie the effect of the drug on memory, possibly also some of the mood changes. The fact that CA1 and 3 pyramidal neurons express the receptor and also are subject to LTP strengthens the link. In addition, the ability of cannabis to produce ataxia and loss of control of movements may be due to actions on CB1 receptors in the substantia nigra and caudate with the receptors likely to be on the latter neurons projecting to the basal ganglia. Cerebellar receptors have also been located. Receptors in the spinal cord and brainstem are likely to be responsible for some of the effects on coordination as well as the analgesic effects. Receptors on neurons in the amygdala may be important in the effects on mood, particularly the sense of well-being and relaxation.

MDMA: 3,4-METHYLENEDIOXYMETHAMPHETAMINE—ECSTASY

General

Ecstasy (Adam, brownies, burgers, Dennis the Menace, disco biscuits, doves, E, Edward, essence, fantasy, love doves, M and Ms, New Yorkers, rhubarb and custard, shamrocks, white doves, X, XTC) is an illegally manufactured drug that comes in tablet or capsule form. The appearance varies considerably ranging from brown, white or pink tablets to yellow, clear, red and black or red and yellow capsules, often with pictures, designs or logos. It is taken orally.

Ecstasy remains a popular drug among young people, mainly those who are into the dance/rave scene, although there are some signs that in the south of the UK at least, Ecstasy might not be quite so fashionable as it was. Although probably the most commonly mentioned drug in the media, use has never been as widespread as cannabis, amphetamine and possibly also LSD.

There have been nearly 80 deaths in the UK related to Ecstasy use. Why this particular group of people died when so many others have also taken the drug is unknown. However, it is clear that many tablets sold as Ecstasy are not what purchasers think they are. The amount of Ecstasy in a tablet can vary greatly and the drug is often mixed with other drugs or a range of adulterants. In Holland, users can submit their pills or tablets to a rough test but this has been criticised by the UK government as condoning drug use.

Despite all the warnings about the dangers of Ecstasy, many young people continue to use it. This has led to 'safer dancing' campaigns which encourage clubs to have 'chill-out' areas, make sure staff are trained in first aid and ensure the water taps in the toilets are working. Some of the deaths from the drug have been due to an overreaction to this advice—water intoxication has been implicated in several of the fatalities.

Ecstasy was first synthesised in 1912 and, like LSD, was used in the 1950s as a potential chemical weapon. The feeling of empathy with others produced by Ecstasy has been used by psychiatrists in therapy sessions. However, animal studies indicating that Ecstasy might damage the brain led to it being banned in the USA in 1985.

As an amphetamine, Ecstasy was already banned in the UK before it became popular in the late 1980s via the House music scene which had developed in America and Ibiza. Ecstasy was used as a stimulant drug to help users stay up all night and to promote empathy and communication between people. It quickly became an important part of the dance scene.

Legal

Ecstasy is a class A drug under the Misuse of Drugs Act. It is illegal to be in possession of or to supply it and the drug cannot be prescribed by doctors. The maximum penalties for possession of Ecstasy is 7 years' imprisonment plus a fine and for supply is life imprisonment plus a fine.

Effects/risks

Ecstasy is a stimulant drug which also has mild hallucinogenic effects. It has been described as being like a mix of amphetamine and a weak form of LSD. This would agree well with the pharmacological effects of the drug which are a combination of the typical amphetamines with some additional effects of 5-HT neurons. Thus Ecstasy releases both dopamine and 5-HT—the former is likely to be responsible for the insomnia, anorexia and elevation of mood while the weak hallucinations and empathy is probably via 5-HT. The latter actions are also likely to lead to the elevated body temperature and blood-clotting abnormalities since 5-HT plays an important role in thermoregulation in the hypothalamus and is present in high amounts in platelets from which it is released by the drug. The effects of taking a moderate dose start after 20–60 min and can last for up to several hours. The pupils become dilated, the jaw tightens and there is often brief nausea, sweating, dry mouth and throat. There is cardiovascular stimulation and a loss of appetite is common. Many users experience an initial rushing feeling followed by a combination of feeling energetic and yet calm. Loss of anger, empathy with other people and an enhanced sense of communication are commonly reported. Some users also report a heightened sense of their surroundings, greater appreciation of music and increased sexual and sensual experiences. Some users have bad experiences. This may include feeling anxious and panicky, confusion and an unpleasant distortion of the senses. The disorientating effect may make accidents more likely. After taking Ecstasy users may feel very tired and low and need a long period of sleep to recover. Regular use may lead to sleep problems, lack of energy, dietary problems (including anorexia nervosa) and feeling depressed. Increased susceptibility to colds and sore throats may follow. While physical dependence is not a problem, psychological dependence on the feelings of euphoria and calmness and the whole scene around the drug can develop.

Little is yet known about the effects of heavy, long-term use of Ecstasy but there are increasing concerns about the possibility of mental health problems, especially chronic depression. Animal studies suggest a long-term depletion of 5-HT can occur and given the known roles of this monoamine in mood (Chapter 9), this is a plausible explanation. It is disturbing that a large number of people may be predisposed to mental problems as a result of this drug use. However, it is probable that major depletions of monoamine systems are needed for overt effects to occur and certainly, the deficits in dopamine levels needed to produce symptoms of Parkinson's disease are likely to be in the order of 75%.

STIMULANTS

General

Amphetamines (speed sulph, sulphate, uppers, wake-ups, billy whizz, whizz, whites, base) are synthetic stimulants which as medicines have been formed into a variety of tablets. Their current medical use is very limited and in fact only dexamphetamine sulphate, Dexedrine, is now available for use solely in the treatment of narcolepsy. The only other amphetamine available for medical use is methylphenidate (Ritalin) for the treatment of attention deficit syndrome in children. As a street drug, amphetamine usually comes as a white, grey, yellowish or pinky powder. The purity rate of street powders is less than 10%, the rest being made up of milder stimulants such as caffeine, other drugs such as paracetamol or substances like glucose, dried baby milk, flour or talcum powder.

The powder form can be snorted up the nose, mixed in a drink or prepared for injection. During the 1990s, amphetamine was a popular drug among young people attending all-night raves and is probably the next most commonly used illegal drug after cannabis. Recent local surveys have shown between 5% and 18% of 16-year-olds claiming to have used it at least once. Amphetamine powder tends to be quite cheap— about £10–12 a gram or £5 for a small 'wrap'.

A new, more concentrated form of amphetamine (known as 'ice') has become common in America. This is a crystallised form of meth (or methyl-) amphetamine that can be smoked or injected. It is very strong and can result in intense paranoia and a very unpleasant come-down. After heroin, amphetamine is probably the most commonly injected street drug in the UK. Amphetamines were first discovered in the 1800s but their medical uses were not recognised until the 1930s. Then they were used to counter low blood pressure, for asthmatics and to suppress appetite. Subsequently, amphetamines were prescribed for a whole range of disorders including inability to sleep, epilepsy, migraine, depression and hyperactivity in children. In the 1950s and 1960s they were widely marketed as slimming tablets. Until 1956 many amphetamine-based drugs could be bought over the counter without a prescription. Use among bored housewives, people who felt low and needed an energy boost ('pep pills' and 'tonics') and people who worked long hours such as long-distance lorry drivers was common. Non-medical use of amphetamines grew in the UK in the 1960s especially among teenage 'mods'. The use of 'purple hearts' (a combination of amphetamine and barbiturate) by thousands of young people led to the first post-war drug craze (and media drug scare) in the UK. Unauthorised possession of amphetamine was banned in 1964. In the 1970s and 1980s street use of amphetamine increased again and centred on a new generation of young people in the all-night club scene of punk rock and Northern Soul. Illicitly manufactured powdered amphetamine and sniffing replaced tablets stolen from factories as the main form of use.

Legal

All amphetamines are prescription only drugs under the Medicines Act. Most are also controlled under the Misuse of Drugs Act. Doctors can prescribe them for patients but it is an offence to be in possession of amphetamines without a prescription. Most amphetamines are controlled as class B drugs under the Misuse of Drugs Act. Maximum penalties for possession are 5 years' imprisonment plus a fine and for supply are 14 years' imprisonment and a fine. If amphetamines are prepared for injection they become class A drugs and increased penalties apply.

Effects/risks

Amphetamines are stimulant drugs. They increase breathing and heart rate, lessen appetite and dilate the pupils. Users tend to feel more alert, energetic, confident and cheerful and less bored or tired. With high doses people often experience a rapid flow of ideas and feel they have increased physical and mental powers although this is usually manifest as talking non-stop. With some people, and especially as the body's energy stores become run down, feelings of anxiety, irritability and restlessness are common. Taking a lot, especially over a few days, can produce a temporary panic and paranoia and with high doses the amphetamine psychosis is like a transient episode of schizophrenia. The effects of a single dose last for about 3–4 h and tend to leave the user feeling tired. Regular amphetamine use can lead to psychological dependence. Users may feel depressed, lethargic, lacking in energy and incredibly hungry without taking the drug. They may be tempted to keep repeating the dose to avoid these feelings. Tolerance also develops with regular use so more is needed to get the same effect.

Heavy, regular use often leads to lack of sleep and food and lowers resistance to disease. Eating disorders, such as anorexia nervosa, may become a problem, especially among women users and work and domestic routines may be disturbed. Many heavy users become very run down and alternate between periods of feeling good and energetic then feeling depressed and low. Delusions, panic attacks, paranoia, a feeling of being 'wired' and possibly hallucinations may also follow. Some users experience violent mood swings and can become very aggressive.

Mode of action

The effects of the amphetamines are discussed in detail in Chapter 7 and are thought to be due to changes in the catecholamines, noradrenaline and dopamine. The peripheral cardiovascular effects probably follow elevated (released) noradrenaline levels in sympathetic neurons while the central effects result from an increased noradrenaline release (anxiety, restlessness) or dopamine (motor stimulation, psychosis). How this is achieved is not absolutely clear but it seems that due to the similarity in structure of amphetamines and catecholamines, amphetamine can enter the nerve terminal by the NA/DA transporter. By so doing, it reduces uptake of the monoamines but more importantly, it causes release of extra NA and DA. This is the result of reverse transport of elevated cytoplasmic monoamines caused by both an inhibition of MAO and a reduction in vesicular uptake of the transmitters. Ecstasy has similar actions but additionally releases 5-HT.

COCAINE

General

Cocaine comes from the Coca plant, grown in the high arid, mountainous areas of South America. It is usually extracted from the leaves of the plant but the leaves themselves can be chewed and a smokable paste made from the leaves is mainly used in countries where the plant grows.

In Britain and America the most common form of cocaine is as a white crystalline powder. Most users sniff it up the nose, often through a rolled banknote or straw, but it can also be made into a solution and injected. Crack is a smokable form of cocaine made into small lumps or 'rocks'. It is usually smoked but can also be prepared for injection. Because it is such a fast-acting drug and the powerful effects wear off quickly, repeated use is common, and since cocaine is a relatively expensive drug it has become closely associated with a rich lifestyle.

Large amounts of cocaine are seized in the UK, but relatively few people present themselves for treatment of dependency. There may be many reasons for this including the fact that those who can afford to have a cocaine problem can afford to attend a private clinic and so are unavailable to researchers and those agencies who collect information about drug use. However, there does seem to be some increase in more general use of the drug. It is appearing in more clubs around the dance/rave scene alongside Ecstasy even though cocaine powder costs more. Crack is around £20–25 for a small rock the size of a sultana, but a rock may have slivers cut from it which are sold for perhaps £10. Although the UK crack problem has not turned out to be as significant as predicted some years ago, crack use has increased in certain inner-city areas bringing with it reports of problems of dependence, drug-related crime and violence.

Coca leaf chewing as an aid to work may have been common among South American Indians as long ago as 250 BC. Cocaine was first extracted from the leaves in 1855 and by the 1870s it was a popular stimulant and tonic and used in a range of patent medicines for all sorts of ailments. Sigmund Freud recommended its use for a range of medical and psychological problems, including alcohol and morphine addiction. Cocaine is a local anaesthetic for eye surgery and in dentistry. Sherlock Holmes, the fictional detective created by Arthur Conan Doyle, was a regular cocaine user while coca-laced wines were enjoyed by popes and royalty in the nineteenth century. Coca-Cola was originally sold as 'a valuable brain tonic and cure for all nervous afflictions' and until 1904 contained small quantities of cocaine. At the turn of the twentieth century doctors began to warn of possible dependence and problems with its use. In America fears developed among white people about 'cocaine-crazed' black people who were rebelling against new discriminatory laws. In Britain concerns arose about the use of cocaine by troops during the First World War. Hysterical press reaction claimed that this was a German plot to destroy the British Empire. In 1916 emergency laws were rushed in to ban possession of cocaine (and opium) and limit its medical use.

Cocaine became more commonly used in the 1960s. Snorting cocaine became fashionable among the 'smart and successful' middle classes and was seen as a glamorous and expensive drug. Meanwhile in America cocaine use was much more widespread and in the mid-1980s, a new, more powerful form of the drug became available, smokable cocaine or crack. This became a major problem for those living in the most deprived areas of inner-city America. Gang warfare, shootings and drug-related crime hit the headlines. In Britain the authorities braced themselves in anticipation of a similar situation but it has turned out to be less of a problem.

Legal

Cocaine and crack are controlled as class A drugs under the Misuse of Drugs Act. It is illegal to be in possession of either crack or cocaine or supply them to other people. Maximum penalties for possession are 7 years' imprisonment plus a fine and for supply life imprisonment plus a fine.

Effects/risks

Cocaine and crack are strong but short-acting stimulant drugs. They tend to make users feel more alert and energetic. Many users say they feel very confident and physically strong and believe they have great mental capacities. Common physical effects include dry mouth, sweating, loss of appetite and increased heart and pulse rate. At higher dose levels users may feel very anxious and panicky. The effects from snorting cocaine start quickly but only last for up to 30 min without repeating the dose. The effects are felt even quicker when smoking crack but are even more short-lived.

Large or quickly repeating doses over a period of hours can led to extreme anxiety, paranoia and even hallucinations. These effects usually disappear as the drug is eliminated from the body. The after-effects of cocaine and crack use may include fatigue and depression as people come down from the high. Excessive doses can cause death from respiratory or heart failure but this is rare.

Neither tolerance nor heroin-like withdrawal symptoms occur with regular use of cocaine. However, regular users may develop a strong psychological dependence on the feelings of physical and mental well-being and may be tempted to keep taking cocaine to avoid feeling tired and depressed. Dependence may be more likely and more severe from smoking crack compared to snorting cocaine. The fact that cocaine and crack are expensive means that people who become dependent may spend vast amounts of money. Those who are not wealthy may find themselves involved in crime or prostitution to fund a habit. With everyday use restlessness, nausea, hyperactivity, insomnia and weight loss may develop. Lack of sleep and weight loss may lead to exhaustion and being very run down. Repeated snorting of cocaine damages the membranes which line the nose. Repeated smoking of crack may cause breathing problems and partial loss of voice. Pregnant women who heavily use cocaine or crack may experience complications and find that their babies are adversely affected. Much has been made in the American press of so-called 'crack babies' and although some babies of crack using mothers may be irritable, difficult to comfort and feed poorly the extent to which this happens has often been exaggerated.

Mode of action

Cocaine, a stimulant, blocks the reuptake of NA (and DA) and so has similar actions to those of the amphetamines which have a number of actions that include the release of NA and DA, and a block of reuptake and metabolism.

BASIS FOR DEPENDENCE

The production of dependence (physical) by the depressant drugs is thought to result from an adaptation of the CNS to the altered environment due to chronic drug use. In the case of the depressant drugs the CNS is believed to establish a new homeostatic state by supersensitivity of the pathways involved, counteracting the drug depression. When the drug is discontinued the depressive effect is removed and dramatic withdrawal symptoms result from this supersensitivity. Since opiates act on specific opiate receptors the withdrawal symptoms are relatively specific, whereas withdrawal from a general depressant such as alcohol produces more marked and generalised symptoms. The supersensitivity has been proposed to result from a number of changes in the depressed pathways that are not mutually exclusive:

(1) Receptor supersensitivity
(2) Unmasking of other neuronal pathways
(3) Synthetic enzyme induction, increasing transmitter levels
(4) Nucleotide changes or coupling to receptor increasing receptor sensitivity
(5) Membrane depolarisation

None of these are mutually exclusive and they may all relate to a common mechanism. Figure 23.2 shows a schematic diagram of how the adaptation of the CNS to a depressant drug can lead to the symptoms of dependence. There is an extensive literature showing that physical withdrawal can be reduced or prevented by drugs acting on related inhibitory systems (e.g. clonidine acting on the G-protein linked α_2-adrenoceptor counters the withdrawal from opioid receptor agonists—the mu opioid receptor has very similar effector mechanisms to the α_2-receptor). Furthermore, withdrawal can be reduced by a large range of drugs that block excitatory systems such as excitatory amino-acid receptor antagonists, calcium channel blockers, etc.

Most of the work has been based on opioids since it is the easiest system to manipulate as administration of the antagonist, naloxone, precipitates withdrawal. Here, the idea that physical dependence results from opposing changes in the neuronal systems depressed by the drug of dependence is borne out by consideration of the acute effects of an opioid and the withdrawal symptoms. They are mirror images of each other:

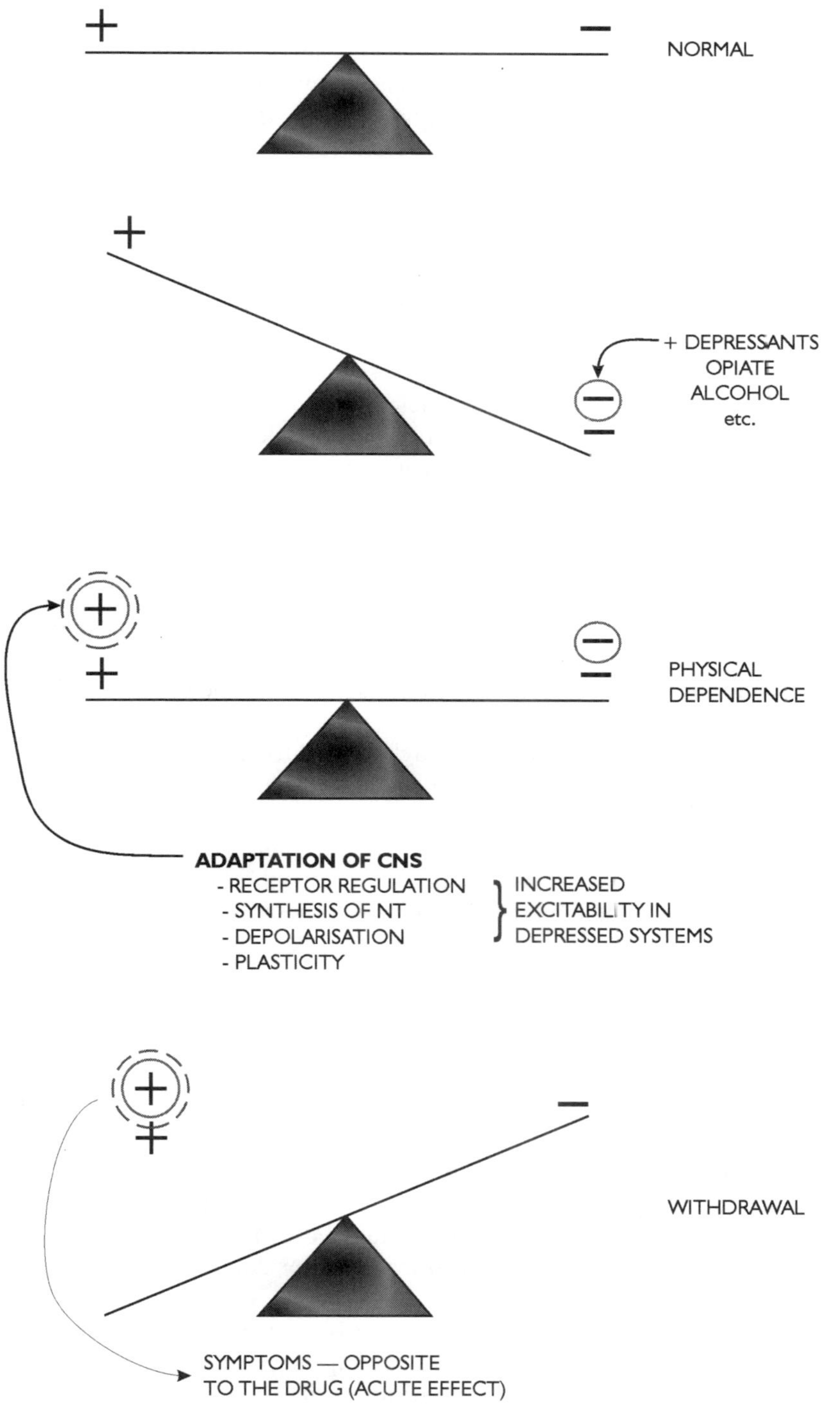

Figure 23.2 A schematic diagram illustrating the ways in which the CNS counters the depressant effects of a drug such as alcohol or an opioid and how this leads to the manifestation of physical dependence when there is abstinence from the drug. These excitatory compensations produce symptoms opposite to the acute effects of the drug

Acute effects	**Symptoms of withdrawal**
Analgesia	Spontaneous aches and pains
Depressed reflexes	Spontaneous twitches (kick the habit)
Feeling of warmth	Feeling of cold (cold turkey)
Anxiolysis	Anxiety and paranoia
Constipation	Diarrhoea
Drying up of secretions	Lacrimation, runny nose, salivation

Although these symptoms last for several days and are not pleasant, they are not that different from a bad cold with influenza yet clearly will be a deterrent to discontinuing the use of a drug. However, a number of people go through withdrawal and yet then go back to the drug. Thus, it is felt that the psychological effects of drugs are critical aspects as are the social issues that interact with continued drug use.

The psychological effects of drugs are poorly understood but involve dopamine systems in the CNS. It is thought that drugs can cause psychological dependence by interactions with dopamine systems that mediate learning so that drug use becomes a learned behaviour. The circuits important in this centre on the nucleus accumbens. The nucleus has inputs from a number of cortical regions and, in turn, projects to the septum, frontal and cingulate cortex and the hypothalamus. The inputs to the accumbens that are thought to be critical for dependence are the dopamine pathways from the ventral tegmental area. Dopamine modulates activity in the nucleus accumbens and these pathways have been implicated in some of the positive symptoms of schizophrenia. In the context of drug dependence of a psychological type, increases in dopamine activity in the VTA are thought to reinforce behaviours occurring at the time. Drug administration becomes associated with environmental cues, such as the paraphenalia associated with the drug and the location where the drug is used. Also the physical and psychological effects of the drug become reinforcing. Thus, electrical stimulation of these areas is rewarding and drug self-administration in animals is reduced by lesions or dopamine receptor antagonists applied to this area. Interestingly, all drugs with psychological dependence liability, despite very different pharmacological actions, produce similar cravings and all increase dopamine activity in the VTA. This is due to release in the case of amphetamine-like drugs and cocaine, via direct depolarisation of the neurons in the case of nicotine. Increased dopamine activity results from disinhibition (of GABA neurons) with alcohol, opioids and cannabinoids although the latter drugs, befitting the mild cravings they produce, only slightly increase activity.

The increased dopamine hypothesis is supported by findings of gene induction in the target areas and the indications that individual differences in dopamine receptors and transporters may underlie impulsive and addictive behaviour in humans. Studies in knock-out mice have, however, provided evidence for complex roles of 5-HT in these processes.

Human data fit well with these ideas since it is very clear that following prolonged drug use the context of the use of the agent has huge importance. Heavily dependent US soldiers in Vietnam during the war, perhaps up to 20% of the troops, were using opium but gave up easily on their return home, where the conditions of war were removed. Many dependent drug users go through physical withdrawal and then re-use the drug when they return to where they took the drug previously, whereas those who move away can do much better in keeping off the drug.

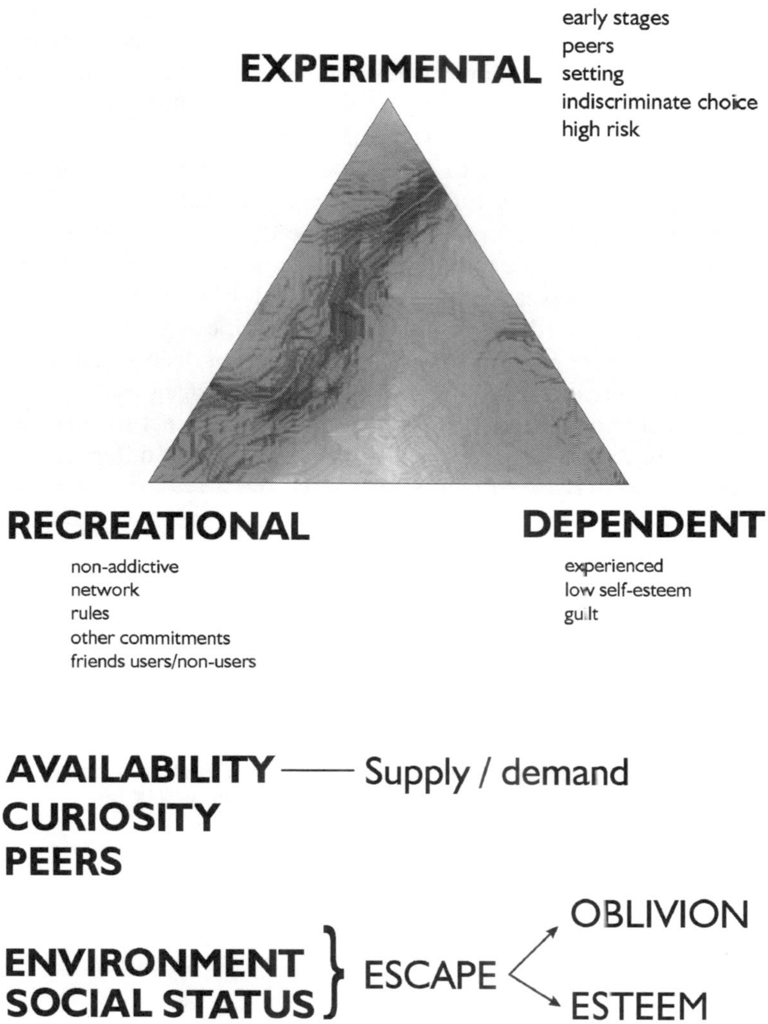

Figure 23.3 Types of drug users and some of the factors that may lead to use of drugs. The triangle represents a simple model whereby three main types of users can be identified—any individual can be at any point on the lines

MAJOR PROBLEMS OF DRUG DEPENDENCE AND ABUSE

(1) Overdose—doses are unknown as is purity
(2) Crime as a result of need to obtain drug
(3) Withdrawal symptoms—may be life-threatening with alcohol
(4) Retreat from society
(5) Acute effects of a particular drug and the chronic pathological effects
(6) AIDS, hepatitis, etc. as a result of injections. Injection of tablets
(7) Drug combinations

The relative importance of these factors will depend on a particular drug, the individual and other factors. Some users are heavily dependent (the prototype addict), others use the drug in very particular circumstances (recreational users) whereas others are only beginners, many of whom will never continue beyond the experimental stage. The physical and psychological effects of the different drugs, individual differences and contextual issues are all interacting to define the nature of drug use and abuse (Fig. 23.3). Finally, social issues are of great importance. The prevalence of serious addiction in areas of social and financial deprivation may be due to the drug being used as a permanent escape from the misery of everyday life with low incomes and housing standards, low job prospects and yet the individual is surrounded by images of affluence. Here drugs are used by an individual to escape from their circumstances, either into oblivion or from modern society into a group of drug users, a society of its own. These types of users are very different from weekend drug users who have strict rules controlling where and when a drug is used and who interact with peers who both use and abstain from drugs. And are these drugs of abuse any different from alcohol and nicotine?

FURTHER READING

Julien, RM (1995) A *Primer of Drug Action*, WH Freeman, New York.

Piomelli, D, Giuffrida, A, Calignano, A and Rodríguez de Fonseca, F (2000) The endocannabinoid system as a target for therapeutic drugs. *Trends Pharm. Sci.* **21**: 218–224.

http://www.streetdrugs.org/ — a US-based site.

http://www.clubdrugs.org/ — another US site.

http://www.drugscope.org.uk/l — a remarkably useful and informative UK site.

Index